QUARK CONFINEMENT AND THE HADRON SPECTRUM VI

Related Titles from AIP Conference Proceedings

QUARK CONFINEMENT AND THE HADRON SPECTRUM VI

6th Conference on Quark Confinement
and the Hadron Spectrum

QCHS 2004

Villasimius, Italy   21 - 25 September 2004

EDITORS

Nora Brambilla
Università degli Studi di Milano

Alberto Devoto
Università degli Studi di Cagliari

Giovanni Maria Prosperi
Università degli Studi di Milano

Umberto D'Alesio
Università degli Studi di Cagliari

Khin Maung
Hampton University

Sergio Serci
Università degli Studi di Cagliari

PONSORING ORGANIZATIONS
nstituto Nazionale di Fisica Nucleare (INFN)
niversità degli Studi di Milano, Dipartimento di Fisica
niversità degli Studi di Cagliari, Dipartmento di Fisica
ezione INFN di Cagliari
egione Sardegna

Melville, New York, 2005
AIP CONFERENCE PROCEEDINGS ■ VOLUME 756

L.C. Catalog Card No 2005924314

ISBN 0-7354-0241-8
ISSN 0094-243X

Printed in the United States of America

PLENARY SESSION PRESENTATIONS

ROUND TABLES AND SUMMARIES

SESSIONS B,C: ROUND TABLE ON PENTAQUARKS AND EXOTICS
(chaired by H. Toki)

SESSION C: SUMMARY AND VISION

SESSION D: ROUND TABLE ON CONFINEMENT/DECONFINEMENT

SESSION E: SUMMARY AND VISION

PARALLEL SESSION CONTRIBUTIONS

SESSION A: THE VACUUM STRUCTURE OF QCD AND THE MECHANISM OF CONFINEMENT

vii

SESSION B: LIGHT QUARKS (AND GLUONIA)

SESSION C: HEAVY QUARKS (AND GLUONIA)

SESSION D: DECONFINEMENT

POSTER SESSION

(Organizer: M. Creutz)

PREFACE

This was the sixth edition of the fortunate series of conferences "Quark Confinement and Hadron Spectrum" that every second year bring together people working on the subject from different points of view. The first conference was held in Como in 1994 and since then it has regularly taken place every two years in different locations, under various chairmen, apparently with increasing success.

This year Villasimius was chosen, an important tourist resort east of the gulf of Cagliari, on this island (the second largest of the Mediterranean sea) steeped in history and incredibly beautiful. There have been practical reasons for this choice, e.g. Gargnano (the site of the preceding edition) being too small for the increasing number of participants. As for me there was also a sentimental reason for choosing Sardinia, since I was born in Cagliari and there I attended school and did most of my University studies.

The present edition has been sponsored by the University of Milano, the University of Cagliari, Istituto Nazionale di Fisica Nucleare, Regione Sardegna, Ente Sardo Industrie Turistiche. In particular the collaboration of the colleagues of the Physics Department of the University of Cagliari and of the local section of INFN have been all important.

The scientific program was organized according to the usual scheme for invited talks, contributed papers and a poster session. The contributed papers were presented in parallel sessions and distributed over five different sections according to the topics. The selection of the papers for oral presentation and the internal organization of each section were entirely devolved to the conveners, who, together with the Organizing Committee and the Advisory Committee, also took active part in the choice of the plenary speakers.

The various sections of the present edition were as follows:

Section A: The vacuum structure of QCD and the Mechanism of confinement.
Conveners M.Faber (Vienna), M. Polikarpov (ITEP), G. Schierholz (DESY/Zeuthen).

Section B: Light Quarks.
Convener: H. Sazdjian (Orsay), M. Testa (Roma), A. Williams (Adelaide).

Section C: Heavy Quarks.
Conveners: G. Bodwin(Argonne), D. Gromes (Heidelberg), J. Soto (Barcellona).

Section D: Deconfinement.
Conveners: Y. Foka (GSI), J. Rafelski (Arizona)

Section E: QCD, New Physics and Experiments.
Conveners: J. Gates (Maryland), R. Mussa (Torino), G. Nardulli (Bari).

As usual the Poster Section was organized by M. Creutz.

On Wednesday 22 we attended an interesting round table on "Pentaquark and Exotic" before dinner, while, after dinner, E. Ribeiro presented a lively "Quark Fantasy", as an example of how abstract ideas on Particle Physics can be made attractive even to young students and the general public.

On the last day, as a synthesis of all subjects, two round tables were organized by the conveners of sections A and D, while for sections B, C and E the most traditional form of summary talks was chosen and very interesting talks were delivered by A. Williams,

J. Soto and G. Nardulli.

More time and attention than usual were dedicated to new experimental results and future perspectives, not only, as natural, in the case section D (deconfinement signature), but also of section C (heavy quark decay or production) and section E (pentaquark search).

On the all, I found scientific works extremely interesting, the discussions were very lively and the opportunities of contact among the participants were enhanced by the very nature of the place.

As regards social events, let me recall the traditional wine sampling on the occasion of the Poster section, during which we could sample various kinds of excellent wines, brought by the participants from all countries in the world. Let me also mention the boat trip, another traditional events in all Italian versions, to a lovely beach and small islands in the vicinity of Villasimius. I think the Thursday dinner in a typical "casa campidanese" was particularly interesting; we were served typical dishes of the local cuisine by hostess and waiters in traditional dress and attended a typical regional singing and dance performance.

And now few words about Sardinia. The island is what remains of a land which existed before the Alps, the Apennines and the entire Italian peninsula formed. It is the oldest geological formation in Italy. Historically, however, it has been permanently inhabited since the IV or the III millennium B.C. Apparently the people were attracted by the rich layers of obsidian, a much appreciated material at the time. The name is believed to come from the Shardana, one of the so called "people of the sea", present in the history of ancient Egypt. A first stage of cultural development was characterized by typical hypogean necropolis carved in the rock (*domus de janas*, fairy homes). Later, between the middle of the II millennium and the Roman times, it was replaced by the more interesting *nuragic culture*, with its typical villages with huts built around fortresses (the *nuraghi*), made up of various rooms and one or more towers covered as tholos. Some of these tholos were possibly even older than the famous Mycenaean tholos. There are about two thousand ruins of such towers, spread over the island, some very interesting to visit. During the VIII century B. C. the Phoenicians established a number of cities on the south-west coast of the island. These colonies (of which Karel, the present Cagliari, was the first) were the gateway to the Punic penetration and Carthage controlled the Island until the III century B. C., when, as a consequence of the wars against Rome, Sardinia became Roman and remained such until the fall of the Western Empire. Then, after a brief period during which it was partially occupied by the Vandals, it became the basis of the Byzantine army during the attempts to restore the Western Empire. Dating back to that time, there remain some simple but very interesting churches. After the conquest of North Africa by the Arabs, due to difficulties of communication, the original Byzantine administration gradually evolved into four independent states, which had close relation with the sea republics of Genoa and Pisa. In XIV century, however, the island was constituted, by a decree of the Pope, as a formally independent kingdom under the king of Aragon, before, and, later, under the king of Spain. For four centuries Sardinia was deeply bound to Spain. It was the time of the American colonization and it is interesting to know that the city of Buenos Ayres is named after the church of Bonaria in Cagliari, in which the Virgin Mary is celebrated as the protector of sailors. Finally in the middle of XVIII century, as a consequence of an international treaty, the island returned under a more direct Italian influence, passing under the sovereignty of the prince of Piemonte and becoming with Piemonte and Liguria the kernel of modern Italy.

I must express my sincere appreciation to all the participants, to the members of the Advisory Committee for their advice, to the conveners and the entire Organizing Committee for all their work. Very special thanks to Sergio Serci, director of the local section of INFN, Alberto Devoto, Umberto D'Alesio, for taking care of the local organization, to Mrs. Barbara Giua of Eurocongress and to all the administrative staff.

I would like to thank again the Rector and the Physics Department of the University of Milano, the Physics Department of the University of Cagliari, INFN and its local Section, Regione Sardegna and ESIT for the financial support that made possible the realization of this Conference and also Eurocongress Secretariat for its excellent organization.

Giovanni M. Prosperi
Chairman of the Conference

International Advisory Board

K. Baker (Jlab/Hampton)
M. Baker (Seattle)
G. Bodwin (Argonne)
M. Creutz (Brookhaven)
H. G. Dosch (Heidelberg)
G. Ecker (Wien)
E. Eichten (Fermilab)
M. Faber (Wien)
H. Georgi (Harvard)
D. Gromes (Heidelberg)
H. Leutwyler (Bern)
W. Lucha (Wien)
M. Luscher (CERN)

A. Manohar (San Diego)
G. Martinelli (Roma)
Y. Ne'eman (Tel Aviv)
M. Neubert (Cornell)
M. G. Olsson (Madison)
E. Predazzi (Torino)
E. Ribeiro (Lisbon)
F. Schöberl (Wien)
M. Shifman (Minnesota)
M. Testa (Roma)
H. Toki (RCNP Osaka)
N. A. Törnqvist (Helsinki)
F. J. Ynduráin (Madrid)

Organizing Committee

G. Prosperi (Milano) [Chair]
N. Brambilla (Milano) [Scientific Secretary]
U. D'Alesio (Cagliari)
A. Devoto (Cagliari)
K. Maung (Hampton and JLAB)
S. Serci (Cagliari)

Conference supported by

Istituto Nazionale di Fisica Nucleare (INFN)
Università degli Studi di Milano, Dipartimento di Fisica (Uni. Milano)
Università degli Studi di Cagliari, Dipartimento di Fisica (Uni. Cagliari)
INFN, Sezione di Cagliari
Regione Sardegna

Quark Fantasy

J. E. F. T. Ribeiro

Centro de Física das Interacções Fundamentais (CFIF),Departamento de Física, Instituto Superior Técnico, Av. Rovisco Pais, P-1049-001 Lisboa, Portugal

PACS: 12.38.Aw, 12.39.Ki, 12.39.Pn

Upon receiving the kind invitation to give the After Dinner Talk at the Conference, I knew immediately that I was in trouble: partly due to the outstanding scientific quality of the audience, partly due to the sheer difficulty to do something interesting enough to hold people in a room after a magnificent dinner in such beautiful surroundings.

Furthermore this talk (organizers-dixit) was supposed to be witty, scientifically interesting and...entertaining. So I decided to make a film ! To this end, I produced-via Flash Mx- a series of short clips and used Macromedia Director, not only to manage memory issues but also to add music.

The main actors of the film had to be the quarks. This is because we have by now a pretty good understanding of their role in strong interactions: exchange, annihilation and unitarity. The film lasts approximately half an hour, has 19474 frames, divided into 12 scenes, some of them in excess of 60 layers, and covers the issues of mass-gap equations; Schwinger-Dyson equations; Valatin-Bogoliubov transformations; quark exchange and annihilation, together with the graphical rules; the Adler zeros as a constraint, imposed by chiral symmetry, between annihilation (meaning hadron-hadron attraction) and exchange (meaning hadronic core repulsion). It ends with the story of the prediction of light scalar resonances as a consequence of the above quark exchange (in this case 3P0-hadron overlaps) and the use of graphical rules.

CP756, *Quark Confinement and the Hadron Spectrum VI*
edited by N. Brambilla, U. D'Alesio, A. Devoto, K. Maung, G.M. Prosperi and S. Serci
© 2005 American Institute of Physics 0-7354-0241-8/05/$22.50

➢ **PLENARY SESSION PRESENTATIONS**

➢ **ROUND TABLES AND SUMMARIES**

SESSIONS B, C: ROUND TABLE ON PENTAQUARKS AND EXOTICS
Chaired by H. Toki
SUMMARY AND VISION (J. Soto)

SESSION D: ROUND TABLE ON CONFINEMENT/DECONFINEMENT

SESSION E: SUMMARY AND VISION (G. Nardulli)

➢ **PARALLEL SESSION CONTRIBUTIONS**

SESSION A: THE VACUUM STRUCTURE OF QCD AND THE MECHANISM OF CONFINEMENT

I)   Topological configurations, monopoles, instantons, dyons, vortex structure; flux-tube formation, and QCD string; duality and strings, confinement and chiral symmetry breaking

II)   The interface between perturbative and nonperturbative QCD; renormalons and power corrections

Conveners: M. Faber (Vienna), M. Polikarpov (Itep), G. Shierholz (Desy/Zeuthen)

SESSION B: LIGHT QUARKS (AND GLUONIA)
Chiral effective theories, sum rules; Schwinger-Dyson equations; masses of light quarks; light-quark loops; phenomenology of light-hadron form factors, spectra and decays; exotics and glueballs; experiments
Conveners: H. Sazdijian (Orsay), M. Testa (Roma), A. Williams (Adelaide)

SESSION C: HEAVY QUARKS (AND GLUONIA)
Effective theories for heavy quarks (HQET, NRQCD, pNRQCD, vNRQCD); sum rules; lattice; heavy quark masses and renormalons; phenomenology of spectra and decays; glueballs; experiments
Conveners: G. Bodwin (Argonne), D. Grommes (Heidelberg), J. Soto (Barcelona)

SESSION D: DECONFINEMENT
QCD at finite temperature; quark-gluon plasma; colour superconductivity; experiments at RHIC and CERN
Conveners: Y. Foka (GSI), J. Rafelski (Arizona)

SESSION E: QCD, NEW PHYSICS AND EXPERIMENTS
Hints on the confinement mechanism from supersymmetric and string theories; precision calculations in QCD with respect to experiments and possible new physics; applications of QCD nonperturbative methods into different fields
Conveners: J. Gates (Maryland), R. Mussa (Torino) G. Nardulli (Bari)

Relativistic Nucleus-Nucleus Collisions and the QCD Phase Diagram

Reinhard Stock

University of Frankfurt, Germany

Abstract. A steep maximum occurs in the Wroblewski ratio between strange and non-strange quarks created in central nucleus-nucleus collisions, of about A=200, at the lower SPS energy $\sqrt{s} \approx$ 7 GeV. By analyzing hadronic multiplicities within the grand canonical statistical hadronization model this maximum is shown to occur at a baryochemical potential of about 450 MeV. In comparison, recent QCD lattice calculations at finite baryochemical potential suggest a steep maximum of the light quark susceptibility, to occur at similar μ_B, indicative of "critical fluctuation" expected to occur at or near the QCD critical endpoint. This endpoint hat not been firmly pinned down but should occur in the 300 MeV $< \mu_B^c <$ 700 MeV interval. It is argued that central collisions within the low SPS energy range should exhibit a turning point between compression/heating, and expansion/cooling at energy density, temperature and μ_B close to the suspected critical point. Whereas from top SPS to RHIC energy the primordial dynamics create a turning point far above in ε and T, and far below in μ_B. And at lower AGS energies the dynamical trajectory stays below the phase boundary. Thus, the observed sharp strangeness maximum might coincide with the critical $\sqrt{s}$ at which the dynamics settles at, or near the QCD endpoint. At RHIC, $\sqrt{s}$=200 GeV, one should probe deep into the deconfined phase.

Bulk hadron production systematics in central nucleus-nucleus collisions at relativistic energy is, overall, well reproduced by a statistical Hagedorn hadronic freeze-out model. A grand canonical version of this model captures the various hadronic species multiplicities, per collision event, from pions to omega hyperons, in terms of a few universal parameters that describe the dynamical stage in which the emerging hadronic matter decays to a quasi-classical gas of free resonances and hadrons [1, 2, 3]. The grand canonical parameters are temperature T, volume V and chemical potential μ. They capture a snapshot of the fireball expansion within the narrow time interval surrounding hadronic chemical freeze-out, which thus appears to populate the hadron/resonance mass and quantum number spectrum, predominantly, by phase space weight [4, 5] thus creating an apparent thermal equilibrium state prevailing in the produced hadron-resonance-population. This chemical equilibrium instantaneously decouples from fireball expansion surviving further (near isentropic) processes. It can thus be retrieved from the finally observed hadronic multiplicities, by state of the art grand canonical model analysis. This analysis succeeds from AGS, via SPS, to RHIC energy [1, 2, 3, 5]. Its main result is that T saturates at 165 ± 10 MeV from top SPS to RHIC energy ($\sqrt{s}$=200 GeV), representing the critical QCD temperature at low μ - in agreement with lattice QCD prediction.

Statistical model analysis is also applicable to elementary collisions, $p + p$, $p + \overline{p}$, and $e^+ e^-$ annihilation as was shown by Hagedorn [6] and, more recently, by Becattini and Collaborators [7, 8]. The canonical version of ensemble analysis is applicable here. Mutatis mutandis the same hadrochemical equilibrium feature is being attested,

CP756, Quark Confinement and the Hadron Spectrum VI
edited by N. Brambilla, U. D'Alesio, A. Devoto, K. Maung, G.M. Prosperi and S. Serci
© 2005 American Institute of Physics 0-7354-0241-8/05/$22.50

emphasizing the statement that the apparent equilibrium does not arise from a thermo-dynamical inelastic rescattering cascade toward equilibrium - there is essentially none in elementary processes - but should stem directly from the QCD hadronization process occuring under phase space dominance [4, 5].

The crucial difference between elementary and central nucleus-nucleus resides, in sta-tistical model view, in a transition from canonical to grand canonical order in the ensu-ing decoupled hadronic state. This transition was studied by Cleymans, Tounsi, Redlich et al. [9]. Its main feature is strangeness enhancement.Comparing the strange to non-strange hadron multiplicities in elementary, and in central nucleus-nucleus collisions at similar energy, one observes an increase of the singly strange hyperons and mesons, relative to pions, of about 2-4, and corresponding higher relative enhancements of mul-tiply strange hyperons [10, 11, 12], ranging up to order-of-magnitude enhancement. In the terminology of Hagedorn statistical models, strangeness is suppressed in the small system, canonical case, of elementary collisions (due to the dictate of local strangeness conservation in a small "fireball" volume), whereas it approaches flavour equipartition in large fireballs due to the occurence of quantum number conservation, on average only, over a large volume - as reflected by the **global** chemical potential featured by the grand canonical ensemble: "strangeness enhancement" occurs as the fading-away of canonical constraints.

From statistical model analysis we obtain a more general view of strangeness relative to non-strangeness production than is provided by considering individual strange to non-strange production ratios, like K/π, Ω/π etc., from $p + p$ to central A+A. The model quantifies strange to non-strange hadron/resonance production by means of Wroblewski quark counting at hadronic freeze-out [13]. It determines the so-called Wroblewski-ratio,

$$\lambda_s = \frac{2(<s> + <\bar{s}>)}{<u> + <\bar{u}> + <d> + <\bar{d}>} \tag{1}$$

which quantifies the overall strangeness to non-strangeness ratio at hadronic freeze-out. Strangeness enhancement (i.e. removal of strangeness suppression in elementary collisions) is quantified, by such an analysis, to proceed from $\lambda_s \approx 0.25$ in elementary collisions, to $\lambda \approx 0.45$ in central nucleus-nucleus collisions [2, 3].

Now we turn to the point of the recent data. From an energy scan conducted at the SPS by NA49, studying hadron multiplicities from $\sqrt{s}$=7 to 17 GeV, a steep maximum was observed [14] in the K^+/π and Λ/π ratios in central Pb+Pb collisions, as shown in Figs.1 and 2. As the K^+ and Λ channels carry most of the total $<s> + <\bar{s}>$ content, this experimental result indicates a kind of "singularity" in the strange to non-strange production ratio, from AGS to RHIC energy. It appears to be unlikely that state of the art hadronic or partonic quasi-classical microscopic transport models should exhibit such non-smooth behaviour which is also absent in $p + p$ collisions. In order to generalize the new NA49 data, away from consideration of individual channel strange to non-strange multiplicities, Becattini et al. [3] analyzed the $\sqrt{s}$ dependence of the Wroblewski-parameter λ_s in the grand canonical statistical hadronization model. Their result is shown in Fig.3 which gives λ_s as a function of the chemical potential μ_B. From top AGS energy (at $\mu_B \approx 550$ MeV) to RHIC energy ($\mu_B \leq 50$ MeV) one perceives an average λ_s of about 0.45 ± 0.08 whereas a steep excursion is seen, to $\lambda_s = 0.6 \pm 0.1$, at

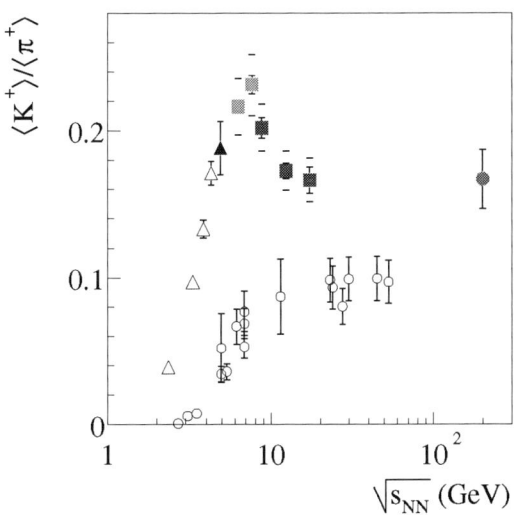

FIGURE 1. Energy dependence of $<K^+>/<\pi^+>$ ratio for central Pb+Pb (Au+Au) collisions (upper points) and p+p interactions (lower points).

μ_B=440 MeV. This point corresponds to the steep maxima observed in Figs.1 and 2, to occur at SPS fixed target energy of 30 GeV/A in central Pb+Pb collisions, corresponding to $\sqrt{s}$=7.3 GeV. The steep singularity of the NA49 K^+/π and Λ/π ratios at this $\sqrt{s}$ thus reflects in a maximum of the Wroblewski λ_s, derived from grand canonical analysis. For the sake of clarity we state here that the latter analysis just takes note, in a snapshot at hadronic freeze-out (to a decoherent hadron/resonance gas) of the general multiplicity order prevailing at the instant of hadronization. The observed λ_s maximum should, therefore, present a hint that hadronization at $\sqrt{s} \approx 7$ GeV should occur under influences, absent at energies above and below. Moreover, NA49 has shown recently [15] that the event-by-event fluctuation of the ratio $(K^+ + K^-)/(\pi^+ + \pi^-)$ measured in central Pb+Pb collisions increases steeply toward $\sqrt{s} = 7$ GeV whereas it was formerly found [16] to amount to be below 4%, at top SPS energy, $\sqrt{s} = 17.3$ GeV.

We thus propose that the dynamical trajectory of central Pb+Pb collisions comes close to the critical point of QCD, at or near $\sqrt{s} = 7$ GeV. This point has been expected to occur on the line in the T, μ_B plane which describes the boundary between the hadronic and partonic QCD phases [17]. Along that line the phase transition is expected to be a crossover at $\mu_B < \mu_B^c$, become second order at $\mu_B = \mu_B^c$, and first order for $\mu_B > \mu_B^c$. At $\mu_B = \mu_B^c$ and T = T_c we thus expect phenomena analogous to critical opalescence. Recent QCD lattice calculations succeeded in an extrapolation to finite chemical potential [18, 19], thus making a first prediction for the phase boundary line and, in particular, the critical point - albeit with considerable uncertainty as was discussed by Redlich at QM04 [20]. This uncertainty stems, firstly, from the uncertainty in the extrapolation to finite μ_B but, secondly, from the unphysical (high) strange quark mass employed in these lattice calculations which, at present, place the critical point somewhere in the interval

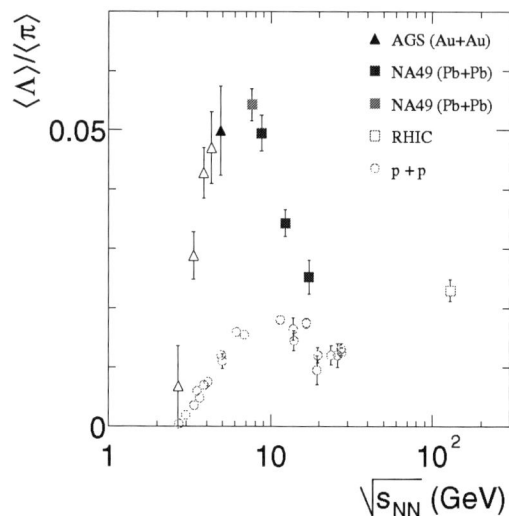

FIGURE 2. Energy dependence of the $< \Lambda > / < \pi > (< \pi >= 1.5 (< \pi^- > + < \pi^+ >))$ ratio for central Pb+Pb (Au+Au) collisions (upper points) and p+p interactions (lower points).

$500 \, \text{MeV} < \mu_B^c \leq 700 \, \text{MeV}$. Redlich argued that it should move to considerably lower μ_b once the s-mass can be chosen closer to the physical quark mass [20]. This expectation was substantiated by recent lattice calculations which show that the critical point might move downward in μ_B once more realistic quark masses are employed [21]. From Fig.3 we see that the strangeness maximum at $\sqrt{s} = 7 \, \text{GeV}$ corresponds to $\mu_B \approx 440 \, \text{MeV}$ and thus quite close to the expected μ_B^c position. Furthermore the energy density at the phase boundary is estimated by lattice QCD to be rather low [22] ($\varepsilon \leq 1 \, \text{GeV/fm}^3$).

Central collisions of heavy nuclei at SPS energy exhibit a general cycle of initial compression and heating which is followed by a maximum energy density stage which then turns into expansion and cooling [23]. The quantities characterizing the overall system dynamics, such as volume and energy-entropy density etc. change very rapidly except during the high density stage which acts analogous to a classical turning point. Also it is only during this stage that relaxation times can be of comparable magnitude to the system evolution time scales (like volume-doubling etc.). One can thus begin only here treating the system dynamical evolution in terms of energy density, temperature and chemical potential - thus defining a dynamical trajectory in the T, μ_B plane. Only if this turning point coincides closely with the QCD critical endpoint one could expect to observe substantial critical phenomena. Now it is well known that the maximum energy density in central collisions of mass 200 nuclei amounts (Bjorken estimate) to above $2 \, \text{GeV/fm}^3$ at top SPS [24], and to about $5 \, \text{GeV/fm}^3$ at RHIC [25] energies, thus overshooting, by far, the critical QCD energy density. The system will thus cross the phase coexistence line, upon re-expansion, whilst already undergoing rapid expansion. Furthermore, the chemical potential is certainly well below 300 MeV at the time of hadronization.The evolution will thus miss the critical point at top SPS, and RHIC energies; and at much lower AGS energies the dynamics falls into the $\mu_B \geq 500 \, \text{MeV}$

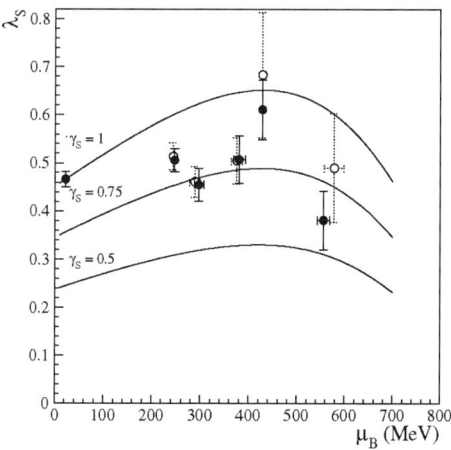

FIGURE 3. Dependence of the λ_s parameter on baryochemical potential extracted from the fits to hadron multiplicities in central Pb+Pb (Au+Au) collisions at AGS, SPS and RHIC energies. The lines show the dependence expected for different values of the γ_s parameter.

domain but the energy might not suffice to reach the phase boundary. In summary we may indeed expect that the dynamical evolution reaches its energy density plateau phase near the expected critical point (i.e. at energy density just below or at 1 GeV/fm^3, and at μ_B between 300 and 500 MeV) somewhere in between maximum AGS and minimum SPS energy.

We now turn to the final point of our line of argument by returning to the lattice results [19, 20] at finite μ_B. They see a steep maximum of the quark number susceptibility

$$\chi_{u,d} \equiv T^2 \left(\frac{d^2}{d(\mu/T)^2} \frac{p}{T^4} \right) \qquad (2)$$

occuring at T=T$_c$= 150 MeV and $\mu_B = 3\mu_Q = 3$ T = 450 MeV. We reproduce the Bielefeld-Swansea results in Fig.4 which also shows the calculations for μ_B = 225 MeV and μ_B = 0 (essentially corresponding to top SPS and RHIC energies, respectively). The latter exhibit no susceptibility peak but a smooth transition from T < T$_c$ to T > T$_c$. As $\chi_{u,d}$ can also be written as

$$\chi_q = T^2 \left(\frac{\delta}{\delta(\mu_{u/T})} + \frac{\delta}{\delta(\mu_{d/T})} \right) \frac{n_u + n_d}{T^3} \qquad (3)$$

we see that the peak in the susceptibility implies a maximum fluctuation of the quark number densities n_u and n_d. We interpret this result as an indication of critical fluctuation occuring in the vicinity of the critical endpoint implicitly present in this calculation. Directly at μ^c the susceptibility would diverge. The critical point in this calculation must thus be near μ_B = 450 MeV and T = 150 MeV. This, in turn, is very close to the

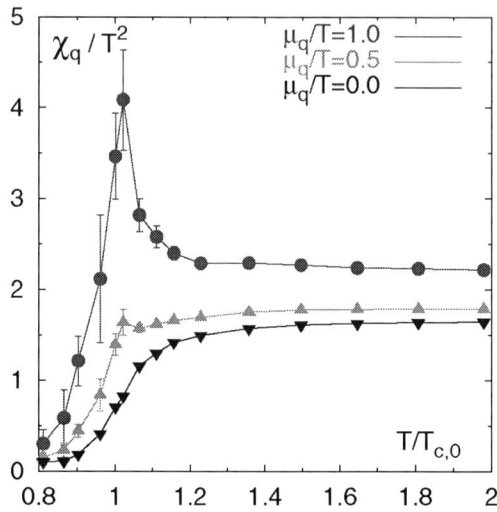

FIGURE 4. The quark number susceptibility calculated within lattice QCD as function of temperature (relative to transition temperature) for different values of quark chemical potential.

parameters of the grand canonical model at the strangeness maximum, $\sqrt{s} = 7$ GeV (Fig.3).

As to the relation between the susceptibility maximum of lattice QCD and the strangeness maximum observed by NA49 (at which the Wroblewski parameter λ_s exhibits an anomaly), Gavai and Gupta [26] have suggested the relationship

$$\lambda_s = \frac{2\chi_s}{\chi_u + \chi_d} \qquad (4)$$

which appears to offer a direct link. In fact they obtain $\lambda_s = 0.48$ from a lattice calculation at zero μ_b: closely coinciding with the value observed at top SPS and RHIC energy (Fig.3). Unfortunately, though, their result refers to $\mu_b = 0$, and the Bielefeld-Swansea calculations at finite μ_b [19] are in two-flavour QCD only. A prediction for χ_s at $\mu_B \approx 450$ MeV, or, more generally, a full three-flavour lattice treatment of the vicinity of the critical point is required to finally assess the line of argument of the present note. If proven correct we would encounter, here, the first **direct** reflection of QCD in nucleus-nucleus collision data.

It has been the aim of the previous sections to discuss hadron production for which we have comprehensive data from Bevalac/SIS to RHIC. The $\sqrt{s}$ dependence exhibits an interesting structure: an initial steep rise up to the lower SPS energies that is followed by indications of a plateau extending over the SPS domain $7 \leq \sqrt{s} \leq 17$ GeV, then by a further steep rise occuring at RHIC energies $\sqrt{s} = 130$ and 200 GeV. This overall pattern has not yet been theoretically understood. Qualitatively one might argue that the plateau structure signals phase coexistence setting in over the SPS energy range, whereas partonic, primordial dynamics becomes dominant at top RHIC energies.

Before turning to this physics I wish to note, however, that the two characteristic temperatures that describe the bulk hadronic dynamical trajectory do indeed reach saturation from top SPS to RHIC energy. The hadronization temperature (inferred from the grand canonical statistical hadronization model) saturates at about 165 ± 10 MeV, in close agreement with lattice QCD estimates of the parton-hadron phase coexistence domain [1, 2, 3, 5, 22]. Furthermore, the second characteristic "freeze-out" temperature, that describes the final bulk hadronic decoupling from strong interaction, appears to saturate at 100 ± 10 MeV. For lack of space I can not discuss here the possible excursions from this universal picture of hadronic phase expansion, as implied by hyperon data. Within such reservation the *thermal* history of bulk hadron expansion may well turn out to be universal from SPS to RHIC, while its hydrodynamical parameters (radial and elliptic flow, as well as mean p_T of m_T) reflect the increasing influence of the pre-hadronic phase, setting the stage for the ensuing hadronic expansion phase.

The above discussion has focused, implicitly, on low p_T physics. The radically new physics, offered by RHIC, stems from expanding our view to p_T up to about 15 GeV. With RHIC we thus turn from soft to hard QCD physics, approaching a situation in which observed high transverse momentum hadrons stem from primordial hard partonic rescattering as described by perturbative QCD. The qualitatively new feature (added to well known pQCD hadron production as it gets imbedded into a large primordial interaction volume) is the in-medium attenuation of the leading partons that are initially emitted in a hard partonic scattering. A colour charge propagating through a colour charged medium suffers induced radiative energy loss thus modifying the well known DGLAP evolution that describes leading parton hadronization in vacuum (in elementary collisions). This energy loss is a QCD analogy to the Landau-Pomeranchuk phenomena of QED that occur once an electric charge traverses an extended electrically charged medium. Due to quantum mechanical interference the net in-medium radiative energy loss becomes proportional to the square of the path-length L over which the propagating charge interacts in-medium. A complication arises as we are not dealing with a homogeneous, infinite volume in central nucleus-nucleus collisions: the simple L^2 law of radiative energy loss gets modified by a transport coefficient that reflects the changing local energy density as experienced by the leading parton traversing the primordial fireball volume while it expands [27].

A vision emerges, for an experimental program of the RHIC experiments (and for further studies at the CERN LHC) that could verify the basic L^2 law of leading parton in-medium energy loss. As a first step the RHIC experiments have demonstrated a universal in-medium suppression of high p_T bulk hadron production yields, as compared to elementary collisions at similar $\sqrt{s}$. The next step results from analysis of the back-to-back production pattern of jets. This primordial hard parton scattering signal offers a distinct geometrical pattern which can be related to the overall geometry of an A+A collision fireball, into which it is embedded. If a jet is created in the periphery of the primordial fireball one of the emerging leading partons may escape into free space essentially unattenuated whereas its opposite side partner traverses the entire radial extent of the reaction volume, thus being maximally attenuated. This opposite side jet quenching phenomenon, as observed by the RHIC experiments, can be quantified, both, versus reaction centrality, and with respect to the location of the impact plane. Such an analysis

7

promises to unravel the two essential parameters of jet attenuation study: the length L of the opposite side jet parton traversal through the dense medium, and the integral of the QCD transport coefficient over the entire trajectory of the emerging opposite-side jet. The first quantity is merely geometric, the second depends on a model of the radial energy density distribution of the primordial fireball, and its evolution during the time interval sampled by the opposite side leading parton while fragmenting into an eventually observed jet. Actually, a multitude of contributions to this Conference show that at RHIC jet energies, in the domain of about 10 GeV studied thus far, the opposite side jet is entirely quenched in central Au+Au collisions, whereas it gradually appears toward smaller L as encountered in semi-peripheral collisions. These, and expected further RHIC data may thus result in verification of the QCD L^2 law, characteristic of a deconfined medium.

REFERENCES

1. F. Becattini, M. Gazdzicki and J. Sollfrank, Nucl. Phys. A638 (1998) 403;
   J. Cleymans and K. Redlich, Phys. Rev. Lett. 81 (1998) 5284.
2. P. Braun-Munzinger, I. Heppe and J. Stachel, Phys. Lett. B465 (1999) 15;
   P. Braun-Munzinger, D. Magestro, K. Redlich and J. Stachel, Phys. Lett B518 (2001) 41;
   P. Braun-Munzinger, K. Redlich and J. Stachel, nucl-th/0304013;
   P. Braun-Munzinger, J. Cleymans, H. Oeschler and K. Redlich, Nucl. Phys. A697 (2002) 902.
3. F. Becattini, M. Gazdzicki, A. Keraenen, J. Manninen and R. Stock, Phys. Rev. C69 (2004) 024905.
4. R. Hagedorn, Nucl. Phys. B24 (1979) 93;
   J. Ellis and K. Geiger, Phys. Rev. D54 (1996) 1967.
5. R. Stock, Phys. Lett B456 (1999) 277;
   R. Stock, hep-ph/0312039.
6. R. Hagedorn, Nuovo Ciemento 35 (1965) 395.
7. F. Becattini, Z. Phys. C69 (1996) 485.
8. F. Becattini and L. Ferroni, hep-ph/0307061.
9. J. Cleymans et al., Phys. Rev. C56 (1997) 2747;
   A. Tounsi and K. Redlich, J. of Physics G28 (2002) 2095.
10. S. V. Afanasiev et al., NA49 Coll., Phys. Rev. C66 (2002) 054902;
    A. Mischke et al., NA49 Coll., Nuc. Phys. A715 (2003) 453.
11. L. Ahle et al., E802 Coll., Phys. Rev. C60 (1999) 064901.
12. F. Antinori et al., NA57 Coll., Nucl. Phys. A698 (2002) 118;
    F. Antinori, talk at QM04.
13. A. K. Wroblewski, Acta Phys. Pol. B16 (1985) 379.
14. V. Friese et al., NA49 Coll., nucl-ex/0305017;
    M. Gazdzicki, talk at QM04, J. Phys. G30 (2004) 701.
15. Ch. Roland et al., NA49 Coll., talk at QM04.
16. S. V. Afanasiev et al., NA49 Coll., Phys. Rev. Lett. 86 (2001) 1965.
17. M. G. Alford, K. Rajagopal and F. Wilczek, Phy. Lett B442 (1998) 247;
    Y. H. Atta and T. Ikeda, Phys. Rev. D67 (2003) 014028.
18. Z. Fodor and S. D. Katz JHEP 0203 (2002) 14.
19. R. V. Gavai and S. Gupta, hep-lat/0303013;
    C. R. Allton et al., hep-lat/0305007.
20. K. Redlich, talk at QM04, J. Phys. G30 (2004) 1271.
21. Z. Fodor and S. D. Katz, hep-lat/0402006;
    F. Karsch et al., Bielefeld-Swansea lattice Collaboration, hep-lat/0309116.
22. F. Karsch, Nucl. Phys. A698 (2002) 199.
23. R. Stock, Phys. Reports, 135 [1986] 259.
24. T. Alber et al., NA49 Coll., Phys. Rev. Lett 75 (1995) 3814.

25. G. Roland, talk at QM04, J. Phys. G30 (2004) 1381.
26. R. V. Gavai and S. Gupta, Phys. Rev. D (2002);
    R. V. Gavai, talk at QM04, J. Phys. G30 (2004) 1333;
    R. V. Gavai and S. Gupta, hep-lat/0303013,
    R. Stock, J. Phys. G30 (2004) 633.
27. A. C. Salgado and U. Wiedemann, Phys. Rev. D68 (2003) 014008.

Direct Photons as a Probe of Deconfinement in Heavy Ion Collisions

Terry C. Awes

Oak Ridge National Laboratory
Oak Ridge, TN 37830, USA

Abstract. We discuss the results and conclusions to be drawn from direct photon measurements in ultra-relativistic heavy ion collisions at the CERN SPS and at RHIC. The direct photon results provide upper limits on the initial temperature of the hot-dense matter produced in central A+A collisions which imply that the system exists in a state with a large number of degrees of freedom, as expected for a deconfined Quark Gluon Plasma. At RHIC energies the preliminary direct photon results indicate that the observed hadron suppression is a final state effect, as expected due to parton energy loss in the produced matter.

INTRODUCTION

Directly radiated thermal photons have long been considered an important penetrating probe with which to study the early phase of the hot and dense matter produced in ultra-relativistic nucleus-nucleus collisions [1, 2, 3]. Single "direct" photons are expected at high transverse momentum, p_T, from well-known hard QCD processes, but also in the p_T region below several GeV/c as a result of thermal radiation from the hot-dense matter. Since the mean free path of the produced photons is considerably larger than the size of the nuclear volume, photons produced throughout all stages of the collision will be observable in the final state. Thus, it is believed that the emitted photons should provide information about the initial conditions of the hot dense system and thereby provide evidence for the possible formation of a system of nuclear dimension of deconfined quarks and gluons, the so-called Quark Gluon Plasma (QGP).

RESULTS AT SPS ENERGIES

In general, experiments which have set out to measure direct photons in ultra-relativistic heavy-ion collisions have been done using a statistical analysis rather than by attempting to identify a direct photon sample using isolation cuts. Due to the desire to pursue the measurement to low transverse momenta and due to the large multiplicity of particles produced in a central heavy ion collision, there is a large probability to falsely attribute a photon to a π^0 due to the many combinatorial possibiliites, while the acceptance to catch both decay photons is small, with the result that the efficiency to correctly identify isolated photons in the p_T region of interest in heavy ion collisions is small.

CP756, *Quark Confinement and the Hadron Spectrum VI*
edited by N. Brambilla, U. D'Alesio, A. Devoto, K. Maung, G.M. Prosperi and S. Serci
© 2005 American Institute of Physics 0-7354-0241-8/05/$22.50

Therefore, the direct photon excess is determined on a statistical basis [4, 5, 6]: the π^0 and η p_T spectra are determined from an invariant mass analysis of photon pairs, and the inclusive photon spectrum is extracted simultaneously. The inclusive photon spectrum is then compared to the spectrum of photons expected from the decay background sources, γ^{bkgd}, which is calculated based on the measured π^0 and η yields (which nominally account for about 98% of γ^{bkgd}) with estimates of the small photon contributions from other radiative decays. In other words, the direct photon excess is extracted as $\gamma^{excess} \equiv \gamma^{obs} - \gamma^{bkgd}$.

Since the low p_T thermal photon excess is expected to be small in comparison to the known background sources, with likely signal/background ratios of 10% or less, it is imperative to minimize and accurately determine possible sources of systematic error. Measurement of the π^0 and η yields simultaneous with the inclusive photon measurement in the same detector allows to eliminate many systematic error sources such as cross section normalization errors and trigger biases, insuring that the π^0, and η yields are measured for the same event selection and in the same acceptance relevant for determination of the decay background contribution to the inclusive photon yield. Furthermore, the data sample itself can be used to estimate and limit most sources of systematic error. For example, the energy dependence of the measured π^0 mass peak may be used to set limits on the error of the energy scale.

Results with Oxygen and Sulphur Beams

The search for direct photons in ultra-relativistic heavy-ion collisions was a major emphasis of the WA80 experiment at the CERN SPS. The first results from WA80 found no significant direct photon signal in central ^{16}O+Au collisions at 60 and 200 A GeV [9]. On the other hand, the preliminary WA80 result for central ^{32}S+Au collisions suggested a direct photon excess at about the 2-sigma level[10], but the excess was found not to be signficant in the final analysis [4]. Nevertheless, using the observed excess and errors on the measurement it was possible to set an upper limit on the direct photon p_T spectrum in the region below about 3.5 GeV/c. The extracted upper limits, at the 90% confidence level, on the invariant yield of excess photons per central ^{32}S+Au collision are shown in Fig. 1 [4].

The WA80 direct photon result generated a great deal of theoretical attention [7, 8, 11, 12, 13, 14]. Most of the model calculations were consistent with the WA80 limit on the photon excess under various, rather standard, QGP formation scenarios [7, 8, 12, 13, 14]. On the other hand, calculations with scenarios in which a QGP did not form over-predicted the observed excess photon yield [7, 8]. The greater amount of thermal radiation predicted in non-QGP scenarios was due to the higher initial temperature resulting from the fewer degrees of freedom in the hadronic matter, which in these initial calculations assumed a simple pion gas. This can be seen in Fig. 1 where the upper limit data are compared with the calculations of Ref. [7] for both the QGP and pure hadron gas scenarios, and with the pure hadron gas calculations of Ref. [8].

The WA80 upper limits on the direct photon production in central ^{32}S+Au collisions allowed to place an upper limit of 250 MeV on the initial temperature [15]. This

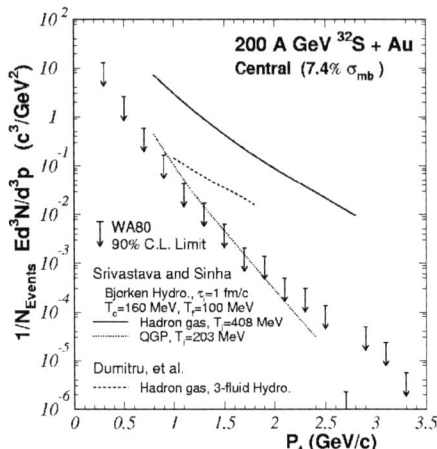

FIGURE 1. Upper limits at the 90% confidence level on the invariant excess photon yield per event for the 7.4% σ_{mb} most central collisions of 200·A GeV ^{32}S+Au. The solid curve is the calculated thermal photon production expected from a hot hadron gas taken from Ref. [7]. The dashed curve is the result of a similar hadron gas calculation taken from Ref. [8]. The dotted curve is the calculated thermal photon production expected in the case of QGP formation also taken from Ref. [7].

temperature limit in conjunction with estimates of the initial energy density implied an Equation of State (EOS) with a large number of degrees of freedom, as in a QGP, or as in a hadronic gas with a rich spectrum of resonances, but ruled out a simple gas of pions [15].

Results with Lead Beams

The WA98 experiment published the first observation of a significant direct photon signal in ultra-relativistic heavy-ion collisions. As shown in Fig. 2 a significant yield of direct photons was observed in central ^{208}Pb+^{208}Pb collisions at 158 A GeV [5, 6] in the region $p_T > 1.5$ GeV/c. In order to determine whether the observed excess is due to thermal radiation from the hot-dense matter produced in the heavy-ion collision, or instead is due to hard scatterings in the initial stage of the collision it is important to compare the observed direct photon yield to that expected for direct photon production in pp collisions.

Unfortunately, no published prompt photon results exist for pp collisions at the $\sqrt{s}$=17.3 GeV of the WA98 measurement. Instead, scaled prompt photon yields for proton-induced reactions on fixed targets at 200 GeV are shown in Fig. 2 for comparison. Results are shown from FNAL experiment E704 [16] for proton-proton reactions, and from FNAL experiment E629 [17] and CERN SPS experiment NA3 [18] for proton-carbon reactions. These results have been divided by the total pp inelastic cross section

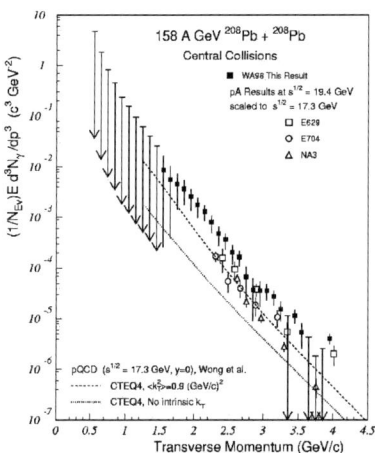

FIGURE 2. The invariant direct photon multiplicity for central 158 A GeV ^{208}Pb+^{208}Pb collisions. The error bars indicate the combined statistical and systematical errors. Data points with downward arrows indicate unbounded 90% CL upper limits. Results of several direct photon measurements for proton-induced reactions have been scaled to central ^{208}Pb+^{208}Pb collisions for comparison. The calculated results shown as curves are described in the text.

($\sigma_{int} = 30$ mb) and by the mass number of the target to obtain the invariant direct photon yield per nucleon-nucleon collision. They have then been multiplied by the calculated average number of nucleon-nucleon collisions (660) for the central Pb+Pb event selection for comparison with the WA98 measurement. This scaling is estimated to have an uncertainty of less than 10%. The proton-induced results have also been scaled from $\sqrt{s} = 19.4$ GeV to the lower $\sqrt{s} = 17.3$ GeV of the WA98 measurement under the assumption that $Ed^3\sigma_\gamma/dp^3 = f(x_T)/s^2$, where $x_T = 2p_T/\sqrt{s}$ [19]. The $\sqrt{s}$-scaling effectively reduces the 19.4 GeV proton-induced results by about a factor of two. This comparison indicates that the observed direct photon production in central ^{208}Pb+^{208}Pb collisions has a shape similar to that expected for proton-induced reactions at the same $\sqrt{s}$ but a yield which is enhanced by a factor of $\sim 2-3$.

QCD predictions of the direct photon production from initial scatterings for $\sqrt{s} = 17.3$ GeV central ^{208}Pb+^{208}Pb collisions are compared to the measured results in Fig. 2. The calculation has been scaled to central ^{208}Pb+^{208}Pb collisions with the number of binary nucleon-nucleon collisions as described above. Results are shown with and without the effects of parton intrinsic k_T by the long-dashed and short-dashed curves, respectively [20]. At this low incident energy, the parton intrinsic k_T is seen to increase the predicted photon yield by a factor which increases with decreasing p_T from about 4 to 8. The predicted direct photon yield with intrinsic k_T effects included is in good agreement with the $\sqrt{s} = 19.4$ GeV proton-induced results scaled to $\sqrt{s} = 17.3$ GeV. It is also in general agreement with the shape of the observed photon spectrum in central ^{208}Pb+^{208}Pb collisions, but underpredicts the observed yield by about a factor of 2.5. This discrepancy could be a result of further deficiencies in the prompt photon calcu-

13

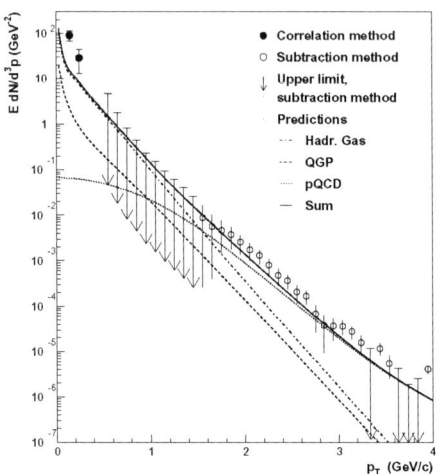

FIGURE 3. Yield of direct photons extracted by the statistical subtraction method (open circles, or arrows indicating upper limits) [5, 6]. Additional points at low p_T have been obtained from the strength of the two-photon correlation ($R_O = 0$ fm: closed circles, $R_O = 6$ fm: closed triangles). Total statistical plus systematical errors are shown. The calculations shown are from Ref. [25].

lations when applied at low incident energy, or it may indicate new effects attributable to nuclear collisions. A possible explanation might be additional p_T broadening of the incoming partons due to soft scatterings prior to the hard scattering which produces the photon [21]. Alternatively, it may be expected that the photon production is enhanced by the additional scatterings which occur as a result of rescattering in nucleus-nucleus collisions.

The central Pb+Pb direct photon results have been compared to a number of model predictions [3, 21, 22, 23, 24, 25]. In Fig. 3 the measured direct photon yield is compared with recent fireball model predictions [25]. The calculated contributions to the total yield from the Quark Gluon Plasma and hadronic stages of the collision are shown. Generally, it is found that the measured excess can be reproduced with modest initial temperatures of about 200 MeV, in which case the thermal contribution from the hadronic phase dominates over the QGP contribution at all p_T, as shown in Fig. 3, but only if an additional nuclear k_T broadening is assumed, which enhances the QCD photon contribution to reproduce the measured results at high p_T. Alternatively, without k_T broadening, the high p_T region can be reproduced but with higher initial temperatures of 250-300 MeV, corresponding to thermalization times of considerably less than 1 fm/c.

An interesting new result from the WA98 experiment is the extraction of the direct photon yield via a photon-photon correlation analysis [26]. It has been shown that photon-photon correlations can provide information about the space-time distribution of the hot matter prior to freeze-out [27]. Moreover, the correlations of direct photons of different transverse momenta will reflect different stages of the collision. Unfortunately, photon interferometry is faced with considerable difficulties compared to hadron interferometry primarily due to the small yield of photons emitted directly from the hot zone in comparison to the huge background of photons produced by the electromagnetic decay of the final hadrons (π^0's).

Under the assumption of a fully chaotic photon source, the direct photon yield N_γ^{Direct} is related to the correlation strength λ and the total inclusive photon yield N_γ^{Total} as [28]

$$N_\gamma^{Direct}/N_\gamma^{Total} = \sqrt{2\lambda} = \sqrt{8\lambda_{inv}K_T R_O/\sqrt{\pi}\,\mathrm{Erf}(2K_T R_O)}$$

where R_O is the outward radius of the emitting source and K_T is the average p_T of the pair. Although the photon R_O has not been measured, a *lower* limit on the yield of direct photons is given by the assumption $R_O = 0$. A *most probable* yield is obtained by assuming a value of $R_O = 6$ fm [29]. The low p_T direct photon yields have been extracted for these two cases and are presented in Fig. 3 (assuming $p_T = \langle K_T \rangle$) [26]. The direct photon yield at high transverse momenta obtained with the subtraction method [5, 6] (Fig. 2) is also shown. In comparison with the calculations it is seen that the contribution from the hadron gas phase dominates the direct photon yield at small p_T, with predicted yields below the measured direct photon yield.

RESULTS FROM RHIC

Pion Production and The Nuclear Suppression

One of the earliest and most exciting results from RHIC was obtained from the first year of RHIC operation. Despite very modest integrated luminosity and consequent limits on the extent of the p_T measurements, the yield of hadrons in central Au+Au collisions at moderately large transverse momenta was observed to be significantly suppressed[30], in qualitative agreement with expectations that it was a consequence of parton energy loss, or jet quenching [31].

Hard processes with small cross section are expected to occur as an incoherent process and therefore should scale with the number of binary nucleon-nucleon collisions in A+A collisions. The nuclear medium effects on particle production can be quantified by the nuclear modification factor, R_{AB}, defined for collisions of A+B as the ratio of invariant yield in A+B to that of p+p, scaled by the number of binary collisions.

$$R_{AB}(p_T) = \frac{(1/N_{AB}^{evt})\,d^2N_{AB}/d\eta dp_T}{\langle N_{coll} \rangle/\sigma_{pp}^{inel} d^2\sigma_{pp}/d\eta dp_T} = \frac{(1/N_{AB}^{evt})\,d^2N_{AB}/d\eta dp_T}{\langle T_{AB} \rangle d^2\sigma_{pp}/d\eta dp_T},$$

where $\langle N_{coll} \rangle$ is the average number of inelastic nucleon-nucleon collisions per event, and $\langle T_{AB} \rangle = \langle N_{coll} \rangle/\sigma_{pp}^{inel}$ is the average of the nuclear overlap function, which is determined purely from the nuclear geometry. Thus $\langle T_{AB} \rangle$ represents the parton luminosity and $\langle T_{AB} \rangle d^2\sigma_{pp}/d\eta dp_T$ gives the expected yield for the experimentally selected nuclear geometry, assuming no nuclear effects.

For heavy ion collisions R_{AB} is expected to be below unity at low p_T where the bulk of the particle production is due to soft processes which scale like the overlap volume,

15

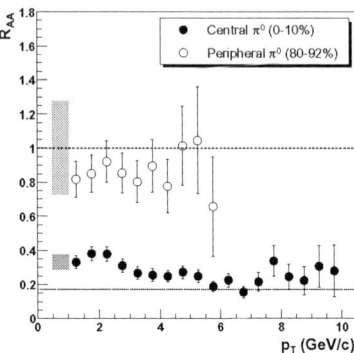

FIGURE 4. Nuclear modification factor $R_{AA}(p_T)$ for π^0 in central (closed circles) and peripheral (open circles) Au+Au at $\sqrt{s_{NN}} = 200$ GeV. The error bars include all point-to-point experimental ($p + p$, Au+Au) errors. The shaded bands represent the fractional uncertainties in $\langle T_{AuAu} \rangle$ and in the π^0 yields normalization added in quadrature, which can move all the points up or down together (in the central case the shaded band shown is the fractional error for the first point).

or number of participant nucleons $\langle N_{Part} \rangle$, rather than as $\langle N_{coll} \rangle$. At high p_T however, R_{AB} should be unity in the absence of nuclear medium effects.

The suppression effect as observed in the PHENIX experiment is shown in Fig. 4 where the neutral pion spectra R_{AA} are shown for peripheral and central Au+Au collisions at $\sqrt{s_{NN}} = 200$ GeV [32]. It is observed that while the peripheral Au+Au results are consistent with the scaled p+p π^0 results ($R_{AA} \approx 1$), the central Au+Au yields are significantly below the scaled expectations, indicating a strong nuclear suppression.

Although the suppression of the high p_T spectra in central A+A collisions was a predicted consequence of energy loss of partons produced in the initial hard scatterings as they traverse the hot-dense matter produced in a central A+A collision, it was alternatively suggested that the suppression could be due to a decreased number of initial hard scatterings due a reduction, or saturation, of the initial gluon density in heavy nuclei [33]. In order to distinguish between an initial state explanation and a final state explanation for the observed suppression, d+Au collisions at $\sqrt{s_{NN}} = 200$ GeV were studied at RHIC during Run 3. While initial state nuclear effects are present in the d+Au system (although to a lesser extent than in the Au+Au system), there will be no large hot-dense medium produced. Thus if the hadron suppression is due to parton energy loss in the produced matter, the suppression should not be observed in d+Au collisions.

In fact, it was observed that the high p_T particles in d+Au collisions are not suppressed [34]. The d+Au results strongly indicate that the suppression in central Au+Au collisions is not an initial state effect, but instead is most likely a final state effect due to the produced dense medium.

Because photons produced in the initial Compton-like quark-gluon scatterings will be sensitive to the initial gluon density, but will hardly be affected by the surrounding dense matter in the central heavy-ion collision, they can provide alternative direct information on whether or not there is saturation of the initial gluon density.

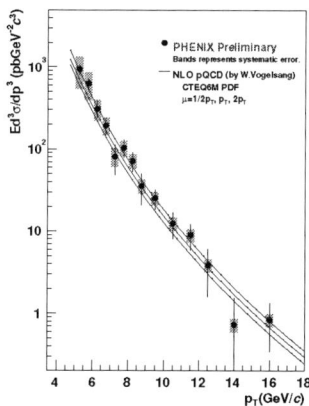

FIGURE 5. Preliminary $p + p \rightarrow \gamma + X$ at $\sqrt{s} = 200$ GeV measured by PHENIX.

Direct Photons at RHIC

The difficulty to interpret nucleus-nucleus collision results without equivalent pp measurements, or reliable QCD predictions, for comparison was made apparent in the above discussion of results from the SPS. At RHIC, pp measurements have been made at $\sqrt{s} = 200$ GeV with similar kinematic coverage for comparison to the A+A results. Moreover, NLO QCD predictions[35] have been shown to provide an accurate description of the π^0 p_T spectra measured in pp collisions[36] as well as preliminary measurements of the direct photon p_T spectrum shown in Fig. 5 [37].

Since the majority of photons observed at a given p_T originate from the decay of π^0's at nearly the same p_T, it is convenient to first investigate the ratios $(\gamma/\pi^0)^{\mathrm{obs}}$ and $(\gamma/\pi^0)^{\mathrm{bkgd}}$, which are less sensitive to systematic error [9]. Furthermore, it is useful to study the ratio $(\gamma/\pi^0)^{\mathrm{obs}}/(\gamma/\pi^0)^{\mathrm{bkgd}}$ which indicates the fraction of photons observed relative to the expected decay background and should have a value of 1 if there are no excess photons. The preliminary PHENIX $(\gamma/\pi^0)^{\mathrm{obs}}/(\gamma/\pi^0)^{\mathrm{bkgd}}$ ratio for the 10% most central Au+Au collisions at $\sqrt{s_{NN}} = 200$ GeV is shown in Fig. 6. The ratio is seen to increase with p_T to much greater than 1 indicating a significant direct photon excess.

The expected ratio in the case that both the direct photon yield and the π^0 yield, with corresponding background decay photon yield, are taken from NLO QCD predictions (Fig. 5), scaled by the number of binary nucleon-nucleon collisions, is shown by the lower solid curve in Fig. 6 and is seen to underpredict the observed ratio. However, this calculation overestimates the decay photon contribution since the π^0 (and η) yield is strongly suppressed in central Au+Au collisions as discussed above (see Fig. 4). If the measured π^0 and η results are instead used in the decay photon background calculation the observed direct photon yield is seen to be in excellent agreement with QCD predictions. This observation strongly indicates that the initial hard-scattering rate is as predicted by QCD, and that the initial parton distributions are not significantly modified in the Au nuclei. This is strong support of the conclusion that the observed

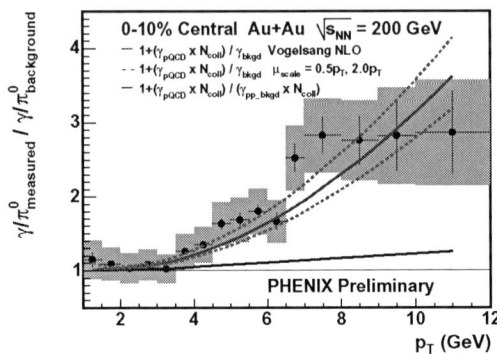

FIGURE 6. Preliminary PHENIX result on the double ratio of measured $(\gamma/\pi^0)_{Measured}$ invariant yield ratio to the background decay $(\gamma/\pi^0)_{Background}$ ratio as a function of p_T for 10% most central Au+Au collisions at $\sqrt{s_{NN}} = 200$ GeV. Statistical and systematical errors are indicated separately. The solid curves are the ratio of QCD predictions described in the text to the background photon invariant yield. The shaded regions indicate the variation of the QCD calculation for scale changes from $p_T/2$ to $2p_T$.

suppression of the hadron spectra is due to the hot dense matter produced in the final state and not due to an initial state effect such as gluon saturation.

The good agreement betwen the measured preliminary direct photon yield and QCD predictions suggests that the thermal photon contributions are small relative to the QCD predictions. It is interesting to note that 30-50% of the photons in this p_T region result from fragmenting partons, rather than from direct scatterings. If the partons do lose energy in the dense matter prior to fragmentation it would be expected that the p_T spectrum of the photons from fragmentation would be reduced, contrary to observation. On the other hand, it has been suggested that there may be additional Bremsstrahlung photons produced as the partons propagate through the dense matter of the final state [38].

CONCLUSION

Direct photon measurements in ultra-relativistic heavy ion collisions have begun to fulfill their promise as sensitive probes with which to diagnose the production of a deconfined state. Already since the first measurements at the SPS for S+Au collisions, the observed low yield of direct photons in central A+A collisions allows to place upper limits on the initial temperature of the hot-dense matter produced. This observation together with the measured large deposited energies allows to infer that the system has access to a large number of degrees of freedom, as for a deconfined QGP state, or a highly complex resonance gas.

The PHENIX preliminary direct photon result for central Au+Au collisions at RHIC is found to be in good agreement with QCD predictions for the yield expected from initial hard-scatterings alone. This result strongly supports the conclusion that the observed large suppression of the hadron spectra is due to parton energy loss as it propagates through the surrounding dense matter prior to fragmentation.

ACKNOWLEDGMENTS

I wish to express my gratitude to my colleagues of the WA80, WA98, and PHENIX collaborations. ORNL is managed by UT-Battelle, LLC, for the U.S. Department of Energy under contract DE-AC05-00OR22725.

REFERENCES

1. Shuryak, E. V., *Phys. Lett.*, **B78**, 150 (1978).
2. McLerran, L. D., and Toimela, T., *Phys. Rev.*, **D31**, 545 (1985).
3. Peitzmann, T., and Thoma, M. H., *Phys. Rept.*, *364*, 175–246 (2002).
4. Albrecht, R., et al., *Phys. Rev. Lett.*, 76, 3506–3509 (1996).
5. Aggarwal, M. M., et al., *Phys. Rev. Lett.*, **85**, 3595–3599 (2000).
6. Aggarwal, M. M., et al. (2000).
7. Srivastava, D. K., and Sinha, B., *Phys. Rev. Lett.*, **73**, 2421–2424 (1994).
8. Dumitru, A., et al., *Phys. Rev.*, **C51**, 2166–2170 (1995).
9. Albrecht, R., et al., *Z. Phys.*, **C51**, 1–10 (1991).
10. Santo, R., et al., *Nucl. Phys.*, **A566**, 61c–68c (1994).
11. Shuryak, E. V., and Xiong, L., *Phys. Lett.*, **B333**, 316–319 (1994).
12. Neumann, J. J., Seibert, D., and Fai, G. I., *Phys. Rev.*, **C51**, 1460–1464 (1995).
13. Arbex, N., Ornik, U., Plumer, M., Timmermann, A., and Weiner, R. M., *Phys. Lett.*, **B345**, 307–312 (1995).
14. Srivastava, D. K., and Sinha, B. C., *Eur. Phys. J.*, **C12**, 109–112 (2000).
15. Sollfrank, J., et al., *Phys. Rev.*, **C55**, 392–410 (1997).
16. Adams, D. L., et al., *Phys. Lett.*, **B345**, 569–575 (1995).
17. McLaughlin, M., et al., *Phys. Rev. Lett.*, **51**, 971 (1983).
18. Badier, J., et al., *Z. Phys.*, **C31**, 341 (1986).
19. Owens, J. F., *Rev. Mod. Phys.*, **59**, 465 (1987).
20. Wong, C.-Y., and Wang, H., *Phys. Rev.*, **C58**, 376–388 (1998).
21. Dumitru, A., Frankfurt, L., Gerland, L., Stocker, H., and Strikman, M., *Phys. Rev.*, **C64**, 054909 (2001).
22. Huovinen, P., Ruuskanen, P. V., and Rasanen, S. S., *Phys. Lett.*, **B535**, 109–116 (2002).
23. Srivastava, D. K., and Sinha, B., *Phys. Rev.*, **C64**, 034902 (2001).
24. Steffen, F. D., and Thoma, M. H., *Phys. Lett.*, **B510**, 98–106 (2001).
25. Turbide, S., Rapp, R., and Gale, C., *Phys. Rev.*, **C69**, 014903 (2004).
26. Aggarwal, M. M., et al., *Phys. Rev. Lett.*, **93**, 022301 (2004).
27. Srivastava, D. K., and Kapusta, J. I., *Phys. Rev.*, **C48**, 1335–1345 (1993).
28. Peressounko, D., *Phys. Rev.*, **C67**, 014905 (2003).
29. Aggarwal, M. M., et al., *Phys. Rev.*, **C67**, 014906 (2003).
30. Adcox, K., et al., *Phys. Rev. Lett.*, **88**, 022301 (2002).
31. Gyulassy, M., and Plumer, M., *Phys. Lett.*, **B243**, 432–438 (1990).
32. Adler, S. S., et al., *Phys. Rev. Lett.*, **91**, 072301 (2003).
33. Kharzeev, D., Levin, E., and McLerran, L., *Phys. Lett.*, **B561**, 93–101 (2003).
34. Adler, S. S., et al., *Phys. Rev. Lett.*, **91**, 072303 (2003).
35. Gordon, L. E., and Vogelsang, W., *Phys. Rev.*, **D50**, 1901–1916 (1994).
36. Adler, S. S., et al., *Phys. Rev. Lett.*, **91**, 241803 (2003).
37. Frantz, J., *J. Phys.*, **G30**, S1003–S1006 (2004).
38. Fries, R. J., Muller, B., and Srivastava, D. K., *Phys. Rev. Lett.*, **90**, 132301 (2003).

Heavy and Light:
New Results from CLEO III and CLEO-c

Jim Napolitano* and the CLEO Collaboration†

*Rensselaer Polytechnic Institute, Troy, NY 12180 USA
†Wilson Laboratory, Cornell University, Ithaca NY, USA

Abstract. We discuss new and developing results from the CLEO III and CLEO-c experiments. The focus of this talk is on hadron spectroscopy and strong interaction dynamics, including various types of decays which bear directly on calculations in Lattice QCD.

INTRODUCTION

The CLEO experiment at Cornell University has been producing results for nearly 25 years [1], utilizing e^+e^- colliding beams at the CESR storage ring. The most recent version of the experiment to take data at high energy, CLEO III, included a new drift chamber [2] and a ring imaging Čerenkov detector for particle identification [3]. Currently, the detector is in a configuration we call CLEO-c [4] to focus on physics in the charm threshold region. This is accompanied by a CESR upgrade to include an array of wiggler magnets which enhance luminosity for running at low energies.

In this talk, we present our most recent results. These include decays of the narrow Υ states with CLEO III, decays of the $\psi(2S)$ with CLEO III and CLEO-c, and decays of charged and neutral D mesons with CLEO-c. We also briefly discuss some plans for physics with a large sample of J/ψ decays.

COLOR SINGLET VERSUS COLOR OCTET IN $\Upsilon(1S) \rightarrow J/\psi + X$

The charmonium production mechanism in hadronic reactions is poorly understood. The first observations by CDF [5, 6, 7] in $p\bar{p}$ collisions indicated a factor of 10 to 50 times higher cross section than first predicted, based on the assumption that the $c\bar{c}$ pair are produced directly as a color singlet. Braaten and Fleming [8] proposed instead a color octet mechanism which could enhance the cross section. In this case, the $c\bar{c}$ is produced by a single gluon, and later sheds soft gluons (nonperturbatively) to evolve eventually in a color singlet state. However, the matrix elements derived from a fit to the data do a poor job predicting the recently observed J/ψ polarization at CDF [9]. Furthermore, predictions of $c\bar{c}$ production in γp [10] and e^+e^- [11, 12, 13] do not agree well with experiments [14, 15, 16]. Finally, at this conference, Mike Leitch has emphasized the importance of understanding the J/ψ hadronic production mechanism for RHIC experiments.

CP756, *Quark Confinement and the Hadron Spectrum VI*
edited by N. Brambilla, U. D'Alesio, A. Devoto, K. Maung, G.M. Prosperi and S. Serci
© 2005 American Institute of Physics 0-7354-0241-8/05/$22.50

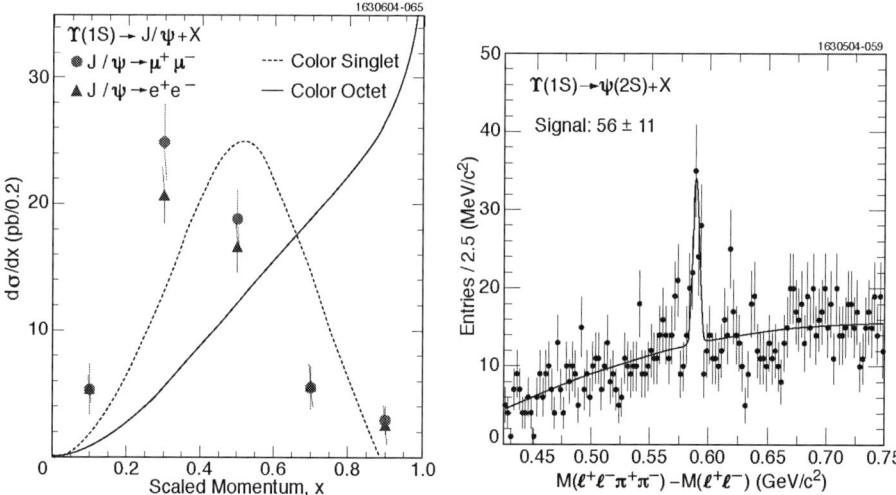

FIGURE 1. Results for inclusive charmonium production in $\Upsilon(1S)$ decay. The momentum spectrum for J/ψ production is shown on the left, where $x \equiv p_{J/\psi}/p_{J/\psi}^{\mathrm{max}}$. The plot on the right shows our signal for $\psi(2S)$ production.

Inclusive charmonium production in $\Upsilon(1S)$ decay is another way to discriminate between color octet and color singlet mechanisms [17, 18, 19, 20]. Because of the additional charmed particles produced in the color singlet mechanism, the J/ψ momentum spectrum should be significantly softer in this case. Using data from CLEO III, we have completed a new measurement [21] of inclusive J/ψ, $\psi(2S)$, and χ_{c_j} in $\Upsilon(1S)$ decay.

Some of these results are shown in Fig. 1. We find a branching ratio for $\Upsilon(1S) \to J/\psi + X$ of $(6.4 \pm 0.4 \pm 0.6) \times 10^{-4}$, which is consistent with *either* color octet or color singlet production. However, the J/ψ momentum spectrum is very soft, even more so than predicted in the color singlet model. Clearly more theoretical work needs to be done in order to understand the mechanism. Our $\Upsilon(1S) \to \psi(2S) + X$ signal is also shown in Fig. 1 yielding a branching ratio of $(41 \pm 11 \pm 8)\%$ relative to J/ψ. We also determine relative branching ratios of $(35 \pm 8 \pm 6)\%$ and $(52 \pm 12 \pm 9)\%$ for χ_{c_1} and χ_{c_2} respectively. These are first observations of $\psi(2S)$ and χ_c production in $\Upsilon(1S)$ decay. We have plans to search for associated production of charmed particles in this data, which would be a necessary signature for the color singlet mechanism.

BRANCHING RATIOS FOR $\Upsilon(NS) \to \mu^+\mu^-$

The total widths of the $\Upsilon(1S)$, $\Upsilon(2S)$, and $\Upsilon(3S)$ resonances, which are below open beauty threshold and therefore narrow, provide useful constraints on strong interaction decay models. These widths are too small to be measured directly. However, by combining a measurement of their branching ratio to $\mu^+\mu^-$ with the partial width to e^+e^-

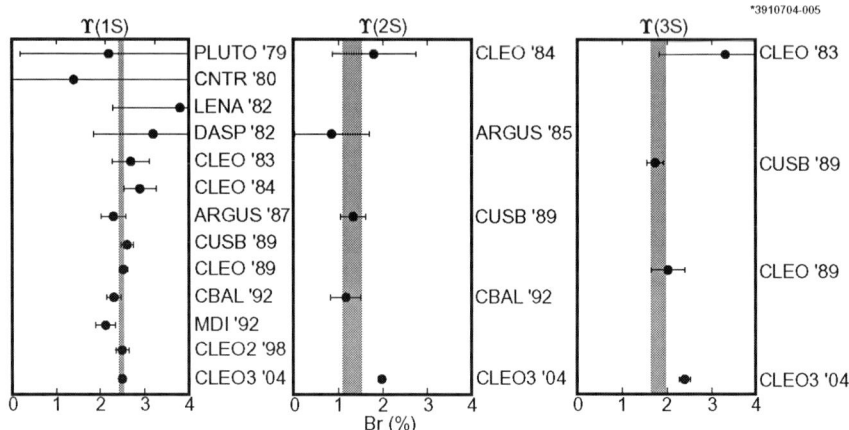

FIGURE 2. Our current measurements (CLEO3 '04) of the $\mu^+\mu^-$ branching ratios for the narrow $\Upsilon(nS)$ resonances, compared with previous results. These values are used to determine the total widths.

(determined from the total cross section in e^+e^- annihilation), the total widths can be found by invoking lepton universality.

CLEO-III has completed a new measurement of these $\mu^+\mu^-$ branching ratios [22]. (A new measurement of the cross section is in progress.) Our results are shown in Fig. 2. We find $\mathcal{B}_{\mu\mu}(1S) = (2.49 \pm 0.02 \pm 0.07)\%$, $\mathcal{B}_{\mu\mu}(2S) = (2.03 \pm 0.03 \pm 0.08)\%$, and $\mathcal{B}_{\mu\mu}(3S) = (2.39 \pm 0.07 \pm 0.10)\%$. The new values are considerably more precise than the existing measurements, and essentially consistent with them. We do, however, see a rather large shift in the value for $\Upsilon(2S)$ decay. Combined with currently available total cross section information, these branching ratios imply total widths $\Gamma(1S) = 52.8 \pm 1.8$ keV, $\Gamma(2S) = 29.0 \pm 1.6$ keV, and $\Gamma(3S) = 20.3 \pm 2.1$ keV.

DECAYS OF THE $\psi(2S)$

We have collected 3.1×10^6 $\psi(2S)$ decays, about half each in the CLEO-III and CLEO-c configurations. This talk mentioned several analyses in progress, including the inclusive photon spectrum [23] and multibody decays [24]. One analysis, the search for the h_c, has since yielded a preliminary result [25].

The talk focussed in particular on $\psi(2S) \rightarrow$ vector $+$ pseudoscalar (VP) [26] and that is what is presented here. An important motivation is the "$\rho\pi$ puzzle", the observation that $\mathcal{B}(\psi(2S) \rightarrow \rho\pi)$ is much smaller than one expects by scaling $\mathcal{B}(J/\psi \rightarrow \rho\pi)$ by the $\mu^+\mu^-$ branching fractions. (This goes by the name "12% rule".) This investigation seeks to not only provide a significant measurement of the $\rho\pi$ branch, but also to investigate other VP decays, including those that are isospin forbidden, and compare them to the corresponding J/ψ decays.

One key aspect of our measurement is that we acquired about four times as much

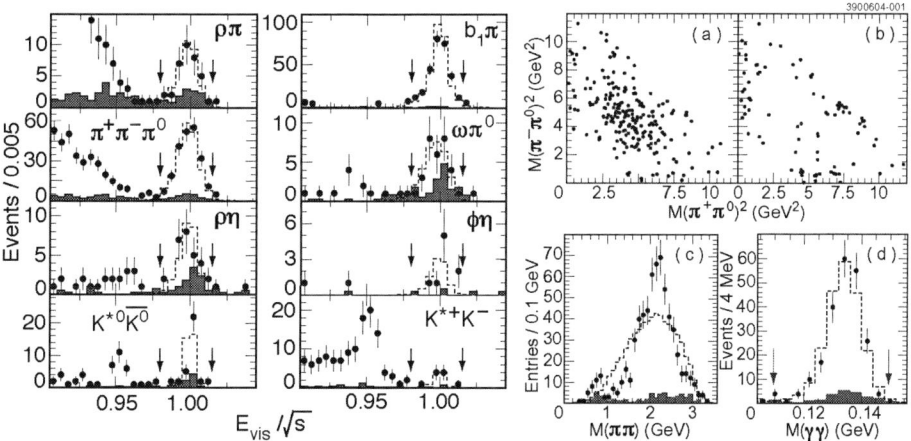

FIGURE 3. *Left:* Signals for various VP decay modes of the $\psi(2S)$, along with the full $\pi^+\pi^-\pi^0$ signal from which $\rho\pi$ is obtained, as well as the $b_1\pi$ "calibration" mode. On-resonance data is plotted as points with error bars, as a function of the total visible energy in the event, scaled by $\sqrt{s}$. Limits of the signal region in this variable are indicated by vertical arrows. The shaded histograms are from data taken slightly below the $\psi(2S)$ resonance, scaled by luminosity. The dashed histogram is a Monte Carlo simulation. *Right:* Focus on $\pi^+\pi^-\pi^0$ including the difference between on-resonance and continuum distributions.

integrated luminosity at an energy just below the $\psi(2S)$ as on resonance. This allowed us to accurately measure the "continuum" background contribution to any signal. We were also able to extract continuum cross sections and momentum distributions.

Various event topologies were allowed in the event selection, including several modes which required $\pi^0 \to \gamma\gamma$ and $\eta \to \gamma\gamma$ detection. We also acquired the $\omega\pi^+\pi^-$ final state so that $\psi(2S) \to b_1^{\pm}\pi^{\mp}$ could be checked for consistency with previous measurements.

The signal for our various final state topologies are shown in Fig. 3. Extracted branching ratios (and continuum cross sections) are given in [26]. This represents the first evidence for nonzero $\psi(2S)$ decay branches to $\pi^+\pi^-\pi^0$, $\rho\pi$, $\rho\eta$, and $K^{*0}\bar{K}^0$. The $b_1\pi$ branch agrees well with previous measurements. Furthermore, although $\rho\pi$ dominates the $\pi^+\pi^-\pi^0$ topology for the continuum, it is a small contribution to $\psi(2S)$ decay.

Based on these measurements, we conclude that isospin allowed VP decays all violated the "12% rule". On the other hand, one observes rough agreement with this guideline for isospin forbidden decays.

STRONG PHYSICS FROM WEAK DECAYS OF D MESONS

Valuable information on strong interaction dynamics can be obtained from the weak decays of D mesons. Much of the CLEO-c program is focussed on acquiring several million events of the type $e^+e^- \to D^+D^-$ and $e^+e^- \to D^0\bar{D}^0$ from running at the $\psi(3770)$ resonance. This will be a particularly clean data set because, when necessary, events can be highly constrained by "tagging" one of the pair of D mesons in order to study the decay of its partner.

Three body decays

We are currently obtaining large samples of three body D meson decays. One area of investigation is the resonant structure in the Dalitz plot which may be due to scalar mesons decaying to $\pi\pi$ or $K\pi$ final states. There was much interest in this topic at the conference. No preliminary results are available at this time, however.

Form factors for $D^0 \to K^- e^+ \nu_e$ and $D^0 \to \pi^- e^+ \nu_e$

The rate for semileptonic D^0 decay[1] is given by

$$\frac{d\Gamma}{dq^2}(D^0 \to h^- e^+ \nu_e) = \frac{G_F^2}{24\pi^3} p_h^3 |V_{cq}|^2 |F(q^2)|^2$$

where, for this discussion, we take $h = K(\pi)$, $q = s(d)$, p_h is the momentum of the recoiling hadron, V_{cq} is the relevant CKM matrix element, q^2 is the invariant mass of the $e^+ \nu_e$ system, and $F(q^2)$ is the hadronic form factor at the $c \to Wq$ vertex. The form factor can be predicted from a number of different theoretical approaches [27, 28, 29] including Lattice QCD [30].

Of all the D^0 semileptonic decays, $D^0 \to \pi^- e^+ \nu_e$ provides the greatest range of q^2. However, this is a challenging measurement because $|V_{cd}|^2 \ll |V_{cs}|^2$ and there is a large potential background from $D^0 \to K^- e^+ \nu_e$. Experimenters must rely on good particle identification and/or kinematic constraints to reduce this background.

High energy ($\sqrt{s} \sim 10$ GeV) data from CLEO-III [31] has been used to complete a recent measurement of the form factors in $D^0 \to K^- e^+ \nu_e$ and $D^0 \to \pi^- e^+ \nu_e$ decay. See Fig. 4. At these energies, D^0 mesons are identified based on their formation through continuum production of $D^{\star+} \to \pi^+_{\text{slow}} D^0$. The neutrino is reconstructed by identifying all other charged and neutral particles in the event and forming the effective $D^{\star+} - D^0$ mass difference ΔM. This is the quantity plotted in the left hand panel of Fig. 4, for three different bins of q^2, for each of the two decays.

In each case "non peaking" backgrounds appear from a variety of sources. Peaking backgrounds appear in the ΔM spectra because of misidentified π^- in $D^0 \to K^- e^+ \nu_e$ decay, which is a small contribution, and because of misidentified hadrons in $D^0 \to \pi^- e^+ \nu_e$, which is a much larger contribution. These relative contributions are determined by simultaneously fitting the distributions, with a good understanding of the Monte Carlo simulations and firm control of the systematic errors. The resulting form factors, for each decay in the three q^2 bins, are shown on the right. Despite the challenging measurement, there is some discrimination between the various

---

[1] Charge conjugation is assumed throughout this paper whenever relevant.

24

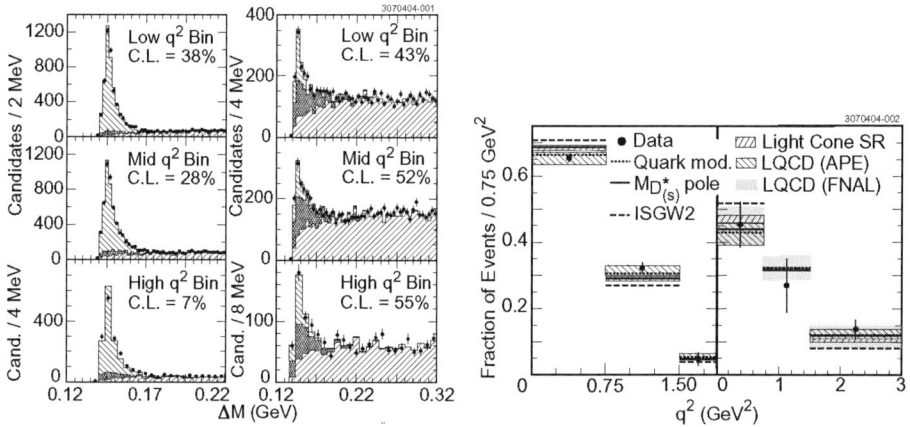

FIGURE 4. CLEO-III results for $D^0 \rightarrow \{K^-, \pi^-\}e^+\nu_e$. See text for details.

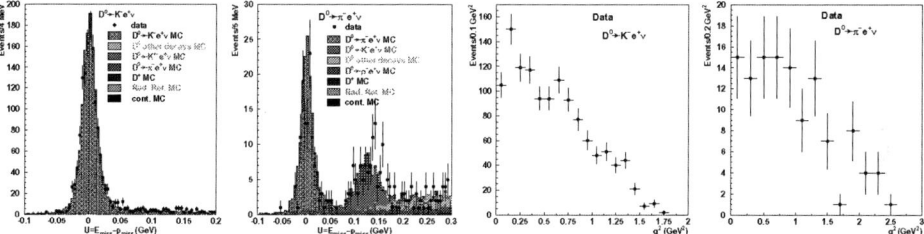

FIGURE 5. CLEO-c preliminary results for $D^0 \rightarrow \{K^-, \pi^-\}e^+\nu_e$. See text for details.

theoretical calculations. The experiment also determines $\mathscr{B}(\pi^-e^+\nu_e)/\mathscr{B}(K^-e^+\nu_e) = 0.082 \pm 0.006 \pm 0.005$.

The power of CLEO-c is nicely demonstrated in our preliminary measurement of these decays [32] using only a few percent of our anticipated data volume. See Fig. 5. The left hand panel shows the kinematic separation from background for each of the two decay modes. No particle identification is in fact necessary. The statistical quality of the form factor measurement is shown on the right, although no corrections have been applied to the spectra in this figure.

The D^+ decay constant in $D^+ \rightarrow \mu^+\nu_\mu$

Meson decay constants, f_X for pseudoscalar meson X, are in a class of measurements that should be calculable at the 1% level Lattice QCD [33, 34]. This quantity essentially measures the overlap of the $q_1\bar{q}_2$ wave function in the meson, and is determined from the purely leptonic decay mode $X^+ \rightarrow \ell^+\nu_\ell$. This decay is helicity suppressed, so the rate is generally quite small, leading to small branching fractions and difficult measurements.

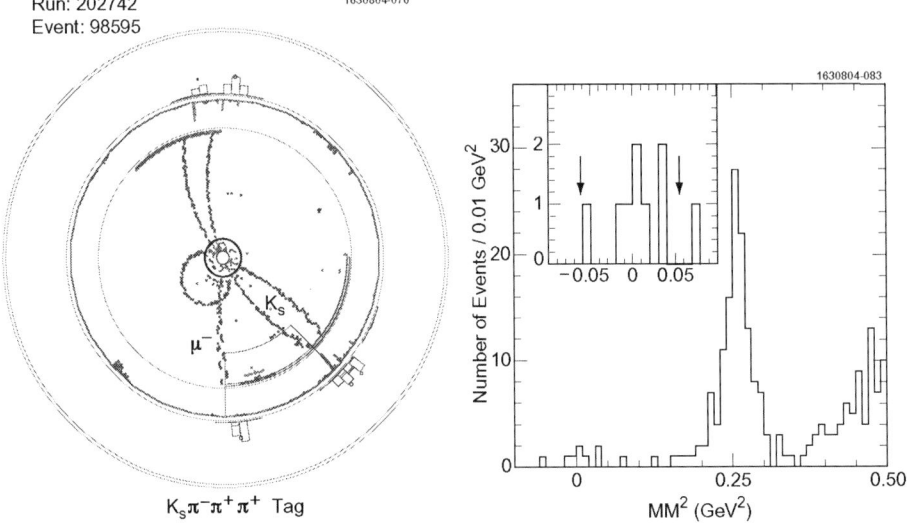

Run: 202742
Event: 98595

1630804-076

1630804-083

$K_S\pi^-\pi^+\pi^+$ Tag

MM^2 (GeV^2)

Number of Events / 0.01 GeV^2

FIGURE 6. CLEO-c measurement of $D^+ \rightarrow \mu^+\nu_\mu$. A typical event showing the decay of a tagged D meson is shown on the left. The histogram on the right shows our signal of eight events, cleanly separated from the large number of $D^+ \rightarrow \pi^+K_L^0$.

The general expression for the decay rate is

$$\Gamma(X^+ \rightarrow \ell^+\nu_\ell) = \frac{G_F^2}{8\pi}f_X^2 m_\ell^2 M_X \left(1 - \frac{m_\ell^2}{M_X^2}\right)^2 |V_{q_1q_2}|^2$$

Using only the small amount of data we have collected so far, CLEO-c has determined a preliminary value for f_{D^+} using the $\mu^+\nu_\mu$ decay of tagged D^+ mesons [35]. An example of such an event tagged with $D^+ \rightarrow K_S\pi^+\pi^+\pi^-$, and observing $D^- \rightarrow \mu^-\bar{\nu}_\mu$, is shown in Fig. 6. The $K_S(\rightarrow \pi^+\pi^-)$ and μ^- are indicated.

CLEO-c employs no particle identification capable of distinguishing $\mu^\pm$ from $\pi^\pm$, so we must rely on other other methods of removing backgrounds. Particularly insidious is $D^+ \rightarrow \pi^+\pi^0$ (0.13% branching ratio) because it is very hard to resolve the low π^0 mass from a massless neutrino. We apply hard cuts on photons from $\pi^0 \rightarrow \gamma\gamma$ (studied thoroughly with Monte Carlo simulation) to remove this background. Another important source of background is $D^+ \rightarrow \tau^+\nu_\tau$ (branching ratio 3.2× that for $\mu^+\nu_\mu$) where $\tau^+ \rightarrow \pi^+\bar{\nu}_\tau$ (11%), and simulations are used to determine its contribution to our signal.

Figure 6 also histograms the missing mass in events with $D^+ \rightarrow \mu^+\nu_\mu$ candidates. The large peak at $MM^2 \approx 0.25$ GeV^2 is due to $D^+ \rightarrow \pi^+K_L^0$, but is well separated from the signal region near $MM^2 = 0$. This signal region is expanded in the inset of the figure, showing a total of eight events. Residual backrounds from $D^+ \rightarrow \pi^+\pi^0$ (0.31 events), $D^+ \rightarrow \tau^+\nu_\tau$ (0.36 events), $D^+ \rightarrow \pi^+K_L^0$ (0.06 events), and others add up to approximately one event. We find $\mathscr{B}(D^+ \rightarrow \mu^+\nu_\mu) = (3.5 \pm 1.4 \pm 0.6) \times 10^{-4}$ and

$f_{D^+} = (201 \pm 41 \pm 17)$ MeV. We emphasize that this is based on only a few percent of our anticipated data sample.

MEASURING THE SCALAR MESON MASS MIXING MATRIX

We conclude with a discussion of an important part of the planned program for CLEO-c, a part which goes to the hard of hadronic physics and confinement.

CLEO-c will collect $\approx 10^9$ J/ψ decays in the latter part of our approved running period. A prime motivation for collecting this data is to shed some light on the prospect of glueball production in the decay $J/\psi \rightarrow \gamma\mathcal{G}$, where $\mathcal{G}$ represents a generic state dominate by a Fock space wave function of two gluons. This idea stimulated much early work in this process [36], and relatively recent calculations have been done to estimate the rate for glueball production in such decays [37].

The experimental situation regarding glueballs is rather murky [38]. Candidates generally are states produced in a "glue rich" reaction, and which appear to overpopulate states predicted by the quark model. The oldest glueball candidate is the pseudoscalar $\eta(1440)$, now recognized as two states $\eta(1405)$ and $\eta(1475)$. It was first observed in J/ψ radiative decay. Also under active study today are the scalar mesons $f_0(1500)$ and $f_0(1710)$ which both show dynamics associated with glueball production and which together also overpopulate the quark model. (The tensor glueball candidate $f_0(2220)$ has not survived recent scrutiny and needs to be confirmed or refuted.) There is no clear experimental consensus on the gluonic content of these mesons.

The theoretical situation is clearer, mainly thanks to advances in Lattice QCD [39, 40, 41]. It seems clear that the lightest glueball should be a scalar and that its *bare* mass should be around 1600 to 1700 MeV. Consequently, the prevailing wisdom at present is that the $f_0(1500)$ and $f_0(1710)$ are mixtures of bare quark model and glueball states, although the nature and amount of that mixing remains the subject of much debate.

CLEO-c seeks to help settle this debate experimentally. There are three well established scalar meson states in the 1300-1700 MeV mass region, whereas the quark model can only account for two of them. (The third state is the wide $f_0(1370)$.) One writes the quark model basis states as $|n\bar{n}\rangle \equiv [u\bar{u} + d\bar{d}]/\sqrt{2}$ and $|s\bar{s}\rangle$. Close, Donnachie, and Kalashnikova [42] have calculated the radiative decay of these scalar states to vector mesons ρ and ϕ, depending on different mixing scenarios. Their results are shown in Table 1. Here "L", "M", and "H" refer to scenarios where the bare glueball is lighter than the bare $n\bar{n}$ state, with mass between the bare $n\bar{n}$ and $s\bar{s}$ states, or heavier than the bare $s\bar{s}$ state, respectively.

The $f_0(1710)$ is clearly observed [43] in J/ψ radiative decay $J/\psi \rightarrow \gamma K\bar{K}$ with a branching ratio of about 10^{-3} [38]. (The $f_0(1500)$ is difficult to observe in this decay because it is hidden underneath the strong $f_2'(1525)$ signal.) Table 1 implies branching ratios somewhere between 10^{-4} and 10^{-2} for radiative decay of the $f_0(1710)$ to vector mesons. With a CLEO-c sample of 10^9 J/ψ decays, there should easily be enough signal to observe $f_0(1710) \rightarrow \gamma\rho$ and $f_0(1710) \rightarrow \gamma\phi$ and discriminate among the mixing scenarios.

This effort would be part of larger problem, namely the determination of the scalar

TABLE 1. Radiative decay widths (in keV) for the decay of a scalar meson f_0 to $\gamma\rho$ and $\gamma\phi$, from a calculation by [42]. See the text for details. The total widths (in MeV) [38] of the physical scalar states are also included for convenience in estimating branching ratios.

| | Radiative Decay Widths (keV) | | | | | | Γ_{TOT} (MeV) |
| | $f_0 \to \gamma\rho(770)$ | | | $f_0 \to \gamma\phi(1020)$ | | | (MeV) |
State	L	M	H	L	M	H	
$f_0(1370)$	443	1121	1540	8	9	32	$\sim$300
$f_0(1500)$	2519	1458	476	9	60	454	109
$f_0(1710)$	42	94	705	800	718	78	125

meson mass mixing matrix $\mathcal{M}$. This matrix mixes the bare states $|n\bar{n}\rangle$, $|s\bar{s}\rangle$, and $|g\rangle$ into the physical states $f_0(1370)$, $f_0(1500)$, and $f_0(1710)$ and would have the form

$$\mathcal{M} = \begin{bmatrix} M_{n\bar{n}} & \Delta_{ns} & \Delta_{ng} \\ \Delta_{ns}^{\star} & M_{s\bar{s}} & \Delta_{sg} \\ \Delta_{ng}^{\star} & \Delta_{sg}^{\star} & M_g \end{bmatrix}$$

Various calculations of this matrix have been carried out. For example, see [44]. These have all been within the context of models, however, and eventually we hope to see Lattice QCD calculations of all elements of the matrix, not just M_g.

Finally, we emphasize that this is only a small portion of the planned program of J/ψ measurements.

ACKNOWLEDGMENTS

It is a pleasure to represent the CLEO collaboration at this conference, and I gratefully acknowledge the opportunity to do so. Much of this work was done while I enjoyed the hospitality of Cornell University and the Wilson Laboratory for Elementary Particle Physics during the 2003/2004 academic year. Finally, I would like to thank the conference organizers, in particular Nora Brambilla, Umberto D'Alesio, and Barbara Giua Marassi for putting together an excellent all around experience.

REFERENCES

1. Berkelman, K., *A Personal History of CESR and CLEO*, World Scientific, 2004.
2. Peterson, D., et al., *Nucl. Instrum. Meth.*, **A478**, 142–146 (2002).
3. Artuso, M., et al., *Nucl. Instrum. Meth.* **A502**, 91–100 (2003).
4. Briere, R. A., et al., *Cornell Laboratory for Nuclear Science Preprint* (2001), CLNS-01-1742.
5. Abe, F., et al., *Phys. Rev. Lett.*, **69**, 3704–3708 (1992).
6. Abe, F., et al., *Phys. Rev. Lett.*, **71**, 2537–2541 (1993).
7. Abe, F., et al., *Phys. Rev. Lett.*, **75**, 1451–1455 (1995).
8. Braaten, E., and Fleming, S., *Phys. Rev. Lett.*, **74**, 3327–3330 (1995).
9. Affolder, T., et al., *Phys. Rev. Lett.*, **85**, 2886–2891 (2000).

10. Cacciari, M., and Kramer, M., *Phys. Rev. Lett.*, **76**, 4128–4131 (1996).
11. Braaten, E., and Chen, Y.-Q., *Phys. Rev. Lett.*, **76**, 730–733 (1996).
12. Fleming, S., Leibovich, A. K., and Mehen, T., *Phys. Rev.*, **D68**, 094011 (2003).
13. Lin, Z.-H., and Zhu, G.-h., *Phys. Lett.*, **B597**, 382–390 (2004).
14. Aubert, B., et al., *Phys. Rev. Lett.*, **87**, 162002 (2001).
15. Abe, K., et al., *Phys. Rev. Lett.*, **88**, 052001 (2002).
16. Abe, K., et al., *Phys. Rev. Lett.*, **89**, 142001 (2002).
17. Trottier, H. D., *Phys. Lett.*, **B320**, 145–151 (1994).
18. Cheung, K.-m., Keung, W.-Y., and Yuan, T. C., *Phys. Rev.*, **D54**, 929–937 (1996).
19. Napsuciale, M., *Phys. Rev.*, **D57**, 5711–5716 (1998).
20. Li, S.-y., Xie, Q.-b., and Wang, Q., *Phys. Lett.*, **B482**, 65–70 (2000).
21. Briere, R. A., et al., *Phys. Rev.*, **D70**, 072001 (2004).
22. Adams, G. S., et al., *hep-ex/0409027* (2004).
23. Athar, S. B., et al., *hep-ex/0408133* (2004).
24. Li, Z., *hep-ex/0408084* (2004).
25. Tomaradze, A., *hep-ex/0410090* (2004).
26. Adam, N. E., et al., *hep-ex/0407028* (2004).
27. Scora, D., and Isgur, N., *Phys. Rev.*, **D52**, 2783–2812 (1995).
28. Melikhov, D., and Stech, B., *Phys. Rev.*, **D62**, 014006 (2000).
29. Khodjamirian, A., Ruckl, R., Weinzierl, S., Winhart, C. W., and Yakovlev, O. I., *Phys. Rev.*, **D62**, 114002 (2000).
30. El-Khadra, A. X., Kronfeld, A. S., Mackenzie, P. B., Ryan, S. M., and Simone, J. N., *Phys. Rev.*, **D64**, 014502 (2001).
31. Huang, G. S., et al., *hep-ex/0407035* (2004).
32. Gao, K. Y., et al., *hep-ex/0408077* (2004).
33. Davies, C. T. H., et al., *Phys. Rev. Lett.*, **92**, 022001 (2004).
34. Shipsey, I., *Nature*, **427**, 591–592 (2004).
35. Besson, D., et al., *hep-ex/0408071* (2004).
36. Godfrey, S., and Napolitano, J., *Rev. Mod. Phys.*, **71**, 1411–1462 (1999).
37. Close, F. E., Farrar, G. R., and Li, Z.-p., *Phys. Rev.*, **D55**, 5749–5766 (1997).
38. Eidelman, S., et al., *Phys. Lett.*, **B592**, 1 (2004).
39. Sexton, J., Vaccarino, A., and Weingarten, D., *Phys. Rev. Lett.*, **75**, 4563–4566 (1995).
40. Morningstar, C. J., and Peardon, M. J., *Phys. Rev.*, **D60**, 034509 (1999).
41. Wada, H., *Nucl. Phys. Proc. Suppl.*, **129**, 432–434 (2004).
42. Close, F. E., Donnachie, A., and Kalashnikova, Y. S., *Phys. Rev.*, **D67**, 074031 (2003).
43. Bai, J. Z., et al., *Phys. Rev.*, **D68**, 052003 (2003).
44. Lee, W.-J., and Weingarten, D., *Phys. Rev.*, **D61**, 014015 (2000).

BARYONS IN THE $1/N_c$ EXPANSION

José L. Goity

Physics Department, Hampton University, Hampton, VA 23668, USA,
and Theory Group, Thomas Jefferson National Accelerator Facility,
Newport News, VA 23606, USA.

Abstract. A brief overview of the $1/N_c$ expansion in the baryon sector is given, with emphasis on recent analyses of masses and decays of excited baryons.

Keywords: 1/N expansion; baryons
PACS: 11.15.Pg, 12.38.Lg,14.20.Gk, 13.30.Eg.

INTRODUCTION

Thirty years ago, 'tHooft [1] discovered that $SU(N_c)$ gauge theory has a natural expansion parameter at large N_c, namely $1/N_c$. He showed that the Feynman diagrams associated with correlations of gauge invariant operators of dimension $\mathcal{O}(N_c^0)$ (which excludes operators carrying non-vanishing baryon number) can be ordered in powers of $1/N_c$ as dictated by their topology. According to this, the class of planar diagrams having the minimal number of quark loops is the dominant one. QCD_2 (QCD in 1+1 dimensions) represents the only known gauge theory example where the the meson spectrum has been worked out exactly in the planar approximation [2]. Only recently, methods based on the AdS/CFT correspondence have allowed to address the large N_c limit in 3+1 dimensions [3], although, strictly speaking, this has been achieved in supersymmetric theory rather than ordinary QCD. These developments are remarkable and will hopefully illuminate our understanding of the non-perturbative workings of QCD. One of the important reasons for considering the large N_c limit is that it is very likely that the key non-perturbative features of QCD, namely confinement and chiral symmetry breaking, can be captured in the planar approximation. The $1/N_c$ expansion can also serve to understand many phenomenological features of QCD. For this purpose the $1/N_c$ expansion can be implemented at the level of effective theory, *i.e.* at the hadronic level. It is at this level that the main impact of the expansion has taken place so far. The $1/N_c$ expansion predicts the narrowness of mesonic resonances ($\Gamma = \mathcal{O}(1/N_c)$), the OZI rule, the suppression of multi-body decays vis-à-vis 2-body decays, the suppression of $qq\bar{q}\bar{q}$ states, etc. In particular, it has been combined with ChPT [4], where the hierarchies implied by the $1/N_c$ expansion are shown to hold. This is strongly suggestive that the features of the $1/N_c$ expansion persist in the real world. An interesting case in point is that of pure gluedynamics, where the expansion is rather in powers of $1/N_c^2$. Lattice calculations for different values of N_c show that the predicted scaling of the corrections holds for N_c as low as 2 [5].

The topic of my talk, the baryon sector, requires a very different approach than the meson/glueball sector in large N_c. Baryons consist of N_c valence quarks as required

CP756, *Quark Confinement and the Hadron Spectrum VI*
edited by N. Brambilla, U. D'Alesio, A. Devoto, K. Maung, G.M. Prosperi and S. Serci
© 2005 American Institute of Physics 0-7354-0241-8/05/$22.50

for having a color singlet state carrying one unit of baryon number. The mass of a baryon is, therefore, $\mathcal{O}(N_c)$. The totally antisymmetric wave function in color makes possible that the N_c quarks be confined within a volume $\sim 1/\Lambda_{QCD}^3$. As Witten pointed out in his seminal work on baryons [6], this fact implies that baryons in large N_c can be treated in the Hartree approximation, and this allows one to pin down the N_c scalings of different quantities rather easily. At a fundamental level, one chief difference with the meson/glueball sector resides in the fact that the $1/N_c$ counting in baryon Feynman diagrams cannot be put in correspondence with some topological index. This may indicate that at the quark-gluon level the study of large N_c baryons will be more difficult than mesons and glueballs. In spite of these differences, the $1/N_c$ expansion for baryons at the effective level can be implemented a shown below and has very successful applications. In my talk I briefly summarized this approach and the results for the ground state (GS) baryon sector and highlighted recent results for excited baryons.

EMERGENT DYNAMICAL SYMMETRY

The $1/N_c$ power counting can be established, following Witten, by considering the valence quark picture in the Hartree approximation. It is expected that the N_c power of the leading contributions to a given quantity is in general independent of the quark masses, and thus this power can be obtained by using the Hartree approximation with non-relativistic quarks. In this picture, the baryon is characterized by a wave function

$$\Psi(x,\xi) = \sum_\sigma \chi(\xi_{\sigma_1},\cdots,\xi_{\sigma_{N_c}})\phi_1(x_{\sigma_1})\cdots\phi_{N_c}(x_{\sigma_{N_c}}), \tag{1}$$

where ξ denote spin and flavor indices, χ is the spin-flavor wave function, and ϕ are spatial wave functions. The permutations assure that the wave function is totally symmetric under the permutations of the positions and spin-flavor indices of the N_c quarks [1]. The content of flavor representations do depend on N_c. For instance, for two flavors ($N_f = 2$) the GS baryons will have all quarks will have a common spatial wave function ϕ, and therefore the spin-flavor wave function will be totally symmetric. The states will then fill a tower where $I = S$ (I being the isospin and S being the spin of the state, $S = \frac{N_c}{2}$, $\frac{N_c}{2} - 1$, etc.). The explicit calculation of matrix elements of operators defined in this picture at the level of the constituent quarks is straightforward (for explicit master formulas see [7]). In particular, the N_c order of the pion-baryon couplings can be determined by calculating matrix elements of the axial-vector currents $A^{ia} = \frac{1}{4}q^\dagger\sigma^i\tau^a q$. Their matrix elements between GS baryons with spin $\mathcal{O}(N_c^0)$ turn out to be $\mathcal{O}(N_c)$ (for this operator, contributions add coherently over the quarks). The π-baryon couplings, once the factor $1/F_\pi = \mathcal{O}(1/\sqrt{N_c})$ is taken into account, are therefore $\mathcal{O}(\sqrt{N_c})$. This strong coupling implies important constraints on the structure of large N_c baryons [8, 9]. In π-baryon scattering, individual diagrams (see Figure 1), are $\mathcal{O}(N_c)$. Unitarity however requires a finite large N_c limit for the scattering amplitude, and the $\mathcal{O}(N_c)$ pieces must

---

[1] The center of mass motion problem is here disregarded. For certain quantities, it ought to be taken into account in order to obtain the correct N_c power counting.

cancel between the two diagrams, leading to consistency relations [8, 9]. Since in large N_c the baryon does not recoil, $k_0 = k_0'$. The cancellation then takes place only if the baryons that give $\mathcal{O}(N_c)$ contributions are degenerate. If such contributions are explicitly displayed, one has:

$$A(\pi B \to \pi B) \propto \frac{g_A^2}{F_\pi^2} k^i k'^j [X^{ia}, X^{jb}] + \mathcal{O}(N_c^0), \tag{2}$$

where $g_A = \mathcal{O}(N_c)$ is the baryon axial-vector coupling, and the π-baryon vertices X^{ia} are $\mathcal{O}(N_c^0)$. Unitarity then leads to the consistency relations (valid for any number of flavors):

$$[X^{ia}, X^{jb}] = \mathcal{O}(1/N_c). \tag{3}$$

In the limit $N_c \to \infty$, the spin operators S^i, flavor operators T^a, and the amplitudes X^{ia} are identified with the generators of a *contracted* spin-flavor group $SU(2N_f)$ [8, 9]. Here the generators X^{ia} become classical as $N_c \to \infty$, and for this reason the contracted group has only infinite dimensional representations. Thus, baryon states in large N_c limit must belong to degenerate towers of states. For instance, the nucleons and Δ resonances would belong into one such a tower, and they should be degenerate for large N_c. Both, the constituent quark picture as well as the Skyrme model lead to the same conclusions.

The contracted spin-flavor symmetry leads to important relations in the baryon sector. These have been studied in much detail in the GS sector [10, 11, 12]. Of particular interest are relations that are violated at $\mathcal{O}(1/N_c^2)$ (gold plated relations) of which there are several, most notably the ratios of π-baryon couplings and of F and D couplings associated with axial currents and with magnetic moments:

$$\frac{g_{\pi N\Delta}}{g_{\pi NN}} = \sqrt{2} + \mathcal{O}(1/N_c^2) \quad (\text{Exp}:1.44)$$

$$\frac{F}{D} = \tfrac{2}{3} + \mathcal{O}(1/N_c^2) \quad (\text{Exp}: \left\{ \begin{array}{l} 0.58 \ \text{from axial couplings} \\ 0.72 \ \text{from magnetic moments} \end{array} \right) \tag{4}$$

(here the π-baryon couplings are defined in terms of reduced matrix elements). The implementation of the $1/N_c$ expansion can be carried out using the framework of the contracted $SU(2N_f)$ with the corresponding infinite towers of states [10], but for practical purposes there is an equivalent method which is easier to work with. In this method one works with large but finite N_c and thus multiplets of states that are finite dimensional, very much as in the valence Hartree picture mentioned earlier. The basis of states is then furnished in terms of irreducible representations of the ordinary $SU(2N_f)$ group [11, 13]. For instance, in this framework, the GS baryons belong to totally symmetric spin-flavor representation with N_c boxes in its Young tableaux.

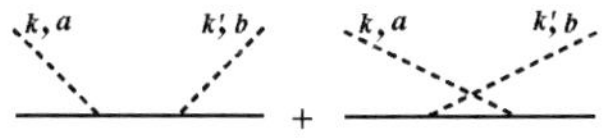

FIGURE 1. π-GS baryon scattering: only GS baryons in the intermediate state give rise to $\mathcal{O}(N_c)$ terms in each diagram.

IMPLEMENTING THE 1/N$_C$ EXPANSION

A key concept for implementing the $1/N_c$ expansion in baryons is that of effective composite operator. A given QCD operator, *e.g.* a current, is represented at the hadronic level by a combination of effective operators with the appropriate quantum numbers. The leading order in $1/N_c$ of each effective operator is generically determined by its n-bodyness and its degree of coherence. An n-body effective operator starts contributing at order $(1/N_c)^{n-1-\kappa}$. The power $n-1$ stems from the fact that in order to generate an $n-body$ operator there must be at least $n-1$ gluon exchanges at the quark level. Figure 2 illustrates how a 2-body operator is generated from a 1-body operator, depicted by a cross.

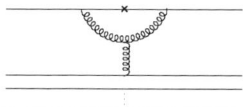

FIGURE 2. Example of diagram that generates effective 2-body operators out of 1-body operators.

Thanks to the Wigner-Eckart theorem, the effective operators con be expressed by means of appropriate products of dynamical symmetry generators. In particular, for the GS baryons the composite operators can be constructed in terms of products of the $SU(2N_f)$ generators. In that case there are simplifications due to reduction relations [10, 11]. In order to illustrate how this works, let us consider the masses and axial currents in the GS sector. The most general mass operator that one can write down, after making use of the corresponding reduction relations for 2-body operators, turns out to be [9]:

$$M = c_0 N_c + c_1 \frac{S^2}{N_c} + \varepsilon \, T^8 + \cdots , \qquad (5)$$

where the term proportional to N_c is the common mass for the tower, there is no flavor singlet $\mathcal{O}(N_c^0)$ mass splittings that one can write down, instead the first operator to furnish such splittings is the hyperfine interaction which is $\mathcal{O}(1/N_c)$. Finally, the third term provides first order $SU(3)$ breaking by the strange quark mass. It is important to remark that there are here flavor singlet $\mathcal{O}(N_c^0)$ contributions, which for instance stem from quark loop corrections. These contributions are encoded in the effective coefficient c_0. Since there are at this point three effective coefficients, and for $N_c = 3$ we have the spin-1/2 octet and spin-3/2 decuplet, entailing a total of eight different masses. There must be, therefore, five parameter-free mass relations: these are the well known Gell-Mann–Okubo and equal spacing relations, plus four relations involving masses of both octet and decuplet masses [10]. Such relations are satisfied up to corrections $\mathcal{O}(1/N_c^2)$, $\mathcal{O}(\varepsilon/N_c)$ and $\mathcal{O}(\varepsilon^2)$. The naturalness of the expansion can be tested by going to higher orders in both $1/N_c$ and $SU(3)$ breaking. By fitting to the masses, the corresponding effective coefficients are determined and found to be a natural size [14], an indication that the expansion seems to work.

Along similar lines the effective operator representation of the axial-vector currents turns out to be:

$$A^{ia} = a_1 G^{ia} + a_2 \frac{1}{N_c} S^i T^a + \cdots, \tag{6}$$

where $G^{ia} = N_c X^{ia}$ are identified with $SU(6)$ generators. Here again the only deviations from the spin-flavor symmetry limit are suppressed by a factor $1/N_c^2$ with respect to the leading term which is here given by G^{ia} and is $\mathscr{O}(N_c)$. This gives rise the gold plated relations depicted in Equation (4). Further results on magnetic moments, electric-quadrupole moments, Hyperon weak decays, etc. have been worked out by several authors.

The constraints implied by the $1/N_c$ expansion have also been implemented at the level of ChPT for baryons [10, 15, 16], where they play a crucial role. In particular, they give rise to partial cancellations in loop diagrams involving virtual spin-1/2 and spin-3/2 baryons.

In summary, the $1/N_c$ expansion within the GS octet-decuplet complex is well established and the hierarchies that it implies are observed to be satisfied remarkably well.

EXCITED BARYONS

The excited baryon sector can be studied in the framework of the $1/N_c$ expansion as well. One key difference with mesons and glueballs is that excited baryon widths are not suppressed in large N_c. A straightforward calculation in the valence quark picture shows that the amplitudes for the decay of an excited baryon to a GS baryon plus a pseudoscalar meson are generically $\mathscr{O}(N_c^0)$ [7, 17]. This fact makes in principle the analysis of excited states considerably more difficult as one would have to carry the baggage of resonance physics. This problem has been recently addressed in the context of the $1/N_c$ expansion [18]. Most of the studies of excited baryons have, however, been carried out by disregarding their finite widths (this would apply strictly in a world where the quark masses are large enough to suppress the decays of the low lying excitations). The chief aspects of the $1/N_c$ expansion are, however, expected to be captured under such an approximation.

In the valence Hartree picture, excited baryons result from having one or more quarks in excited states, leaving the rest in the ground state (the core). Since excited quarks can carry non-vanishing orbital angular momentum, a convenient basis of excited states can be built in terms of multiplets of the $O(3) \times SU(2N_f)$ group. The convenience of this choice can be readily seen in the observed excited states which nicely fit into such multiplets, e.g. (for three flavors) the Roper $[0^+, 56]$, $[1^-, 70]$, $[2^+, 56]$, etc. Although the dynamical $O(3) \times SU(2N_f)$ symmetry is not exact in large N_c limit, being broken at the level of the excited baryon masses by zeroth order couplings of spin-orbit type [19, 20, 21], it is still very useful as an approximate symmetry as shown below. The $1/N_c$ expansion is implemented along similar lines as in the case of the GS baryons by means of effective operators [19, 22, 23] as it is illustrated in the applications given below. The excited baryons that have been extensively analyzed are those where $\ell \leq 4$ and belong to the $SU(2N_f)$ totally symmetric (S) or the mixed symmetric (MS) $(N_c - 1, 1)$ (in

Young tableaux language) representations. In what follows we address the main issues in excited baryons where the $1/N_c$ expansion provides important insights.

Configuration mixings: The first important issue to be addressed with excited baryons is that of configuration mixing: the physical states will in general be admixtures of different $O(3) \times SU(2N_f)$ multiplets. An analysis at the level of the non-relativistic Hartree picture gives the hierarchy of mixings. Denoting by GS the states where all quarks have the same spatial wave function, one finds [7]:

- The mixing angle between $[\ell = 0, S]_{GS}$ and:
  - $[\ell = 0, MS]$ is $\mathcal{O}(1/\sqrt{N_c})$,
  - $[\ell > 0, S]$ is $\mathcal{O}(1/N_c^{\frac{3}{2}})$,
  - $[\ell > 0, MS]$ is $\mathcal{O}(1/\sqrt{N_c})$,
- Mixing angles between excited states are as follows:
  - mixing between $\ell = 0$ S and MS states is $\mathcal{O}(1/N_c)$,
  - $\Delta \ell = 0$ mixing between $\ell > 0$ S and MS states is $\mathcal{O}(N_c^0)$,
  - $\Delta \ell \neq 0$ mixing between S and S states is $\mathcal{O}(1/N_c^2)$,
  - $\Delta \ell \neq 0$ mixing between $\ell > 0$ S and MS states is $\mathcal{O}(1/N_c)$,
  - $\Delta \ell \neq 0$ mixing between $\ell > 0$ MS and MS states is $\mathcal{O}(N_c^0)$.

As expected, the mixings with other configurations affect the properties of GS baryons only at sub-leading order in $1/N_c$. One example is the electric-quadrupole amplitude in $\Delta \to N\gamma$ transition, which can be shown to be driven by the $[\ell = 2, MS]$ content of the GS wave functions, which according to the above is $\mathcal{O}(1/\sqrt{N_c})$. The only zeroth order mixings are driven by coupling to orbital angular momenta and occur between excited states with non-vanishing ℓ's. This seems to be a problem, as such a mixing would make the use of $O(3) \times SU(2N_f)$ multiplets meaningless as a good starting point for the analysis. There is plenty of evidence, however, that all couplings of orbital degrees of freedom are anomalously weak for dynamical reasons yet to be understood. This is shown clearly in the analysis of masses discussed below, where such effects are substantially smaller than the dominant $\mathcal{O}(1/N_c)$ effects.

Masses: Analyses of the masses of several excited multiplets have been carried out in several works [19, 23, 24, 25, 26, 27]. Here, the case of the negative parity $SU(6)$ 70-plet is presented as example. This multiplet has $\ell = 1$, and belongs to the MS spin-flavor representation. In the MS representation one box of the Young tableaux is distinguished. The $N_c - 1$ symmetrized boxes are known as the core, and the spin-flavor generators acting on the corresponding indices are denoted with a label c. The generators acting on the distinguished index are denoted with lower cases. The most general mass operator will then be given by a combination of effective operators built in terms of products of the $O(3)$ and the $SU(6)$ generators. A basis of operators to $\mathcal{O}(1/N_c)$ and to linear order in $SU(3)$ breaking is shown in the Table [25]. The mass operator to that order is:

$$M_{70} = \sum_{n=1}^{11} c_n O_n + \sum_{n=1}^{4} d_n B_n .$$ (7)

35

Here there are three $\mathcal{O}(N_c^0)$ operators that break the spin-flavor symmetry, all of them involving the coupling of the orbital angular momentum. One is the $\ell \cdot s$ operator while the other two involve flavor exchange. At linear order in $SU(3)$ breaking there are four different spin-flavor structures. The inputs for the fit include two mixing angles (there are two pairs of octets, one with $J = 1/2$ and one with $J = 3/2$, and the mixing angles refer to mixing in each pair of the basis states with $S = 1/2$ and $S = 3/2$). In total there are thirty five observables in the multiplet of which twenty one are masses and fourteen are mixing angles (the two just mentioned plus those induced by $SU(3)$ breaking). Since there are a total of fifteen operators in the basis, there are twenty parameter free relations. Of special interest are relations that do not involve mixing angles, of which there are thirteen. Among them there are four novel relations involving states in the different $SU(3)$ multiplets. Although these relations cannot be tested due to the fact that some of the states in the 70-plet are poorly know or unknown, they provide parameter free predictions for the masses of all these unknown states.

TABLE 1. Basis of operators. The operators O_n are $SU(3)$ singlets and the operators B_n are octets. The coefficients were obtained from the best fit to the known **70**-plet masses and mixings.

Operator	Order in N_c	Coefficient [MeV]
$O_1 = N_c\, 1$	1	$c_1 = 449 \pm 2$
$O_2 = l_i\, s_i$	0	$c_2 = 52 \pm 15$
$O_3 = \frac{3}{N_c} l_{ij}^{(2)}\, g_{ia}\, G_{ja}^c$	0	$c_3 = 116 \pm 44$
$O_4 = \frac{4}{N_c+1} l_i\, t_a\, G_{ia}^c$	0	$c_4 = 110 \pm 16$
$O_5 = \frac{1}{N_c} l_i\, S_i^c$	-1	$c_5 = 74 \pm 30$
$O_6 = \frac{1}{N_c} S_i^c\, S_i^c$	-1	$c_6 = 480 \pm 15$
$O_7 = \frac{1}{N_c} s_i\, S_i^c$	-1	$c_7 = -159 \pm 50$
$O_8 = \frac{2}{N_c} l_{ij}^{(2)} s_i\, S_j^c$	-1	$c_8 = 3 \pm 55$
$O_9 = \frac{3}{N_c^2} l_i\, g_{ja}\{S_j^c, G_{ia}^c\}$	-1	$c_9 = 71 \pm 51$
$O_{10} = \frac{2}{N_c^2} t_a\{S_i^c, G_{ia}^c\}$	-1	$c_{10} = -84 \pm 28$
$O_{11} = \frac{3}{N_c^2} l_i\, g_{ia}\{S_j^c, G_{ja}^c\}$	-1	$c_{11} = -44 \pm 43$
$B_1 = t_8 - \frac{1}{2\sqrt{3}N_c}O_1$	0	$d_1 = -81 \pm 36$
$B_2 = T_8^c - \frac{N_c-1}{2\sqrt{3}N_c}O_1$	0	$d_2 = -194 \pm 17$
$B_3 = \frac{10}{N_c} d_{8ab}\, g_{ia}\, G_{ib}^c + \frac{5(N_c^2-9)}{8\sqrt{3}N_c^2(N_c-1)}O_1$ $+ \frac{5}{2\sqrt{3}(N_c-1)}O_6 + \frac{5}{6\sqrt{3}}O_7$	0	$d_3 = -15 \pm 30$
$B_4 = 3\, l_i\, g_{i8} - \frac{\sqrt{3}}{2}O_2$	0	$d_4 = -27 \pm 19$

The chief approximation in this analysis is that configuration mixing has been neglected. We argue below that this is likely to be a good approximation. The results from the fit are very revealing. With the normalization chosen for each operator, the natural sizes of the coefficients c_n is 500 MeV and of the coefficients b_n is 200 MeV. The first important observation is that $c_{2,3,4} \ll 500$ MeV, implying that the zeroth order spin-flavor symmetry breaking is in practice rather small. The weakness of these spin-orbit interactions is a dynamical issue yet to be understood in QCD. These operators are however important for various smaller effects: O_2 drives the splitting between the two $SU(3)$

singlet Λ's, while the operators O_3 and O_4 are necessary for resolving the so-called spin-orbit puzzle in the quark model represented by the incompatibility between the spin-orbit splittings of the various spin-orbit partners. On the other hand, O_3 is the driving operator for the $J = 1/2$ octet-octet mixing angle. The dominant spin-flavor symmetry breaking is due to the $\mathcal{O}(1/N_c)$ hyperfine interaction operator O_6. This is the hyperfine operator naturally generated in the quark picture by one-gluon exchange. All other operators of that order are suppressed by the QCD dynamics. Finally, the $SU(3)$ breaking operators that dominate are the B_1 and B_2 that represent the simple constituent quark mass splitting between the s and the $u - d$ quarks. It is indeed remarkable that this analysis gives back a picture that resembles in its main aspects the constituent quark model. The weakness of the spin-orbit couplings observed here is also observed in other multiplets [28, 27]. The neglected configuration mixing we ought to worry about is the zeroth order one, which is driven by spin-orbit type operators. However, the observed weakness of such type of operators in the mass fits suggests that also the operators driving configuration mixing will be rather weak. This is the strongest argument favoring our main approximation, together with the fact that there are no multiplets in the vicinity of the $\ell = 1$ 70-plet with which it can have zeroth order mixing. Finally, if the number of colors in the real world would have been truly large, the hyperfine interaction would be suppressed, and the analysis could proceed with a different choice of state basis, now dictated by the zeroth order spin-flavor breaking [21].

Decays: The decays of excited baryons have been addressed in several works [22, 24, 28]. The direct decays into GS baryons are the dominant channels [7] with partial widths $\mathcal{O}(N_c^0)$. As illustration, we briefly discuss the decays of the non-strange members of the 70-plet where the details can be found in [28]. The decays proceed via the emission of a single π or η meson in S- or D-wave. The partial widths are given by:

$$\Gamma^{[\ell_P, I_P]} = f_{ps} \frac{|\sum_q C_q^{[\ell_P, I_P]} \mathcal{B}_q(\ell_P, I_P, S, I, J^*, I^*, S^*)|^2}{\sqrt{(2J^*+1)(2I^*+1)}} \tag{8}$$

$$f_{ps} = \frac{k_P^{1+2\ell_P}}{8\pi^2 \Lambda^{2\ell_P}} \frac{M_{B^*}}{M_B},$$

where M_B and M_{B^*} are respectively GS and excited baryon masses, I and I^* their respective isospins, etc., ℓ_P is the pseudoscalar meson angular momentum, I_P its isospin, k_P its momentum, and Λ is a scale of the order of the QCD scale. The $\mathcal{B}_q(\ell_P, I_P, S, I, J^*, I^*, S^*)$ are effective operator matrix elements, and the $C_q^{[\ell_P, I_P]}$ are the effective coefficients to be fitted. The effective operators are given in terms of spin-flavor operators $\mathcal{G}_q^{[j,I]}$ of spin j and isospin I that connect the MS with the S representations. Their matrix elements are then given in terms of reduced matrix elements of the spin-flavor operators according to:

$$\mathcal{B}_q(\ell_P, I_P, S, I, J^*, I^*, S^*) = (-1)^{j+J^*+\ell_P+J+1} \sqrt{(2J^*+1)(2\ell_P+1)}$$

$$\times \begin{Bmatrix} J^* & S^* & 1 \\ j & \ell_P & S \end{Bmatrix} s\langle S; I \| \left(\mathcal{G}^{[j,I_P]} \right)_q \| S^*; I^* \rangle_{MS}, \tag{9}$$

where the labels S and MS in the reduced matrix elements indicate the corresponding spin-flavor representation. For the π-channel S-wave decays the basis of operators consists of one zeroth order operator and three $\mathscr{O}(1/N_c)$ operators, while in the D-wave decays one has respectively two and six operators. In the η-channel there is one zeroth order operator for both partial waves, and at $\mathscr{O}(1/N_c)$ there are one and two for S- and D-wave respectively. The 1-body operators, which are of zeroth order, have the natural structure of the operators in the chiral quark model or Skyrme model. The rest of the operators are 2- and 3-body operators.

The partial widths are determined from the analysis of π-N scattering data. These analyses have included some model dependence whose effects are not easy to asses. For the channels used in the fit the errors used are those estimated by the Particle Data Group. The errors in most cases are at the level of 30% or more, which means that the data is not precise enough to give clear understanding of the effects of the different sub-leading operators. Another fact that contributes to this is that the number of sub-leading operators is rather large, comparable to the number of available data, and thus there is significant correlations among the coefficients. The fit shows, however, some clear-cut features. The first feature is that the sub-leading order coefficients have natural size (taking into account the fact that their errors are large). The second observation is that in the D-wave π channels the 2-body zeroth order operator has a coefficient that is about a factor 1/3 smaller than the coefficient of the 1-body operator. Thus, the natural mechanism of decay predicted in the chiral quark model is dominant. A third feature is that it is necessary to include sub-leading order contributions in order to explain the S-wave η-channel decay of the N^* states: the ratio $N^*(1535) \to \eta N$ vs. $N^*(1650) \to \eta N$ requires those corrections to agree with observation. Finally, it is through the decays that the two octet-octet mixing angles are obtained. There is some disagreement between the results of the $1/N_c$ analyses [22, 28]. However, this is mostly related to the large errors in the inputs and to the fact that in [22] the whole 70-plet decays were considered (this work did not use a complete basis of operators at $\mathscr{O}(1/N_c)$).

SUMMARY AND OUTLOOK

The implementation of the $1/N_c$ expansion in baryons provides a means to understand numerous key observations in GS as well as excited baryons. The main consequence of the expansion is the existence of an approximate dynamical spin-flavor symmetry that turns out to play a central role. Baryon states are organized in multiplets as dictated by the spin-flavor group, providing a very useful framework in which one can implement the $1/N_c$ expansion via the effective operator analysis. The expansion leads at each order to parameter free relations, which will naturally be violated only by higher order corrections. All such relations, where there is enough empirical information for testing them, are shown to hold remarkably well to the corresponding order of accuracy. Other relations, e.g. mass relations for excited multiplets, that have not been tested because of lack of data, become predictions that will hopefully to be tested in the future.

The application to excited baryons clearly exposes some important characteristics of the QCD dynamics that we do not understand yet. An example of that is the weakness of the $\mathscr{O}(N_c^0)$ spin-flavor breaking effects that result in an enhancement of the role of that symmetry for excited baryons. Moreover, in the analysis of baryon masses it is observed that only a few effective operators have effects of natural magnitude as dictated by their

$1/N_c$ order, while the rest are suppressed, leading to a picture of baryons that in its main characteristics is similar to that of the constituent quark model. One of the great challenges in baryon physics is the understanding of how and why those suppressions occur. The merit of the $1/N_c$ expansion is that it provides the ordering principle in which to frame that challenge. In a similar vein, the current lattice QCD studies of excited baryons can draw great benefit from the $1/N_c$ expansion for interpreting the results.

ACKNOWLEDGMENTS

I thank the organizers for the kind invitation to give a talk about large N_c baryons. I also thank the organization of the conference for financial support. I am indebted to my collaborators Carlos Schat and Norberto Scoccola for many insights on the topics of my talk. This work was supported by DOE contract DE-AC05-84ER40150 under which SURA operates the Thomas Jefferson National Accelerator Facility, and by the National Science Foundation (USA) through grant # PHY-0300185 (JLG).

REFERENCES

1. G. 'tHooft, *Nucl. Phys.* **B 72**, 461 (1974).
2. G. 'tHooft, *Nucl. Phys.* **B 75**, 461 (1974).
3. See for instance M. Kruczenski, D. Mateos, R. C. Myers and D. J. Winters, *JHEP* **07**, 049 (2003).
4. P. Herrera-Siklódy, J. I. Latorre, P. Pascual and J. Taron, Nucl. Phys. **B 497** (1997) 345.
   R. Kaiser and H. Leutwyler, Eur. Phys. J. **C17** (2000) 623.
5. M. Teper, *hep-th/0412005*, and references therein.
6. E. Witten, *Nucl. Phys.* **B 160**, 57 (1979).
7. J. L. Goity, *hep-ph/0405304*.
8. J. -L. Gervais and B. Sakita, *Phys. Rev. Lett.* **52**, 87 (1984), and *Phys. Rev.* **D 30**, 1795 (1984)
9. R. F. Dashen and A. V. Manohar, *Phys. Lett.* **B 315**, 425 (1993), and **B 315**, 438 (1993).
10. R. F. Dashen, E. Jenkins and A. V. Manohar, *Phys. Rev.* **D 49**, 4713 (1994).
11. R. F. Dashen, E. Jenkins and A. V. Manohar, *Phys. Rev.* **D 51**, 3697 (1995).
12. J. Dai, R. F. Dashen, E. Jenkins and A. V. Manohar, *Phys. Rev.* **D 53**, 273 (1996).
13. M. A. Luty and J. M. March-Russell, *Nucl. Phys.* **B 426**, 71 (1994).
    C. D. Carone, H. Georgi and S. Osofsky, *Phys. Lett.* **B 322**, 227 (1994).
14. E. Jenkins and R. F. Lebed, *Phys. Rev.* **D 52**, 282 (1995).
15. R. Flores-Mendieta, Ch. P. Hofmann, E. Jenkins and A. V. Manohar, *Phys. Rev.* **D 62**, 034001 (2000).
16. R. Flores-Mendieta, *hep-ph/0410171*.
17. T. D. Cohen, D. C. Dakin, A. Nellore and R. F. Lebed, *Phys. Rev.* **D 69**, 056001 (2004).
18. T. D. Cohen, D. C. Dakin, A. Nellore and R. F. Lebed, *Phys. Rev.* **D 70**, 056004 (2004).
19. J. L. Goity, *Phys. Lett.* **B 414**, 140 (1997).
20. D. Pirjol and T.-M. Yan, *Phys. Rev.* **D 57**, 1449 (1998), and **D 57**, 5434 (1998).
21. D. Pirjol and C. L. Schat, *Phys. Rev.* **D 67**, 096009 (2003).
22. C. D. Carone, H. Georgi, L. Kaplan and D. Morin, *Phys.Rev.* **D50**, 5793 (1994).
23. C. E. Carlson, Ch. Carone, J. L. Goity and R. F. Lebed, *Phys. Lett.* **B 438**, 327 (1998), and *Phys. Rev.* **D 59** 114008 (1999).
24. C. E. Carlson and C. D. Carone, *Phys. Lett.* **B 484**, 260(2000).
25. J. L. Goity, C. L. Schat and N. N. Scoccola, *Phys. Rev. Lett.* **88**, 102002 (2002), and *Phys. Rev.* **D 66**, 114014 (2002).
26. J. L. Goity, C. L. Schat and N. N. Scoccola, *Phys. Lett.* **B 564**, 83 (2003).
27. N. Matagne and Fl. Stancu, *hep-ph/0409261*.
28. J. L. Goity, C. L. Schat and N. N. Scoccola, *hep-ph/0411092*.

Relevance of the strange quark sector in chiral perturbation theory

H. Sazdjian

Groupe de Physique Théorique, Institut de Physique Nucléaire, Université Paris XI,
F-91406 Orsay Cedex, France

Abstract. Results obtained in recent years in the strange quark sector of chiral perturbation theory are reviewed and the theoretical relevance of this sector for probing the phase structure of QCD at zero temperature with respect to the variation of the number of massless quarks is emphasized.

Keywords: Spontaneous symmetry breaking, Chiral Lagrangians, Chiral perurbation theory, Strange quark, Phase transition.
PACS: 11.30.Rd, 12.39.Fe, 13.75.Lb.

CHIRAL PERTURBATION THEORY

Chiral perturbation theory (ChPT) is an effective theory of QCD at low energies, in which the dynamical degrees of freedom are those of the pseudoscalar Goldstone bosons (π, K, η) of the chiral group $SU(3) \times SU(3)$. That theory was developed by Weinberg [1] and applied in more detail to the QCD case by Gasser and Leutwyler [2, 3]. The effective lagrangian $\mathscr{L}_{eff}$ is the most general locally chiral invariant lagrangian in the presence of external source terms. It contains an infinite series of terms constructed out of the pseudoscalar meson fields and the external sources. The series is arranged according to an increasing power of derivatives and quark masses. At low energies, the Goldstone boson interactions are weak (they are of second order in the derivatives and of first order in the quark masses) and therefore a perturbative calculation of their transition amplitudes is meaningful. The expansion parameter is essentially the order of magnitude of the external momenta or of the quark masses divided by the hadronic mass scale (which is of the order of 1 GeV).

The counting rules of the dimensionalities of various diagrams are based on the counting of the numbers of external momenta, of the mass terms and of loops, each quark mass being equivalent to a momentum to the power 2. Furthermore, because of the weakness of the interactions at the tree level at low momenta, each loop introduces two additional powers of momenta, thus contributing to nonleading orders [1]. Generally, there are terms of order $O(p^2)$, then terms of order $O(p^4)$, $O(p^6)$, etc. The effective lagrangian takes the following corresponding expansion:

$$\mathscr{L}_{eff} = \mathscr{L}_2 + \mathscr{L}_4 + \mathscr{L}_6 + \dots , \tag{1}$$

each index indicating the power of the momenta that will emerge from a tree-level calculation of a process. It is evident that at low energies it is the terms with smallest indices that will be the dominant ones.

CP756, *Quark Confinement and the Hadron Spectrum VI*
edited by N. Brambilla, U. D'Alesio, A. Devoto, K. Maung, G.M. Prosperi and S. Serci
© 2005 American Institute of Physics 0-7354-0241-8/05/$22.50

The effective lagrangian is renormalizable order by order in perturbation theory. The term $\mathcal{L}_2$ contributes, through its one-loop effects, to the $O(p^4)$ terms, which are generally ultra-violet divergent. Those divergences are then absorbed in redefinitions of the coupling constants contained in the terms of the lagrangian $\mathcal{L}_4$. The latter, in turn, produces, with $\mathcal{L}_2$, one-loop divergences of order $O(p^6)$ that are absorbed, together with the two-loop divergences of the lagrangian $\mathcal{L}_2$, by the coupling constants of the lagrangian $\mathcal{L}_6$, and so forth.

At each order of the perturbation series there are a certain number of coupling constants, called low energy constants (LEC), that are order parameters of spontaneous chiral symmetry breaking. They are expected to be determined from experimental data. In addition, one also encounters the quark masses m_u, m_d, m_s.

At order $O(p^2)$, one has two LECs: F_0, which is the pion weak decay constant F_π in the chiral limit, and $B_0 = - <0|\bar{u}u|0>_0 / F_0^2$, which is proportional to the quark condensate in the vacuum in the chiral limit [3, 4]. At order $O(p^4)$, one has 10 (observable) LECs, called L_i, $i = 1,\ldots,10$ [3]. At order $O(p^6)$, one has 90 (observable) LECs, called C_i, $i = 1,\ldots,90$ [5]. It is worthwhile to emphasize that not all LECs enter in a definite process.

The theory relatively simplifies if one sticks to processes related to the nonstarnge sector of the quarks (u, d). The chiral group now becomes $SU(2) \times SU(2)$. Here, the strange quark can be considered as heavy and the corresponding field integrated out.

The $SU(2) \times SU(2)$ version of ChPT contains less LECs than its $SU(3) \times SU(3)$ version. At order $O(p^2)$ one still has two LECs, F and B, the analogs of F_0 and B_0, but now considered in the $SU(2) \times SU(2)$ chiral limit. At order $O(p^4)$, there are 7 LECs, called ℓ_i, $i = 1,\ldots,7$ [2]. At order $O(p^6)$, there are 53 LECs, called c_i, $i = 1,\ldots,53$ [5].

A detailed study of the elastic $\pi\pi$ scattering amplitude up to order $O(p^6)$ was done by several groups [6, 7, 8, 9]. The rate of convergence of ChPT seems rather satisfactory: $O(p^4)$ effects represent approximately 20-25% of the global quantities under consideration, while $O(p^6)$ effects represent 7-8% of the contributions. Recent experimental data from the E865 experiment at Brookhaven about K_{e4} decay [10], which provides information about the $\pi\pi$ scattering amplitude near threshold through the final state interaction, have confirmed the hypothesis that the quark condensate parameter B is the leading order parameter of chiral symmetry breaking [4, 9, 11]. This means that the QCD vacuum is very similar to a ferromagnetic medium, as far as chiral symmetry breaking is concerned.

Extension of ChPT to $SU(3) \times SU(3)$ allows the study of sectors involving the K and η mesons. But here, the strange quark mass m_s is not as small as the non-strange ones, m_u and m_d. From the tree-level relation [4] $2m_s/(m_u + m_d) = 2m_K^2/m_\pi^2 - 1 \simeq 25$ one guesses that ChPT might converge more slowly than in the $SU(2) \times SU(2)$ case. Apart from that aspect, which by itself leads to unavoidable complications, it was emphasized by Descotes, Girlanda and Stern [12] that the type of dependence of physical quantities on the strange quark mass m_s might reveal some important theoretical features of QCD.

PHASE TRANSITION IN THE NUMBER OF MASSLESS QUARKS

It is known that $SU(N_c)$ gauge theories might undergo a zero temperature chiral phase transition when the number N_f of massless fermions (in the fundamental representation) reaches some critical value $N_f^* < 11N_c/2$ [13]. The argument goes as follows. For $N_f < 11N_c/2$, one has an asymptotically free theory, while for $N_f > 11N_c/2$ asymptotic freedom is lost. From the first two terms of the beta-function one infers the existence of an infra-red stable fixed point appearing when N_f reaches from below a critical value $N_{f0}(< 11N_c/2)$. In perturbation theory, one has $N_{f0} \simeq 34N_c/13$. When the value of N_f further increases and reaches the vicinity of $11N_c/2$ from below, then the domain of variation of the effective coupling constant becomes tiny and the theory becomes fully perturbative, in the infra-red and in the ultra-violet, reducing to a conformal theory. Such theories, because of the smallness of the coupling constant, do not undergo spontaneous chiral symmetry breaking, neither display confinement [14]. On the other hand, at small values of N_f, one has chiral symmetry breaking and confinement [15, 16, 17, 18]. Therefore, there should exist a critical value of N_f, N_f^*, such that $N_{f0} < N_f^* < 11N_c/2$, where the theory undergoes a chiral phase transition (Fig. 1).

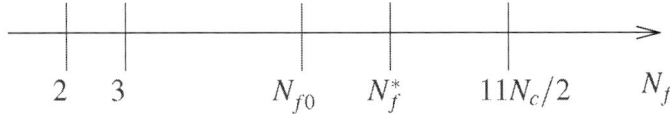

FIGURE 1. Domains of N_f. At N_{f0}, reached from below, an infra-red fixed point appears. At N_f^* a chiral phase transition from the Goldstone mode ($N_f < N_f^*$) to the Wigner mode ($N_f^* < N_f < 11N_c/2$) occurs. Above $11N_c/2$ asymptotic freedom is lost.

It is evident that at N_f^* all chiral order parameters should have vanished. Among those, F_0 would be the last order parameter to vanish, since it is the fundamental order parameter of chiral symmetry breaking, directly related to the Goldstone theorem.

The precise value of N_f^* is dependent on the dynamical models that are used to evaluate it. Appelquist, Terning and Wijewardhana [14], combining perturbation theory and gap equation calculations find $N_f^* \simeq 4N_c = 12$. Lattice calculations give a rather wide range of values. Kogut and Sinclair [19], Brown *et al.* [20] find $8 \le N_f^* \le 12$; Iwasaki *et al.* [21] find $N_f^* \simeq 6$, while Mawhinney [22] finds $N_f^* \simeq 4$. Within the instanton liquid model, Velkovsky and Shuryak [23] find $N_f^* \simeq 6$.

From another viewpoint, Descotes, Girlanda and Stern have studied the dependence on N_f of various chiral order parameters [12]. Using properties of the Dirac operator in the background gluon field in euclidean space (placed in a box) and bounds derived by Vafa and Witten [24] concerning its eigenvalues, they obtain the following inequalities when the number of massless fermions changes from N_f to $N_f + 1$:

$$F_0^2[N_f + 1] < F_0^2[N_f], \tag{2}$$

$$|<\overline{u}u>|_{(N_f+1)} < |<\overline{u}u>|_{N_f}. \tag{3}$$

These inequalities, which do not hinge on any hypothesis about the existence of a chiral phase transition, are manifestly compatible, at least locally, with the behaviors of order parameters as expected from such an hypothesis.

In summary, the eventual existence of a chiral phase transition in N_f would have the tendancy to decrease the values of order parameters with increasing N_f. The slope of the variation would strongly depend on the value of the critical point N_f^*. It would be stronger for smaller N_f^*. In the real world one does not have much freedom to vary N_f. The only possibility that we have is to vary N_f from 2 ($SU(2) \times SU(2)$) to 3 ($SU(3) \times SU(3)$). (For $N_f = 1$, chiral symmetry is destroyed by the $U_A(1)$-anomaly.) Possibly small values of N_f^* (4-6, say) would induce rather strong variations of order parameters in passing from $N_f = 2$ to $N_f = 3$. Therefore, phenomenological studies of such effects would represent indirect tests about the vicinity of N_f^*. In particular, quantities that are Zweig-rule suppressed in the large-N_c limit (almost true for $N_f = 2$ and $N_c = 3$), should be enhanced for $N_f = 3$ [12]. Those mainly concern scalar meson sectors and the LECs L_4 and L_6. Simultaneously, loops of the strange quark might provide important contributions and destabilize certain results obtained in the large-N_c limit [12].

PHENOMENOLOGY

The above problem was first studied by Moussallam [25]. He calculated the ratio of the quark condensate evaluated in a theory with three massless quarks to the condensate evaluated in a theory with two massless quarks:

$$R_{32} \equiv \frac{<\overline{u}u>_{N_f=3}}{<\overline{u}u>_{N_f=2}}. \tag{4}$$

In the $N_f = 2$ case, the mass of the strange quark is fixed at its "physical" value; but since the latter is still small compared to the massive hadron masses, one can use perturbation theory for it and keep only the leading contribution in m_s. One thus obtains, at the one-loop level:

$$R_{32} = 1 - \frac{m_s B_0}{F_\pi^2} \left[32 L_6(\mu) - \frac{1}{16\pi^2} \left(\frac{11}{9} \ln(\frac{m_s B_0}{\mu^2}) + \frac{2}{9} \ln(\frac{4}{3}) \right) \right] + O(m_s^2), \tag{5}$$

where μ is the renormalization mass. The quantity $m_s B_0$ can be replaced by its tree-level expression, $(m_K^2 - m_\pi^2/2)$. L_6 is then evaluated from the correlator of scalar-isoscalar densities, $(\overline{u}u + \overline{d}d)$ and $\overline{s}s$:

$$\int d^4x e^{ip \cdot x} \langle T[(\overline{u}u(x) + \overline{d}d(x))\overline{s}s(0)] \rangle_c. \tag{6}$$

This is precisely a Zweig-rule violating term. It is evaluated by saturating the intermediate states with $\pi\pi$ and $K\overline{K}$ states, yielding the pion and kaon scalar form factors. One obtains coupled Muskelishvili–Omnès equations. Use of experimental values of phase shifts and phases leads to:

$$L_6(m_\eta) = (0.6 \pm 0.2) \times 10^{-3}, \qquad (7)$$
$$R_{32} = 0.46 \pm 0.27. \qquad (8)$$

The last result indicates a strong variation of the quark condensate when passing from two massless quarks to three.

A similar study, by a different method, was also done in Ref. [26], confirming the above conclusions. $O(p^6)$ effects, estimated by means of a resonance model and the sigma-model, do not seem to qualitatively change the above results [27].

An important quantity in the strange quark sector is the πK elastic scattering amplitude. The tree-level (current algebra), $O(p^2)$, values of the S-wave isospin 1/2 and 3/2 scattering lengths had been calculated by Weinberg [28]:

$$a_0^{1/2} = 0.14, \qquad a_0^{3/2} = -0.07. \qquad (9)$$

(In units of m_π^{-1}.)

The scattering amplitude at the one-loop level, $O(p^4)$, was calculated by Bernard, Kaiser and Meissner [29]. The scattering lengths become:

$$a_0^{1/2} = 0.19, \qquad a_0^{3/2} = -0.05. \qquad (10)$$

Until recently, experimental knowledge of the scattering lengths was very poor. As for $\pi\pi$, it is not possible to realize direct scattering experiments at low energies, because pions and kaons decay. One must then use extrapolations of high energy data to low energies. Recently, a detailed evaluation of the low-energy πK elastic scattering amplitude was done by Büttiker, Descotes-Genon and Moussallam [30], by means of Roy and Steiner type equations [31, 32]. Those equations use dispersion relations, crossing symmetry, unitarity and partial-wave analysis, together with high-energy data, to reconstruct the elastic scattering amplitude at low energies. The method was already used for $\pi\pi$ scattering [33, 34, 35, 36, 9]. In πK, one ends up with six coupled integral equations. Solutions with rather small uncertainties have been obtained for the scattering lengths [30]:

$$a_0^{1/2} = 0.224 \pm 0.022, \qquad a_0^{3/2} = -0.045 \pm 0.008. \qquad (11)$$

$O(p^6)$ effects in $\pi\pi$ and πK scattering were evaluated (for $N_f = 3$) by Bijnens, Dhonte and Talavera [37, 38]. The corresponding LECs are calculated by resonance saturation methods. They do an overall fit to all existing data (scattering amplitudes, form factors, masses, etc.), leaving the Zweig-rule violating LECs L_4 and L_6 as free parameters. They obtain several sets of results, depending on which experimental quantities optimization is imposed by varying slightly the resonance parameters. For the set producing the best fit with the scattering amplitudes, they find for the πK scattering lengths:

$$a_0^{1/2} = 0.220, \qquad a_0^{3/2} = -0.047. \qquad (12)$$

The relevant LECs are:

$$L_4 = 0.2 \times 10^{-3}, \qquad L_6 = 0.0 \times 10^{-3}. \qquad (13)$$

The above values of the scattering lengths match, within the allowed uncertainties, those obtained from the Roy–Steiner extrapolation method of high energy data [Eq. (11)]. The values of the LECs are also compatible with a small violation of the Zweig rule. At this point one might conclude that ChPT is rapidly converging, without sizable Zweig-rule violating effects. However, the overall fit of Refs. [37, 38] displays in some instances contradictory aspects. One notices that in other sectors (mainly the scalar ones) $O(p^6)$ effects are more important than $O(p^4)$ effects, indicating bad convergence (in particular in the pionic sector, which, in the $SU(2) \times SU(2)$ case had a rapid convergence). In such cases, the meaning of the $O(p^4)$ LECs L_4 and L_6 becomes questionable. Perhaps optimization with respect to global convergence of $SU(3) \times SU(3)$ should be tried.

In this respect, Descotes-Genon, Fuchs, Girlanda and Stern have proposed a different method of evaluation of high-order effects [39]. They suggest to isolate those terms which might be sensitive to Zweig-rule violating effects (four in all) and to treat them nonperturbatively, while treating the rest perturbatively. The method was already applied to the $\pi\pi$ scattering case; its application to the other sectors could still reduce the existing uncertainties.

Finally, we mention here some future useful experiments about the πK system.

The process

$$D \longrightarrow \pi K e \nu, \tag{14}$$

could be analyzed in the FOCUS experiment at FermiLab. It would give information, through the final state interaction, about the elastic πK phase shifts. It plays an analogous role as the $K_{\ell 4}$ decay for $\pi\pi$ scattering.

The process

$$\tau \longrightarrow \pi K \nu, \tag{15}$$

could be analyzed in the CLEOIII experiment at Cornell. It would give information about the $K_{\ell 3}$ form factors.

The observation and measurement of the properties of the hadronic atom $(K^+\pi^-)_{at.}$ would also give complementary informations about the scattering lengths. Hadronic atoms are Coulomb bound states of charged hadrons, which generally decay under the effect of the strong interactions into neutral isospin partners. Thus the above atom would mainly decay as

$$(K^+\pi^-)_{at.} \longrightarrow K^0\pi^0. \tag{16}$$

The lifetime of the atom depends essentially on the combination $(a_0^{1/2} - a_0^{3/2})$ of the πK scattering lengths, while the energy level splittings depend upon the combination $(2a_0^{1/2} + a_0^{3/2})$ [40, 41, 42, 43]. The experimental study of the pionium (the $\pi^+\pi^-$-atom) is currently done at CERN in the DIRAC experiment. For the $K^+\pi^-$-atom, projects are being prepared. To precisely reconstruct the strong interaction scattering lengths from the hadronic atom properties, one needs to take into account isospin breaking and electromagnetic radiative corrections, as well as relativistic corrections. A recent theoretical study of the $K\pi$-atom was done by Schweizer [44].

In conclusion, the study of the strange quark sector up to order $O(p^6)$ offers the possibility of a full test of $SU(3) \times SU(3)$ ChPT and at the same time of an indirect probe of a possible phase transition in the number of massless flavors in QCD.

ACKNOWLEDGMENTS

Institut de Physique Nucléaire is Unité Mixte de Recherche 8608. This work was supported in part by the European Community network EURIDICE under contract No. HPRN-CT-2002-00311

REFERENCES

1. S. Weinberg, *Physica A* **96**, 327 (1979).
2. J. Gasser and H. Leutwyler, *Ann. Phys. (N.Y.)* **158**, 142 (1984).
3. J. Gasser and H. Leutwyler, *Nucl. Phys.* **B250**, 465 (1985).
4. M. Gell-Mann, R. Oakes and B. Renner, *Phys. Rev.* **175**, 2195 (1968).
5. J. Bijnens, G. Colangelo and G. Ecker, *JHEP* **9902**, 020 (1999) [hep-ph/9902437]; *Ann. Phys. (N.Y.)* **280**, 100 (2000) [hep-ph/9907333].
6. J. Bijnens, G. Colangelo, G. Ecker, J. Gasser and M. E. Sainio, *Phys. Lett. B* **374**, 210 (1996) [hep-ph/9511397].
7. M. Knecht, B. Moussallam, J. Stern and N. H. Fuchs, *Nucl. Phys. B* **457**, 513 (1995) [hep-ph/9507319]; *ibid. B* **471**, 445 (1996) [hep-ph/9512404].
8. L. Girlanda, M. Knecht, B. Moussallam and J. Stern, *Phys. Lett. B* **409**, 461 (1997) [hep-ph/9703448].
9. G. Colangelo, J. Gasser and H. Leutwyler, *Phys. Lett. B* **488**, 261 (2000) [hep-ph/0007112]; *Nucl. Phys. B* **603**, 125 (2001) [hep-ph/0103088].
10. S. Pislak *et al.* [E865 Collaboration], *Phys. Rev. D* **67**, 072004 (2003) [hep-ex/0301040].
11. G. Colangelo, J. Gasser and H. Leutwyler, *Phys. Rev. Lett.* **86**, 5008 (2001) [hep-ph/0103063].
12. S. Descotes, L. Girlanda and J. Stern, *JHEP* **0001**, 041 (2000) [hep-ph/9910537].
13. Banks and A. Zaks, *Nucl. Phys. B* **196**, 189 (1982).
14. T. Appelquist, J. Terning and L. C. R. Wijewardhana, *Phys. Rev. Lett.* **77**, 1214 (1996) [hep-ph/9602385].
15. G. 't Hooft, in *Recent Developments in Gauge Theories*, edited by G. 't Hooft *et al.*, Plenum Press, New York, 1980.
16. S. Coleman and E. Witten, *Phys. Rev. Lett.* **45**, 100 (1980).
17. Y. Frishman, A. Schwimmer, T. Banks and S. Yankielowicz, *Nucl. Phys. B* **177**, 157 (1981).
18. S. Coleman and B. Grossman, *Nucl. Phys. B* **203**, 205 (1982).
19. J. B. Kogut and D. K. Sinclair, *Nucl. Phys. B* **295 [FS21]** 465 (1988).
20. F. Brown *et al.*, *Phys. Rev. D* **46**, 5655 (1992) [hep-lat/9206001].
21. Y. Iwasaki *et al.*, *Prog. Theor. Phys.* **131**, 415 (1994) [hep-lat/9804005].
22. R. D. Mawhinney, *Nucl. Phys.* **60A** *(Proc. Suppl.)*, 306 (1998) [hep-lat/9805031].
23. M. Velkovsky and E. Shuryak, *Phys. Lett. B* **437**, 398 (1998) [hep-ph/9703345].
24. C. Vafa and E. Witten, *Comm. Math. Phys.* **95**, 257 (1984).
25. B. Moussallam, *Eur. Phys. J. C* **14**, 111 (2000) [hep-ph/9909292].
26. S. Descotes-Genon and J. Stern, *Phys. Lett. B* **488**, 274 (2000) [hep-ph/0007082].
27. B. Moussallam, *JHEP* **0008**, 005 (2000) [hep-ph/0005245].
28. S. Weinberg, *Phys. Rev. Lett.* **17**, 616 (1966).
29. V. Bernard, N. Kaiser and U.-G. Meissner, *Nucl. Phys. B* **357**, 129 (1991).
30. P. Büttiker, S. Descotes-Genon and B. Moussallam, *Eur. Phys. J. C* **33**, 409 (2004) [hep-ph/0310283].
31. S. M. Roy, *Phys. Lett. B* **36**, 353 (1971).
32. F. Steiner, *Fortsch. Phys.* **19**, 115 (1971).
33. J.-L. Basdevant, J.-C. Le Guillou and H. Navelet, *Nuovo Cimento A* **7**, 363 (1972).
34. J.-L. Basdevant, C. D. Froggatt and J. L. Petersen, *Phys. Lett. B* **41**, 173 (1972); *ibid. B* **41**, 178 (1972); *Nucl. Phys. B* **74**, 413 (1974).
35. M. R. Pennington and S. D. Protopopescu, *Phys. Rev. D* **7**, 1429 (1973); *ibid. D* **7**, 2591 (1973).
36. C. D. Froggatt and J. L. Petersen, *Nucl. Phys. B* **129**, 89 (1977).
37. J. Bijnens, P. Dhonte and P. Talavera, *JHEP* **0401**, 050 (2004) [hep-ph/0401039]; *ibid.* **0405**, 036 (2004) [hep-ph/0404150].
38. J. Bijnens, *Chiral meson physics at two loops*, hep-ph/0409068.

39. S. Descotes-Genon, N. H. Fuchs, L. Girlanda and J. Stern, *Eur. Phys. J. C* **34**, 201 (2004) [hep-ph/0311120].
40. S. Deser, M. L. Goldberger, K. Baumann and W. Thirring, *Phys. Rev.* **96**, 774 (1954).
41. J. L. Uretsky and T. R. Palfrey, Jr., *Phys. Rev.* **121**, 1798 (1961).
42. T. L. Trueman, *Nucl. Phys.* **26**, 57 (1961).
43. S. M. Bilen'kii, Nguyen Van Hieu, L. L. Nemenov and F. G. Tkebuchava, *Sov. J. Nucl. Phys.* **10**, 469 (1969).
44. J. Schweizer, *Phys. Lett. B* **587**, 33 (2004) [hep-ph/0401048]; *Eur. Phys. J. C* **36**, 483 (2004) [hep-ph/0405034].

Consistency checks of pion-pion scattering data and chiral dispersive calculations

J. R. Peláez* and F. J. Ynduráin†

*Departamento de Física Teórica, II, Facultad de Ciencias Físicas, Universidad Complutense de Madrid, E-28040, Madrid, Spain
†Departamento de Física Teórica, C-XI Universidad Autónoma de Madrid, Canto Blanco, E-28049, Madrid, Spain.

Abstract. We have evaluated forward dispersion relations for scattering amplitudes that follow from direct fits to several sets of $\pi\pi$ scattering experiments, together with the precise K decay results, and high to energy data. We find that some of the most commonly used experimental sets, as well as some recent theoretical analyses based on Roy equations, do not satisfy these constraints by several standard deviations. Finally, we provide a consistent $\pi\pi$ amplitude by improving a global fit to data with these dispersion relations.

INTRODUCTION

A precise knowledge of the $\pi\pi$ scattering amplitude has become increasingly important since it provides crucial tests for one and two loop Chiral Perturbation Theory (ChPT), as well as crucial information on three topics under intensive experimental and theoretical investigation: light meson spectroscopy, pionic atom decays and CP violation in kaons. Unfortunately, these precision studies are very cumbersome due to the poor quality of the data which is affected by large systematic errors. Here we review our recent work where we checked the fulfillment of dispersion relations by different sets of data commonly used in the literature, and provided parametrizations consistent with such requirements.

Recently, Ananthanarayan, Colangelo, Gasser and Leutwyler (ACGL)[1] and Colangelo, Gasser and Leutwyler (CGL)[2] have used data, analyticity and unitarity through Roy equations, and ChPT, to build a $\pi\pi$ amplitude. They provide phase shifts up to 0.8 GeV, scattering lengths and effective ranges claiming an outstanding precision. While the methods of CGL constitute a substantial improvement over previous ones, their analysis has to rely on some input, part of which we have recently questioned. First of all, their Regge high energy representation does not describe the high energy data, and does not satisfy well certain sum rules. Second, their D2 wave is incompatible with a number of requirements. Finally some of their input has remarkably small errors and relies precisely on some data sets that do not satisfy well forward dispersion relations. All this is discussed in the present note, which is based our recent works [3, 4, 5].

---

0 Presented by F. J. Ynduráin at "Quark confinement and the Hadron Spectrum", Sardinia, Sept. 2004.

CP756, *Quark Confinement and the Hadron Spectrum VI*
edited by N. Brambilla, U. D'Alesio, A. Devoto, K. Maung, G.M. Prosperi and S. Serci
© 2005 American Institute of Physics 0-7354-0241-8/05/$22.50

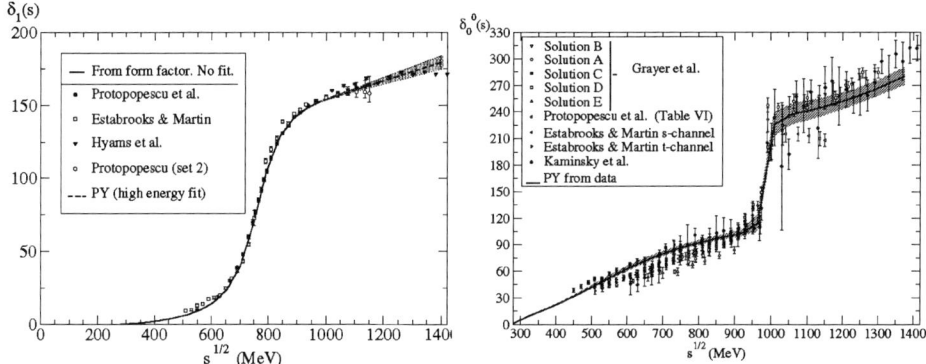

FIGURE 1. a) P wave phase shifts from $\pi\pi$ scattering [9, 10] compared with the prediction with the parameters (2.1) (solid line below 1 GeV). Note that this is *not* a fit to these data, but is obtained from the pion form factor [6]. The error here is like the thickness of the line. Above 1 GeV, the dotted line and error (PY) are as follows from the fit in [5]. b) S0 phase shifts and error band as given by Eqs.(5,6) below 1 GeV, and from [5] above. The K_{l4} and $K_{2\pi}$ decay data are not shown. (see our [5] for details).

LOW ENERGY PARTIAL WAVES FROM FITS TO DATA

The S0, S2 and P partial waves at low energy, $s^{1/2} \lesssim 1$ GeV

We first consider wave-by-wave *fits to data* for the S0, S2, P waves, as in [5], which improve our "tentative solution" in [3]. To fit the phase shifts, $\delta(s)$, we parametrize $\cot\delta(s)$ taking into account its analytic properties, as well as its zeros (associated with resonances) and poles (when the phase shift crosses $n\pi$, $n =$integer).

For the P wave, up to $\simeq 1\,\mathrm{GeV}$ we use the results from a fit to the pion form factor as given in [6]. The comparison with $\pi\pi$ scattering data can be seen in Fig.1. We take s_0 as the point at which inelasticity begins to be nonegligible, and we write

$$\cot\delta_1(s) = \frac{s^{1/2}}{2k^3}(M_\rho^2 - s)\left\{B_0 + B_1\frac{\sqrt{s}-\sqrt{s_0 - s}}{\sqrt{s}+\sqrt{s_0 - s}}\right\}; \quad s_0^{1/2} = 1.05 \text{ GeV}. \quad (1)$$

$$B_0 = 1.069 \pm 0.011, \quad B_1 = 0.13 \pm 0.05, \quad M_\rho = 773.6 \pm 0.9,$$
$$a_1 = (37.6 \pm 1.1) \times 10^{-3}M_\pi^{-3}, \quad b_1 = (4.73 \pm 0.26) \times 10^{-3}M_\pi^{-5}; \quad (2)$$

For the S2 wave we fit data where two like charge pions are produced:[7] although these pions are not all on their mass shell, at least there is no problem of interference among various isospin states. At low energies, we fix the Adler zero at $z_2 = M_\pi$ and fit only the low energy data, $s^{1/2} < 1.0\,\mathrm{GeV}$; later on we allow z_2 to vary. We have

$$\cot\delta_0^{(2)}(s) = \frac{s^{1/2}}{2k}\frac{M_\pi^2}{s - 2z_2^2}\left\{B_0 + B_1\frac{\sqrt{s}-\sqrt{s_0 - s}}{\sqrt{s}+\sqrt{s_0 - s}}\right\}, \quad s_0^{1/2} = 1.05 \text{ GeV}, \quad (3)$$

$$B_0 = -80.4 \pm 2.8, \quad B_1 = -73.6 \pm 12.6;$$

$$a_0^{(2)} = (-0.052 \pm 0.012)\,M_\pi^{-1}; \quad b_0^{(2)} = (-0.085 \pm 0.011)\,M_\pi^{-3}. \tag{4}$$

The S0 wave experimental situation is somewhat confusing, and we consider two methods of data selection. In both, we fit K_{l4} and $K \to 2\pi$ decay data, in which pions are on the *mass shell*. In the first method, called *global fit*, we include some points at $0.81\,\mathrm{GeV} \le s^{1/2} \le 0.97\,\mathrm{GeV}$, where the various experiments agree within $\lesssim 1.5\sigma$. Care is exercised to compose errors realistically, see details in [5], Subsect 2.2.2. In this case we fix the Adler zero at $z_0 = M_\pi$ and find

$$\cot \delta_0^{(0)}(s) = \frac{s^{1/2}}{2k} \frac{M_\pi^2}{s - \frac{1}{2}z_0^2} \frac{M_\sigma^2 - s}{M_\sigma^2} \left\{ B_0 + B_1 \frac{\sqrt{s} - \sqrt{s_0 - s}}{\sqrt{s} + \sqrt{s_0 - s}} \right\}, \tag{5}$$

$$B_0 = 21.04, \quad B_1 = 6.62, \quad M_\sigma = 782 \pm 24\,\mathrm{MeV}; \quad \delta_0^{(0)}(m_K) = 41.0° \pm 2.1°;$$

$$a_0^{(0)} = (0.230 \pm 0.010)M_\pi^{-1}, \quad b_0^{(0)} = (0.268 \pm 0.011)M_\pi^{-3};$$

this fit (shown in Fig 1.b as PY) is valid for $s^{1/2} \le 0.95\,\mathrm{GeV}$. The B_i errors are strongly correlated; uncorrelated errors are obtained if using the parameters x, y with

$$B_0 = y - x; \quad B_1 = 6.62 - 2.59x; \quad y = 21.04 \pm 0.70, \quad x = 0 \pm 2.6. \tag{6}$$

The other method is to fit only K_{l4} and $K \to 2\pi$ data, or to add to this, individually, data from the various experimental analyses. The results can be found in Table 1.

The S0, S2 and P partial waves at $1\,\mathrm{GeV} \lesssim s^{1/2} \lesssim 1.42\,\mathrm{GeV}$

The D and F data are scanty, and have large errors. To stabilize the fits we impose the values of the scattering lengths that follow from the Froissart–Gribov representation. This is not circular reasoning since their Froissart–Gribov representation depends mostly on the S0, S2 and P waves, and very little on the D0, D2, F waves themselves. We do not discuss here the D0 and F waves (see [5]) as they do not present special features.

For D2 we only expect important inelasticity when the $\pi\pi \to \rho\rho$ channel opens up, so that $s_0 = 1.45^2\,\mathrm{GeV}^2 \sim 4M_\rho^2$. A pole term is necessary here, since we expect $\delta_2^{(2)}$ to change sign near threshold: the data [7] give negative and small values for $\delta_2^{(2)}$ above some $500\,\mathrm{MeV}$, while, from the Froissart–Gribov representation, it is known[11] that the scattering length must be positive. Indeed we include in the fit the value $a_2^{(2)} = (2.72 \pm 0.36) \times 10^{-4}\,M_\pi^{-5}$. In addition, the clear inflection seen in data around $1\,\mathrm{GeV}$ asks for a third order conformal expansion. So we write

$$\cot \delta_2^{(2)}(s) = \frac{s^{1/2}}{2k^5} \left\{ B_0 + B_1 w(s) + B_2 w(s)^2 \right\} \frac{M_\pi^4 s}{4(M_\pi^2 + \Delta^2) - s}, \quad w(s) = \frac{\sqrt{s} - \sqrt{s_0 - s}}{\sqrt{s} + \sqrt{s_0 - s}}.$$

And we find $B_0 = (2.4 \pm 0.3) \times 10^3$, $B_1 = (7.8 \pm 0.8) \times 10^3$, $B_2 = (23.7 \pm 3.8) \times 10^3$, $\Delta = 196 \pm 20\,\mathrm{MeV}$. The fit, which may be found in Fig 2, returns reasonable numbers for the scattering length and for the effective range parameter, $b_2^{(2)}$:

$$a_2^{(2)} = (2.5 \pm 0.9) \times 10^{-4}\,M_\pi^{-5}; \quad b_2^{(2)} = (-2.7 \pm 0.8) \times 10^{-4}\,M_\pi^{-7}. \tag{7}$$

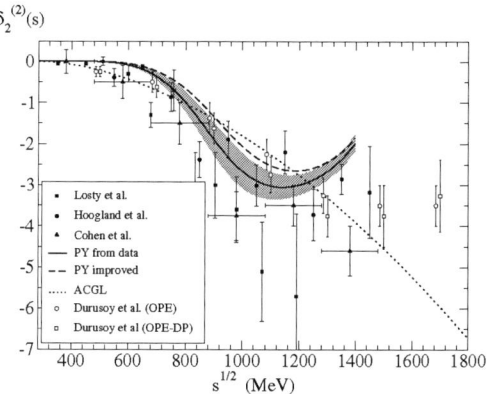

FIGURE 2. Continuous line: The $I = 2$, D-wave phase shift, obtained by only fitting the experimental data. Broken line: with the parameters improved using dispersion relations. Dotted line: the fit, valid between $s^{1/2} = 0.625$ GeV and 1.375 GeV, of Martin, Morgan and Shaw which ACGL and CGL, however, use from threshold to $s^{1/2} = 2$ GeV. The experimental points are from [7].

THE HIGH ENERGY ($S^{1/2} \geq 1.42$ GEV) INPUT

In order to test dispersion relations we also need the imaginary part of the scattering amplitude at $s^{1/2} \geq 1.42$ GeV, that we take from a Regge fit to data [4] (and the slightly improved rho residue of [5]). We note that, in the early 1970s, when $\pi\pi$ phase shifts were poorly known and, above all, when it was still not clear that the standard Regge picture is a QCD feature, Regge factorization was questioned [12] using crossing sum rules and then-existing low energy phase shift data. This was adopted later by ACGL and CGL, assuming a too large rho residue and a Pomeron a *third* of what factorization and the experimental data on the total $\pi\pi$ cross section implies, as well as unconventional slopes. Unfortunately this has been also used in subsequent Roy equation analyses. As discussed in [4, 5], however, standard Regge factorization describes experiment [13, 14] and is perfectly consistent with crossing sum rules if assumed to hold above 1.42 GeV.

In Fig 3 we show our Regge description of the imaginary parts of $\pi\pi$ scattering amplitudes [4, 5, 14] together with the data[13], compared with that used by ACGL[1], CGL[2] above 2 GeV.

Between 1.42 GeV $\leq s^{1/2} \leq 2$ GeV, these authors use the scattering amplitude reconstructed from one Cern–Munich phase shift analysis, and, in particular for S0, the re-elaboration of Au, Morgan and Pennington [10]. Unfortunately, in this region the inelasticity is large and the Cern–Munich experiments, which only measure the *differential* cross section for $\pi\pi \to \pi\pi$ are insufficient to reconstruct without ambiguity the full imaginary part. In addition, in [3, 5] we showed that Cern–Munich phases fail to pass a number of consistency tests. This is also seen clearly in Fig 3, where we plot the total cross section for $\pi^+\pi^-$ that follows Hyams et al.,[10], which is incompatible with other experimental data [13], as well as with Regge factorization.

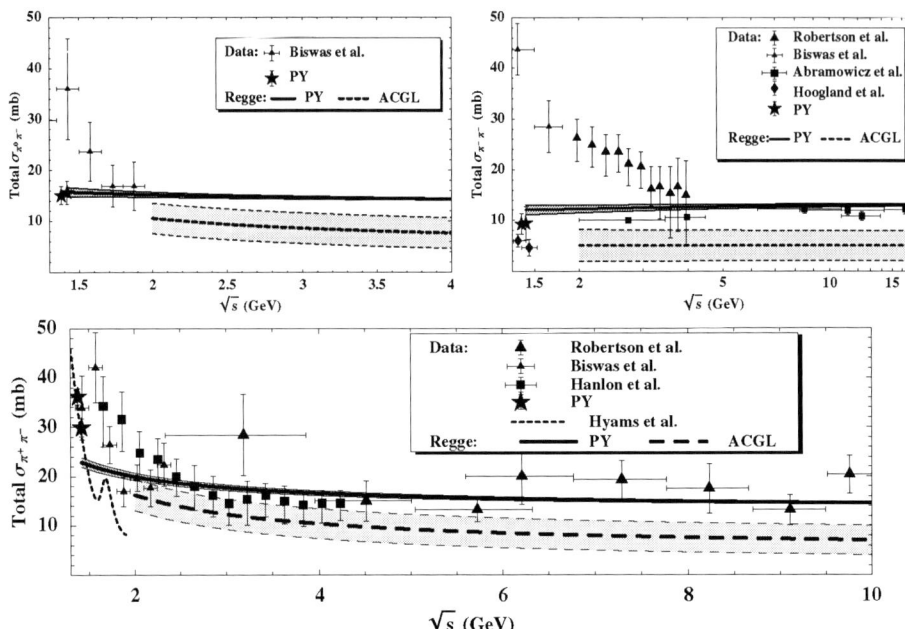

FIGURE 3. The $\pi\pi$ cross sections. Experimental points from [13]. The stars at 1.38 and 1.42 GeV (PY) are from the phase shift analysis of experimental data given in [5]. Continuous lines, from 1.42 GeV (PY): Regge formula, with parameters as in [4] (the three lines per fit cover the error in the theoretical values of the Regge residues). Dashed lines, above 2 GeV: the cross sections following from ACGL;[1] the gray band covers their error band. Below 2 GeV, the dotted line corresponds to the $\pi^+\pi^-$ cross section from the Cern–Munich analysis; cf. Fig 7 in the paper of Hyams et al.[10]

In a recent paper Caprini, Colangelo, Gasser and Leutwyler,[17] to be denoted by CCGL, review our work in [3] and conclude that, still, they consider the CGL solution consistent. They also raised the contention that our Reggeistics could not be correct because it violates certain sum rules. In view of Fig.3 this contention is meaningless since the PY cross sections are perfectly compatible with high energy ($s^{1/2} \geq 1.42\,\text{GeV}$) *experimental* data, while the ACGL ones are not. In [3, 4, 5] we also checked that our representation satisfies two crossing sum rules.

Concerning D2, ACGL and CGL borrow an old fit in the book of Martin, Morgan and Shaw,[15] where only intermediate energy data were fitted.

$$\delta_2^{(2)}(s) = -0.003(s/4M_\pi^2)\left(1 - 4M_\pi^2/s\right)^{5/2}, \tag{8}$$

which fails at threshold (it gives a negative scattering length) and does not fit well data below 1.42 GeV, as shown in Fig 2. Above 1 GeV, this D2 phase grows quadratically with the energy, while Regge theory predicts all phases to go to a multiple of π. In particular D2 should go to zero; see Appendix C of [5] for details. It is true that this D2 wave is small but, given the accuracy claimed by CGL, it is certainly not negligible.

52

CHECKING FORWARD DISPERSION RELATIONS

In the present Section we study how well the previous amplitudes obtained from fits to different sets of data satisfy forward dispersion relations. We consider three independent scattering amplitudes in t-symmetric or antisymmetric combinations, that form a complete set: $\pi^0\pi^0 \to \pi^0\pi^0$, $\pi^0\pi^+ \to \pi^0\pi^+$, and the t channel isospin one amplitude, $I_t = 1$. The reason is that the two first depend only on two isospin states, and have positivity properties: their imaginary parts are sums of positive terms, thus reducing the final uncertainties. Hence, for $\pi^0\pi^0$, we have

$$\operatorname{Re} F_{00}(s) - F_{00}(4M_\pi^2) = \frac{s(s - 4M_\pi^2)}{\pi} \text{P.P.} \int_{4M_\pi^2}^{\infty} ds' \frac{(2s' - 4M_\pi^2)\operatorname{Im} F_{00}(s')}{s'(s' - s)(s' - 4M_\pi^2)(s' + s - 4M_\pi^2)}. \quad (9)$$

In particular, for $s = 2M_\pi^2$, which will be important for the Adler zeros, we have

$$F_{00}(4M_\pi^2) = F_{00}(2M_\pi^2) + D_{00}, \quad D_{00} = \frac{8M_\pi^4}{\pi} \int_{4M_\pi^2}^{\infty} ds \frac{\operatorname{Im} F_{00}(s)}{s(s - 2M_\pi^2)(s - 4M_\pi^2)}. \quad (10)$$

For the $\pi^0\pi^+$ channel, which does not depend on S0:

$$\operatorname{Re} F_{0+}(s) - F_{0+}(4M_\pi^2) = \frac{s(s - 4M_\pi^2)}{\pi} \text{P.P.} \int_{4M_\pi^2}^{\infty} ds' \frac{(2s' - 4M_\pi^2)\operatorname{Im} F_{0+}(s')}{s'(s' - s)(s' - 4M_\pi^2)(s' + s - 4M_\pi^2)}.$$

At the point $s = 2M_\pi^2$, this becomes

$$F_{0+}(4M_\pi^2) = F_{0+}(2M_\pi^2) + D_{0+}, \quad D_{0+} = \frac{8M_\pi^4}{\pi} \int_{8M_\pi^2}^{\infty} ds \frac{\operatorname{Im} F_{0+}(s)}{s(s - 2M_\pi^2)(s - 4M_\pi^2)}. \quad (11)$$

Finally, for isospin unit exchange, which does not require subtractions,

$$\operatorname{Re} F^{(I_t=1)}(s,0) = \frac{2s - 4M_\pi^2}{\pi} \text{P.P.} \int_{4M_\pi^2}^{\infty} ds' \frac{\operatorname{Im} F^{(I_t=1)}(s',0)}{(s' - s)(s' + s - 4M_\pi^2)}. \quad (12)$$

at threshold this is known as the Olsson sum rule.

Depending on the method we use to fit the S0 wave we find the results in Table 1, where, we have separated on top those fits to data with a total $\chi^2/d.o.f. < 6$ for the $\pi^0\pi^0$ and $I_t = 1$ dispersion relations up to 0.925 GeV, a fairly reasonable $\chi^2/d.o.f.$ since these fits were obtained independently of the dispersive approach.

However, in Table 1 we also list the very frequently used t and s-channel solutions of Estabrooks and Martin [10], those of Protopopescu et al.[9], from Table VI, VIII and table XII, as well as the solution A of Grayer et al. [10]. Their $I_t = 1$ plus $\pi^0\pi^0$ dispersion relation total $\chi^2/d.o.f.$ is surprisingly poor: 11.3, 10.1, 7, 6, 7.5, 9.9, respectively. *Therefore, any result that relies heavily on these sets should be taken very cautiously.*

TABLE 1. PY: our global fit, Eqs.(5,6). We do not give its B_0 and B_1 uncertainties as they are strongly correlated, see Eq.(6) for the uncorrelated ones. Grayer B, C, E: different solutions in Grayer et al.[10]. Kaminski: [10]. In [5] we have also studied fits to the data in Tables VI, XII and VIII in [9], to Solution A in[10], as well as fits to the theoretical outcome in Estabrooks and Martin.[10]. They all give a total $\chi/\text{d.o.f.} \geq 6$

	B_0	B_1	M_σ (MeV)	$\dfrac{I_t=1}{\chi^2}$ d.o.f.	$\dfrac{\pi^0\pi^0}{\chi^2}$ d.o.f.	$\delta_0^{(0)}(0.8^2)$
PY, Eqs.(5,6)	21.04	6.62	782 ± 24	0.3	3.5	$91.9°$
K decay only	18.5 ± 1.7	$\equiv 0$	766 ± 95	0.2	1.8	$93.2°$
K decay data + Grayer, B	22.7 ± 1.6	12.3 ± 3.7	858 ± 15	1.0	2.7	$84.0°$
K decay data + Grayer, C	16.8 ± 0.85	-0.34 ± 2.34	787 ± 9	0.4	1.0	$91.1°$
K decay data + Grayer, E	21.5 ± 3.6	12.5 ± 7.6	1084 ± 110	2.1	0.5	$70.6°$
K decay data + Kaminski	27.5 ± 3.0	21.5 ± 7.4	789 ± 18	0.3	5.0	$91.6°$
K decay data + Grayer, A	28.1 ± 1.1	26.4 ± 2.8	866 ± 6	2.0	7.9	$81.2°$
K decay data + EM, s−channel	29.8 ± 1.3	25.1 ± 3.3	811 ± 7	1.0	9.1	$88.3°$
K decay data + EM, t−channel	29.3 ± 1.4	26.9 ± 3.4	829 ± 6	1.2	10.1	$85.7°$
K decay data + Protopopescu VI	27.0 ± 1.7	22.0 ± 4.1	855 ± 10	1.2	5.8	$82.9°$
K decay data + Protopopescu XII	25.5 ± 1.7	18.5 ± 4.1	866 ± 14	1.2	6.3	$82.2°$
K decay data + Protopopescu 3	27.1 ± 2.3	23.8 ± 5.0	913 ± 18	1.8	4.2	$76.7°$

IMPROVED FITS USING DISPERSION RELATIONS

We now improve the previous low energy fits parameters by fitting also the dispersion relations up to 0.925 GeV, thus obtaining parametrizations more compatible with analyticity and $s - u$ crossing. This is an alternative method to Roy equations; it is better in that we do not need as input the scattering amplitude for $|t|$ up to $30M_\pi^2$, where the Regge fits existing in the literature disagree strongly (see [5], Appendix B) and also in that we can test all energies[5], whereas Roy equations are valid for $s^{1/2} < \sqrt{60}M_\pi \sim 1.1$ GeV (and only applied up to 0.8 GeV). Starting from Eqs.(5,6), we find, in M_π units,

$$\text{S0; } s^{1/2} \leq 2m_K: \quad B_0 = 17.4 \pm 0.5; \; B_1 = 4.3 \pm 1.4;$$
$$M_\sigma = 790 \pm 21 \text{ MeV}; \; z_0 = 195 \text{ MeV [Fixed]};$$
$$a_0^{(0)} = 0.230 \pm 0.015; \; b_0^{(0)} = 0.312 \pm 0.014.$$

$$\text{S2; } s^{1/2} \leq 1.0: \quad B_0 = -80.8 \pm 1.7; \; B_1 = -77 \pm 5; \; z_2 = 147 \text{ MeV [Fixed]};$$
$$a_0^{(2)} = -0.0480 \pm 0.0046; \; b_0^{(2)} = -0.090 \pm 0.006.$$

$$S2;\ 1.0 \le s^{1/2} \le 1.42: \quad B_0 = -125 \pm 6;\ B_1 = -119 \pm 14;\ \varepsilon = 0.17 \pm 0.12.$$

$$P;\ s^{1/2} \le 1.05: \quad B_0 = 1.064 \pm 0.11;\ B_1 = 0.170 \pm 0.040;\ M_\rho = 773.6 \pm 0.9 \text{ MeV};$$
$$a_1 = (38.7 \pm 1.0) \times 10^{-3};\ b_1 = (4.55 \pm 0.21) \times 10^{-3}.$$

$$D0;\ s^{1/2} \le 1.42: \quad B_0 = 23.5 \pm 0.7;\ B_1 = 24.8 \pm 1.0;\ \varepsilon = 0.262 \pm 0.030;$$
$$a_2^{(0)} = (18.4 \pm 3.0) \times 10^{-4};\ b_2^{(0)} = (-8.6 \pm 3.4) \times 10^{-4}.$$

$$D2;\ s^{1/2} \le 1.42: \quad B_0 = (2.9 \pm 0.2) \times 10^3;\ B_1 = (7.3 \pm 0.8) \times 10^3;$$
$$B_2 = (25.4 \pm 3.6) \times 10^3;\ \Delta = 212 \pm 19;$$
$$a_2^{(2)} = (2.4 \pm 0.7) \times 10^{-4};\ b_2^{(2)} = (-2.5 \pm 0.6) \times 10^{-4}.$$

$$F;\ s^{1/2} \le 1.42: \quad B_0 = (1.09 \pm 0.03) \times 10^5;\ B_1 = (1.41 \pm 0.04) \times 10^5;$$
$$a_3 = (7.0 \pm 0.8) \times 10^{-5}. \tag{13}$$

In Fig.4 we show the improved curves for S0 and S2, and that of D2 in Fig.2.

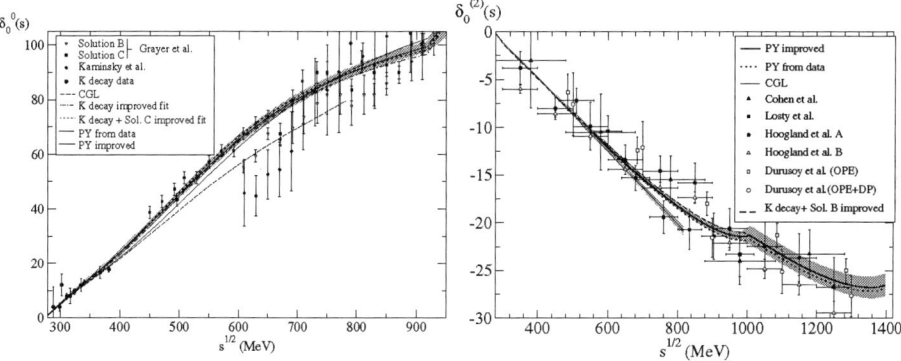

FIGURE 4. a) The improved S0 phase shift (PY improved, Eq.13), the global fit (PY from data, he S0 Eqs. (5,6)), and the *improved* solutions "K decay only" and "Grayer C" of Table 2 (almost on top of PY improved). The solution CGL[2] (dashed line) is also shown. b) S2 improved Phase shift (PY improved, Eq. (13)); global fit (PY from data, Eq. (4)); the solution CGL [2] (thin continuous line) and the improved parametrization with K decays and So. B of Grayer et al.[10].

Concerning the improved fits to individual sets of data, we get somewhat different results for S0, listed in Table 2. In Table 2 we also show the $\chi^2/d.o.f.$ of each forward dispersion relation and the standard deviations for the sum rule in Eq.10, which are more than four for K decay plus the Grayer B or E or Kaminski improved solutions. Concerning the other waves, no matter what set of parameters from data fits we start from, we end up with very similar values to those given in Eq.13. This can be checked in Fig.4.b, where we show the improved "K decay + Grayer Sol. B" S2 wave. Even though it is the one for which we obtained the most different central values for the S0 wave compared with those given in Eq.13, it falls perfectly within the uncertainty of our improved solution.

TABLE 2. Improved fits. Names are as in Table 1. Although errors are given for the Adler zero, we fix it when evaluating other errors, to break the otherwise very large correlations

Improved fits:	Improved PY, Eq.13	K decay only	K decay +grayer C	K decay +Grayer B	K decay +Grayer E	K decay + Kamiński
B_0	17.4 ± 0.5	16.4 ± 0.9	16.2 ± 0.7	20.7 ± 1.0	20.2 ± 2.2	20.8 ± 1.4
B_1	4.3 ± 1.4	$\equiv 0$	0.5 ± 1.8	11.6 ± 2.6	8.4 ± 5.2	13.6 ± 43.7
M_σ (MeV)	790 ± 30	809 ± 53	788 ± 9	861 ± 14	982 ± 95	798 ± 17
z_0 (MeV)	195 ± 30	182 ± 34	182 ± 39	233 ± 30	272 ± 50	245 ± 39
$I_t = 1$ $\chi^2/d.o.f.$	0.40	0.30	0.37	0.37	0.60	0.43
$\pi^0\pi^0$ $\chi^2/d.o.f.$	0.66	0.29	0.32	0.83	0.09	1.08
$\pi^+\pi^-$ $\chi^2/d.o.f.$	1.62	1.77	1.74	1.60	1.40	1.36
Eq.(10)	1.6σ	1.5σ	1.5σ	4.0σ	6.0σ	4.5σ
$\delta_0^{(0)}(0.8^2\text{GeV}^2)$	$91.3°$	$91.3°$	$91.0°$	$85.1°$	$78.0°$	$91.8°$

DISPERSION RELATIONS AND THE CGL SOLUTION

We have also checked the fulfillment of forward dispersion relations for the CGL solution for the S0, S2 and P waves al low energy. This is depicted in Fig 5, where we show, both for CGL and our improved fit, Eq.(13), the mismatch between the real part and the dispersive evaluations, that is to say, the differences Δ_i,

$$\Delta_1 \equiv \operatorname{Re} F^{(I_t=1)}(s,0) - \frac{2s - 4M_\pi^2}{\pi} \,\text{P.P.}\! \int_{4M_\pi^2}^\infty \mathrm{d}s' \, \frac{\operatorname{Im} F^{(I_t=1)}(s',0)}{(s'-s)(s'+s-4M_\pi^2)}, \quad (14)$$

$$\Delta_{00} \equiv \operatorname{Re} F_{00}(s) - F_{00}(4M_\pi^2) \quad (15)$$
$$- \frac{s(s-4M_\pi^2)}{\pi} \,\text{P.P.}\! \int_{4M_\pi^2}^\infty \mathrm{d}s' \, \frac{(2s'-4M_\pi^2)\operatorname{Im} F_{00}(s')}{s'(s'-s)(s'-4M_\pi^2)(s'+s-4M_\pi^2)},$$

$$\Delta_{0+} \equiv \operatorname{Re} F_{0+}(s) - F_{0+}(4M_\pi^2) \quad (16)$$
$$- \frac{s(s-4M_\pi^2)}{\pi} \,\text{P.P.}\! \int_{4M_\pi^2}^\infty \mathrm{d}s' \, \frac{(2s'-4M_\pi^2)\operatorname{Im} F_{0+}(s')}{s'(s'-s)(s'-4M_\pi^2)(s'+s-4M_\pi^2)}.$$

These quantities would vanish, $\Delta_i = 0$, if the dispersion relations were exactly satisfied.

We include in the comparison of Fig 5 the uncertainties; in the case of CGL, these errors are as follow from the parametrizations given by these authors in [2], for $s^{1/2} \lesssim 0.8$ GeV. At higher energies they are taken from data via our parametrizations. It can be clearly seen that the CGL parametrizations do not satisfy these forward dispersion relations by several standard deviations.

One might wonder why in Table 2, the S0 wave improved "K decay+solution B" yields $\chi^2/d.o.f.$ of order one for the forward dispersion relations, being so similar to what CGL get for that wave. The reason is that, as we show in Fig.4.b. the improved solution B, requires an S2 wave that is even farther from the CGL S2 wave than our global improved solution.

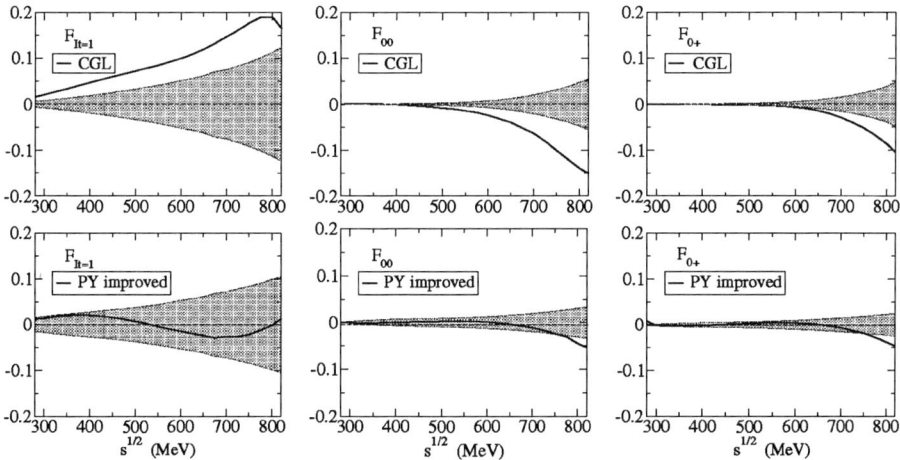

FIGURE 5. Dispersion relations for the $\pi\pi$ amplitudes of [2] (CGL) and for our improved global fit (PY improved, Eq.13. We plot the differences Δ_i, Eqs. 14, between real parts calculated directly from the parametrizations, or from the dispersive formulas. Consistency within one sigma occurs within the shaded bands. The progressive deterioration of the CGL results as the energy increases is apparent here.

LOW ENERGY PARAMETERS IN THE LITERATURE

We here present, in Table 3, the low energy parameters obtained from Roy equations by CGL, by Descotes et al.[15], that we denote by DFGS, and by Kamiński, Leśniak and Loiseau[15], denoted by KLL. This is compared with what we find fitting experimental data, improved with dispersion relations (see [5] for details).

The mismatches between many of the parameters of CGL and PY are apparent here; not surprisingly, they affect mostly parameters sensitive to high energy:

- The S wave parameters are compatible, mainly due to the large PY uncertainties.
- The mismatch between CGL and PY for $a_2^{(I)}$ and $b_2^{(I)}$ is roughly 2.5σ. In ref.[3] we pointed out that the ACGL and CGL results did not satisfy the Froissart Gribov sum rules. This happens to more than four standard deviations for the difference between the CGL calculation using Wanders sum rules minus the Froissart-Gribov representation. This larger mismatch, as pointed out in [17], does not involve the S and P waves, and is due to the Regge and $L \geq 2$ wave input, evidently very different from the beginning for CGL and PY, but that certainly affects the values of $a_2^{(I)}$ and $b_2^{(I)}$.
- The b_1 calculation differs by more than 4 standard deviations for PY and CGL.

TABLE 3. Units of M_π. The numbers in the CGL column are as given by CGL in Table 2 and elsewhere in their text. In PY, the values for the D, F waves parameters are from the Froissart–Gribov representation. The rest are from the fits, improved with dispersion relations, except for a_1 and b_1 that have been taken as in [5].

	DFGS	KLL	CGL	PY
$a_0^{(0)}$	0.228 ± 0.032	0.224 ± 0.013	0.220 ± 0.005	0.230 ± 0.015
$a_0^{(2)}$	-0.0382 ± 0.0038	-0.0343 ± 0.0036	-0.0444 ± 0.0010	-0.0480 ± 0.0046
$b_0^{(0)}$		0.252 ± 0.011	0.280 ± 0.001	0.312 ± 0.014
$b_0^{(2)}$		-0.075 ± 0.015	-0.080 ± 0.001	-0.090 ± 0.006
$a_1 \times 10^3$		39.6 ± 2.4	37.9 ± 0.5	38.4 ± 0.8
$b_1 \times 10^3$		2.83 ± 0.67	5.67 ± 0.13	4.75 ± 0.16
$a_2^{(0)} \times 10^4$			17.5 ± 0.3	18.70 ± 0.41
$a_2^{(2)} \times 10^4$			1.70 ± 0.13	2.78 ± 0.37
$b_2^{(0)} \times 10^4$			-3.55 ± 0.14	-4.16 ± 0.30
$b_2^{(2)} \times 10^4$			-3.26 ± 0.12	-3.89 ± 0.28
$a_3 \times 10^5$			5.6 ± 0.2	6.3 ± 0.4

THE ACGL, CGL PHASE AT $S^{1/2} = 0.8\,\mathrm{GEV}$

In the ACGL, CGL analyses, by the input phases for the S0, S2 and P waves at the point, $s^{1/2} = 0.8$ GeV, where they match the solutions to the Roy equations to the experimental amplitude. Indeed it is dominant for their Olsson sum rule calculation, which involves the $I_t = 1$ channel.

The quantity $\delta_0^{(0)}((0.8\,\mathrm{GeV})^2)$ is in fact given in Eq. (7.3) of ACGL as

$$\delta_0^{(0)}((0.8\,\mathrm{GeV})^2) = 82.3 \pm 3.4^\circ. \qquad (17)$$

whose error may be contrasted with the estimates of [5], which vary, for the data above 0.8 GeV, between 6° and 18°, or with the $\delta_0^{(0)}((0.8\,\mathrm{GeV})^2)$ values we obtain from fits to different sets of data in Table 1, or the improved fits in Table 2. Small errors could be expected from theoretical analysis including many data but the above small error was used as an *input*.

The reason to consider such an small error is that ACGL consider the difference $\delta_1 - \delta_0^{(0)}$ at 0.8 GeV, in the hope that some of the uncertainties will cancel. Then they interpolate and then average the points from a choice of three analysis of the CERN/Munich experiment[10]

$$\delta_1((0.8\,\mathrm{GeV})^2) - \delta_0^{(0)}((0.8\,\mathrm{GeV})^2) = \begin{cases} 23.4 \pm 4.0^\circ & \text{[Hyams et al.]} \\ 24.8 \pm 3.8^\circ & \text{[Estabrooks and Martin, } s\text{-channel]} \\ 30.3 \pm 3.4^\circ & \text{[Estabrooks and Martin, } t\text{-channel].} \end{cases}$$

and set

$$\delta_0^{(0)}((0.8\,\mathrm{GeV})^2) - \delta_1((0.8\,\mathrm{GeV})^2) = 26.6 \pm 2.8^\circ.$$

However, this error does not include systematics. All numbers here stem from the *same experiment*, and differ only on the method of analysis. Their spread is an indication of the *systematic* uncertainties, roughly an additional $\pm 4°$. In addition, the Hyams et al. value above is only one of *five* solutions in Grayer et al.[10], and considering also data of Protopopescu et al.,[9] the systematic error would increase to $10°$. Remarkably, Estabrooks and Martin themselves, point out in their section 4 (first paragraph) that different D wave input "lead to systematic changes in δ_S^0 of the order of $10°$".

ACKNOWLEDGMENTS

F.J. Ynduráin is grateful to the organizing committee for the opprtunity to talk at the meeting and for financial support.

REFERENCES

1. Ananthanarayan, B., Colangelo, G., Gasser, J., and Leutwyler, H., *Phys. Rep.*, **353**, 207, (2001).
2. Colangelo, G., Gasser, J., and Leutwyler, H., *Nucl. Phys.* **B603**, 125, (2001).
3. Peláez, J. R., and Ynduráin, F. J., *Phys. Rev.* **D68**, 074005 (2003).
4. Peláez, J. R., and Ynduráin, F. J., *Phys. Rev.* **D69**, 114001 (2004).
5. Peláez, J. R., and Ynduráin, F. J., FTUAM 04-14 hep-ph/0411334.
6. de Trocóniz, J. F., and Ynduráin, F. J., *Phys. Rev.*, **D65**, 093001, (2002) and hep-ph/0402285. When quoting numbers, we will quote from this last paper.
7. Losty, M. J., et al. *Nucl. Phys.*, **B69**, 185 (1974); Hoogland, W., et al. *Nucl. Phys.*, **B126**, 109 (1977); Cohen, D. et al., *Phys. Rev.* **D7**, 661 (1973); Durusoy, N. B., et al., *Phys. Lett.* **B45**, 517 (19730.
8. Rosselet, L., et al. *Phys. Rev.* **D15**, 574 (1977); Pislak, S., et al. *Phys. Rev. Lett.*, **87**, 221801 (2001).
9. Protopopescu, S. D., et al., *Phys Rev.* **D7**, 1279, (1973).
10. Cern-Munich experiment: Hyams, B., et al., *Nucl. Phys.* **B64**, 134, (1973); Grayer, G., et al., *Nucl. Phys.* **B75**, 189, (1974). See also the analysis of the same data in Estabrooks, P., and Martin, A. D., *Nucl. Physics*, **B79**, 301, (1974); Kamiński, R., Lesniak, L, and Rybicki, K., *Z. Phys.* **C74**, 79 (1997) and *Eur. Phys. J. direct* **C4**, 4 (2002); Au, K. L., Morgan, D., and Pennington, M. R. *Phys. Rev.* **D35**, 1633 (1987).
11. Palou, F. P., and Ynduráin, F. J., *Nuovo Cimento*, **19A**, 245, (1974); Palou, F. P., Sánchez-Gómez, J. L., and Ynduráin, F. J., *Z. Phys.*, **A274**, 161, (1975).
12. Pennington, M. R., *Ann. Phys.* (N.Y.), **92**, 164, (1975).
13. Biswas, N. N., et al., *Phys. Rev. Letters*, **18**, 273 (1967) [$\pi^-\pi^-$, $\pi^+\pi^-$ and $\pi^0\pi^-$]; Cohen, D. et al., *Phys. Rev.* **D7**, 661 (1973) [$\pi^-\pi^-$]; Robertson, W. J., Walker, W. D., and Davis, J. L., *Phys. Rev.* **D7**, 2554 (1973) [$\pi^+\pi^-$]; Hoogland, W., et al. *Nucl. Phys.*, **B126**, 109 (1977) [$\pi^-\pi^-$]; Hanlon, J., et al, *Phys. Rev. Letters*, **37**, 967 (1976) [$\pi^+\pi^-$]; Abramowicz, H., et al. *Nucl. Phys.*, **B166**, 62 (1980) [$\pi^+\pi^-$]. These references cover the region between 1.35 and 16 GeV, and agree within errors in the regions where they overlap (with the exception of $\pi^-\pi^-$ below 2.3 GeV, discussion in [5]).
14. Pelaez, J. R. Proceedings of MESON 2004, Cracow, Poland, 4-8 Jun 2004. arXiv:hep-ph/0407213.
15. Martin, B. R., Morgan, D., and and Shaw, G. *Pion-Pion Interactions in Particle Physics*, Academic Press, New York (1976).
16. Descotes, S. *et al.*, Eur. Phys. J. C, **24**, 469, (2002); Kamiński, R., Leśniak, L., and Loiseau, B. *Phys. Letters*, **B551**, 241 (2003).
17. Caprini, I., Colangelo, G., Gasser, J., and Leutwyler, H. *Phys. Rev.***D68**,074006 (2003)

$\pi\pi$ scattering, pion form factors and chiral perturbation theory

Gilberto Colangelo

Institut für Theoretische Physik der Universität Bern
Sidlerstr. 5 3012 Bern Switzerland

Abstract. I discuss recent progress in our understanding of the $\pi\pi$ scattering amplitude at low energy thanks to the combined use of chiral perturbation theory and dispersion relations. I also comment on the criticism raised by Peláez and Yndduráin on this work.

1. INTRODUCTION

In the previous conference of this series I was invited to present results [1] concerning the experimental determination of the $\bar{q}q$ condensate in the SU(2) chiral limit [2], based on an analysis of the Brookhaven E865 data [3] on the low-energy $\pi\pi$ phase shift as extracted from K_{e4} decays. These experimental results and their analysis closed a long-standing discussion about the size of this order parameter of chiral symmetry breaking in QCD (cf. [4] and references therein): the scenario in which the SU(2) $\bar{q}q$ condensate is unexpectedly small, or even vanishing, though interesting, is now experimentally excluded. Only the dependence of this condensate on the number of massless flavours still remains an open issue (cf. [5] and references therein). In the yet earlier conference [6], within a general discussion of recent progress in chiral perturbation theory (CHPT), I had already presented our predictions for the two S-wave scattering lengths [7].

This work on $\pi\pi$ scattering has a few features which are worth stressing:

- the precision reached (at the level of a few percent) is quite unusual for hadronic physics;
- this precision concerns a prediction – experiments have not yet reached the same level of accuracy ("theory is ahead of experiment" as Heiri Leutwyler puts it [8]);
- despite a rather heavy machinery which is necessary to obtain this prediction, the latter does follow from QCD, and indeed the experimental tests tell us something about QCD, as the conclusion about the size of the $\bar{q}q$ condensate shows.

The precision obtained in our theoretical understanding of the $\pi\pi$ scattering amplitude at low energy is not only important *per se*, but has also important consequences for a number of other processes. In almost every low energy hadronic process the interaction among pions plays an important role, and being able to treat this accurately may lead to relevant improvements. An example of this is the anomalous magnetic moment of the muon, where one can make good use of the accurate knowledge of the $\pi\pi$ P-wave phase shift [9].

CP756, Quark Confinement and the Hadron Spectrum VI
edited by N. Brambilla, U. D'Alesio, A. Devoto, K. Maung, G.M. Prosperi and S. Serci
© 2005 American Institute of Physics 0-7354-0241-8/05/$22.50

An essential role in this improved understanding of the $\pi\pi$ scattering amplitude at low energy has been played by the combined use of CHPT and dispersion relations. In CHPT, even after a two-loop calculation of the $\pi\pi$ scattering amplitude, one finds out that only close to the center of the Mandelstam triangle the series converges rather fast, whereas at threshold the convergence is surprisingly slow: a direct evaluation of the scattering lengths in CHPT would not have reached the same precision level [10]. On the other hand a purely dispersive analysis of $\pi\pi$ scattering, as performed in the seventies (for a review of this early work cf. [11]) using Roy equations [12] did not lead to precise predictions either, because of lack of information on the subtraction constants. If one uses CHPT to pin down the latter, the scheme becomes predictive and accurate. In our work we first had to redo the Roy equation analysis [13] and then matched the dispersive representation to the chiral one [14].

Some of the input used in the Roy equation analysis in [13] has been criticised by Peláez and Ynduráin [15], and doubts have been cast on the level of precision reached in our analysis. This criticism has been immediately answered [16]. Also, the more recent objections raised in [17] on the dispersive determination of the scalar radius discussed in [14] were shown to be unfounded [18].

In the present edition of the conference a session has been devoted to a discussion of these issues. In this contribution I will present my view on these issues and on the ongoing discussion – of course, the current view of Peláez and Ynduráin can also be found in these proceedings [19]. Rather then concentrating on the reply to the criticism raised by Peláez and Ynduráin, which is rather technical and can anyway be found in the original papers [16, 18], I will review the work we did on the $\pi\pi$ scattering amplitude and discuss its importance also in view of future experimental tests, as well as tests and comparisons with lattice calculations. I will also briefly discuss the criticism and our reply, but I wish to stress right away that the points raised in [15] have all been answered in [16], and that the claimed violation of a "robust lower bound" [17] in our calculation of the scalar radius has been shown to be a non-issue because this lower bound does not exist [18]. The discussion on these issues is closed. In a more recent paper [20] Peláez and Ynduráin claim that our representation for the $\pi\pi$ scattering amplitude fails to satisfy some dispersion relations. We have not yet evaluated these, and I can therefore not comment on this claimed failure. Moreover, in their contribution to these proceedings they criticize our choice of one of the input parameters in our Roy analysis, the value of the S-wave isoscalar phase shift at 0.8 GeV – I will comment on this point below.

2. $\pi\pi$ SCATTERING: ROY EQUATIONS AND CHPT

In SU(2) CHPT the expansion parameter is $\hat{m}/M_\rho$ and one expects higher order corrections to be of the order of a few percents. There are many known examples in which this naive expectation is violated and corrections are substantially larger. A well known example is the $\pi\pi$ S-wave isoscalar scattering length which has been first evaluated by Weinberg [21] to leading order in the chiral expansion. Numerically this gives $a_0^0(\text{LO}) = 0.16$, but the next-to-leading order corrections, first calculated by Gasser and Leutwyler [22] shift this value by about 25%, $a_0^0(\text{NLO}) = 0.20$. The next-to-next-to-leading order corrections have also been evaluated [10] and have been found to be not yet negligible, shifting the value by another 10% up to $a_0^0(\text{NNLO}) = 0.22$. The error estimate for this quantity is not trivial: first of all one has to determine a number of low-energy constants (LEC) which appear in the chiral expansion of this quantity and estimate the corresponding error. Second, one has to estimate the size of yet higher order corrections. At first sight going below the 10% level for the total error appears to be difficult. The reason for the large size of the higher order corrections for this quantity, however, is well understood and is due to the strong interaction of the pions in the $I = 0$ S wave: the perturbative expansion for these unitarity corrections converges slowly. In a dispersive framework, on the other hand, these unitarity corrections can be treated exactly.

If one combines the dispersive and the chiral approaches one can make a quantum jump in accuracy: the dispersive treatment can be used to evaluate the unitarity corrections which are problematic in the chiral expansion, and CHPT can be used to fix the subtraction constants which represent the only true degrees of freedom in the dispersive treatment at low energy. The crucial point is that if one chooses the subtraction constants properly, the chiral expansion for these does indeed follow the naive expectations about the size of the higher order corrections. Moreover, in the low energy region, the dispersive treatment does lead to very precise results.

I will now illustrate in some more detail this program and first discuss the Roy equations and their numerical solution and then the matching to the chiral representation and the numerical prediction for the scattering lengths.

2.1. Roy equations

In 1971 Roy [12] showed that using crossing one can write a set of dispersion relations for the $\pi\pi$ scattering amplitude which involve only physical region singularities. When projected onto partial waves these equations take the form of an infinite set of integral equations in which the real part of any partial wave is given by an integral over the imaginary parts of all partial waves in the physical region. At low energy (say below 1 GeV) the S and P waves dominate, and it suffices to consider the equations only for these lowest partial waves. For example, the Roy equation for the S, $I = 0$ wave reads as follows

$$\text{Re}\, t_0^0(s) = k_0^0(s) + \int_{4M_\pi^2}^{E_0^2} ds' K_{00}^{00}(s, s') \, \text{Im}\, t_0^0(s') + \int_{4M_\pi^2}^{E_0^2} ds' K_{01}^{01}(s, s') \, \text{Im}\, t_1^1(s')$$

$$+ \int_{4M_\pi^2}^{E_0^2} ds' K_{00}^{02}(s, s') \, \text{Im}\, t_0^2(s') + f_0^0(s) + d_0^0(s) \,, \tag{1}$$

where k_0^0 is the contribution of the subtraction polynomial, f_0^0 the contribution from the intermediate energy region and d_0^0 the so-called driving term containing both the contribution from the high-energy region as well as that from the higher partial waves:

$$k_0^0(s) = a_0^0 + \frac{s - 4M_\pi^2}{12M_\pi^2} \left(2a_0^0 - 5a_0^2\right)$$

62

$$f_0^0(s) = \sum_{I'=0}^{2} \sum_{\ell'=0}^{1} \int_{E_0^2}^{E_1^2} ds' K_{0\ell'}^{0I'}(s,s') \operatorname{Im} t_{\ell'}^{I'}(s')$$

$$d_0^0(s) = \text{all the rest .} \tag{2}$$

The two energies $E_{0,1}$ were chosen in [13] to be $E_0 = 0.8$ GeV and $E_1 = 2$ GeV. With this choice of E_0 the solution has been shown to be unique [23]. The f_0^0 term is evaluated using the imaginary parts $\operatorname{Im} t_\ell^I$ measured in $\pi N \to \pi\pi N$ experiments, whereas the driving terms are evaluated using experimental information on the lowest lying resonances in D and higher partial waves, and Regge representations for the high-energy $\pi\pi$ scattering amplitude. The kernels $K_{II'}^{\ell\ell'}(s,s')$ are all known explicitly [13].

The upshot of the analysis in [13] is that if the input above E_0 is given, the solution of the equations below E_0 is uniquely fixed by the two scattering lengths a_0^0 and a_0^2. Actually only one of the two is a true free parameter if the solution has to be physical, i.e. if it has no cusps at E_0. Since the input above E_0 is not known with infinite accuracy, this correlation among the two input parameters is not a line, but gets broadened into a band known as the Universal Band. A similar correlation among the two scattering lengths is also given by the Olsson sum rule [24].

2.2. Matching the chiral and the dispersive representation

In the early literature on Roy equations the subtraction constants were taken as free parameters – in the region below the matching point E_0 these were the main sources of uncertainty and it was difficult to turn the Roy equation machinery into a predictive scheme. The best use one could make of Roy equation solutions was to analyze data in the low-energy region, like those from K_{e4} decays, in order to *determine* the scattering lengths – in much the same way as the E865 collaboration [3] has used our Roy equation solutions [13]. The point of view has changed drastically once it has become clear that CHPT can provide rather accurate predictions for the scattering lengths [22].

As mentioned above, however, the scattering lengths are not the quantities that CHPT can most accurately predict. Since the choice of the subtraction point is arbitrary, one can exploit this freedom and subtract below threshold, close to the center of the Mandelstam triangle, in order to be far from the singularities that make the chiral series converge slowly. By doing so one can optimize the accuracy of the whole scheme as is well illustrated by the following breakdown of a_0^0 into various contributions:

$$a_0^0 = \frac{7M_\pi^2}{32\pi F_\pi^2} C_0 + M_\pi^4 \alpha_0 + O(M_\pi^8) \tag{3}$$

where

$$C_0 = 1 + \frac{M_\pi^2}{3} \langle r^2 \rangle_s - \frac{5M_\pi^2}{224\pi^2 F_\pi^2} \left\{ \bar{\ell}_3 - \frac{563}{525} \right\} + O(M_\pi^4) \tag{4}$$

(here given only at next-to-leading order, for simplicity) is the part of the scattering length which only depends on the quark masses, whereas α_0 is the part which is due

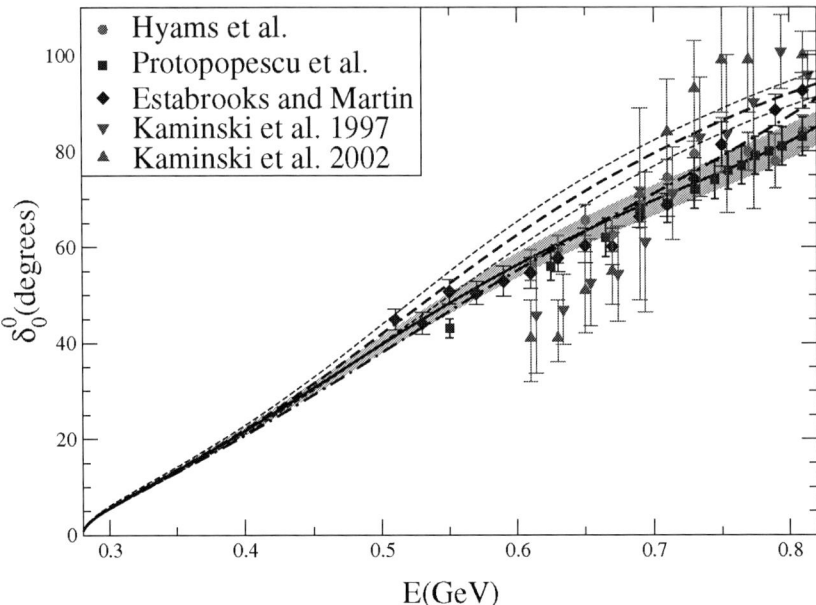

FIGURE 1. Phase shift for the $I = 0$ S wave. The shaded band is the result of the analysis in [14] and the solid line is the Roy equation solution obtained with the input above 1.4 GeV proposed in [15]. The dashed curve and the corresponding uncertainty band are the "tentative alternate solution" proposed in [15]. The dotdashed curve is the Roy solution fit to the Kaminski et al. 1997 [25] data obtained in [26]. The other data sets shown are from Refs. [27, 28, 29]

to the momentum dependent part of the amplitude, here evaluated at threshold. The constant C_0 is to be chosen as subtraction constant in the dispersive treatment, because for this the chiral expansion converges fast: the numerical evaluation at the two-loop level gives

$$C_0 = 1.096 \pm 0.021 \tag{5}$$

amounting to a 10% shift evaluated with 20% of relative uncertainty. This correction shifts a_0^0 from 0.16 to 0.17. The bulk of the correction, however, comes from the momentum dependent part of the amplitude, the α_0 term which can be accurately evaluated through a dispersive integral. The final result for both S-wave scattering lengths reads

$$a_0^0 = 0.220 \pm 0.005 \qquad a_0^2 = -0.0444 \pm 0.0010 \ , \tag{6}$$

with an accuracy at the level of a few percent for both. Notice that the situation is completely different for a_0^2: here the tree level result is -0.0454 and the momentum-dependent part of the correction is not particularly large. As is well known the $\pi\pi$ interaction in the $I = 2$ channel is weak.

If one uses the dispersive representation for the $\pi\pi$ scattering amplitude, this high level of accuracy obtained for the scattering lengths is reflected in the whole energy region below E_0: having fixed the scattering lengths all other sources of noise generate

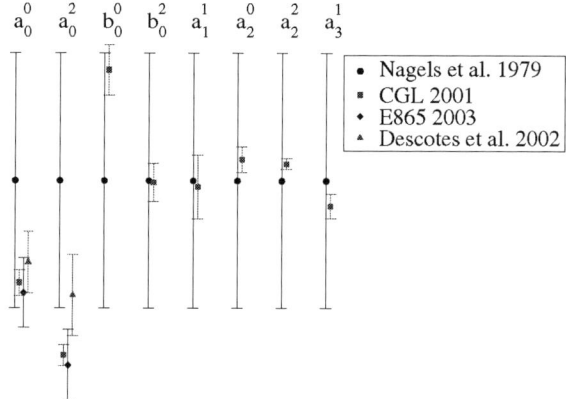

FIGURE 2. Accuracy improvement in the evaluation of various threshold parameters due to the use of CHPT for the subtraction constants.

remarkably little uncertainty as illustrated for the case of the $S0$ wave in Fig. 1. Notice that the band is a prediction which relies on experimental input only at 0.8 GeV and above – it is *not* a fit to any of the data sets shown in the same plot. The plot contains various data sets as well as the "tentative alternate solution" proposed by Peláez and Ynduráin in [15] and the Roy solution fit of Kaminski et al. [26]. These will be commented upon in the next section. That CHPT is mostly responsible for the improvement in the accuracy in the Roy treatment is well illustrated in Fig. 2, where the values for a number of threshold parameters obtained from Roy equation analyses as reported in the compilation of data [30] are given in arbitrary units chosen such that all errors are normalized to the same size. The errors obtained after matching the dispersive and the chiral representation are about an order of magnitude smaller.

3. THE CRITICISM OF PELÁEZ AND YNDURÁIN

In [15] Peláez and Ynduráin have criticized our work [13, 14] and made the following claims: that the input used in the Roy equation analysis above 1.4 GeV was incorrect and that as a consequence the solutions we obtained below 0.8 GeV were also incorrect. The claim about the incorrectness of the input has two aspects: between 1.4 GeV and 2 GeV we used experimental information and they claim that this is unreliable – above 2 GeV, where we relied on a Regge representation, they claimed that the one we used is not orthodox because it does not respect factorization. According to the latter property, the residues of the Regge poles which appear in the $\pi\pi$ scattering amplitude must be given by the square of the residue for πN divided by the NN residue. As explicitly stated in [13] the Regge representation we used served the purpose of giving us a fair account of the contributions to the dispersive integrals from the regions between 2 and 3 GeV – the contributions from yet higher energies are negligible because the Roy equations are twice subtracted. In fact even the contributions from the region above 1.4 GeV are rather small and play a minor role in the Roy equations. For this reason we have not made our own analysis of this part of the input and took what was available in the literature.

The observation that contributions from above 1.4 GeV do not matter much for the Roy solutions below 0.8 GeV appears to be in contradiction with the second claim made by Peláez and Ynduráin, namely that the solutions we obtained were "spurious" because of the incorrect input used. It is important to stress here that Peláez and Ynduráin made this claim without supporting it with a calculation, but only with indirect arguments. I will come back to these indirect arguments later but I first must say that the Roy equation solutions for the input proposed by Peláez and Ynduráin as the correct one have been calculated in [16]. The outcome is the solid line in Fig. 1 and is indistinguishable from the solution obtained with the input originally used in [13]. The calculation shows that the second claim of Peláez and Ynduráin is wrong.

This takes us back to the indirect arguments they had used to support their claim. They gave three arguments:

1. a mismatch in the Olsson sum rule;
2. a discrepancy among two different determinations of the P-wave effective range;
3. a discrepancy among two different evaluations of the D and F wave threshold parameters.

In [16] we have discussed in detail all these indirect arguments and shown that either the discrepancy is not there (as in the case of the P wave effective range) or that the conclusion that the discrepancy can be cured by changing the Roy solution below 0.8 GeV is incorrect. The interested reader is referred to [16] for a detailed discussion of all these points.

As explicitly stated also in [16], a better input than the Regge representation that we have used in [13] is certainly possible and can obtained with a thorough analysis of all the available information (like high-energy total cross section data, sum rules etc.) [31]. In this perspective, also the data on total $\pi\pi$ cross sections which Peláez and Ynduráin have pointed out [19] are useful information which has to be taken into account – the representation used in [13] does not describe these very well and can be improved. It is however a fact that the influence of improvements in the high-energy input on the scattering lengths and the whole low-energy scattering amplitude will be negligible. Such improvements may be of interest in applications which rely on the $\pi\pi$ scattering amplitude close to 1 GeV, like the dispersive representation of hadronic contributions to a_μ which relies on the P wave phase shifts [9].

3.1. The input phase $\delta_0^0(0.8\text{GeV})$

In the most recent papers Peláez and Ynduráin have moved on to discuss other points and to raise further criticism to our analysis. In particular they have criticized our value and error for $\delta_0^0(E_0)$ which, as discussed at length in [13], is one of the most important input parameters in the Roy analysis. In this paper we had observed that if one looks at the data on this particular wave it is difficult to draw any conclusion because different data sets (in fact different analyses of the same $\pi N \to \pi\pi N$ scattering

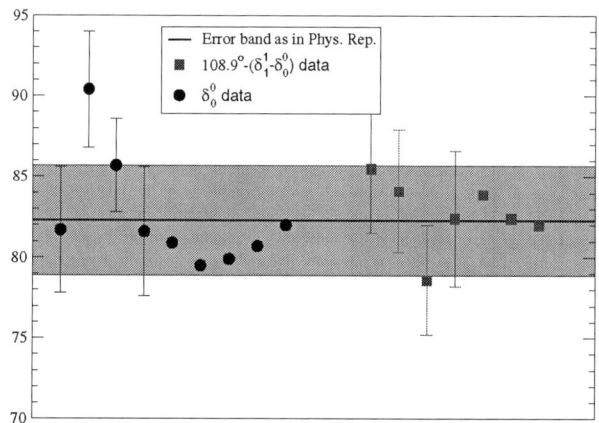

FIGURE 3. Graphical representation of the data on $\delta_0^0(0.8\text{GeV})$ as given in Table 2, p. 226 of Ref. [13]: on the left-hand side of the plot the plain data for $\delta_0^0(0.8\text{GeV})$ are shown. On the right-hand side those for the difference $\delta_0^0(0.8\text{GeV}) - \delta_1^1(0.8\text{GeV})$ shifted by $108.9°$, the value of the P wave phase shift as extracted from form factor data.

data) are mutually incompatible. The situation improves dramatically if one looks at the difference $\delta_1^1(E_0) - \delta_0^0(E_0)$, for which different data sets give a coherent picture. The fact that the δ_1^1 phase is now known much better thanks to the data on the vector form factor then leads to a rather good determination of $\delta_0^0(E_0)$ which we estimated to be $82.3° \pm 3.4°$. This is illustrated in Fig. 3 which compares the direct determinations of $\delta_0^0(E_0)$ to those that exploit the phase difference. In their recent discussions Peláez and Ynduráin have given particular emphasis to the recent analyses by Kaminski et al. [25]. This is indeed important new work on $\pi\pi$ scattering because it analyzes a large body of polarized $\pi N \to \pi\pi N$ data, which are certainly very interesting and useful. The data of these two analyses are shown in Fig. 1: around 0.8 GeV these data are substantially higher than those of Hyams et al. or Protopopescu et al. and our band. In that region the "tentative alternate solution" proposed by Peláez and Ynduráin in [15] is in better agreement with them. Below 0.7 GeV, however, the Kaminski et al. data become lower than the other data sets and also lower than the two bands shown: even the "tentative alternate solution" is now in flat disagreement with these data. The problem is seen also in a recent paper by Kaminski et al. [26] where they solve Roy equations and fit the data with the solutions: the overall fit is not particularly good precisely because this peculiar shape of the data in this energy region (low between 0.6 and 0.7 GeV, rising steeply at 0.7 GeV and then high until 0.8 GeV) cannot be followed by Roy equation solutions. This best fit to their data is shown in Fig. 1 as dotdashed curve, and is evidently a trade-off between the high and the low data. This curve lies almost everywhere inside our Roy solution band and has $\delta_0^0(0.8\text{GeV}) = 87°$, about 1.5 σ higher than the range we had used as input.

This most recent analysis does not clarify the experimental situation concerning the $S0$ wave. Since no information is provided on the phase difference $\delta_1^1 - \delta_0^0$ we could not make use of these data when we fixed the input for the Roy equations. Moreover the results of the Roy equation analysis of Kaminski et al. [26] we just discussed show that there is no reason to modify our central value for $\delta_0^0(0.8\text{GeV})$ or to stretch its error.

3.2. Scalar radius

Another point on which criticism has been raised against our analysis concerns the scalar radius $\langle r^2 \rangle_s$ which appears in the chiral expansion of both S-wave scattering lengths, cf. Eq. (4). The input value we used [14]:

$$\langle r^2 \rangle_s = 0.61 \pm 0.04 \text{fm}^2 \tag{7}$$

had been determined through a dispersive analysis of the scalar form factor, following [32] after updating the $\pi\pi$ phase shifts which are used as input. Ynduráin has recently claimed that the outcome of this calculation violates a "robust lower bound" he derives [17]. The same bound is violated also by other calculations of the same quantity. One of them, performed by Moussallam [33], makes a thorough analysis of the different phenomenological inputs available in the literature and comes to a conclusion which is in perfect agreement with Eq. (7): the values he finds for the radius lie between 0.58 and $0.65\,\text{fm}^2$.

This apparent puzzle has been recently clarified in [18]: the "robust lower bound" does not exist because its derivation relies on an untenable assumption, namely that the phase of the scalar form factor in the region above 1.1 GeV must be close to the $S0$ phase shift. The correct conclusion is that above 1.1 GeV where the elasticity is again close to one, the difference between the phase of the scalar form factor and the phase shift has to be close to a multiple of π: the various available calculations all agree that this difference is close to π rather than to zero. Also this criticism is unjustified.

4. OUTLOOK

The predictions for the S wave $\pi\pi$ scattering lengths (6) discussed here are of unusual precision in hadronic physics. They are derived under the assumption that the quark condensate is the leading order parameter of the spontaneous symmetry breaking in QCD. Experimental tests have confirmed this hypothesis and have put the standard picture of the QCD vacuum on a solid experimental basis. The predictions are however still more precise than the experimental measurements of the same quantities, and there is still room for a more stringent test of QCD at low energy. Experiments aiming at performing these tests are currently underway: the DIRAC experiment at CERN [34] is measuring the lifetime of pionic atoms which is proportional to the square of the difference of the two scattering lengths and plans to reach a 5% precision in the measurement of this difference. The NA48II experiment [35] will gather twice the statistics of K_{e4} decays of the E865 experiment, thus reaching an improvement of a factor $\sqrt{3}$ in the final uncertainty. The phase shift difference measured in these decays is mostly sensitive to a_0^0. The prospects for precision low-energy hadronic physics are at the moment particularly good. We look forward to the experimental results.

ACKNOWLEDGMENTS

It is a pleasure to thank the organizers for the invitation and the perfect organization of the conference in a wonderful environment. I also thank B. Ananthanarayan, I. Caprini, J. Gasser and H. Leutwyler for the longstanding and very pleasant collaboration on the topics discussed here and for useful comments and suggestions on the manuscript.

REFERENCES

1. G. Colangelo, *Proc. of the 5th International Conference on Quark Confinement and the Hadron Spectrum, Gargnano, Italy, 10-14 Sep 2002*
2. G. Colangelo, J. Gasser and H. Leutwyler, Phys. Rev. Lett. **86** (2001) 5008 [arXiv:hep-ph/0103063].
3. S. Pislak *et al.*, Phys. Rev. D **67** (2003) 072004 [arXiv:hep-ex/0301040].
4. S. Descotes-Genon, N. H. Fuchs, L. Girlanda and J. Stern, Eur. Phys. J. C **24** (2002) 469 [arXiv:hep-ph/0112088].
5. S. Descotes-Genon, L. Girlanda and J. Stern, Eur. Phys. J. C **27** (2003) 115 [arXiv:hep-ph/0207337].
6. G. Colangelo, arXiv:hep-ph/0011025.
7. G. Colangelo, J. Gasser and H. Leutwyler, Phys. Lett. B **488** (2000) 261 [arXiv:hep-ph/0007112].
8. H. Leutwyler, AIP Conf. Proc. **670** (2003) 45 [arXiv:hep-ph/0212323].
9. G. Colangelo, arXiv:hep-ph/0312017.
10. M. Knecht, B. Moussallam, J. Stern and N. H. Fuchs, Nucl. Phys. B **457** (1995) 513 [arXiv:hep-ph/9507319], and Nucl. Phys. B **471** (1996) 445 [arXiv:hep-ph/9512404]. J. Bijnens, G. Colangelo, G. Ecker, J. Gasser and M. E. Sainio, Phys. Lett. B **374** (1996) 210 [arXiv:hep-ph/9511397], and Nucl. Phys. B **508** (1997) 263 [Err.-ibid. B **517** (1998) 639] [arXiv:hep-ph/9707291].
11. M. R. Pennington, Annals Phys. **92** (1975) 164.
12. S. M. Roy, Phys. Lett. B **36** (1971) 353.
13. B. Ananthanarayan, G. Colangelo, J. Gasser and H. Leutwyler, Phys. Rept. **353** (2001) 207 [arXiv:hep-ph/0005297].
14. G. Colangelo, J. Gasser and H. Leutwyler, Nucl. Phys. B **603** (2001) 125 [arXiv:hep-ph/0103088].
15. J. R. Pelaez and F. J. Yndurain, Phys. Rev. D **68** (2003) 074005 [arXiv:hep-ph/0304067].
16. I. Caprini, G. Colangelo, J. Gasser and H. Leutwyler, Phys. Rev. D **68** (2003) 074006 [arXiv:hep-ph/0306122].
17. F. J. Yndurain, Phys. Lett. B **578** (2004) 99 [Err.-ibid. B **586** (2004) 439] [arXiv:hep-ph/0309039].
18. B. Ananthanarayan et al. Phys. Lett. B **602** (2004) 218 [arXiv:hep-ph/0409222].
19. J. R. Pelaez and F. J. Yndurain arXiv:hep-ph/0411334.
20. J. R. Pelaez and F. J. Yndurain arXiv:hep-ph/0412320, these proceedings.
21. S. Weinberg Phys. Rev. Lett. **17** (1966) 616.
22. J. Gasser and H. Leutwyler, Phys. Lett. B **125** (1983) 325, and Annals Phys. **158** (1984) 142.
23. J. Gasser and G. Wanders, Eur. Phys. J. C **10** (1999) 159 [arXiv:hep-ph/9903443].
24. M. G. Olsson, Phys. Lett. **B410** (1997) 311 [hep-ph/9703247].
25. R. Kaminski, L. Lesniak and K. Rybicki, Z. Phys. C **74** (1997) 79 [arXiv:hep-ph/9606362], Eur. Phys. J. directC **4** (2002) 4 [arXiv:hep-ph/0109268].
26. R. Kaminski, L. Lesniak and B. Loiseau, Phys. Lett. B **551** (2003) 241 [arXiv:hep-ph/0210334].
27. B. Hyams *et al.*, Nucl. Phys. **B64** (1973) 134.
28. S. D. Protopopescu *et al.*, Phys. Rev. **D7** (1973) 1279.
29. P. Estabrooks and A. D. Martin, Nucl. Phys. **B79** (1974) 301.
30. M. M. Nagels *et al.*, Nucl. Phys. B **147** (1979) 189.
31. I. Caprini, G. Colangelo and H. Leutwyler, work in progress.
32. J. F. Donoghue, J. Gasser and H. Leutwyler, Nucl. Phys. B **343** (1990) 341.
33. B. Moussallam, Eur. Phys. J. C **14** (2000) 111 [arXiv:hep-ph/9909292].
34. Preliminary results can be found at http://dirac.web.cern.ch/DIRAC/.
35. See: http://na48.web.cern.ch/NA48/NA48-2/NA48_2.html

Photoproduction of Gluonic Hybrids and Exotics

Paul Eugenio *for the CLAS Collaboration*

Florida State University

Abstract.
Motivated by recent experimental results for gluonic hybrid meson candidates and from recent theoretical Lattice QCD and Flux-tube model calculations, photoproduction should provide an ideal hunting ground for gluonic matter. Jefferson Lab offers an excellent opportunity to undertake the study of meson spectroscopy at intermediate energies. Current studies are underway at CLAS which are showing the feasibility of using CLAS as a meson spectrometer.

Keywords: photoproduction, exotic, mesons, hybrids
PACS: 11.80.Et

MOTIVATIONS

Discoveries of new phenomena in nuclear and particle physics have provided insight into the fundamental constituents of matter. In the past few decades we have seen a new picture emerge in which quarks form the building blocks of nearly all matter. Yet the gluon, which carries the force which binds quarks, can interact with other gluons to form a bound state, or interact as a fundamental constituent of matter along with the quarks. Thus new forms of gluonic or hybrid matter should exist.

The search for hybrids in recent years has resulted in considerable excitement. Theoretical predictions from both gluonic flux-tube models and recent lattice gauge theory results predict the lightest hybrid at a mass of 1.9 GeV for the exotic $J^{PC} = 1^{-+}$ $q\bar{q}g$-hybrid(see references [1] and [2]). Exotic meson states are those with quantum numbers not accessible to conventional $q\bar{q}$ bound states. Recent experimental results find two very promising 1^{-+} exotic candidates. The $\pi_1(1400)$ seen decaying to $\eta\pi^-$ at Brookhaven has a mass somewhat too low for the theory prediction for a gluonic hybrid [3]. A higher mass observed state, the $\pi_1(1600)$ is tantalizing as a gluonic hybrid, but its decay to $\rho\pi$ was unexpected [4]. Even though the existence of both states appears very clear, these states have had a history of controversy, particular those produced via pion beams [5].

It has been pointed out by Close and Page [6] that in the case of photoproduction, where the photon can be effectively replaced by a ρ interacting with an exchange π, ρ, or ω, the production strength for producing gluonic hybrids could be considerable. Furthermore, Szczepaniak and Swat [7] concluded that in the case of photoproduction, the π_1 exotic and the well known a_2 should be produced on an equal footing, whereas in pion production the exotic is suppressed by a factor of 10.

As supported by recent lattice gauge calculations using excited adiabatic potentials, it is a good approximation to decouple the quark degrees of freedom from the gluonic degrees of freedom. This is based on the idea that quarks in the systems react much slower than the gluonic fields responsible for strong confinement. As in ordinary $q\bar{q}$

CP756, *Quark Confinement and the Hadron Spectrum VI*
edited by N. Brambilla, U. D'Alesio, A. Devoto, K. Maung, G.M. Prosperi and S. Serci
© 2005 American Institute of Physics 0-7354-0241-8/05/$22.50

mesons where the addition of one unit of orbital angular momentum costs about 1 GeV, calculations show that gluonic excitations are also of the order of 1 GeV in energy, and that there is a kind of orthogonality decoupling the quarks and flux-tube (gluon) degrees of freedom. In lattice QCD calculations and in flux-tube models, excited flux-tubes can have

$$_{flux-tube}J^{PC} = 1^{+-} \text{ or } 1^{-+} \text{ [8].}$$

Therefore with pseudoscalar probes, such as pion beams, coupling the quark degrees of freedom to those of an excited flux-tube results in

$$_{quarks}J^{PC} \otimes _{flux-tube}J^{PC} = 1^{--}, 1^{++}.$$

On the other hand, for vector probes, such as the photon viewed as a vector meson, coupling the quark degrees of freedom with that of an excited flux-tube results in

$$_{quarks}J^{PC} \otimes _{flux-tube}J^{PC} = 0^{-+}, 1^{-+}, 2^{-+}, 0^{+-}, 1^{+-}, 2^{+-}.$$

It is interesting to note that the vector probe has access to manifestly exotic quantum numbers (exotic $J^{PC} = 0^{+-}, 1^{-+}, 2^{+-}$). Since there exists a wealth of data with pseudoscalar hadronic probes and very little data with vector probes like the photon, this may explain the lack of observations of gluonic hybrids.

Flux-tube model calculations of gluonic hybrid decays prefer decay channels to (L=0) + (L=1) meson pairs. For example, according to these calculations the lowest lying exotic state ($J^{PC} = 1^{-+}$) should have typical partial widths [9]

$$b_1\pi : f_1\pi : \rho\pi = 170\ MeV : 60\ MeV : 10\ MeV.$$

Figure 1 shows the $b_1\pi$ invariant mass for a recent analysis of SLAC photoproduction data [10]. The sparse data clearly demonstrates the need for better photoproduction experiments.

Studies are underway to pursue gluonic matter at nuclear physics labs around the world. At JLab there are plans to upgrade the accelerator by doubling its maximum beam energy. A linchpin for the Lab's upgrade plans[2] is the addition of a new experimental program called GlueX[3]. The goal of GlueX is to unambiguously discover and study gluonic matter. Although GlueX will be the definitive experiment for studying gluonic matter, it will be several years until this experimental program begins to acquire data. In the meantime, there is timely and complementary experimental program in meson spectroscopy to search for new and unusual mesons via photoproduction using the CLAS detection facility at Jefferson Lab.

RECENT RESULTS FROM CLAS PHOTOPRODUCTION AT HIGH ENERGIES

The CLAS collaboration took data at the end of August 2001 with a real photon beam with an energy range of 4.8 GeV to 5.47 GeV. Since the CLAS detector is not well designed to study forward-going systems, this effort used a modified orientation of the

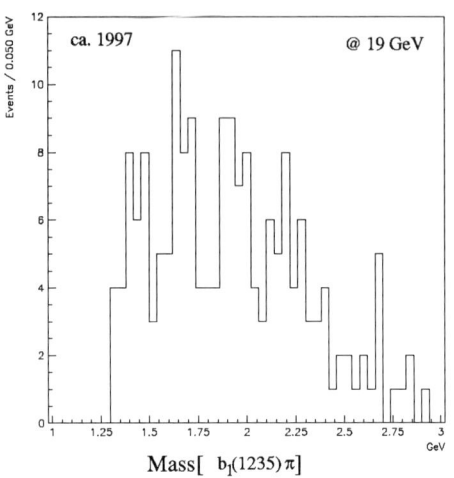

FIGURE 1. The $b_1(1235)\pi$ mass spectrum from $\gamma p \rightarrow p\pi^+\pi^+\pi^-\pi^-\pi^0$ at 16.5 -21.0 GeV.

CLAS detector configuration (torus magnet at half maximum field and target moved one meter upstream of its standard position) in order to maximize acceptance for the reaction $\gamma p \rightarrow n\pi^+\pi^+\pi^-$ in the low-t region. The experiment was limited to a total of 8 days run time. The analysis of these data is now underway, and early results have shown the feasibility of using CLAS for meson spectroscopy. With a photon beam energy in the range of 5 to 6 GeV, the experiment readily photo-produce states of masses up to about 2 GeV/c^2. This allows the experiment to access the exotic 1^{-+} candidates states that have been recently observed [3, 4].

The first CLAS partial wave analysis on a small subset of the data are shown for the benchmark reaction $\gamma p \rightarrow p\pi^+\pi^-$ in Figure 2. The accepted $\pi^+\pi^-$ invariant mass is shown in Figure 2a. The results of the partial wave analysis are shown in Figure 2b-d. In Figure 2b the $\rho(770)$ shows up clearly in the $J^{PC}|M| = 1^{--}|1|$ partial wave as expected for helicity conservation in the t-channel. Two other partial wave intensities are shown: in Figure 2c, the non-helicity conserving ρ partial wave; and in Figure 2d, the isotropic $J^{PC} = 0^{++}$ partial wave. Note that there appears to be a slight ambiguity in the analysis between the $1^{--}|1|$ and 0^{++} in the mass range above 1 GeV, that is events with hard zeros in the 0^{++} distribution leak into the 1^{--} distribution.

The problem has been attributed to backgrounds from Δ's and N^*'s which are produced at one of the πp vertexes. Figure 3 exhibits the $\pi^- p$ invariant mass for these data. The $\Delta^0(1232)$, $N(1520)/N(1535)$, and $\Delta^0(1620)$ are clearly visible in the mass spectrum. The $\Delta^{++}(1232)$ (not shown) is also cleanly produced. While it is relatively easy to cut out the $\Delta(1232)$ without much loss in statistics, it is much more difficult to cleanly cut the higher mass Δ's and N^*'s. The effects of this baryon background on the phi (ϕ_{TY})

TABLE 1. Partial waves included in the fit of $\gamma p \rightarrow n\pi^+\pi^+\pi^-$

J^{PC}	m^ε	L	Isobar	# Waves
0^{-+}	0^+	0	σ	1
0^{-+}	0^+	1	$\rho(770)$	1
1^{++}	$0^+, 1^\pm$	0,2	$\rho(770)$	6
1^{++}	$0^+, 1^\pm$	1	σ	3
1^{-+}	$0^-, 1^\pm$	1	$\rho(770)$	3
2^{++}	$0^-, 1^\pm, 2^\pm$	2	$\rho(770)$	5
2^{-+}	$0^+, 1^\pm$	1,3	$\rho(770)$	6
2^{-+}	$0^+, 1^\pm$	2	σ	3
2^{-+}	$0^+, 1^\pm$	0,2	$f_2(1270)$	6
Background				

angular distribution[1] in the $\pi^+\pi^-$ Gottfried-Jackson Frame can be seen in the "V" plot of Mass($p\pi$) vs ϕ_{TY}. Figure 4a shows the π^-p "V" plot for accepted $\gamma p \rightarrow p\pi^+\pi^-$ data. The corresponding ϕ_{TY} distribution is shown in Figure 4b. A clear linear relationship is exhibited. This figure shows that events from different Δ's and N^*'s add to the structure of the angular distribution. This is a pure kinematic effect. It is interesting to note that one can turn this situation around, that is, a $\pi\pi$ resonance of high spin can have narrow structures in the ϕ_{TY} angular distributions which could result in a kinematic reflection of resonance-like structure in the $p\pi$ invariant mass distribution.

$$\gamma p \rightarrow n\pi^+\pi^+\pi^-$$

From the same data run the reaction $\gamma p \rightarrow \pi^+\pi^+\pi^- n$ was studied. The three charged pions were measured in CLAS, while the neutron was identified and measured from missing four-momentum. Laboratory angle cuts on the pions, as well as the selection of low t events, greatly reduce the baryon resonance background. Figure 5 shows various distributions of the data: the $t\prime$ distribution, the missing neutron mass, and di-pion effective masses.

The preliminary PWA results are shown in Figure 6; waves included in the fit are listed in Table 1. While these results are very preliminary, there is a very clear signal for the 2^{++} $a_2(1320)$. There is some evidence for photoproduction of the 1^{++} $a_1(1260)$ and also the 2^{-+} $\pi_2(1670)$. There is some strength in the exotic $J^{PC} = 1^{-+}$ partial wave near 1600 MeV/c^2, but it is not conclusive.

---

[1] The Treiman-Yang angle is the ϕ in the Gottfried-Jackson Frame.

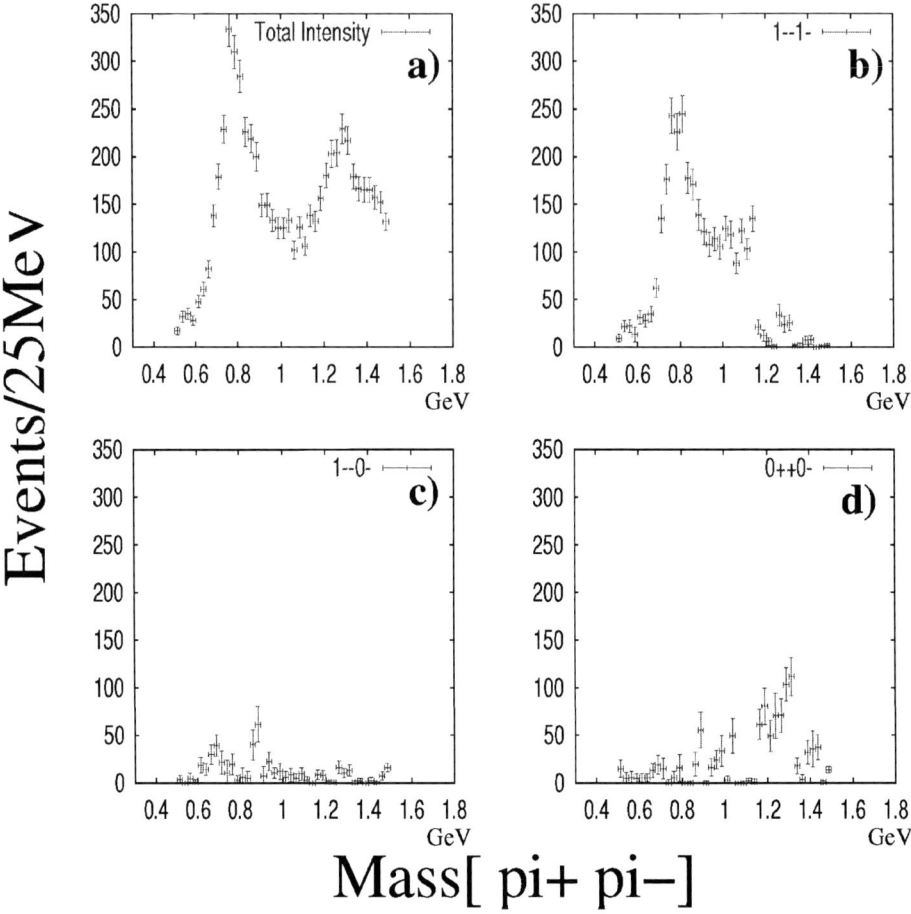

FIGURE 2. The reaction $\gamma p \rightarrow p\pi^+ + \pi^-$. Preliminary mass independent partial wave analysis results from E01-017: a) The total accepted $\pi^+\pi^-$ mass spectrum, b) the partial wave intensity for $J^{PC}|M| = 1^{--}1$, c) the partial wave intensity for $J^{PC}|M| = 1^{--}0$, d) the partial wave intensity for $J^{PC}|M| = 0^{++}0$. Each data point in mass represents an independent partial wave analysis.

$$\gamma p \rightarrow p p \bar{p}$$

The proton-antiproton system has had a rich history spanning more than thirty years. Initially, the $p\bar{p}$ system attracted much interest due to theoretical predictions of exotic matter. These predictions included: nucleon-antinucleon states that are loosely bound in a molecule-like structure called quasi-nuclear baryonium, and tightly-bound multi-quark baryonium ($qq - \bar{q}\bar{q}$) which have favored decays to nucleon-antinucleon final states. In

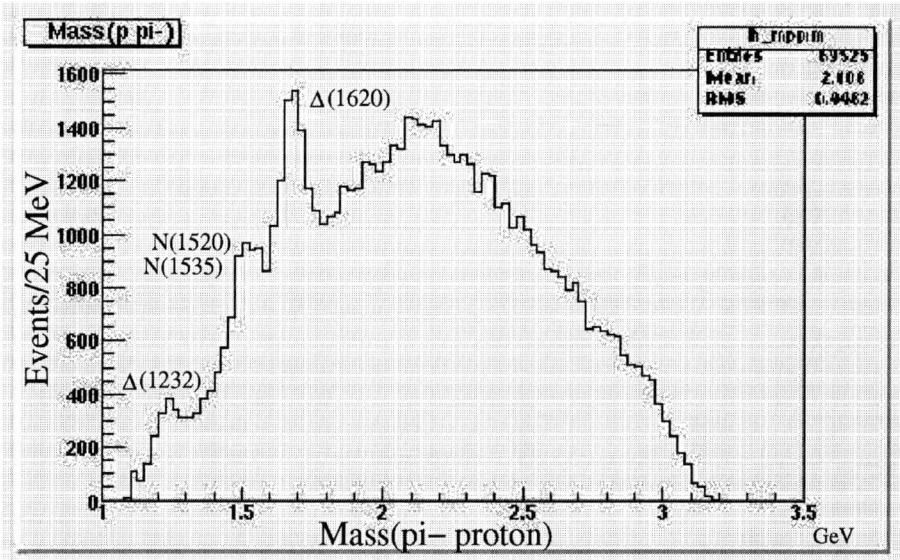

FIGURE 3. The accepted $\pi^- p$ invariant mass from the reaction $\gamma p \rightarrow p\pi^+\pi^-$.

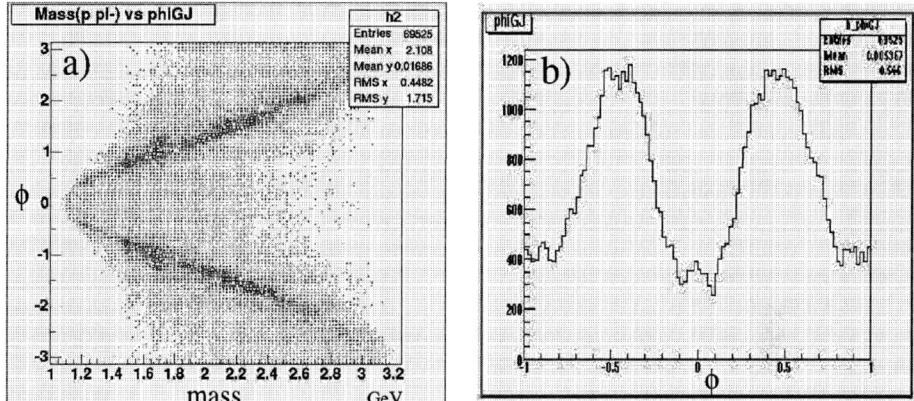

FIGURE 4. Experimental results for $\gamma p \rightarrow p\pi^+\pi^-$: a) ϕ_{TY} vs Mass($\pi^- p$), b) ϕ_{TY}/π. Not corrected for acceptance.

the early 1970's, there were claims of a unusually-narrow meson resonance with a mass of 1.93 GeV/c^2 [11, 12]. It was believed that this particle was not an ordinary meson, and that it would couple to the proton-antiproton system. There were then claims that experiments found the narrow resonance in proton-antiproton scattering experiments

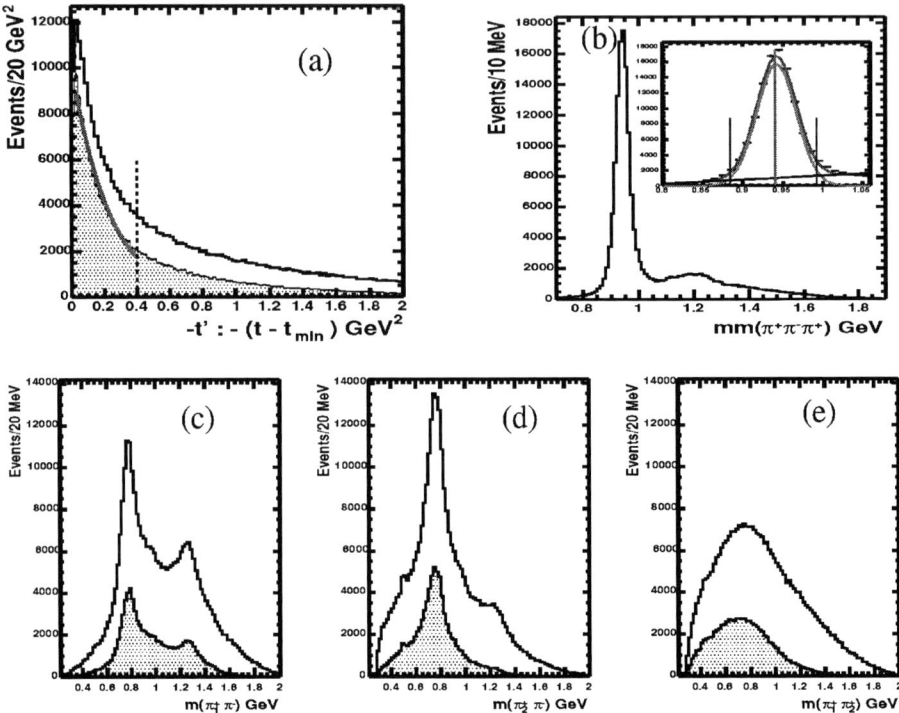

FIGURE 5. Various distributions of the reaction $\gamma p \rightarrow \pi^+\pi^+\pi^- n$ at 5.2 GeV. (a) $t\prime$ distribution. Events to the left of the dotted line were selected for the PWA. The shaded area has cuts on the pion laboratory angles described in the text. (b) Missing mass off of $\pi^+\pi^+\pi^-$ showing the missing neutron. (c) Mass of the π^- and slow π^+ showing both the ρ and $f_2(1270)$ isobars. (d) Mass of the π^- and fast π^+. (e) Mass of $\pi^+\pi^+$. (c),(d), and (e) shaded areas are events selected for the final partial wave analysis.

[13, 14, 15, 16]. Also, in the late 1970s there were claims of additional higher mass narrow resonances at 2.02 and 2.20 GeV/c^2 in the proton-antiproton system [17, 16, 18]. However, follow up experiments did not make such claims[19, 20], and until recently, the debate had died out. In 1997, CERN refuted their earlier claims of the 1.93 and 2.02 GeV/c^2 resonances, yet in 1999, a reanalysis of the CERN data confirmed the existence of the 2.02 and 2.2 GeV/c^2 resonances. Recently the BES collaboration has claimed to observe a narrow baryonium state with a mass near the proton-antiproton mass threshold, which decays to proton-antiproton[21]. Presently, the only well-known particle that decays to proton-antiproton is the J/ψ particle, with a mass of 3.097 GeV/c^2 [22]. Most of the past experiments involved proton-antiproton scattering or pion production. Recently Jefferson Laboratory has provided the first look at the proton-antiproton system through photoproduction.

In the recent CLAS high-energy photon data run, nearly five thousand exclusive

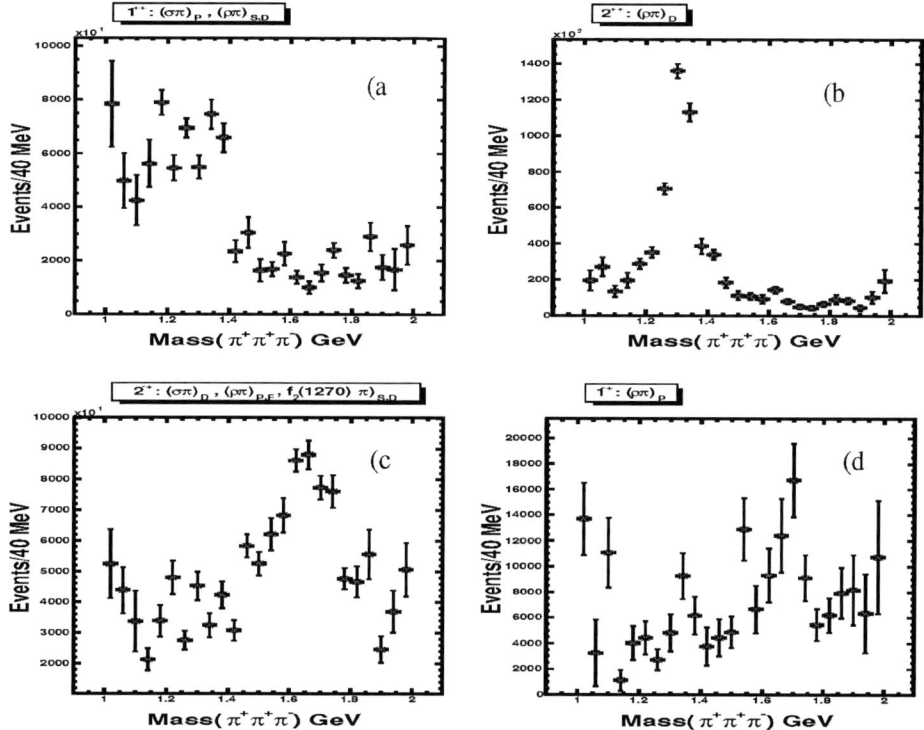

FIGURE 6. Preliminary mass independent partial wave decomposition of the reaction $\gamma p \to \pi^+ \pi^+ \pi^- n$. (a) $J^{PC} = 1^{++}$ partial wave. (b) 2^{++} partial wave. (c) 2^{-+} wave. (d) Exotic 1^{-+} wave.

$\gamma p \to p p \bar{p}$ events were observed where all final state particles were identified in the CLAS spectrometer. However, in CLAS there are regions of the detector where particles can go unmeasured. For example, the CLAS toroidal magnetic field bends negatively charged particles back toward the beam. Quite often, these particles end up going back into the beam-line, and are lost. To increase the exclusive data yield, the anti-proton was allowed to be identified via the missing mass.

Figure 7 shows the missing-mass-squared of events containing two identified protons. There is a prominent peak at a mass squared of $0.880 \; (GeV/c^2)^2$, which is consistent with a missing antiproton. Selecting the events consistent with a missing antiproton $[0.85 \; (GeV/c^2)^2 \leq MM^2 \leq 0.91 \; (GeV/c^2)^2]$ yields approximately 17,100 $\gamma p \to p p (\bar{p})$ events.

Possible mechanisms which could describe the photoproduction of a proton-antiproton pair are diffraction/meson exchange, baryon exchange, and anti-baryon exchange. In each process, an intermediate resonance may be produced. In meson exchange the photon transfers very little momentum to the target, but interacts with the

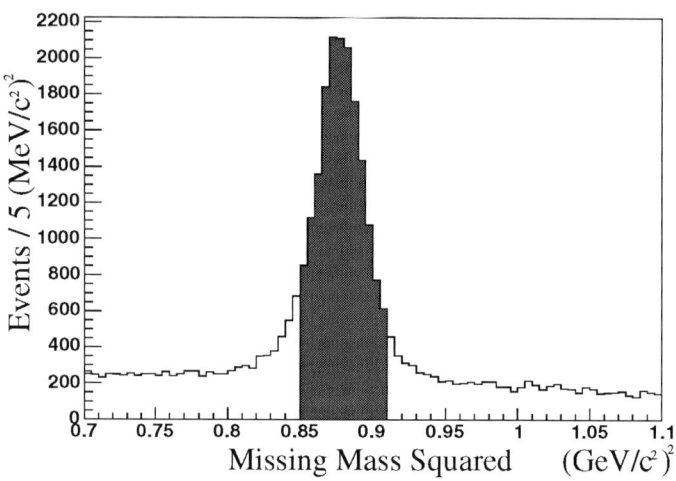

FIGURE 7. The reaction $\gamma p \rightarrow pp\bar{p}$: missing mass squared off two protons.

exchanged meson to produce a resonance that decays to a fast forward-going proton-antiproton pair. In baryon exchange, the photon interacts with an exchanged baryon converting it to a fast forward-going proton, leaving behind a slow moving meson resonance at the target vertex which decays to a proton-antiproton pair. For anti-baryon exchange, the photon interacts with an exchange anti-baryon, converting it to a fast forward-going antiproton and leaving behind a resonance at the target vertex which decays to two protons.

The distinction of meson exchange and baryon exchange production is clouded by the two identical protons. Without information identifying which is which, the two mechanisms are nearly indistinguishable. A simple way to distinguish protons is to sort on the proton momentum. In the cases of meson and baryon exchange, one proton should be moving fast and in the forward direction, while the other proton is produced at or near the target vertex, receives very little momentum transfer from the beam, and is expected to be slow. Therefore, one can use the momentum of the two protons on an event by event basis and associate a $p_{fast}\bar{p}$ resonance with meson exchange and a $p_{slow}\bar{p}$ resonance with baryon exchange.

No obvious features are observed in the two proton invariant mass(not shown). In the invariant mass distribution of $p_{fast}\bar{p}$ there are no obvious structures indicating resonant nature. The invariant mass of $p_{slow}\bar{p}$ is shown in Figure 8. The distribution has some interesting structures, with a sharp rise at threshold and a possible narrow peak or dip near 2.0 GeV and broader peak at 2.04 GeV. While it is possible that these feature could be due to acceptance, preliminary Monte Carlo studies suggest otherwise and that the acceptance is smoothly varying as a function of the $p\bar{p}$ invariant mass. Current analysis plans include performing a partial wave analysis to search for resonant behavior in $p\bar{p}$ system.

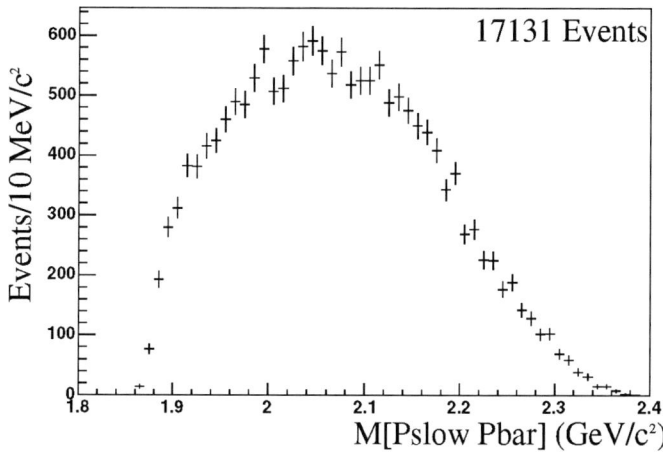

FIGURE 8. Preliminary results of the reaction $\gamma p \rightarrow pp\bar{p}$: the invariant mass of the slow proton with the antiproton.

REFERENCES

1. N. Isgur, R. Kokoski, and J. Paton, *Phys. Rev. Lett.* **54**, 869 (1985).
2. P. Lacock, et.al.(UKQCD Collaboration), *Phys. Lett.* **B401**, 308 (1997).
3. D. R. Thompson et. al. (BNL-E852 Collaboration,) *Phys. Rev. Lett.* **79**, 1630 (1997).
4. G. S. Adams et. al. (BNL-E852 Collaboration), *Phys. Rev. Lett.* **81**, 5760 (1998).
5. See contributions from both A. V. Popov and D. V. Amelin, *9th International Conference on Hadron Spectroscopy: HADRON01* Protvino, Russia 2001.
6. F. Close and P. Page, *Phys. Rev.* **D52**, 1706 (1995).
7. A. Szczepaniak and M. Swat, *Phys. Lett.* **B516**, 72 (2001).
8. N. Isgur and J. Paton, *Phys. Rev.* **D31**, 2910 (1985).
9. T. Barnes, F. E. Close, and E.S. Swanson, *Phys.Rev.* **D52** 5242 (1995).
10. G. R. Blackett, et al, *arXiv.org e-Print Archive* **hep-ex/9708032** (1997).
11. M.N. Focacci *et. al.*, Phys. Rev. Lett. 17, 890(1966).
12. D. Cline *et. al.*, Phys. Rev. Lett. 17, 1268(1968).
13. A.S. Carroll *et. al.*, Phys. Rev. Lett. 32, 247(1974).
14. T.E. Kalogeropoulos and G.S. Tzanakos Phys. Rev. Lett. 34, 1047(1975)
15. V. Chaloupka *et. al.*, Phys. Lett. 61 B, 487(1976).
16. P. Benkheiri *et. al.*, Phys. Lett. 68 B, 483(1977).
17. J. Bodenkamp *et. al.*, Phys. Lett. 133 B, 275(1983).
18. B.G. Gibbard *et. al.*, Phys. Rev. Lett. 42, 1593(1979).
19. R. Bizzarri *et. al.*, Phys. Rev. D 6, 160(1972).
20. J. Bensinger *et. al.*, Phys. Rev. D 23, 1417(1983).
21. J. Z. Bai *et al.* [BES Collaboration], Phys. Rev. Lett. **91**, 022001 (2003) [arXiv:hep-ex/0303006].
22. M.W. Eaton *et. al.*, Phys. Rev. D 29, 805(1984).

Pentaquarks – Status Report

Marek Karliner

Cavendish Laboratory
University of Cambridge, UK
and
School of Physics and Astronomy
Raymond and Beverly Sackler Faculty of Exact Sciences
Tel Aviv University, Tel Aviv, Israel
e-mail: marek@proton.tau.ac.il

Abstract. I discuss the recent experimental and theoretical developments following the discovery of the Θ^+ pentaquark – an exotic $uudd\bar{s}$ baryon resonance observed in the KN channel by several experiments, and an exotic Ξ^{*--} ($ddss\bar{u}$) reported by NA49 at CERN. I focus on the theoretical interpretation of the data, both in terms of quark and chiral degrees of freedom, on the predictions for related exotic states, and on several unresolved questions raised by the experimental data, such as why some experiments observe the pentaquarks and other don't, the apparently extremely narrow width of the Θ^+ and the determination of its parity. I also describe the likely properties of the proposed heavy-quark pentaquarks – an anticharmed exotic baryon Θ_c ($uudd\bar{c}$) and Θ_b^+, ($uudd\bar{b}$), which are expected to be extremely narrow or even stable against strong decays. H1 recently reported observation of a possible Θ_c candidate in $D^{*-}p$ channel. Pentaquarks are also being searched for in e^+e^- annihilation and $\gamma\gamma$ collisions in the LEP data and at B-factories.

Keywords: QCD, exotics, quark model
PACS: 12.38.-t,12.39.-x,12.39.Mk

INTRODUCTION

In the course of the last year we have witnessed a remarkable renaissance of QCD spectroscopy, with several new surprising experimental results: two new extremely narrow mesons containing c and $\bar{s}$ quarks (BaBar, CLEO, Belle); a new very narrow resonance precisely at at $D^{0*}D^0$ threshold (Belle, CDF, D0); enhancements near $\bar{p}p$ thresholds (BES, Belle); a $\Lambda_c\bar{p}$ resonance (Belle), and exotic 5-quark resonances: Θ^+ ($uudd\bar{s}$), Ξ^{*--} ($ddss\bar{u}$), Θ_c ($uudd\bar{c}$). The existence of these states provides a serious challenge to the traditional picture of hadrons made either of three quarks or a quark and an antiquark. Clearly, QCD bound-state dynamics is still an open problem. In this brief review I will focus on the pentaquarks.

THE EXPERIMENTAL STATUS OF THE Θ^+ PENTAQUARK

By now there is a large number of experimental reports on observing the Θ^+ pentaquark [1] as either K^+n or K_sp resonance, as shown in Fig. 1. One experiment (ZEUS) reported also observing the anti particle, $\bar{\Theta}^-$. All experiments report relatively narrow widths, but so far these are all consistent with the experimental resolution. The true width is likely to be much more narrow $\lesssim$ 1-4 MeV, as suggested by several indirect but

CP756, Quark Confinement and the Hadron Spectrum VI
edited by N. Brambilla, U. D'Alesio, A. Devoto, K. Maung, G.M. Prosperi and S. Serci
© 2005 American Institute of Physics 0-7354-0241-8/05/$22.50

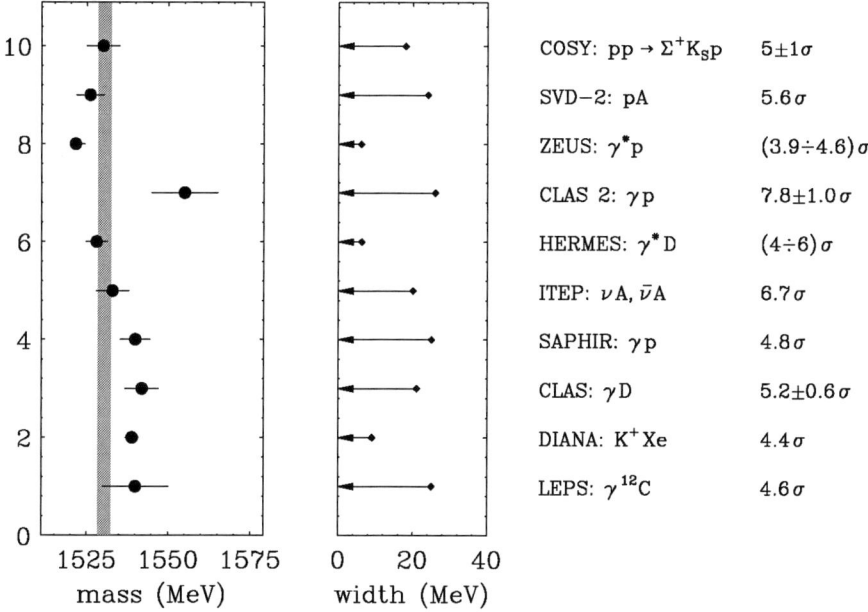

FIGURE 1. Summary of experiments which reported Θ^+ observation.

quite robust arguments. Such a narrow width of a resonance at 100 MeV above threshold is a puzzle in itself.

Despite a large number of experiments reporting observation of the Θ^+, the experimental situation is not clear for several reasons.

First, there is a substantial scatter of the Θ^+ mass values, indicating possible systematic effects, or presence of additional resonances.

Second, the relevant cross sections are very small, probably on the order of μb, while the non-exotic processes are $\sim mb$. Therefore, in order to extract the signal from the background, sophisticated cuts are needed, depending on the specific experimental setup. The systematic effects introduced by these cuts continue to be studied.

Third, several experiments (HERA-B, PHENIX, DELPHI and ALEPH) looked for the Θ^+ and did not see it. At present, it is an open question why some experiments see the Θ^+ and others do not. Two of these are LEP experiments which see a lot of protons, but no deuterons. H1 reports $\bar{d}/\bar{p} \approx 5.0^{-4}$, so it is puzzling why the LEP experiments do not see antideuterons. This has to be resolved before we know if we should worry that they do not see the Θ^+.

One possible resolution [20] of the contradiction between the various experiments is that a specific production mechanism is present in the experiments that see the Θ^+ and is absent in those that do not see it. The CLAS data on $\gamma p \rightarrow \pi^+ K^- K^+ n$, and in particular the $(K^+ K^- n)$ mass distribution which shows a peak at the mass of 2.4 GeV suggest that there might be a cryptoexotic N^* resonance with hidden strangeness.

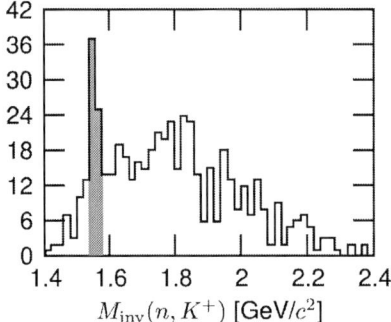 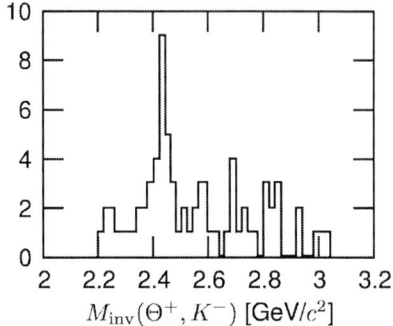

Fig. 2. Left panel: $(K^+ n)$ invariant mass distribution in CLAS $\gamma p \to \pi^+ K^- K^+ n$ experiment, hep-ex/0311046; right panel: $(K^+ K^- n)$ invariant mass distribution corresponding to the data under the Θ^+ peak in the left panel.

Searches for such baryon resonances with hidden strangeness have indicated possible candidates but these searches did not go up to 2.4 GeV.

DEVELOPMENT OF THE PENTAQUARK THEORY

The possible existence of pentaquarks was suggested as early as 1977 by Jaffe. In the early 1980's a negative-parity $\bar{c}suud$ pentaquark was considered by Lipkin, but the specific story of the Θ^+ really began with the revival of the Skyrme model at the end of 1983. The Skyrme model is a low-energy approximation to large-N_c QCD, which in turn shares many features of real-world QCD. The fundamental degrees of freedom in the Skyrme model are nonlinearly coupled quasi-Goldstone $SU(3)_f$ pseudoscalars and the baryons emerge as solitons. A somewhat more general class of similar models is collectively referred to as chiral soliton models ($\mathcal{X}$SM). When properly quantized, the ground state soliton is a $J^P = \frac{1}{2}^+$ $SU(3)_f$ octet. The first excited state is a $J^P = \frac{3}{2}^+$ $SU(3)_f$ decuplet. The next one is a $J^P = \frac{1}{2}^+$ $SU(3)_f$ antidecuplet which cannot be constructed out of 3 quarks. This was realized early on and several research groups estimated the mass of the lightest member of the $\overline{10}$ at around 1540 MeV. But most people viewed this as a problem for the model, since it was well known and documented by the Review of Particle Properties that such states did not exist [4]. Then in 1997 Diakonov, Petrov and Polyakov (DPP) did two things [5]: (a) they took the prediction seriously, effectively declaring that $\overline{10}$ is not a bug, but a feature (b) they estimated that the state is less than 15 MeV wide, which made its detection seem feasible. Later on it turned out that certain results in their paper needed to be revised, including the specific values of the width and the predictions for masses of other states in $\overline{10}$. In particular, they predicted $M(\Xi^{--}) = 2070$ MeV, vs. the NA49 result [6] 1862 MeV (more on this below). But the paper triggered the first experiment in Japan and this really got the ball rolling.

Recently we re-analyzed [11] the predictions of chiral-soliton models for the masses and decay widths of baryons in the $\overline{\mathbf{10}}$. We found 1430 MeV $< M(\Theta^+) < 1660$ MeV and 1790 MeV $< M(\Xi^{--}) < 1970$ MeV. These are consistent with the masses reported recently, but more precise predictions rely on ambiguous identifications of non-exotic baryon resonances. The overshoot in the original DPP prediction for $M(\Xi^{--})$ is mainly due to an outdated value of the πN Σ-term. Parametrically $\Gamma(\overline{\mathbf{10}}) \sim \mathscr{O}(1/N_c^2)$, but with realistic couplings it is hard to get $\Gamma(\overline{\mathbf{10}}) < 10$ MeV. A key prediction is a light $\mathbf{27}$ with $J^P = \frac{3}{2}^+$, i.e. a Θ-like $I = 1$ state within 100 MeV above $\Theta^+(I = 0)$.

One remarkable prediction of the χSM is that the $SU(3)$ breaking in $\overline{\mathbf{10}}$ is linear in hypercharge. This is similar to the baryon decuplet, where it amounts to counting the number of strange quarks. But for $\overline{\mathbf{10}}$, it is seemingly counterintuitive, as it implies that Θ^+ with one antistrange quark is *lighter* than the nonstrange $N^* \in \overline{\mathbf{10}}$. To understand where this comes from, it is best to rederive the result in the quark language, by carefully constructing the $\overline{\mathbf{10}}$ quark wave functions.

Starting from $|\Theta^+\rangle = |uudd\bar{s}\rangle$, we can generate the other states in $\overline{\mathbf{10}}$ by repeatedly applying a U-spin lowering operator which replaces d by s: $U_-|d\rangle = |s\rangle$, $U_-|\bar{s}\rangle = -|\bar{d}\rangle$. Thus:

$$|p^*\rangle = U_-|uudd\bar{s}\rangle = -\sqrt{\frac{1}{3}}\,|uud\,d\bar{d}\rangle + \sqrt{\frac{2}{3}}\,|uud\,s\bar{s}\rangle \tag{1}$$

So p^* is heavier than Θ^+ because one of the components in its wave function contains *two* heavier quarks in the form of an $\bar{s}s$ pair. It has no net strangeness and is *cryptoexotic*. The leading $SU(3)$ breaking effect is proportional to the total number of strange *plus* antistrange quarks, $\langle \#s + \#\bar{s}\rangle_{p^*} = 2 \times (\sqrt{2/3})^2 = 4/3$, so $\Delta\langle \#s + \#\bar{s}\rangle = 1/3$ and $\Delta M \sim m_s/3$. There are also subleading effects, having to do with the color hyperfine interaction $\sim 1/m_q$, but these depend on the specific form of the wave function.

A definitive full QCD analysis of the pentaquarks will be eventually provided by lattice gauge theory (LGT). It will probably take a while, as dealing with unstable resonances in LGT is notoriously difficult. The main problem is the need to separate the resonance from scattering states with the same quantum numbers. In our case, one needs to make sure that the Θ^+ two-point function on the lattice is not contaminated by contributions from KN. This requires careful measurement of finite-volume effects and is extremely costly in computer time.

The quark model and the χSM provide complementary descriptions of the pentaquarks properties. If we want to "peek inside", the quark model is clearly the way to go.

CORRELATED QUARKS – DIQUARKS AND TRIQUARKS

Most quark model treatments of multiquark spectroscopy use the color-magnetic short-range hyperfine interaction as the dominant mechanism for possible binding. The hyperfine interaction between two quarks denoted by i and j is then written as

$$V_{hyp} = -V(\vec{\lambda}_i \cdot \vec{\lambda}_j)(\vec{\sigma}_i \cdot \vec{\sigma}_j) \tag{2}$$

where $\vec{\lambda}$ and $\vec{\sigma}$ denote the generators of $SU(3)_c$ and the Pauli spin operators, respectively. The interaction is attractive in states symmetric in color × spin and repulsive in antisymmetric states. Because of Pauli principle the interaction is always repulsive between same-flavor quarks. This flavor antisymmetry suggests that the bag or single-cluster models commonly used to treat normal hadrons may not be adequate for multi-quark systems. In such a state, with identical pair correlations for all pairs in the system, all same-flavor quark pairs are necessarily in a higher-energy configuration, due to the repulsive nature of their hyperfine interaction. The $uudd\bar{s}$ pentaquark is really a complicated five-body system where the optimum wave function to give minimum color-magnetic energy can require flavor-dependent spatial pair correlations for different pairs in the system; *e.g.*, that keep the like-flavor uu and dd pairs apart, while minimizing the distance and optimizing the color couplings within the other pairs.

We consider here a possible model for a strange pentaquark that implements these ideas by dividing the system into two color non-singlet clusters which separate the pairs of identical flavor quarks. The two clusters, a ud diquark and a $ud\bar{s}$ triquark, are in a relative P-wave, are separated by a distance larger than the range of the color-magnetic force and are kept together by the color electric force. Therefore the color hyperfine interaction operates only within each cluster, but is not felt between the clusters, as shown schematically in Fig. 3.

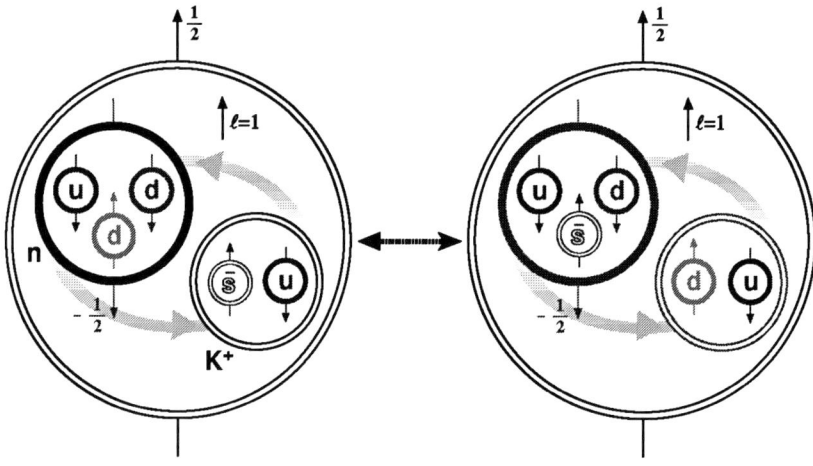

Kn configuration   diquark – triquark configuration of the $uudd\bar{s}$ pentaquark

Fig. 3. K^+n and the diquark-triquark configuration of the $uudd\bar{s}$ pentaquark.

An unusual aspect of the $uudd\bar{s}$ pentaquark is that the S-wave has higher energy than the P-wave. This is because in the S-wave there is no angular momentum barrier to prevent repulsive interaction between same-flavor quarks. Therefore this correlated quark picture predicts a positive parity pentaquark, in agreement with the χSM. It is extremely important to measure the parity in an experiment. If it turns out to be negative, you can throw away all my papers on the subject, together with most of the other theoretical papers!

Using the diquark-triquark configuration as the starting point, we can extract some specific properties of the Θ^+. The $|ud\,du\bar{s}\rangle$ configuration contains a ud diquark, which is an isosinglet, has $S=0$ and is a color antitriplet. The $ud\bar{s}$ triquark contains another isosinglet ud pair, but this time with $S=1$. The triquark has $S=\frac{1}{2}$ and is a color triplet. Since ud and $ud\bar{s}$ are in a relative P-wave, Θ^+ has $J^P=\frac{1}{2}^+$, $I=0$ and is in $\overline{10}$ of $SU(3)_f$.

A very similar structure was proposed in Ref. [21], shortly after Ref. [9] appeared. The only difference is that the second ud pair is assumed to have $S=0$, rather than $S=1$. This means that there is no hyperfine interaction between $\bar{s}$ and the light quarks, and so the hyperfine binding is somewhat weaker than in Ref. [9]. In Ref. [7] it was pointed out that the ud-$ud\bar{s}$ and $(ud)^2$-$\bar{s}$ configurations mix strongly, so the true ground state has a somewhat lower energy that either of the two.

In the diquark-triquark configuration the hyperfine binding turns out [9, 10] to be about 50 MeV stronger than the total hyperfine interaction in the KN system. But this does not mean that the state is below KN threshold, because there is the additional cost of putting the $\{ud\}\{du\bar{s}\}$ system in a P-wave. The latter can be estimated by noticing that the cost of such excitations only depends on the reduced mass of the system. The reduced mass of the $\{ud\}\{du\bar{s}\}$ is quite close to that of $c\bar{s}$ in the D_s, where the P-wave excitation energy is about 200 MeV. Putting it all together, one obtains [9, 10] an estimate $M(\Theta^+) \approx 1592 \pm 50$ MeV, reasonably close to the experimental value of 1530 ± 10 MeV.

This looks encouraging, but one must also deal with the other member of the $\overline{10}$, the Ξ^{--} which was observed by NA49 at 1862 ± 2 MeV [6]. This is to be contrasted with a triquark-diquark configuration prediction of 1720 ± 50 MeV. This difference of roughly 100 MeV is generic for all correlated quark models [21]. One should note however, that Ξ^{--} is 400 MeV above the $\Xi\pi$ threshold, while Θ^+ is only 100 MeV above the KN threshold. This is an open challenge for the theory. Moreover, one can derive a variational mass inequality [13] relating the Ξ^{--} and Θ^+ masses: $M(\Xi^{--}) - M(\Theta^+) \lesssim 300$ MeV, vs. the experimental value of 300 MeV. This puts strong constraints on models of 5-quark structure and indicates a urgent need for experimental confirmation of the NA49 results.

The existence of strongly mixed ud-$ud\bar{s}$ $(ud)^2$-$\bar{s}$ configurations for the Θ^+ provides a possible explanation of its narrow width [12]. It is a standard feature of quantum mechanics that in the case of an exact degeneracy the two configurations have equal weight in the mixed state, and the relative phase is such that the two decay amplitudes into KN destructively interfere, exactly cancelling each other and decoupling the mixed state from the KN decay channel. When the two configurations are almost degenerate, the two decay amplitudes almost cancel, yielding a very narrow width of the mixed state, on the order of a few MeV.

There is an associated new experimental prediction: the destructive interference mechanism suppresses only the coupling of the Θ^+ to KN, but not the coupling to K^*N. The latter channel is above threshold, so it cannot be seen in a decay, but the lack of suppression in Θ^+K^*N coupling should be observable in the baryon-exchange K^-p reactions where the kaon is observed going backward in the center-of-mass system:

$$K^-p \to \bar{K}^o n; \quad K^-p \to \bar{K}^{*o} n; \quad K^-p \to \bar{K}^o N^{*o}; \quad K^-p \to \bar{K}^{*o} N^{*o} \qquad (3)$$

where N^{*o} denotes any $I = 1/2$ electrically neutral baryon resonance.

These reactions shown in Fig. 4 can only proceed via the t-channel exchange of an exotic positive-strangeness baryon. But if the Θ^+ couples only weakly to KN, the $K^-\Theta^+ \to n$ and $p \to \bar{K}^o\Theta^+$ vertices are also weak by crossing and the $K^-p \to \bar{K}^{*o}N^{*o}$ reaction should be much stronger than the other three which require a Θ^+KN coupling [12].

The $\Theta^+K\Delta$ coupling is forbidden by isospin if the Θ^+ is an isoscalar. Therefore the presence of the Δ in these baryon exchange reactions is a test for the presence of exotic positive strangeness baryons with higher isospin.

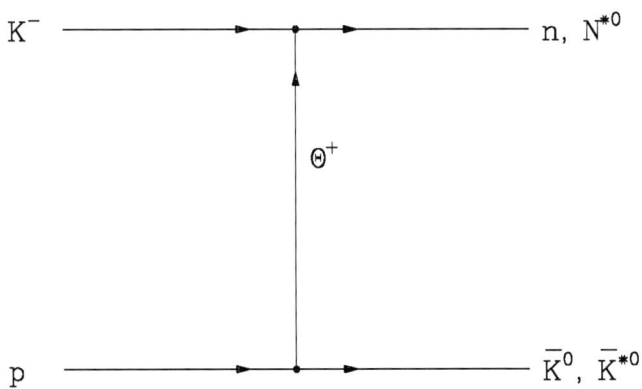

Fig. 4. Baryon-exchange diagram corresponding to the reactions (3).

EXPERIMENTAL CHALLENGES AND FUTURE DIRECTIONS

In my opinion, the most pressing issues in pentaquark research at the moment are experimental. The most important among these is the confirmation of Θ^+ and Ξ^{*--}.

After that, it is essential to measure the parity. Several methods have been proposed. Most of them rely on identifying some measurable asymmetry which depends on the pentaquark parity. All these proposals are quite challenging. A detailed discussion of their relative merits is outside the scope of this talk, both because of space limitations and because of the limited competence of the speaker in such matters.

In parallel it is important to search for additional exotic states:
(a) those obtained by replacing $\bar{s} \to \bar{c}, \bar{b}$: Θ_c ($uudd\bar{c}$)and Θ_b^+ ($uudd\bar{b}$). More about these shortly.
(b) $\overline{10}$ with $J = \frac{3}{2}$ [22]: assuming that Θ^+ and other members of the $\overline{10}$ have $J = \frac{1}{2}$ which results from $S = \frac{1}{2}$ and $L = 1$, it is natural to look for partners with $J = \frac{3}{2}$. Current estimates [22] indicate that such states could be within $\lesssim 50$ MeV of their $J = \frac{1}{2}$ counterparts.

(c) exotics in higher representations: **27, 35**, etc. There are indications from the $\mathcal{X}$SM [11] that such states could be within $\lesssim 100$ MeV of the $\overline{\mathbf{10}}$. Clearly, a whole new spectroscopy waits to be explored!

Predictions: Θ_c and Θ_b^+

If the existence of Θ^+ is confirmed, the case for the existence of its heavy cousins will be quite strong. The basic idea is quite simple [14]: assuming we have a reasonable approximate quark wave function for the Θ^+, replace $\bar{s}$ by $\bar{c}$ or $\bar{b}$ and compute the properties of the resulting state. At present we do not know how far this strategy can be pushed, because the strength of the color-magnetic hyperfine interaction is inversely proportional to the quark mass. So the configuration which is optimal for $\bar{s}$ might not be optimal for a heavy quark. Still, it is worthwhile to explore the consequences of this approach.

A rough prediction of this approach is that Θ_c has $J^P = \frac{1}{2}^+$, $I = 0$, mass of about 3000 ± 50 MeV and a width $(1 \div 2) \times 20$ MeV. Similarly, Θ_b^+ has $J^P = \frac{1}{2}^+$, $I = 0$, mass of about 6.4 GeV and a very narrow width, $(1 \div 2) \times 4$ MeV.

There are two basic methods in searching for such states:
(a) look for unexpectedly narrow peaks in DN, D^*N and BN B^*N, invariant mass distributions where the mesons contain a heavy antiquark;
(b) look for a proton coming out from a vertex which is known to carry anti-charm or anti-bottom flavor. This approach is particularly well suited to B factories where the flavor of a secondary vertex can easily be tagged.

Recently H1 published evidence for a narrow anticharmed baryon resonance [15] in the $D^{*-} p$ and $D^{*+} \bar{p}$ channels, i.e. $uudd\bar{c}$ and $\bar{u}\bar{u}\bar{d}\bar{d}c$, with a mass $3099 \pm 3 \pm 5$ MeV and width 12 ± 3 MeV, and estimated statistical significance of 5.4σ, as shown in Fig. 5.

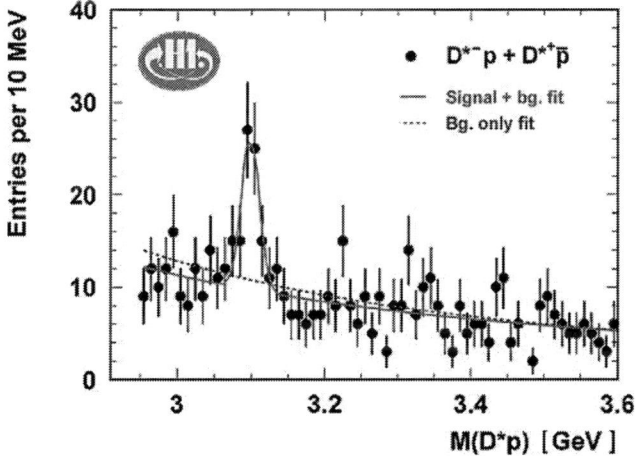

Fig. 5. H1 data for a resonance in $D^{*-} p$ and $D^{*+} \bar{p}$ invariant mass spectra.

This is of course very exciting, but the sister ZEUS experiment sees no sign of such a resonance in their data, despite a somewhat larger data sample. In addition, there are conference reports from FOCUS [17], ALEPH [18], CLEO, BaBar and CDF [19] who looked for this resonance in their data and did not see it. At present no one understands the reason for this disagreement between the various experiments. However, since H1 sees the resonance at the same mass in both $D^{*-}p$ and $D^{*+}\bar{p}$, one can safely rule out a statistical fluctuation.

Recently, with Bryan Webber we have estimated the probability of producing Θ_c in LEP and the Tevatron, taking the H1 data as input and assuming formation through D^*p coalescence [23]. In our model the cross section for Θ_c formation is proportional to the rate of production of pD^{*-} (or $\bar{p}D^{*+}$) pairs in close proximity both in momentum space and in coordinate space. The constant of proportionality is determined from the Θ_c cross section in deep inelastic scattering as reported by the H1. The HERWIG Monte Carlo is used to generate simulated DIS events and also to model the space-time structure of the final state, as shown in Fig. 6.

Requiring the proton and the D^* be within a 100 MeV mass window and separated by a spacelike distance of no more than 2 fm, we find that a "coalescence enhancement factor" $F_{co} \sim 4$ is required to account for the H1 signal. The same approach is then applied in order to estimate the number of Θ_c events produced at LEP and the Tevatron.

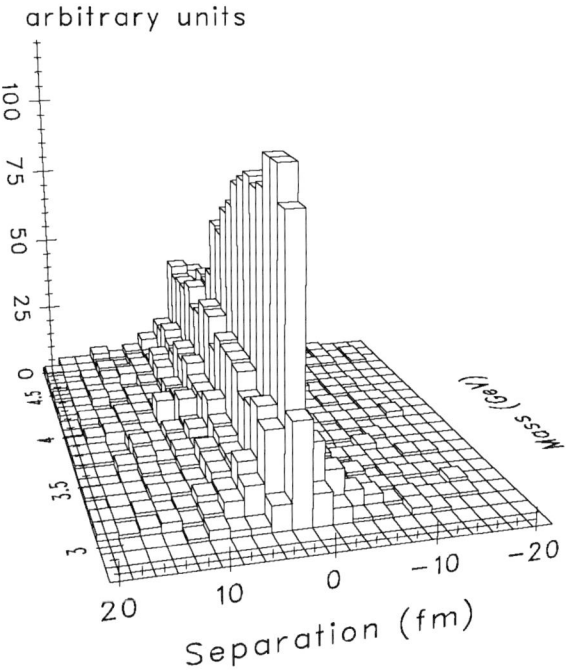

Fig. 6. Predicted pD^* joint mass–separation distribution in DIS.

For each of the four LEP experiments the model predicts between 25 and 40 H1-like Θ_c events. For the Tevatron a signal of many thousands of events would have been expected.

Since both LEP and Tevatron experiments reported null results, our analysis implies that the either the H1 signal is spurious and due to an unknown systematic effect, or alternatively that it corresponds to a real resonance, whose production mechanism in Tevatron. DIS is substantially different from the production mechanism in e^+e^- and the Yet another possibility is that either the theoretical or experimental analysis is missing an essential ingredient.

Search for pentaquarks at B factories

In B factories one expects a reasonable branching ratio for $B \to$ baryon + antibaryon, somewhere below 10^{-6}. Producing a pentaquark with an antibaryon requires production of an additional $\bar{q}q$ pair, as shown in Fig. 7.

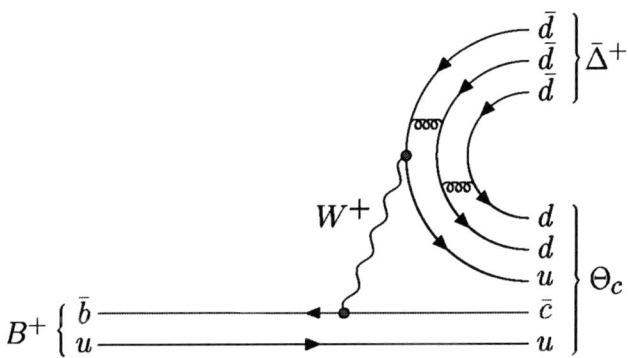

Fig. 7. Pentaquark production in B decays.

Making such an extra $\bar{q}q$ pair carries a penalty in the BR. It is hard to make a precise estimate of this penalty, but it is probably at least an order of magnitude. So with enough data the $B \to$ pentaquark + antibaryon decay should be attainable. This decay has a particularly striking signature. Since it is a two-body decay where the mass of the initial state is exactly known, energy and momentum conservation in the B CM frame ensure that unlike in the hadron reactions, there are no kinematical ambiguities. Moreover, for modes with $\Theta^+ \to K_s p$ decay, the K_s flavor is tagged by the antibaryon.

ACKNOWLEDGMENTS

The original work described here was done in collaboration with Harry Lipkin, John Ellis and Michał Praszałowicz and Bryan Webber. This research was supported in part by a grant from the United States-Israel Binational Science Foundation (BSF), Jerusalem.

REFERENCES

1. T. Nakano *et al.* [LEPS Coll.], Phys.Rev.Lett. **91**,012002(2003), hep-ex/0301020; V.V. Barmin *et al.* [DIANA Coll.], Phys. Atom. Nucl. **66**,1715(2003), hep-ex/0304040; S. Stepanyan *et al.* [CLAS Coll.], Phys. Rev. Lett. **91**, 252001 (2003), hep-ex/0307018; J. Barth *et al.* [SAPHIR Coll.], hep-ex/0307083; A. E. Asratyan *et al.*, hep-ex/0309042; V. Kubarovsky *et al.* [CLAS Coll.], Phys. Rev. Lett. **92**,032001(2004); E. – ibid. **92**,049902(2004), hep-ex/0311046; R. Togoo *et al.*, Proc. Mongolian Acad. Sci., **4**(2003)2; A. Airapetian *et al.* [HERMES Coll.], Phys. Lett. B **585**,213(2004), hep-ex/0312044; A. Aleev *et al.* [SVD Coll.], hep-ex/0401024; M. Abdel-Bary *et al.* [COSY-TOF Coll.], hep-ex/0403011; P. Z. Aslanyan *et al.*, hep-ex/0403044; S. Chekanov *et al.* [ZEUS Coll.], hep-ex/0403051; T. Nakano, talk at NSTAR 2004, March 24-27, Grenoble, France, http://lpsc.in2p3.fr/congres/nstar2004/talks/nakano.pdf ; Y. A. Troyan *et al.*, hep-ex/0404003.
2. S. Nussinov, hep-ph/0307357; R. W. Gothe and S. Nussinov, hep-ph/0308230; R. A. Arndt, I. I. Strakovsky and R. L. Workman, Phys. Rev. C **68** (2003) 042201 [Erratum-ibid. C **69**, 019901 (2004)], nucl-th/0308012 and nucl-th/0311030; J. Haidenbauer and G. Krein, hep-ph/0309243; R. N. Cahn and G. H. Trilling, hep-ph/0311245; A. Casher and S. Nussinov, Phys. Lett. B **578** (2004) 124, hep-ph/0309208.
3. J. Z. Bai *et al.* [BES Coll.], hep-ex/0402012; K. T. Knopfle *et al.* [HERA-B Coll.], hep-ex/0403020; C. Pinkenburg [PHENIX Coll.], nucl-ex/0404001; P. Hansen [for ALEPH Coll.], talk at DIS 2004, http://www.saske.sk/dis04/talks/C/hansen.pdf ; T. Wengler [DELPHI Coll.], Moriond '04 QCD, hep-ex/0405080.
4. Particle Data Group, Review of Particle Properties, 1984, p. S243.
5. D. Diakonov, V. Petrov and M. V. Polyakov, Z. Phys. A **359** (1997) 305, hep-ph/9703373.
6. C. Alt *et al.* [NA49 Coll.], hep-ex/0310014.
7. For a review of pentaquark models, see B.K. Jennings and K. Maltman, Phys. Rev. D **69**, 094020 (2004), hep-ph/0308286.
8. V. Burkert, at NSTAR 2004, March 24-27, Grenoble, France, lpsc.in2p3.fr/congres/nstar2004/talks/burkert_2.pdf .
9. M. Karliner and H. J. Lipkin, hep-ph/0307243.
10. M. Karliner and H.J. Lipkin, Phys. Lett. B **575** (2003) 249.
11. J. R. Ellis, M. Karliner and M. Praszałowicz, JHEP **05**(2004)002, hep-ph/0401127.
12. M. Karliner and H. J. Lipkin, Phys. Lett. B **586**, 303 (2004) hep-ph/0401072.
13. M. Karliner and H. J. Lipkin, Phys. Lett. B **594**, 273 (2004), hep-ph/0402008.
14. M. Karliner and H. J. Lipkin, hep-ph/0307343.
15. A. Aktas *et al.* [H1 Coll.], Phys. Lett. B **588**, 17 (2004), hep-ex/0403017.
16. ZEUS Coll., hep-ex/0409033.
17. *"A Search for a State which Decays to a Charged D* and Proton at FOCUS"*, http://www-focus.fnal.gov/penta/penta_charm.html .
18. S. Schael *et al.*, [ALEPH Coll.] Phys. Lett. B **B599** 1 (2003).
19. Yeon Sei Chung [for the CDF collaboration], Chicago Flavor Seminar, http://b0urpc.fnal.gov/~yschung/Chicago-Flavor.pdf .
20. M. Karliner and H. J. Lipkin, Phys. Lett. B **597**, 309 (2004), hep-ph/0405002.
21. R. L. Jaffe and F. Wilczek, Phys. Rev. Lett. **91**, 232003 (2003), hep-ph/0307341.
22. J. J. Dudek and F. E. Close, Phys. Lett. B **583**, 278 (2004), hep-ph/0311258.
23. M. Karliner and B. R. Webber, hep-ph/0409121.

Soft-Collinear Factorization and the Calculation of the $B \to X_s \gamma$ Rate

Matthias Neubert[1]

Institute for High-Energy Phenomenology
Newman Laboratory for Elementary-Particle Physics, Cornell University
Ithaca, NY 14853, U.S.A.

Abstract. Using results on soft-collinear factorization for inclusive B-meson decay distributions, a systematic study of the partial $B \to X_s \gamma$ decay rate with a cut $E_\gamma \geq E_0$ on photon energy is performed. For values $E_0 \leq 1.9$ GeV the rate can be calculated without reference to shape functions. The result depends on three large scales: m_b, $\sqrt{m_b \Delta}$, and $\Delta = m_b - 2E_0$. The sensitivity to the scale $\Delta \approx 1.1$ GeV (for $E_0 \approx 1.8$ GeV) introduces significant uncertainties, which have been ignored in the past. Our new prediction for the $B \to X_s \gamma$ branching ratio with $E_\gamma \geq 1.8$ GeV is $\mathrm{Br}(B \to X_s \gamma) = (3.44 \pm 0.53 \pm 0.35) \times 10^{-4}$, where the errors refer to perturbative and parameter uncertainties, respectively. The implications of larger theory uncertainties for New Physics searches are explored with the example of the type-II two-Higgs-doublet model.

Keywords: B physics, resummation, rare decays, searches for New Physics
PACS: 12.38.Cy,12.39.Hg,12.39.St,13.20.He

INTRODUCTION

Given the prominent role of $B \to X_s \gamma$ decay in searching for physics beyond the Standard Model, it is of great importance to have a precise prediction for its inclusive rate and CP asymmetry in the Standard Model. The total inclusive $B \to X_s \gamma$ decay rate can be calculated using a conventional operator-product expansion (OPE) based on an expansion in logarithms and inverse powers of the b-quark mass. However, in practice experiments can only measure the high-energy part of the photon spectrum, $E_\gamma \geq E_0$, where typically $E_0 = 2$ GeV or slightly below (measured in the B-meson rest frame) [1, 2]. With E_γ restricted to be close to the kinematic endpoint at $M_B/2$, the hadronic final state X_s is constrained to have large energy $E_X \sim M_B$ but only moderate invariant mass $M_X \sim (M_B \Lambda_{QCD})^{1/2}$. In this kinematic region, an infinite number of leading-twist terms in the OPE need to be resummed into a non-perturbative shape function, which describes the momentum distribution of the b-quark inside the B meson [3, 4].

Conventional wisdom based on phenomenological studies of shape-function effects says these effects are important near the endpoint of the photon spectrum, but they can be ignored as soon as the cutoff E_0 is lowered below about 1.9 GeV. In other words, there should be an instantaneous transition from the "shape-function region" of large non-perturbative corrections to the "OPE region", in which hadronic corrections to the rate are suppressed by at least two powers of Λ_{QCD}/m_b. Below, we argue that this notion

---

[1] Work supported by the National Science Foundation under Grant PHY-0355005

is based on a misconception. While it is correct that once the cutoff E_0 is chosen below 1.9 GeV the decay rate can be calculated using a local short-distance expansion, we show that this expansion involves three "large" scales. In addition to the hard scale m_b, an intermediate scale $\sqrt{m_b \Delta}$ corresponding to the typical invariant mass of the hadronic final state X_s, and a low scale $\Delta = m_b - 2E_0$ related to the width of the energy window over which the measurement is performed, become of crucial importance. The precision of the theoretical calculations is ultimately determined by the value of the lowest short-distance scale Δ, which in practice is of order 1 GeV or only slightly larger. The theoretical accuracy that can be reached is therefore not as good as in the case of a conventional heavy-quark expansion applied to the B system. More likely, it is similar to (if not worse than) the accuracy reached, say, in the description of the inclusive hadronic decay rate of the τ lepton.

While we are aware that this conclusion may come as a surprise to many practitioners in the field of flavor physics, we believe that it is an unavoidable consequence of our analysis. Not surprisingly, then, we find that the error estimates for the $B \to X_s \gamma$ branching ratio that can be found in the literature are, without exception, too optimistic. Since there are unknown $\alpha_s^2(\Delta)$ corrections at the low scale $\Delta \sim 1$ GeV, we estimate the present perturbative uncertainty in the $B \to X_s \gamma$ branching ratio with E_0 in the range between 1.6 and 1.8 GeV to be of order 10–15%. In addition, there are uncertainties due to other sources, such as the b- and c-quark masses. The combined theoretical uncertainty is of order 15–20%, about twice as large as what has been claimed in the past. While this is a rather pessimistic conclusion, we stress that the uncertainty is limited by unknown higher-order perturbative terms, not by non-perturbative effects, which we find to be under good control. Therefore, there is room for a reduction of the uncertainty by means of well-controlled perturbative calculations.

QCD FACTORIZATION THEOREM

Using recent results on the factorization of inclusive B-meson decay distributions [5, 6], it is possible to derive a QCD factorization formula for the integrated $B \to X_s \gamma$ decay rate with a cut $E_\gamma \geq E_0$ on photon energy. In the region of large E_0, the leading contribution to the rate can be factorized in the form [7]

$$\Gamma^{\text{leading}}_{\bar{B} \to X_s \gamma}(E_0) = \frac{G_F^2 \alpha}{32 \pi^4} |V_{tb} V_{ts}^*|^2 \, \overline{m}_b^2(\mu_h) \, |H_\gamma(\mu_h)|^2 \, U_1(\mu_h, \mu_i) \tag{1}$$

$$\times \int_0^{\Delta_E} dP_+ \, (M_B - P_+)^3 \int_0^{P_+} d\hat{\omega} \, m_b \, J\big(m_b(P_+ - \hat{\omega}), \mu_i\big) \, \hat{S}(\hat{\omega}, \mu_i),$$

where $\Delta_E = M_B - 2E_0$ is twice the width of the window in photon energy over which the measurement of the decay rate is performed. The variable $P_+ = E_X - |\vec{P}_X|$ is the "plus component" of the 4-momentum of the hadronic final state X_s, which is related to the photon energy by $P_+ = M_B - 2E_\gamma$. The endpoint region of the photon spectrum is defined by the requirement that $P_+ \leq \Delta_E \ll M_B$, in which case P^μ is called a hard-collinear momentum [8].

In the factorization formula, $\mu_h \sim m_b$ is a hard scale, while $\mu_i \sim \sqrt{m_b \Lambda_{\text{QCD}}}$ is an intermediate hard-collinear scale of order the invariant mass of the hadronic final state.

The precise values of these matching scales are irrelevant, since the rate is formally independent of μ_h and μ_i. The hard corrections captured by the function $H_\gamma(\mu_h)$ result from the matching of the effective weak Hamiltonian of the Standard Model (or any of its extensions) onto a leading-order current operator of soft-collinear effective theory (SCET) [9]. At tree level, $H_\gamma(\mu_h) = C_{7\gamma}^{\text{eff}}(\mu_h)$ is equal to the "effective" coefficient $C_{7\gamma}^{\text{eff}} = C_{7\gamma} - \frac{1}{3}C_5 - C_6$. The expression valid at next-to-leading order can be found in [7]. The function $H_\gamma(\mu_h)$ is multiplied by the running b-quark mass $\overline{m}_b(\mu_h)$ defined in the $\overline{\text{MS}}$ scheme, which is part of the electromagnetic dipole operator $Q_{7\gamma}$.

The jet function $J(m_b(P_+ - \hat{\omega}), \mu_i)$ in (1) describes the physics of the final-state hadronic jet. An expression for this function valid at next-to-leading order in perturbation theory has been derived in [5, 6]. The perturbative expansion of the jet function can be trusted as long as $\mu_i^2 \sim m_b \Delta$ with $\Delta = m_b - 2E_0 \ll M_B$. Note that the "natural" choices $\mu_h \propto m_b$ and $\mu_i^2 \equiv m_b \tilde{\mu}_i$ with $\tilde{\mu}_i$ independent of m_b remove all reference to the b-quark mass (other than in the arguments of running coupling constants) from the factorization formula.

The shape function $\hat{S}(\hat{\omega}, \mu_i)$ parameterizes our ignorance about the soft physics associated with bound-state effects inside the B meson [3, 4]. Its naive interpretation is that of a parton distribution function, governing the distribution of the light-cone component k_+ of the residual momentum of the b quark inside the heavy meson. Once radiative corrections are included, however, a probabilistic interpretation of the shape function breaks down [5]. For convenience, the shape function is renormalized in (1) at the intermediate hard-collinear scale μ_i rather than at a hadronic scale μ_{had}. This removes any uncertainties related to the evolution from μ_i to μ_{had}. Since the shape function is universal, all that matters is that it is renormalized at the same scale when comparing different processes.

The last ingredient in the factorization formula is the function $U_1(\mu_h, \mu_i)$, which describes the renormalization-group (RG) evolution of the hard function $|H_\gamma|^2$ from the high matching scale μ_h down to the intermediate scale μ_i, at which the jet and shape functions are renormalized. The exact expression for this quantity and its perturbative expansion valid at next-to-next-to-leading logarithmic order can be found in [7].

As written in (1), the decay rate is sensitive to non-perturbative hadronic physics via its dependence on the shape function. This sensitivity is unavoidable as long as the scale $\Delta = m_b - 2E_0$ is a hadronic scale, corresponding to the endpoint region of the photon spectrum above, say, 2 GeV. Here we are interested in a situation where E_0 is lowered out of the shape-function region, such that Δ can be considered large compared with Λ_{QCD}. For orientation, we note that with $m_b = 4.7$ GeV and the cutoff $E_0 = 1.8$ GeV employed in a recent analysis by the Belle Collaboration [2] one gets $\Delta = 1.1$ GeV. The values of the three relevant physical scales as functions of the photon-energy cutoff E_0 are shown in Figure 1. This plot illustrates the fact that the transition from the shape-function region to the region where a conventional OPE can be applied is not abrupt but proceeds via an intermediate region, in which a short-distance analysis based on a multi-scale OPE (MSOPE) can be performed. The transition from the shape-function region into the MSOPE region occurs when the scale Δ becomes numerically large compared with Λ_{QCD}. Then terms of order $\alpha_s^n(\Delta)$ and $(\Lambda_{\text{QCD}}/\Delta)^n$, which are non-perturbative in the shape-function region, gradually become decent expansion parameters. Only for very

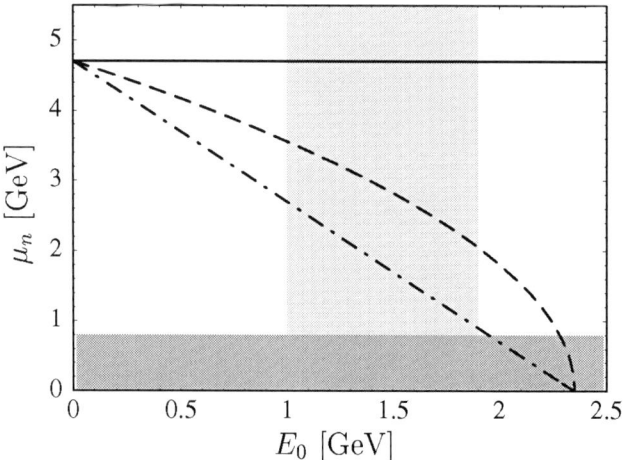

FIGURE 1. Dependence of the three scales $\mu_h = m_b$ (solid), $\mu_i = \sqrt{m_b \Delta}$ (dashed), and $\mu_0 = \Delta$ (dash-dotted) on the cutoff E_0, assuming $m_b = 4.7\,\text{GeV}$. The gray area at the bottom shows the domain of non-perturbative physics. The light gray band in the center indicates the region where the MSOPE must be applied.

low values of the cutoff ($E_0 < 1\,\text{GeV}$ or so) it is justified to treat Δ and $\sqrt{m_b \Delta}$ as scales of order m_b.

Separating the contributions associated with these scales requires a sophisticated multi-step procedure. The first step, the separation of the hard scale from the intermediate scale, has already been achieved in (1). To proceed further, we use that integrals of smooth weight functions with the shape function $\hat{S}(\hat{\omega}, \mu)$ can be expanded in a series of forward B-meson matrix elements of local operators in heavy-quark effective theory (HQET) [10], provided that the integration domain is large compared with Λ_{QCD} [5, 6]. The perturbative expansions of the associated Wilson coefficient functions can be trusted as long as $\mu \sim \Delta$. In order to complete the scale separation, it is therefore necessary to evolve the shape function in (1) from the intermediate scale $\mu_i \sim \sqrt{m_b \Delta}$ down to a scale $\mu_0 \sim \Delta$. This can be achieved using the analytic solution to the integro-differential RG evolution equation for the shape function in momentum space [5, 11].

As a final comment, we stress that the main purpose of performing the scale separation using the MSOPE is not that this allows us to resum Sudakov logarithms. Indeed, the "large logarithm" $\ln(m_b/\Delta) \approx 1.5$ is only parametrically large, but not numerically. What is really important is to disentangle the physics at the low scale $\mu_0 \sim \Delta$, which is "barely perturbative", from the physics associated with higher scales, where a short-distance treatment is on much safer grounds. The MSOPE allows us to distinguish between the three coupling constants $\alpha_s(m_b) \approx 0.22$, $\alpha_s(\sqrt{m_b \Delta}) \approx 0.29$, and $\alpha_s(\Delta) \approx 0.44$ (for $\Delta = 1.1\,\text{GeV}$), which are rather different despite the fact that there are no numerically large logarithms in the problem. Given these values, we expect that scale separation between Δ and m_b is as important as that between m_b and the weak scale M_W.

CALCULATION OF THE SHAPE-FUNCTION INTEGRAL

The scale dependence of the renormalized shape function is governed by an integro-differential RG evolution equation, whose exact solution in momentum space can be found using a technique developed in [11]. The result takes the remarkably simple form

$$\hat{S}(\hat{\omega}, \mu_i) = U_2(\mu_i, \mu_0) \frac{e^{-\gamma_E \eta}}{\Gamma(\eta)} \int_0^{\hat{\omega}} d\hat{\omega}' \frac{\hat{S}(\hat{\omega}', \mu_0)}{\mu_0^\eta (\hat{\omega} - \hat{\omega}')^{1-\eta}} . \tag{2}$$

The exact expression for the evolution function $U_2(\mu_i, \mu_0)$ can be found in [7], and

$$\eta = \int_{\alpha_s(\mu_i)}^{\alpha_s(\mu_0)} d\alpha \frac{2\Gamma_{\text{cusp}}(\alpha)}{\beta(\alpha)} = \frac{\Gamma_0}{\beta_0} \ln \frac{\alpha_s(\mu_0)}{\alpha_s(\mu_i)} + \dots \tag{3}$$

is given in terms of the cusp anomalous dimension [12].

Relation (2) accomplishes the evolution of the shape function from the intermediate scale down to the low scale $\mu_0 \sim \Delta$. The remaining task is to expand the integral over the shape function in (1) in a series of forward B-meson matrix elements of local HQET operators of increasing dimension, multiplied by perturbative coefficient functions. This can be done whenever $\Delta = \Delta_E - \bar{\Lambda} = m_b - 2E_0$ is large compared with Λ_{QCD} [5]. The result is

$$
\begin{aligned}
\Gamma_{\bar{B} \to X_s \gamma}^{\text{leading}}(E_0) &= \frac{G_F^2 \alpha}{32\pi^4} |V_{tb} V_{ts}^*|^2 m_b^3 \overline{m}_b^2(\mu_h) |H_\gamma(\mu_h)|^2 U_1(\mu_h, \mu_i) U_2(\mu_i, \mu_0) \\
&\times \frac{e^{-\gamma_E \eta}}{\Gamma(1+\eta)} \left(\frac{\Delta}{\mu_0}\right)^\eta \left[1 + \frac{C_F \alpha_s(\mu_i)}{4\pi} \mathcal{J}(\Delta) + \frac{C_F \alpha_s(\mu_0)}{4\pi} \mathcal{S}(\Delta) \right] \\
&\times \left[1 + \frac{\eta(\eta-1)}{2} \frac{(-\lambda_1)}{3\Delta^2} + \dots \right],
\end{aligned}
\tag{4}
$$

where

$$
\begin{aligned}
\mathcal{J}(\Delta) &= 2\ln^2 \frac{m_b \Delta}{\mu_i^2} - [4h(\eta) + 3] \ln \frac{m_b \Delta}{\mu_i^2} + 2h^2(\eta) + 3h(\eta) - 2h'(\eta) + 7 - \frac{2\pi^2}{3}, \\
\mathcal{S}(\Delta) &= -4\ln^2 \frac{\Delta}{\mu_0} + 4[2h(\eta) - 1] \ln \frac{\Delta}{\mu_0} - 4h^2(\eta) + 4h(\eta) + 4h'(\eta) - \frac{5\pi^2}{6},
\end{aligned}
\tag{5}
$$

and $h(\eta) = \psi(1+\eta) + \gamma_E$ is the harmonic function generalized to non-integer argument. Even though it is parametrically larger than ordinary power corrections of order $(\Lambda_{\text{QCD}}/m_b)^2$, the "enhanced" λ_1/Δ^2 correction in (4) remains small in the region of "perturbative" values of Δ, where the MSOPE can be trusted. The net effect amounts to a reduction of the decay rate by less than 5%.

The rate in (4) is formally independent of the three matching scales, at which we switch from QCD to SCET (μ_h), from SCET to HQET (μ_i), and finally at which the shape-function integral is expanded in a series of local operators (μ_0). In practice, a

95

residual scale dependence arises because we have truncated the perturbative expansion. Varying the three matching scales about their default values provides some information about unknown higher-order terms. In the limit where the intermediate and low matching scales μ_i and μ_0 are set equal to the hard matching scale μ_h, our result reduces to the conventional formula used in previous analyses of the $B \to X_s\gamma$ decay rate. However, this choice cannot be justified on physical grounds.

In (4) we have accomplished a complete resummation of (parametrically) large logarithms at next-to-next-to-leading logarithmic order in RG-improved perturbation theory, which is necessary in order to calculate the decay rate with $\mathcal{O}(\alpha_s)$ accuracy. Specifically, it means that terms of the form $\alpha_s^n L^k$ with $k = (n-1), \ldots, 2n$ and $L = \ln(m_b/\Delta)$ are correctly resummed to all orders in perturbation theory. To the best of our knowledge, a complete resummation at next-to-next-to-leading order has never been achieved before. Finally, we stress that the various next-to-leading order terms in the expression for the decay rate should be consistently expanded to order α_s before applying our results to phenomenology.

Up to this point, the b-quark mass m_b entering the formula (4) for the decay rate is defined in the on-shell scheme. While this is most convenient for performing calculations using heavy-quark expansions, it is well known that HQET parameters defined in the pole scheme suffer from infra-red renormalon ambiguities. It is necessary to replace them in favor of some physical, short-distance parameters. For our purposes, the "shape-function scheme" provides for a particularly suitable definition of the heavy-quark mass [5]. The idea is that a good estimate of a shape-function integral can be obtained using the mean-value theorem, replacing $\hat{\omega}$ with

$$\langle \hat{\omega} \rangle_\Delta = \frac{\int_0^{\Delta_E} d\hat{\omega}\, \hat{\omega}\, \hat{S}(\hat{\omega}, \mu_0)}{\int_0^{\Delta_E} d\hat{\omega}\, \hat{S}(\hat{\omega}, \mu_0)} \equiv \bar{\Lambda}(\Delta, \mu_0) = M_B - m_b(\Delta, \mu_0). \tag{6}$$

Here $m_b(\Delta, \mu_0)$ is the running shape-function mass, which depends on a hard cutoff Δ in addition to the renormalization scale μ_0. The quantity Δ in the shape-function scheme is defined by the implicit equation $\Delta = \Delta_E - \bar{\Lambda}(\Delta, \mu_0) = m_b(\Delta, \mu_0) - 2E_0$. The shape-function scheme provides a physical definition of m_b, which can be related to any other short-distance definition using perturbation theory. Based on various sources of phenomenological information including Υ spectroscopy and moments of inclusive B-meson decay spectra, the value of the shape-function mass at a reference scale $\mu_* = 1.5\,\text{GeV}$ has been determined as $m_b(\mu_*, \mu_*) = (4.65 \pm 0.07)\,\text{GeV}$ [5].

The results discussed so far provide a complete description of the $B \to X_s\gamma$ decay rate at leading order in the $1/m_b$ expansion, where the two-step matching $\text{QCD} \to \text{SCET} \to \text{HQET}$ is well understood. For practical applications, however, it is necessary to include corrections arising at higher orders in the heavy-quark expansion. Most important are "kinematic" power corrections of order $(\Delta/m_b)^n$, which are not associated with new hadronic parameters. Unlike the non-perturbative corrections, these effects appear already at first order in Δ/m_b, and they are numerically dominant in the region where $\Delta \gg \Lambda_{\text{QCD}}$. Technically, the kinematic power corrections correspond to subleading jet functions arising in the matching of QCD onto higher-dimensional SCET operators, as well as subleading shape functions arising in the matching of SCET onto HQET operators. The corresponding terms are known in fixed-order perturbation theory, without scale separation and RG resummation [13, 14]. To perform a complete RG analysis of even the first-order terms in Δ/m_b is beyond the scope of our discussion. Since for typical values of E_0 the power corrections only account for about 15% of the $B \to X_s \gamma$ decay rate, an approximate treatment suffices at the present level of precision. Details of how these corrections are implemented can be found in [7].

RATIOS OF DECAY RATES

The contributions from the three different short-distance scales entering the central result (4) and the associated theoretical uncertainties can be disentangled by taking ratios of decay rates. Some ratios probe truly short-distance physics (i.e., physics above the scale $\mu_h \sim m_b$) and so remain unaffected by the new theoretical results presented above. For some other ratios, the short-distance physics associated with the hard scale cancels to a large extent, so that one probes physics at the intermediate and low scales, irrespective of the short-distance structure of the theory.

Ratios insensitive to low-scale physics. Physics beyond the Standard Model may affect the theoretical results for the $B \to X_s \gamma$ branching ratio and CP asymmetry only via the Wilson coefficients of the various operators in the effective weak Hamiltonian. As a result, the ratio of the $B \to X_s \gamma$ decay rate in a New-Physics model relative to that in the Standard Model remains largely unaffected by the resummation effects studied in the present work. From (4), we obtain

$$\frac{\Gamma_{\bar{B} \to X_s \gamma}|_{\mathrm{NP}}}{\Gamma_{\bar{B} \to X_s \gamma}|_{\mathrm{SM}}} = \frac{|H_\gamma(\mu_h)|^2_{\mathrm{NP}}}{|H_\gamma(\mu_h)|^2_{\mathrm{SM}}} + \text{power corrections.} \tag{7}$$

The power corrections would introduce some mild dependence on the intermediate and low scales μ_i and μ_0, as well as on the cutoff E_0.

Another important example is the direct CP asymmetry in $B \to X_s \gamma$ decays, for which we obtain

$$A_{\mathrm{CP}} = \frac{\Gamma_{\bar{B} \to X_s \gamma} - \Gamma_{B \to X_{\bar{s}} \gamma}}{\Gamma_{\bar{B} \to X_s \gamma} + \Gamma_{B \to X_{\bar{s}} \gamma}} = \frac{|H_\gamma(\mu_h)|^2 - |\overline{H}_\gamma(\mu_h)|^2}{|H_\gamma(\mu_h)|^2 + |\overline{H}_\gamma(\mu_h)|^2} + \text{power corrections,} \tag{8}$$

where $\overline{H}_\gamma$ is obtained from H_γ by CP conjugation, which in the Standard Model amounts to replacing the CKM matrix elements by their complex conjugates. It follows that predictions for the CP asymmetry in the Standard Model and various New Physics scenarios [15] remain largely unaffected by our considerations.

Ratios sensitive to low-scale physics. The multi-scale effects studied in this work result from the fact that in practice the $B \to X_s \gamma$ decay rate is measured with a restrictive cut on the photon energy. These complications would be absent if it were possible to measure the fully inclusive rate. It is convenient to define a function $F(E_0)$ as the ratio of the $B \to X_s \gamma$ decay rate with a cut E_0 divided by the total rate,

$$F(E_0) = \frac{\Gamma_{\bar{B} \to X_s \gamma}(E_0)}{\Gamma_{\bar{B} \to X_s \gamma}(E_*)}. \tag{9}$$

Because of a logarithmic soft-photon divergence for very low energy, it is conventional [14] to define the "total" inclusive rate as the rate with a very low cutoff $E_* = m_b/20$. The denominator in the expression for $F(E_0)$ can be evaluated using the standard OPE, which corresponds to setting all three matching scales equal to μ_h. The numerator is given by our expression in (4), supplemented by power corrections.

NUMERICAL RESULTS

We are now ready to present the phenomenological implications of our findings. A complete list of the relevant input parameters and their uncertainties is given in [7], where we also explain our strategy for estimating the perturbative uncertainty as well as the uncertainty due to parameter variations.

We begin by presenting predictions for the CP-averaged $B \to X_s\gamma$ branching fraction with a cutoff $E_\gamma \geq E_0$ applied on the photon energy measured in the B-meson rest frame. Lowering E_0 below $2\,\text{GeV}$ is challenging experimentally. The first measurement with $E_0 = 1.8\,\text{GeV}$ has recently been reported by the Belle Collaboration [2]. It yields[2]

$$\text{Br}(B \to X_s\gamma)\Big|_{E_\gamma > 1.8\,\text{GeV}} = (3.38 \pm 0.30 \pm 0.29) \cdot 10^{-4}. \qquad (10)$$

For $E_0 = 1.8\,\text{GeV}$ we have $\Delta \approx 1.1\,\text{GeV}$, which is sufficiently large to apply the formalism developed in the present work. (For comparison, the value $E_0 = 2.0\,\text{GeV}$ adopted in the CLEO analysis [1] implies $\Delta \approx 0.7\,\text{GeV}$, which we believe is too low for a short-distance treatment.) We find

$$\text{Br}(B \to X_s\gamma)\Big|_{E_0 = 1.8\,\text{GeV}} = (3.44 \pm 0.53\,[\text{pert.}] \pm 0.35\,[\text{pars.}]) \times 10^{-4}, \qquad (11)$$

where the first error refers to the perturbative uncertainty and the second one to parameter variations. The largest parameter uncertainties are due to the b- and c-quark masses. Our result is in excellent agreement with the experimental value shown in (10). Comparing the two results, and naively assuming Gaussian errors, we conclude that

$$\text{Br}(B \to X_s\gamma)_{\text{exp}} - \text{Br}(B \to X_s\gamma)_{\text{SM}} < 1.4 \cdot 10^{-4} \quad (95\%\ \text{CL}). \qquad (12)$$

Mainly as a result of the enlarged theoretical uncertainty, this bound is much weaker than the one derived in [17], where this difference was found to be less than $0.5 \cdot 10^{-4}$. Hence, we obtain a much weaker constraint on New Physics parameters. For instance, for the case of the type-II two-Higgs-doublet model, we may use the analysis of [18] to deduce

$$m_{H^+} > (\text{slightly below})\ 200\,\text{GeV} \quad (95\%\ \text{CL}), \qquad (13)$$

which is significantly weaker than the constraints $m_{H^+} > 500\,\text{GeV}$ (at 95% CL) and $m_{H^+} > 350\,\text{GeV}$ (at 99% CL) found in [17].

---

[2] To obtain this result, we had to undo a theoretical correction accounting for the effects of the cut $E_\gamma > 1.8\,\text{GeV}$, which had been applied to the experimental data.

The function $F(E_0)$ provides us with an alternative way to discuss the effects of imposing the cutoff on the photon energy. In contrast to the branching ratio, it is independent of several input parameters (e.g., $\overline{m}_b(\overline{m}_b)$, $|V_{ts}^* V_{tb}|$, τ_B, $\lambda_{1,2}$), and it shows a very weak sensitivity to variations of the remaining parameters. We obtain

$$F(1.8\,\text{GeV}) = (92^{+7}_{-10}\,[\text{pert.}] \pm 1\,[\text{pars.}])\%\,. \tag{14}$$

This is the first time that this fraction has been computed in a model independent way. The result may be compared with the values $(95.8^{+1.3}_{-2.9})\%$ and $(95 \pm 1)\%$ obtained from two studies of shape-function models [14, 16], in which perturbative uncertainties have been ignored. We obtain a significantly smaller central value with a much larger uncertainty.

CONCLUSIONS AND OUTLOOK

We have performed the first systematic analysis of the inclusive decay $B \to X_s\gamma$ in the presence of a photon-energy cut $E_\gamma \geq E_0$, where E_0 is such that $\Delta = m_b - 2E_0$ can be considered large compared to Λ_{QCD}, while still $\Delta \ll m_b$. This is the region of interest to experiments at the B factories. The first condition ($\Delta \gg \Lambda_{\text{QCD}}$) ensures that a theoretical treatment without shape functions can be applied. However, the second condition ($\Delta \ll m_b$) means that this treatment is *not* a conventional heavy-quark expansion in powers of $\alpha_s(m_b)$ and Λ_{QCD}/m_b. Instead, we have shown that three distinct short-distance scales are relevant in this region. They are the hard scale m_b, the hard-collinear (or jet) scale $\sqrt{m_b\Delta}$, and the low scale Δ. To separate the contributions associated with these scales requires a multi-scale operator product expansion (MSOPE).

Our approach allows us to study analytically the transition from the shape-function region, where $\Delta \sim \Lambda_{\text{QCD}}$, into the MSOPE region, where $\Lambda_{\text{QCD}} \ll \Delta \ll m_b$, and further into the region $\Delta = \mathcal{O}(m_b)$, where a conventional heavy-quark expansion applies. This is a significant improvement over previous work. For instance, it has sometimes been argued that exactly where the transition to a conventional heavy-quark expansion occurs is an empirical question, which cannot be answered theoretically. Our formalism provides a precise, quantitative answer to this question. In particular, for $B \to X_s\gamma$ with a realistic cut on the photon energy, one is *not* in a situation where a short-distance expansion at the scale m_b can be justified. The analysis makes it evident that the precision that can be achieved in the prediction of the $B \to X_s\gamma$ branching ratio is, ultimately, determined by how well perturbative and non-perturbative corrections can be controlled at the lowest relevant scale Δ, which in practice is of order $1\,\text{GeV}$. Consequently, we estimate significantly larger theoretical uncertainties than previous authors. These uncertainties are dominated by yet unknown higher-order perturbative effects. Non-perturbative, hadronic effects at the scale Δ appear to be small and under control.

This is not the first time in the history of $B \to X_s\gamma$ calculations that issues of scale setting have changed the prediction and error estimate for the branching ratio (see, e.g., the discussion in [17]). In our case, however, the change in perspective about the theory of $B \to X_s\gamma$ decay is more profound, as it imposes limitations on the very validity of a short-distance treatment. If the short-distance expansion at the scale Δ fails, then the rate *cannot* be calculated without resource to non-perturbative shape functions, which introduces an irreducible amount of model dependence. In practice, while $\Delta \approx 1.1$ GeV (for $E_0 \approx 1.8$ GeV) is probably sufficiently large to trust a short-distance analysis, it would be unreasonable to expect that yet unknown higher-order effects should be less important than in the case of other low-scale applications of QCD.

Obtaining a precise prediction for the $B \to X_s\gamma$ decay rate in the Standard Model is an important target of heavy-flavor theory. The present work shows that the ongoing effort to calculate the dominant parts of the next-to-next-to-leading corrections in the conventional heavy-quark expansion is only part of what is needed to achieve this goal. Equally important will be to compute the dominant higher-order corrections proportional to $\alpha_s^2(\Delta)$ and $\alpha_s^2(\sqrt{m_b\Delta})$, and to perform a renormalization-group analysis of the leading kinematic power corrections of order Δ/m_b. In fact, our error analysis suggests that these effects are potentially more important that the hard matching corrections at the scale m_b.

ACKNOWLEDGMENTS

I am grateful to the organizers for the invitation to deliver this talk. It is a pleasure to thank Alex Kagan and Björn Lange for useful discussions.

REFERENCES

1. S. Chen *et al.* [CLEO Collaboration], Phys. Rev. Lett. **87**, 251807 (2001) [hep-ex/0108032].
2. P. Koppenburg *et al.* [Belle Collaboration], Phys. Rev. Lett. **93**, 061803 (2004) [hep-ex/0403004].
3. M. Neubert, Phys. Rev. D **49**, 3392 (1994) [hep-ph/9311325]; Phys. Rev. D **49**, 4623 (1994) [hep-ph/9312311].
4. I. I. Y. Bigi, M. A. Shifman, N. G. Uraltsev and A. I. Vainshtein, Int. J. Mod. Phys. A **9**, 2467 (1994) [hep-ph/9312359].
5. S. W. Bosch, B. O. Lange, M. Neubert and G. Paz, Nucl. Phys. B **699**, 335 (2004) [hep-ph/0402094].
6. C. W. Bauer and A. V. Manohar, Phys. Rev. D **70**, 034024 (2004) [hep-ph/0312109].
7. M. Neubert, hep-ph/0408179.
8. S. W. Bosch, R. J. Hill, B. O. Lange and M. Neubert, Phys. Rev. D **67**, 094014 (2003) [hep-ph/0301123].
9. C. W. Bauer, S. Fleming, D. Pirjol and I. W. Stewart, Phys. Rev. D **63**, 114020 (2001) [hep-ph/0011336]; C. W. Bauer, D. Pirjol and I. W. Stewart, Phys. Rev. D **65**, 054022 (2002) [hep-ph/0109045].
10. For a review, see: M. Neubert, Phys. Rept. **245**, 259 (1994) [hep-ph/9306320].
11. B. O. Lange and M. Neubert, Phys. Rev. Lett. **91**, 102001 (2003) [hep-ph/0303082].
12. G. P. Korchemsky and A. V. Radyushkin, Nucl. Phys. B **283**, 342 (1987); I. A. Korchemskaya and G. P. Korchemsky, Phys. Lett. B **287**, 169 (1992).
13. C. Greub, T. Hurth and D. Wyler, Phys. Rev. D **54**, 3350 (1996) [hep-ph/9603404].
14. A. L. Kagan and M. Neubert, Eur. Phys. J. C **7**, 5 (1999) [hep-ph/9805303].
15. A. L. Kagan and M. Neubert, Phys. Rev. D **58**, 094012 (1998) [hep-ph/9803368].
16. I. Bigi and N. Uraltsev, Int. J. Mod. Phys. A **17**, 4709 (2002) [hep ph/0202175].
17. P. Gambino and M. Misiak, Nucl. Phys. B **611**, 338 (2001) [hep-ph/0104034].
18. F. M. Borzumati and C. Greub, Phys. Rev. D **58**, 074004 (1998) [hep-ph/9802391].

Open Problems in Heavy Quarkonium Physics

Antonio Vairo

Dipartimento di Fisica dell'Università di Milano and INFN via Celoria 16, 20133 Milan, Italy

Abstract. Some recent progress and a personal selection of open problems in heavy quarkonium physics (spectroscopy, decay and production) inspired by the activity of the Quarkonium Working Group are reviewed.

Keywords: Heavy quarkonia, spectra, decays, production
PACS: 12.38.-t,14.40.Lb,14.40.Nd

INTRODUCTION

The wealth and quality of new experimental findings and the theoretical progress in the construction and use of Effective Field Theories (EFTs) of QCD are among the reasons of the heavy quarkonium physics *renaissance* witnessed during the last years. In order to keep track and make immediately available to a larger community the progress in the field, experimental and theoretical physicists have gathered in the last three years to form a Quarkonium Working Group [1]. The offspring of this collaboration have been three workshops, a school and a newly issued CERN Yellow Report [2]. In the following we will review some recent results and open problems in heavy quarkonium physics. We refer the reader to [2] for exhaustive presentations.

SPECTROSCOPY

There has been in recent years a renewed interest and a noteworthy progress in the calculation of the heavy quarkonium spectrum from perturbative QCD. The progress comes essentially from three directions:

(1) progress in the construction of EFTs of QCD suitable to describe non-relativistic bound state systems. These have helped to organize higher-order calculations and to factorize high-energy perturbative and low-energy non-perturbative contributions. For a recent review we refer to [3].

(2) fixed order calculations [4, 5, 6, 7, 8, 9, 10, 11, 12, 13].

(3) resummation of large contributions (large logs or large contributions associated to renormalon singularities) [14, 15, 16, 17, 18, 19].

By definition, a perturbative treatment of the bound state is possible if the momentum transfer scale p of the heavy quarks in the bound state is much larger than Λ_{QCD}. Two situations may occur under this condition [20]. Let us call E the typical kinetic energy of the heavy quark and antiquark in the centre-of-mass reference frame: in a non-relativistic bound state $p \sim mv \gg E \sim mv^2$, v being the heavy-quark velocity in the bound state.

CP756, *Quark Confinement and the Hadron Spectrum VI*
edited by N. Brambilla, U. D'Alesio, A. Devoto, K. Maung, G.M. Prosperi and S. Serci
© 2005 American Institute of Physics 0-7354-0241-8/05/$22.50

The first situation corresponds to quarkonium states for which $E \gtrsim \Lambda_{\rm QCD}$. Under this circumstance the heavy-quarkonium potential is purely perturbative; we may call this case Coulombic. In the case $E \gg \Lambda_{\rm QCD}$ non-perturbative contributions are encoded into local condensates, in the case $E \sim \Lambda_{\rm QCD}$ into non-local ones (see also [21] and references therein). The second situation corresponds to quarkonium states for which $p \gg \Lambda_{\rm QCD} \gg E$. Under this circumstance the potential contains a perturbative part and short-range non-perturbative contributions. We may call quasi Coulombic the case in which non-perturbative contributions to the potential turn out to be small and may be treated perturbatively in the calculation of observables.

Clearly, perturbative calculations applied to heavy-quarkonium ground states are on a more solid ground that those to higher resonances, and those applied to bottomonium more than those to charmonium, since the (quasi-)Coulombic case is more likely to be realized. Perturbative determinations of the $\Upsilon(1S)$ and J/ψ masses have been used to extract the b and c masses (see *Masses*). The main uncertainty in these determinations comes from non-perturbative contributions. Information on their size can come from other ground-state observables, like the hyperfine splittings (see *Hyperfine splittings*), and the B_c mass (see *B_c mass*). Recent studies of higher quarkonium states in perturbation theory may be found in [4, 22, 23, 5, 24, 25, 12]. Likely only some of the lowest bottomonium resonances may be treated consistently as Coulombic or quasi-Coulombic bound states. It is surprising, therefore, that some of the gross features of the bottomonium spectrum, like the equal spacing of the radial excitations, may be qualitatively reproduced by pure perturbative calculations (see *Higher bottomonium states*).

TABLE 1.   Collection of recently obtained values of $m_b^{\overline{\rm MS}}(m_b^{\overline{\rm MS}})$ and $m_c^{\overline{\rm MS}}(m_c^{\overline{\rm MS}})$ from the $\Upsilon(1S)$ and J/Ψ masses.

Ref.	order	$m_b^{\overline{\rm MS}}(m_b^{\overline{\rm MS}})$ (GeV)
[27]	NNLO	4.24 ± 0.09
[28]	NNLO	4.21 ± 0.09
[29]	NNLO	$4.210 \pm 0.090 \pm 0.025$
[25]	NNLO	$4.190 \pm 0.020 \pm 0.025$
[26]	NNNLO	4.349 ± 0.070
[30]	NNNLO	4.20 ± 0.04
[31]	NNNLO	4.241 ± 0.070

Ref.	order	$m_c^{\overline{\rm MS}}(m_c^{\overline{\rm MS}})$ (GeV)
[24]	NNLO	1.24 ± 0.020

Masses

Table 1 shows some recent determinations of the b and c masses from the quarkonium ground-state masses. Finite charm-mass effects have been included in the analyses of [25, 30, 31]. With the exception of [27] the conversion from the threshold (or pole) masses to the $\overline{\rm MS}$ masses has been performed at three-loop accuracy. All NNNLO analyses are only partially complete, since the full three-loop static potential has not been calculated yet. Different schemes have been used to implement the leading renormalon cancellation. This may explain some of the differences between the results. The most substantial discrepancy, which is between the result of [26] and all the other ones, may be possibly ascribed to the use of the on-shell scheme, as well as to the way non-perturbative effects have been implemented. In [24] the charm $\overline{\text{MS}}$ mass has been extracted from the J/Ψ mass.

In the above analyses non-perturbative effects constitute the major source of uncertainty together with possible effects due to subleading renormalons. Indeed, in the situation $E \sim \Lambda_{\text{QCD}}$ non-perturbative corrections to the spectrum ($\sim \Lambda_{\text{QCD}}^3/p^2 \sim mv^4$) and the subleading renormalon in the mass ($\sim \Lambda_{\text{QCD}}^2/m \sim mv^4$) are parametrically of the same order as NNLO corrections in the perturbative expansion. As long as non-perturbative effects will not be incorporated in a quantitative manner, perturbative calculations beyond NNLO will not improve the determination of the heavy-quark masses from the quarkonium system.

The dominant non-perturbative correction to the mass of a quarkonium state $|n\rangle$ of LO binding energy $E_n^{(0)}$ is encoded in the expression

$$\delta E_n = -i\frac{g^2}{9}\int_0^\infty dt\, \langle n|\mathbf{r}e^{it(E_n^{(0)}-h_o)}\mathbf{r}|n\rangle \, \langle \mathbf{E}(t)\,\mathbf{E}(0)\rangle(\mu),$$

$h_o = p^2/m + \alpha_s/(6r)$ being the LO octet Hamiltonian and μ a factorization scale [8, 20]. A precise determination of this formula would need having an accurate determination of the chromoelectric correlator. In this respect the available data are puzzling. Quenched lattice calculations of gluelump masses indicate an inverse correlation length for the correlator $\langle \mathbf{E}(t)\,\mathbf{E}(0)\rangle$ of about 1.25 GeV [32, 33]. This scale is not only larger than the scale E but also of the same order as the momentum-transfer scale p, if p is identified with $m_b/2 \times 4/3 \times \alpha_s(p)$ (e.g. $m_b \approx 4.7$ GeV and $\alpha_s(p) \approx 0.4$). This may potentially question the whole perturbative approach to the ground state of bottomonium. On the other hand this approach is supported by the fact that it provides a value of the b mass consistent with those obtained from low moments sum rule calculations or from the B system [34]. Only a better determination of the chromoelectric correlator can solve the conundrum. A possible solution could come, for instance, from

- unquenching, if it lowers the value of the inverse correlation length as calculations with cooling techniques seem to indicate [35].
- a particular parametrization of the chromoelectric correlator that sets in the non-perturbative behaviour at a scale μ lower than the inverse correlation length.

The assumption on the Coulombic behaviour of ground-state quarkonia may be also tested on other observables, like the B_c mass and the hyperfine splittings.

B_c **mass**

Table 2 shows some recent determinations of the B_c mass in perturbation theory at NNLO accuracy compared with the value of the B_c mass in [36] and in a very recent lattice study [37]. The b and c pole masses are expanded in terms of the $\Upsilon(1S)$ and

J/ψ masses respectively in [38], and in terms of the $\overline{\text{MS}}$ masses in [24, 25]. In [25] finite charm-mass effects are included. Since the charm quark effectively decouples in the scheme of Ref. [38], we may consider the results of Refs. [38] and [25] at the same level of theoretical accuracy and the difference between them as due to higher-order corrections beyond NNLO accuracy.

TABLE 2. Different perturbative determinations of the B_c mass compared with the experimental value and a recent lattice determination.

B_c mass (MeV)					
State	[36] (expt)	[37] (lattice)	[38] (NNLO)	[24] (NNLO)	[25] (NNLO)
1^1S_0	6400(400)	$6304 \pm 12^{+12}_{-0}$	6326(29)	6324(22)	6307(17)

In *Fermilab Today*, December 3, 2004, the CDF collaboration has announced the preliminary results of a search for the B_c, using decays into a J/ψ and a charged pion. Their preliminary result is $M_{B_c} = 6287 \pm 5$ MeV. This value, if confirmed, would support the assumption that non-perturbative contributions to the quarkonium ground state are of the same magnitude as NNLO or even NNNLO corrections, which would be consistent with a $E \gtrsim \Lambda_{\text{QCD}}$ power counting.

Hyperfine splittings

Charmonium and bottomonium ground state hyperfine splittings have been recently calculated at NLL in [18]. The result in the charmonium case is consistent with the experimental value. This, again, supports the assumption that non-perturbative contributions for the quarkonium ground state are consistent with a $E \gtrsim \Lambda_{\text{QCD}}$ power counting. In the bottomonium case there are not yet experimental data.

We note, however, that NNLO corrections may be potentially important for this observable, also considering that at this order new type of corrections will appear for the first time:

- non-perturbative corrections, which are parametrically of NNLO in the situation $\Lambda_{\text{QCD}} \sim E$.
- corrections to the mass. In order to show the possible impact of this type of corrections, in Fig. 1 we display the hyperfine splittings at NLO accuracy calculated using $1/2$ of the bottomonium and charmonium vector ground state masses and the $\overline{\text{MS}}$ masses. The difference between the two determinations (taken around the maximum in μ) is about 70% in the charmonium case and about 30% in the bottomonium case.

Higher bottomonium states

Higher bottomonium resonances have been investigated in the framework of perturbative QCD most recently in [24, 25, 39]. The calculation of [24] is accurate at NNLO, that one of [25] includes finite charm-mass effects, while [39] is a numerical calculation

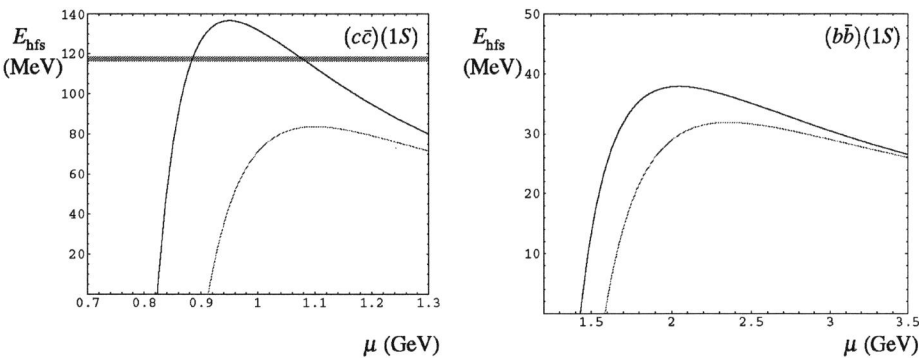

FIGURE 1. Hyperfine splittings of the $1S$ charmonium (left) and bottomonium (right) states at NLO in the $1S$ mass scheme, $m_c = M_{J/\psi}/2$, $m_b = M_{\Upsilon(1S)}/2$, (lower curves) and in the $\overline{\text{MS}}$ mass scheme, $m_c = m_c^{\overline{\text{MS}}}(m_c^{\overline{\text{MS}}})(1 + 4\alpha_s/(3\pi))$, $m_b = m_b^{\overline{\text{MS}}}(m_b^{\overline{\text{MS}}})(1 + 4\alpha_s/(3\pi))$, (upper curves) as a function of the normalization scale μ. α_s runs at 4-loop accuracy with 3 massless flavours. Left plot: $m_c^{\overline{\text{MS}}}(m_c^{\overline{\text{MS}}}) = 1.28$ GeV, $\alpha_s(M_{J/\psi}/2) = 0.352$, $\alpha_s(m_c^{\overline{\text{MS}}}(m_c^{\overline{\text{MS}}})) = 0.405$. Right plot: $m_b^{\overline{\text{MS}}}(m_b^{\overline{\text{MS}}}) = 4.22$ GeV, $\alpha_s(M_{\Upsilon(1S)}/2) = 0.209$, $\alpha_s(m_b^{\overline{\text{MS}}}(m_b^{\overline{\text{MS}}})) = 0.218$.

TABLE 3. Masses of $b\bar{b}$ states. The result of Ref. [24] comes from a full perturbative calculation up to $O(m\alpha_s^4)$ without finite charm-mass corrections; the result of Ref. [25] comes from a full perturbative calculation up to $O(m\alpha_s^4)$ including finite charm-mass corrections; the result of [39] incorporates full corrections up to $O(m\alpha_s^4)$ in the individual levels and full corrections up to $O(m\alpha_s^5)$ in the fine splittings, includes finite charm-mass corrections and also depends on some assumptions about the long-range non-perturbative behaviour of the static potential. Numbers without errors are those without explicit or reliable error estimates in the corresponding works.

		$b\bar{b}$ states		
State	expt	[24]	[25]	[39]
1^3S_1	9460	9460	9460	9460
1^3P_2	9913	9916(59)	10012(89)	9956
1^3P_1	9893	9904(67)	10004(86)	9938
1^3P_0	9860	9905(56)	9995(83)	9915
2^3S_2	10023	9966(68)	10084(102)	10032
2^3P_2	10269		10578(258)	10270
2^3P_1	10255		10564(247)	10260
2^3P_0	10232	10268	10548(239)	10246
3^3S_1	10355	10327(208)	10645(298)	10315

of the spectrum that includes also NLO spin-dependent potentials. This last calculation also relies on some assumptions about the long-range behaviour of the static potential. The results are summarized in Table 3.

The surprising result of these studies is that some gross features of the lowest part of

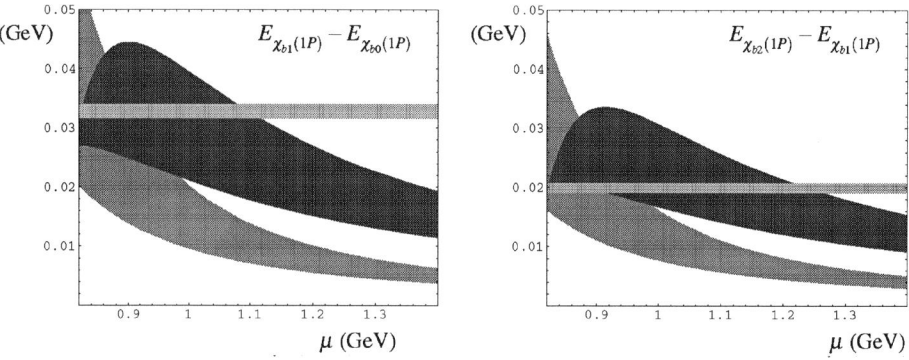

FIGURE 2. $E_{\chi_{b1}(1P)} - E_{\chi_{b0}(1P)}$ (left) and $E_{\chi_{b2}(1P)} - E_{\chi_{b1}(1P)}$ (right) versus the normalization scale μ. The light band shows the LO expectation, the dark one the NLO one. The widths of the bands account for the uncertainty in $\alpha_s(M_Z) = 0.1187 \pm 0.002$ [36]. The horizontal bands show the experimental values [36]. From [13].

the bottomonium spectrum, like the approximate equal spacing of the radial levels, is reproduced by a perturbative calculation that implements the leading-order renormalon cancellation. If this is coincidental or reflects the (quasi-)Coulombic nature of the states will be decided by further studies. A recent NLO calculation of the $1P$ bottomonium fine splittings has been performed in [13]. The results are summarized in Fig. 2. Figure 3 plots ρ, the ratio of the fine splittings considered in Fig. 2, as a function of the normalization scale μ (see [40] and references therein for early studies of this quantity). It seems to indicate either the existence of large NLL/NNLO corrections (as it happens in the hyperfine splittings of the $1S$ levels) or sizeable non-perturbative corrections, somehow hidden in the error bands of Fig. 2.

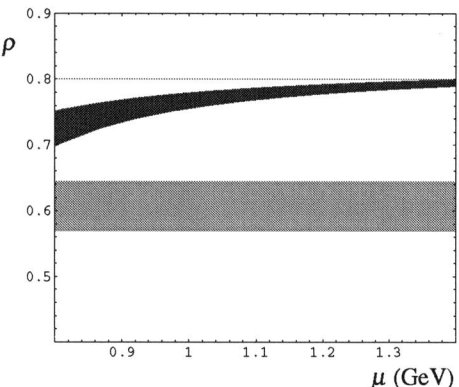

FIGURE 3. $\rho = (E_{\chi_{b2}} - E_{\chi_{b1}})/(E_{\chi_{b1}} - E_{\chi_{b0}})$ versus the normalization scale μ. The horizontal line at 0.8 corresponds to the LO expectation, the curve to the NLO one. The band accounts for the uncertainty in $\alpha_s(M_Z) = 0.1187 \pm 0.002$ [36]. The horizontal band shows the experimental value [36]. From [13].

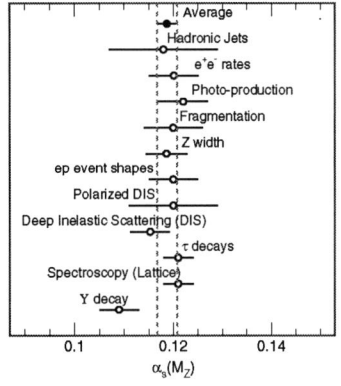 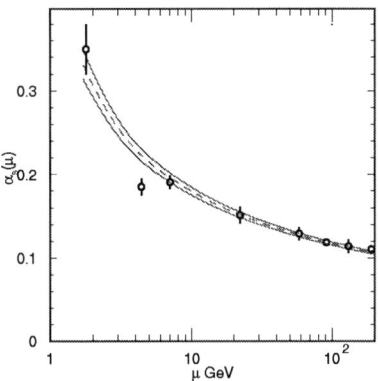

FIGURE 4. Different determinations of $\alpha_s(M_Z)$ (left) and of $\alpha_s(\mu)$ at different scales μ (right). In both figures the determinations outside the band come from Υ decay data. From [36].

DECAY

Υ inclusive decays

The NRQCD factorization formulas state that inclusive decay widths to light hadrons (*l.h.*) may be written as sums of products of matrix elements and imaginary parts of Wilson coefficients of 4-fermion operators [41]. In the case of the Υ system they read:

$$\Gamma(\Upsilon \to l.h.) = \frac{2}{m^2}\left(\mathrm{Im}\, f_1(^3S_1)\langle O_1(^3S_1)\rangle_\Upsilon \right.$$

$$+ \mathrm{Im}\, f_8(^3S_1)\langle O_8(^1S_0)\rangle_{\eta_b} + \frac{\mathrm{Im}\, f_8(^1S_0)}{3}\langle O_8(^3S_1)\rangle_{\eta_b}$$

$$\left. + \mathrm{Im}\, g_1(^3S_1)\frac{\langle \mathscr{P}_1(^1S_0)\rangle_{\eta_b}}{m^2} + \frac{\sum_J(2J+1)\mathrm{Im}\, f_8(^3P_J)}{9}\frac{\langle O_8(^1P_1)\rangle_{\eta_b}}{m^2} + \cdots \right).$$

The use of these formulas (and similar ones for the electromagnetic decay widths) to extract α_s at the b-mass scale has turned out problematic, as shown by Fig. 4. It seems that uncertainties have been underestimated [2]. The reasons may be the following:

- poor knowledge of the matrix elements. Note that usually octet matrix elements have been neglected in the analyses. This may not be justified considering that $\mathrm{Im}\, f_1(^3S_1)$ and $\mathrm{Im}\, g_1(^3S_1)$ are suppressed by an extra α_s with respect to most of the $\mathrm{Im}\, f_8$ coefficients (see [42] and references therein) that may compensate, under some circumstances, the suppression in the velocity expansion of the octet operator matrix elements. The poor knowledge of the matrix element $\langle \mathscr{P}_1 \rangle$ has been indicated as one of the major source of uncertainty in [43]. Studies of decay matrix elements on the lattice can be found in [44, 45, 46]. Factorization formulas for decay matrix elements in pNRQCD have been derived in [47, 48].

- large higher-order corrections to the Wilson coefficients [49].

TABLE 4. Comparison of ratios of χ_{cJ} partial widths. The experimental values PDG 2004 and PDG 2000 are obtained from the world averages of [36] and [50] respectively. The chosen ratios do not depend at leading order in the velocity expansion on octet or singlet matrix elements. The LO and NLO columns refer to a leading and next-to-leading order calculation done at the renormalization scale $2m_c$ with the following choice of parameters: $m_c = 1.5$ GeV and $\alpha_s(2m_c) = 0.245$. From [2].

Ratio	[36] (PDG 2004)	[50] (PDG 2000)	LO	NLO
$\dfrac{\Gamma(\chi_{c0} \to \gamma\gamma)}{\Gamma(\chi_{c2} \to \gamma\gamma)}$	5.1±1.1	13±10	≈ 3.75	≈ 5.43
$\dfrac{\Gamma(\chi_{c2} \to l.h.) - \Gamma(\chi_{c1} \to l.h.)}{\Gamma(\chi_{c0} \to \gamma\gamma)}$	410±100	270±200	≈ 347	≈ 383
$\dfrac{\Gamma(\chi_{c0} \to l.h.) - \Gamma(\chi_{c1} \to l.h.)}{\Gamma(\chi_{c0} \to \gamma\gamma)}$	3600±700	3500±2500	≈ 1300	≈ 2781
$\dfrac{\Gamma(\chi_{c0} \to l.h.) - \Gamma(\chi_{c2} \to l.h.)}{\Gamma(\chi_{c2} \to l.h.) - \Gamma(\chi_{c1} \to l.h.)}$	7.9±1.5	12.1±3.2	≈ 2.75	≈ 6.63
$\dfrac{\Gamma(\chi_{c0} \to l.h.) - \Gamma(\chi_{c1} \to l.h.)}{\Gamma(\chi_{c2} \to l.h.) - \Gamma(\chi_{c1} \to l.h.)}$	8.9±1.1	13.1±3.3	≈ 3.75	≈ 7.63

χ_c inclusive decays

In the case of χ_c inclusive decay widths the NRQCD factorization formulas read:

$$\Gamma(\chi_{cJ} \to l.h.) = 9 \operatorname{Im} f_1(^3P_J) \frac{|R'_{\chi_{cJ}}(0)|^2}{\pi m^4} + \frac{2 \operatorname{Im} f_8(^3S_1)}{m^2} \langle O_8(^1S_0) \rangle_{\chi_{cJ}} + \cdots,$$

$$\Gamma(\chi_{cJ} \to \gamma\gamma) = 9 \operatorname{Im} f_{\gamma\gamma}(^3P_J) \frac{|R'_{\chi_{cJ}}(0)|^2}{\pi m^4} + \cdots \qquad J = 0, 2,$$

where $R'_{\chi_{cJ}}(0)$ is the derivative of the radial part of the χ_{cJ} wave function at the origin. The displayed terms are accurate at leading order in the velocity expansion. The Wilson coefficients are known at NLO accuracy in α_s.

The experimental determination of the χ_c decay widths has dramatically improved in the last four years mainly due to the measurements of the E835 experiment at the Fermilab Antiproton Accumulator. A way to see the impact of these measurements is provided by Tab. 4, where we compare the PDG 2000 [50] with the PDG 2004 [36] determinations of different ratios of hadronic and electromagnetic widths. There have been sizable shifts in some central values and considerable reductions in the errors. In particular, the error on the ratio of the electromagnetic χ_{c0} and χ_{c2} widths has been reduced by about a factor 10, while in all other ratios the errors have been reduced by a factor 2 or 3. The considered ratios of hadronic and electromagnetic widths do not depend at leading order in the velocity expansion on any non-perturbative parameter. Therefore, they can be determined in perturbation theory only. The last two columns of Tab. 4 show the result of a LO and NLO calculation respectively. Clearly the data have become sensitive to NLO corrections and may be used, in principle, to determine α_s at the charm-mass scale. Before this, a necessary step is the calculation of the decay widths at next-to-leading order in the velocity expansion, since these contributions are potentially of the same magnitude as NLO corrections in the Wilson coefficients (see Ref. [51] for a calculation in the electromagnetic case).

PRODUCTION

In this section we briefly mention two of the main open problems in our understanding of charmonium production. We refer to [2] for a proper treatment.

Charmonium polarization

Quarkonium production at large p_T is dominated by gluon fragmentation. NRQCD predicts that the dominant gluon-fragmentation process is gluon fragmentation into a quark-antiquark pair in a color octet $^3S_1$ state. At large p_T the fragmenting gluon is transversely polarized. In the standard NRQCD power counting it is expected that the octet quark-antiquark pair keeps the transverse polarization as it evolves into a S-wave quarkonium. Different power countings [52] or higher-order corrections may somehow dilute the polarization, which, however, is expected to show up in high p_T data. The present Tevatron data [53] do not seem to confirm this expectation. The uncertainties are, however, too large to make any definite claim. The issue here is mainly experimental and more precise Tevatron data are eagerly awaited.

Double charmonium production

The most challenging open problem in charmonium production concerns double charmonium production in e^+e^- annihilations. The Belle data of inclusive $J/\psi + c\bar{c}$

production,

$$\frac{\sigma(e^+e^- \to J/\psi + c\bar{c})}{\sigma(e^+e^- \to J/\psi + X)} = 0.82 \pm 0.15 \pm 0.14, \quad \text{Belle [54]},$$

and of exclusive $J/\psi + \eta_c$ production,

$$\sigma(e^+e^- \to J/\psi + \eta_c)\,\text{Br}(c\bar{c}_{\text{res}} \to> 2 \text{ ch. track}) = 25.6 \pm 2.8 \pm 3.4 \text{ fb}, \quad \text{Belle [55]},$$

are far by almost an order of magnitude from the available theoretical predictions:

$$\frac{\sigma(e^+e^- \to J/\psi + c\bar{c})}{\sigma(e^+e^- \to J/\psi + X)} \approx 0.1, \quad [56, 57, 58],$$

$$\sigma(e^+e^- \to J/\psi + \eta_c) = 2.31 \pm 1.09 \text{ fb}, \quad [59, 60].$$

On the other hand, the present upper bound of Belle on $J/\psi + J/\psi$ production is consistent with theoretical expectations:

$$\sigma(e^+e^- \to J/\psi + J/\psi)\,\text{Br}(c\bar{c}_{\text{res}} \to> 2 \text{ ch. track}) < 9.1 \text{ fb}, \quad \text{Belle [55]},$$

$$\sigma(e^+e^- \to J/\psi + J/\psi) = 8.70 \pm 2.94 \text{ fb}, \quad [61, 62].$$

Even if independent data by BaBar would be most welcome, the issue here seems to be mainly theoretical and the data may be the signal of some new production mechanism.

REFERENCES

1. http://www.qwg.to.infn.it
2. N. Brambilla *et al.*, arXiv:hep-ph/0412158.
3. N. Brambilla, A. Pineda, J. Soto and A. Vairo, arXiv:hep-ph/0410047.
4. S. Titard and F. J. Yndurain, Phys. Rev. D **49**, 6007 (1994) [arXiv:hep-ph/9310236].
5. A. Pineda and F. J. Yndurain, Phys. Rev. D **58**, 094022 (1998) [arXiv:hep-ph/9711287].
6. N. Brambilla, A. Pineda, J. Soto and A. Vairo, Phys. Rev. D **60**, 091502 (1999) [arXiv:hep-ph/9903355].
7. B. A. Kniehl and A. A. Penin, Nucl. Phys. B **563**, 200 (1999) [arXiv:hep-ph/9907489].
8. N. Brambilla, A. Pineda, J. Soto and A. Vairo, Phys. Lett. B **470**, 215 (1999) [arXiv:hep-ph/9910238].
9. D. Eiras and J. Soto, Phys. Lett. B **491**, 101 (2000) [arXiv:hep-ph/0005066].
10. A. H. Hoang, arXiv:hep-ph/0008102.
11. M. Melles, Phys. Rev. D **62**, 074019 (2000) [arXiv:hep-ph/0001295].
12. B. A. Kniehl, A. A. Penin, V. A. Smirnov and M. Steinhauser, Nucl. Phys. B **635**, 357 (2002) [arXiv:hep-ph/0203166].
13. N. Brambilla and A. Vairo, arXiv:hep-ph/0411156.
14. M. E. Luke, A. V. Manohar and I. Z. Rothstein, Phys. Rev. D **61**, 074025 (2000) [arXiv:hep-ph/9910209].
15. A. H. Hoang, M. C. Smith, T. Stelzer and S. Willenbrock, Phys. Rev. D **59**, 114014 (1999) [arXiv:hep-ph/9804227].
16. M. Beneke, Phys. Lett. B **434**, 115 (1998) [arXiv:hep-ph/9804241].
17. A. Pineda, Phys. Rev. D **65**, 074007 (2002) [arXiv:hep-ph/0109117].
18. B. A. Kniehl, A. A. Penin, A. Pineda, V. A. Smirnov and M. Steinhauser, Phys. Rev. Lett. **92**, 242001 (2004) [arXiv:hep-ph/0312086].
19. A. A. Penin, A. Pineda, V. A. Smirnov and M. Steinhauser, Phys. Lett. B **593**, 124 (2004) [arXiv:hep-ph/0403080].

20. N. Brambilla, A. Pineda, J. Soto and A. Vairo, Nucl. Phys. B **566**, 275 (2000) [arXiv:hep-ph/9907240].
21. N. Brambilla and A. Vairo, arXiv:hep-ph/0004192.
22. S. Titard and F. J. Yndurain, Phys. Rev. D **51**, 6348 (1995) [arXiv:hep-ph/9403400].
23. S. Titard and F. J. Yndurain, Phys. Lett. B **351**, 541 (1995) [arXiv:hep-ph/9501338].
24. N. Brambilla, Y. Sumino and A. Vairo, Phys. Lett. B **513**, 381 (2001) [arXiv:hep-ph/0101305].
25. N. Brambilla, Y. Sumino and A. Vairo, Phys. Rev. D **65**, 034001 (2002) [arXiv:hep-ph/0108084].
26. A. A. Penin and M. Steinhauser, Phys. Lett. B **538**, 335 (2002) [arXiv:hep-ph/0204290].
27. M. Beneke and A. Signer, Phys. Lett. B **471**, 233 (1999) [arXiv:hep-ph/9906475].
28. A. H. Hoang, Phys. Rev. D **59**, 014039 (1999) [arXiv:hep-ph/9803454].
29. A. Pineda, JHEP **0106**, 022 (2001) [arXiv:hep-ph/0105008].
30. T. Lee, JHEP **0310**, 044 (2003) [arXiv:hep-ph/0304185].
31. C. Contreras, G. Cvetic and P. Gaete, Phys. Rev. D **70**, 034008 (2004) [arXiv:hep-ph/0311202].
32. M. Foster and C. Michael [UKQCD Collaboration], Phys. Rev. D **59**, 094509 (1999) [arXiv:hep-lat/9811010].
33. G. S. Bali and A. Pineda, Phys. Rev. D **69**, 094001 (2004) [arXiv:hep-ph/0310130].
34. A. X. El-Khadra and M. Luke, Ann. Rev. Nucl. Part. Sci. **52**, 201 (2002) [arXiv:hep-ph/0208114].
35. M. D'Elia, A. Di Giacomo and E. Meggiolaro, Phys. Lett. B **408**, 315 (1997) [arXiv:hep-lat/9705032].
36. S. Eidelman *et al.* [Particle Data Group Collaboration], Phys. Lett. B **592**, 1 (2004).
37. I. F. Allison, C. T. H. Davies, A. Gray, A. S. Kronfeld, P. B. Mackenzie and J. N. Simone [HPQCD Collaboration], arXiv:hep-lat/0411027.
38. N. Brambilla and A. Vairo, Phys. Rev. D **62**, 094019 (2000) [arXiv:hep-ph/0002075].
39. S. Recksiegel and Y. Sumino, Phys. Rev. D **67**, 014004 (2003) [arXiv:hep-ph/0207005].
40. N. Brambilla and A. Vairo, arXiv:hep-ph/9904330.
41. G. T. Bodwin, E. Braaten and G. P. Lepage, Phys. Rev. D **51**, 1125 (1995) [Erratum-ibid. D **55**, 5853 (1997)] [arXiv:hep-ph/9407339].
42. A. Vairo, Mod. Phys. Lett. A **19**, 253 (2004) [arXiv:hep-ph/0311303].
43. A. Pich, *talk at the 1st QWG Workshop, CERN, 2002*, http://www.qwg.to.infn.it/WS-nov02/index.html.
44. G. T. Bodwin, J. Lee and D. K. Sinclair, arXiv:hep-lat/0412006.
45. G. T. Bodwin, D. K. Sinclair and S. Kim, Phys. Rev. D **65**, 054504 (2002) [arXiv:hep-lat/0107011].
46. G. T. Bodwin, D. K. Sinclair and S. Kim, Phys. Rev. Lett. **77**, 2376 (1996) [arXiv:hep-lat/9605023].
47. N. Brambilla, D. Eiras, A. Pineda, J. Soto and A. Vairo, Phys. Rev. Lett. **88**, 012003 (2002) [arXiv:hep-ph/0109130].
48. N. Brambilla, D. Eiras, A. Pineda, J. Soto and A. Vairo, Phys. Rev. D **67**, 034018 (2003) [arXiv:hep-ph/0208019].
49. A. A. Penin, A. Pineda, V. A. Smirnov and M. Steinhauser, Nucl. Phys. B **699**, 183 (2004) [arXiv:hep-ph/0406175].
50. D. E. Groom *et al.* [Particle Data Group Collaboration], Eur. Phys. J. C **15**, 1 (2000).
51. J. P. Ma and Q. Wang, Phys. Lett. B **537**, 233 (2002) [arXiv:hep-ph/0203082].
52. S. Fleming, I. Z. Rothstein and A. K. Leibovich, Phys. Rev. D **64**, 036002 (2001) [arXiv:hep-ph/0012062].
53. T. Affolder *et al.* [CDF Collaboration], Phys. Rev. Lett. **85**, 2886 (2000) [arXiv:hep-ex/0004027].
54. K. Abe *et al.* [Belle Collaboration], BELLE-CONF-0331, contributed paper, International Europhysics Conference on High Energy Physics (EPS 2003), Aachen, Germany, 2003.
55. K. Abe *et al.* [Belle Collaboration], Phys. Rev. D **70**, 071102 (2004) [arXiv:hep-ex/0407009]; S. Olsen, *talk at the 3rd QWG Workshop, IHEP, 2004*, http://www.qwg.to.infn.it/WS-oct04/index.html.
56. P. L. Cho and A. K. Leibovich, Phys. Rev. D **54**, 6690 (1996) [hep-ph/9606229].
57. S. Baek, P. Ko, J. Lee, and H. S. Song, Phys. Lett. B **389**, 609 (1996) [hep-ph/9607236].
58. F. Yuan, C. F. Qiao, and K. T. Chao, Phys. Rev. D **56**, 321 (1997) [hep-ph/9703438].
59. E. Braaten and J. Lee, Phys. Rev. D **67**, 054007 (2003) [hep-ph/0211085].
60. K. Y. Liu, Z. G. He, and K. T. Chao, Phys. Lett. B **557**, 45 (2003) [hep-ph/0211181].
61. G. T. Bodwin, J. Lee, and E. Braaten, Phys. Rev. Lett. **90**, 162001 (2003) [hep-ph/0212181].
62. G. T. Bodwin, J. Lee, and E. Braaten, Phys. Rev. D **67**, 054023 (2003) [hep-ph/0212352].

Non-perturbative Heavy Quark Effective Theory: a test and its matching to QCD

Rainer Sommer

DESY, Platanenallee 6, 15738 Zeuthen, Germany

Abstract. We give an introduction to the special problems encountered in a treatment of HQET beyond perturbation theory in the gauge coupling constant. In particular, we report on a recent test of HQET as an effective theory for QCD and discuss how HQET can be implemented on the lattice including the non-perturbative matching of the effective theory to QCD.

Keywords: Heavy quark effective theory; Non-perturbative renormalization; Matching; Lattice QCD; Static approximation; Mass of the b-quark
PACS: 11.10.Gh; 11.15.Ha; 12.15.Hh; 12.38.Gc; 12.39.Hg; 14.65.Fy

1. INTRODUCTION

Heavy quark effective theory is routinely used in phenomenology. In these applications, the matching to QCD is achieved perturbatively and matrix elements of the operators in the effective theory are determined from experiment and models. However, HQET took its origin as an effective theory in the lattice regularization, where it was designed as a so-lution to the problem of treating quarks which are heavy compared to the inverse lattice spacing and thus do not propagate properly in the standard relativistic framework [1].

Unfortunately, after considerable initial activity (see e.g. [2, 3, 4] and references therein) the non-perturbative treatment of the effective theory on the lattice had been somewhat dormant for a while. The reason is that it was realized [5] that a non-perturbative matching to QCD is needed; otherwise the continuum limit does not exist. A practicable solution of this problem was only found recently [6, 7, 8].

Here we point out that a non-perturbative matching is necessary on *and off* the lattice, the problem being most severe on the lattice and we review a non-perturbative test of HQET. We then explain a recent strategy to perform fully non-perturbative computations in HQET and discuss the status and perspectives of this approach.

2. HQET AS AN ASYMPTOTIC EXPANSION OF QCD

Consider QCD at energies low enough such that the top-quark may be neglected alto-gether. In the QCD Lagrangian

$$\mathscr{L}_{\text{QCD}} = -\frac{1}{2g_0^2} \text{tr}\{F_{\mu\nu}F_{\mu\nu}\} + \sum_f \overline{\psi}_f[D_\mu\gamma_\mu + m_f]\psi_f \tag{1}$$

the sum over flavors then extends over $f = \text{u,d,s,c,b}$. An effective theory, HQET is expected to provide the asymptotic expansion of a certain (large) set of energies (e.g.

CP756, *Quark Confinement and the Hadron Spectrum VI*
edited by N. Brambilla, U. D'Alesio, A. Devoto, K. Maung, G.M. Prosperi and S. Serci
© 2005 American Institute of Physics 0-7354-0241-8/05/$22.50

mass splittings) and matrix elements of QCD in terms of the inverse of the mass of the b-quark.[1] We restrict our discussion to the energies and matrix elements of states which contain a single b-quark at rest and refer to reviews such as [9, 10] for the general case of finite velocity. The Lagrangian of the theory, which may be obtained by a formal $1/m_b$ expansion (see e.g. [11]), is then given by the replacement

$$\overline{\psi}_b[D_\mu \gamma_\mu + m_b]\psi_b \quad \rightarrow \quad \mathscr{L}_{stat} + \mathscr{L}^{(1)} + \ldots, \quad \mathscr{L}_{stat} = \overline{\psi}_h[D_0 + \delta m]\psi_h. \tag{2}$$

Here $\mathscr{L}^{(1)}$ is of order $1/m_b$ and the mass term of the b-quark has been removed from the Lagrangian such that observable quantities in the b-sector have a finite limit as $m_b \rightarrow \infty$ (with a suitable counter term δm). The effective heavy quark field ψ_h has only two degrees of freedom as appropriate for a non-relativistic spin 1/2 particle. Still it is notationally convenient to keep ψ_h as a 4-component spinor but impose the constraint

$$P_+\psi_h = \psi_h, \, P_+ = (1+\gamma_0)/2; \tag{3}$$

i.e. the lower components vanish in the Dirac representation. In order to discuss matrix elements, such as the B-meson decay constant, also the composite fields involving b-quarks are translated to the effective theory, for example:

$$A_0 = Z_A \overline{\psi}_1 \gamma_0 \gamma_5 \psi_b \quad \rightarrow \quad A_0^{stat} = Z_A^{stat} \overline{\psi}_1 \gamma_0 \gamma_5 \psi_h. \tag{4}$$

Here Z_A, Z_A^{stat} are the renormalization constants of the composite fields.

The effective theory is valid for the low-lying energy levels as well as their matrix elements, the simplest one being

$$\Phi^{QCD} \equiv F_B \sqrt{m_B} = Z_A \langle 0|A_0|B \rangle. \tag{5}$$

It is scale independent, due to the chiral symmetry of QCD in the massless limit (including $m_b = 0$). In the effective theory this symmetry is absent and Z_A^{stat} depends on the energy scale, μ, used in the renormalization condition which defines the finite current. Instead of $\Phi^{stat}(\mu) \equiv Z_A^{stat}(\mu)\langle 0|A_0|B \rangle_{stat}$ it is therefore better to consider the renormalization group invariant matrix element

$$\Phi_{RGI} = \lim_{\mu \rightarrow \infty} \left[2b_0 \bar{g}^2(\mu) \right]^{-\gamma_0/2b_0} \Phi^{stat}(\mu). \tag{6}$$

It is both μ and renormalization scheme independent, as is easily seen using $\Phi_{scheme}^{stat}(\mu) = \Phi_{scheme'}^{stat}(\mu)(1 + O(\bar{g}^2(\mu)))$. In eq. (6), the coefficients b_0, γ_0 defined by

$$\beta(\bar{g}) \equiv \mu \frac{d}{d\mu}\bar{g} = -b_0 \bar{g}^3 + O(\bar{g}^5), \quad \gamma(\bar{g}) \equiv \frac{\mu}{Z_A^{stat}} \frac{d}{d\mu} Z_A^{stat} = -\gamma_0 \bar{g}^2 + O(\bar{g}^4) \tag{7}$$

---

[1] Powers of $1/m_b$ are understood to be accompanied by slowly (logarithmically) varying functions of m_b.

enter. We can now write down the HQET-expansion of the QCD matrix element

$$\Phi^{\text{QCD}} = C_{\text{PS}}(M_{\text{b}}/\Lambda_{\overline{\text{MS}}}) \times \Phi_{\text{RGI}} + O(1/M_{\text{b}}), \tag{8}$$

$$M_{\text{b}} = \lim_{\mu \to \infty} \left[2b_0 \bar{g}(\mu)\right]^{-d_0/2b_0} \bar{m}(\mu), \quad \tau(\bar{g}) \equiv \frac{\mu}{\bar{m}} \frac{\text{d}}{\text{d}\mu} \bar{m} = -d_0 \bar{g}^2 + \dots \tag{9}$$

$$\Lambda_{\overline{\text{MS}}} = \lim_{\mu \to \infty} \mu \left(b_0 \bar{g}_{\overline{\text{MS}}}^2(\mu)\right)^{-b_1/(2b_0^2)} e^{-1/(2b_0 \bar{g}_{\overline{\text{MS}}}^2(\mu))}. \tag{10}$$

Let us dicuss the somewhat unfamiliar form of eq. (8) and the conversion function $C_{\text{PS}}(M_{\text{b}}/\Lambda_{\overline{\text{MS}}})$. In a more conventional form we have

$$\Phi^{\text{QCD}} = C_{\text{match}}(m_{\text{b}}/\mu) \times \Phi_{\overline{\text{MS}}}(\mu) + O(1/m_{\text{b}}) \tag{11}$$

with a matrix element $\Phi_{\overline{\text{MS}}}(\mu)$ renormalized in the effective theory in the $\overline{\text{MS}}$ scheme and the matching coefficient $C_{\text{match}}(m_{\text{b}}/\mu)$ depending on the b-quark mass m_{b} in the $\overline{\text{MS}}$ scheme at scale m_{b}, i.e. $\bar{m}_{\overline{\text{MS}}}(m_{\text{b}}) = m_{\text{b}}$. The factor C_{match} is determined (usually in perturbation theory) such that eq. (11) holds for some particular matrix element of the current and will then be valid *for all matrix elements*. Contact to eq. (8) is easily made by using

$$\frac{\Phi_{\text{RGI}}}{\Phi_{\overline{\text{MS}}}(\mu)} = \left[2b_0 \bar{g}^2(\mu)\right]^{-\gamma_0/2b_0} \exp\left\{ -\int_0^{\bar{g}(\mu)} \text{d}g \left[\frac{\gamma_{\overline{\text{MS}}}(g)}{\beta_{\overline{\text{MS}}}(g)} - \frac{\gamma_0}{b_0 g}\right] \right\}, \tag{12}$$

setting the arbitrary renormalization point μ to m_{b} and identifying

$$C_{\text{PS}}\left(\frac{M_{\text{b}}}{\Lambda_{\overline{\text{MS}}}}\right) = C_{\text{match}}(1) \frac{\Phi_{\overline{\text{MS}}}(m_{\text{b}})}{\Phi_{\text{RGI}}} \tag{13}$$

$$= \left[2b_0 \bar{g}^2(m_{\text{b}})\right]^{\gamma_0/2b_0} \exp\left\{ \int_0^{\bar{g}(m_{\text{b}})} \text{d}g \left[\frac{\gamma_{\text{match}}(g)}{\beta_{\overline{\text{MS}}}(g)} - \frac{\gamma_0}{b_0 g}\right] \right\},$$

where $\bar{g}$ is taken in the $\overline{\text{MS}}$ scheme. The last equation may be taken as a definition of the anomalous dimension γ_{match} in the "matching scheme". It has contributions from $\gamma_{\overline{\text{MS}}}$ as well as from $C_{\text{match}} = 1 + c_1 \bar{g}^2 + \dots$, namely

$$\gamma_{\text{match}}(\bar{g}) = -\gamma_0 \bar{g}^2 - [\gamma_1^{\overline{\text{MS}}} + 2b_0 c_1] \bar{g}^4 + \dots. \tag{14}$$

Note that replacing the $\overline{\text{MS}}$ coupling by a non-perturbative one, γ_{match} may also be defined beyond perturbation theory through eqs. (13,8).[2] Another advantage of eq. (13) is that C_{PS} is independent of the arbitrary choice of renormalization scheme for the effective operators in the effective theory. Apart from the choice of the QCD coupling, the "convergence" of the series eq. (14) is dictated by the physics, nothing else. Note further that (at leading order in $1/M$) the conversion function C_{PS} contains the full

---

[2] Clearly the r.h.s. of eq. (13) is a function of $\bar{g}^2(m_{\text{b}})$, i.e. a function of $m_{\text{b}}/\Lambda_{\overline{\text{MS}}}$. We prefer to write it as a function of the ratio of renormalization group invariants, $M_{\text{b}}/\Lambda_{\overline{\text{MS}}}$.

(logarithmic) mass-dependence. The non-perturbative effective theory matrix elements are mass independent numbers. Conversion functions such as C_{PS} are universal for all (low energy) matrix elements of their associated operator. Thus

$$Z_A^2 \langle A_0^\dagger(x)A_0(0)\rangle_{QCD} \overset{x^2 \gg 1/M_b^2}{\sim} [C_{PS}(\tfrac{M_b}{\Lambda_{\overline{MS}}})]^2 \langle A_0^{stat}(x)^\dagger A_0^{stat}(0)\rangle_{RGI} + O(\tfrac{1}{M_b}) \quad (15)$$

is a straight forward generalization of eq. (8).

Analogous expressions for the conversion functions are valid for the time component of the axial current replaced by other composite fields, for example the space components of the vector current. Based on the work of [12, 13, 14] and recent efforts their perturbative expansion is known including the 3-loop anomalous dimension γ_{match} obtained from the 3-loop anomalous dimension $\gamma_{\overline{MS}}$ [15] and the 2-loop matching function C_{match} [16, 17, 18]. Figure 1, taken from [19], illustrates that the remaining $O(\bar{g}^6(m_b))$ errors in C_{PS} seem to be relatively small.

Although it is generally accepted that HQET is an effective theory of QCD in the sense that was just described, tests of this equivalence are rare and mostly based on phenomenological analysis of experimental results. A pure theory test can be performed if QCD including a heavy enough quark can be simulated on the lattice at lattice spacings which are small enough to be able to take the continuum limit. This has recently been achieved [19] and will be summarized below.

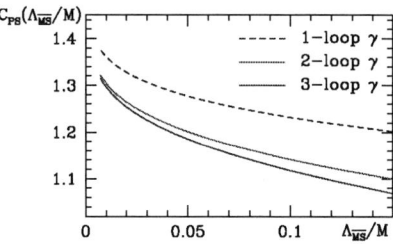

1: C_{PS} estimated in perturbation theory.

2.1. Tests of HQET in a finite volume

We start with the QCD side of such a test. Lattice spacings such that $am_b \ll 1$ can be reached if one puts the theory in a finite volume, $L^3 \times T$ with L, T not too large. We shall use $T = L$. For various practical reasons, so-called Schrödinger functional boundary conditions are chosen, i.e. Dirichlet in time (at $x_0 = 0, T$) and periodic in space [20, 21]. Equivalent boundary conditions are easily imposed in the effective theory [22]. We then form correlation functions of boundary quark fields ζ (located at $x_0 = 0$) and composite fields such as the time component of the axial current in the bulk ($0 < x_0 < T$), as illustrated in Fig. 2 and given for example by

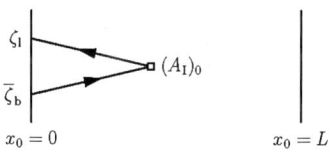

2: The correlation function f_A.

$$f_A(x_0) = -\frac{a^6}{2} \sum_{y,z} \langle (A_I)_0(x)\, \bar{\zeta}_b(y)\gamma_5\zeta_l(z)\rangle . \quad (16)$$

(The current A_I represents the $O(a)$-improved version of the axial current for which lattice artifacts linear in the lattice spacing are absent.) Another correlation function,

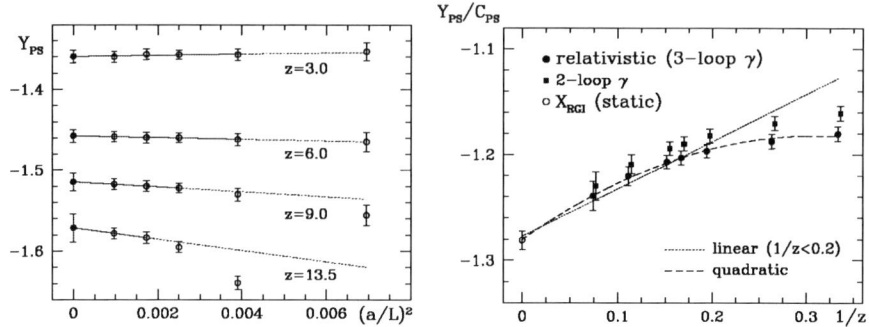

FIGURE 3. Testing eq. (19) through numerical simulations in the quenched approximation and for $L \approx 0.2\,\mathrm{fm}$ [19]. The physical mass of the b-quark corresponds to $z \approx 5$.

f_1, describes the propagation of a quark-antiquark pair from the $x_0 = 0$ boundary to the $x_0 = T$ boundary. For details we refer to [19].

We then take a ratio for which the renormalization factors of the boundary fields cancel,

$$Y_{PS}(L,M) \equiv Z_A \left.\frac{f_A(L/2)}{\sqrt{f_1}}\right|_{T=L} = \frac{\langle \Omega(L)|A_0|B(L)\rangle}{||\,|\Omega(L)\rangle\,||\;||\,|B(L)\rangle\,||}, \tag{17}$$

$$|B(L)\rangle = \mathrm{e}^{-LH/2}|\varphi_B(L)\rangle\,,\quad |\Omega(L)\rangle = \mathrm{e}^{-LH/2}|\varphi_0(L)\rangle\,. \tag{18}$$

As shown in the above equations, Y_{PS} can be represented as a matrix element of the axial current between a normalized state $|B(L)\rangle$ with the quantum numbers of a B-meson and $|\Omega(L)\rangle$ which has vacuum quantum numbers. The time evolution $\mathrm{e}^{-LH/2}$ ensures that both of these states are dominated by energy eigenstates with energies around $2/L$ and less. In other words, HQET is applicable if $1/L \ll M$ (and of course $\Lambda \ll M$).

One then expects (for fixed $L\Lambda$)

$$Y_{PS}(L,M)/C_{PS}(M/\Lambda) = X_{RGI} + \mathrm{O}(1/z)\,,\quad z = ML, \tag{19}$$

where the $1/M$ corrections are written in the dimensionless variable $1/z$ and X_{RGI} is defined as Y_{PS} but at lowest order in the effective theory and renormalized as in eq. (6). Of course such relations are expected after the continuum limit of both sides have been taken separately. For the case of $Y_{PS}(L,M)$, this is done by the following steps:

- Fix a value u_0 for the renormalized coupling $\bar{g}^2(L)$ (in the Schrödinger functional scheme) at vanishing quark mass. In [19] u_0 is chosen such that $L \approx 0.2$fm.
- For a given resolution L/a, determine the bare coupling from the condition $\bar{g}^2(L) = u_0$. This can easily be done since the relation between bare and renormalized coupling is known [23].
- Fix the bare quark mass $\tilde{m}_q$ of the heavy quark such that $LM = z$ using the known renormalization factor Z_m in $M = Z_m\tilde{m}_q$ [23].
- Evaluate Y_{PS} and repeat for better resolution a/L.
- Extrapolate to the continuum as shown in Fig. 3, left.

As can be seen in the figure, the continuum extrapolation becomes more difficult as the mass of the heavy quark is increased and $O((aM)^2)$ discretization errors become more and more important. In contrast the continuum extrapolation in the static effective theory (Fig. 4) is much easier (once the renormalization factor relating bare current and RGI current is known [6]). After the continuum limit has been taken, the finite mass QCD observable $Y_{PS}(L,M)$ turns smoothly into the prediction from the effective theory as illustrated in the r.h.s. figure. Indeed, several such successful tests were performed in [19], one of them free of the perturbative uncertainty in the conversion function (due to reparametrization invariance [24, 25, 26]) and two others with the static $(M \to \infty)$ limit known from the spin symmetry of HQET. For lack of space we do not show more examples. We only note that the coefficient of the $1/z^n$ terms in naive fits to the finite mass results are roughly of order unity.

Of course, finite mass lattice QCD results have been compared to the static limit over the years, see for example [27, 3, 4, 28, 29, 30, 31, 32, 33, 34, 35, 36, 37] and references therein. The new quality of the tests just discussed is that the composite fields were renormalized non-perturbatively throughout and that, by considering a small volume, the continuum limit could be taken at large quark masses.

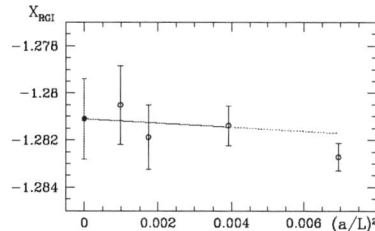

4: Continuum extrapolation of X_{RGI} [19].

2.2. Beyond the leading order: the need for non-perturbative conversion functions

Both from looking at Fig. 3 and from just a naive estimate of Λ/M_b, one expects that the effective theory has to be implemented beyond the leading order in $1/M$ to reach an acceptable precision in this expansion. However, if one wants to do this consistently, i.e. one wants to obtain the true coefficients in the expansion, the leading order conversion functions such as C_{PS} have to be known non-perturbatively. This general problem in the determination of power corrections in QCD is seen by considering the error made in eq. (13) (or eq. (11)) when the anomalous dimension has been computed at l loops and C_{match} at $l-1$ loop order. The conversion function C_{PS} is then known up to an error

$$\Delta(C_{PS}) \propto [\bar{g}^2(m_b)]^l \sim \left\{ \frac{1}{2b_0 \ln(M_b/\Lambda)} \right\}^l \overset{M_b \to \infty}{\gg} \frac{\Lambda}{M_b}. \tag{20}$$

As m_b is made large, this error becomes dominant. Taking a perturbative conversion function and adding power corrections to the leading order effective theory is thus to be regarded as a phenomenological approach, where one assumes that the coefficient of the $[\bar{g}^2(m_b)]^l$ term is small, such that the $(\Lambda/M_b)^n$ corrections dominate over a certain mass interval. Indeed, returning to our example, Fig. 3 indicates that the power corrections are larger than the perturbative ones at $1/z = 0.1 \ldots 0.2$. Nevertheless, it remains a fact that a theoretically consistent evaluation of power corrections requires a fully non-perturbative formulation of the theory including a non-perturbative matching to QCD.

3. NON-PERTURBATIVE FORMULATION OF HQET

The discussion in this section summarizes the main points of [8]. We regularize the theory on a space-time lattice. In the $1/m_b$ part of the Lagrangian,

$$\mathcal{L}^{(1)}(x) = \sum_i \omega_i^{(1)} \mathcal{L}_i^{(1)}(x), \tag{21}$$

$$\mathcal{L}_1^{(1)} = \overline{\psi}_h(-\sigma \cdot \mathbf{B})\psi_h, \quad \mathcal{L}_2^{(1)} = \overline{\psi}_h(-\tfrac{1}{2}\mathbf{D}^2)\psi_h \tag{22}$$

the chromo-magnetic field $\mathbf{B}$ and the 3-d Laplacian $\mathbf{D}^2$ are then discretized in the standard way. Details will be irrelevant for our discussion. The coefficients $\omega_i^{(1)}$ are functions of the bare coupling g_0 as well as the mass of the heavy quark. They have to be determined such as to match the effective theory to QCD. Matching at the classical level fixes

$$\omega_1^{(1)} = 1/m_b + O(g_0^2) = \omega_2^{(1)}. \tag{23}$$

Furthermore, we note that also in eq. (4) a dimension 4 composite field with coefficient $\propto 1/m_b + O(g_0^2)$ has to be added on the r.h.s. when $1/m_b$ corrections are considered. As an additional essential ingredient in the formulation of the effective theory we always expand the formal weight in the path integral, $\exp\left(\sum_x -(\mathcal{L}_{\text{stat}}(x) + \mathcal{L}^{(1)}(x) + \ldots)\right)$, in a power series in $1/m_b$. The correlation functions are then defined by

$$\langle \mathcal{O} \rangle = \tfrac{1}{Z} \int D[\varphi]\, \mathcal{O}[\varphi] \exp\left(-a^4 \sum_x \mathcal{L}_{\text{stat}}(\mathbf{x})\right) \times \left\{ 1 - a^4 \sum_x \mathcal{L}^{(1)}(x) + \ldots \right\}, \tag{24}$$

where φ denotes collectively the fields of the theory and the denominator Z insures $\langle 1 \rangle = 1$. The higher order terms in the Lagrangian then appear only as insertions into the correlation functions of the static effective theory. The latter is renormalizable by power counting and as a result also the effective theory truncated *at any finite order in* $1/m_b$ *is renormalizable*. With higher dimensional operators in the exponential, as in NRQCD, this would not be the case. For the lattice theory renormalizability is important because it means that the continuum limit exists and is independent of the details of the lattice formulation (universality).

3.1. Power divergencies

The coefficients $\delta m, \omega_i^{(1)}$ in eq. (2) and eq. (21) have a regular expansion in the bare coupling g_0^2. Still, perturbative precision is in general insufficient, since operators of higher dimensions mix with those of lower dimension, e.g.

$$\mathcal{O}_R^{d=5} = \sum_k z_k \mathcal{O}_k^{d=5} + \sum_k c_k \mathcal{O}_k^{d=4}, \quad c_k = \frac{c_k^{(0)} + c_k^{(1)} g_0^2 + \ldots}{a}. \tag{25}$$

Since the lattice spacing decreases as $a \sim \exp(-1/(2b_0 g_0^2))$ for small bare gauge coupling g_0, a truncation of the perturbative series leaves terms undetermined which diverge as the lattice spacing goes to zero. The origin of this problem is the same as the need for non-perturbative conversion functions, but the consequence is more drastic due to the presence of the hard cutoff in the lattice theory. Without non-perturbative precision for $\delta m, \omega_i^{(1)}$, the continuum limit does not exist.

3.2. Matching strategy

The definition of the effective theory is essentially given by eq. (24), supplemented by a definition of the effective composite fields. The only missing piece is a practical strategy for ascertaining how the parameters in the Lagrangian and in the effective fields can be determined beyond perturbation theory. At a given order n in the $1/m_b$-expansion, we denote the parameters in the effective theory by $c_k, k = 1, \ldots, N_n$. Observables, e.g. renormalized correlation functions or energies are denoted by $\Phi_k^{\mathrm{HQET}}(L,M)$ $(\Phi_k^{\mathrm{QCD}}(L,M))$ in the effective theory (in QCD), with the argument M referring to the mass of the heavy quark and L the linear extent of the finite volume. It is then sufficient to impose

$$\Phi_k^{\mathrm{HQET}}(L_0,M) = \Phi_k^{\mathrm{QCD}}(L_0,M), \quad k = 1, \ldots, N_n. \tag{26}$$

to determine all parameters $\{c_k, k = 1, \ldots, N_n\}$ in the effective theory. Observables used originally to fix $\{c_k, k = 1, \ldots, N_f\}$, the parameters of QCD, may be amongst these Φ_k^{QCD}. The matching conditions, eq. (26), define the set $\{c_k\}$ for any value of the lattice spacing (or equivalently bare coupling). Here, a typical choice is $L_0 \approx 0.2 \ldots 0.4 \,\mathrm{fm}$, since in such a volume the r.h.s. of the equation can be evaluated, see Sect. 2.1. In practice, the parameters of the effective theory are then determined at rather small lattice spacings in a range of $a = 0.01\,\mathrm{fm}$ to $a = 0.04\,\mathrm{fm}$. Large volumes as they are needed to compute the physical mass spectrum or matrix elements then require very large lattices $(L/a > 50)$. A further step is needed to bridge the gap to practicable lattice spacings. A well-defined procedure is as follows: First we assume that all observables $\Phi_k^{\mathrm{HQET}}(L,M)$ have been made dimensionless by multiplication with appropriate powers of L. Next, we define step scaling functions [38], F_k, by

$$\Phi_k^{\mathrm{HQET}}(sL,M) = F_k(\{\Phi_j^{\mathrm{HQET}}(L,M), j = 1, \ldots, N_n\}), \quad k = 1, \ldots, N_n, \tag{27}$$

where typically one uses scale changes of $s = 2$. These dimensionless functions describe the change of the complete set of observables $\{\Phi_k^{\mathrm{HQET}}\}$ under a scaling of $L \to sL$. In order to compute them one selects a lattice with a certain resolution a/L. The specification of $\Phi_j^{\mathrm{HQET}}(L,M)$, $j = 1, \ldots, N_n$, then fixes all (bare) parameters of the theory. The l.h.s. of eq. (27) is now computed, keeping the bare parameters fixed while changing $L/a \to L'/a = sL/a$. The values for the continuum F_k can then be be reached by extrapolating the resulting lattice numbers to $a/L \to 0$.

After repeating this step two or three times with $s = 2$, lattice spacings appropriate for infinite volume computations will be reached.

3.3. Example: the mass of the b-quark at lowest order

For illustration purposes we consider a simple example here, the computation of the b-quark mass, starting from the observed B-meson mass. Already at the lowest order in $1/m_b$ the mixing of operators of different dimensions is relevant in this case: $\overline{\psi}_h D_0 \psi_h$ mixes with $\overline{\psi}_h \psi_h$. Hence $\delta m = (c_1 g_0^2 + c_2 g_0^4 + \ldots)/a$, the coefficient of $\overline{\psi}_h \psi_h$ in the Lagrangian eq. (2), has to be determined (or eliminated) non-perturbatively. In eq. (26) we have $n = 0$, $N_0 = N_f + 1$ and we omit the discussion of the choices for Φ_k, $k = 1, \ldots, N_f$ which fix the bare light quark masses and coupling (both in QCD and HQET). Obviously any energy in the b-sector will do to fix δm. We choose[8, 39] [3]

$$\Gamma(L,M) = \tfrac{1}{2a} \ln \left[f_A(x_0 - a)/f_A(x_0 + a) \right] \qquad (x_0/L \text{ fixed}) \tag{28}$$

and we require eq. (26) where for $k = N_f + 1$ we identify

$$\Phi_{N_f+1}^{\text{QCD}}(L,M) \equiv L\Gamma(L,M), \quad \Phi_{N_f+1}^{\text{HQET}}(L,M) \equiv L(\Gamma_{\text{stat}}(L) + m), \tag{29}$$

Here m represents the quark mass that was removed from all energies in defining the effective theory such that the $m \to \infty$ limit exists and Γ_{stat} refers to eq. (28) at the lowest order in $1/M$. The relevant part of eq. (27) can then be written in the simple form,

$$\Phi_{N_f+1}^{\text{HQET}}(2L,M) = 2\Phi_{N_f+1}^{\text{HQET}}(L,M) + \sigma_m\left(\bar{g}^2(L)\right), \tag{30}$$

$$\sigma_m\left(\bar{g}^2(L)\right) \equiv 2L[\Gamma_{\text{stat}}(2L) - \Gamma_{\text{stat}}(L)]. \tag{31}$$

In σ_m the divergent δm (as well as the mass shift m) cancel. It is independent of the mass. We now see immediately that

$$m_B = \underbrace{E_{\text{stat}} - \Gamma_{\text{stat}}(L_2)}_{a \to 0 \text{ in HQET}} + \underbrace{\Gamma_{\text{stat}}(L_2) - \Gamma_{\text{stat}}(L_0)}_{a \to 0 \text{ in HQET}} + \underbrace{\Gamma(L_0, M_b)}_{a \to 0 \text{ in QCD}} + O(\Lambda^2/M_b). \tag{32}$$

Here, $E_{\text{stat}} = \lim_{L \to \infty} \Gamma_{\text{stat}}(L)$ is the infinite volume energy of a B-meson in static approximation. It is often called the static binding energy. The whole strategy is illustrated in Fig. 5. As indicated in eq. (32), the continuum limit can be taken in each individual step; a numerical example is shown in Fig. 6.

After obtaining all pieces in eq. (32), the equation is numerically solved for $z_b = M_b L_0$. Since also the size of L_0 in units of $r_0 \approx 0.5\,\text{fm}$ [40] is known, one can quote (remember m_b is in the $\overline{\text{MS}}$ scheme at scale m_b)

$$r_0 M_b = 16.12(24)(15) \rightarrow M_b = 6.36(10)(6)\,\text{GeV}, \quad m_b = 4.12(7)(4)\,\text{GeV}. \tag{33}$$

We emphasize that this result is in the quenched approximation but includes the lowest non-trivial order in $1/m_b$. An estimate of the associated $O(\Lambda^2/M_b)$ uncertainty is *not* included in the errors shown. Our discussion mainly serves to illustrate the potential of the approach in an example where the power divergent mixing needed to be solved non-perturbatively.

---

[3] In practice Γ is replaced by the spin averaged energy [8, 39].

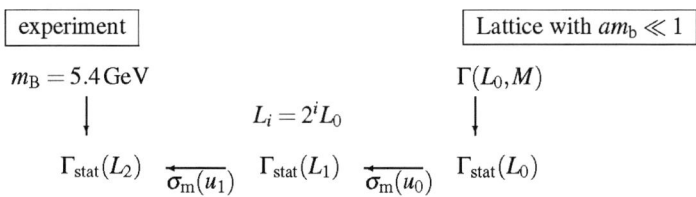

FIGURE 5. Connecting experimental observables to renormalized HQET. Γ_{stat} is a renormalized quantity in HQET and $\sigma_{\text{m}}(\bar{g}^2(L))$ connects $\Gamma_{\text{stat}}(L)$ and $\Gamma_{\text{stat}}(2L)$.

4. PERSPECTIVES

Non-perturbative HQET at the leading order in $1/m$ has reached a satisfactory status, with the b-quark mass [8] and the B_s-meson decay constant [6, 41] known in the continuum limit of the quenched approximation. Their precision can and will still be improved. Applying these methods to the theory with dynamical fermions is straight forward; "only" the usual problems with the light quarks have to be solved. By themselves such lowest order results are not expected to have an interesting precision for phenomenological applications, but

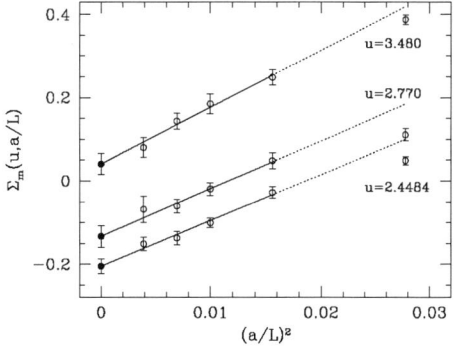

6: Extrapolation $\Sigma_{\text{m}}(u,a/L) = \sigma_{\text{m}}(u) + c\,a^2/L^2$

certainly they can constrain the large mass behavior computed with other methods [27, 3, 4, 28, 29, 30, 31, 32, 33, 34, 35, 36, 37].

More interestingly, also $1/m_b$ corrections can, in principle, be computed in the effective theory. Here, details of the necessary numerical steps have not yet been implemented but the very first tests have been encouraging [42]. Another relevant technical advance has been the realization that a change of the regularization details allows to achieve much better statistical errors in HQET, while keeping the discretization errors small [41]. In summary, we believe that all ingredients exist which are needed to apply HQET beyond the leading order in $1/m_b$.

ACKNOWLEDGMENTS

I would like to thank my colleagues S. Dürr, M. Della Morte, A. Jüttner, J. Rolf, and in particular J. Heitger for a pleasant collaboration in developing non-perturbative HQET. I am grateful to H. Simma for useful comments on this manuscript.

REFERENCES

1. E. Eichten, *Nucl. Phys. Proc. Suppl.*, **4**, 170 (1988).
2. E. Eichten, *Nucl. Phys. Proc. Suppl.*, **20**, 475–487 (1991).
3. R. Sommer, *Phys. Rept.*, **275**, 1–47 (1996), hep-lat/9401037.
4. H. Wittig, *Int. J. Mod. Phys.*, **A12**, 4477–4538 (1997), hep-lat/9705034.
5. L. Maiani, G. Martinelli, and C. T. Sachrajda, *Nucl. Phys.*, **B368**, 281–292 (1992).
6. J. Heitger, M. Kurth, and R. Sommer, *Nucl. Phys.*, **B669**, 173–206 (2003), hep-lat/0302019.
7. J. Heitger, and R. Sommer, *Nucl. Phys. Proc. Suppl.*, **106**, 358–360 (2002), hep-lat/0110016.
8. J. Heitger, and R. Sommer, *JHEP*, **02**, 022 (2004), hep-lat/0310035.
9. M. Neubert, *Phys. Rept.*, **245**, 259–396 (1994), hep-ph/9306320.
10. T. Mannel (1996), hep-ph/9606299.
11. B. A. Thacker, and G. P. Lepage, *Phys. Rev.*, **D43**, 196–208 (1991).
12. D. J. Broadhurst, and A. G. Grozin, *Phys. Lett.*, **B267**, 105–110 (1991), hep-ph/9908362.
13. M. A. Shifman, and M. B. Voloshin, *Sov. J. Nucl. Phys.*, **45**, 292 (1987).
14. H. D. Politzer, and M. B. Wise, *Phys. Lett.*, **B206**, 681 (1988).
15. K. G. Chetyrkin, and A. G. Grozin (2003), hep-ph/0303113.
16. X. Ji, and M. J. Musolf, *Phys. Lett.*, **B257**, 409 (1991).
17. D. J. Broadhurst, and A. G. Grozin, *Phys. Rev.*, **D52**, 4082–4098 (1995), hep-ph/9410240.
18. V. Gimenez, *Nucl. Phys.*, **B375**, 582–624 (1992).
19. J. Heitger, A. Jüttner, R. Sommer, and J. Wennekers, *JHEP*, **11**, 048 (2004), hep-ph/0407227.
20. M. Lüscher, R. Narayanan, P. Weisz, and U. Wolff, *Nucl. Phys.*, **B384**, 168–228 (1992), hep-lat/9207009.
21. S. Sint, *Nucl. Phys.*, **B421**, 135–158 (1994), hep-lat/9312079.
22. M. Kurth, and R. Sommer, *Nucl. Phys.*, **B597**, 488–518 (2001), hep-lat/0007002.
23. S. Capitani, M. Lüscher, R. Sommer, and H. Wittig, *Nucl. Phys.*, **B544**, 669 (1999), hep-lat/9810063.
24. M. E. Luke, and A. V. Manohar, *Phys. Lett.*, **B286**, 348–354 (1992), hep-ph/9205228.
25. W. Kilian, and T. Ohl, *Phys. Rev.*, **D50**, 4649–4656 (1994), hep-ph/9404305.
26. R. Sundrum, *Phys. Rev.*, **D57**, 331–336 (1998), hep-ph/9704256.
27. C. Alexandrou, et al., *Z. Phys.*, **C62**, 659–668 (1994), hep-lat/9312051.
28. A. X. El-Khadra, A. S. Kronfeld, P. B. Mackenzie, S. M. Ryan, and J. N. Simone, *Phys. Rev.*, **D58**, 014506 (1998), hep-ph/9711426.
29. S. Aoki, et al., *Phys. Rev. Lett.*, **80**, 5711–5715 (1998).
30. C. W. Bernard, et al., *Phys. Rev. Lett.*, **81**, 4812–4815 (1998), hep-ph/9806412.
31. D. Becirevic, et al., *Phys. Rev.*, **D60**, 074501 (1999), hep-lat/9811003.
32. A. Ali Khan, et al., *Phys. Rev.*, **D64**, 034505 (2001), hep-lat/0010009.
33. K. C. Bowler, et al., *Nucl. Phys.*, **B619**, 507–537 (2001), hep-lat/0007020.
34. L. Lellouch, and C. J. D. Lin, *Phys. Rev.*, **D64**, 094501 (2001), hep-ph/0011086.
35. S. M. Ryan, *Nucl. Phys. Proc. Suppl.*, **106**, 86–97 (2002), hep-lat/0111010.
36. G. M. de Divitiis, M. Guagnelli, R. Petronzio, N. Tantalo, and F. Palombi, *Nucl. Phys.*, **B675**, 309–332 (2003), hep-lat/0305018.
37. G. M. de Divitiis, M. Guagnelli, F. Palombi, R. Petronzio, and N. Tantalo, *Nucl. Phys.*, **B672**, 372–386 (2003), hep-lat/0307005.
38. M. Lüscher, P. Weisz, and U. Wolff, *Nucl. Phys.*, **B359**, 221–243 (1991).
39. J. Heitger, and J. Wennekers, *JHEP*, **02**, 064 (2004), hep-lat/0312016.
40. R. Sommer, *Nucl. Phys.*, **B411**, 839 (1994), hep-lat/9310022.
41. M. Della Morte, et al., *Phys. Lett.*, **B581**, 93–98 (2004), hep-lat/0307021.
42. S. Dürr, A. Jüttner, J. Rolf, and R. Sommer (2004), hep-lat/0409058.

Lattice QCD Study for the Interquark Force in Three-Quark and Multi-Quark Systems

H. Suganuma[*], T.T. Takahashi[†], F. Okiharu[**] and H. Ichie[*]

[*]Faculty of Science, Tokyo Institute of Technology, Tokyo 152-8551, Japan
[†]Yukawa Institute for Theoretical Physics, Kyoto University, Kyoto 606-8502, Japan
[**]Department of Physics, Nihon University, Kanda-Surugadai 1-8, Chiyoda, Tokyo 101, Japan

Abstract. We study three-quark and multi-quark potentials in SU(3) lattice QCD. From accurate calculations for more than 300 different patterns of 3Q systems, the static ground-state 3Q potential $V_{3Q}^{g.s.}$ is found to be well described by the Coulomb plus Y-type linear potential (Y-Ansatz) within 1%-level deviation. As a clear evidence for Y-Ansatz, Y-type flux-tube formation is actually observed on lattices in maximally-Abelian projected QCD. For about 100 patterns of 3Q systems, we perform accurate calculations for the 1st excited-state 3Q potential $V_{3Q}^{e.s.}$ by diagonalizing the QCD Hamiltonian in presence of three quarks, and find a large gluonic-excitation energy $\Delta E_{3Q} \equiv V_{3Q}^{e.s.} - V_{3Q}^{g.s.}$ of about 1 GeV, which gives a physical reason on success of the quark model. ΔE_{3Q} is found to be reproduced by "inverse Mercedes Ansatz", which indicates a complicated bulk excitation for the gluonic-excitation mode. We study also tetra-quark and penta-quark potentials in lattice QCD, and find that they are well described by the OGE Coulomb plus multi-Y type linear potential, which supports the flux-tube picture even for multi-quarks. Finally, narrow decay width of low-lying penta-quark baryons is discussed in terms of the QCD string theory.

Keywords: lattice QCD, inter-quark potential, multi-quark, gluonic excitation, confinement, string
PACS: 12.38.Gc, 12.38.Aw, 12.39.Mk, 12.39.Jh, 12.39.Pn, 13.30.Eg, 14.20.Jn, 11.25.Wx

1. INTRODUCTION

Quantum chromodynamics (QCD), an SU(3) gauge theory, was first proposed by Yoichiro Nambu [1] in 1966 as a candidate for the fundamental theory of strong interaction, just after introduction of a "new" quantum number, "color" [2]. In spite of its simple form, QCD creates thousands of hadrons and leads to various interesting nonperturbative phenomena such as color confinement [3, 4, 5] and dynamical chiral-symmetry breaking [6]. Even at present, it is very difficult to deal with QCD analytically due to its strong-coupling nature in an infrared region. Instead, lattice QCD has been applied as a direct numerical analysis for nonperturbative QCD.

In 1979, the first application [7] of lattice QCD Monte Carlo simulations was done for the inter-quark potential between a quark and an antiquark using the Wilson loop. Since then, the study of inter-quark forces has been one of the important issues in lattice QCD [8]. Actually, in hadron physics, the inter-quark force can be regarded as an elementary quantity to connect the "quark world" to the "hadron world", and plays an important role to hadron properties.

In this paper, we perform detailed and high-precision analyses for inter-quark forces in three-quark and multi-quark systems with SU(3) lattice QCD [9, 10, 11, 12, 13, 14, 15, 16, 17], and try to extract the proper picture for hadrons including multi-quarks.

CP756, *Quark Confinement and the Hadron Spectrum VI*
edited by N. Brambilla, U. D'Alesio, A. Devoto, K. Maung, G.M. Prosperi and S. Serci
© 2005 American Institute of Physics 0-7354-0241-8/05/$22.50

2. THE THREE-QUARK POTENTIAL IN LATTICE QCD

In general, three-body forces are regarded as residual interaction in most fields in physics. In QCD, however, the three-body force among three quarks is a "primary" force reflecting SU(3) gauge symmetry. In fact, the three-quark (3Q) potential is directly responsible for structure and properties of baryons, similar to the relevant role of the $Q\bar{Q}$ potential for meson properties. Furthermore, the 3Q potential is a key quantity to clarify quark confinement in baryons. However, in contrast to the $Q\bar{Q}$ potential [8], there were only a few pioneering lattice studies [18] done in 80's for the 3Q potential before our study in 1999 [9], in spite of its importance in hadron physics.

As for the functional form of the inter-quark potential, we note two theoretical arguments at short and long distance limits.

1. At short distances, perturbative QCD is applicable, and therefore the inter-quark potential is expressed as the sum of two-body OGE Coulomb potentials.
2. At long distances, the strong-coupling expansion of QCD is plausible, and it leads to the flux-tube picture [19].

Then, we theoretically conjecture the functional form of the inter-quark potential as the sum of OGE Coulomb potentials and a linear potential based on the flux-tube picture,

$$V = \frac{g^2}{4\pi} \sum_{i<j} \frac{T_i^a T_j^a}{|\mathbf{r}_i - \mathbf{r}_j|} + \sigma L_{\min} + C, \tag{1}$$

where $L_{\min}$ is the minimal value of the total length of flux-tubes linking static quarks. Of course, it is highly nontrivial that these simple arguments on UV and IR limits of QCD hold for intermediate distances. Nevertheless, lattice QCD indicates that the $Q\bar{Q}$ potential $V_{Q\bar{Q}}(r)$ is well described with this form as [8, 10, 11]

$$V_{Q\bar{Q}}(r) = -\frac{A_{Q\bar{Q}}}{r} + \sigma_{Q\bar{Q}} r + C_{Q\bar{Q}}. \tag{2}$$

For 3Q systems, there appears a junction which connects three flux-tubes from three quarks, and Y-type flux-tube formation is expected [10, 11, 19, 20, 21]. Therefore, the (ground-state) 3Q potential is expected to be the Coulomb plus Y-type linear potential, i.e., Y-Ansatz,

$$V_{3Q}^{\text{g.s.}} = -A_{3Q} \sum_{i<j} \frac{1}{|\mathbf{r}_i - \mathbf{r}_j|} + \sigma_{3Q} L_{\min} + C_{3Q}, \tag{3}$$

where $L_{\min}$ is Y-shaped flux-tube length.

For more than 300 different patterns of spatially-fixed 3Q systems, we calculate the ground-state 3Q potential $V_{3Q}^{\text{g.s.}}$ from the 3Q Wilson loop W_{3Q} using SU(3) lattice QCD [10, 11, 12, 13] with the standard plaquette action at the quenched level on various lattices, i.e., (β=5.7, $12^3 \times 24$), (β=5.8, $16^3 \times 32$), (β=6.0, $16^3 \times 32$) and ($\beta = 6.2, 24^4$). For accurate measurements, we construct ground-state-dominant 3Q operators using the smearing method [10, 11]. Note that the lattice QCD calculation is completely independent of any Ansatz for the potential form.

To conclude, we find that the static ground-state 3Q potential $V_{3Q}^{g.s.}$ is well described by the Coulomb plus Y-type linear potential (Y-Ansatz) within 1%-level deviation [10, 11]. To demonstrate this, we show in Fig.1(a) 3Q confinement potential V_{3Q}^{conf}, i.e., the 3Q potential subtracted by its Coulomb part, plotted against Y-shaped flux-tube length L_{min}. For each β, clear linear correspondence is found between 3Q confinement potential V_{3Q}^{conf} and L_{min}, which indicates Y-Ansatz for the 3Q potential.

Recently, as a clear evidence for Y-Ansatz, Y-type flux-tube formation is actually observed in maximally-Abelian (MA) projected lattice QCD from measurements of the action density in spatially-fixed 3Q systems [14, 22]. (See Figs.1 (b) and (c).) In this way, together with recent several analytical studies [23, 24], Y-Ansatz for the static 3Q potential seems to be almost settled.

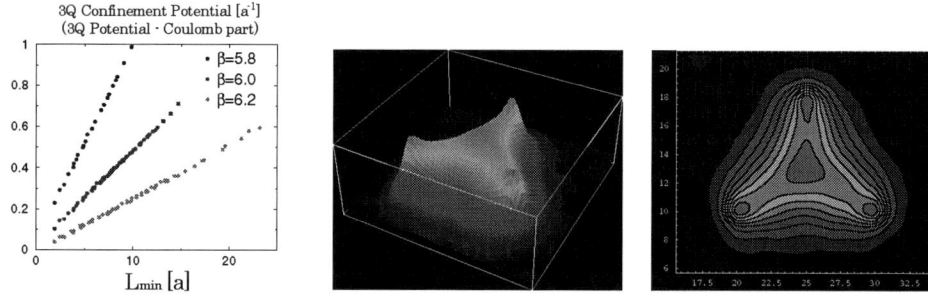

FIGURE 1. (a) 3Q confinement potential V_{3Q}^{conf}, i.e., the 3Q potential subtracted by its Coulomb part, plotted against Y-shaped flux-tube length L_{min} at β=5.8, 6.0 and 6.2 in the lattice unit. (b) A bird's-eye view and (c) a contour map of the lattice QCD result for Y-type flux-tube formation in a spatially-fixed 3Q system in MA projected QCD. The distance between the junction and each quark is about 0.5 fm.

3. GLUONIC EXCITATIONS IN 3Q SYSTEMS

In this section, we study excited-state 3Q potentials and gluonic excitations in 3Q systems using lattice QCD [12, 13]. The excited-state 3Q potential $V_{3Q}^{e.s.}$ is the energy of the excited state in the static 3Q system. The energy difference $\Delta E_{3Q} \equiv V_{3Q}^{e.s.} - V_{3Q}^{g.s.}$ between $V_{3Q}^{g.s.}$ and $V_{3Q}^{e.s.}$ is called as the gluonic-excitation energy, and physically means the excitation energy of the gluon-field configuration in the static 3Q system. In hadron physics, the gluonic excitation is one of interesting phenomena beyond the quark model, and relates to hybrid hadrons such as $q\bar{q}G$ and $qqqG$ in the valence picture.

For about 100 different patterns of 3Q systems, we calculate the excited-state potential in SU(3) lattice QCD with $16^3 \times 32$ at β=5.8 and 6.0 at the quenched level by diagonalizing the QCD Hamiltonian in presence of three quarks. In Fig.2, we show the 1st excited-state 3Q potential $V_{3Q}^{e.s.}$ and the ground-state potential $V_{3Q}^{g.s.}$. The gluonic excitation energy $\Delta E_{3Q} \equiv V_{3Q}^{e.s.} - V_{3Q}^{g.s.}$ in 3Q systems is found to be about 1GeV in hadronic scale as $0.5\text{fm} \leq L_{min} \leq 1.5\text{fm}$. Note that the gluonic excitation energy of about 1GeV is rather large compared with excitation energies of quark origin. This result predicts that the lowest hybrid baryon $qqqG$ has a large mass of about 2 GeV.

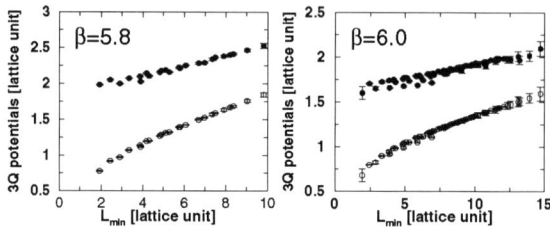

FIGURE 2. The 1st excited-state 3Q potential $V_{3Q}^{e.s.}$ and the ground-state 3Q potential $V_{3Q}^{g.s.}$. The lattice results at $\beta = 5.8$ and $\beta = 6.0$ well coincide apart from an irrelevant overall constant. The gluonic excitation energy $\Delta E_{3Q} \equiv V_{3Q}^{e.s.} - V_{3Q}^{g.s.}$ is about 1GeV in hadronic scale as $0.5\text{fm} \leq L_{min} \leq 1.5\text{fm}$.

Inverse Mercedes Ansatz for Gluonic Excitations in 3Q Systems

Next, we investigate the functional form of $\Delta E_{3Q} \equiv V_{3Q}^{e.s.} - V_{3Q}^{g.s.}$, where the Coulomb part is expected to be canceled between $V_{3Q}^{g.s.}$ and $V_{3Q}^{e.s.}$. After some trials, as shown in Fig.3, we find that the lattice data of the gluonic excitation energy $\Delta E_{3Q} \equiv V_{3Q}^{e.s.} - V_{3Q}^{g.s.}$ are relatively well reproduced by "inverse Mercedes Ansatz" [13],

$$\Delta E_{3Q} = \frac{K}{L_{\bar{Y}}} + G, \quad L_{\bar{Y}} \equiv \sum_{i=1}^{3} \sqrt{x_i^2 - \xi x_i + \xi^2} = \frac{1}{2} \sum_{i \neq j} \overline{P_i Q_j} \quad (x_i \equiv \overline{PQ_i}, \ \xi \equiv \overline{PP_i}), \quad (4)$$

where $L_{\bar{Y}}$ denotes "modified Y-length" defined by a half perimeter of "Mercedes form" as shown in Fig.3(a). As for (K, G, ξ), we find $(K \simeq 1.43, G \simeq 0.77 \text{ GeV}, \xi \simeq 0.116 \text{ fm})$ at $\beta = 5.8$, and $(K \simeq 1.35, G \simeq 0.85 \text{ GeV}, \xi \simeq 0.103 \text{ fm})$ at $\beta = 6.0$.

Inverse Mercedes Ansatz indicates that the gluonic-excitation mode is realized as a complicated bulk excitation of the whole 3Q system.

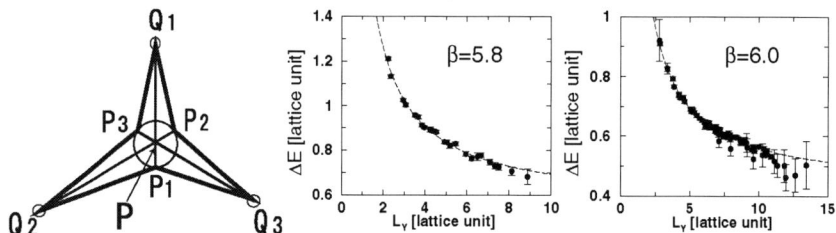

FIGURE 3. (a) Mercedes form for a 3Q system. (b) Lattice QCD results of the gluonic excitation energy $\Delta E_{3Q} \equiv V_{3Q}^{e.s.} - V_{3Q}^{g.s.}$ in 3Q systems plotted against modified Y-length $L_{\bar{Y}}$ at $\beta = 5.8$. (c) The same at $\beta=6.0$. The dashed curve denotes inverse Mercedes Ansatz.

Behind Success of the Quark Model

Here, we consider connection between QCD and the quark model in terms of gluonic excitations [12, 13, 14, 15]. While QCD is described with quarks and gluons, the simple quark model successfully describes low-lying hadrons even without explicit gluonic modes. In fact, gluonic excitations seem invisible in low-lying hadron spectra, which is rather mysterious.

On this point, we find the gluonic-excitation energy to be about 1GeV or more, which is rather large compared with excitation energies of quark origin. Therefore, contribution of gluonic excitations is considered to be negligible and dominant contribution is brought by quark dynamics such as spin-orbit interaction for low-lying hadrons. Thus, the large gluonic-excitation energy of about 1GeV gives a physical reason for the invisible gluonic excitation in low-lying hadrons, which would play a key role for success of the quark model without gluonic modes [12, 13, 14, 15].

In Fig.4, we present a possible scenario from QCD to the massive quark model in terms of color confinement and dynamical chiral-symmetry breaking (DCSB).

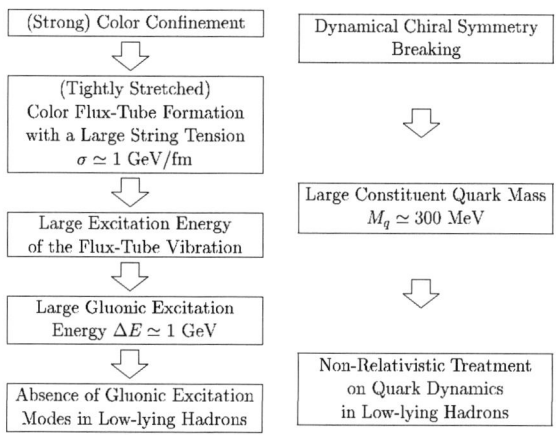

FIGURE 4. A possible scenario from QCD to the quark model in terms of color confinement and DCSB. DCSB leads to a large constituent quark mass of about 300 MeV, which enables non-relativistic treatment for quark dynamics approximately. Color confinement results in color flux-tube formation among quarks with a large string tension of $\sigma \simeq 1$ GeV/fm. In the flux-tube picture, gluonic excitations are described as flux-tube vibrations, which are expected to be large in hadronic scale. Indeed, the large gluonic-excitation energy of about 1 GeV observed in lattice QCD leads to absence of gluonic modes in low-lying hadrons, which plays a key role to success of the quark model without gluonic excitation modes.

4. TETRA-QUARK AND PENTA-QUARK POTENTIALS

In this section, we perform the first study of multi-quark potentials in SU(3) lattice QCD, motivated by recent experimental discoveries of multi-quark hadrons, i.e., X(3872) and $D_s(2317)$ as candidates of tetra-quark (QQ-$\bar{Q}\bar{Q}$) mesons, and $\Theta^+(1540)$, $\Xi^{--}(1862)$ and $\Theta_c(3099)$ as penta-quark ($4Q$-$\bar{Q}$) baryons [25]. As unusual features of multi-quark hadrons, their decay widths are extremely narrow, e.g., $\Gamma(X(3872)) < 2.3$MeV (90% C.L.). For physical understanding of multi-quark hadrons, theoretical analyses are necessary as well as experimental studies. In particular, for realistic model calculations of multi-quark hadrons, it is required to clarify inter-quark forces such as the quark confinement force in multi-quark systems based on QCD.

OGE Coulomb plus Multi-Y Ansatz

As a theoretical form of the multi-quark potential, we present one-gluon-exchange (OGE) Coulomb plus multi-Y Ansatz [15, 16, 17] based on Eq.(1), i.e., the sum of OGE Coulomb potentials and the linear confinement potential proportional to multi-Y-shaped flux-tube length L_{min}.

On the 4Q potential V_{4Q}, we investigate QQ-$\bar{\text{Q}}\bar{\text{Q}}$ systems where two quarks locate at $(\mathbf{r}_1, \mathbf{r}_2)$ and two antiquarks at $(\mathbf{r}_3, \mathbf{r}_4)$ as shown in Fig.5. For connected 4Q systems, a plausible form of V_{4Q} is OGE plus multi-Y Ansatz [17],

$$V_{c4Q} \equiv -A_{4Q}\{(\frac{1}{r_{12}} + \frac{1}{r_{34}}) + \frac{1}{2}(\frac{1}{r_{13}} + \frac{1}{r_{14}} + \frac{1}{r_{23}} + \frac{1}{r_{24}})\} + \sigma_{4Q}L_{min} + C_{4Q}, \quad (5)$$

while V_{4Q} for disconnected 4Q systems would be approximated by "two-meson" Ansatz as $V_{2Q\bar{Q}} \equiv V_{Q\bar{Q}}(r_{13}) + V_{Q\bar{Q}}(r_{24})$.

On the 5Q potential V_{5Q}, we investigate QQ-$\bar{\text{Q}}$-QQ systems where the two quarks at $(\mathbf{r}_1, \mathbf{r}_2)$ and those at $(\mathbf{r}_3, \mathbf{r}_4)$ form $\bar{\mathbf{3}}$ representation of SU(3) color, respectively, and the antiquark locates at $\mathbf{r}_5$, as shown in Fig.5. For the 5Q system, OGE Coulomb plus multi-Y Ansatz is expressed as $V_{5Q} = V_{5Q}^{Coul} + \sigma_{5Q}L_{min} + C_{5Q}$ with the Coulomb part as

$$V_{5Q}^{Coul} = -A_{5Q}\{(\frac{1}{r_{12}} + \frac{1}{r_{34}}) + \frac{1}{2}(\frac{1}{r_{15}} + \frac{1}{r_{25}} + \frac{1}{r_{35}} + \frac{1}{r_{45}}) + \frac{1}{4}(\frac{1}{r_{13}} + \frac{1}{r_{14}} + \frac{1}{r_{23}} + \frac{1}{r_{24}})\}.(6)$$

We theoretically set (A_{4Q}, σ_{4Q}) and (A_{5Q}, σ_{5Q}) to be $(A_{3Q}, \sigma_{3Q}) \simeq (0.1366, 0.046a^{-2})$ in the 3Q potential [11]. Note that there is no adjustable parameter in the theoretical Ansätze apart from an irrelevant constant.

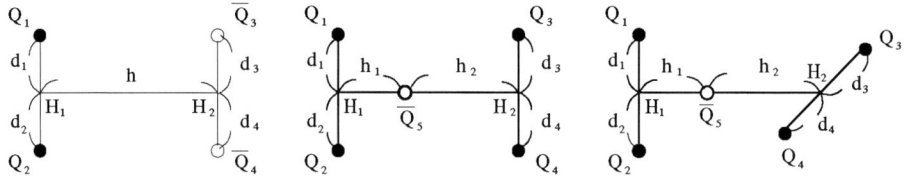

FIGURE 5. (a) A planar tetra-quark (QQ-$\bar{\text{Q}}\bar{\text{Q}}$) configuration. (b) A planar penta-quark (4Q-$\bar{\text{Q}}$) configuration. (c) A twisted penta-quark configuration with $Q_1Q_2 \perp Q_3Q_4$. We here take $d_1 = d_2 = d_3 = d_4 \equiv d$.

Multi-quark Wilson loops and multi-quark potentials

In QCD, static multi-quark potentials can be obtained from the corresponding multi-quark Wilson loops. As shown in Fig.6, we define the 4Q Wilson loop W_{4Q} and the 5Q

Wilson loop W_{5Q} [15, 16, 17] by

$$W_{4Q} \equiv \frac{1}{3}\text{tr}(\tilde{M}_1 \tilde{L}_{12} \tilde{M}_2 \tilde{R}_{12}), \quad W_{5Q} \equiv \frac{1}{3!}\varepsilon^{abc}\varepsilon^{a'b'c'}\tilde{M}^{aa'}(\tilde{L}_3\tilde{L}_{12}\tilde{L}_4)^{bb'}(\tilde{R}_3\tilde{R}_{12}\tilde{R}_4)^{cc'}, \quad (7)$$

where $\tilde{L}_i, \tilde{R}_i, \tilde{M}, \tilde{M}_j$ $(i = 1,2,3,4, j = 1,2)$ are given by

$$\tilde{L}_i, \tilde{R}_i, \tilde{M}, \tilde{M}_j \equiv P\exp\{ig\int_{L_i, R_i, M, M_j} dx^\mu A_\mu(x)\} \in SU(3)_c, \quad (8)$$

i.e., $\tilde{L}_i, \tilde{R}_i, \tilde{M}, \tilde{M}_j$ $(i = 3,4, j = 1,2)$ are line-like variables and $\tilde{L}_i, \tilde{R}_i$ $(i = 1,2)$ are staple-like variables, and $\tilde{L}_{12}, \tilde{R}_{12}$ are defined by

$$\tilde{L}_{12}^{a'a} \equiv \frac{1}{2}\varepsilon^{abc}\varepsilon^{a'b'c'}\tilde{L}_1^{bb'}\tilde{L}_2^{cc'}, \quad \tilde{R}_{12}^{a'a} \equiv \frac{1}{2}\varepsilon^{abc}\varepsilon^{a'b'c'}\tilde{R}_1^{bb'}\tilde{R}_2^{cc'}. \quad (9)$$

Note that both the 4Q Wilson loop W_{4Q} and the 5Q Wilson loop W_{5Q} are gauge invariant.

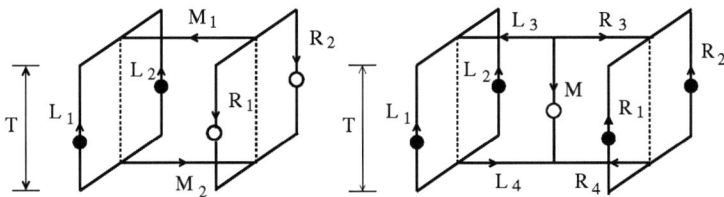

FIGURE 6. (a) The tetra-quark (4Q) Wilson loop W_{4Q}. (b) The penta-quark (5Q) Wilson loop W_{5Q}.

We calculate the multi-quark potentials (V_{4Q}, V_{5Q}) from the multi-quark Wilson loops (W_{4Q}, W_{5Q}) in SU(3) lattice QCD with $\beta = 6.0$ (i.e., $a \simeq 0.1$fm) and $16^3 \times 32$ at the quenched level [15, 16, 17], using the smearing method to reduce excited-state components. In this paper, we investigate planar and twisted configurations for multi-quark systems as shown in Fig.5, and show the results for $d_1 = d_2 = d_3 = d_4 \equiv d$ and $h_1 = h_2 \equiv h/2$.

Figure 7 shows the 4Q potential V_{4Q} [17]. For large h, V_{4Q} coincides with the energy $V_{c4Q}(d, h)$ of the connected 4Q system. For small h, V_{4Q} coincides with the energy $V_{2Q\bar{Q}} = 2V_{Q\bar{Q}}(h)$ of the "two-meson" system composed of two flux-tubes. Thus, we get the relation of $V_{4Q} = \min\{V_{c4Q}(d, h), 2V_{Q\bar{Q}}(h)\}$, and find the "flip-flop" between the connected 4Q system and the "two-meson" system around the level-crossing point where these two systems are degenerate as $V_{c4Q}(d, h) = 2V_{Q\bar{Q}}(h)$.

In Fig.8, we show the 5Q potential V_{5Q}. The lattice data denoted by the symbols are found to be well reproduced by the theoretical curve of OGE plus multi-Y Ansatz [15, 16, 17] with (A_{5Q}, σ_{5Q}) fixed to be (A_{3Q}, σ_{3Q}) in the 3Q potential [11].

As a remarkable fact, we find universality of the string tension and the OGE result among Q$\bar{Q}$, 3Q, 4Q and 5Q systems as [10, 11, 12, 13, 14, 15, 16, 17]

$$\sigma_{Q\bar{Q}} \simeq \sigma_{3Q} \simeq \sigma_{4Q} \simeq \sigma_{5Q}, \quad \frac{1}{2}A_{Q\bar{Q}} \simeq A_{3Q} \simeq A_{4Q} \simeq A_{5Q}. \quad (10)$$

This result supports the flux-tube picture on the confinement mechanism even for multi-quark systems [15, 16, 17].

129

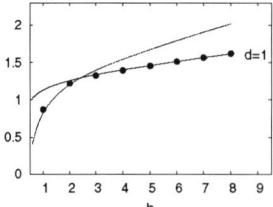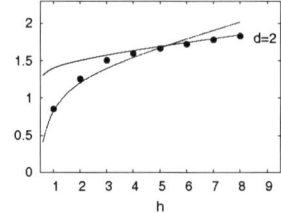

FIGURE 7. The 4Q potential V_{4Q} in the lattice unit for planar 4Q configurations with $d = 1$ (left) and $d = 2$ (right) as shown in Fig.5(a). The symbols denote lattice QCD results. We add the theoretical curves for the connected 4Q system (the solid curve) and for the "two-meson" system (the dashed curve).

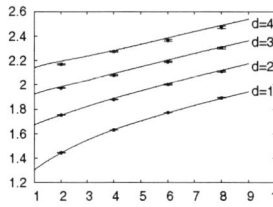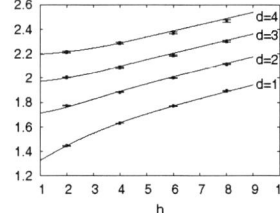

FIGURE 8. The 5Q potential V_{5Q} in the lattice unit for planar configurations (left) and twisted configurations (right) as shown in Figs.5(b) and (c). The symbols denote lattice QCD results. We add the theoretical curve of OGE plus multi-Y Ansatz with (A_{5Q}, σ_{5Q}) fixed to be (A_{3Q}, σ_{3Q}).

5. QCD STRING THEORY FOR PENTA-QUARK DECAY

Our lattice QCD studies on various inter-quark potentials indicate the flux-tube picture for hadrons, which is idealized as the QCD string model. In this section, we consider penta-quark dynamics, especially for its extremely narrow width, in terms of the QCD string theory.

The ordinary string theory mainly describes open and closed strings corresponding to mesons and glueballs, and has only two types of reaction process as shown in Fig.9:

1. String breaking (or fusion) process.
2. String recombination process.

On the other hand, the QCD string theory describes also baryons and antibaryons as Y-shaped flux-tubes, which is different from the ordinary string theory. Note that appearance of Y-type junctions is peculiar to the QCD string theory. Accordingly, the QCD string theory includes third reaction process as shown in Fig.10:

3. Junction (J) and anti-junction ($\bar{\text{J}}$) pair creation (or annihilation) process.

Through this J-$\bar{\text{J}}$ pair creation process, baryon and anti-baryon pair creation can be described.

As a remarkable fact in the QCD string theory, decay process (or creation process) of penta-quark baryons inevitably accompanies J-$\bar{\text{J}}$ creation [26] as shown in Fig.11.

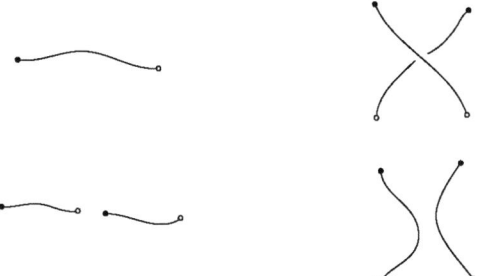

FIGURE 9.   Reaction process in the ordinary string theory: string breaking (or fusion) process (left) and string recombination process (right).

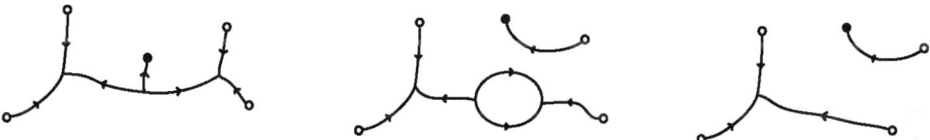

FIGURE 10.   Junction (J) and anti-junction ($\bar{\text{J}}$) pair creation (or annihilation) process peculiar to the QCD string theory.

Here, the intermediate state is considered as a gluonic-excited state, since it clearly corresponds to a non-quark-origin excitation.

Our lattice QCD study indicates that such a gluonic-excited state is a highly-excited state with an excitation energy above 1GeV. Then, in the QCD string theory, the decay process of penta-quark baryons near the threshold can be regarded as a quantum tunneling, and therefore the penta-quark decay is expected to be strongly suppressed. This leads to a very small decay width of penta-quark baryons.

FIGURE 11.   A decay process of penta-quark baryons in the QCD string theory. The penta-quark decay process inevitably accompanies J-$\bar{\text{J}}$ creation, which is a kind of gluonic excitation.

Now, we try to estimate the decay width of penta-quark baryons near the threshold in the QCD string theory. In the quantum tunneling as shown in Fig.11, the barrier height corresponds to the gluonic excitation energy ΔE of the intermediate state, and can be estimated as $\Delta E \simeq 1\text{GeV}$. The time scale T for this tunneling process is expected to be hadronic scale as $T = 0.5 \sim 1\text{fm}$, since T cannot be smaller than the spatial size of the reaction area due to causality. Then, the suppression factor for this penta-quark decay can be roughly estimated as

$$|\exp(-\Delta E T)|^2 \simeq |e^{-1\text{GeV}\times(0.5\sim1)\text{fm}}|^2 \simeq 10^{-2} \sim 10^{-4}. \qquad (11)$$

Note that this suppression factor $|\exp(-\Delta E T)|^2$ appears in the decay process of low-lying penta-quark baryons for both positive-parity and negative-parity states.

For the decay of $\Theta^+(1540)$ into N and K, the Q-value Q is $Q = M(\Theta^+) - M(\mathrm{N}) - M(\mathrm{K}) \simeq (1540 - 940 - 500)\mathrm{MeV} \simeq 100\mathrm{MeV}$. In ordinary sense, the decay width is expected to be controlled by $\Gamma_{\mathrm{hadron}} \simeq Q \simeq 100\mathrm{MeV}$. Considering the extra suppression factor of $|\exp(-\Delta ET)|^2$, we get a rough order estimate for the decay width of $\Theta^+(1540)$ as $\Gamma[\Theta^+(1540)] \simeq \Gamma_{\mathrm{hadron}} \times |\exp(-\Delta ET)|^2 \simeq 1 \sim 10^{-2}\mathrm{MeV}$.

ACKNOWLEDGMENTS

H.S. would like to thank Profs. G.M. Prosperi and N. Brambilla for their kind hospitality at Confinement VI. H.S. is also grateful to Profs. T. Kugo and A. Sugamoto for useful discussions on the QCD string theory. The lattice QCD Monte Carlo simulations have been performed on NEC-SX5 at Osaka University and on HITACHI-SR8000 at KEK.

REFERENCES

1. Y. Nambu, in *Preludes in Theoretical Physics*, (North-Holland, Amsterdam, 1966).
2. M.Y. Han and Y. Nambu, *Phys. Rev.* **139**, B1006 (1965).
3. Y. Nambu, in *Symmetries and Quark Models* (Wayne State University, 1969); *Lecture Notes at the Copenhagen Symposium* (1970).
4. Y. Nambu, *Phys. Rev.* **D10**, 4262 (1974).
5. For instance, articles in *Color Confinement and Hadrons in Quantum Chromodynamics*, edited by H. Suganuma *et al.* (World Scientific, 2004).
6. Y. Nambu and G. Jona-Lasinio, *Phys. Rev.* **122**, 345 (1961); *ibid.* **124**, 246 (1961).
7. M. Creutz, *Phys. Rev. Lett.* **43**, 553 (1979); *ibid.* **43**, 890 (1979); *Phys. Rev.* **D21**, 2308 (1980).
8. H.J. Rothe, *Lattice Gauge Theories*, 2nd edition (World Scientific, 1997) p.1.
9. T.T. Takahashi, H. Matsufuru, Y. Nemoto and H. Suganuma, in *Dynamics of Gauge Fields*, Tokyo, Dec. 1999, edited by A. Chodos *et al.*, (Universal Academy Press, 2000) 179; H. Suganuma, Y. Nemoto, H. Matsufuru and T.T. Takahashi, *Nucl. Phys.* **A680**, 159 (2000).
10. T.T. Takahashi, H. Matsufuru, Y. Nemoto and H. Suganuma, *Phys. Rev. Lett.* **86**, 18 (2001).
11. T.T. Takahashi, H. Suganuma, Y. Nemoto and H. Matsufuru, *Phys. Rev.* **D65**, 114509 (2002).
12. T.T. Takahashi and H. Suganuma, *Phys. Rev. Lett.* **90**, 182001 (2003).
13. T.T. Takahashi and H. Suganuma, *Phys. Rev.* **D70**, 074506 (2004).
14. H. Suganuma, T.T. Takahashi and H. Ichie, in *Color Confinement and Hadrons in Quantum Chromodynamics*, (World Scientific, 2004) p.249; *Nucl. Phys.* **A737** (2004) S27.
15. H. Suganuma, T. T. Takahashi, F. Okiharu and H. Ichie, in *QCD Down Under*, March 2004, Adelaide, *Nucl. Phys.* **B** (Proc. Suppl.) (2004) in press.
16. F. Okiharu, H. Suganuma and T. T. Takahashi, "First study for the pentaquark potential in SU(3) lattice QCD", hep-lat/0407001.
17. F. Okiharu, H. Suganuma, T. T. Takahashi, in *Pentaquark04*, July 2004, SPring-8, Japan (World Sci.).
18. R. Sommer and J. Wosiek, *Phys. Lett.* **149B**, 497 (1984); *Nucl. Phys.* **B267**, 531 (1986).
19. J. Kogut and L. Susskind, *Phys. Rev.* **D11**, 395 (1975); J. Carlson, J. Kogut and V. Pandharipande, *Phys. Rev.* **D27**, 233 (1983); *ibid.* **D28**, 2807 (1983).
20. M. Fable de la Ripelle and Yu. A. Simonov, *Ann. Phys.* **212**, 235 (1991).
21. N. Brambilla, G.M. Prosperi and A. Vairo, *Phys. Lett.* **B362**, 113 (1995).
22. H. Ichie, V. Bornyakov, T. Streuer and G. Schierholz, *Nucl. Phys.* **A721**, 899 (2003); *Nucl. Phys.* **B** (Proc. Suppl.) **119**, 751 (2003).
23. D.S. Kuzmenko and Yu. A. Simonov, *Phys. Atom. Nucl.* **66**, 950 (2003).
24. J.M. Cornwall, *Phys. Rev.* **D69**, 065013 (2004).
25. For a recent review, S. L. Zhu, *Int. J. Mod. Phys.* **A19**, 3439 (2004) and references therein.
26. M. Bando, T. Kugo, A. Sugamoto and S. Terunuma, *Prog. Theor. Phys.* **112**, 325 (2004).

Confinement versus Bose-Einstein condensation

Kurt Langfeld

Institut für Theoretische Physik, Universität Tübingen, Germany.

Abstract. The deconfinement phase transition at high baryon densities and low temperatures evades a direct investigation by means of lattice gauge calculations. In order to make this regime of QCD accessible by computer simulations, two proposal are made: (i) A Lattice Effective Theory (LET) is designed which incorporates gluon and diquark fields. The deconfinement transition takes place when the diquark fields undergo Bose-Einstein condensation. (ii) Rather than using eigenstates of the particle number operator, I propose to perform simulations for a fixed expectation value of the baryonic Noether current. This approach changes the view onto the finite density regime, but evades the sign and overlap problems. The latter proposal is exemplified for the LET: Although the transition from the confinement to the condensate phase is first order in the coupling constant space at zero baryon densities, the transition at finite densities appears to be a crossover.

INTRODUCTION

Lattice gauge simulations leave no doubt that Quantum Chromodynamics (QCD) exhibits a transition from the baryonic regime to the Quark Gluon Plasma (QGP) at high temperatures and small baryon densities. It is this high temperature regime which is currently under investigation at RHIC, Brookhaven [1] and which will be a major target of LHC, Cern. It is believed that a transition to the QGP also appears at high baryon densities and small temperatures. Very little is known about the latter transition from first principle simulations: lattice simulations at finite values of the baryonic chemical potential μ encounter a severe sign/overlap problem which limits their scope to the range of small μ [2, 3].

At asymptotic baryon densities it is assumed that the quarks form a Fermi surface. In this case, perturbative gluon interactions support the existence of a diquark BCS state known as color superconductor [4, 5]. At the present stage of research, no first principle results are available for the region of the QCD phase diagram where the transition from the baryonic phase to the QGP occurs at small temperatures and intermediate densities. Here, I will argue that the transition is driven by the Bose-Einstein condensation of diquarks. At high densities, the Bose-Einstein condensate gradually develops to a diquark BCS state.

In the present paper, two proposal are put forward to provide access to the finite density transition of QCD: (i) It is argued that the transition is within the reach of a Lattice Effective Theory (LET) which incorporates gluons and diquarks as dynamical degrees of freedom. At the stage of the present model, the baryonic current is entirely supported by the diquarks. (ii) As in the case of QCD, the LET also suffers from a sign problem at finite baryon chemical potential. In order to get first insights, I propose

CP756, *Quark Confinement and the Hadron Spectrum VI*
edited by N. Brambilla, U. D'Alesio, A. Devoto, K. Maung, G.M. Prosperi and S. Serci
© 2005 American Institute of Physics 0-7354-0241-8/05/$22.50

to change the point of view: Rather than considering only eigenstates of the particle density operator, simulations are performed for a given expectation value of the baryonic Noether density. In the case of the LET, we will find that the finite density transition is a crossover rather than of first order.

LATTICE EFFECTIVE THEORY

Model building

The central assumption for describing the finite density transition is that only gluon and diquark degrees of freedom are relevant for the intermediate density region of the phase diagram. In the hadronic phase, diquarks are confined to a length scale of $\approx 1\,\text{fm}$. Even if the transition is first order, the correlation length might become much larger than $\approx 1\,\text{fm}$ before the system is disordered by bubble nucleation. Therefore, the working hypothesis of the present approach is that the degrees of freedom relevant at the transition are gluons and point-like scalar (diquark) fields, i.e.,

$$\phi^a(x) = \varepsilon^{abc}\,\varepsilon_{AB}\,\bar{q}^b_{\text{ch}A}(x)\,\gamma_5\,q^c_B(x), \tag{1}$$

where $q^b_{\text{ch}A}(x)$ are the charge conjugated quark fields, $A,B = 1\ldots 2$ are flavor- and $a,b,c = 1\ldots 3$ are color indices, respectively. The Effective Action, which should describe physics at the transition scale, is a SU(3) pure gauge theory supplemented with a scalar (Higgs, diquark) field which belongs to the the fundamental representation of the gauge group. The Lattice Effective Theory is modeled by the action

$$S = -\frac{\beta}{3}\sum_{\mu<\nu,x} \text{tr}\,U_\mu(x)U_\nu(x+\mu)U^\dagger_\mu(x+\nu)U^\dagger_\nu(x) \tag{2}$$
$$+ \sum_x \phi^a(x)\phi^a(x) - \kappa\sum_{\mu,x}\phi(x)U_\mu(x)\phi^\dagger(x+\mu) - \text{h.c.} + \sum_x \lambda\,[\phi^2(x)]^2.$$

Thereby, the gluon degrees of freedom are encoded by the link fields $U_\mu(x)$, β is the usual prefactor of the Wilson action, which largely describes the gluon-dynamics, and κ is the Higgs hopping parameter which is related to the (bare) Higgs mass m by

$$m^2a^2 = 1 - 8\kappa, \tag{3}$$

where a is the lattice spacing. Turning off the gluon interaction ($U_\mu = 1$), a Bose-Einstein condensate (BEC) is formed for $\kappa > 1/8$. Thereby, the BEC regime is stabilized by the Higgs quartic term ($\lambda > 0$).

The $SU(3)$ Higgs mechanism and residual confinement

Confinement effects in the gauged SU(2) Higgs model were extensively studied in [6, 7, 8]. Here, the SU(3) Higgs system will be investigated for the first time.

For a sufficiently large Higgs hopping parameter κ, we expect that the system passes into the phase of condensed diquarks. In order to realize the formation of a scalar expectation value, we are forced to fix the gauge degree of freedom: Since a gauge transformation $\Omega(x) \in SU(3)$ acts on the fields as

$$U_\mu^\Omega(x) = \Omega(x) U_\mu(x) \Omega^\dagger(x+\mu) , \qquad \phi^\Omega(x) = \Omega(x) \phi(x) ,$$

any residual gauge degree of freedom would wipe out the expectation value $\langle \phi(x) \rangle$. Here we choose the Minimal Landau Gauge, i.e.,

$$\sum_{\mu,x} \text{tr}\, U_\mu^\Omega(x) \xrightarrow{\Omega} \text{maximal} . \tag{4}$$

Note that the gauge constraint (4) leaves a global gauge transformation $\Omega(x) = \Omega$ unfixed. The spontaneous breaking of this residual global gauge symmetry is signaled by a non-vanishing value $\langle \phi^\Omega(x) \rangle$ and marks the occurrence of the Higgs phase [7].

Which fields gain a mass due to the formation of the BEC of diquarks? What is the fate of the Would-Be Goldstone bosons? What is different for SU(3) compared with the familiar SU(2) Higgs mechanism? In order to answer these questions, let us invoke a semi-classical approach for the moment. Thereby, the scalar field is decomposed into a classical part and fluctuations, $\phi^\Omega(x) = \phi_c + \varphi(x)$. After (minimal) Landau gauge fixing, we may choose without a loss of generality

$$\phi_c = v\,(1,0,0)^T ; \qquad SU(3) \longrightarrow SU(2) . \tag{5}$$

This implies that only a part of the global SU(3) color group is broken by the condensate, and that a SU(2) color symmetry remains intact. From the Higgs kinetic term, we detect the masses of the gluon fields A_μ^a, $a = 1 \ldots 8$, i.e.,

$$\left[D_\mu \phi(x) \right]^\dagger D_\mu \phi(x) = \ldots + A_\mu^a(x) M^{ab} A_\mu^b(x) , \qquad M^{ab} = \frac{1}{2} \phi_c^\dagger \{ t^a, t^b \} \phi_c ,$$

where D_μ is the gauge covariant derivative, and t^a, $a = 1 \ldots 8$ are the generators of the SU(3) algebra. Using (5), a direct calculation of the mass matrix M^{ab} reveals that

$$A_\mu^1, A_\mu^2, A_\mu^3 + \sqrt{3} A_\mu^8, A_\mu^4, A_\mu^5 : \text{massive}; \quad A_\mu^6, A_\mu^7, A_\mu^3 - \sqrt{3} A_\mu^8 : \text{massless}.$$

Unless in the case of SU(2), there are not enough Higgs fields to give a mass to *all* gluons. As expected, the gluons corresponding to the unbroken global SU(2) color symmetry remain massless. The interesting question which solely arises in the context of SU(3) (and which will be partially answered below) is whether color charges which transform under the invariant SU(2) subgroup are still confined. Since the diquark field ϕ^1 is built up from quarks of color 2 and 3 (see 1), the integrity of ϕ^1 as point-like particle would be preserved by confining effects throughout the BEC transition.

In order to explore the phase diagram of the Lattice Effective Theory of gluons and diquarks as function of β and κ (λ will be a given number in the studies below), we need an order parameter which detects the BEC phase. After installing the gauge condition (4), one might think to use $\langle \phi_c \rangle$. The point is that in an ergodic lattice simulation each lattice configurations would generate a different direction for ϕ_c implying that $\langle \phi_c \rangle = 0$ by virtue of the residual *global* gauge degree of freedom. Let us define

$$ v^a = \frac{1}{N_x} \sum_x \phi^{\Omega a}(x) , \tag{6} $$

where N_x is the number of space-time points. The crucial observation is that in the BEC phase the fields $\phi^{\Omega}(x)$ are (almost) uniquely oriented throughout space-time, i.e., $v^2 = \mathcal{O}(1)$, while in the color unbroken phase $v^2 \approx 0$. This suggests to use

$$ \Phi^2 = \left\langle \frac{1}{N_x^2} \sum_a \left[\sum_x \phi^{\Omega a}(x) \right] \left[\sum_y \phi^{\Omega a}(y) \right] \right\rangle \tag{7} $$

as the Litmus paper for the BEC transition. In the (global) color unbroken phase, the correlation length ξ of the gauged scalar fields is defined from the (disconnected) Green function by

$$ \langle \phi^{\Omega a}(x) \phi^{\Omega a}(y) \rangle \propto \exp\{-|x-y|/\xi\} , \qquad \xi \text{ finite} . \tag{8} $$

We therefore find that Φ^2 vanishes in the infinite volume limit, i.e.,

$$ \lim_{N_x \to \infty} \Phi^2 \approx \lim_{N_x \to \infty} \xi/N_x = 0 , $$

while $\Phi^2 = \mathcal{O}(1)$ in the BEC phase.

Numerical results

The Lattice Effective Theory, corresponding to the action (2), can be simulated on a computer using a generalized version of the Cabibbo Marinari algorithm. Microcanonical reflections concerning both, the scalar fields and the link fields, are employed to reduce autocorrelations. In this first investigation, neither a scaling analysis nor a study of the line of constant physics is pursued. The aim of the simulations was to reveal the underlying physics at a qualitative level. The numerical results were obtained for $\lambda = 0.1$ on 8^4 and 12^4 lattices. Landau gauge, see (4), is implemented using a standard iteration over-relaxation procedure.

The findings for the "diquark condensate" Φ (7) are summarized in figure 1. The open symbols have been obtained on a 8^4 lattice, while the full symbols correspond to a 12^4 lattice.

We find that diquark condensation sets in if κ exceeds a critical value. This finding is a highly non-trivial result for the following reason: Landau gauge fixing is performed at the level of the link fields and implies the maximization of the gauge functional (4). In particular the iterative procedure used here only locates a *local* maximum, and repeating

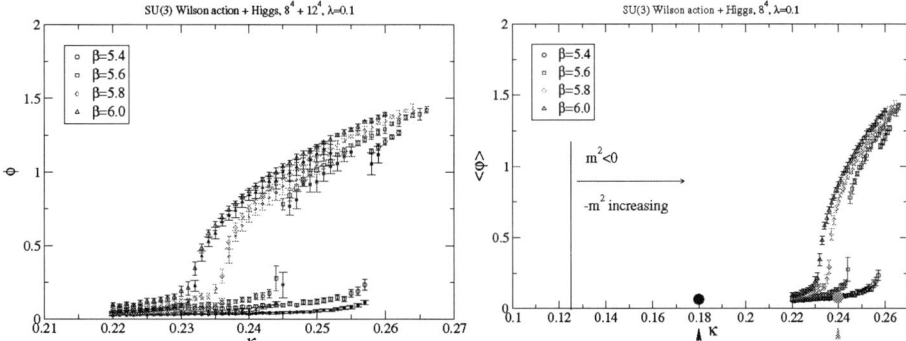

FIGURE 1. The diquark condensate defined in (7) as function of the Higgs hopping parameter κ for several values β (for a definition of the parameters see (2)). The colored symbols indicate the positions in parameter space where finite density simulations will be carried out.

the maximization on the same link configurations generically yields different gauge transformations $\Omega(x)$, each of which generates a different Gribov copy of the links and scalar fields, respectively. This implies that an average over the Gribov copies within the first Gribov horizon is performed when the expectation value Φ (7) is calculated. Since the scalar field transforms homogeneously, the first guess would be that the correlation of the gauged scalar fields is destroyed by the average over the first Gribov horizon. The non-trivial result is that this is not the case: The scalar correlation length ξ (8) is *insensitive* to the Gribov noise.

Having in mind that in the case without gluonic interactions Bose-Einstein condensation sets in for $\kappa > 0.125$, see (3), we here observe that the onset of the condensation is postponed to much larger values of κ due to gluonic interactions. A coexistence of the hadronic phase and the diquark BEC phase is not observed.

For β values as large as 5.8 (and for a 8^4 lattice), the system is in the deconfinement phase by virtue of temperature and volume effects. Figure 1 shows that the BEC transition changes form first order to second order (or higher) when β is increased. A possible explanation for this observation is: If the gluons are deconfined due to volume effects, only five of them acquire a mass by virtue of the SU(3) Higgs mechanism. Three gluons remain massless and give rise to the critical phenomenon with infinite correlation length. At small β values, the system is in the "hadronic" phase and the gluons possess a mass gap. If confinement persists for color states corresponding to the unbroken SU(2) subgroup, the three SU(2) gluons possess a mass gap due to confinement. The remaining five gluons are massive because of the Higgs mechanism. This would imply that there is no massless excitation which could give rise to a second order transition. This line of arguments favors the picture that color states of the residual SU(2) subgroup are still confined after the BEC transition.

FINITE DENSITY DECONFINEMENT TRANSITION

Lattice gauge theory results

The generic approach to Yang-Mills thermodynamics at finite baryon densities is based on the introduction of a non-zero chemical potential. In the case of a SU(2) gauge group, the fermion determinant is real and can be included in the probabilistic measure. Numerical simulations can be performed by using standard algorithms, although this numerical approach consumes a lot of computer time due to the non-local nature of the action [9]. In the case of a SU(3) gauge group, the fermion determinant acquires imaginary parts for a non-vanishing chemical potential and cannot be considered to be part of the probabilistic measure. The most prominent example to circumvent this conceptual difficulty considers the fermion determinant as part of the correlation function to be calculated. Thereby, the probabilistic measure of zero-density Yang-Mills theory is used to generate the gauge field configurations. However, it turns out that this approach suffers from the so-called "overlap" problem implying that for realistic lattice sizes an unrealistic number of Monte-Carlo steps is necessary to achieve reliable results [10].

At the present stage, the scope of lattice QCD simulations is limited to the regime of small baryon densities. Two approaches have been proven to be fruitful: (i) the approach based upon a Taylor expansion with respect to the chemical potential μ around $\mu = 0$ [11]; (ii) the method employing simulations at imaginary chemical potential and finally seeking a continuation to real chemical potential [2]. Finally, I would like to mention a recently proposed technique where multi-parameter re-weighting is used in order to reduce the severeness of the overlap problem [12].

It is fair to say that a direct lattice study of the QCD phase transition at intermediate baryon densities (and small temperature) is not feasible at the present level of investigations.

A change of view

The solution of the sign/overlap problem in finite density lattice QCD probably requires new type of algorithms such as cluster algorithms or D-theory [13, 14]. Here, I would like to suggest to change the question of interest in a way which makes the problem solvable and which nevertheless sheds light onto the region of the QCD phase diagram where the finite density deconfinement transition takes place.

In order to put my proposal into the proper context, let me briefly review the origin of the sign problem. The standard question which we used to ask is: what is the ground state energy of the system if we only consider field configurations which are eigen states of the particle number operator $\hat{N}$:

$$E(B) = \langle \phi | \hat{H} | \phi \rangle , \qquad \hat{N} | \phi \rangle = B | \phi \rangle , \qquad (9)$$

where $\hat{H}$ is the Hamilton operator. A conversion of the latter formulation to a functional integral setup usually involves the introduction of a chemical potential and generically leads to the sign problem.

Here, I propose to consider, instead of (9), the ground state energy where the fields *on average* possess a given particle number, i.e.,

$$E(B) = \langle \phi | \hat{H} | \phi \rangle \,, \qquad \langle \phi | \hat{N} | \phi \rangle = B \,. \tag{10}$$

This approach certainly disregards certain features of the multi-fermion system; the hope is, however, that the approach sketches the deconfinement transition at high densities at least qualitatively. It is well known how to formulate the approach (10) in the functional integral language [15]. The quantity of interest is the effective action Γ, which originates from the partition function by means of a Legendre transformation:

$$Z[\mu] = \int \mathcal{D}\phi \, \exp \left\{ -S + \int d^4x \, \mu(x) \, \rho(x) \right\} \,, \tag{11}$$

$$\Gamma[\rho_c] = -\ln Z[\mu] + \int d^4x \, \mu(x) \, \rho_c(x) \,, \qquad \rho_c(x) = \frac{\delta \ln Z[\mu]}{\delta \mu(x)} \,, \tag{12}$$

where $\rho(x)$ is the zeroth component of U(1) Noether current corresponding to conserved baryon charge. Using a constant external source, i.e., $\mu(x) = \mu$, the effective action is turned into the effective potential for a constant baryon density ρ_c. The technical advantage of the approach to systems of finite (classical) density is that the additional factor $exp \int d^4x \, \mu(x) \, \rho(x)$ in (11) can be included to the action during the Monte-Carlo Updates leading to significant overlap with the finite density configurations.

LET study of the finite density transition

Let me stress that the classical density approach, outlined in the previous subsection, is applicable to lattice QCD. Nevertheless, such simulations involve the inclusion of dynamical quarks and are very time consuming. For this reason, we will here explore the method resorting to the Lattice Effective Theory (LET) discussed in the previous section.

In the present case, the baryon density is solely supported by the diquarks. The corresponding Noether density $\rho(x)$ is given by the zeroth component of the current

$$j_\mu(x) = \frac{1}{2i} \left[\phi^\dagger(x) \, U_\mu(x) \, \phi(x+\mu) - \phi^\dagger(x+\mu) \, U_\mu^\dagger(x) \, \phi(x) \right] \,.$$

The action of the LET is then given by

$$S_{\text{den}} = S - \mu \sum_x \rho(x) \,, \tag{13}$$

where the zero density part S is defined in (2). It is still feasible to modify the Cabibbo Marinari algorithm to include the finite μ term making the lattice simulation straightforward.

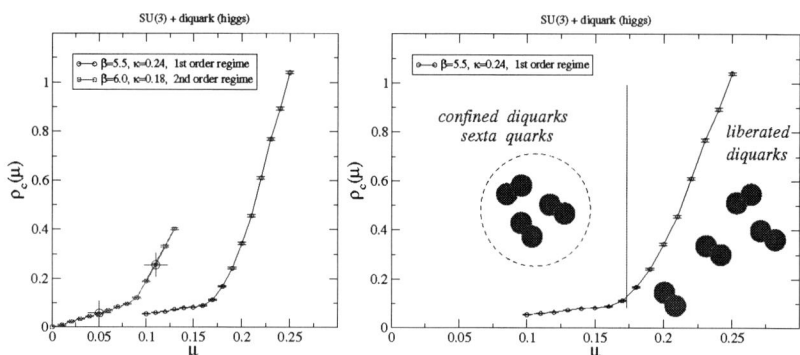

FIGURE 2. Left Panel: The expectation value of the baryonic Noether current as function of the external source μ. The blue crosses indicate the parameters with which the probability distribution of ρ will be studied below. Right panel: Illustration of the baryonic matter contributing to ρ_c.

For the actual simulation, two regimes are of particular interest: (i) The large volume regime, where the gluons which belong to the residual unbroken SU(2) subgroup are confined. (ii) The small volume regime, where the Higgs transition (as function of the Higgs hopping parameter κ) is 2nd order. Here, I used the parameters (see figure 1, right panel for an illustration)

$$\beta = 5.5, \quad \kappa = 0.24 \qquad \text{(large volume regime)}$$
$$\beta = 6.0, \quad \kappa = 0.18 \qquad \text{(small volume regime)} .$$

The results are presented in figure 2. We observe that the behavior of the expectation value of the baryonic Noether current, $\rho_c = \langle \rho \rangle$, as function of the overlap enhancing factor μ is qualitatively the same for both scenarios: $\langle \rho \rangle$ increases linearly with μ until a critical value μ_c is reached. For $\mu > \mu_c$, the linear dependence continues with a bigger slope. There is an intuitive understanding for this behavior (for an illustration see figure 2, right panel): At small μ, we are in a "confined phase" where small color electric flux tubes connect static color sources until string breaking occurs at large distances. In this phase, the diquarks are bound to color singlet states which are in the present theory 3-diquark states or, equivalently, sexta-quark states. For $\mu > \mu_c$, the theory looses its confining capabilities; diquarks are no longer bound to color singlet sexta-quarks. The baryon number susceptibility has increased due to deconfinement. It is interesting to note that even the Higgs transition is of first order when the Higgs hopping term is varied, the finite density transition is more like a crossover which describes sexta-quarks dissolving into liberated diquarks.

Let us discuss how a particular expectation value $\rho_c = \langle \rho \rangle$ is realized in a lattice simulation for a given external parameter μ. For this purpose, we ask: How big is the probability, $P(\rho)\,d\rho$, of finding ρ in the interval $[\rho, \rho + d\rho[$ for an actual lattice configuration? The normalized distribution $dP/d\rho$ is shown in figure 3 for $\mu = 0.05$ (confined phase) and for $\mu = 0.12$ (deconfined phase). It is intuitive that the distribution is broader in the deconfined regime. Finally, I point out that simulations for a given value

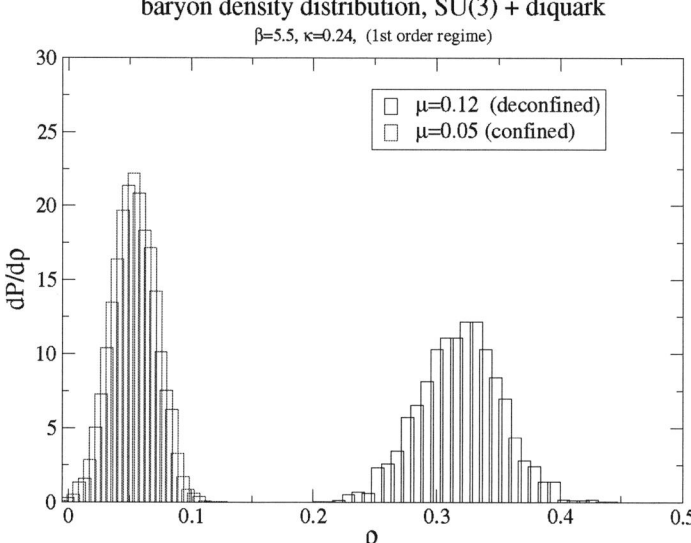

FIGURE 3. Probability distribution of the baryonic Noether density.

of ρ_c can be done without to much of a loss of statics: In this case, we would confine us to configurations with ρ belonging to a bin where $dP/d\rho$ peaks.

CONCLUSIONS

In this paper, the proposal [7] to describe the deconfinement phase transition at high baryon density by an effective theory of diquarks and gluons is thoroughly investigated by means of lattice simulations. The dynamics of the Lattice Effective Theory (LET) is dictated by the action of a gauged SU(3) Higgs model where the scalar Higgs field simulates the point-like diquark.

In a first step the regime of vanishing baryon density is explored. For this purpose, a detailed lattice simulation of the SU(3) Yang-Mills theory with a scalar field in fundamental representation was performed for the first time. In the large volume limit a first order deconfinement transition occurs, if the Higgs hopping parameter exceeds a critical strength. The diquarks undergo Bose-Einstein condensation. It turns out that only five of the eight gluons acquire a mass by the SU(3) Higgs mechanism, and preliminary evidence was found that confinement with respect to the unbroken SU(2) subgroup is intact.

In order to gain first insights into the regime of intermediate baryon densities, I here propose to perform lattice simulations at a fixed expectation value ρ_c of the baryonic Noether current. This procedure is outline for the LET designed above. I stress, however, that this approach is applicable to full QCD simulations. The practical benefit of this change of view onto the density regime is that severe overlap and sign problems are avoided. Overlap is ensured by an external parameter μ which couples to the Noether density.

At small μ, the linear rise of ρ_c with increasing μ is due to the population of the vacuum with color singlet bound states consisting of three diquarks (sexta-quarks). Above a critical value, deconfinement occurs, and the vacuum is populated by liberated diquarks. The transition appears to be a crossover. A significant signal of deconfinement is only visible in the baryon number susceptibility which rapidly increases at the transition.

More realistic LETs also incorporate the "valence quark" and give rise to baryons as bound states of diquarks and valence quarks. The investigation of these LETs are left to future work.

ACKNOWLEDGMENTS

This work is supported in parts by the Ministry of Science, Research and the Arts of Baden Würtemberg under Az: 24-7532.23-19-18/1 and by the *Virtual Institute* of the Helmholtz Association no. VH-VI-041.

REFERENCES

1.   U. W. Heinz, Nucl. Phys. A **721**, 30 (2003) [arXiv:nucl-th/0212004].
2.   P. de Forcrand and O. Philipsen, Nucl. Phys. B **642**, 290 (2002) [arXiv:hep-lat/0205016].
3.   Z. Fodor and S. D. Katz, Phys. Lett. B **534**, 87 (2002) [arXiv:hep-lat/0104001].
4.   D. Bailin and A. Love, Phys. Rept. **107**, 325 (1984).
5.   M. G. Alford, K. Rajagopal and F. Wilczek, Nucl. Phys. B **537**, 443 (1999) [arXiv:hep-ph/9804403].
     [6]
6.   K. Langfeld, arXiv:hep-lat/0109033.
7.   K. Langfeld, Talk given at Workshop on Strong and Electroweak Matter (SEWM 2002), Heidelberg, Germany, 2-5 Oct 2002, arXiv:hep-lat/0212032.
8.   R. Bertle, M. Faber, J. Greensite and S. Olejnik, Phys. Rev. D **69**, 014007 (2004) [arXiv:hep-lat/0310057].
9.   S. Hands, J. B. Kogut, M. P. Lombardo and S. E. Morrison, Nucl. Phys. B **558**, 327 (1999) [arXiv:hep-lat/9902034].
10. I. M. Barbour [UKQCD Collaboration], Nucl. Phys. A **642**, 251 (1998).
11. S. Ejiri, C. R. Allton, S. J. Hands, O. Kaczmarek, F. Karsch, E. Laermann and C. Schmidt, Prog. Theor. Phys. Suppl. **153**, 118 (2004) [arXiv:hep-lat/0312006].
12. Z. Fodor and S. D. Katz, JHEP **0404**, 050 (2004) [arXiv:hep-lat/0402006].
13. M. G. Alford, S. Chandrasekharan, J. Cox and U. J. Wiese, Nucl. Phys. B **602**, 61 (2001) [arXiv:hep-lat/0101012].
14. S. Chandrasekharan and U. J. Wiese, Nucl. Phys. B **492**, 455 (1997) [arXiv:hep-lat/9609042].
15. S. Coleman, *Aspects of Symmetry*, Cambridge University Press, 1985.

Spontaneous CP violation and quark mass ambiguities

Michael Creutz

Physics Department, Brookhaven National Laboratory, Upton, NY 11973, USA

Abstract. I explore the regions of quark masses where CP will be spontaneously broken in the strong interactions. The boundaries of these regions are controlled by the chiral anomaly, which manifests itself in ambiguities in the definition of non-degenerate quark masses. In particular, the concept of a single massless quark is ill defined.

INTRODUCTION

In this talk I discuss two apparently distinct but deeply entwined topics. First, I ask for what quark masses is CP spontaneously broken. Second, I investigate whether a vanishing up quark mass is a physically meaningful concept. I should emphasize at the outset that I will be exploring rather unphysical regions of parameter space. This in some sense is a theorists fantasy, with no direct experimental relevance. Indeed, my real goal is to understand better how chiral symmetry works in the strong interactions. The main content of this talk is contained in two recent papers [1, 2]. The basic ideas have roots in a talk I gave in Como at the 1996 edition of this conference [3].

I require a few assumptions. First, the continuum limit of QCD should exist and confine, with the only relevant parameters being the coupling and the quark masses. Then I will assume that chiral symmetry is spontaneously broken in the usual way and that effective chiral Lagrangians are qualitatively correct. Finally I assume that the anomaly removes any flavor singlet chiral symmetry. In particular, this implies that a single massless quark gives no exact Goldstone boson.

The underlying concepts are all quite old. In 1971 Dashen [4] showed how CP symmetry could be spontaneously broken in the strong interactions. DiVecchia and Veneziano [5] observed the CP violation in chiral Lagrangians at negative quark masses. Georgi and McArthur [6] showed that non-perturbative effects could give a non-multiplicative shift to the up quark mass. This motivated Kaplan and Manohar [7] in their classic studies of ambiguities in the up quark mass in the context of effective chiral Lagrangians. Banks, Nir and Seiberg [8] discussed the fact that the concept of a vanishing up quark mass is not so clean.

CP756, *Quark Confinement and the Hadron Spectrum VI*
edited by N. Brambilla, U. D'Alesio, A. Devoto, K. Maung, G.M. Prosperi and S. Serci
© 2005 American Institute of Physics 0-7354-0241-8/05/$22.50

THE EFFECTIVE MESON THEORY

I base my initial discussion on the effective theory for pseudoscalar mesons in terms of a field taking values in the group $SU(3)$

$$\Sigma = \exp(i\pi_\alpha \lambda_\alpha / f_\pi) \in SU(3) \tag{1}$$

I work with the three flavor theory, i.e. I include the up, down, and strange quarks. The standard generators of the group SU(3) are given by λ_α, and the pseudoscalar octet fields are denoted as π_α.

Chiral symmetry is manifested in independent left or right global rotations on this field

$$\Sigma \to g_L^\dagger \Sigma g_R \tag{2}$$

This symmetry is explicitly broken by the quark masses. The lowest order effective Lagrangian including the masses takes the form

$$L = \frac{f_\pi^2}{4} \text{Tr}(\partial_\mu \Sigma^\dagger \partial_\mu \Sigma) - v \, \text{Re} \, \text{Tr}(\Sigma M) \tag{3}$$

with the mass matrix

$$M = \begin{pmatrix} m_u & 0 & 0 \\ 0 & m_d & 0 \\ 0 & 0 & m_s \end{pmatrix} \tag{4}$$

Expanding this density to quadratic order in meson fields and then diagonalizing the resulting non-derivative term gives the usual result that the meson masses squared are proportional to the quark masses, including $m_{\pi^\pm}^2 \sim m_u + m_d$. For my purposes, I am particularly interested in the isospin violation arising from the up-down mass difference $m_d - m_u$. This results in a mixing of the π^0 and η mesons, giving somewhat complicated formulae for their masses

$$\begin{aligned} m_{\pi^0}^2 &\sim \tfrac{2}{3}\left(m_u + m_d + m_s - \sqrt{m_u^2 + m_d^2 + m_s^2 - m_u m_d - m_u m_s - m_d m_s}\right) \\ m_\eta^2 &\sim \tfrac{2}{3}\left(m_u + m_d + m_s + \sqrt{m_u^2 + m_d^2 + m_s^2 - m_u m_d - m_u m_s - m_d m_s}\right) \end{aligned} \tag{5}$$

Note in particular that it is possible to tune the parameters such that $m_{\pi^0}^2$ vanishes. This occurs when

$$m_u = \frac{-m_s m_d}{m_s + m_d}. \tag{6}$$

This vanishing mass does not require chiral symmetry. It does, however, occur at a somewhat unphysical location, requiring at least one of the quark masses to be negative.

SPONTANEOUS CP VIOLATION

Going to negative quark masses at first seems a bit strange, but in such a regime unusual things do happen. Note that because of the anomaly, the signs of the quark masses can

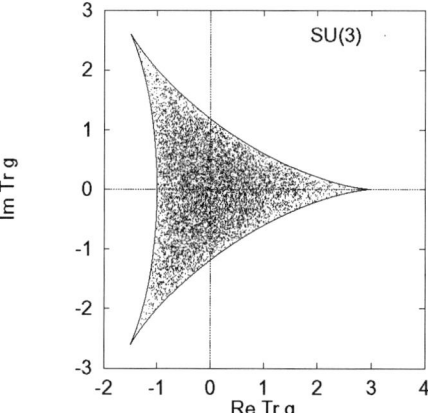

FIGURE 1. The real and imaginary parts of the traces of 10,000 randomly chosen $SU(3)$ matrices. Note that the minimum real part occurs at two distinct cube roots of unity.

become significant. Flavored chiral rotations can move the signs around, but the overall sign of the determinant of the mass matrix is invariant.

The effective Lagrangian is useful for clarifying the expected behavior with negative masses. In the usual case with positive masses, the vacuum involves quantum fluctuations about the maximum of $\mathrm{ReTr}\Sigma$. This occurs at $\Sigma = I$. However, now consider the case of degenerate negative masses. The vacuum instead should occur at the minimum of $\mathrm{ReTr}\Sigma$. The important point is that $-I$ does not lie in the group $SU(3)$. A simple analysis shows that the minimum is doubly degenerate, occuring at $\Sigma = \exp(\pm 2\pi i/3)$. Fig. 1 plots the traces of 10,000 random matrices to illustrate this result. Note that CP symmetry takes Σ to Σ^*; thus, either of these solutions involves a spontaneous breaking of CP.

This CP violating phase can be approached continuously by passing through the values of the quark masses in Eq. 6 where $m_{\pi^0}^2$ vanishes. Indeed, this equation represents the boundary for the occurance of a pion condensed phase with $\langle \pi^0 \rangle \neq 0$. Similar boundaries occur at the appropriate branches of

$$m_u = \frac{-m_s m_d}{\pm m_s \pm m_d} \tag{7}$$

The full phase diagram is sketched in Fig. 2.

In the CP violating phases the vacuum fluctuates about non-trivial complex matrices of form

$$\Sigma = \begin{pmatrix} e^{i\phi_1} & 0 & 0 \\ 0 & e^{i\phi_2} & 0 \\ 0 & 0 & e^{-i\phi_1 - i\phi_2} \end{pmatrix} \tag{8}$$

where the angles satisfy

$$m_u \sin(\phi_1) = m_d \sin(\phi_2) = -m_s \sin(\phi_1 + \phi_2) \tag{9}$$

145

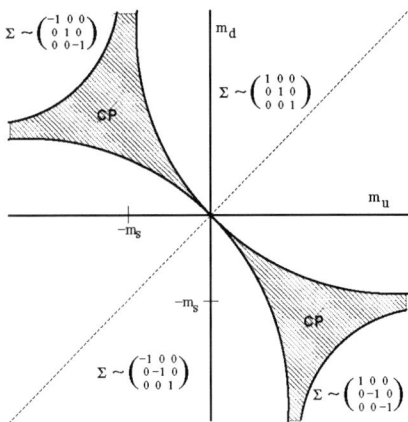

FIGURE 2. The phase diagram of QCD as a function of the up and down quark masses with a fixed positive strange quark mass.

The CP violating phases are separated from the conserving ones by second order transition lines occuring when $m_{\pi^0} = 0$. The former have two degenerate vacua related by $\phi_i \leftrightarrow -\phi_i$.

AMBIGUITIES IN THE UP QUARK MASS

I now take a section through this diagram. Phenomenologically, the down and strange quark masses appear to definitely not vanish. Consider fixing them at some positive values and study the dependence of the theory as a function of the up quark mass alone. Thus follow a horizontal line at fixed m_d in Fig. 2. To enable continuation around the boundary of the CP violating phase, extend the m_u dependence into the complex plane. This gives the qualitative structure sketched in Fig. 3.

The expectation is a first order transition along negative Re m axis. This ends at a second order critical point at non-zero Re $m < 0$. Along the first order line there is a spontaneous breaking of CP. The transition has a simple order parameter $\langle \pi_0 \rangle$. The presence of the gap below $m_u = 0$ and the CP violating phase were noted some time ago by Di Vecchia and Veneziano [5].

Note that nothing significant occurs at $m_u = 0$ when $m_d \neq 0$. This raises an interesting question: Does $m_u = 0$ have any physical significance? I now argue that this is not a well posed question if $m_d \neq 0$ and $m_s \neq 0$. One consequence of this observation is that a vanishing up quark mass is an unacceptable solution to the strong CP problem.

A crucial message here is that the concept of an "underlying basic Lagrangian" does not exist. Field theory is full of divergences that must be regulated. It is only the underlying symmetries of the theory that remain significant. The case of a single massless quark gives no special symmetry because of the anomaly. Unlike the multiple degenerate quark case, no exact Goldstone bosons should appear at $m_u = 0$.

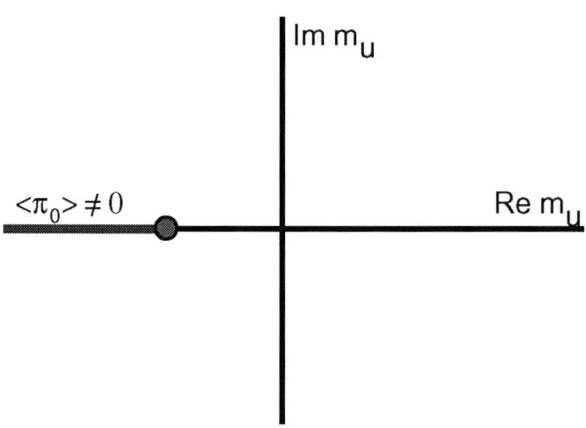

FIGURE 3. The qualitative phase diagram as a function of a complex up quark mass with fixed strange and down quark masses. (Here I ignore the reappearance of a CP conserving phase at large negative up quark mass.)

The renormalization group trajectory

A continuum field theory is defined as a limit of a cutoff theory. For QCD the bare parameters are the coupling g and quark masses m_i. These must be renormalized for the continuum limit; indeed both bare parameters renormalize to zero in a well defined way given by the renormalization group equations. For this discussion I denote the cutoff as a minimum length a. This corresponds to the inverse of a large momentum scale. To simplify the discussion I will consider a single quark mass. Then the well known flow equations in the small coupling limit take the form

$$a\frac{d}{da}g = \beta(g) = \beta_0 g^3 + \beta_1 g^5 + \ldots + \text{non-perturbative}$$
$$a\frac{d}{da}m = m\gamma(g) = m(\gamma_0 g^2 + \gamma_1 g^4 + \ldots) + \text{non-perturbative} \tag{10}$$

The first few coefficients β_0, β_1, and γ_0 are scheme independent. In these equations the "non-perturbative" parts fall faster than any power of g as $g \to 0$. Q crucial point, to which I will return, is that these contributions are not proportional to the quark mass.

The solution to these equations is standard

$$a = \frac{1}{\Lambda}e^{-1/2\beta_0 g^2} g^{-\beta_1/\beta_0^2}(1 + O(g^2))$$
$$m = M g^{\gamma_0/\beta_0}(1 + O(g^2)) \tag{11}$$

Rewriting shows how the coupling and mass go to zero in the continuum limit $a \to 0$

$$g^2 \sim \frac{1}{\log(1/\Lambda a)} \to 0$$
$$m \sim M \left(\frac{1}{\log(1/\Lambda a)}\right)^{\gamma_0/2\beta_0} \to 0 \tag{12}$$

The first part of this equation represents the famous phenomenon of "asymptotic freedom."

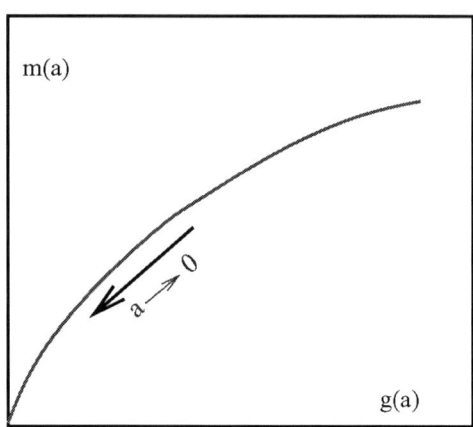

FIGURE 4. The continuum limit involves following a renormalization group trajectory in the space of bare parameters.

At a basic level these equations arise from holding a few physical quantities fixed along the "renormalization group trajectory," as sketched in Fig. 4. For this discussion it is convenient to use the lightest baryon and the lightest meson masses, m_p, m_π, as physical quantities to hold fixed. With multiple quark flavors one would hold several meson masses fixed.

The parameters Λ and M represent "integration constants" of the renormalization group equations. Λ is conventionally interpreted as the "QCD scale," while M defines a the "renormalized quark mass." Their values follow from limits along the renormalization group trajectory

$$\Lambda = \lim_{a \to 0} \frac{e^{-1/2\beta_0 g^2} g^{-\beta_1/\beta_0^2}}{a} \tag{13}$$

$$M = \lim_{a \to 0} m g^{-\gamma_0/\beta_0} \tag{14}$$

The precise numerical values of Λ and M depend on the renormalization scheme.

The physical masses map directly onto the integration constants, $\Lambda = \Lambda(m_p, m_\pi)$ and $M = M(m_p, m_\pi)$. Inverting, gives the physical masses as functions of Λ and M: $\longrightarrow$ $m_i = m_i(\Lambda, M)$. Simple dimensional analysis implies this relationship must take the form $m_i = \Lambda f_i(M/\Lambda)$, with $f(x)$ some *a priori* unknown function.

In the case of multiple degenerate fermions more is known about the behavior of f. In particular, Goldstone bosons should appear as the quark mass goes to zero: $m_\pi^2 \sim m_q$. This implies the existence of a square root singularity $f_\pi(x) \sim x^{1/2}$. The location of this singularity defines what is meant by zero mass quarks, thus removing any additive ambiguity in defining M.

The single massless flavor case, however, is somewhat special. Then the meson mass $m_\pi = \Lambda f_\pi(M/\Lambda)$ does not vanish at $M = 0$. The anomaly precludes chiral symmetry and

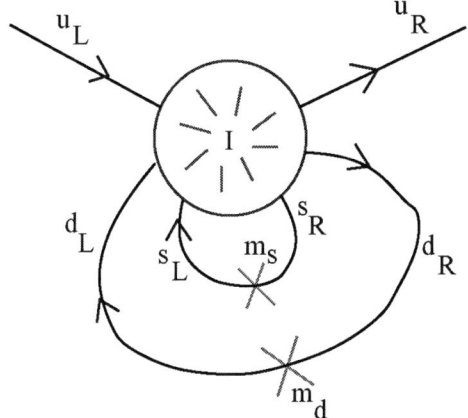

FIGURE 5. Non-perterbative effects can flip all quark spins through the anomaly. Tying the heavier quark lines together with mass terms generates a shift in the lightest quark mass.

Goldstone bosons. While the meson mass can be forced to vanish, the earlier discussion shows that this requires a special tuning to a negative quark mass. For the function $f_\pi(x)$ one expects a smooth and non-vanishing behavior at $x = 0$.

The shift of the singularity in x away from zero is due to non-perturbative effects. Indeed, non-perturbative contributions to the mass flow are expected which are not proportional to quark mass. As shown some time ago by t'Hooft [9], there are nonperturbative classical effects called "instantons" that flip all quark spins simultaneously. Tying the heavier quarks together with their mass terms as sketched in Fig. 5 gives rise to a contribution to the up quark mass flow

$$\Delta m_u \sim \frac{m_d m_s}{\Lambda_{qcd}}, \quad \Lambda_{qcd} \tag{15}$$

These effects indicate that $m_u = 0$ is not renormalization group invariant. If the up quark mass vanishes as some point on the trajectory, it will not for other points.

Matching between schemes

This non-invariance of the quark mass under the renormalization group raises the specter of scheme dependence. On changing renormalization schemes one should preserve the lowest order perturbative limit as $g \to 0$ at fixed scale a

$$\tilde{g} = g + O(g^3) + \text{non-perturbative}$$
$$\tilde{m} = m(1 + O(g^2)) + \text{non-perturbative} \tag{16}$$

Here "non-perturbative" terms should vanish faster than any power of g. As mentioned above, the integration constants Λ and M in general depend on the chosen scheme.

149

It is important to recognize that this perturbative matching at fixed a is not the continuum limit. Indeed, taking $g \to 0$ at fixed a gives perturbation theory on free quarks, while taking $a \to 0$ at fixed g gives the divergences of field theory. For the confining physics of the real world, one must go between these limits and take a and g together to zero along the renormalization group trajectory.

To dramatically illustrate the issue, consider a particularly cooked up new scheme

$$\tilde{a} = a$$
$$\tilde{g} = g \tag{17}$$
$$\tilde{m} = m - Mg^{\gamma_0/\beta_0} \times \frac{e^{-1/2\beta_0 g^2} g^{-\beta_1/\beta_0^2}}{\Lambda a}$$

Here I have crafted the last factor in the mass expression to approach unity. This non-perturbative redefinition of the bare parameters makes

$$\tilde{M} \equiv \lim_{a \to 0} \tilde{m} \tilde{g}^{-\gamma_0/\beta_0} = M - M = 0 \tag{18}$$

While this may be somewhat artificial, it shows that some scheme always exists where the renormalized quark mass vanishes. Of course doing this for the top quark will insert ridiculously large non-perturbative effects, but it is possible in principle. Because of this ambiguity, $M = 0$ is scheme dependent and thus is not a physical concept. Of course with degenerate quarks one can precisely define masslessness by the location of the square root singularity in $f(x)$ as defined above.

On the lattice

As in the general case, on the lattice the renormalization flows depend on details of the lattice action. Various gauge actions as well as the fermion formulation need to be considered. Recent discussions have concentrated on overlap/Ginsparg-Wilson fermion operators [10, 11], which bring a remnant of chiral symmetry to the lattice. However even these operators are not unique. The overlap operator relies on a projection, but it depends on the particular Dirac operator being projected. When starting with a Wilson Dirac operator, the input negative mass is adjustable over a finite range.

The one flavor theory dynamically generates a gap which will appear in the spectrum of the final Dirac operator. The overlap projection does not protect the size of this gap. With the gap present, the Ginsparg-Wilson condition is not sufficient to guarantee the preservation of $M = 0$ between schemes.

With a Ginsparg-Wilson action the concept of a massless quark is synonymous with zero topological susceptibility. If a single massless quark is an ill defined concept, this raises the question of whether the topological susceptibility is uniquely defined for $N_f < 2$. This question has also been asked in the context of making the Witten-Veneziano formula [12, 13] for the η' mass precise [14, 15]. From a perturbative point of view infinities are not a problem [16, 17]. However, to give a unique winding number to a gauge configuration requires a degree of smoothness for the gauge fields. One popular condition that ensures a well defined winding number forbids plaquettes further than a finite distance δ from the identity [18]. However this condition is rather strong and leaves unresolved issues. In particular this "admissibility condition" has recently been shown to be in conflict with reflection positivity [19].

CONCLUSIONS

Effective chiral Lagrangians show that the strong interactions will spontaneously violate CP for large regions of parameter space. This phenomenon requires negative quark masses, a concept made physical by the anomaly. Based on this qualitative picture, I have argued that $m_u = 0$ is not a meaningful concept. As such, it cannot be regarded as a possible solution to the strong CP problem. These effects are entirely non-perturbative. As a corollary, the topological susceptibility is not uniquely defined for $N_f < 2$. Finally, I note that available simulation algorithms cannot explore this fascinating physics because it involves regions where the fermion determinant is not positive, i.e. Monte Carlo methods have a sign problem.

ACKNOWLEDGMENTS

This manuscript has been authored under contract number DE-AC02-98CH10886 with the U.S. Department of Energy. Accordingly, the U.S. Government retains a non-exclusive, royalty-free license to publish or reproduce the published form of this contribution, or allow others to do so, for U.S. Government purposes.

REFERENCES

1. M. Creutz, Phys. Rev. Lett. **92**, 201601 (2004) [arXiv:hep-lat/0312018].
2. M. Creutz, Phys. Rev. Lett. **92**, 162003 (2004) [arXiv:hep-ph/0312225].
3. M. Creutz, arXiv:hep-ph/9608216.
4. R. F. Dashen, Phys. Rev. D **3**, 1879 (1971).
5. P. Di Vecchia and G. Veneziano, Nucl. Phys. B **171**, 253 (1980).
6. H. Georgi and I. N. McArthur, HUTP-81/A011
7. D. B. Kaplan and A. V. Manohar, Phys. Rev. Lett. **56**, 2004 (1986).
8. T. Banks, Y. Nir and N. Seiberg, arXiv:hep-ph/9403203.
9. G. 't Hooft, " Phys. Rev. D **14**, 3432 (1976) [Erratum-ibid. D **18**, 2199 (1978)].
10. H. Neuberger, Phys. Lett. B **417**, 141 (1998) [arXiv:hep-lat/9707022].
11. P. H. Ginsparg and K. G. Wilson, Phys. Rev. D **25**, 2649 (1982).
12. E. Witten, Nucl. Phys. B **156** (1979) 269.
13. G. Veneziano, Nucl. Phys. B **159** (1979) 213.
14. E. Seiler and I. O. Stamatescu, MPI-PAE/PTh 10/87
15. E. Seiler, Phys. Lett. B **525** (2002) 355 [arXiv:hep-th/0111125].
16. L. Giusti, G. C. Rossi and M. Testa, Phys. Lett. B **587**, 157 (2004) [arXiv:hep-lat/0402027].
17. M. Luscher, arXiv:hep-th/0404034 (unpublished).
18. M. Luscher, Commun. Math. Phys. **85**, 39 (1982).
19. M. Creutz, Phys. Rev. D, in press (2004) [arXiv:hep-lat/0409017].

Running coupling constant and masses in QCD, the meson spectrum

M. Baldicchi and G. M. Prosperi

Dipartimento di Fisica, Università di Milano
I.N.F.N., sezione di Milano
via Celoria 16, I20133 Milano, Italy

Abstract. In line with some previous works, we study in this paper the meson spectrum in the framework of a second order quark-antiquark Bethe-Salpeter formalism which includes confinement. An analytic one loop running coupling constant $\alpha_s(Q)$, as proposed by Shirkov and Sovlovtsov, is used in the calculations. As for the quark masses, the case of a purely phenomenological running mass for the light quarks in terms of the c. m. momentum is further investigated. Alternatively a more fundamental expression $m_P(Q)$ is introduced for light and strange quarks, combining renormalization group and analyticity requirements with an approximate solution of the Dyson-Schwinger equation. The use of such running coupling constant and masses turns out to be essential for a correct reproduction of the the light pseudoscalar mesons.

Keywords: Running coupling constant, masses, QCD, meson spectrum
PACS: 12.38.Aw, 11.10.St, 12.38.Lg, 12.39.Ki

1. INTRODUCTION

In a series of papers [1, 2] we have applied a second order Bethe-Salpeter formalism [1], previously established [3], to the evaluation of the quark-antiquark spectrum, in the context of QCD. Taking advantage of a Feynman-Schwinger representation for the quark propagator in an external field, the kernels of the Bethe-Salpeter and the Dyson-Schwinger equations were obtained, starting from an appropriate ansatz on the Wilson loop correlator. Such an ansatz consisted in adding an area term to the lowest perturbative expression of $\ln W$. By a 3D reduction of the original 4D BS equation a mass operator was obtained and applied to the determination of the $q\bar{q}$ bound states [4].

In that way, using a fixed strong coupling constant α_s and appropriate values for the other variables, the entire spectrum was reasonably well reproduced with, however, the relevant exception of the light pseudoscalar mesons (π, K, η_s). Agreement even for the latter states could be obtained using an analytic running coupling constant $\alpha_s(Q)$ proposed by Shirkov and Sovlovtsov, which is modified in the infrared region with respect to the ordinary purely perturbative expression [5, 6]. In conjunction it was also necessary to use a phenomenological running constituent mass for the light quarks u and d, written as a polynomial in the center of the mass quark momentum $\mathbf{k}$ [2].

In this paper we reconsider and improve the above procedure from two aspects:

---

[1] second order in the sense of the differential equations

CP756, *Quark Confinement and the Hadron Spectrum VI*
edited by N. Brambilla, U. D'Alesio, A. Devoto, K. Maung, G.M. Prosperi and S. Serci

a) we evaluate the hyperfine $^3S_1 - {}^1S_0$ separation for the light pseudoscalar mesons to the second rather than to the first order perturbation theory,

b) we use running constituent masses for u, d and s quarks, obtained by an approximate solution of the appropriate DS equations and analytic running current masses.

As a consequence of a) a significant improvement is obtained in agreement with the data already with a phenomenological running mass for the light quarks. As for case b), preliminary calculations seem to provide results numerically similar to the above ones, but more satisfactory from the conceptual point of view.

The plan of the remaining part of the paper is as follows. In Sect. 2 we briefly recall the second order BS formalism to establish notations. In Sect. 3 we discuss the DS equation and the 3D reduction of the BS equation. In Sect. 4 we consider the infrared behavior of the running coupling constant and obtain the corresponding running masses. In sect. 5 and 6 we report our results and draw some conclusions.

2. SECOND ORDER BETHE-SALPETER FORMALISM

In the QCD framework a *second order* four point quark-antiquark function and a full quark propagator can be defined as

$$H^{(4)}(x_1, x_2; y_1, y_2) = -\frac{1}{3} \text{Tr}_{\text{color}} \langle \Delta_1(x_1, y_1; A) \Delta_2(y_2, x_2; A) \rangle \tag{1}$$

and

$$H^{(2)}(x - y) = \frac{i}{\sqrt{3}} \text{Tr}_{\text{color}} \langle \Delta(x, y : A) \rangle, \tag{2}$$

where

$$\langle f[A] \rangle = \int DA \, M_F[A] \, e^{iS_G[A]} f[A], \tag{3}$$

$M_F[A] = \text{Det} \, \Pi^2_{j=1}[1 + g\gamma^\mu A_\mu (i\gamma^\nu_j \partial_{j\nu} - m_j)^{-1}]$ and $\Delta(x, y; A)$ is the *second order* quark propagator in an external gauge field.

The quantity Δ is defined by the second order differential equation

$$(D_\mu D^\mu + m^2 - \frac{1}{2} g \, \sigma^{\mu\nu} F_{\mu\nu}) \Delta(x, y; A) = -\delta^4(x - y), \tag{4}$$

$(\sigma^{\mu\nu} = \frac{i}{2}[\gamma^\mu, \gamma^\nu]$ and $D_\mu = \partial_\mu + igA_\mu)$ and it is related to the corresponding first order propagator by $S(x, y; A) = (i\gamma^\nu D_\nu + m)\Delta(x, y; A)$.

The advantage of considering second order quantities is that the spin terms are more clearly separated and it is possible to write for Δ a generalized Feynman-Schwinger representation, *i. e.* to solve eq. (4) in terms of a quark path integral [3, 1]. Using the latter in (1) or (2) a similar representation can be obtained for $H^{(4)}$ and $H^{(2)}$.

The interesting aspect of this representation is that the gauge field appears only through a Wilson line correlator W. In the limit $x_2 \to x_1$, $y_2 \to y_1$ or $y \to x$ the Wilson lines close in a single Wilson loop Γ and if Γ stays on a plane, $i \ln W$ can be written in a first approximation as the sum of its lowest perturbative expression and an area term

$$i \ln W = \frac{4}{3} g^2 \oint dz^\mu \oint dz^{\nu\prime} D_{\mu\nu}(z - z') + \tag{5}$$

$$\sigma \oint dz^0 \oint dz^{0\prime} \delta(z^0 - z^{0\prime}) |\mathbf{z} - \mathbf{z}'| \int_0^1 d\lambda \left\{ 1 - [\lambda \frac{d\mathbf{z}_\perp}{dz^0} + (1 - \lambda) \frac{d\mathbf{z}'_\perp}{dz^{0\prime}}]^2 \right\}^{\frac{1}{2}}.$$

The area term here is written as the algebraic sum of successive equal time strips and $d\mathbf{z}_\perp = d\mathbf{z} - (d\mathbf{z} \cdot \mathbf{r}) \mathbf{r}/r^2$ denotes the transversal component of $d\mathbf{z}$. The basic assumption now is that in the center of mass frame (5) remains a good approximation even in the general case, *i. e.* for non flat curves and when $x_2 \neq x_1, y_2 \neq y_1$ or $y \neq x$.

Then, by appropriate manipulations on the resulting expressions, an inhomogeneous Bethe-Salpeter equation for the 4-point function $H^{(4)}(x_1, x_2; y_1, y_2)$ and a Dyson-Schwinger equation for $H^{(2)}(x - y)$ can be derived in a kind of generalized ladder and rainbow approximation. This should appear plausible, even from the point of view of graph resummation, for the analogy between the perturbative and the confinement terms in (5). We may refer to such terms as a *gluon exchange* and a *string connection*.

In momentum representation, the corresponding homogeneous BS-equation becomes

$$\Phi_P(k) = -i \int \frac{d^4 u}{(2\pi)^4} \hat{I}_{ab} \left(k - u; \frac{1}{2} P + \frac{k + u}{2}, \frac{1}{2} P - \frac{k + u}{2} \right)$$

$$\hat{H}_1^{(2)} \left(\frac{1}{2} P + k \right) \sigma^a \Phi_P(u) \sigma^b \hat{H}_2^{(2)} \left(-\frac{1}{2} P + k \right), \tag{6}$$

where we have set $\sigma^0 = 1$; $a, b = 0, \mu \nu$; the center of mass frame has to be understood, $P = (m_B, \mathbf{0})$; $\Phi_P(k)$ denotes an appropriate *second order* wave function [2].

Similarly, in terms of the irreducible self-energy, defined by $\hat{H}^{(2)}(k) = \frac{i}{k^2 - m^2} + \frac{i}{k^2 - m^2} i \Gamma(k) \hat{H}^{(2)}(k)$, the DS-equation can be written

$$\hat{\Gamma}(k) = \int \frac{d^4 l}{(2\pi)^4} \hat{I}_{ab} \left(k - l; \frac{k + l}{2}, \frac{k + l}{2} \right) \sigma^a \hat{H}^{(2)}(l) \sigma^b. \tag{7}$$

The kernels in (6) and (7) are the same in the two equations, consistently with the requirement of chiral symmetry limit [7], and are given by

$$\hat{I}_{0;0}(Q; p, p') = 16\pi \frac{4}{3} \alpha_s p^\alpha p'^\beta \hat{D}_{\alpha\beta}(Q) +$$

$$+ 4\sigma \int d^3 \zeta e^{-i\mathbf{Q}\cdot\zeta} |\zeta| \varepsilon(p_0) \varepsilon(p'_0) \int_0^1 d\lambda \{ p_0^2 p_0'^2 - [\lambda p'_0 \mathbf{p}_T + (1 - \lambda) p_0 \mathbf{p}'_T]^2 \}^{\frac{1}{2}}$$

---

[2] In terms of the second order field $\phi(x) = (i\gamma^\mu D_\mu + m)^{-1} \psi(x)$ this wave function is defined by

$$\langle 0 | \phi(\frac{\xi}{2}) \bar{\psi}(-\frac{\xi}{2}) | P \rangle = \frac{1}{(2\pi)^2} \Phi_P(k) e^{-ik\xi}.$$

$$\hat{I}_{\mu\nu;0}(Q;p,p') = 4\pi i \frac{4}{3}\alpha_s (\delta_\mu^\alpha Q_\nu - \delta_\nu^\alpha Q_\mu) p'_\beta \hat{D}_{\alpha\beta}(Q) -$$

$$-\sigma \int d^3\zeta \, e^{-i\mathbf{Q}\cdot\zeta} \varepsilon(p_0) \frac{\zeta_\mu p_\nu - \zeta_\nu p_\mu}{|\zeta|\sqrt{p_0^2 - \mathbf{p}_T^2}} p'_0$$

$$\hat{I}_{0;\rho\sigma}(Q;p,p') = -4\pi i \frac{4}{3}\alpha_s p^\alpha (\delta_\rho^\beta Q_\sigma - \delta_\sigma^\beta Q_\rho) \hat{D}_{\alpha\beta}(Q) +$$

$$+\sigma \int d^3\zeta \, e^{-i\mathbf{Q}\cdot\zeta} p_0 \frac{\zeta_\rho p'_\sigma - \zeta_\sigma p'_\rho}{|\zeta|\sqrt{p_0'^2 - \mathbf{p}_T'^2}} \varepsilon(p'_0)$$

$$\hat{I}_{\mu\nu;\rho\sigma}(Q;p,p') = \pi \frac{4}{3}\alpha_s (\delta_\mu^\alpha Q_\nu - \delta_\nu^\alpha Q_\mu)(\delta_\rho^\alpha Q_\sigma - \delta_\sigma^\alpha Q_\rho) \hat{D}_{\alpha\beta}(Q), \qquad (8)$$

where in the second and in the third equation $\zeta_0 = 0$ has to be understood. Notice that, due to the privileged role given to the c. m. frame, the terms proportional to σ in (8) are not formally covariant.

3. DS EQUATION AND MASS OPERATOR

Concerning eq. (7), let us observe that the unity matrix, $\sigma^{\mu\nu}$ and γ^5 form a subalgebra of the Dirac algebra. Consequently $\Gamma(k)$ can be assumed to depend only on this set of matrices and, since it must be a three dimensional scalar, only on terms like $k_j \sigma^{0j}$. In fact, it can be checked that $\Gamma(k)$ can be consistently assumed to be completely spin independent and eq. (7) can be written in the form

$$\Gamma(k) = i \int \frac{d^4 l}{(2\pi)^4} \frac{R(k,l)}{l^2 - m^2 + \Gamma(l)}, \qquad (9)$$

with

$$R(k,l) = 4\pi \frac{4}{3}\alpha_s \left[4\frac{p^2 l^2 - (pl)^2}{(k-l)^2} + \frac{3}{4} \right] +$$

$$+ \sigma \int d^3 r \, e^{-i(k-l)\cdot r} r (k_0 + l_0)^2 \sqrt{1 - \frac{(\mathbf{k}_\perp + \mathbf{l}_\perp)^2}{(k_0 + l_0)^2}}, \qquad (10)$$

$\mathbf{k}_\perp$ and $\mathbf{l}_\perp$ denoting as above the transversal part of $\mathbf{k}$ and $\mathbf{l}$.

Notice that, once (10) is solved, the pole or constituent mass m_P, to be used in bound states problems, is given by the equation

$$m_P^2 - m^2 + \Gamma(m_P^2) = 0. \qquad (11)$$

We can try to solve eq. (9) iteratively and we have at the first step

$$\Gamma(k) = i \int \frac{d^4 l}{(2\pi)^4} \frac{R(k,l)}{l^2 - m^2}. \qquad (12)$$

155

In a preliminary calculation we omit altogether the perturbative contribution to $R(k,l)$ (notice the overplacing of curves b and c in Fig. 1) and neglect the term in $(\mathbf{k}_\perp + \mathbf{l}_\perp)^2$ in the string part. Strictly, the second approximation is justified only for S bound states (classically $k_\perp r$ is the angular momentum of the bound state) but it is necessary in order to make the integral analytically calculable.

Then introducing a cut off μ, we obtain

$$\Gamma(k) = \frac{\sigma}{\pi}[k_0^2 A(m, |\mathbf{k}|) - B_\mu - B(m, |\mathbf{k}|), \tag{13}$$

where $B_\mu = 2\ln\frac{\mu}{m} - 1$,

$$A(m, |\mathbf{k}|) = \frac{1}{\mathbf{k}^2 + m^2}\left[1 + \frac{m^2}{2|\mathbf{k}|\sqrt{\mathbf{k}^2 + m^2}}\ln\frac{\sqrt{\mathbf{k}^2 + m^2} + |\mathbf{k}|}{\sqrt{\mathbf{k}^2 + m^2} - |\mathbf{k}|}\right] \tag{14}$$

and $B(m, |\mathbf{k}|)$ is a more complicated expression that we do not report explicitly here for lack of space. The resulting pole mass is

$$\overline{m}_P^2(m, |\mathbf{k}|) = \frac{m^2 + \frac{\sigma}{\pi}[B_\mu + B(m, |\mathbf{k}|) - k^2 A(m, |\mathbf{k}|)]}{1 + \frac{\sigma}{\pi}A(m, |\mathbf{k}|)}, \tag{15}$$

The above expression depends on the current mass m and on the quark c. m. momentum $|\mathbf{k}|$, (see Fig. 1a). Notice that such dependence on $|\mathbf{k}|$ is clearly an artifact of the schematic way we have introduced confinement in eq. (5) and that the curve is rather flat in the region of interest. Correspondingly it seems reasonable to chose as true mass $m_P(m)$ the value of $\overline{m}_P(m, |\mathbf{k}|)$ at its stationary point in $|\mathbf{k}|$.

Then, in a neighbor of its singularity $k^2 = m_P^2$, the full propagator can be written as $\hat{H}^{(2)}(k) = \frac{iZ}{k^2 - m_P^2}$, where the residuum Z differs from 1 only for terms proportional to α_s or σ. Consistently in (6) we can simply take $Z = 1$ and are left with the free propagator with a constituent mass. If, in addition, we replace $\hat{I}_{ab}$ with its so called instantaneous approximation $\hat{I}_{ab}^{\text{inst}}(\mathbf{k}, \mathbf{u})$, we can explicitly perform the integration in u_0 and arrive at a three dimensional reduced equation.

Such a reduced equation takes the form of the eigenvalue equation for a squared mass operator [3], $M^2 = M_0^2 + U$, with $M_0 = w_1 + w_2$, $w_{1,2} = \sqrt{m_{1,2}^2 + \mathbf{k}^2}$ and

$$\langle \mathbf{k}|U|\mathbf{k}'\rangle = \frac{1}{(2\pi)^3}\sqrt{\frac{w_1 + w_2}{2w_1 w_2}}\,\hat{I}_{ab}^{\text{inst}}(\mathbf{k}, \mathbf{k}')\sqrt{\frac{w_1' + w_2'}{2w_1' w_2'}}\,\sigma_1^a \sigma_2^b \tag{16}$$

(for an explicit expression we refer to [2, 1]). The quadratic form of the above equation obviously derives from the second order formalism we have used.

Alternatively, in more usual terms, one can look for the eigenvalue of the mass operator or center of mass Hamiltonian $H_{\text{CM}} \equiv M = M_0 + V$ with V defined by $M_0 V + VM_0 + V^2 = U$. Neglecting the term V^2 the linear form potential V can be obtained from U by the kinematic replacement $\sqrt{\frac{(w_1 + w_2)(w_1' + w_2')}{w_1 w_2 w_1' w_2'}} \to \frac{1}{2\sqrt{w_1 w_2 w_1' w_2'}}$. The resulting expression is particularly useful for a comparison with models based on potential. In particular, in the static limit V reduces to the Cornell potential

$$V_{\text{stat}} = -\frac{4}{3}\frac{\alpha_s}{r} + \sigma r; \tag{17}$$

in the semirelativistic limit (up to $\frac{1}{m^2}$ terms after an appropriate Foldy-Wouthuysen transformation) it equals the potential discussed in ref. [8], if full relativistic kinematics is kept, but the spin dependent terms are neglected, it becomes identical to the potential of the relativistic flux tube model [3].

4. RUNNING COUPLING CONSTANT AND MASSES

As we said, diagonalizing M^2 or H_{CM} with fixed coupling constant and quark masses, a general good fit of the data was obtained. Actually a serious problem was represented by the masses of the light pseudo scalar mesons that turned out too large. The results obtained in ref. [2] suggest, however, that the situation can be greatly improved using an appropriate running coupling constant and running quark masses.

At one loop, the running coupling constant is usually written

$$\alpha_s(Q) = \frac{4\pi}{\beta_0 \ln(Q^2/\Lambda^2)}, \tag{18}$$

with $\beta_0 = 11 - \frac{2}{3}N_f$ and N_f the number of 'active' quarks. However, the singularity occurring in such expression is an artifact of perturbation theory and it contradicts general analyticity properties, therefore the expression must be somewhat modified in the infrared region [9]. Notice that this is particularly important for the quark-antiquark bound state problem, where the variable Q^2 is usually identified with the squared momentum transfer $Q^2 = (\mathbf{k} - \mathbf{k}')^2$, which ranges typically from $(0.1\,\text{GeV})^2$ to $(1\,\text{GeV})^2$ for different quark masses and states.

The most naive modification of eq. (18) would consist in freezing $\alpha_s(Q^2)$ to a certain maximum value $\bar{\alpha}_s$ as Q^2 decreases and in treating this value as a phenomenological parameter (truncation prescription). However, various more sensible proposals have been made on different bases [5, 6].

In particular Shirkov and Solovtsov [6] suggest to replace (18) with

$$\alpha_s(Q) = \frac{4\pi}{\beta_0} \left(\frac{1}{\ln(Q^2/\Lambda^2)} + \frac{\Lambda^2}{\Lambda^2 - Q^2} \right). \tag{19}$$

This remains regular for $Q^2 = \Lambda^2$ and has a finite Λ independent limit $\alpha_s(0) = 4\pi/\beta_0$ for $Q^2 \to 0$. Eq. (19) is obtained assuming a dispersion relation for $\alpha_s(Q)$ with a cut for $-\infty < Q^2 < 0$ and applying (18) to the evaluation of the spectral function.

The running mass expression corresponding to (19) can be written in the form

$$m(Q) = \hat{m} \left(\frac{Q^2/\Lambda^2 - 1}{Q^2/\Lambda^2 \ln(Q^2/\Lambda^2)} \right)^{\gamma_0/2\beta_0}, \tag{20}$$

where in the $\overline{MS}$ scheme $\gamma_0 = 8$. Eq. (20) is obtained integrating the one loop renormalization group equation

$$\frac{Q}{m(Q)}\frac{dm(Q)}{dQ} = -\gamma_0 \frac{\alpha_s(Q)}{4\pi}, \qquad (21)$$

where (19) has been used and $\hat{m}$ denotes an integration constant. Notice that $m(Q)$ is singular for $Q \to 0$, contrary to $\alpha_s(Q)$.

Finally, if we replace the running mass (20) in (15) after maximizing we obtain a running constituent mass $m_P(Q)$ of the type reported in Fig. 1 d that can be used together with the running coupling constant (19) in the expression of the operator M^2 (see Sec. 3). [3]

5. CALCULATIONS AND RESULTS

The calculations we report in this paper follow a similar line to those of Ref. [1]. The general strategy for solving the eigenvalue equation for M^2 and the numerical treatment are basically the same.

We neglect spin-orbit terms, but include the hyperfine terms in U (see eq. (16)). We solve first the eigenvalue equation for $M_{stat} = M_0 + V_{stat}$ (see Eq. (17)) by the Rayleigh-Ritz method with an harmonic oscillator basis and then treat $M^2 - M_{stat}^2$ as a perturbation (up to the first order this is obviously equivalent to taking $m_B^2 = \langle M^2 \rangle$).

In the above general framework, in Fig. 2 we graphically report and compare with the data [10] three different type of results, corresponding to different choices for the strong coupling constant α_s, the string tension σ and the constituent masses.

Diamonds correspond to results already reported in [1]. A running coupling constant $\alpha_s(Q)$ was assumed equal to the one loop perturbative expression (18) frozen at the maximum value $\overline{\alpha}_s = 0.35$, with $N_f = 4$ and $\Lambda = 200$ MeV. Q was identified with $|\mathbf{k} - \mathbf{k}'|$ and σ was set equal to 0.2 GeV^2. Fixed masses $m_u = m_d = 10$ MeV, $m_s = 200$ MeV, $m_c = 1.394$ GeV, $m_b = 4.763$ GeV were adopted. The results do not differ essentially from the fixed coupling constant case; the spectrum is reasonably well reproduced on the whole with the exception of the light pseudoscalar mesons π, η_s and K (the η_s mass is derived from the masses of η and η' with the usual assumptions).

Circlets correspond to results of the type reported in [2], but in which the hyperfine separation for the 1S and 2S states has been evaluated up to the second order of perturbation theory. In this case as running coupling constant we have taken the Shirkov-Solovtsov expression (19). We have set again $N_f = 4$ and $\Lambda = 200$ MeV, but $\sigma = 0.18$ GeV^2 and $m_s = 0.39$ GeV, $m_c = 1.545$ GeV, $m_b = 4.898$ GeV. On the contrary for the light quarks we have taken a phenomenological running mass in terms of the modulus of the c.m. quark momentum, $m_u^2 = m_d^2 = 0.17|\mathbf{k}| - 0.025|\mathbf{k}|^2 + 0.15|\mathbf{k}|^4$ GeV^2

---

[3] At first sight it could seem strange that we should talk of a Q dependence for a quantity like the constituent mass that should have a definite physical value. The point is that we are using $m_P(Q)$ in the context of certain approximations and it is the accuracy of such approximations that depend on the scale Q.

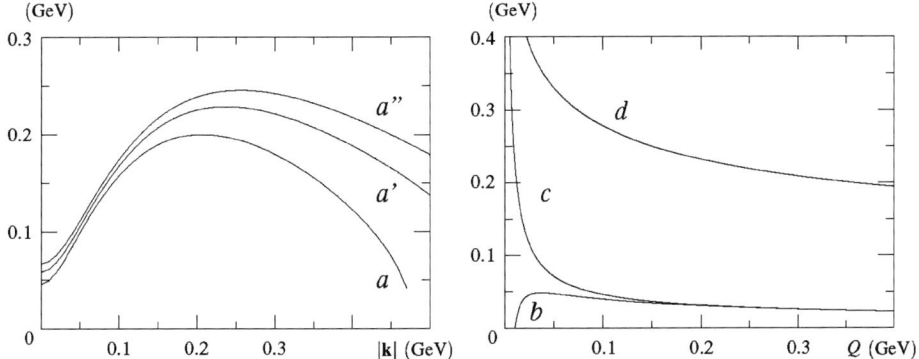

FIGURE 1. *a), a'), a")* $\bar{m}_P = \bar{m}_P(m, |\mathbf{k}|)$ for increasing value of m (16); *d)* running constituent mass $m_{\mathrm{p}}(Q)$; *c)* analytic running current mass (21); *b)* perturbative running pole masses corresponding to *c)*.

with $|\mathbf{k}|$ in GeV. We can see that in this way even the light pseudoscalar mesons turn out correctly, with possibly some problems for the $q\bar{b}$ states and some other highly excited states for which coupling with other channels are probably important.

Squares correspond to preliminary completely new calculations, made using the analytic running constant (19), running constituent masses $m_{\mathrm{p}}(Q)$ (as described in Sec. 4) for the light and strange quarks, fixed masses for the charm and the beauty quarks. Inside $\alpha_{\mathrm{s}}(Q)$ the quantity Q has been again identified with $|\mathbf{k} - \mathbf{k}'|$. On the contrary, for computational difficulties, inside $m_{\mathrm{p}}(Q)$ we have taken

$$Q = \frac{1}{e^{\gamma_{\mathrm{E}}} \langle r \rangle}, \tag{22}$$

where γ_{E} is the Euler constant $\gamma_{\mathrm{E}} = 0.5772\ldots$ and $\langle r \rangle$ is the radius of the unperturbed bound state [11]. We have chosen $N_f = 3$, $\Lambda = 180$ MeV, $\sigma = 0.18\,\mathrm{GeV}^2$, $\frac{\sigma}{\pi}B_\mu = 0.48$ GeV in (15), both for the light and the strange quarks, and then $\hat{m}_u = \hat{m}_d = 25.0$ MeV, $\hat{m}_s = 87.3$ MeV in (20) (in order to reproduce correctly the ρ and the ϕ masses). Finally we have taken $m_c = 1.508$ GeV and $m_b = 4.842$ for c and b quarks. The results are not of a better quality than those obtained in the preceding calculation but obviously conceptually more satisfactory.

As an example, in the table numerical values for the three types of calculations are reported in the order for the light-light channel. For the third case in the last column the pertinent values of running constituent light quark mass are also reported for the various states.

6. CONCLUSIONS

In conclusion we can confirm what already noticed in references [2] that our reduced second order formalism together with ansatz (5) can reproduce reasonably well the general structure of the entire meson spectrum, light-light, heavy-heavy and light-heavy sectors included. In order to obtain the masses of light pseudo scalar mesons π, η_s and

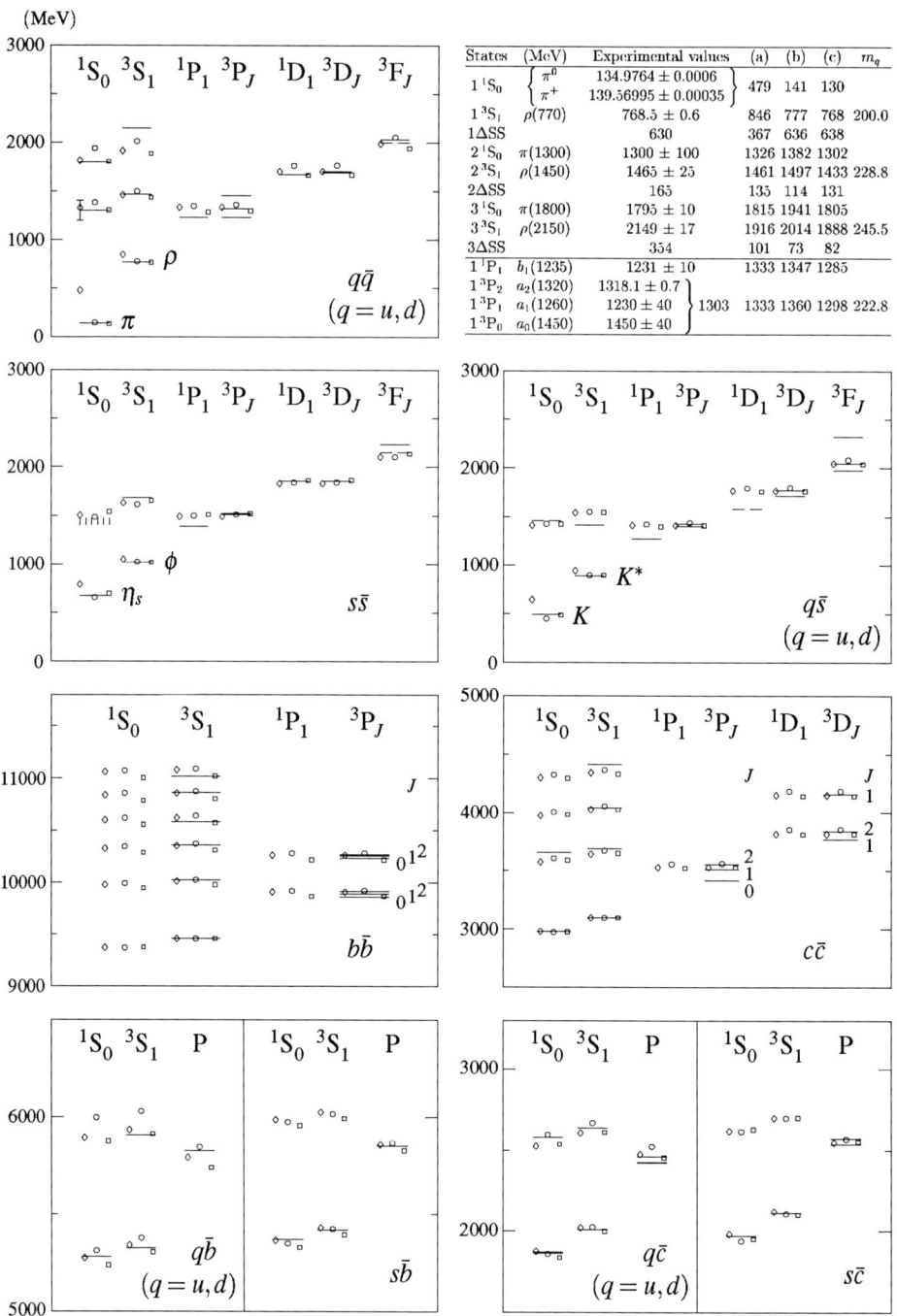

States	(MeV)	Experimental values	(a)	(b)	(c)	m_q
$1\,{}^1S_0$	π^0	134.9764 ± 0.0006	479	141	130	
	π^+	139.56995 ± 0.00035				
$1\,{}^3S_1$	$\rho(770)$	768.5 ± 0.6	846	777	768	200.0
$1\Delta SS$		630	367	636	638	
$2\,{}^1S_0$	$\pi(1300)$	1300 ± 100	1326	1382	1302	
$2\,{}^3S_1$	$\rho(1450)$	1465 ± 25	1461	1497	1433	228.8
$2\Delta SS$		165	135	114	131	
$3\,{}^1S_0$	$\pi(1800)$	1795 ± 10	1815	1941	1805	
$3\,{}^3S_1$	$\rho(2150)$	2149 ± 17	1916	2014	1888	245.5
$3\Delta SS$		354	101	73	82	
$1\,{}^1P_1$	$b_1(1235)$	1231 ± 10	1333	1347	1285	
$1\,{}^3P_2$	$a_2(1320)$	1318.1 ± 0.7	1303	1333	1360	1298	222.8
$1\,{}^3P_1$	$a_1(1260)$	1230 ± 40				
$1\,{}^3P_0$	$a_0(1450)$	1450 ± 40				

FIGURE 2. Quarkonium spectrum (lines represent experimental data).

K, however, a correct consideration of the infrared behavior of the running coupling constant and of some kinds of running constituent mass for the light quarks is essential. The analytic Shirkov-Solovtsov coupling constant seems to provide such a behavior.

What is new in this paper is the inclusion of second order perturbative corrections to the hyperfine splitting in the case of phenomenological running masses considered in [2] (circlets in Fig. 2) and the use of a running mass obtained combining renormalization group and analyticity requirements with an approximate solution of the quark Dyson-Schwinger equation (squares in Fig. 2).

ACKNOWLEDGMENTS

We are indebted to C. Simolo for various calculations on the running masses.

REFERENCES

1. M. Baldicchi, G.M. Prosperi, *Phys. Rev.* **D 62** (2000) 114024; *Fizika* **B 8** (1999) 2, 251; *Phys. Lett.* **B 436** (1998) 145.
2. M. Baldicchi and G. M. Prosperi, *Phys. Rev.* **D 66** (2002) 074008; *Color confinement and hadrons Quantum Chromodynamics*, Page. 183, H. Suganuma, *et al.* eds. World Scientific 2004, hep-ph/0310213.
3. N. Brambilla, E. Montaldi, G.M. Prosperi, *Phys. Rev.* **D 54** (1996) 3506; G.M. Prosperi, *Problems of Quantum Theory of Fields*, Pag. 381, B.M. Barbashov, G.V. Efimov, A.V. Efremov Eds. JINR Dubna 1999, hep-ph/9906237.
4. For a different approach to the relativistic bound state problem see : Yu.S. Kalashnikova, A.V. Nefediev, Yu.A. Simonov, *Phys. Rev.* **D 64** (2001) 014037; and references therein.
5. A.C. Mattingly, P.M. Stevenson, *Phys. Rev.* **D 49** (1994) 437; S. J. Brodsky, hep-ph/0412101; S. J. Brodsky, S. Menke, C. Merino, *Phys. Rev.* **D 67** (2003), 055008; P. Boucaud *et al.*, JHEP **04** (2000) 006; A. Ringwald and F. Schrempp *Phys. Lett.* **B 459** (1999) 24; Yu.L. Dokshitzer, A. Lucenti, G. Marchesini, G.P. Salam, *JHEP* 9805 (1998) 003; Yu.L. Dokshitzer, V.A. Khoze, S.I. Troyan, *Phys. Rev.* **D 53** (1996) 89; N.V. Krasnikov, A.A. Pivovarov, *Phys. Atom. Nucl.* **64** (2001) 1500; G. Grunberg, *Phys. Rev.* **D 46** (1992) 2228; A.I. Sanda *Phys. Rev. Lett.* **42** (1979) 1658.
6. D.V. Shirkov, I.L. Solovtsov, *Phys. Rev. Lett.* **79** (1997) 1209; *Theor. Math. Phys.* **120** (1999) 1220; D.V. Shirkov, hep-ph/0408272; A.I. Alekseev, B.A. Arbuzov, *Mod. Phys. Lett.* **A 13** (1998) 1747; A.I. Alekseev, *Few Body Syst.* **32** (2003) 193; see also A. V. Nesterenko, *Int. Journ. Mod. Phys.* **A 18** (2003) 5475.
7. M.B. Hecht, C.D. Roberts, S.M. Schmidt, *Phys. Rev.* **C 63** (2001) 025213, and references therein.
8. A. Barchielli, E. Montaldi, G.M. Prosperi, *Nucl. Phys.* **B 296** (1988) 625; Erratum-ibid. **B 303** (1988) 752; A. Barchielli, N. Brambilla, G.M. Prosperi, *Il Nuovo Cimento* **103 A** (1990) 59; N. Brambilla, P. Consoli, G.M. Prosperi, *Phys. Rev.* **D 50** (1994) 5878.
9. I.F. Ginzburg, D.V. Shirkov, *Sov. Phys. JEPT* **22** (1966) 234.
10. The Review of Particle Physics, S. Eidelman *et al.*, *Phys. Lett.* **B 592** (2004) 1.
11. W. Lucha, F. Schöberl, D. Gromes, *Phys. Rep.* **200** (1991) 127.

Relations between the Gribov-horizon and center-vortex confinement scenarios[1]

Jeff Greensite*, Štefan Olejník† and Daniel Zwanziger**

*Physics and Astronomy Dept., San Francisco State University, San Francisco, CA 94117, USA
†Institute of Physics, Slovak Academy of Sciences, SK–845 11 Bratislava, Slovakia
**Physics Department, New York University, New York, NY 10003, USA

Abstract. We review numerical evidence on connections between the center-vortex and Gribov-horizon confinement scenarios.

> So oft in theologic wars,
> The disputants, I ween,
> Rail on in utter ignorance
> Of what each other mean,
> And prate about an Elephant
> Not one of them has seen!
>
> *John Godfrey Saxe* (1816-1887) [1]

INTRODUCTION

Once upon a time, several blind men from a Hindustan village touched different parts of an elephant's body and, judging from their perceptions, quarreled about what the whole creature might look like. In a similar way, we explore color confinement by isolating different aspects of the phenomenon, and build our limited experience into various model schemes. The moral from the ancient parable teaches us that it is vitally important to unify all different views before we can appreciate the whole beauty of the beast.

The aspirations of this contribution are more modest. I will concentrate on common points of two seemingly unrelated pictures of color confinement. The former assumes that confinement arises due to the condensation of a particular type of topological excitations, so called center vortices (for a review see [2]), in the QCD vacuum; while the latter is the Gribov-horizon scenario in Coulomb gauge, which has been advocated by Gribov [3] and Zwanziger [4].

I will first briefly introduce the idea of the Gribov-horizon scenario, then formulate a simple criterion of confinement for static color charges in Coulomb gauge, discuss how the fulfillment of this criterion depends on presence/absence of center vortices, and finally present results for the Coulomb energy of a pair of static charges. I will present a subset of our numerical results, details, as well as some analytic insights on

---

[1] Plenary talk presented by Š. Olejník.

CP756, *Quark Confinement and the Hadron Spectrum VI*
edited by N. Brambilla, U. D'Alesio, A. Devoto, K. Maung, G.M. Prosperi and S. Serci
© 2005 American Institute of Physics 0-7354-0241-8/05/$22.50

the connections between center vortices and the Gribov horizon, can be found in recent publications [5, 6, 7].

CONFINEMENT SCENARIO IN COULOMB GAUGE

In Coulomb gauge, the Hamiltonian of QCD has the following form [8]:

$$H = H_{glue} + H_{coul}, \tag{1}$$

$$H_{glue} = \frac{1}{2} \int d^3\mathbf{x} \left(\mathscr{J}^{-1/2} \mathbf{E}^a \mathscr{J} \cdot \mathbf{E}^a \mathscr{J}^{-1/2} + \mathbf{B}^a \cdot \mathbf{B}^a \right), \tag{2}$$

$$H_{coul} = \frac{1}{2} \int d^3\mathbf{x} d^3\mathbf{y} \, \mathscr{J}^{-1/2} \rho^a(x) \mathscr{J} K^{ab}(x,y;A) \rho^b(y) \mathscr{J}^{-1/2}, \tag{3}$$

$$\rho^a = \rho^a_{matter} - g f^{abc} \mathbf{A}^b \cdot \mathbf{E}^c. \tag{4}$$

A prominent role in this expression is played by the Faddeev–Popov operator

$$M(A) \equiv -\nabla \cdot \mathscr{D}(A), \qquad \text{where} \qquad \mathscr{D}^{ac}_i(A) = \partial_i \delta^{ac} + f^{abc} A^b_i(x), \tag{5}$$

which enters both the interaction kernel K and the Jacobian factor $\mathscr{J}$

$$K^{ab}(x,y;A) \equiv \left[M^{-1} \left(-\nabla^2 \right) M^{-1} \right]^{a,b}_{x,y}, \qquad \mathscr{J} \equiv \det \left[M(A) \right]. \tag{6}$$

It is well-known since the seminal paper of Gribov [3], that Coulomb-gauge fixing in a non-abelian theory is a difficult problem. The transversality condition $\nabla \cdot \mathbf{A} = 0$ does not fix the gauge completely. Gribov [3] suggested to restrict integration over configurations to the so-called Gribov region (GR), defined as the subspace of transverse gauge fields for which the Faddeev–Popov operator is positive, and which are therefore local minima of the functional

$$I[\mathbf{A},g] = \int dx \left[{}^g \mathbf{A}^a(x) \right]^2, \qquad \text{where} \qquad {}^g A_i = g^{-1} A_i g + g^{-1} \partial_i g. \tag{7}$$

The boundary of this region is called the Gribov horizon.

However, there exist gauge orbits which intersect the Gribov region more than once; the next step [9] then is to restrict fields to the fundamental modular region (FMR), the set of absolute minima of the functional (7). Both the GR and the FMR are bounded in every direction and convex.

The essence of the Gribov-horizon confinement scenario can be phrased in a simple way: The dimension of gauge-field configuration space is huge, so it is reasonable to expect that most configurations are located close to its boundary (the horizon; in a similar way, the volume measure $r^{d-1} dr$ of a d-dimensional sphere is peaked at its surface). The interaction kernel K, which determines the interaction energy of static color sources, contains the inverse of the Faddeev–Popov operator, which is strictly zero on the horizon and near-zero close to the horizon. A high density of configurations near the horizon can thus lead to a strong enhancement of the Coulomb interaction energy, and hopefully cause color confinement.

A CONFINEMENT CONDITION IN TERMS OF
FADDEEV–POPOV EIGENSTATES

Let us consider a single static color charge, which can be written in Coulomb gauge as

$$\Psi_C^\alpha[A;x] = \psi^\alpha(x)\Psi_0[A],\qquad(8)$$

where α is the color index for a point charge in color group representation r, and Ψ_0 is the Coulomb-gauge ground state. The excitation energy of this state, above the ground state, is given by

$$\mathcal{E}_r = \frac{\langle\Psi_C^\alpha|H_{coul}|\Psi_C^\alpha\rangle}{\langle\Psi_C^\alpha|\Psi_C^\alpha\rangle} - \langle\Psi_0|H_{coul}|\Psi_0\rangle,\qquad(9)$$

and is proportional to

$$\mathcal{E} = \frac{1}{N^2-1}\langle K^{aa}(x,x;A)\rangle.\qquad(10)$$

The quantity $\mathcal{E}_r$ is the color Coulomb self-energy of unscreened color charge and is expected to be both ultraviolet and infrared divergent in a confining theory. The UV divergence can be regulated by a lattice cut-off, however, the quantity must still be divergent at infinite volume, even after lattice regularization, due to IR effects.

On a lattice, one can express (10) through eigenstates of the Faddeev–Popov operator

$$\sum_{b,y}M_{xy}^{ab}\,\phi_y^{(n)b} = \lambda_n\phi_x^{(n)a}\qquad(11)$$

simply as (assuming that M is invertible, i.e. excluding its zero modes)

$$\mathcal{E} = \frac{1}{3V_3}\sum_n\left\langle\frac{F_n}{\lambda_n^2}\right\rangle,\qquad\text{where}\qquad F_n \equiv \sum_{a,xy}\phi_x^{(n)a}(-\nabla^2)_{xy}\phi_y^{(n)a*},\qquad(12)$$

and V_3 is the lattice 3-volume. In SU(2) lattice gauge theory, the link variables can be expressed as

$$U_\mu(x) = b_\mu(x) + i\sigma^c a_\mu^c(x),\qquad b_\mu(x)^2 + \sum_c a_\mu^c(x)^2 = 1,\qquad(13)$$

and the lattice Faddeev–Popov operator

$$\begin{aligned}M_{xy}^{ab} =\ & \delta^{ab}\sum_k\left\{\delta_{xy}\left[b_k(x)+b_k(x-\hat{k})\right] - \delta_{x,y-\hat{k}}b_k(x) - \delta_{y,x-\hat{k}}b_k(y)\right\}\\ & - \varepsilon^{abc}\sum_k\left\{\delta_{x,y-\hat{k}}a_k^c(x) - \delta_{y,x-\hat{k}}a_k^c(y)\right\}\end{aligned}\qquad(14)$$

(x,y denote lattice sites at fixed time) is a $3V_3 \times 3V_3$ sparse matrix with $3V_3$ linearly independent eigenstates. If we denote $N(\lambda,\lambda+\Delta\lambda)$ the number of eigenvalues in the range between λ and $\lambda+\Delta\lambda$, we can introduce, on a large lattice, the density of states $\rho(\lambda) \equiv N(\lambda,\lambda+\Delta\lambda)/(3V_3\Delta\lambda)$. Then, as $V_3 \to \infty$,

$$\mathcal{E} = \int_0^{\lambda_{max}}\frac{d\lambda}{\lambda^2}\,\langle\rho(\lambda)F(\lambda)\rangle.\qquad(15)$$

A (necessary) condition for confinement can now be formulated: *The excitation energy $\mathcal{E}_r$ of a static, unscreened color charge is divergent if, at infinite volume,*

$$\lim_{\lambda \to 0} \frac{\langle \rho(\lambda) F(\lambda) \rangle}{\lambda} > 0. \qquad (16)$$

CENTER VORTICES AND THE CONFINEMENT CONDITION

It is interesting to check whether the above condition is fulfilled in various ensembles of lattice configurations. First, at zero-th order in the gauge coupling, the Faddeev–Popov operator is simply a lattice Laplacian and its eigenstates are just plane waves. One can easily verify that, in this case,

$$\rho(\lambda) = \frac{1}{4\pi^2}\sqrt{\lambda}, \qquad F(\lambda) = \lambda, \qquad \mathcal{E} = \frac{1}{2\pi^2}\sqrt{\lambda_{max}}, \qquad (17)$$

the excitation energy is IR finite and the confinement criterion is not met. One needs some mechanism of enhancement of $\rho(\lambda)$ and $F(\lambda)$ in the region of small λ values. I will argue that such an enhancement exists in full lattice configurations, and is provided by center vortices.

Center vortices are identified by fixing to an adjoint gauge, and then projecting link variables to the Z_N subgroup of SU(N) [10]. The excitations of the projected theory are known as P-vortices. In the direct maximal center gauge (DMCG) in SU(2) [11] one fixes to the maximum of

$$\mathcal{R}_{MCG} = \sum_{x,\mu} \left| \tfrac{1}{2} \mathrm{Tr}[U_\mu(x)] \right|^2, \qquad (18)$$

and center projects by

$$U_\mu(x) \longrightarrow Z_\mu(x) = \mathrm{sign}\, \mathrm{Tr}[U_\mu(x)]. \qquad (19)$$

A lot of evidence has been accumulated that center vortices alone reproduce much of confinement physics, for a review see [2].

We have determined the density of states $\rho(\lambda)$ and the mean value of the lattice Laplacian $F(\lambda)$ in three ensembles of lattice configurations:

1. full configurations, $\{U_\mu(x)\}$,
2. "vortex-only" configurations, $\{Z_\mu(x) = \mathrm{sign}\, \mathrm{Tr}[U_\mu(x)]\}$, and
3. "vortex-removed" configurations, $\{U_\mu^{(R)}(x) = Z_\mu^\dagger(x)U_\mu(x)\}$.

The procedure of vortex removal, first introduced by de Forcrand and D'Elia [12], is known to remove the string tension, eliminate chiral symmetry breaking, and send the topological charge of lattice configurations to zero.

Each of the three ensembles was brought to Coulomb gauge by maximizing, on each time-slice,

$$\mathcal{R}_{coul}(t) = \sum_{\mathbf{x}} \sum_{k=1}^{3} \tfrac{1}{2}\mathrm{Tr}[U_k(\mathbf{x},t)]. \qquad (20)$$

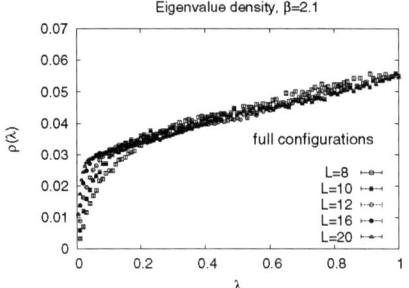
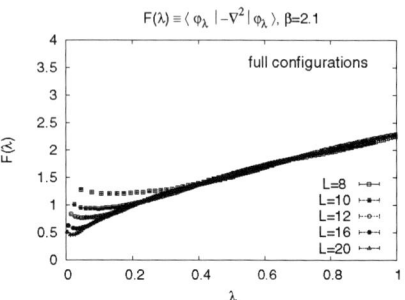

FIGURE 1. The density $\rho(\lambda)$ and $F(\lambda)$ for full lattice configurations.

Our results for full configurations at $\beta = 2.1$ are exemplified in Figure 1, for a variety of lattice volumes. One can observe a sharp "bend" near $\lambda \to 0$. Both quantities behave near 0 like λ^p, λ^q, with p, q small numbers. A scaling analysis along the lines of random matrix theory (see Appendix B in [7] for details and references) gives the estimates

$$\rho(\lambda) \sim \lambda^{0.25}, \qquad F(\lambda) \sim \lambda^{0.38}, \tag{21}$$

with a subjective error of about 20% in the exponents. The Coulomb gauge confinement condition, Eq. (16), is thus clearly fulfilled. This result provides a confirmation of the mechanism envisaged in the Gribov-horizon scenario.

To pinpoint the mechanism which might be responsible for the enhancement of eigenvalues near the horizon, we considered the Faddeev–Popov observables in the vortex-only configurations. Our data are displayed in Figure 2. The enhancement near 0 is even more pronounced; it appears that both quantities converge to a non-zero value in the infinite volume limit

$$\rho(0) \sim 0.06, \qquad F(0) \sim 1.0, \tag{22}$$

which is confirmed also by a RM scaling analysis. Once again, the confinement criterion (16) is obviously satisfied.

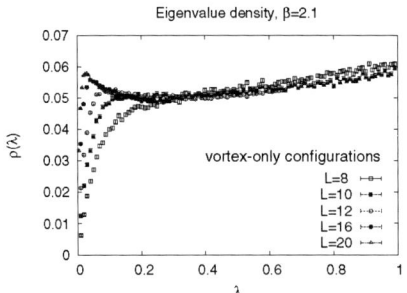
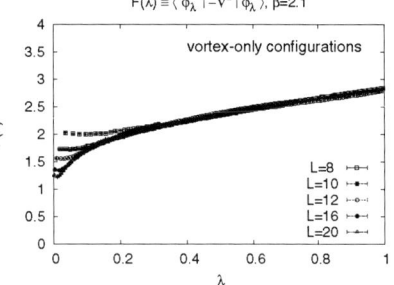

FIGURE 2. The density $\rho(\lambda)$ and $F(\lambda)$ for vortex-only configurations.

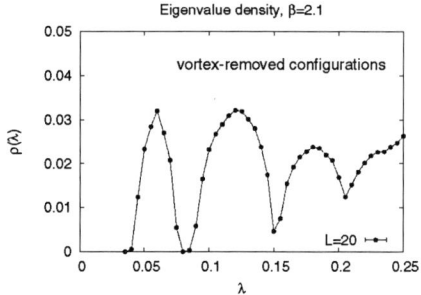
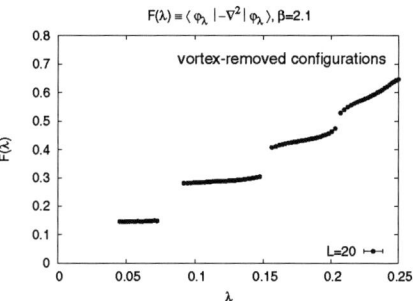

FIGURE 3. The density $\rho(\lambda)$ and $F(\lambda)$ for vortex-removed configurations.

Finally, results for "vortex-removed" configurations are shown in Figure 3, for the largest available, 20^4 lattice volume. The eigenvalue spectrum is strikingly different in this case. The eigenvalue density consists of a series of distinct peaks, while values of $F(\lambda)$ are organized into separate bands, the lowest few of them clearly separated by gaps. This result can easily be understood by inspection of the eigenvalue spectrum of the Laplacian operator (equal to M at zero-th order in the coupling). The eigenvalue density, at finite volume, is a sum of delta-functions, and each eigenvalue λ_k is multiply (N_k-times) degenerate. The quantity $F(\lambda_k)$ is equal to λ_k. Now if one compares the degeneracy N_k of the k-th eigenvalue with the number of eigenvalues inside the k-th "band" of $F(\lambda)$ (the right panel of Figure 3), one finds a precise match. This can be simply interpreted: the vortex-removed configuration is just a small perturbation of the zero-field limit $U_\mu = \mathbf{1}$. This perturbation lifts the degeneracy of degenerate eigenvalues and spreads them into bands of finite width. The estimate of the Coulomb self-energy in this case indicates that it remains IR finite in the infinite volume limit.

In pure gauge theory, we used the procedure of de Forcrand and D'Elia [12] to study effects of vortex removal on the density of Faddeev–Popov eigenvalues in the small λ region. However, in some lattice models, confining vortex configurations can be

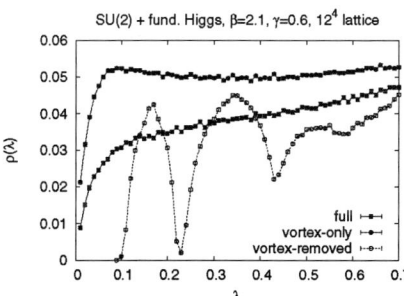
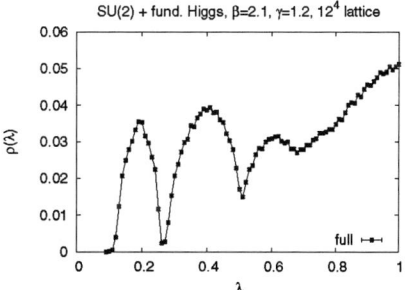

FIGURE 4. The density $\rho(\lambda)$ for a gauge-Higgs system in the "confined" phase (left) and the "Higgs" phase (right).

suppressed by changing the coupling constants. A prototype example is the gauge field coupled to a scalar field of unit modulus in the fundamental representation of the gauge group [13]. The action of the model is

$$S = S_W + \frac{\gamma}{2} \sum_{x,\mu} \text{Tr}[\phi^\dagger(x) U_\mu(x) \phi(x+\hat{\mu})], \tag{23}$$

where S_W is the usual Wilson gauge action and ϕ is an SU(2) group-valued field. In the strict sense, this theory is non-confining for all values of couplings β, γ, due to the well-known Osterwalder-Seiler–Fradkin-Shenker theorem [14, 15]. Even though there is no thermodynamic phase transition separating the pseudo-Higgs phase from the pseudo-confinement phase of the model, there is a symmetry breaking transition between the phases (a "Kertész" line), and center vortices percolate in the "confinement" phase, and cease to percolate in the "Higgs" phase [6, 16, 17, 18]. By varying γ at fixed gauge coupling β, one can modify the vortex content and study its effects on Faddeev–Popov eigenvalues.

In Figure 4 I show the eigenvalue density at $\beta = 2.1$ for two values of gauge-Higgs coupling: $\gamma = 0.6$, deep in the "confinement" phase, and $\gamma = 1.2$ corresponding to the "Higgs" phase. In the former, the densities for full, vortex-only, and vortex-removed configurations clearly resemble those in pure gauge theory, while the density in the Higgs phase for full configurations looks almost identical to the vortex-removed data in the confinement phase.

COULOMB ENERGY

We have seen that the Coulomb self-energy of a color non-singlet state is IR divergent, due to the enhanced density of Faddeev–Popov eigenvalues near zero. Another question is whether the color Coulomb potential of a charge-anticharge pair grows linearly and, if so, whether it is also sensitive to the presence/absence of center vortices. This question was addressed in Ref. [5], and I will summarize here briefly the main results.

Let $|\Psi_{q\bar{q}}\rangle = \bar{q}^a(0) q^a(R) |\Psi_0\rangle$ denote a physical heavy static quark-antiquark state in Coulomb gauge. Then

$$\mathscr{E}_{q\bar{q}} = \langle \Psi_{q\bar{q}} | H | \Psi_{q\bar{q}} \rangle - \langle \Psi_0 | H | \Psi_0 \rangle = E_{se} + V_{coul}(R) \tag{24}$$

is a sum of self-energy contributions and the R-dependent color Coulomb potential. It can be computed from the correlator of two timelike Wilson lines in Coulomb gauge:

$$G(R,T) = \langle \tfrac{1}{2} \text{Tr}[L^\dagger(\mathbf{x},T) L(\mathbf{y},T)] \rangle \Big|_{R=|\mathbf{x}-\mathbf{y}|} \; ; \qquad L(\mathbf{x},T) - \exp\left[i \int_0^T dt \, A_0(\mathbf{x},t) \right]. \tag{25}$$

It is easy to show that

$$\mathscr{E} = E_{se} + V_{coul}(R) = \lim_{T \to 0} V(R,T); \qquad \mathscr{E}_{min} = E'_{se} + V(R) = \lim_{T \to \infty} V(R,T), \tag{26}$$

where

$$V(R,T) = -\tfrac{\partial}{\partial T} \log[G(R,T)]. \tag{27}$$

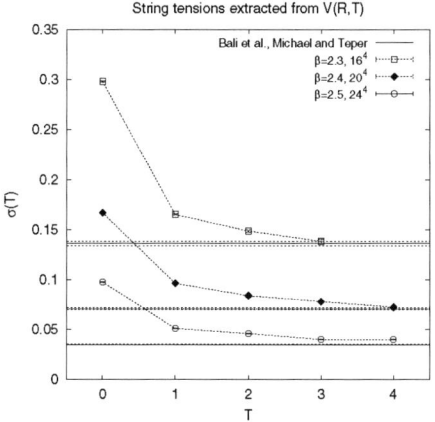

FIGURE 5. $\sigma(T)$ vs. T. Solid lines indicate the accepted values of the asymptotic string tension at each β value, with dashed lines indicating the error bars.

Above, $\mathcal{E}_{min}$ is the minimal energy of a state containing two static charges, and $V(R)$ is the static interquark potential. Since $\mathcal{E} > \mathcal{E}_{min}$, it is clear that if $V(R)$ is confining, then so is $V_{coul}(R)$, and $V(R)$ is bounded from above by $V_{coul}(R)$, as first proven by Zwanziger [19].

On a lattice we introduce

$$L(\mathbf{x}, T) = U_0(\mathbf{x}, a) U_0(\mathbf{x}, 2a) \dots U_0(\mathbf{x}, T), \tag{28}$$

and

$$V(R, T) = -\tfrac{1}{a} \log \left[\frac{G(R, T+a)}{G(R, T)} \right], \quad V(R, 0) = -\tfrac{1}{a} \log[G(R, a)], \tag{29}$$

from which we obtain an estimate of $V_{coul}(R)$ (exact in the continuum limit).

Our results, for SU(2), are shown in Figures 5 and 6.[2] The former figure represents a consistency check of our procedure. It verifies that the string tension $\sigma(T)$ extracted form $V(R, T)$ approaches the accepted value of the asymptotic string tension at large T. In Figure 6 we plot the potential $V(R, T)$ for $T = 0$ (our estimate of the color Coulomb potential) and for $T = 4$ (approaching the static quark-antiquark potential $V(R)$). Upper lines of data points in both figures clearly demonstrate that both $V_{coul}(R)$ and $V(R)$ rise linearly with distance. However, the slope of this linear rise is larger in the color Coulomb potential, $\sigma_{coul} \approx 3\sigma$.[3] The fact that the color Coulomb potential overconfines does not contradict Zwanziger's bound [19], nor is really surprising. There is no reason

---

[2] First preliminary results of the determination of the color Coulomb potential in SU(3) lattice gauge theory were presented by Nakamura [20] at this conference.

[3] The color Coulomb potential was measured using different methods by Cucchieri and Zwanziger [21], and Langfeld and Moyaerts [22], with lower values for the ratio σ_{coul}/σ. The origin of this discrepancy has not been clarified yet.

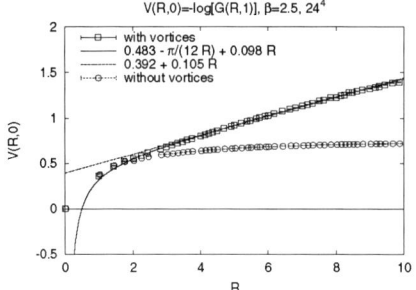

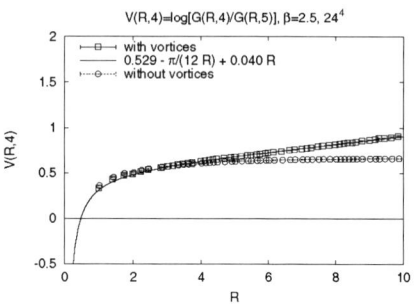

FIGURE 6. $V(R,0)$ and $V(R,4)$ at $\beta = 2.5$. The former tends to the color Coulomb potential in the continuum limit, while the latter approximates the static asymptotic interquark potential.

to believe that the quark-antiquark state in the Coulomb gauge is the true QCD flux tube state, the minimal energy state containing a static quark and antiquark. The Coulombic force can be lowered to the true asymptotic one e.g. by constituent gluons present in the QCD flux tube, as was suggested in the gluon chain model of Greensite and Thorn [23].

The effect of vortex removal is illustrated in lower lines of data points of Figure 6. Removing center vortices also removes the confining property of the color Coulomb potential. Since the potential, in Coulomb gauge, is sensitive to the interaction kernel K in the Hamiltonian (1), and the kernel in turn to the density of near-zero eigenvalues of the Faddeev–Popov operator, this result again confirms the observation of the previous section: Removal of center vortices alters the density quite drastically.

We also note that the confining property of the color Coulomb potential is tied to the unbroken realization of a remnant global gauge symmetry in Coulomb gauge. This connection was studied in detail in Ref. [6]. It was demonstrated there on a few examples (deconfined phase in pure gauge theory, pseudo-confinement phase of the gauge–fundamental-Higgs theory) that confinement in the color Coulomb potential is not identical to confinement in the static interquark potential, and center symmetry breaking, spontaneous or explicit, does not necessarily imply remnant symmetry breaking.

CONCLUSIONS

Results of our numerical simulations suggest an appealing picture: The low-lying eigenvalues of the Faddeev–Popov operator in Coulomb gauge tend towards zero as the lattice volume increases. The density of the eigenvalues goes as a small power of λ, and this, together with a similar behavior of the average Laplacian, $F(\lambda)$, assures the infrared divergence of the energy of an unscreened color charge. Also, due to the enhancement of near-zero modes of the Faddeev–Popov operator, the Coulomb energy of a pair of color charges rises linearly with their separation. Both facts support the ideas of the Gribov-horizon confinement scenario.

The constant density of low-lying eigenvalues can be attributed to the vortex component of gauge-field configurations. A thermalized configuration in a pure gauge theory factors into a confining piece (the vortex-only part), and a piece which closely resembles the lattice of a gauge–Higgs theory in the Higgs phase (the vortex-removed configuration). This establishes a firm connection between the center-vortex picture and the Gribov-horizon scenario. This connection is exemplified also by the fact that vortex removal removes the color Coulomb string tension of the color Coulomb potential. It is also consistent with recent investigations of Gattnar et al. [24] in Landau gauge.

In this talk, I covered results of our numerical investigations. Related analytical developments were omitted: (i) thin center vortices lie on the Gribov horizon; (ii) the Gribov horizon is a convex manifold in lattice configuration space and thin center vortices are conical singularities on that manifold; (iii) the Coulomb gauge is an attractive fixed point of a more general gauge condition, interpolating between the Coulomb and Landau gauges. Interested readers are invited to find them in our recent publications [6, 7].

ACKNOWLEDGMENTS

Our research is supported in part by the U.S. Department of Energy under Grant No. DE-FG03-92ER40711 (J.G.), the Slovak Grant Agency for Science, Grant No. 2/3106/2003 (Š.O.), and the National Science Foundation, Grant No. PHY-0099393 (D.Z.). Š.O. is grateful to the organizers of the conference for invitation to present this talk and for creating a stimulating, yet relaxed atmosphere.

REFERENCES

1. Saxe, J. G. (1816–1887), *The Blind Men and the Elephant*.
2. Greensite, J., *Prog. Part. Nucl. Phys.*, **51**, 1 (2003), hep-lat/0301023.
3. Gribov, V. N., *Nucl. Phys.*, **B139**, 1 (1978).
4. Zwanziger, D., *Nucl. Phys.*, **B518**, 237 (1998).
5. Greensite, J., and Olejník, Š., *Phys. Rev.*, **D67**, 094503 (2003), hep-lat/0302018.
6. Greensite, J., Olejník, Š., and Zwanziger, D., *Phys. Rev.*, **D69**, 074506 (2004), hep-lat/0401003.
7. Greensite, J., Olejník, Š., and Zwanziger, D. (2004), hep-lat/0407032.
8. Christ, N. H., and Lee, T. D., *Phys. Rev.*, **D22**, 939 (1980).
9. Dell'Antonio, G., and Zwanziger, D., *Commun. Math. Phys.*, **138**, 291 (1991).
10. Del Debbio, L., Faber, M., Greensite, J., and Olejník, Š., *Phys. Rev.*, **D55**, 2298 (1997), hep-lat/9610005.
11. Del Debbio, L., Faber, M., Giedt, J., Greensite, J., and Olejník, Š., *Phys. Rev.*, **D58**, 094501 (1998), hep-lat/9801027.
12. de Forcrand, P., and D'Elia, M., *Phys. Rev. Lett.*, **82**, 4582 (1999), hep-lat/9901020.
13. Lang, C. B., Rebbi, C., and Virasoro, M., *Phys. Lett.*, **B104**, 294 (1981).
14. Osterwalder, K., and Seiler, E., *Ann. Phys.*, **110**, 440 (1978).
15. Fradkin, E. H., and Shenker, S. H., *Phys. Rev.*, **D19**, 3682 (1979).
16. Chernodub, M. N., Gubarev, F. V., Ilgenfritz, E. M., and Schiller, A., *Phys. Lett.*, **B434**, 83 (1998), hep-lat/9805016.
17. Langfeld, K. (2001), hep-lat/0109033.
18. Bertle, R., Faber, M., Greensite, J., and Olejník, Š., *Phys. Rev.*, **D69**, 014007 (2004), hep-lat/0310057.
19. Zwanziger, D., *Phys. Rev. Lett.*, **90**, 102001 (2003), hep-lat/0209105.
20. Nakamura, A. (2004), talk at this conference.
21. Cucchieri, A., and Zwanziger, D., *Nucl. Phys. Proc. Suppl.*, **119**, 727 (2003), hep-lat/0209068.
22. Langfeld, K., and Moyaerts, L., *Phys. Rev.*, **D70**, 074507 (2004), hep-lat/0406024.
23. Greensite, J., and Thorn, C. B., *JHEP*, **02**, 014 (2002).
24. Gattnar, J., Langfeld, K., and Reinhardt, H., *Phys. Rev. Lett.*, **93**, 061601 (2004), hep-lat/0403011.

The dual Meissner effect in SU(2) Landau gauge

Tsuneo Suzuki*, Katsuya Ishiguro†, Yoshihiro Mori† and Toru Sekido†

*Institute for Theoretical Physics, Kanazawa University, Kanazawa 920-1192, Japan
and
RIKEN, Radiation Laboratory, Wako 351-0158, Japan
†Institute for Theoretical Physics, Kanazawa University, Kanazawa 920-1192, Japan

Abstract. The dual Meissner effect is observed without monopoles in quenched $SU(2)$ QCD with Landau gauge-fixing. Abelian as well as non-Abelian electric fields are squeezed. Magnetic displacement currents which are time-dependent Abelian magnetic fields play a role of solenoidal currents squeezing Abelian electric fields. Monopoles are not always necessary to the dual Meissner effect. The squeezing of the electric flux means the dual London equation and the massiveness of the Abelian electric fields as an asymptotic field. The mass generation of the Abelian electric fields is related to a gluon condensate $< A_\mu^a A_\mu^a > \neq 0$ of mass dimension 2.

INTRODUCTION

To understand color confinement mechanism is still an unsolved important problem. The dual Meissner effect is believed to be the mechanism[1, 2]. However what causes the dual Meissner effect is not clarified. A possible candidate is magnetic monopoles which appear after projecting $SU(3)$ QCD to an Abelian $U(1)^2$ theory by a partial gauge fixing[3]. If such monopoles condense, color confinement could be understood due to the dual Meissner effect. Actually an Abelian projection adopting a special gauge called Maximally Abelian gauge (MA)[4, 5] leads us to interesting results[6, 7] which support the idea of monopoles after an Abelian projection. See Fig.1 in which the dual Meissner effect due to monopole currents as a solenoidal current is seen beautifully[8].

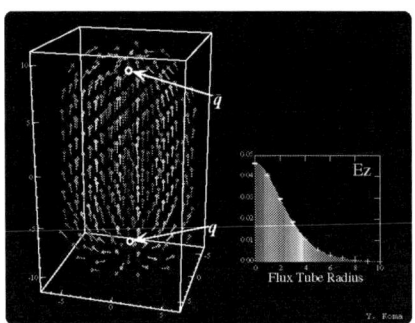 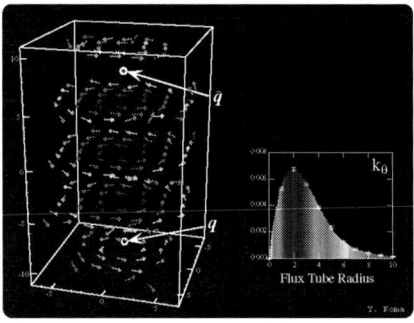

FIGURE 1. Abelian electric field flux and monopole currents in MA gauge[8].

Now a natural question arises. What happens in other general Abelian projections

CP756, *Quark Confinement and the Hadron Spectrum VI*
edited by N. Brambilla, U. D'Alesio, A. Devoto, K. Maung, G.M. Prosperi and S. Serci
© 2005 American Institute of Physics 0-7354-0241-8/05/$22.50

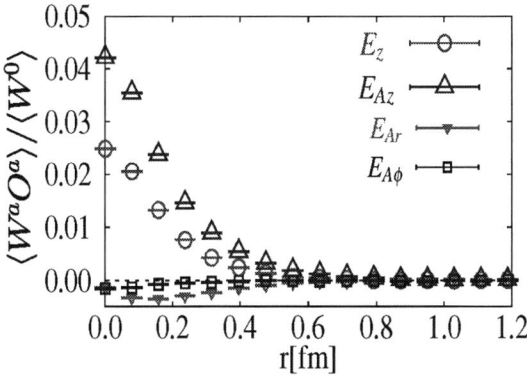

FIGURE 2. Abelian $\vec{E}_A$ and non-Abelian $\vec{E}$ electric field profiles in Landau gauge. $W(R \times T = 6 \times 6)$ is used.

or even without any Abelian projection? For example, consider an Abelian projection diagonalizing Polyakov loops. Monopoles exist in the continuum limit at a point where eigenvalues of Polyakov loops are degenerate[3]. But it is easy to show that such a point runs only in the time-like direction. This means only time-like monopoles which do not contribute to the string tension exist in the Abelian projection[9]. Discuss another simple case of Landau gauge. There vacuum configurations are so smooth and it is easy to check numerically that no monopoles coming from singularities exist. Without space-like monopoles or monopoles themselves, monopole condensation could not occur. We have to find another confinement mechanism.

In this note, we show that the dual Meissner effect in an Abelian sense works good even when monopoles do not exist, performing Monte-Carlo simulations of quenched $SU(2)$ QCD with Landau gauge fixing. Instead of monopoles, time-dependent Abelian magnetic fields regarded as magnetic displacement currents are squeezing Abelian electric fields. The dual Meissner effect leads us to the dual London equation and the mass generation of the Abelian electric fields which suggests the existence of a dimension 2 gluon condensate. Our present numerical results are not perfect, since the continuum limit, the infinite-volume limit and the gauge-independence are not studied yet. Moreover our discussions use Abelian components only on the basis of not yet clarified assumption that Abelian components are dominant in the infrared QCD (Abelian dominance[6, 7, 10, 11]). Nevertheless authors think the present results are very interesting to general readers, since they show for the first time the Abelian dual Meissner effect is working in lattice non-Abelian QCD without resorting to monopoles coming from a singular gauge transformation[12]. The gauge adopted here is the simplest one only to get smooth configurations. The present results hence suggest the Abelian dual Meissner effect is the real universal mechanism of color confinement which has been sought for many years. Moreover the relation of the Abelian dual Meissner effect with the dimension 2 gluon condensate sheds new light on the importance of the gluon condensate [13, 14, 15, 16, 17]. Detailed studies, a direct proof of Abelian dominance and extensions to $SU(3)$ and full QCD in zero and finite-temperature cases are in progress.

SIMULATION DETAILS

As a lattice quenched $SU(2)$ QCD, we use an improved gluonic action found by Iwasaki[18].

$$S = \beta \left\{ c_0 \sum Tr(plaquette) + c_1 \sum Tr(rectangular) \right\}, \tag{1}$$

with which better scaling behaviors of physical quantities are expected. The mixing parameters are fixed as $c_0 + 8c_1 = 1$ and $c_1 = -0.331$.

To measure correlations of gauge-variant electric and magnetic fields directly, we adopt a simplest gauge called Landau gauge which maximizes $\sum_{s,\mu} Tr[U_\mu(s) + U_\mu^\dagger(s)]$. After the gauge fixing, we try to measure electric and magnetic flux distributions by evaluating correlations of Wilson loops and field strengths. For comparison, we also use MA gauge where $\sum_{s,\mu} Tr[U_\mu(s)\sigma_3 U_\mu^\dagger(s)\sigma_3]$ is maximized. To get a good signal to noise ratio, the APE smearing technique[19] is used when evaluating Wilson loops $W(R,T) = W^0 + iW^a\sigma^a$:

$$U_i(s) \longrightarrow N \left\{ U_i(s) + \alpha \sum_{j \neq i} U_j(s)U_i(s+\hat{j})U_j^\dagger(n+\hat{i}) \right\}, \tag{2}$$

where N is normalization factor and α is a free parameter. We have used $\alpha = 0.2$ and $N = 80$.

Measurements of the string tension make us fix the scale when we use $\sqrt{\sigma} = 440$MeV. We adopt a coupling constant $\beta = 1.2$ in which the lattice distance $a(\beta = 1.2)$ is $0.07921(22)$[fm]. The scale is chosen only because we compare our results with those studied extensively in MA gauge[20, 8]. The lattice size is 32^4 and after 5000 thermalization, we have prepared 5000 thermalized configurations per each 100 sweeps for measurements.

Non-Abelian electric and magnetic fields are defined from 1×1 plaquette $U_{\mu\nu}(s) = U_{\mu\nu}^0 + iU_{\mu\nu}^a\sigma^a$ as done in Ref.[21]:

$$
\begin{aligned}
E_k^a(s) &\equiv \frac{1}{2}(U_{4k}^a(s-\hat{k}) + U_{4k}^a(s)) \\
B_k^a(s) &\equiv \frac{1}{8}\varepsilon_{klm}(U_{lm}^a(s-\hat{l}-\hat{m}) \\
&\quad + U_{lm}^a(s-\hat{l}) + U_{lm}^a(s-\hat{m}) + U_{lm}^a(s))
\end{aligned}
$$

We also define Abelian electric (E_{Ai}^a) and magnetic fields (B_{Ai}^a) similarly using Abelian plaquettes $\theta_{\mu\nu}^a(s)$ defined through link variables $\theta_\mu^a(s)$:

$$\theta_{\mu\nu}^a(s) \equiv \theta_\mu^a(s) + \theta_\nu^a(s+\hat{\mu}) - \theta_\mu^a(s+\hat{\nu}) - \theta_\nu^a(s) \tag{3}$$

where $\theta_\mu^a(s)$ is given by $U_\mu(s) = \exp(i\theta_\mu^a(s)\sigma^a)$. In MA gauge, the Abelian link variables $\theta_\mu^{MA}(s)$ are defined by a phase of the diagonal part of the non-Abelian link field.

$$U_\mu^0(s) = \sqrt{1 - |c_\mu(s)|^2}\cos\theta_\mu^{MA}(s),$$
$$U_\mu^3(s) = \sqrt{1 - |c_\mu(s)|^2}\sin\theta_\mu^{MA}(s).$$

Since the off-diagonal part $|c_\mu(s)|$ is small[6], $\theta_\mu^{MA}(s) \sim \theta_\mu^3(s)$ in MA gauge. As a source corresponding to a static quark and antiquark pair, we use here only non-Abelian Wilson loops.

RESULTS

First we show in Fig.2 Abelian and non-Abelian electric flux profiles around a pair of static quark and antiquark in Landau gauge. The profiles are mainly studied on the perpendicular plane at the midpoint between the quark pair. Note that electric fields perpendicular to the $Q\bar{Q}$ axis are found to be negligible. It is very interesting to see from Fig.2 that Abelian electric field E_{Az} simply defined here is squeezed also. Moreover the squeezing of E_{Az} is stronger than that of non-Abelian one E_z. To know how squeezing of the Abelian flux occurs seems hence essential.

Let us discuss from now on flux distributions of Abelian fields alone. It is checked numerically that there are no DeGrand-Toussaint monopoles[22]. See Fig.3 in which histograms of Abelian field strength $\theta_{\mu\nu}$ are plotted in Landau gauge (Left) and MA gauge (Right).

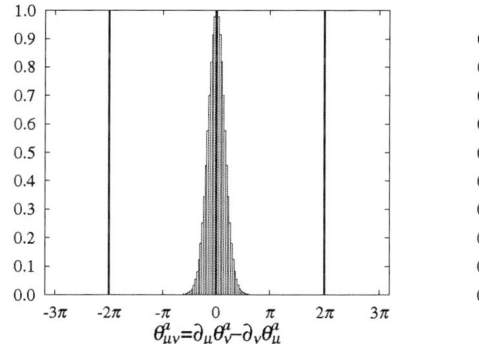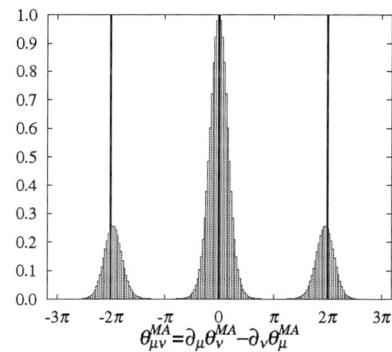

FIGURE 3. Histograms of $\theta_{\mu\nu}$ in Landau (Left) and MA (Right) gauges

Hence the Abelian fields satisfy kinematically the simple Abelian Bianchi identity:

$$\vec{\nabla} \times \vec{E}_A^a = \partial_4 \vec{B}_A^a, \qquad \vec{\nabla} \cdot \vec{B}_A^a = 0. \tag{4}$$

In the case of MA gauge, there are additional monopole current $(\vec{k}, k_4)$ contributions:

$$\vec{\nabla} \times \vec{E}^{MA} = \partial_4 \vec{B}^{MA} + \vec{k}, \qquad \vec{\nabla} \cdot \vec{B}^{MA} = k_4. \tag{5}$$

175

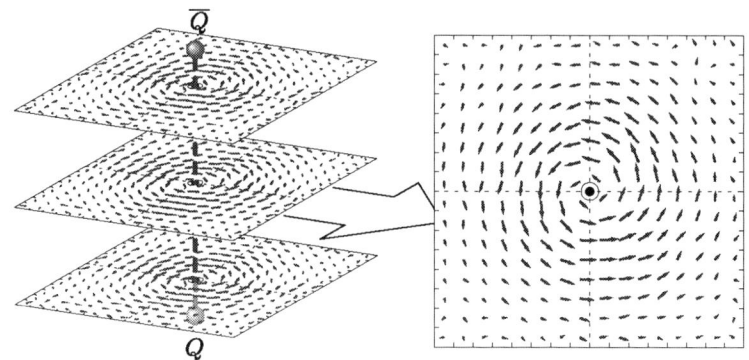

FIGURE 4. Magnetic displacement currents in Landau gauge as a solenoidal current.

Here $\vec{E}^{MA}$ and $\vec{B}^{MA}$ are defined in terms of plaquette variables $\theta_{\mu\nu}^{MA}(s)$ (mod 2π) which are constructed by $\theta_{\mu}^{MA}(s)$.

The Coulombic electric field coming from the static source is written in the lowest perturbation theory in terms of the gradient of a scalar potential. Hence it does not contribute to the curl of the Abelian electric field nor to the Abelian magnetic field in the above Abelian Bianchi identity Eq.(4). The dual Meissner effect says that the squeezing of the electric flux occurs due to cancellation of the Coulombic electric fields and those from solenoidal magnetic currents. In the case of MA gauge, magnetic monopole currents $\vec{k}$ play the role of the solenoidal current[23, 20, 8].

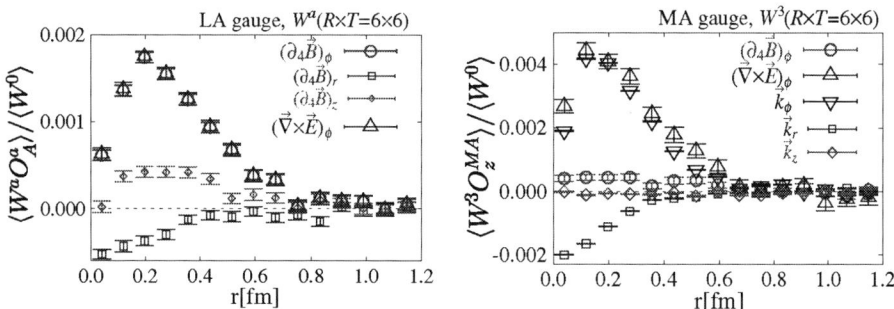

FIGURE 5. Curl of Abelian electric fields and magnetic displacement currents around a static quark pair in Landau (Left) and in MA (Right) gauges. Monopole currents are also plotted in MA gauge.

Now what happens in a smooth gauge like the Landau gauge where monopoles do not exist? From Eq.(4), only $\partial_4 \vec{B}_A$ regarded as a magnetic displacement current could play the role of the solenoidal current. It is very interesting to see Fig.4 in which this happens actually in Landau gauge. Note that the solenoidal current has a direction squeezing the Coulombic electric field. Let us see also the detailed distributions shown in Fig.5. The other components of the magnetic displacement current $\partial_4 B_{Ar}$ and $\partial_4 B_{Az}$ are not vanishing but they are much suppressed consistently with Fig.2. In comparison,

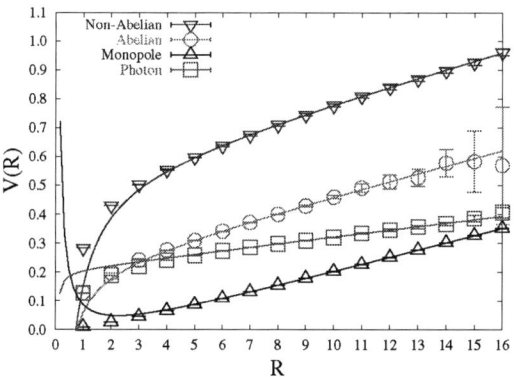

FIGURE 6. Abelian, monopole and photon static potentials.

we show the case of MA gauge also in Fig.5. Here $\partial_4 B_{A\phi}$ is found to be negligible numerically as already expected from the works[23, 11, 20]. Instead monopole currents are circulating[23, 20, 8]. In this case, k_r is non-vanishing, although it is also suppressed in comparison with k_ϕ. k_z is almost zero. The authors think that non-vanishing of the radial and z components of $\partial_4 \vec{B}$ in Landau gauge and $\vec{k}$ in MA gauge is due to lattice artifacts and also due to the smallness of the Wilson loop size adopted here. It is interesting that the shapes of $\partial_4 B_{A\phi}$ in Landau gauge and k_ϕ in MA gauge look similar, although the strengths are different. They have a peak at almost the same distance around 0.2[fm] and almost vanish around 0.7[fm].

ABELIAN DOMINANCE TEST

The reader may wonder if the above consideration of the Abelian fields alone is enough. Some people believe that the non-perturbative confinement problem in infrared QCD could be understood in terms of Abelian quantities[6, 7, 24].

First we evaluate the Abelian and monopole contributions to the string tension in MA gauge using the Iwasaki improved action (1). It is plotted in Fig.6. Our results are as follows:

$$\frac{\sigma_{ab}}{\sigma} = 0.94 \pm 0.05, \quad \frac{\sigma_{mo}}{\sigma_{ab}} = 0.98 \pm 0.02, \quad \frac{\sigma_{mo}}{\sigma} = 0.92 \pm 0.01.$$

This is compared with the result obtained using Wilson gauge action in Ref.[25]:

$$\frac{\sigma_{mo}}{\sigma} = 0.87 \pm 0.02$$

Hence substantial improvement is obtained with the use of Iwasaki impoved action. This suggests the abelian monopole part could reproduce the full string tension in the continuum limit.

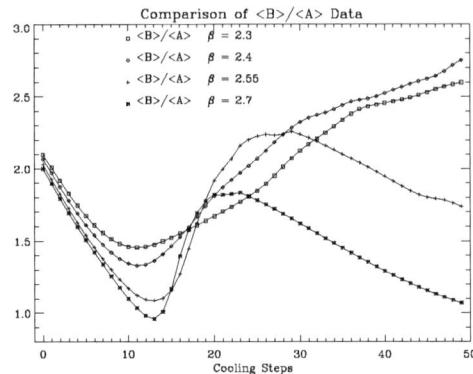

FIGURE 7.   Non-abelianicity Q vs. Cooling Step, for several values of β. This figure is from Ref.[10].

Abelian dominance in infrared QCD is also checked in some works with the use of controlled cooling. Giedt and Greensite [10] have measured the ratio B/A where

$$B = -\frac{1}{V}\sum_x \frac{1}{n_p(n_p-1)}\sum_{i>j}\sum_{m>n}\mathrm{Tr}\{[F_{ij}(x),F_{mn}(x)]^2\}$$

$$A = \frac{1}{n_p V}\sum_x\sum_{i>j}\mathrm{Tr}\{F_{ij}^4\}$$

and have shown that it decreases as the cooling steps as far as the string tension is kept non-vanishing. See Fig.7 taken from Ref.[10]. Cooling is expected to reduce the big Coulombic interaction in the non-Abelian case while keeping infrared confinement property.

Cea and Cosmai[11] have measured connected correlation operators of Wilson loops and 1×1 plaquette using the controlled cooling and have obtained the penetration length in non-Abelian case. It is almost equal to the penetration length determined by Abelian Wilson loop and Abelian electric fields in MA gauge. See Fig.8 taken from Ref.[11].

We also try to check Abelian dominance in Landau gauge using a controlled cooling [26] under which the string tension remains non-vanishing. For reader convenience let us, briefly, illustrate our cooling procedure. The lattice gauge configurations are cooled by replacing the matrix $U_\mu(s)$ associated to each link $l \equiv (s,\hat{\mu})$ with a new matrix $U'_\mu(s)$ in such a way that the local contribution to the lattice action

$$S(s) = 1 - \frac{1}{2}\mathrm{tr}\left\{U_\mu(s)k(s)F(s)\right\} \tag{6}$$

is minimized. $\widetilde{F}(s) = k(s)F(s)$ is the sum over the "U-staples" involving the link l and $k(s) = \sqrt{\det\left(\widetilde{F}(s)\right)}$, so that $F(s) \in$ SU(2). In a "controlled" or "smooth" cooling step we have

$$U_\mu(s) \to U'_\mu(s) = V(s)U_\mu(s)\,, \tag{7}$$

178

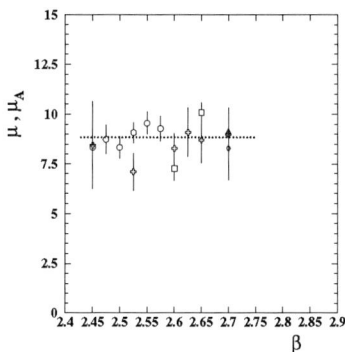

FIGURE 8. μ and μ_A (in units of $\Lambda_{\overline{MS}}$) versus β for square Wilson loops. Circles, squares, and triangle refer to $L = 16, 20, 24$ respectively. Crosses and diamond refer to the Abelian projected correlator ρ_W^A with $L = 16, 20$ respectively. This figure is from Ref. [11].

where $V(s)$ is the SU(2) matrix which maximizes

$$\text{tr}\left\{V(s)U_\mu(s)F(s)\right\} \tag{8}$$

subjected to the following constraint on the SU(2) distance between $U_\mu(s)$ and $U'_\mu(s)$:

$$\frac{1}{4}\text{tr}\left[\left(U_\mu^\dagger(s) - U_\mu'^\dagger(s)\right)\left(U_\mu(s) - U'_\mu(s)\right)\right] \le \delta^2. \tag{9}$$

We adopt $\delta = 0.3$. A complete cooling sweep consists in the replacement Eq. (7) at each lattice site.

We find the profile of the non-Abelian electric field E_z^a tends to that of the Abelian one of $E_{A_z}^a$ in the long-range region as shown in Fig.9. This is consistent with the above result[11] in a different approach.

DIMENSION 2 GLUON CONDENSATE

Now we have shown that the magnetic displacement currents are important in the dual Meissner effect when there are no monopoles. Then how can we understand the origin of the dual Meissner effect without monopoles? The Abelian dual Meissner effect indicates the massiveness of the Abelian electric field as an asymptotic field:

$$(\partial_\rho^2 - m^2)\vec{E}_A \sim 0. \tag{10}$$

This leads us to a dual London equation which is a key to the dual Meissner effect. Let us evaluate the curl of the magnetic displacement current. Using Eq.(4), we get

$$\vec{\nabla} \times \partial_4 \vec{B}_A = \vec{\nabla}(\vec{\nabla} \cdot \vec{E}_A) - \vec{\nabla}^2 \vec{E}_A.$$

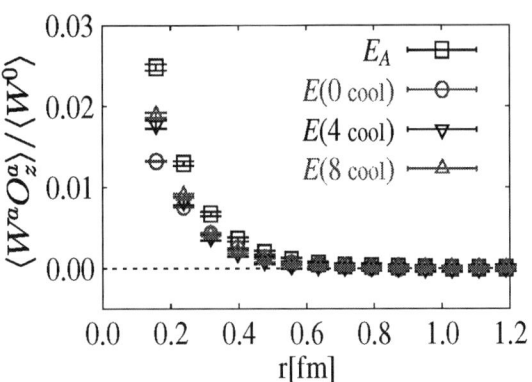

FIGURE 9. Profiles of non-Abelian electric field E_z^a after cooling. E_{Az}^a stands for the Abelian electric field. $W(R \times T = 6 \times 6)$ is used.

From Eq.(10), we get the dual London equation:

$$\vec{\nabla} \times \partial_4 \vec{B}_A \sim (\partial_4^2 - m^2)\vec{E}_A. \tag{11}$$

Let us remember a simple mean-field approach developed by Fukuda[27]. Neglecting gauge-fixing and Fadeev-Popov terms, we have equations of motion $D_\mu^{ab} F_{\mu\nu}^b = 0$ and the (non-Abelian) Bianchi identity $D_\mu^{ab*} F_{\mu\nu}^b = 0$. Applying D operator to the Bianchi identity and using the Jacobi identity and the equations of motion, we get

$$(D_\rho^2)^{ab} F_{\mu\nu}^b = 2g\varepsilon^{abc} F_{\mu\alpha}^b F_{\nu\alpha}^c. \tag{12}$$

Notice $(D_\rho^2)^{ab} = \partial_\rho^2 \delta^{ab} + g\varepsilon^{acb}(\partial_\rho A_\rho^c) + g^2(A_\rho^a A_\rho^b - \delta^{ab}(A_\rho^c)^2)$. Hence if $< A_\mu^a A_\nu^b >= \delta^{ab}\delta_{\mu\nu}v^2 \neq 0$, we see asymptotically that the electric fields become massive $(\partial_\rho^2 - m^2)E_k^a \sim 0$ with $m^2 = 8g^2v^2$[28]. Now the Abelian electric field is also massive asymptotically $(\partial_\rho^2 - m^2)E_{Ak}^a \sim 0$. Hence the dual London equation (11) is obtained.

The importance of the dimension 2 gluon condensate has been stressed by Zakharov and his collaborators[13] and Refs.[14, 15]. Recent discussions on the value of the gluon condensate are seen in Ref.[16]. Some of them discuss the mass generation of the gluon propagator. But as pointed out by Fukuda[27], the non-vanishing dimension 2 gluon condensate leads us to a conclusion that field strengths $F_{\mu\nu}^a(s)$ instead of gluon fields $A_\mu^a(s)$ become good canonical variables having a finite mass, whereas gluon propagators have a $(p^2)^{-2}$ behavior showing confinement.

Although the operator of the gluon condensate is gauge-variant, it is proved recently the expectation value is gauge invariant[17]. Hence the gluon condensate has a physical importance, if the proof is correct.

ACKNOWLEDGMENTS

The numerical simulations of this work were done using RSCC computer clusters in RIKEN. The authors would like to thank RIKEN for their support of computer facilities. T.S. is supported by JSPS Grant-in-Aid for Scientific Research on Priority Areas 13135210 and (B) 15340073. T.S. is thankful to Valentine Zakharov, Mikhail Polikarpov and Manfred Faber for useful discussions.

REFERENCES

1. G. 't Hooft, in Proceedings of the EPS International Conference, edited by A. Zichichi, p. 1225 (1976).
2. S. Mandelstam, Phys. Rept. **23C**, 245 (1976).
3. G. 't Hooft, Nucl. Phys. **B190**, 455 (1981).
4. T. Suzuki, Prog. Theor. Phys. **69**, 1827(1983).
5. A. S. Kronfeld, M. L. Laursen, G. Schierholz and U. J. Wiese, Phys. Lett. B **198**, 516 (1987); A. S. Kronfeld, G. Schierholz and U. J. Wiese, Nucl. Phys. B **293**, 461 (1987).
6. T. Suzuki and I. Yotsuyanagi, Phys. Rev. **D42**, R4257 (1990);
7. T. Suzuki, Nucl. Phys. Proc. Suppl. **30**, 176 (1993); M. N. Chernodub and M. I. Polikarpov, in "Confinement, duality, and nonperturbative aspects of QCD", Ed. by P. van Baal, Plenum Press, p. 387 (1997).
8. Y. Koma, M. Koma, E.-M. Ilgenfritz, T. Suzuki, and M. I. Polikarpov, Phys. Rev. **D68**, 094018 (2003).
9. M.N. Chernodub, Phys.Rev. **D69**, 094504 (2004).
10. J. Giedt and J. Greensite, Phys. Rev. **D55**, 4484 (1997).
11. P. Cea and L. Cosmai, Phys. Rev. **D52**, 5152 (1995).
12. The dual London equation was discussed gauge-invariantly with a definition of a non-abelian monopole current in P. Skala, M. Faber and M. Zach, Nucl. Phys. **B494**, 293 (1997); Phys. Lett. **B424**, 335 (1998).
13. L. Stodolsky, Pierre van Baal and V.I. Zakharov, Phys. Lett. **B552**, 214 (2003) and references therein.
14. K-I. Kondo, Phys. Lett. **B514**, 335 (2001); Phys.Lett. **B572**, 210 (2003).
15. D. Dudal et al., JHEP **0401**, 044 (2004) and references therein.
16. E.R. Arriola, P.O. Bowman and W. Bronioski, "Landau-gauge condensate from the quark propagator on the lattice", hep-ph/0408309 and references therein.
17. A. A. Slavnov, " Gauge invariance of dimension two condensate in Yang-Mills theory", hep-th/0407194
18. Y.Iwasaki, Nucl. Phys. **B258**, 141 (1985); Univ. Tsukuba Preprint UTHEP-118(1983) unpublished.
19. APE Collaboration: M. Albanese et al., Phys. Lett. **B192** 163 (1987).
20. G. S. Bali, V. Bornyakov, M. Muller-Preussker and K. Schilling, Phys. Rev. **D54**, 2863 (1996).
21. G.S. Bali, K.Schilling and Ch. Schlichter, Phys. Rev. **D51**, 5165 (1995).
22. T. A. DeGrand and D. Toussaint, Phys. Rev. **D22**, 2478 (1980).
23. V. Singh D. A. Browne, and R. W. Haymaker, Phys. Lett. **B306**, 115 (1993).
24. K-I. Kondo, "Magnetic condensation, Abelian dominance, and instability of Savvidy vacuum", hep-th/0404252 and references therein.
25. G.S. Bali, V. Bornyakov, M. Mueller Preussker and K. Schilling, Phys.Rev. **D54**, 2863 (1996).
26. M.Campostrini et al., Phys. Lett. **B225**, 403 (1989).
27. R. Fukuda, Prog. Theor. Phys. **67**, 648 (1982); ibid. 655 (1982).
28. A mass counter term is needed rigorously.

Strings from lattice Yang-Mills theory

V. I. Zakharov

Max-Planck Institut für Physik
Werner-Heisenberg Institut
Föhringer Ring 6, 80805, Munich
E-mail: xxz@mppmu.mpg.de

Abstract. We review properties of lower-dimension vacuum defects observed in lattice simulations of SU(2) Yang-Mills theories. One- and two-dimensional defects are associated with ultraviolet divergent action. The action is the same divergent as in perturbation theory but the fluctuations extend over submanifolds of the whole 4d space. The ultraviolet action is self tuned to a divergent entropy and the 2d defects can be thought of as strings populated with particles. The newly emerging 3d defects are closely related to the confinement mechanism. Namely, there is a kind of holography so that information on the confinement is encoded in a 3d submanifold. Furthermore, there are data indicating that spontaneouis symmetry breaking is related to submanifolds. We introduce an SU(2) invariant classification scheme which would allows for a unified description of $d = 1, 2, 3$ defects. The scheme fits known data and predicts that 3d defects are related to chiral symmetry breaking. Relation to stochastic vacuum model is briefly discussed as well.

Keywords: monopoles, strings, confinement, lattice
PACS: 11.15.Ha, 11.25.Tq, 12.38.Aw

INTRODUCTION

Studies of the confinement mechanism have become since long a prerogative of the lattice simulations, for a recent review see [1]. The continuum theory provided in fact little guidance for the search of the confinement mechanism. Equations which one borrows from the continuum physics refer mostly to U(1) Higgs models or instantons, see, e.g., [2]. However, these hints from the continuum theory could be used at a qualitative level at best. Indeed, the $U(1)$ models do not observe the original symmetry properties of QCD while instantons do not confine.

Painstaking analysis of the lattice simulations did allow to extract vacuum fluctuations which are actually responsible for the confinement. These are so called monopoles and central vortices, for reviews see, e.g., [3] and [1, 4]. By construction, monopoles are infinitely thin closed trajectories while the central vortices are infinitely thin closed 2d surfaces [1]. Separation of the two types of the defects is actually superficial. Rather, one observes vortices populated with monopoles. Monopoles live on 2d surfaces, not in the whole 4d space and there can be no vortices without monopoles, see in particular [5, 6]. Infinitely thin (with size of the lattice spacing a), percolating trajectories and surfaces look very different from, say, instantons which are bulky fields, with size of order Λ_{QCD}^{-1}. Thus, one is tempted to say that lattice simulations uncovered existence of objects of

---

[1] both trajectories and surfaces are defined actually on the dual lattice.

lower dimensions in the vacuum state of Yang-Mills theories. However, a prevailing viewpoint, for a recent presentation see, e.g., [2], is that apparent point-likeness of the monopoles and vortices is an artifact of their definition and in fact they only mark corresponding bulky field fluctuations. There are no models for the conjectured thick monopoles and vortices in terms of the original Yang-Mills fields and interpretation of the lattice data remains then obscure.

It is only rather recently that it was understood that the monopoles and vortices might be physical lower-dimensional defects. The basic observation which brings about such a conclusion is the ultraviolet divergence in the corresponding *non-Abelian* action [2] associated with the monopoles (see [8] and references therein) and vortices [5]. To explain the survival of the monopoles and vortices– despite of their ultraviolet divergent action– one is forced to postulate [9] self-tuning of the ultraviolet divergent action and of ultraviolet divergent entropy [3].

Observation of the ultraviolet divergent action associated with confining field con- figurations seems to be most dramatic evidence in favor of relevance of hard fields to confinement. The d=2 defects appear to be nothing else but strings with excitations of scalar field living on them. There exist, however, other observations [10, 11, 12] which indicate that lower dimensional defects are of physical significance [4].

When the lattice studies were undertaken first, there was no theory of extended objects at all. However, more recently the idea that strings are relevant to QCD has become quite common. More specifically, one expects that if a dual formulation of YM theories exists, it would be a string theory [13]. Usually, however, one believes that it is only in the limit of infinite number of colors, $N_c \to \infty$, that one might find a dual formulation.

Thus, there appears a possibility that the languages of lattice and continuum theories would get unified again, this time in terms of theory of extended objects. The author of the present review does believe that this unification may become one of major advances in gauge theories. One of possible feedback from lattice studies to the continuum theory is that topological excitations observed within a 'direct' formulation might become fundamental variables of the dual formulation of the same theory, see, e.g., [14]. Thus, if the strings are indeed observed as excitations in lattice simulations of YM theories, this can be considered as indication that there exists a dual formulation in terms of fundamental strings [15].

It goes without saying that there is no well defined strategy yet to achieve this synthesis of the lattice and continuum-theory studies. Here we address a modest problem of reformulating some of the lattice results in terms of the continuum theory. The point is that many results, especially on the lattice strings, are obtained originally in terms of so called projected fields, see, e.g., [1]. We will discuss a classification scheme of the defects in explicitly SU(2) invariant terms. Also we will comment on possible relation to the stochastic picture of the vacuum [16, 17].

---

[2] It is worth emphasizing that the non-perturbative ultraviolet divergent fields are no more divergent than perturbation theory, for details see [7].
[3] Visible entropy of the 2d and 1d defects explodes indeed exponentially with $a \to 0$ [6].
[4] Note that for lower-dimension defects to be relevant the corresponding fields are to be singular.

LATTICE STRINGS

We have reviewed recently the properties of the two-dimensional defects, or lattice strings [18] and will be brief here.

Magnetic monopoles

Theoretically, the most difficult point about the monopoles is their definition on the lattice. Monopoles are topological excitations of a compact $U(1)$ [19]. To define them in non-Abelian case one uses projection of the original YM fields on the 'closest' Abelian configuration. The physical idea behind considering the monopoles is that confinement is mostly due to Abelian degree of freedom [20].

While the definition of the monopoles is not so transparent, many observed properties are beautiful and formulated in perfectly SU(2) invariant way. Monopoles are observed as clusters of trajectories. Infinite, or percolating cluster corresponds to the classical expectation value, $< \phi_M >$ of a magnetically charged field ϕ_M. Short, or ultraviolet clusters correspond to quantum fluctuations of the field ϕ_M. The total length of the clusters is trivially proportional to the total volume of the lattice, V_4 and one conventionally introduces corresponding densities:

$$L_{tot} = 4\rho_{tot} \cdot V_4 = 4(\rho_{perc} + \rho_{finite}) \cdot V_4 . \tag{1}$$

According to the data [21]:

$$\rho_{tot} \approx 1.6(fm)^{-3} + 1.5(fm)^{-2} \cdot a^{-1} , \tag{2}$$

where a is the lattice spacing. The a^{-1} term is entirely due to the finite clusters. For the percolating cluster the density is a constant in the physical units.

One can translate (2) into the standard filed theoretic language by observing that [9]:

$$\langle |\phi_M|^2 \rangle = (const)a \cdot \rho_{tot} . \tag{3}$$

Thus, we have

$$\langle |\phi_M|^2 \rangle \sim \Lambda_{QCD}^2 . \tag{4}$$

Point-like facet of the monopoles

Eq (4) looks as if monopoles were bulky fields like instantons. However, the monopoles action diverges with $a \to 0$, the same as for point-like particles. Namely [8]:

$$S_{mon} \equiv M \cdot L_{mon} , \tag{5}$$

and

$$M(a) \approx \ln 7 \cdot a^{-1} \tag{6}$$

where $M(a)$ corresponds to the radiative mass and is found by measuring extra *non-Abelian* action associated with the monopoles. Note that we quote the data in a way which allows a straightforward theoretical interpretation. Namely, the propagating mass is no longer $M(a)$ but [22]:

$$m_{prop}^2 = \frac{(const)}{a}\left(M(a) - \frac{\ln 7}{a}\right) , \qquad (7)$$

where the constants $const, \ln 7$ are of pure geometrical origin and depend on the lattice used. In particular, $\ln 7$ corresponds to the hypercubic lattice. Note that in Euclidean space a physical mass of a point-like particle can appear only as a result of tuning between divergent action and entropy.

Thus, the data (6) correspond to a small monopole mass. However, data (2) imply that globally monopoles live on a 2d surface. For ordinary point-like particles $\rho_{tot} \sim a^{-3}$.

Closed strings

Closed surfaces are topological defects of the Z_2 gauge theory. In simulations of SU(2) theory these surfaces are defined in terms of Z_2 projection which replaces the original YM fields with $Z_\mu(x) = \pm 1$. The central vortices are defined as unification of all the plaquettes on the dual lattice which pierce negative plaquettes in the Z_2 projection, for review see [1, 4].

Two most striking properties of the central vortices is that their total area scales in physical units, for review see [1, 4] while non-Abelian action is ultraviolet divergent [5]:

$$A_{tot} \approx 4 \, (fm)^{-2} V_4 , \quad S_{tot} \approx 0.54 \frac{A_{tot}}{a^2} . \qquad (8)$$

Moreover, the excess of the action disappears on the plaquettes next to those belonging to the vortex. In other words, the vortices are infinitely thin, at least on the presently available lattices.

It is worth emphasizing that the properties (8) amount to observing an elementary string. Indeed, the data on the total area imply that the tension is of order Λ_{QCD}^2 while the ultraviolet divergence in the action assumes vanishing thickness. The suppression due to the action is to be compensated by enhancement due to the entropy. Fine tuning of the entropy and action is a generic feature of any consistent theory of an elementary string in Euclidean space.

Another striking feature of the lattice strings is that the monopole trajectories, discussed in the preceding subsection, lie in fact on the central vortices. Thus, the two types of defects merge with each other.

THREE DIMENSIONAL DOMAINS

The 3d defects are much more recent than the strings and have been studied in less detail. Moreover, there are a few independent pieces of evidence in favor of existence of 3d defects which are, in fact, not necessarily related to each other.

'Strong' potentials

The central vortices are defined in terms of negative plaquettes in Z_2 projection. In Z_2 projection links take on values $\pm$. Generically, the values $(+1)$ and (-1) are the same frequent. One can, however, minimize the number of negative links. using remaining Z_2 invariance. Physicswise, one fixes the gauge by localizing large potentials on as small number of links as possible. Since link values correspond to potentials and are gauge dependent, one can wonder what is the objective meaning of such minimization. The point is that minimizing, say, potential squared one can arrive at gauge invariant quantities [23]. Minimizing number of negative links as a variation of such a procedure.

And, indeed, one finds [11] that volume of negative links scales as a physical 3d defect:

$$V_3 = c_3 \Lambda_{QCD} \cdot V_4 \ . \tag{9}$$

Note that by construction the volume are bound by the central vortices. This volume can be called Dirac volume [1]. Eq (9) then states that the minimal Dirac volume scales in physical units, or, alternatively, has a zero fractal dimension.

Holography and confinement

Relation of the volume discussed above to the confinement is revealed through a remarkable observation of the authors of Ref [24]. One replaces the original link matrices $U_\mu(x)$ by $\tilde{U}_\mu(x)$ where

$$\tilde{U}_\mu(x) \equiv U_\mu(x) \cdot Z_\mu(x) \ , \tag{10}$$

where $Z_\mu(x)$ are projected values of the same links. Next, one evaluates the Wilson loop and quark condensate $\langle \bar{q}q \rangle >$ in terms of the modified links $\tilde{U}$. The result [24] is that both the confining potential and spontaneous breaking of the chiral symmetry disappear.

Originally [24] the change (10) affected approximately half of the total number of links. Now, we see that it is enough to perform the change (10) on a 3d submanifold of the whole 4d space to kill the confinement and quark condensate. Thus, we observe a kind of holography when information on the confinement is encoded on a submanifold of the whole space.

In more detail, consider a plane on which we will draw a Wilson line. Consider, furthermore, a particular configuration of the gauge fields generated with the standard action. Determine then the 3d volume described in the preceding section. Intersections of this volume with the plane considered are segments of 1d lines. Now, we can draw any Wilson line on the plane. The statement is that the sign of the Wilson line can be determined by counting the number of intersections with segments of 1d defects. It is a highly non-trivial observation, difficult to interpret. Note that there is no logical contradiction, though. Indeed, there are gauges where the confining fields are soft, of order $A_\mu \sim \Lambda_{QCD}$. Apparently, one can use gauge invariance to choose a gauge where the confining fields are of order $A_\mu \sim g/a$ but occupy a 3d volume [5].

Chiral symmetry breaking

There is a series of observations, not directly related to each other that indicate relevance of some 3d defects to the spontaneous breaking of the chiral symmetry [6]:

(a) procedure of Ref [12] described above makes also the quark condensate vanish:

$$\langle \bar{q}q \rangle_{\tilde{U}} \approx 0 \qquad (11)$$

Now, we know [11] that the change (10) affects not a finite part of the 4d space but only a 3d submanifold.

(b) there is evidence in favor of long range topological structure in QCD vacuum which is related to chiral symmetry breaking [10]. The search process for the topological structure is formulated in terms of eigenfunctions of the Dirac operator and explicitly gauge invariant.

(c) One introduces the so called inverse participation ratio [26] defined in terms of eigenfunctions of the Dirac operator:

$$I = N\Sigma_x \rho_i^2(x) \ , \qquad (12)$$

where N is the number of lattice sites x,

$$\rho_i(x) = \psi_i^\dagger \psi_i(x) \ ,$$

and $\psi_i(x)$ is the i-th normalized $(\Sigma_x \rho_i(x) = 1)$ lowest eigenvector of the Dirac operator.

Dependence of the inverse participation ratio on the lattice spacing a was studied in Ref [12]. The result is:

$$\langle I \rangle = c_1 + c_2 \cdot a^{-\gamma} \ , \qquad (13)$$

with a non-vanishing exponent γ:

$$1 \leq \gamma \leq 2 \ .$$

Note that the value $\gamma = 1$ would correspond, in the limit $a \to 0$ to localization of the eigenfunctions on a 3d volume. It is worth emphasizing that the a dependence observed refers to an explicitly gauge invariant quantity.

To summarize, there are indications that the chiral symmetry breaking is determined by gauge fields living on a subspace. Since the confinement itself also seems to be related to a 3d volume (see above), it is not clear whether we deal with a phenomenon specific for chiral symmetry breaking or with an effect common to confinement.

---

[5] Some considerations on possible relation between gauge invariance and holography in the gravitational case can be found in [25]

[6] We are considering the quenched approximation

CLASSIFICATION SCHEME

Invariants

There is no theory of the defects in the non-Abelian case. However, even in the absence of such a theory one can try to find a SU(2) invariant classification scheme. Generically, the first example of such a scheme for monopoles was proposed long time ago [27]. In pure YM theory, there are no classical monopole solutions. However, imagine that there exists a scalar field, vector in the color space H^a, $a = 1, 2, 3$. Then one could fix the gauge by rotating vector H^a to the third direction at each point. This fixation of the gauge would fail however at the point where

$$H^a = 0 . \tag{14}$$

Condition (14) can be viewed as three equations defining 1d defects in the 4d space which can be identified with monopole trajectories. It is crucial that (14) is SU(2) invariant.

For various reasons, this idea does not seem to work in the realistic case, for review and references see [28]. Rather, monopoles are associated with singular non-Abelian fields (see above). Let us try to adjust the classification scheme to this set up [18, 28, 29].

Begin with YM theory in three dimensions and assume that monopoles violate the Bianchi identities. The Bianchi identities read

$$D\tilde{G} = 0 , \tag{15}$$

and can be used then to express the gauge potential A in terms of the field strength tensor, see, e.g., [30]. In somewhat symbolic form,

$$A = \frac{1}{g} \partial \tilde{G} \tilde{G}^{-1} . \tag{16}$$

The inverse matrix exists unless the determinant constructed on the components of $\tilde{G}$ vanishes. Denoting $\tilde{G}_{ik}^a \equiv \varepsilon_{ikl} B_l^a$ we have the following condition for the Bianchi identities to be violated:

$$\varepsilon_{ikl} \varepsilon_{abc} B_i^a B_k^b B_l^c = 0 . \tag{17}$$

Note that the condition (17) is perfectly gauge and rotational invariant.

Let us now consider the 4d Euclidean case. The rotational group in 4d splits into a product of two SU(2) groups. Since we do not introduce spinors, SU(2) can be replaced by $O(3)$, $O(4) = O(3) \times O(3)$. The corresponding representations of the $O(3)$ groups are chiral fields $(H_i^a \pm E_i^a)$.

Looking for a generalization of (17) we notice that there are now two possibilities:

$$det(E_i^a + H_i^a) = 0 , \ or \ det(E_i^a - H_i^a) = 0 . \tag{18}$$

Imposing either of them we specify a 3d defect. On this 3d submanifold one can use as independent three fields of a certain chirality but not of the opposite one. Association of

3d defects with chiral symmetry breaking arises as a consequence of the symmetry of the problem.

The boundary of these 3d defects is determined by conditions;

$$det(E_i^a + H_i^a) = 0, \; and \; det(E_i^a - H_i^a) = 0. \tag{19}$$

which determine 2d defects. Moreover, if both conditions (19) are satisfied, there is no inversion of the Bianchi identities similar to (16).

Finally, zeros of a second order of the determinant would define 1d defects.

Classification scheme vs data

The classification scheme proposed above is based on symmetry alone and is not unique. But, nevertheless, let us try to identify the 2d and 1d defects arising within this scheme with the central vortices and monopoles. There are a few quite remarkable confirmations of such an identification:

(a) the central vortices are associated, according to the scheme, with singular fields and, possibly, the Bianchi identities are violated on these 2d defects. As we know, indeed, the central vortices are associated with singular action. Moreover, monopoles live on the surface and may well signify violation of the Bianchi identities;

(b) non-Abelian fields associated with the 2d defects are aligned with the surface. This is confirmed by the measurements, according to which the excess of the action vanishes already on the plaquettes next to the central vortices [5];

(c) the monopole trajectories are predicted to lie on the central vortices, in agreement with the data [31, 5];

(d) 'monopoles' appear to be Abelian fields since zero of second order of the determinant constructed on three independent (within a 3d defect) fields implies that there is only a single independent color vector;

(e) on the other hand, the non-Abelian field of the monopoles is not spherically symmetrical but rather aligned with the surface. This collimation of the field was observed in measurements, [31].

It is worth emphasizing that all the properties (a) - (e) are gauge invariant. Thus, the data so far do confirm that through projections one detects gauge invariant objects.

Finally, the scheme predicts that breaking of the chiral symmetry is associated with 3d defects. The corresponding lattice data were summarized in the preceding section.

STOCHASTIC FIELDS

Let us notice that all our definitions of the defects assumed finite lattice spacing a. In the continuum limit, the correlators of the confining fields become singular: For example,

$$\langle G_{\mu\nu}^a(x), G_{\mu\nu}^a(y) \rangle_{strings} = \Lambda_{UV}^2 \Lambda_{QCD}^2 f(x-y) + (regular \; terms), \tag{20}$$

where

$$f(0) = 1, \; f(x \neq 0) = 0 .$$

For a finite a the function $f(x)$ has a resolution of order a. Stochastic model of vacuum was used in Ref.[17] to derive the area law for the Wilson line. One can readily see that the derivation survives in the singular limit we are considering.

Furthermore, denote by $\bar{A}$ the confining potential obtained in the gauge minimizing the number of negative links (see above). Then

$$\langle\, \bar{A}(x),\, \bar{A}(y)\, \rangle \;=\; \Lambda_{QCD}\cdot\Lambda_{UV}\; f(x-y)\; +(regular\ terms)\,. \tag{21}$$

The singular nature of the confining potential could explain observed dependence of the localization of zero modes on the lattice spacing, see above.

It is worth emphasizing that reduction of the confining potential to the 'white noise' would be a great oversimplification [7]. Indeed, the 3d nature of the domains assumes also non-trivial correlators for the derivatives of the potential. The issue deserves further consideration.

CONCLUSIONS

Physics of confinement might undergo quite a dramatic change soon. There have been emerging data indicating relevance to confinement of lower-dimension defects, or singular fields. Two-dimensional defects with divergent action and entropy, which selftuned to each other are naturally interpreted as the dual string, observed as a vacuum excitation. The string is detected through projection but possesses many SU(2) invariant properties. Other emerging phenomena, a kind of holography and localization of modes on a sub-manifold shrinking to zero with $a \to 0$, are observed in explicitly SU(2) invariant terms. The price is that the structure of the fields responsible for these observational phenomena is less transparent.

ACKNOWLEDGMENTS

This review is devoted to interpretation of the lattice data obtained mostly by colleagues from ITEP: P.Yu. Boyko, M.N. Chernodub, F.V. Gubarev, A.N. Kovalenko, M.I. Polikarpov, S.N. Syritsyn. I am thankful to all of them for common work, used in these notes, and numerous discussions. I am thankful to W. Bardeen, A. DiGiacomo, A. Gorsky, I. Horvath, J. Greensite, A. Polyakov, L. Stodolsky, L. Susskind, A. Vainshtein for enlightening discussions.

I am thankful to organizers of the Conference 'Confinement VI" for the invitation and hospitality.

REFERENCES

1. J. Greensite, *Prog. Part. Nucl. Phys.* **51** 1 (2003) (hep-lat/0301023).

---

[7] Actually, the 'white noise' would not confine.

2. M. Engelhardt, *"Generation of confinement and other nonperturbative effects by infrared gluonic degrees of freedom"*, hep-lat/0409023.
3. M.N. Chernodub, F.V. Gubarev, M.I. Polikarpov, A.I. Veselov, *Progr. Theor. Phys. Suppl.* **131**, 309 (1998);
   A. Di Giacomo, *Progr. Theor. Phys. Suppl.* **131**, 161 (1998) (hep-lat/9802008);
   T. Suzuki, *Progr. Theor. Phys. Suppl.* **131**, 633 (1998).
4. K. Langfeld, *et. al.*, *"Vortex induced confinement and the IR properties of Green functions"*, hep-lat/0209173.
5. F.V. Gubarev, et al., *Phys. Lett.* **B574** (2003) 136, (hep-lat/0212003)
6. A.V. Kovalenko, M.I. Polikarpov, S.N. Syritsyn, V.I. Zakharov, *"Interplay of monopoles and P-vortices"*, hep-lat/0309032.
7. V.I. Zakharov, *"Nonperturbative match of ultraviolet renormalon"*, hep-ph/0309178.
8. V.G. Bornyakov, *et al.*, *Phys. Lett.* **B537**, 291 (2002).
9. V.I. Zakharov, *"Hidden mass hierarchy"*, (hep-ph/0202040);
   M.N. Chernodub, V.I. Zakharov, *Nucl. Phys.* **B669**, 233 (2003) (hep-th/0211267);
   V.I. Zakharov, *Usp. Fiz. Nauk* **47** 39 (2004).
10. I. Horvath, et al., *Phys. Rev.* **D68** (2003) 114505 (hep-lat/0302009);
    I. Horvath, *" The analysis of space-time structure in QCD vacuum: localization vs global behaviour in local observables and Dirac eigenmodes"*, hep-lat/0410046.
11. M.I. Polikarpov, S.N. Syritsyn, V.I. Zakharov, *"A novel probe of the vacuum of the lattice gluodynamics"*, hep-lat/0402018;
    A.V. Kovalenko, M.I. Polikarpov, S.N. Syritsyn, V.I. Zakharov, *"Three dimensional vacuum domains in four dimensional SU(2) gluodynamics"*, hep-lat/0408014.
12. MILC Collaboration (C. Aubin et al.), *"The scaling dimension of low lying Dirac eigenmodes of the topological charge density"* hep-lat/0410024
13. J.M. Maldacena, *Adv. Theor. Math.* **2**, 231 (1998), (hep-th/9711200);
    A.M. Polyakov, *Int. J. Mod. Phys.* **A14**, 645 (1999), (hep-th/9809057).
14. R. Savit, *Rev. Mod. Phys.* **52** (1980) 453.
15. V.I. Zakharov, *"Hints on dual variables from lattice SU(2) gluodynamics"*, hep-ph/0309301.
16. J. Ambjorn, P. Olesen, *Nucl. Phys.* **B170** (1980) 60;
    T.I. Belova, Yu.M. Makeenko, M.I. Polikarpov, A.I. Veselov, *Nucl. Phys.* **B230** (1984) 473;
    H.G. Dosch, *Phys. Lett.* **B190** (1987) 177;
    H.G. Dosch, Yu.A. Simonov, *Phys. Lett.* **B205** (1988) 339.
17. Yu.A. Simonov, *"Nonperturbative QCD: confinement and deconfinement"*, hep-ph/0211330;
    A. Di Giacomo, H.G. Dosch, V.I. Shevchenko, Yu.A. Simonov, *Phys. Rept.*, 372 (2002) (hep-ph/0007223)
18. V.I. Zakharov, *"Lower-dimansion vacuum defects in lattice Yang-Mills theory"*, hep-ph/0410034.
19. A.M. Polyakov, *Phys. Lett.* **B59** (1975) 82.
20. T. Suzuki, I. Yotsuyanagi, *Phys. Rev.* **D42** (1990) 4257.
21. P.Yu. Boyko, M.I. Polikarpov, V.I. Zakharov, *Nucl. Phys. Proc. Suppl.* **119** 724 (2003), (hep-lat/0209075);
    V.G. Bornyakov, P.Yu. Boyko, M.I. Polikarpov, V.I. Zakharov, *Nucl. Phys.* **B672** 222 (2003) (hep-lat/0305021);
22. J. Ambjorn, *"Quantization of geometry"*, hep-th/9411179.
23. F.V. Gubarev, L. Stodolsky, V.I. Zakharov, *Phys. Rev. Lett.* **86**, 2220 (2001) (hep-ph/0010057)
24. P. de Forcrand, M. D'Elia, *Phys. Rev. Lett.* **82**, 458 (1999) (hep-lat/9901020).
25. G. 't Hooft, *"The hidden information in the standard model"*, ITP-UU-02/68.
26. Ch. Gattringer, et al., *Nucl. Phys.* **B617** (2001) 101,(hep-lat/0107016).
27. G. t'Hooft, *Nucl. Phys.*, **B190**, 455 (1981).
28. F.V. Gubarev, V.I. Zakharov, *"Interpreting the lattice monopoles in the continuum terms"* hep-lat/0211033.
29. F.V. Gubarev, V.I. Zakharov, *Int. J. Mod. Phys.* **A17** (2002) 157 (hep-th/0004012).
30. M.B. Halpern,*Phys. Rev.* **D19** (1979) 517;
    O. Ganor, J. Sonnenschein, *Int. J. Mod. Phys.* **A11** (1996) 5701, (hep-th/9507036).
31. J. Ambjorn, J. Giedt, J. Greensite, *JHEP* **0002**, 033 (2000) (hep-lat/9907021).

191

Recent SELEX Results on the Properties of Charmed Hadrons

Jürgen Engelfried

Instituto de Física, Universidad Autónoma de San Luis Potosí, Mexico
for the SELEX Collaboration

Abstract. The SELEX [1] Fixed Target Experiment (Fermilab E781) employs beams of Σ^-, pions and protons to study the production and decay properties of charmed Mesons and Baryons. Here we present recent results on doubly-charmed baryons and charmed-strange mesons.

Keywords: Charmed baryons, Charmed mesons
PACS: 14.20.Lq, 14.40.Lb

INTRODUCTION

In 2002 the SELEX collaboration reported the first observation of a candidate for a double charm baryon, decaying as $\Lambda_c^+ K^- \pi^+$ [2]. The state had a mass of $3519 \pm 2\,\mathrm{MeV}/c^2$, and its observed width was consistent with experimental resolution, less than $5\,\mathrm{MeV}/c^2$. The final state contained a charmed baryon and negative strangeness (Λ_c^+ and K^-), consistent with the Cabbibo-allowed decay of a Ξ_{cc}^+ configuration. Starting from a Λ_c^+ sample used previously [4, 5], we looked for an additional vertex between the primary vertex and the Λ_c^+ decay point. We used all tracks not assigned to the single charm candidate in the search. The new secondary vertex had to have an acceptable fit χ^2 and a separation of at least 1σ from the new primary, where σ denotes the combined error of the primary and (new) secondary vertex position in the direction of flight of the Ξ_{cc}^+ candidate. The cuts were developed and fixed in previous searches for short-lived single-charm baryon states. We have applied them here without change. The number of events in the signal region is 22 events. We estimate the number of expected background events in the signal region from the sidebands as 6.1 ± 0.5 events. The single-bin significance of this signal is 6.3σ. The Poisson probability of observing at least this excess, including the Gaussian uncertainty in the background, is 1.0×10^{-6}. The overall probability of observing an excess at least as large as the one we see anywhere in the search interval is 1.1×10^{-4}. For more details see references [2, 3].

In order to confirm the interpretation of this state as a double charm baryon, it is essential to observe the same state in some other way. Other experiments with large charm baryon samples, e.g., the FOCUS [10] and E791 fixed target charm experiments at Fermilab or the B-factories, have not confirmed the double charm signal. This is not inconsistent with the SELEX results. The report in Ref. [2] emphasized that this new state was produced by the baryon beams (Σ^-, proton) in SELEX, but not by the π^- beam. It also noted that the apparent lifetime of the state was significantly shorter than that of the Λ_c^+, which was not expected in model calculations [6, 7, 8].

CP756, *Quark Confinement and the Hadron Spectrum VI*
edited by N. Brambilla, U. D'Alesio, A. Devoto, K. Maung, G.M. Prosperi and S. Serci
© 2005 American Institute of Physics 0-7354-0241-8/05/$22.50

Another way to confirm the Ξ_{cc}^{+} is to observe it in a different decay mode that also involves a final state with baryon number and charm (not anti-charm). One such mode involving only stable charged particles is the channel $pD^{+}K^{-}$, another one $\Xi_{c}^{+}\pi^{+}\pi^{-}$.

In addition, one has to look for the other members of the double-charm baryon family, namely the Ξ_{cc}^{++} and the Ω_{cc}^{+}.

SECOND DCB OBSERVATION: $\Xi_{cc}^{+} \to pD^{+}K^{-}$

After the discovery and publication of the $\Lambda_c K^- \pi^+$ signal we sought to confirm the discovery in another decay mode which was likely to have a significant branching ratio. Obvious choices were $\Xi_c^+ \pi^+ \pi^-$ and pD^+K^-. Since the SELEX D^+ signal is large and well studied we began with it.

With a similar analysis technique we observe [9] in this new decay mode an excess of 5.4 events over an expected background of 1.6 ± 0.35 events. The Poisson probability that a background fluctuation can produce the apparent signal is less than $1.5 \cdot 10^{-5}$. The observed mass of this state is $3518 \pm 3 \,\mathrm{MeV}/c^2$ consistent with the previous result. Averaging the two results gives a mass of $3518.7 \pm 1.7 \,\mathrm{MeV}/c^2$. The observation of this new weak decay mode confirms the previous suggestion that this state is a double charm baryon.

The relative branching ratio $\Gamma(pD^+K^-)/\Gamma(\Lambda_c^+K^-\pi^+) = 0.078 \pm 0.045$. The lifetime of this state in both decay modes is very short; less than 33 fsec at 90 % confidence. The properties of these two signals are consistent with each other. Even though the number of observed events seems high compared to predictions, the signals have a large significance, and the wrong sign combination reproduce the background level and show no peaks.

THE SELEX $D_{sJ}^{\star}(2632) \to D_s^+ \eta$ AND $D^0 K^+$

Motivated by the 2003 observations by BaBar [11], CLEO [12], and Belle [13] of narrow D_s resonances, decaying into $D_s \pi^0$ at $2319 \,\mathrm{MeV}/c^2$ and $D_s^* \pi^0$ at $2.46 \,\mathrm{GeV}/c^2$, SELEX search for similar resonances. Recently, SELEX announced [14] the observation of yet another narrow D_s resonance, this time above the strong decay threshold, in two decay modes, $D_s \eta$ and $D^0 K^+$, at $2632.6 \pm 1.6 \,\mathrm{MeV}/c^2$. In the $D^0 K^+$ we see the previously reported $D_s(2573)$ in addition to the new peak at the same mass as the peak observed in $D_s \eta$. This new excited D_s state shows two surprising properties: 1) The width of the 2632 state ($< 17 \,\mathrm{MeV}/c^2$) is clearly narrower than the width of the $D_s(2573)$; this is surprising, because the Q value of the decay is a factor of 1.5 larger. 2) The branching ratio $\Gamma(D^0K^+)/\Gamma(D_s^+\eta) = 0.14 \pm 0.06$ is opposite to what one would expect from an excited D_s. A long list of theory and phenomenology papers try to explain this state and its properties.

SUMMARY AND OUTLOOK

The SELEX collaboration observes for the first time a doubly charmed baryon, the $\Xi_{cc}^+(3520)$, in two decay modes: $\Xi_{cc}^+ \to \Lambda_c^+ K^- \pi^+$ and $\Xi_{cc}^+ \to p D^+ K^-$. These states are produced by baryons (Σ^-, p), and have a lifetime at resolution limit: $\tau_{cc} < 33\,\text{fsec}$.

SELEX reports also the observation of a new D_s resonance in two decay modes: $D_{sJ}(2632) \to D_s^+ \eta$ and $D^0 K^+$.

SELEX is now working on extending the measurements on the doubly charm baryon family, by searching for additional decay modes of the Ξ_{cc}^+, the Ξ_{cc}^{++}, and the Ω_{cc}^+.

SELEX is now working on extending the measurement of the doubly charmed baryon family by search for the Ξ_{cc}^{++} and the Ω_{cc}^+ and additional decay modes of the Ξ_{cc}^+.

ACKNOWLEDGMENTS

This work was supported in part by the Consejo Nacional de Ciencia y Tecnología (CONACyT), Mexico.

REFERENCES

1. The SELEX (Fermilab E781) Collaboration: Ball State University, Bogazici University, Carnegie-Mellon University, Centro Brasileiro de Pesquisas Fisicas, Fermilab, Institute For High Energy Physics (Protvino), Institute of High Energy Physics (Beijing), Institute of Theoretical and Experimental Physics (Moscow), Max-Planck-Institute for Nuclear Physics, Moscow State University, Petersburg Nuclear Physics Institute, Tel Aviv University, Universidad Autónoma de San Luis Potosí, Universidade Federal da Paraíba, H. H. Wills Physics Laboratory, University of Bristol, University of Iowa, University of Michigan-Flint, University of Rochester, University of Rome La Sapienza and INFN, University of São Paulo, University of Trieste and INFN.
   http://www-selex.fnal.gov
2. SELEX Collaboration, M. Mattson et al.: *First observation of the doubly charmed baryon* Ξ_{cc}^+. Phys. Rev. Letters **89** (2002) 112001, [arXiv:hep-ex/0208014].
3. M. Mattson, Ph.D. thesis, Carnegie Mellon University, 2002, FERMILAB-THESIS-2002-03.
4. SELEX Collaboration, A. Kushnirenko, et al.: Phys. Rev. Lett. **86** (2001) 5243, [arXiv:hep-ex/0010014].
5. SELEX Collaboration, F. Garcia et al.: Phys. Lett. **B528** (2002) 49, [arXiv:hep-ex/0109017].
6. B. Guberina, B. Melic, and H. Stefancic: Phys. Lett. **B484** (2000) 43.
7. M. Moinester: Zeit. Phys. **A355** (1996) 349.
8. V. Kiselev and A. Likhoded: Phys. Usp. **45** (2002) 455-506, Usp.Fiz.Nauk **172** (2002) 497-550, [arXiv:hep-ex/0103169], and references therein.
9. SELEX Collaboration, A. Ocherashvili, et al.:, *Confirmation of the double charm baryon* $\Xi_{cc}^+(3520)$ *via its decay to* $p D^+ K^-$. Submitted to Phys. Rev. Letters, [arXiv:hep-ex/0406033].
10. S. Ratti, BEACH2002, Vancouver, B.C., see also
    www.hep.vanderbilt.edu/~stenson/xicc/xicc_focus.html
11. BaBar Collaboration, B. Aubert, et al., Phys. Rev. Lett. **90** (2003) 242001, [arXiv:hep-ex/0304021].
12. CLEO Collaboration, D. Besson, et al., Phys. Rev. **D68** (2003) 032002, [arXiv:hep-ex/0305100].
13. Belle Collaboration, P. Krokovny, et al.: Phys. Rev. Lett. **91** (2003) 262002, [arXiv:hep-ex/0308019].
14. SELEX Collaboration, A.V. Evdokimov, et al.: Phys. Rev. Lett. **93** (2004) 242001, [arXiv:hep-ex/0406045].

Review of the experimental evidence on pentaquarks and critical discussion

Sonia Kabana

Laboratory for High Energy Physics, University of Bern, Sidlerstrasse 5, 3012 Bern, Switzerland

Abstract. We review and discuss the experimental evidence on predicted baryonic states made by 4 quarks and one antiquark, called pentaquarks. Theoretical and experimental advances in the last few years led to the observation of pentaquark candidates by some experiments, however with relatively low individual significance. Other experiments did not observed those candidates. Furthermore, the masses of the $\theta^+(1540)$ candidates exhibit a large variation in different measurements. We discuss to which extend these contradicting informations may lead to a consistent picture.

1. INTRODUCTION

Pentaquarks is a name devoted to describe baryons made by 4 quarks and one antiquark. These states, predicted long time ago to exist [1, 2, 3], were searched for already in the 60'ies but few candidates found have not been confirmed [4]. Recent advances in theoretical [3] and experimental work [5] led to a number of new candidates in the last 2 years of searches [5, 6, 7, 8, 9, 10, 11, 12, 13, 14, 15, 16, 17, 18, 19, 20]. For recent reviews on pentaquarks see [21, 22, 23]. The current theoretical description of pentaquarks (e.g. [3, 24, 25, 26, 27, 28, 29, 30]) does not lead to a unique picture on the pentaquark existence and characteristics, reflecting the complexity of the subject. In the following, we review and discuss the experimental observations of pentaquark candidates, as well as their lack of observation by some experiments.

2. EXPERIMENTAL EVIDENCE ON PENTAQUARK CANDIDATES

$\theta_{\bar{s}}^+$: $uudd\bar{s}$ The first observation of a candidate for the θ^+ pentaquark has been reported by the LEPS collaboration [5] in reactions $\gamma + A$ with γ energy 1.5-2.4 GeV and in the decay channel nK^+. Recent preliminary analysis of new data taken recently by LEPS lead to a confirmation of the seen peak with about 90 entries in the peak above background, as compared to 19 measured previously [31]. This first observation were followed by a number of experiments which have seen the θ^+ candidate peak [6, 7, 8, 9, 10, 11, 12, 13, 14, 15]. Figure 1, left, shows the masses of all θ^+ candidate peaks measured. The θ^+ peak has been observed in two decay channels. The open points correspond to the decay channel $K_s^0 p$, while the closed points to the decay channel $K^+ n$. The candidate θ^+ peak is seen in different reactions namely of $\gamma + A$, $\nu + A$, $p + p$, $K + Xe$, $e + d$, $e + p$, $K + Xe$. All of these reactions involved at least a baryon in the initial state. The energies are small (few GeV range) for all $\gamma + A$ reactions and vary

CP756, *Quark Confinement and the Hadron Spectrum VI*
edited by N. Brambilla, U. D'Alesio, A. Devoto, K. Maung, G.M. Prosperi and S. Serci

for the rest up to $\sqrt{s}$=300 GeV for $e + p$. In the experiments measuring the decay channel K^+n the neutron was not directly measured. Even though the θ^+ candidate peak has been observed by several experiments, the individually achieved statistical significance of the signal and this is mostly not large. The largest significance was $S/\sqrt{B} = 7.8 \pm 1$ [8].

A remarkable observation has been made by the CLAS collaboration in the same publication [8], namely they observed that the θ^+ candidate seem to be preferably produced through the decay of a possible new narrow resonance $N^0(2400)$. A preliminary analysis of CLAS [32] showed also a second peak in the invariant mass (K^+n) at 1573 ± 5 MeV with a significance of about 6 σ. The second peak is a candidate for an excited θ^+ state which is expected to exist with about ~ 50 MeV higher mass than the ground state, in agreement with the observation. A preliminary cross section estimate gives 5-12 nb for the low mass peak and 8-18 nb for the high mass peak. Cross sections have been reported also from the COSY-TOF collaboration [13] (proton beam 2.95 GeV on protons) which observed a θ^+ peak in the invariant mass pK_s^0. They measure a cross section of $0.4 \pm 0.1 \pm 0.1$ (syst) μb which is in rough agreement with predictions of 0.1-1 μb for p+p, p+n near threshold. The ZEUS collaboration [12] (e^+p $\sqrt{s}$=300-318 GeV) is the only experiment which observed for the first time the $\overline{\theta}^-$ state decaying in $\overline{p}K_s^0$ (fig. 1, right). Most experiments measure a θ^+ width consistent with the experimental resolution, while Zeus and Hermes give a measurement of width somewhat larger than their resolution. A measurement with a much improved resolution would be important. Non-observation of θ^+ in previous experiments lead to an estimate of its width to be of the order of 1 MeV or less [33]. This limit would gain in significance, once the lack of observation of the θ^+ peak by several experiments will be better understood.

Study of the θ^+ mass variation

Figure 1, left, as previously mentioned shows a compilation of the masses of θ^+ candidate peaks observed by several experiments. The statistical and systematic errors (when given) have been added in quadrature. For GRAAL we assume an error of 5 MeV as no error has been given in [19]. For the two preliminary peaks of CLAS we assume the systematic error of 10 MeV quoted previously by CLAS. The lines indicate the mean value of the mass among the $\theta^+ \rightarrow pK_s^0$ and the $\theta^+ \rightarrow nK^+$ observations.

It appears that the mass of θ^+ from $\theta^+ \rightarrow nK^+$ observations is systematically higher than the one from $\theta^+ \rightarrow pK_s^0$ observations. This may be related to the special corrections needed for the Fermi motion and/or to details of the analysis with missing mass instead of direct measurement of the decay products.

All observations together give a mean mass of 1.533 ± 0.023 GeV and they deviate from their mean with a χ^2/DOF of 3.92. The χ^2/DOF for the deviation of the $\theta^+ \rightarrow pK_s^0$ observations from their mean of 1.529 ± 0.011 GeV is 3.76. The χ^2/DOF for the deviation of the $\theta^+ \rightarrow nK^+$ observations from their mean of 1.540 ± 0.020 GeV is 0.94.

The bad χ^2/DOF for the $\theta^+ \rightarrow pK_s^0$ observations maybe due to an underestimation of the systematic errors. In particular in some cases no systematic errors are given, sometimes because the results are preliminary. If we add a systematic error of 0.5% of the measured mass (therefore of about 8 MeV) on all measurements for which no systematic error was given by the experiments, we arrive to a χ^2/DOF for the $\theta^+ \rightarrow pK_s^0$

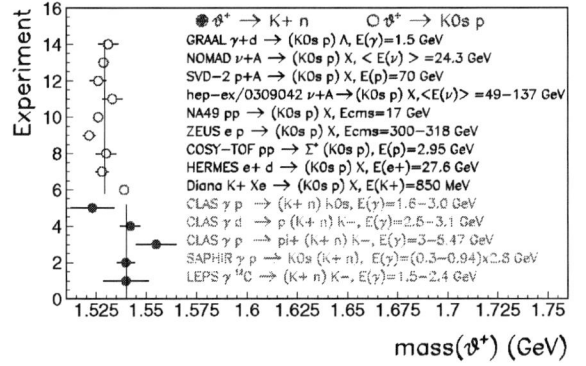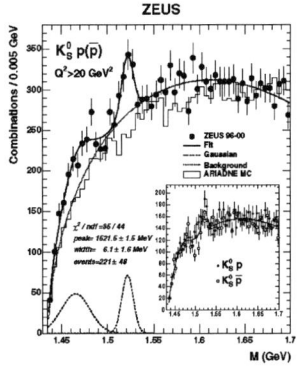

FIGURE 1. Left: Compilation of measured masses of θ^+ candidates. Right: Zeus results on the θ^+ candidate peak and its antiparticle.

observations of 0.95 and a mean mass of 1.529 ± 0.022 GeV. The χ^2/DOF for the $\theta^+ \rightarrow nK^+$ observations almost don't change by this, (mean mass = 1.540 ± 0.022 GeV, χ^2/DOF=0.91), because the experiments mostly give the systematic errors for this decay channel. All observations together give then a mean mass of 1.533 ± 0.031 GeV and they deviate from their mean with a χ^2/DOF of 2.1, reflecting mainly the difference of masses between the two considered decay channels. It is important to understand the origin of this discrepancy. This can be studied measuring $\theta^+ \rightarrow K^+ n$ in experiments with direct detection of the neutron or the antineutron for the $\overline{\theta^-}$ like PHENIX and GRAAL.

θ^{++} A preliminary peak is quoted by CLAS [32] for the candidate $\theta^{++} \rightarrow pK^+$ produced in the reaction $\gamma p \rightarrow \theta^{++} K^- \rightarrow pK^+ K^-$ at 1579 ± 5 MeV. A previous peak observed by CLAS in the invariant mass pK^+ has been dismissed as due to ϕ and hyperon resonance reflexion [34]. The STAR collaboration quoted a preliminary peak in the pK^+ and $\overline{p}K^-$ invariant masses at 1.530 GeV in d+Au collisions at $\sqrt{s}$=200 GeV [35].

Ξ, N^0 The NA49 experiment has observed in p+p reactions at $\sqrt{s}$=17 GeV the pentaquark candidates $\Xi^{--}(1862 \pm 2MeV) \rightarrow \Xi^- \pi^-$, the $\Xi^0(1864 \pm 5MeV) \rightarrow \Xi^- \pi^+$ and their antiparticles [15]. They measure a width consistent with their resolution of about 18 MeV. They also observe preliminary results of the decay $\Xi^-(1850) \rightarrow \Xi^0(1530)\pi^-$ with simarly narrow width as the other candidates [16].

The experiment STAR has shown preliminary results on a N^0 ($udsd\overline{s}$) or Ξ ($udss\overline{d}$) I=1/2 candidate [18]. STAR uses minimum bias Au+Au collisions at $\sqrt{s}$=200 geV and observes a peak in the decay channel ΛK_s^0 at a mass 1734 ± 0.5 (stat) ± 5 (syst) MeV with width consistent with the experimental resolution of about 6 MeV and $S/\sqrt{B}$ between 3 and 6 depending on the method used [18]. STAR also observe signs of the known narrow states $\Xi(1690)$ and $\Xi(1820)$ with a lower significance.

The GRAAL experiment has shown preliminary results on two narrow N^0 candidates. One candidate is observed at a mass of 1670 MeV in the invariant mass of ηn from the reaction $\gamma d \to \eta n X$. The neutron has been directly detected. The other is observed at a mass of 1727 MeV in the invariant masses of ΛK_s^0 as well as in the invariant masses of $\Sigma^- K^+$ at the same mass and with the same width [20]. The second reaction allow to establish the strange quark content and therefore to exclude the Ξ hypothesis. The difference of 7 MeV between the STAR and GRAAL measured masses of 1727 and 1734 MeV, should be compared to the systematic errors. STAR quotes a systematic error of 5 MeV while GRAAL quotes no systematic error.

The mass of the peaks at 1670 and at (1727,1734) MeV is in good agreement with the N masses suggested by Arndt et al [36]. In this paper a modified Partial Wave analysis allows to search for narrow states and presents two candidate N masses, 1680 and/or 1730 MeV with width below 30 MeV.

θ_c^0 The H1 collaboration at DESY used $e^- p$ collisions at $\sqrt{s}$=300 and 320 GeV and have observed a peak in the invariant masses $D^{*-} p$ and $D^{*+}\overline{p}$ at a mass 3099 ± 3 (stat) ± 5 (syst) MeV and width of 12 ± 3 MeV [17]. This peak is a candidate for the state θ_c^0 = $uudd\overline{c}$ and is the first charmed pentaquark candidate seen.

Lack of observation of pentaquark candidates

Several experiments have reported preliminary or final results on the non-observation of pentaquarks e.g. e^+e^-: Babar, Belle, Bes, LEP experiments, $p\overline{p}$: CDF, D0 pA:E690, γp: FOCUS, pA: HERA-B, ep: Zeus (for the θ_c^0) μ^+ ^6LiD: COMPASS, Hadronic Z decays: LEP, π, K, p on A: HyperCP, $\gamma\gamma$: L3, π, p, Σ on p: SELEX, pA: SPHINX, $\Sigma^- A$: WA89 $K^+ p$: LASS, [37]. HERMES has reported the non-observation of a θ^{++} candidate peak in the pK^+ invariant masses [10]. No other experiment has observed the candidates for the Ξ and the θ_c pentaquarks seen by NA49 and H1. Especially Zeus has searched for the θ_c under similar conditions as H1 and with similar statistics, without observing a peak [38]. Many of the experiments reporting non observation of pentaquarks have a very high statistics and good mass resolution.

It has been argued that the non-observation of pentaquark states in the above experiments could be due to an additional strong suppression factor for pentaquark production in e^+e^- collisions, as well as in B decays which is lifted in reactions like γA in which $s\overline{s}$ and a baryon are present in the initial state [39]. The constituents of the θ^+ are already present in the initial state of e.g. low energy photoproduction experiments, while in other experiments baryon number and strangeness must be created from gluons [39]. It is important to try to assess the expected cross sections.

The non observation of pentaquarks in high energy interactions of hadrons (CDF ($p\overline{p}$), E690 (pA) etc) can be a consequence of the decrease of the pentaquark cross section with increasing energy [40, 41]. This depends however on the kinematic region considered, and it is suggested to look for pentaquarks in the central rapidity region [40, 41].

In addition, if the θ^+ is produced preferably through the decay of a new resonance $N^0(2400) \to \theta^+ K^-$ as suggested by CLAS and NA49 and as discussed in [39, 42], neglecting this aspect maybe a further cause of its non-observation in some experiments. Some authors pointed out the importance to exclude kinematic reflexions as reason behind the θ^+ peak [43]. This known source of systematic errors is under investigation by the experiments which observe pentaquark candidates.

It is clear that a higher statistic is desirable in order to confirm the pentaquark observations reported so far. New data taken in 2004 and planned to be taken in 2005 will lead to enhancements in statistics of experiments up to a factor of 15 allowing to test the statistical significance and make more systematic studies. Experiments searching for pentaquarks should test also the production mechanisms proposed in the literature e.g. the θ^+ production through the $N^0(2400)$ decay. For example Phenix could search for the final state $\overline{\theta^-}K^+$ or $\overline{\theta^-}K_s^0$ demanding the invariant mass of $\overline{\theta^-}K_s^0$ and $\overline{\theta^-}K^+$ to be in the range 2.3 to 2.5 GeV, and study the option to trigger online on this channel.

3. SUMMARY AND CONCLUSIONS

Recent theoretical and experimental advances led to the observation of candidates for a number of pentaquarks states. In particular candidate signals have been observed for the $\theta^+(1533)$, $\theta^{++}(1530/1579)$, $\theta_{\bar{c}}^0(3099)$, $\Xi^{--}(1862)$, $\Xi^0(1864)$, $\Xi^-(1850)$, $N/\Xi^0(1734/1727)$, $N^0(2400)$, $N^0(1670)$ states as well as a possible excited $\theta_{\bar{s}}^+(1573)$ state. These observations are promising, however despite the large number of e.g. the $\theta_{\bar{s}}^+$ observations, they all suffer from a low individual statistical significance. A much higher statistics is needed to support and solidify the existing evidence. Several other high statistics experiments have reported lack of observation of those candidates. The inconsistency among experiments waits to be clarified through high statistics measurements of pentaquark candidates, their characteristics (cross sections, quantum numbers) and upper limits in the case of non-observation. Furthermore, systematic studies are needed as well as advanced theoretical understanding of the observations, in particular of the narrow width and the production mechanism of the observed candidates as well as the possible reasons behind their non-observations by some experiments. Combined theoretical and experimental efforts should be able to answer soon the question if pentaquarks exist.

REFERENCES

1. H. Walliser, Nucl. Phys. A 548 (1984) 649. A. Manohar, *Nucl. Phys.* **B 248** (1984) 19. H. J. Lipkin, Phys. Lett. 45 B, (1973), 267. R. L. Jaffe, K. Johnson, Phys. Lett. B 60, (1976), 201. R. Jaffe, talk given in the topological conference on Baryon Resonances, Oxford July 5-9, 1976, Oxford Top. Conf. 1976, p. 455.
2. M Preszalowicz, *World Scientific* (1987) 112, hep-ph/0308114.
3. D. Diakonov, M. Polyakov, V. Petrov, *Z. Phys. C* **359** (1997) 305.
4. M. Aguilar-Benitez et al., (Particle Data Group), Phys. Lett. B 170, (1986), 289. J. Amirzadeh et al., Phys. Lett. 89B (1979) 125.
5. LEPS Collaboration, (T. Nakano et al.), *Phys. Rev. Lett.* **91** (2003) 012002, hep-ex/0301020.
6. CLAS Collaboration, (S. Stepanyan et al.), *Phys. Rev. Lett.* **91**, (2003) 252001.
7. SAPHIR Collaboration, (J. Barth et al.), *Phys. Lett.* **B 572**, (2003) 127, hep-ex/0307083.
8. CLAS Collaboration, (V. Kubarovsky et al.), *Phys. Rev. Lett.* **92**, (2004) 032001, hep-ex/0311046.
9. DIANA Collaboration, (V. Barmin et al), *Phys. Atom. Nucl.* **66**, (2003) 1715, and Yad. Fiz. 66, (2003), 1763, hep-ex/0304040.
10. HERMES Collaboration, Phys. Lett. B 585, (2004), 213, [hep-ex/0312044].
11. A. E. Asratyan et al., Phys. Atom. Nucl. 67, (2004), 682, Yad. Fiz. 67, (2004) 704, hep-ex/0309042.
12. ZEUS Collaboration, Phys. Lett. B 591, (2004), 7, [hep-ex/0403051].
13. COSY-TOF Collaboration, Phys. Lett. B 595, (2004), 127, hep-ex/0403011.

14. L. Camilleri, NOMAD Collaboration, contribution to the '21st International Conference on Neutrino Physics and Astrophysics', Paris, 14-19 June, 2004.
15. NA49 Collaboration, (C. Alt et al.), *Phys. Rev. Lett.* **92** (2004) 042003.
16. K. Kadija, *talk presented in the Pentaquark 2003 Workshop*, Jefferson Lab, November 6-8, 2003, Virginia, USA.
17. H1 Collaboration, A. Aktas et al., accepted for publication in Phys. Lett. B, hep-ex/0403017.
18. S. Kabana et al, STAR Coll., Proceedings of the 20th Winter Workshop in Nuclear Dynamics, Jamaica, March 2004, hep-ex/0406032.
19. C. Schaerf et al., GRAAL collaboration, contribution to the conference Pentaquarks 2004, Spring-8, July 20, 2004.
20. V. Kouznetsov et al, GRAAL Collaboration, contribution to the workshop Pentaquarks, Trento, 10-12 February, 2004. http://www.tp2.ruhr-uni-bochum.de/talks/trento04/
21. R Jaffe, F. Wilczek, Phys. World 17, (2004), 25. K. Hicks, hep-ph/0408001. S.L. Zhu, hep-ph/0406204.
22. K. Hagiwara et al., *Phys. Rev.* **D 66** (2002) 010001. S. Eidelman et al., (Particle Data Group), Phys. Lett. B 592, 1 (2004).
23. A. Dzierba et al., hep-ex/0412077. K. Hicks, hep-ex/0412048. S. Kabana, proceedings of the International Workshop "IX Hadron Physics and VII Relativistic Aspects of Nuclear Physics: a Joint Meeting on QCD and QGP", (HADRON-RANP 2004), 28 March - 03 April, 2004, Rio de Janeiro, Brasil. S. Kabana, proceedings of the the International Conference on 'Strangeness in Quark Matter', 15-21 Sept. 2004, Cape Tawn, South Africa (J. of Phys. G: Nuclear and Particle Physics).
24. R. L. Jaffe, F. Wilczek, *Phys. Rev. Lett.* **91** (2003) 232003, hep-ph/0307341.
25. R. L. Jaffe, F. Wilczek, Phys. Rev. D 69, (2004) 114017, hep-ph/0312369.
26. D. Diakonov, V. Petrov, [hep-ph/0310212].
27. J. Ellis et al, JHEP 0405:002, 2004, hep-ph/0401127.
28. Fl Stancu, hep-ph/0408042.
29. M. Bleicher et al., Phys. Lett. B 595, (2004), 595. F. Csikor et al., hep-lat/0407033.
30. F. M. Liu et al, Phys. Lett. B 597, (2004), 333, hep-ph/0404156.
31. T. Nakano, LEPS Collaboration, contribution to the conference Pentaquarks 2004, Spring-8, July 20, 2004. http://www.rcnp.osaka-u.ac.jp/ penta04/talk/program_new.html
32. M. Battaglieri, CLAS Collaboration, contribution to the conference Pentaquarks 2004, Spring-8, July 20, 2004.
33. R. A. Arndt et al., Phys. Rev. C 68, (2003), 042201, nucl-th/0308012.
34. H. G. Juengst, CLAS Collaboration, nucl-ex/0312019.
35. J. Ma, APS meeting, 5 Jan. 2004. S. Kabana, RHIC and AGS user's meeting, BNL, 10-15 July 2004.
36. R. A. Arndt et al., *Phys. Rev.* **C69** (2004) 035208, hep-ph/0312126.
37. BABAR Collaboration, Contributions to the Intern. Conference on High Energy Physics (ICHEP 2004), Beijing, 16-22 August 2004, hep-ex/0408037 and hep-ex/0408064. I. Abt et al., HERA-B Collaboration, hep-ex/0408048. T. Wengler et al., LEP results (Aleph, Delphi, L3, Opal), Contribution to the Intern. Conference 39th Rencontres de Moriond on QCD and High-Energy hadronic interactions, 28 March-4 April 2004, La Thuile, hep-ex/0405080. I. Gorelov et al., CDF Collaboration, hep-ex/0408025. B. T. Huffman et al, D0 Collaboration, Fermilab-Conf-04/074-E, June 2004. R. Mizuk, Belle Collaboration, contribution to the conference Pentaquarks 2004, Spring-8, July 20, 2004. E. Gottschalk et al, E690 Collaboration, contribution to the conference Pentaquarks 2004, Spring-8, July 20, 2004. G. Brona et al., COMPASS Coll., 2004 wwwcompass.cern.ch/compass/notes/2004-5. K. Stenson et al., FOCUS Coll., 2004, hep-ex/0412021. M. Longo et al, HyperCP coll., hep-ex/0410027. J Napolitano et al, hep-ex/0412031. L3 Coll., ICHEP 2004 Conference, ichep04.ihep.ac.cn S. Armstrong, hep-ex/0410080. J Engelfried et al., SELEX Coll., these proceedings. Y. Antipov et al., SPHINX Coll., Eur. Phys. J. A21 2004 455, hep-ex/0407026. M Antamovich et al., WA89 Coll., hep-ex/0405042.
38. L. Gladilin, these proceedings, hep-ex/0412014.
39. M. Karliner, H. Lipkin, Phys. Lett. B 597, (2004), 309, hep-ph/0405002.
40. D. Diakonov, hep-ph/0406043.
41. A. Titov at al, nucl-th/0408001.
42. Y. I. Azimov, I. I. Strakovsky, hep-ph/0406312.
43. A. R. Dzierba et al., Phys. Rev. D 69, (2004), 051901, hep-ph/0311125.

A narrow "peanut" pentaquark

Dmitri Melikhov

Nuclear Physics Institute, Moscow State University, 119992, Moscow, Russia

Abstract. We analyse the decay $\Theta_s(1/2^+) \to NK$ in a non-relativistic Fock space description using three and five constituent quarks for the nucleon and the pentaquark, respectively. Following Jaffe and Wilczek [1], we assume that quark-quark correlations in spin-zero state play an important role for the pentaquark internal structure. Within this scenario, a strong dynamical suppression of the decay width is shown to be possible only if the pentaquark has an asymmetric "peanut" structure with the strange antiquark in the center and the two extended composite diquarks rotating around. In this case a decay width of $\simeq 1$ MeV may be a natural possibility.

The existence of pentaquarks is not yet undoubtedly established. But if these particles exist, the exotic members of the pentaquark multiplet must have a very small decay width of order 1 MeV or even lower. For the possible origin of the small pentaquark width many qualitative suggestions have been put forward. In a scenarios proposed by Jaffe and Wilczek [1] the positive-parity spin-1/2 pentaquark consists of an antiquark and two scalar diquarks in a relative P-wave state. In this talk I present the results of a fully dynamical quark-model calculation of the pentaquark width done together with B. Stech and S. Simula [2] using a non-relativistic Fock space representation for the $J^P = \frac{1}{2}^+$ pentaquark in the Jaffe-Wilczek scenario.

The decay amplitude $T(\Theta \to KN)$ is related to the matrix element

$$
\begin{aligned}
\langle N(p')|\bar{s}\gamma_\mu\gamma_5 d|\Theta(p)\rangle = & \; g_A(q^2)\bar{u}_N(p')\gamma_\mu\gamma_5 u_\Theta(p) + g_P(q^2)q_\mu\bar{u}_N(p')\gamma_5 u_\Theta(p) \\
& + g_T(q^2)\bar{u}_N(p')\sigma_{\mu\nu}q^\nu\gamma_5 u_\Theta(p), \qquad q = p - p'.
\end{aligned}
$$

Here the form factors g_i contain poles at $q^2 > 0$ due to strange meson resonances with the appropriate quantum numbers. The residue of the pole in g_P at $q^2 = M_K^2$ is related to the amplitude of interest $T(\Theta \to NK)$: $(M_K^2 - q^2)g_P(q^2)\bar{u}_N(p')\gamma_5 u_\Theta(p) \to f_K T(\Theta \to NK)$ for $q^2 \to M_K^2$, where $f_K = 160$ MeV is the kaon decay constant. The form factor g_A contains the pole at $q^2 = (K_A^*)^2$, but at $q^2 = M_K^2$ it is a regular function. Making use of the relationship between the form factors g_A and g_P emerging in the limit of spontaneously broken chiral symmetry [2] gives $T(\Theta \to NK) = \frac{M_\Theta + M_N}{f_K} g_A(M_K^2) \cdot \bar{u}_N(p')i\gamma_5 u_\Theta(p)$ and

$$
\Gamma(\Theta) = \Gamma(\Theta \to K^+ n) + \Gamma(\Theta \to K^0 p) \simeq \frac{1}{\pi}\frac{|\vec{q}|^3}{f_K^2} g_A^2(M_K^2).
$$

For $M_\Theta = 1540$ MeV one finds $|\vec{q}| = 270$ MeV and $\Gamma(\Theta) = 240\, g_A^2$ MeV. For transitions between hadrons of the same quark structure $g_A \simeq 1$ (e.g. for the nucleon $g_A \simeq 1.23$). So for a normal resonance one would expect $\Gamma(\Theta) \simeq 200$ MeV. To obtain a width of ≤ 10 MeV one needs a strongly suppressed value $g_A \leq 0.2$.

CP756, *Quark Confinement and the Hadron Spectrum VI*
edited by N. Brambilla, U. D'Alesio, A. Devoto, K. Maung, G.M. Prosperi and S. Serci
© 2005 American Institute of Physics 0-7354-0241-8/05/$22.50

In [2] we calculated the amplitude $\langle N|\bar{s}\gamma_\mu\gamma_5 d|\Theta\rangle$ and the form factor $g_A(q^2)$ using a non-relativistic equal-time Fock space representation. **The nucleon** in this framework is described by its radial wave function depending on the relative coordinates $\vec{\rho}_N = \vec{r}_2 - \vec{r}_3$ and $\vec{\lambda}_N = \frac{1}{2}(\vec{r}_2 + \vec{r}_3) - \vec{r}_1$, for which we take the Gaussian function

$$\Psi_N(r_1|r_2,r_3) \sim \exp\left(-\frac{1}{2\alpha_{\rho N}^2}\vec{\rho}_N^2 - \frac{2}{3\alpha_{\lambda N}^2}\vec{\lambda}_N^2\right).$$

The pentaquark wave function depends on the relative coordinates $\vec{r}_{23} = \vec{r}_2 - \vec{r}_3$, $\vec{R}_{23} = \frac{1}{2}(\vec{r}_2 + \vec{r}_3)$, $\vec{r}_{45} = \vec{r}_4 - \vec{r}_5$, $\vec{R}_{45} = \frac{1}{2}(\vec{r}_4 + \vec{r}_5)$, $\vec{\rho}_\Theta = \vec{R}_{23} - \vec{R}_{45}$, $\vec{\lambda}_\Theta = \frac{1}{2}(\vec{R}_{23} + \vec{R}_{45}) - \vec{r}_1$, where $\vec{r}_1$ is the position of the strange particle, $\vec{R}_{23}$ and $\vec{R}_{45}$ are the positions of the two diquarks. As required by the quark-diquark scenario, the pentaquark wave function factorizes into the diquark wave functions and the wave function of the three-particle quark-diquar-diquark system, for which we take again Gaussian parameterizations

$$\Psi_\Theta(r_1|r_2,r_3|r_4,r_5) \sim \exp\left(-\frac{1}{2\alpha_{\rho\Theta}^2}\vec{\rho}_\Theta^2 - \frac{2}{3\alpha_{\lambda\Theta}^2}\vec{\lambda}_\Theta^2\right)\exp\left(-\frac{\vec{r}_{23}^2}{2\alpha_D^2}\right)\exp\left(-\frac{\vec{r}_{45}^2}{2\alpha_D^2}\right).$$

The form factor g_A can be expressed through the following vector overlap amplitude

$$\frac{24}{\sqrt{3}}\int d\vec{r}_2 d\vec{r}_4 d\vec{r}_5 \exp\left(i\vec{q}\,\frac{\vec{r}_2 + \vec{r}_4 + \vec{r}_5}{3}\right)\vec{\rho}_\Theta\,\Psi_\Theta(r_s|r_2,r_d|r_4,r_5)$$

$$\times\{2\Psi_N(r_2|r_4,r_5) + \Psi_N(r_4|r_2,r_5)\},\qquad \vec{\rho}_\Theta = \frac{1}{2}(\vec{r}_2 + \vec{r}_d - \vec{r}_4 - \vec{r}_5).$$

Details of this calculation can be found in our paper [2].

Numerical estimates. We present now numerical results for the pentaquark width. Two assumptions reduce the number of parameters:

1. The structure of the diquark in the nucleon and in the pentaquark coincide, i.e. the size-parameter α_D of the diquark wave function Φ_D is equal to the parameter $\alpha_{\rho N}$ of the nucleon wave function, $\alpha_D = \alpha_{\rho N}$.

2. The parameters of the nucleon wave function are chosen such that the experimental nucleon electromagnetic form factor is reproduced for small momentum transfers, $\alpha_{\lambda N}^2/16 + \alpha_{\rho N}^2/48 = 1/M_\rho^2$. We first take a symmetric wave function $\alpha_{\lambda N} = \alpha_{\rho N} = 0.9$ fm. The diquark size parameter is then $\alpha_D = \alpha_{\rho N} = 0.9$ fm. Now only the two free parameters of the pentaquark wave function $\alpha_{\rho\Theta}$ and $\alpha_{\lambda\Theta}$ remain to be fixed. Recall that $\alpha_{\rho\Theta}$ determines the average distance between the two extended diquarks, and $\alpha_{\lambda\Theta}$ determines the average distance between the s-antiquark and the center-of-mass of the two diquarks. Little is known about the details of the pentaquark structure. Therefore we allow the parameters $\alpha_{\lambda\Theta}$ and $\alpha_{\rho\Theta}$ to vary in a broad range $0.6\,fm < \alpha_{\lambda\Theta},\,\alpha_{\rho\Theta} < 1.6\,fm$ and study the dependence of g_A and the width on these parameters.

Fig. 1(a) shows $\Gamma(\Theta)$ vs the pentaquark size parameters $\alpha_{\lambda\Theta}$ and $\alpha_{\rho\Theta}$. If both parameters are $\simeq 1$ fm, then $g_A \simeq 0.8$ and the width is 150 MeV. No suppression due to a possible mismatch of color and flavour quantum numbers in the initial and final states takes place. However, a strong dynamical suppression occurs if the structure of the pentaquark is asymmetric: For instance, for $\alpha_{\lambda\Theta} = 0.6\ fm, \alpha_{\rho\Theta} = 1.4\ fm$, we get $g_A = 0.05$ and $\Gamma(\Theta) = 1$ MeV.

Fig. 1(b) presents $\Gamma(\Theta)$ vs the diquark size α_D for fixed values of the pentaquark size-parameters $\alpha_{\rho\Theta} = \alpha_{\lambda\Theta} = 1$ fm. A sizeable reduction of the pentaquark width occurs only for a very small diquark size which corresponds to implausibly large deviations from a symmetric nucleon wave function. Such compact diquarks are not supported by a successful description of the nucleon properties with a symmetric wave function.

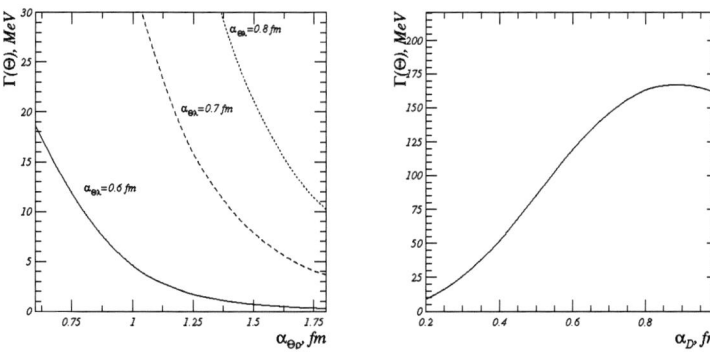

FIGURE 1. *Left (a):* $\Gamma(\Theta)$ vs the pentaquark size parameters $\alpha_{\rho\Theta}$ and $\alpha_{\lambda\Theta}$. *Right (b):* $\Gamma(\Theta)$ vs the diquark size parameter α_D for a symmetric pentaquark $\alpha_{\Theta\rho} = \alpha_{\lambda\Theta} = 1$ fm.

Summing up, the pentaquark decay width $\Gamma(\Theta)$ is found to depend strongly on the pentaquark configuration: when all size-parameters of the pentaquark wave function are close to 1 fm, one obtains a width of about 150 MeV, i.e. a typical hadronic value. *The color-flavour structure of the pentaquark causes no suppression of the width.*

A strong dynamical suppression of the amplitude occurs for a "peanut"-shaped pentaquark, i.e. when it has an asymmetric structure with $\alpha_{\lambda\Theta} \ll \alpha_{\rho\Theta}$. For instance, $\alpha_{\lambda\Theta} = 0.6$ fm and $\alpha_{\rho\Theta} = 1.4$ fm brings the width down to 1 MeV.

We therefore conclude that *if the pentaquark can be described as a five-quark system, in which two composite spin-zero diquarks are in the relative P-wave state, the small width requires a rather asymmetric "peanut" structure with two extended diquarks rotating about the strange antiquark localized near the center.*

ACKNOWLEDGMENTS

I would like to thank my friends Berthold Stech and Silvano Simula for the most enjoyable collaboration, and Nora Brambilla for the invitation to participate in this exciting Conference.

REFERENCES

1. R.Jaffe, F.Wilczek, *Phys. Rev. Lett.* **91**, 232003 (2003).
2. D.Melikhov, S.Simula, B.Stech, *Phys. Lett.* **B594**, 265 (2004).

Heavy Quarks

Joan Soto

Universitat de Barcelona, Departament d'Estructura i Constituents de la Matèria, Diagonal 647,
08028 Barcelona, Catalonia, Spain

Abstract. I summarize the contributions to the Heavy Quark Section presented in the conference. The topics discussed include beauty production, heavy-light systems, heavy quarkonium, doubly heavy baryons and Higgs-bound heavy quarkonium.

Keywords: heavy quarks, heavy quark effective theory, non-relativistic QCD.

PACS: 12.38.Bx, 12.38.Cy, 12.38.Gc, 12.39.Hg, 12.39.Jh, 12.39.St, 13.20.Gd, 13.20.He, 13.30.Eg, 14.20.Lq, 14.40.Gx, 14.40.Lb, 14.40.Nd .

INTRODUCTION

The scope of the Heavy Quark Section of this conference was somewhat detailed in the summary talk of the previous edition [1]. Let me then start here by comparing the gross features of the contributions to this conference to the ones of the previous one . This time we had two experimental contributions [2, 3], two lattice contributions [4, 5], and seven phenomenological ones [6, 7, 8, 9, 10, 11, 12]. This means a decrease in the number of experimental contributions, which was exceptionally large two years ago, in favor of an increase in the number of phenomenological and lattice ones. Heavy-light systems [3, 4, 8, 9], heavy quarkonium [5, 6, 7, 10, 11, 12] and doubly heavy baryons [3] were discussed in both conferences, with a larger number of talks on heavy quarkonium this time. It was also noticeable the increase of talks which were reporting on results obtained using effective field theory (EFT) techniques of various sorts [4, 5, 6, 7, 9].

The paper will be distributed as follows. In the first section beauty production at Hera-B will be briefly discussed. The next three sections will be devoted to heavy-light systems, heavy quarkonium and doubly heavy baryons respectively. In the fift section, the possibility of having heavy quarkonium states bound by Higgs exchange (rather than by gluon exchange) will be briefly discussed. Within each section I will first mention related plenary talks in the conference (if any) and next focus on experimental contributions (if any), then on contributions using phenomenological models (if any) and finally on contributions based on QCD (if any), in particular those using EFTs. I will finish up with a brief sumary and outlook.

BEAUTY PRODUCTION

We had an experimental talk by Maaijke Mevius on bottom production at HERA-B [2]. This experiment studies collisions of $920 GeV$ protons on different nuclear targets. The reported open beauty production was $\sigma(b\bar{b}) = 9.8 \pm 14 \pm 20$nb/nucleon. This measurement is normalized to the J/ψ production cross section obtained in the E789 [14]

CP756, *Quark Confinement and the Hadron Spectrum VI*
edited by N. Brambilla, U. D'Alesio, A. Devoto, K. Maung, G.M. Prosperi and S. Serci
© 2005 American Institute of Physics 0-7354-0241-8/05/$22.50

and E771 [15] experiments scaled to HERA-B energies ($\sigma(J/\psi) = 357 \pm 8 \pm 27$). Results from hidden beauty production were also reported ($BR(\Upsilon \rightarrow l^+ l^-) \times \frac{d\sigma_\Upsilon}{dy}|_{y=0} = 3.4 \pm 0.8$pb/nucleon, $l^+ l^- = e^+ e^-, \mu^+ \mu^-$).

HEAVY-LIGHT MESONS

Heavy-light systems were pretty well represented in the plenary talks. Experimental results from CLEO were reported by John Napolitano [16]. Spectroscopy was discussed by Estia Eichten [17]. Factorization in semi-inclusive and exclusive decays was discussed by Matthias Neubert [18] within the framework of Soft-Collinear Effective Theory [19]. Finally the lattice treatment of Heavy Quark Effective Theory [20, 27, 28] was discussed by Rainer Sommer [21].

Concerning the contributions in the parallel session, we had an interesting experimental talk [3] reporting on recent results from the SELEX collaboration. Concerning heavy-light mesons, the highlight was the discovery of a narrow ($\Gamma < 17 MeV$) charmed-strange state $D_{sJ}^+(2632)$, which has been observed in two decay modes, namely $D_s^+ \eta$ and $D^0 K^+$, with surprising branching fractions $\Gamma(D_s^+ \eta)/\Gamma(D^0 K^+) \sim 6$ [13].

The model [22], which includes coupled channel effects, was presented by George Rupp [8]. This model is not specially designed for heavy-light mesons but for mesons in general and hence it may be of interest to specialist of Section B [23]. It also appears to accomodate well heavy-light states, in particular $D_{s0}^*(2317)$.

Fulvia de Fazio gave a talk on B-meson decays to two light vector particles [9]. In order to understand the experimental measurements on the polarization fractions, a model, which incorporates chiral and heavy quark symmetries [25, 26], is used to analize the rescattering of D, D_s-mesons [24]. It was argued that these effects can modify the helicity amplitudes in penguin dominated processes.

Anthonny Green reported on latice results for excited states of heavy-light mesons [4, 29]. The heavy quark is treated as a static source and the mass of the light quark is close to the strange quark. UKQCD configurations, quenched and dynamical, are used. Predictions for yet-to-be-discovered B-states are given using linear interpolations between known D-states and the static limit.

HEAVY QUARKONIUM

Heavy Quarkonium was also present in the plenary session talks by John Napolitano on CLEO results [16], by Estia Eichten on spectroscopy and threshold effects [17] and by Antonio Vairo [31], who reported on selected results from the Quarkonium Working Group [32].

Concerning the contributions in the parallel sessions, Orlando Oliveira gave a talk in which using classical solutions of Yang-Mills theories as a background for propagating heavy quarks, the spectrum of heavy quarkonium could be obtained [10, 30].

Dieter Gromes presented preliminary results on a calculation of the $^1P_1^{+-}$ quarkonium annihilation decay in the weak coupling regime using the Bethe-Salpeter equation [11, 33]. Such a calculation would allow to extract the Non-Relativistic QCD (NRQCD)

[34, 35, 36] matrix element $\langle {}^1P_1|O_8({}^1S_0)|{}^1P_1\rangle$ in the weak coupling regime [35]. It would be interesting to see if modern EFTs techniques [36] provide a more economical way to obtain such a matrix element in this regime.

Geoffrey Bodwin presented preliminary results on a lattice evaluation of a number of NRQCD decay matrix elements [5]. This is important in order to check the velocity scaling rules proposed in [35], which have been challenged by several authors. These rules are crucial to predict transverse polarization of ψs and Υs at large $\mathbf{p}_\perp$ [37]. The preliminary results, which are obtained in the quenched approximation and have not been translated to $\overline{MS}$ scheme, are in general compatible with the original scaling rules.

Matthias Steinhauser reported on a number of results for heavy quarkonium which were obtained using Potential NRQCD (pNRQCD) [38, 36] in the weak coupling regime [7]. These included (i) NNNLO calculations of the spectrum [39], which can be used to extract m_b and eventually m_t at future linear colliders; (ii) renormalization group improved calculations, which lead to a NLL prediction of the η_b mass [40] and of the B_c^* mass [41], and to a NNLL prediction of $\Gamma(\eta_b \to \gamma\gamma)$ [42].

Xavier Garcia i Tormo presented results on the end-point region of the photon spectrum in semi-inclusive radiative decays of $\Upsilon(1S)$ [6]. They are obtained by combining techniques of SCET [19, 43], NRQCD [34, 35, 36] and pNRQCD [38, 36]. They include a calculation of the so-called color octet shape functions using pNRQCD in the weak coupling regime [44] and lead to a very good description of experimental data [45].

DOUBLY HEAVY BARYONS

Part of the talk by Jürgen Engelfried on recent results by the SELEX collaboration [3] was devoted to double charm baryons. The highlight was the confirmation of the Ξ_{cc}^+ state [46] through the recently observed decay to pD^+K^- [47]. Isospin partners of various candidate states were also discussed.

ELECTROWEAK HEAVY QUARKONIUM

Quarks of masses similar or larger than the top quark mass are not expected to form QCD bound states since they rapidly decay due to electroweak interactions. Namely, the corresponding hadrons have decay widths similar or larger than the QCD binding energies. For would-be toponium only a bump is expected to be seen in $t\bar{t}$ production cross section [48]. However, for masses large enough, the Higgs exchange between the quark and the anti-quark produces an atractive Yukawa potential, which overtakes the strength of the one-gluon exchange potential. If such quarks, for instance belonging to a fourth generation, existed in nature, a new type of heavy quarkonium state could be realized. The talk by Gennady Kozlov entertained this possibility [12, 49]. Constrains on the masses of such a quark and the Higgs were obtained for such a new heavy quarkonium state to exists. Also, an exhaustive analysis was presented on its experimental signatures both at the Tevatron and at the LHC.

SUMMARY AND OUTLOOK

Let me try to give the flavor of the evolution of the Heavy Quark Section in time by comparing to it again to the one of the previous conference. There were some talks which show a certain continuity in a line of research. The clearest example is the one by Anthonny Green [4], who continues reporting on lattice investigations of heavy-light mesons in the static limit [50], this time on excited states, which shows that progress is steadily being made. A second line which was also present in the last conference concerns hadronic effects in B-meson decays. We had a talk by T.N. Pham last time [51], and one by Fulvia de Fazio this one [9]. A third line following the same pattern are models for heavy-light spectroscopy. We had a talk by last time by D. Ebert [52] and one by George Rupp this time [8], although Rupp's model is claimed to hold for light meson spectroscopy as well. Experimental talks on heavy-light spectroscopy were also given last [53] and this [3] time. There is also some sort of continuity in talks which have more to do with electroweak physics than with QCD heavy quark bound states. We had an experimental talk by S. Jain last time [56] on the possibility of producing new particles in the $t\bar{t}$ channel at the Tevatron, and, this time, we had an interesting talk by Gennady Kozlov on a new kind (Higgs-bound) of heavy quarkonium state [12]. Continuity also exists in contributions which, being related to heavy quark bound states, do not belong to the main lines of research that people follow nowadays: we had a talk by A. Weber [57] last time and a talk by Orlando Oliveira this one [10] which fall into this class.

Doubly heavy baryons were also present in the last conference in the talk on a potential model description of them by D. Ebert [52] and have also been present in this one by the interesting experimental talk by Juergen Engelfried [3]. As briefly discussed in [1], a QCD based EFT description of these sytems has not been worked out in detail yet (see [60, 61] for early contributions), and would be an interesting developement for the future.

A qualitative change has occurred for contributions related to heavy quarkonium. We had quite a few experimental contributions last time [54, 53, 55], one contribution on NRQCD sum rules [58], and one contribution on potential models [59]. This time, however, we had no experimental contribution, which was somewhat compensated for by the plenary talk of John Napolitano [16], one talk on Bethe-Salpeter treatment by Dieter Gromes [11], one on lattice NRQCD by Geoffrey Bodwin [5] and two talks on pNRQCD by Xavier Garcia i Tormo and Matthias Steinhauser respectively [6, 7]. Clearly, QCD based approaches, mainly using Effective Field Theories, are taking over. In particular, I would like to single out two things: in the talk by Matthias Steinhauser the higher order calculations carried out so far in the weak coupling regime were presented, and in the talk by Xavier Garcia i Tormo, SCET was first used in a talk of the heavy quark section (see also the plenary talk by Matthias Neubert [18]). I expect more talks using SCET in next conference.

ACKNOWLEDGMENTS

I acknowledge financial support from a CICYT-INFN 2004 collaboration contract, Acciones Integradas (Spain-Italy) HI2003-0362, the MCyT and Feder (Spain) grant FPA2001-3598, the CIRIT (Catalonia) grant 2001SGR-00065 and the network EU-RIDICE (EU) HPRN-CT2002-00311.

REFERENCES

1. J. Soto, arXiv:hep-ph/0301138.
2. M. Mevius, *in these proceedings*.
3. J. Engelfried, *in these proceedings*.
4. A. M. Green, *in these proceedings*.
5. G. T. Bodwin, *in these proceedings*.
6. X. Garcia i Tormo, *in these proceedings*, arXiv:hep-ph/0410052.
7. M. Steinhauser, *in these proceedings*.
8. G. Rupp, *in these proceedings*.
9. F. de Fazio, *in these proceedings*.
10. O. Oliveira, *in these proceedings*.
11. D. Gromes, *in these proceedings*.
12. G. A. Kozlov,, *in these proceedings*.
13. A. V. Evdokimov *et al.* [SELEX Collaboration], arXiv:hep-ex/0406045.
14. M. H. Schub *et al.* [E789 Collaboration], Phys. Rev. D **52** (1995) 1307 [Erratum-ibid. D **53** (1996) 570].
15. T. Alexopoulos *et al.* [E-771 Collaboration], Phys. Rev. D **55** (1997) 3927.
16. J. Napolitano, *in these proceedings*.
17. E. Eichten, *in these proceedings*.
18. M. Neubert, *in these proceedings*.
19. C. W. Bauer, S. Fleming, D. Pirjol and I. W. Stewart, Phys. Rev. D **63** (2001) 114020 [arXiv:hep-ph/0011336].
20. M. A. Shifman and M. B. Voloshin, Sov. J. Nucl. Phys. **45** (1987) 292 [Yad. Fiz. **45** (1987) 463]; Sov. J. Nucl. Phys. **47** (1988) 511 [Yad. Fiz. **47** (1988) 801]; E. Eichten and B. Hill, Phys. Lett. B **234** (1990) 511; B. Grinstein, Nucl. Phys. B **339** (1990) 253. H. Georgi, Phys. Lett. B **240** (1990) 447.
21. R. Sommer, *in these proceedings*.
22. E. van Beveren and G. Rupp, Mod. Phys. Lett. A **19**, 1949 (2004) [arXiv:hep-ph/0406242].
23. A. Williams, *in these proceedings*.
24. P. Colangelo, F. De Fazio and T. N. Pham, Phys. Lett. B **597** (2004) 291 [arXiv:hep-ph/0406162].
25. G. Burdman and J. F. Donoghue, Phys. Lett. B **280** (1992) 287.
26. R. Casalbuoni, A. Deandrea, N. Di Bartolomeo, R. Gatto, F. Feruglio and G. Nardulli, Phys. Rept. **281** (1997) 145 [arXiv:hep-ph/9605342].
27. N. Isgur and M. B. Wise, Phys. Lett. B **232** (1989) 113; Phys. Lett. B **237** (1990) 527.
28. M. Neubert, Phys. Rept. **245** (1994) 259 [arXiv:hep-ph/9306320].
29. A. M. Green, J. Koponen, C. McNeile, C. Michael and G. Thompson [UKQCD Collaboration], Phys. Rev. D **69** (2004) 094505 [arXiv:hep-lat/0312007].
30. O. Oliveira and R. A. Coimbra, arXiv:hep-ph/0305305.
31. A. Vairo, *in these proceedings*.
32. http://www.qwg.to.infn.it/
33. D. Gromes, Eur. Phys. J. C **36** (2004) 169 [arXiv:hep-ph/0403290].
34. W. E. Caswell and G. P. Lepage, Phys. Lett. B **167** (1986) 437; B. A. Thacker and G. P. Lepage, Phys. Rev. D **43** (1991) 196.
35. G. T. Bodwin, E. Braaten and G. P. Lepage, Phys. Rev. D **51** (1995) 1125 [Erratum-ibid. D **55** (1997) 5853] [arXiv:hep-ph/9407339].
36. N. Brambilla, A. Pineda, J. Soto and A. Vairo, arXiv:hep-ph/0410047.

37. P. L. Cho and M. B. Wise, Phys. Lett. B **346** (1995) 129 [arXiv:hep-ph/9411303].
38. A. Pineda and J. Soto, Nucl. Phys. Proc. Suppl. **64** (1998) 428 [arXiv:hep-ph/9707481]; N. Brambilla, A. Pineda, J. Soto and A. Vairo, Nucl. Phys. B **566** (2000) 275 [arXiv:hep-ph/9907240].
39. A. A. Penin and M. Steinhauser, Phys. Lett. B **538** (2002) 335 [arXiv:hep-ph/0204290].
40. B. A. Kniehl, A. A. Penin, A. Pineda, V. A. Smirnov and M. Steinhauser, Phys. Rev. Lett. **92** (2004) 242001 [arXiv:hep-ph/0312086].
41. A. A. Penin, A. Pineda, V. A. Smirnov and M. Steinhauser, Phys. Lett. B **593** (2004) 124 [arXiv:hep-ph/0403080].
42. A. A. Penin, A. Pineda, V. A. Smirnov and M. Steinhauser, Nucl. Phys. B **699** (2004) 183 [arXiv:hep-ph/0406175].
43. C. W. Bauer, C. W. Chiang, S. Fleming, A. K. Leibovich and I. Low, Phys. Rev. D **64** (2001) 114014 [arXiv:hep-ph/0106316]; S. Fleming and A. K. Leibovich, Phys. Rev. D **67** (2003) 074035 [arXiv:hep-ph/0212094].
44. X. Garcia i Tormo and J. Soto, Phys. Rev. D **69**, 114006 (2004) [arXiv:hep-ph/0401233].
45. B. Nemati *et al.* [CLEO Collaboration], Phys. Rev. D **55** (1997) 5273 [arXiv:hep-ex/9611020].
46. M. Mattson *et al.* [SELEX Collaboration], Phys. Rev. Lett. **89** (2002) 112001 [arXiv:hep-ex/0208014].
47. A. Ocherashvili *et al.* [SELEX Collaboration], arXiv:hep-ex/0406033.
48. A. H. Hoang *et al.*, Eur. Phys. J. directC **2** (2000) 3 [arXiv:hep-ph/0001286].
49. G. A. Kozlov, A. N. Sisakian, Z. I. Khubua, G. Arabidze, G. Khoriauli and T. Morii, J. Phys. G **30** (2004) 1201.
50. A. M. Green, J. Koponen, P. Pennanen and C. Michael [UKQCD Collaboration], arXiv:hep-lat/0212017.
51. arXiv:hep-ph/0301160.
52. D. Ebert, R. N. Faustov and V. O. Galkin, *Prepared for 5th International Conference on Quark Confinement and the Hadron Spectrum, Gargnano, Brescia, Italy, 10-14 Sep 2002*
53. *Prepared for 5th International Conference on Quark Confinement and the Hadron Spectrum, Gargnano, Brescia, Italy, 10-14 Sep 2002*
54. Y. S. Zhu [BES Collaboration], *Prepared for 5th International Conference on Quark Confinement and the Hadron Spectrum, Gargnano, Brescia, Italy, 10-14 Sep 2002*
55. J. Tseng [CDF collaboration], FERMILAB-CONF-02-348-E *Presented at 5th International Conference on Quark Confinement and the Hadron Spectrum, Gargnano, Brescia, Italy, 10-14 Sep 2002*
56. S. Jain, N. K. Mondal and D. Chakraborty [D0 Collaboration], *Presented at 5th International Conference on Quark Confinement and the Hadron Spectrum, Gargnano, Brescia, Italy, 10-14 Sep 2002*; Pramana **62** (2004) 561 [arXiv:hep-ex/0302037].
57. A. Weber, *Presented at 5th International Conference on Quark Confinement and the Hadron Spectrum, Gargnano, Brescia, Italy, 10-14 Sep 2002*
58. M. Eidemuller, *Presented at 5th International Conference on Quark Confinement and the Hadron Spectrum, Gargnano, Brescia, Italy, 10-14 Sep 2002*; Nucl. Phys. Proc. Suppl. **121** (2003) 213 [arXiv:hep-ph/0209022].
59. S. Godfrey and J. L. Rosner, arXiv:hep-ph/0210399.
60. M. J. Savage and M. B. Wise, Phys. Lett. B **248** (1990) 177.
61. M. A. Sanchis-Lozano, Phys. Lett. B **321** (1994) 407; Nucl. Phys. B **440** (1995) 251 [arXiv:hep-ph/9502359].

Hadronization and Quark Probes of Deconfinement at RHIC

Huan Z. Huang* and Johann Rafelski†

*Department of Physics and Astronomy, University of California, Los Angeles, CA 90095-1547
†Department of Physics, University of Arizona, Tucson, AZ 85721-0081

Abstract. [1] We discuss experimental features of identified particle production from nucleus-nucleus collisions. These features reflect hadronization from a deconfined partonic matter whose particle formation scheme is distinctly different from fragmentation phenomenology in elementary collisions. Multi-parton dynamics, such as quark coalescences or recombinations, appear to be essential to explain the experimental measurements at the intermediate transverse momentum of 2–5 GeV/c. Constituent quarks seem to be the dominant degrees of freedom at hadronization. Heavy quark production should help quantify deconfined matter properties.

Keywords: Quark Gluon Plasma, Deconfinement, Relativistic heavy-ion collisions, Strangeness, Hadronization, Collective Flow, Elliptic Flow, Charm production
PACS: 25.75.Nq,25.75.-q,24.10.Pa,25.75.Dw

INTRODUCTION

The advent of the Relativistic Heavy Ion Collider (RHIC) at Brookhaven National Laboratory (BNL) has turned the next page in search for, and study of, the Quark Gluon Plasma (QGP). Most recent dAu baseline reaction measurements by all four RHIC experiments [1, 2, 3, 4] confirmed that a dense strongly interacting medium has been created in central AuAu collisions at RHIC. The QCD nature of the dense matter created at RHIC and whether the current experimental evidence proves the discovery of the QGP have been since under debate within the heavy ion physics community [5, 6, 7, 8]. We will address here primarily the physics 'soft' hadron experimental results and the related evidence for deconfinement. The major topics we address are: 1) features of hadronization and other evidence for a color deconfined bulk partonic matter; 2) the QCD properties of the matter at hadronization; and 3)charm production and related future experimental measurements capable to further quantify the properties of deconfined phase.

Among topics we discuss in depth are: mechanisms of hadronization, experimental status of strange particle production enhancement, azimuthal parton flow anisotropy, bulk properties of dense hadron matter at parton fireball breakup, charm experimental status at RHIC.

---

[1] This report constitutes the combined contribution of the authors to the "VI Quark Confinement and the Hadron Spectrum" conference, held 21–25 September 2004 in Sardinia, Italy covering their oral presentations and the 'Deconfinement' discussion session; proceedings to be published by the American Institute of Physics.

CP756, *Quark Confinement and the Hadron Spectrum VI*
edited by N. Brambilla, U. D'Alesio, A. Devoto, K. Maung, G.M. Prosperi and S. Serci
© 2005 American Institute of Physics 0-7354-0241-8/05/$22.50

FEATURES OF HADRONIZATION OBSERVED AT RHIC

What we know about hadronization

We observe, in the final state, a multitude of hadrons, irrespective of what happened and which reaction system is observed. The paradigm emerged that QCD color charges are confined and hadrons exist in color singlet state. Individual reactions occur between leptons and/or quarks. However, the conversion, *i.e.*, hadronization of quarks and gluons (partons) has not been understood based on first principles. There are several widely studied models:

1. Hadron formation in elementary e^+e^- and qq (that is nucleon–nucleon) collisions has been described in the pQCD domain in terms of several components. The particle production process is factorized into parton distribution functions, parton interaction processes and in final step, fragmentation functions for hadron formation.

   The fragmentation function is assumed to be universal and can be obtained phenomenologically from e^+e^- collision measurments. A typical Feynmen-Field [9] fragmentation process involves a leading parton of momentum p, which fragments into a hadron of momentum p_h whose properties are mostly determined by the leading parton. The fragmentation function is a function of variable $z = p_h/p$, $z \in (0,1)$. Note that baryon production is found to be significantly suppressed compared to the production of light mesons: the baryon to pion ratio increases with z, but never exceeds 20% [10], see also the pp-STAR results in figure 2 below.

2. In the soft (low $p_\perp$) particle production region, the pQCD framework and factorization break down, particles are believed not to be from fragmentation of partons. In the elementary interactions physics, string fragmentation models were inspired by the QCD description of the quark and anti-quark interaction. The Lund string model is one of the popular hadron formation models which has been successfully implemented in Monte Carlo description of e^+e^-, pp, and nuclear collisions [11]. Note that, in the string fragmentation models, the baryon production is also suppressed because baryon formation requires the clustering of three quarks [12].

3. Soft multi particle production in particular in the AA (nuclear) collision domain is described within the (Fermi) Statistical Hadronization (SH) model. SH is a model of particle production in which the birth process of each particle fully saturates (maximizes) the quantum mechanical probability amplitude, and thus, all hadron yields are determined by the appropriate integrals of the accessible phase space.

   The statistical hadronization model introduced in 1950 [13, 14] has matured today to a full fledged tool for soft strongly interacting particle production, capable to describe in detail hadron abundances once the mass spectrum of hadron resonances is included [15]. The key SH parameters within the grand canonical formulation of particle phase space are the temperature T and, at a finite baryon density present in a AA reaction system, the baryochemical potential μ_B. It is generally accepted

211

that as the energy of the colliding nuclei varies, a wide domain of T and μ_B is explored, see figure 2 in [16]. Note that the ratio of baryons to mesons is, in chemical equilibrium, also suppressed, since $e^{\frac{m_{\pi,K} - m_{\text{baryon}}}{T}} \ll 1$

4. The reaction picture of a soup of quarks in a dense expanding fireball inevitably triggers development of recombination and coalescence models [17, 18]. Recent experimental RHIC results triggered further developments on quark coalescence [19, 20] and recombination [21, 22] which all have the essence of multiparton dynamics for hadron formation, despite significant differences in details.

(Anti)Baryon yield

Strangeness plays a particularly important role as a characteristic QGP observable. The enhanced production of (strange) (anti)baryons has been an expected feature in recombinant hadronization of QGP. In nuclear collisions, the strange hadron yield derives from two independent reaction steps following each other in time:

1. the establishment of a ready supply of strange quark pairs $s, \bar{s}$ which occurs predominantly in the initial hot phase of bulk partonic matter by the *thermal* pQCD gluon fusion processes $gg \to s\bar{s}$ [23], in a manner independent of the production of final state hadrons;

2. the high initial $s, \bar{s}$ yield survives the process of fireball expansion evolution [17], and contributes in hadronization of pre-formed $s, \bar{s}$ quarks to an unusually high multistrange (anti)baryon abundance [24].

This is illustrated in figure 1.

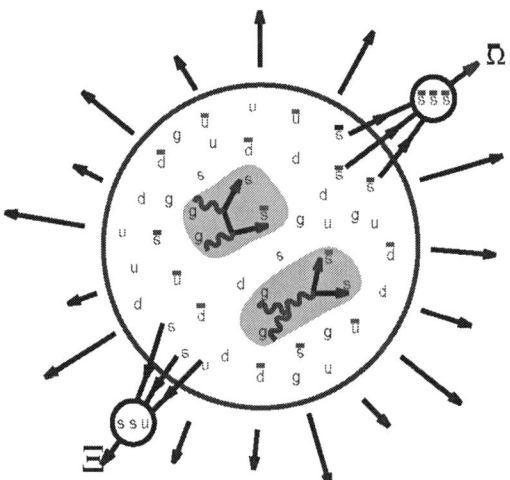

FIGURE 1. Illustration of the two step mechanism of strange hadron formation from QGP: inserts show gluon fusion into strangeness, followed by QGP recombinant hadronization.

Experiments indeed show very significant new features of hadron formation in AA interactions compared to elementary e^+e^- and qq (that is pp) collisions.

- Baryon production from AA collisions, especially for multi-strange hyperons, has been measured to be much larger than theoretical (superposition, cascade) model calculations. The hyperon production per number of participant pairs from AA collisions at the SPS is significantly enhanced in comparison with the value from pp collisions. The enhancement factor increases from Λ to Ω hyperons and with the collision centrality [25, 26].
- The increase in the baryon production from AA collisions has become much more prominent at RHIC energies. Figure 2 shows the ratios of $\bar{p}/\pi$ and $\bar{\Lambda}/K_S$ from central AuAu collisions at $\sqrt{s_{NN}} = 130$, and 200 GeV measured by PHENIX [27] and STAR [28, 29, 30].

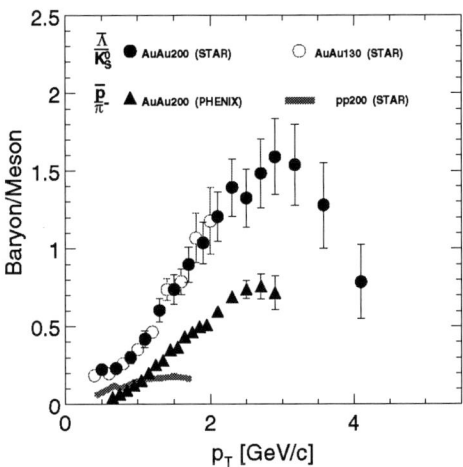

FIGURE 2. Ratios of $\bar{\Lambda}$ to K_S from AuAu and pp collisions (STAR) and $\bar{p}$ to π from AuAu collisions (PHENIX) as a function of transverse momentum $(p_\perp)$. In addition to resonance contributions in all hadrons, the $\bar{\Lambda}$ includes contributions from Σ^0 decays.

The apparent difference, in figure 2, between the STAR and PHENIX ratios can be understood since the STAR $\bar{\Lambda}$ data include the electromagnetic decay contribution from Σ^0 and both comprise different 'towers' of hadron resonance decays. Note also that the PHENIX data are corrected for post-reaction weak decay contributions $Y, \bar{Y} \to N\bar{N}$. Similarly, the STAR $\bar{\Lambda}$ data have been corrected for feed-down contributions from multi-strange hyperon decays. Some early data, for example [31], are not shown because these data were not corrected for weak decay contributions.

The large baryon to meson ratio, seen in figure 2, cannot be accommodated by the traditional, $e.g.$, fragmentation scheme. The large ratio at the intermediate $p_\perp$ region provide clear evidence that particle formation dynamics in AA collisions at RHIC are distinctly different from the traditional hadron formation mechanism via fragmentation processes developed for the elementary e^+e^- and nucleon-nucleon collisions.

The recombination models [21, 22] provided a satisfactory description of the particle yields, in particular, the large production of baryons in the intermediate $p_\perp$ region. The formation of a dense partonic system provides a parton density dependent increase in baryon yield as a function of collision centrality through the coalescence mechanism.

Comparing AuAu with pp reactions at $\sqrt{s_{NN}} = 200$ GeV, see figure3, we see another large baryon production enhancement for strange (anti)baryons $\Lambda, \overline{\Lambda}, \Xi^-, \overline{\Xi}^+$ reported by STAR [32]. This enhancement increases with centrality [33, 34] and with greater strangeness content as found in strangeness recombination model [17]. These feature were recognized early on as a characteristic signature of QGP [24].

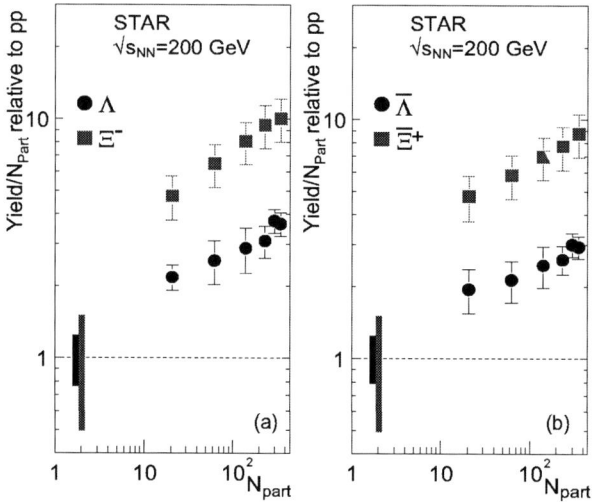

FIGURE 3. The STAR yields per participant for production of Λ and Ξ^- on left (a) and $\overline{\Lambda}$ and $\overline{\Xi}^+$ on right (b) in AuAu collisions at $\sqrt{s_{NN}} = 200$ GeV. Error bars are statistical. Ranges for pp reference data indicate the systematic uncertainty.

This pattern of baryon enhancement has been observed by the WA97 [25] and NA57 [26] experiments at lower reaction energy available at the CERN-SPS. There is a gradual increase in the strange antibaryon yield with reaction energy [26]. The large difference between baryon and antibaryon yields at SPS energy range are due to the presence of a significant baryon density at lower reaction energy. For purpose of comparison with RHIC results, this baryon density effect can be removed by considering the geometric mean of the baryon and antibaryon yield.

In our opinion, the results presented above for the systematics of strange hadron enhancement demonstrate that same novel mechanism operates in the entire collision energy interval spanned by these data. Only a deconfined quark-gluon plasma is a copious source of these often rarely produced hadrons. When the plasma fireball breaks up into final hadrons, the high abundance of strange quarks and antiquarks manifests itself yielding high abundances of multi strange hadrons. In conventional reaction schemes, the production of particles containing two or more strange quarks is suppressed by the rarity of the required reaction processes.

Nuclear modification factor

The nuclear modification factor is defined in a collision number scaled comparison of peripheral with central head-on collisions:

$$R_{CP} = \frac{[yield/N_{bin}]^{central}}{[yield/N_{bin}]^{peripheral}}. \quad (1)$$

The nuclear modification factor has also been defined by,

$$R_{AA} = \frac{[yield]^{AA}}{N_{bin} \times [yield]^{pp}}, \quad (2)$$

where N_{bin} is the number of binary nucleon–nucleon collisions. The $[yield]^{AA}$ and $[yield]^{pp}$ are particle yields ($d^2n/dp_\perp dy$) from AA and pp collisions, respectively.

A R_{AA} or R_{CP} of unity implies that particle production from AA collision is equivalent to a superposition of independent nucleon–nucleon collisions. Hard scattering processes within the pQCD framework are believed to follow approximately the binary scaling in the kinematic region where the nuclear shadowing of parton distribution function and the Cronin effect are not significant.

Measurements of charged hadrons and neutral pions have revealed a strong suppression at high $p_\perp$ region in central AuAu collisions [35, 36, 37, 38]. Recent dAu measurements [1, 2, 3, 4] have demonstrated that the large suppression of high $p_\perp$ particles in central AuAu collisions is mainly due to energy loss, presumably of partons traversing the dense matter created in these collisions.

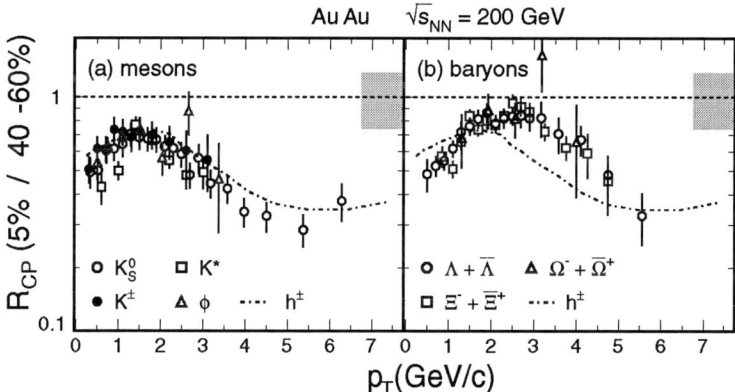

FIGURE 4. R_{CP} of $K^\pm$, K_S, K^*, ϕ, Λ, Ξ and Ω in comparison with charged hadron in dashed line. Distinct meson and baryon groups are observed.

Figure 4 shows the R_{CP} of $K^\pm$, K_S, K^*, ϕ, Λ, Ξ and Ω as a function of $p_\perp$ from AuAu collisions at $\sqrt{s_{NN}} = 200$ GeV measured by the STAR collaboration, where the ratio is derived from the most central 5% to the peripheral $40--60\%$ collisions. The dashed line is the R_{CP} of charged hadrons for reference. In the low $p_\perp$ region, soft particle production is dominated by the number of participant scaling.

The particle type dependence can be described using hydrodynamic flow which predicts a mass dependence for the low $p_\perp$ spectra. In the intermediate $p_\perp$ region of 2 to 5.5 GeV/c the $p_\perp$ dependence of R_{CP} falls into two groups, one for mesons and one for baryons. Despite the large mass differences between K_S and K^*/ϕ, and between Λ and Ξ little difference among the mesons and among the baryons has been observed within statistical errors. Particle dependence in the nuclear modification factor disappears only above a $p_\perp$ of 6 GeV/c, consistent with the expectation from conventional fragmentation processes. The unique meson and baryon dependence in the intermediate $p_\perp$ region indicates the onset of a production dynamics very different from both fragmentation at high $p_\perp$ and hydrodynamic behavior at low $p_\perp$.

Azimuthal anisotropy

The azimuthal angular particle distribution can be described by a Fourier expansion,

$$\frac{d^2 n}{p_\perp dp_\perp d\phi} \propto (1 + 2\sum_n v_n \cos n(\phi - \Psi_R)), \tag{3}$$

where Ψ_R is the reaction plane angle and v_2 has been called elliptic flow [39]. In a non-central AA collision the overlapping participants form an almond shaped particle emission source, see figure 5. The reaction plane is defined by the vectors x (impact parameter direction) and z (beam direction). The pressure gradient is greater and particles would experience larger expansion force along the short axis direction (in plane) than along the long axis (out plane), see figure 5 left, resulting in the final state in an ellipsoid in transverse momentum space. Theoretical, (typically hydrodynamic) model calculations find that this $p_\perp$ dependent component in v_2 is generated mostly at the early stage of the nuclear collision.

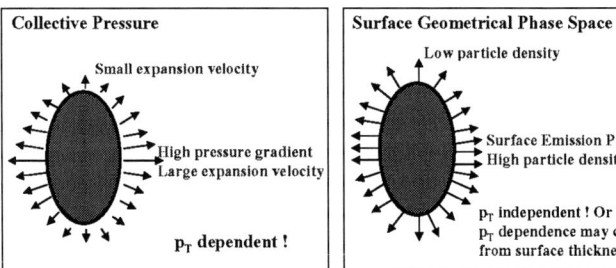

FIGURE 5. Schematic diagram to show two dynamical origins of angular anisotropy, one from hydrodynamical expansion and the other from surface geometrical phase space.

There is another dynamical mechanisms responsible for v_2 due to the spatial geometrical component in the phase space of emitting source. Particle production from a dense matter can be squeezed more in the reaction plane than that out of the reaction plane.

More generally, a freeze-out geometry will impose itself on particle abundance ellipticity. However, any $p_\perp$ dependence would be a 2nd order effect associated *e.g.* with the optical depth of the freeze-out system. Figure 5 shows a schematic diagram for these two mechanisms of generating azimuthal angular anisotropy v_2 in non-central collisions.

Figure 6 shows the v_2 as a function of $p_\perp$ for π, K, p, Λ and Ξ from PHENIX [40] and STAR [41] measurements. The angular anisotropy v_2 reveals three salient features:

1. particles exhibit hydrodynamic behavior in the low $p_\perp$ region — a common expansion velocity may be established from the pressure of the system and the heavier the particle the larger the momentum from the hydrodynamic motion leading to a decreasing ordering of v_2 from π, to K and p for a given $p_\perp$;
2. v_2 values do not depend strongly on $p_\perp$ at the intermediate $p_\perp$ region in contrast to the strong $p_\perp$ dependence in the yield of particles;
3. the saturated v_2 values for baryons are higher than those for mesons and there is a distinct grouping among mesons and baryons.

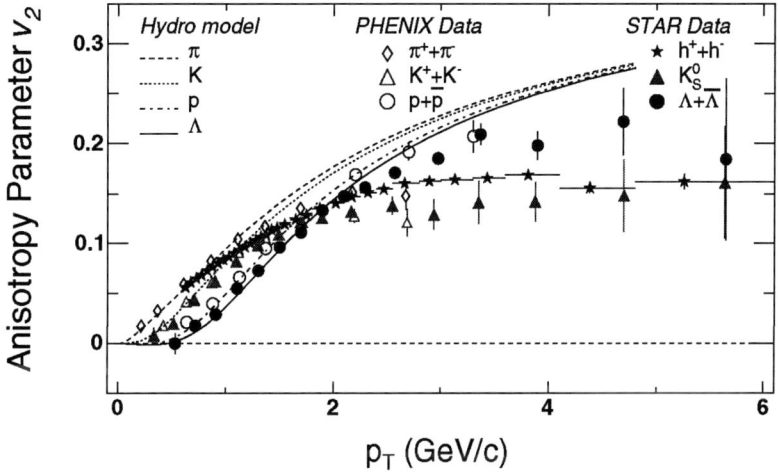

FIGURE 6. Azimuthal angular anisotropy v_2 as a function of $p_\perp$ for identified particles. The hydrodynamic calculation results are by P. Huovinen *et al.* [42].

The absence of a strong $p_\perp$ dependence at the intermediate $p_\perp$ region is an intriguing phenomenon for the angular anisotropy of particle emission. Parton energy loss in the dense medium created in *AA* collisions was proposed as a possible mechanism for generating an angular anisotropy v_2. High energy partons are quenched inside the dense medium and lead to an effective particle emission just from a shell area of the participating volume [43]. Jet quenching scenario [44] (parton energy loss scenario) cannot explain the particle dependence in both the nuclear modification factor and the angular anisotropy v_2 at the intermediate $p_\perp$ region. Within the parton energy loss scenario the larger v_2 of baryons implies a higher energy loss than that of mesons; the larger nuclear modification factor of baryons, however, is only consistent with a smaller energy loss than that of mesons. This scenario may be important for the considerable

v_2 magnitude for charged hadrons at a $p_\perp$ greater than 6 GeV/c or so measured by STAR [45].

Indeed, the magnitude of the measured v_2 is significantly larger than what can be accommodated based on particle emission from a geometrical ellipsoid source within an energy loss scenario [46]. Using a more realistic Wood–Saxon description of the colliding nuclei for the ellipsoid source, the predicted theoretical v_2 is much smaller than that from a hard-sphere model of the colliding nuclei, leading to a greater discrepancy between the measurement and the theoretical expectation.

We conclude that the magnitude and the particle dependence of v_2 at the intermediate $p_\perp$ region cannot have a dynamical origin either from hydrodynamic flow or from parton energy loss alone. Another plausible procedure would be to relate the v_2 in the intermediate $p_\perp$ region to the geometrical shape of the emitting particle source, see figure 5. A surface emission pattern from the almond shaped participant volume should naturally lead to a saturation of v_2. A surface emission scenario is possible if particles are produced by surface related dynamical instabilities and the hadronization duration is relatively short [47].

Quark Flow Anisotropy

An empirical quark number n scaling of v_2 has been noted [48]. n is the number of valance quarks and antiquarks in a hadron. Figure 7 presents the v_2/n as a function of $p_\perp/n$ for K, p, Λ and Ξ from AuAu200 collisions, where the line is a polynomial fit to the data points. The bottom panel shows the ratio of data points to the fit. At the intermediate $p_\perp$ region ($0.6 < p_\perp/n < 2$ GeV/c), all meson and baryon data points fall onto an uniform curve. The π data (not shown) are significantly above the v_2 of other mesons. The large fraction of resonance decay contribution to the π yield is known to enhance the v_2 of π at a given $p_\perp$ [49, 50].

For $p_\perp/n < 0.6$ GeV, there appears a small residual but systematic particle dependent deviation from the fit curve, which agrees best with the K_S results. The residual mass and/or strangeness and/or meson–baryon dependence of this deviation can have many causes involving microscopic transport and/or collective flow phenomena. Its detailed understanding could play an important role in the demonstration of physics processes that lead to the constituent quark level azimuthal flow at hadronization.

We believe that the particle dependence on v_2, which is largley explained by n-scaling, requires a v_2 distribution at the constituent quark level. The scaled azimuthal angular anisotropy (v_2/n) may be interpreted as the constituent quark anisotropy just prior to the hadron formation. The n-scaling works both for strange and nonstrange quarks, indicating that the azimuthal flow anisotropy is the same for all three 'light' flavors, and by extension, that the collective quark flow is the same for the three flavors. The precision of the n scaling leaves very little space for the presence of gluon (fragmentation) participation in the intermediate momentum particle production process. This suggests that semi-hard gluons have effectively disapperad as independent partonic degrees of freedom in the final hadron formation.

We have shown in this section that the hadron formation dynamics, at RHIC at

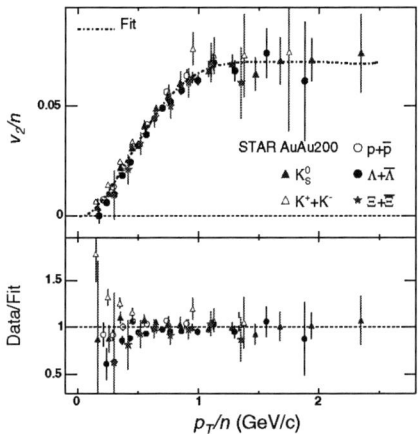

FIGURE 7. Azimuthal angular anisotropy v_2/n as a function of $p_\perp/n$ for identified particles where n is the number of constituent quarks. The line is a polynomial fit to the data points excluding the π data. The bottom panel shows the (multiplicative) deviation from the fit line.

$0.6 < p_\perp/n < 2$, GeV/c is very different from the parton fragmentation picture where the leading parton plays a dominant role in determining properties of the final state hadron. The measured features for these intermediate $p_\perp$ hadrons produced at RHIC require that all quark ingredients ($n = 2$ for mesons and $n = 3$ for baryons) play an approximately equal role in the hadron formation and that the hadron properties are determined by the sum of contributing partons.

DECONFINED BULK MATTER

Bulk properties as function of centrality

The global study of soft hadron yields and spectra implies that we can fully characterize the properties of bulk matter at time of hadronization by adding up strangeness, entropy, etc., contained in the final state hadrons [51]. This is done particularly easily within the statistical hadronization scheme, which have been successfully applied to describe stable particle production.

As a first step, this leads to a quantitative understanding of the variation of the bulk properties with centrality [52]. In such an analysis of $\sqrt{s_{NN}} = 200$ GeV STAR and PHENIX results, the contents in entropy, baryon number, strangeness, and energy grows linearly with the participant number for $A > 20$. This implies that the conversion of colliding nuclear matter into bulk parton matter is at RHIC rather independent of the size of nuclear overlap and complete, including all matter. In our above discussion of v_2, we have tacitly assumed that the properties of bulk partonic matter formed in non central collisions are not significantly dependent on centrality, as the present study confirms

In most central 5% of the reactions, one finds that the net baryon density per unit of rapidity is at mid-rapidity $d(B-\bar{B})/dy = 14\pm2$, the strangeness yield $ds/dy = 135\pm10$ and the entropy yield $dS/dy = 4900\pm400$. The errors comprise the uncertainty of the data and the chemical scheme used. The strangeness per entropy yield $s/S = (2.9\pm0.3)10^{-3}$ is more than 4 times enhanced compared to the AGS energy scale. For the most peripheral reactions, this ratio is $s/S = (1.9\pm0.3)10^{-3}$ which shows the influence of the fireball expansion dynamics on production of strangeness.

Given the experimental yields of particles at central rapidity, the intensive properties of bulk matter at hadronization, such as P pressure, ε energy density, σ entropy density, $E/TS = \varepsilon/T\sigma$, seen in figure 8, follow summing the properties of individual particle fractions. They are given for the two chemical models in figure 8 on left. The lines guide the eye, the actual results are the squares and circles centered at the mean trigger centrality of the data considered. These results are for the full chemical non-equilibrium (squares) and for strangeness chemical non-equilibrium (circles) respectively.

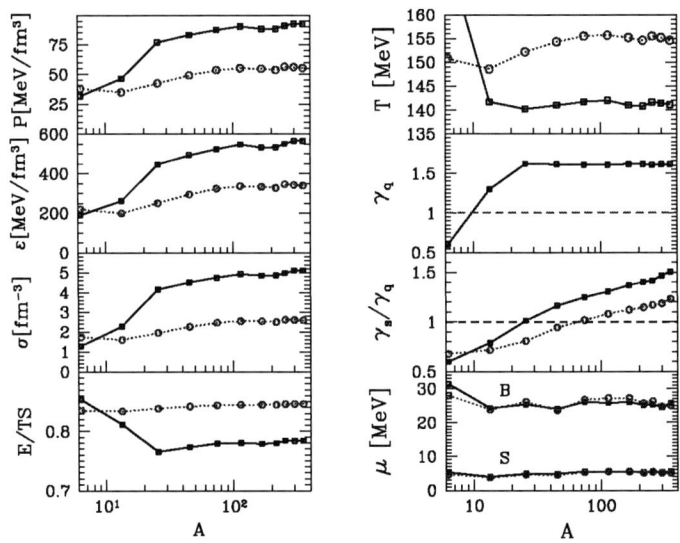

FIGURE 8. Left: pressure P, energy density $\varepsilon = E/V$, entropy density S/V and E/TS and; Right: temperature T, light quark phase space occupancy γ_q, the ratio of strange to light quark phase space occupancies γ_s/γ_q and the chemical potentials (B for baryochemical μ_B and S for strangeness μ_S) as a function of centrality. After Ref. [52].

To obtain the physical properties, a fit of the statistical hadronization model parameters had been performed and the results are shown on right: the hadronization temperature T, phase space occupancies $\gamma_q, \gamma_s/\gamma_q$, and the chemical potential $\mu_{B,S}$ of baryon number and strangeness, respectively. We see the expected rise of strangeness occupancy yield with size (centrality) of the system, related to the extended lifespan during which strangeness is produced [33, 34]. Otherwise there is remarkable stability of the bulk properties with centrality of the collision. The bulk matter created at RHIC at $\sqrt{s_{NN}} = 200$ GeV, small or large in size and independently of models of chemical (non)equilibrium has a baryochemical potential $\mu_B = 25\pm1$ MeV.

The Phase Boundary

The hadronization temperature $T = 140$–155 ± 8 MeV is near but below the phase transformation boundary for QCD with 2+1 flavors. The infinite matter static equilibrium transformation is expected at $T = 164 \pm 4$ MeV, for more detail about the structure of the cross-over region, see figure 9.

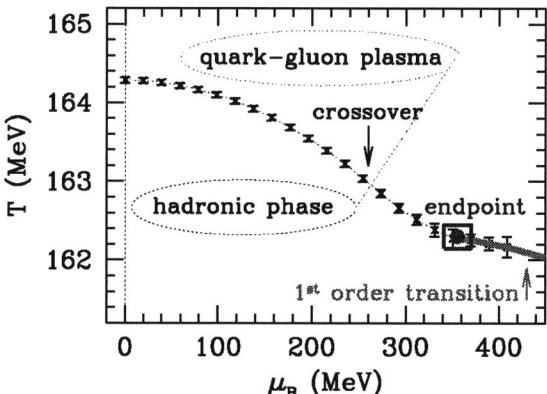

FIGURE 9. The phase diagram at low baryon density in physical units, results of Ref. [53]. Phase crossover ends in the square which shows the endpoint where 1st order phase transition sets in.

The lower RHIC freeze-out temperature we reported above is associated with a rapid expansion of the RHIC fireball: the flow of color charge pushes the vacuum and the bulk breakup occurs suddenly from an over expanded supercooled condition. Quantitatively the supercooling is of the required magnitude [47]. The sudden hadronization mechanism also explains why single freeze-out analysis of particle spectra at SPS and RHIC succeed [54, 55].

HEAVY QUARKS

Theoretical remarks

Heavy quarks (charm, in future bottom) are produced mostly in initial parton scattering (gluon fusion) and in any case during the initial primary and/or very high temperature phase of the collision. Therefore, heavy quark measurement can probe the initial parton flux, the dynamical evolution and the quark energy loss in dense medium. If heavy quarks are found to participate in the collective motion of the medium (radial or elliptic flow), this will lend further confirmation for parton collectivity.

At RHIC, one expects that multiple heavy quark–antiquark pairs will be produced in an individual AA collision. Therefore one can expect a novel mechanism of J/Ψ formation by recombination [56]. Then, there should be a contribution to heavy quarkonium formation which utilizes combinations of initially uncorrelated quark and antiquark, leading to a *quadratic* increase with the total number of heavy flavor quarks in the event. This can occur within the deconfined phase, considering that recent Lattice QCD (LQCD) calculations indicate that spectral functions of pseudo scalar and vector mesons have non-trivial shapes at a temperature above the critical T_c [57]. In particular, heavy quarkonium such as J/ψ may survive at a temperature above $1.6T_c$ [58, 59].

Charm quark transport dynamics in dense nuclear medium will provide unique probes to the QCD properties of the medium. If the initial temperature is very high, $T \sim 500$ MeV or higher, the yield of total charm quarks can also be increased through thermal gluon–gluon scatterings. Possible suppression of charm mesons at high $p_\perp$ will test the energy loss dynamics of charm quark propagation in a QCD medium. Theoretical calculations have predicted a reduced medium induced energy loss for heavy quarks and the high $p_\perp$ suppression of charmed hadrons should not be as strong as light hadrons [60, 61]. For a recent review of theoretical charm quark situation see Ref. [62].

RHIC experimental status

Both the STAR and PHENIX collaborations at RHIC have been pursuing vigorous heavy quark physics programs, both in analysis of current data and in future detector upgrade plans, to provide better capabilities for heavy quark measurements.

The total charm quark pair production cross section ($\sigma_{c\bar{c}}$) is an important constraint on the collision dynamics and the heavy quark evolution. Both STAR and PHENIX have presented results on the $\sigma_{c\bar{c}}$ measurement of pp collisions from charm semi-leptonic decays. In addition, STAR has also derived an equivalent $\sigma_{c\bar{c}}$ for pp collisions based on direct reconstruction of hadronic decays of charm mesons from dAu collisions. Figure 10 shows the $p_\perp$-spectra of electron and D^0 from STAR. The PHENIX preliminary non-photonic electron data are represented by the fitted line with a reported measurement $\sigma_{c\bar{c}} = 709 \pm 85(\text{stat})+332 - 281(\text{sys}) \, \mu b$ [63]. STAR has measured $\sigma_{c\bar{c}} = 1300 \pm 200 \pm 400 \, \mu b$ and a mean transverse momentum for D^0 of $\langle p_\perp \rangle = 1.32 \pm 0.08$ GeV/c from direct D^0 reconstruction [64]. A next-to-leading order pQCD calculation of the charm quark production cross section [65] has yielded $\sigma_{c\bar{c}} = 300 - 450 \, \mu b$, significantly below the STAR measurement and at the lower end of the PHENIX range of uncertainty.

Several comments on the cross section measurements are in order. When measuring the charm cross section through semi-leptonic decay, the quality of the electron data for $p_\perp \lesssim 1$ GeV/c is significantly deteriorated because of a large combinatorial background. An electron of $p_\perp \sim 1$ GeV/c typically comes from the decay of a D meson with $p_\perp \sim 2$ GeV/c, which is significantly beyond the average $p_\perp$ of 1.32 ± 0.08 GeV/c reported by STAR. Therefore, one has to extrapolate to the low $p_\perp$ region by over a factor of two to obtain the total charm cross section. Such an extrapolation is often model dependent and has a large uncertainty. The semi-leptonic decay branching ratios for D^0, D^*, $D^\pm$ and D_S are different. The electron yield from decays of these D mesons depends on the relative yield which is another important contribution to the uncertainties of the charm production cross section derived from electron measurement. The direct reconstruction of the D decay kinematics provides a broad coverage of $p_\perp$ and does not suffer from the same uncertainties as the leptonic decay electron measurement. However, present

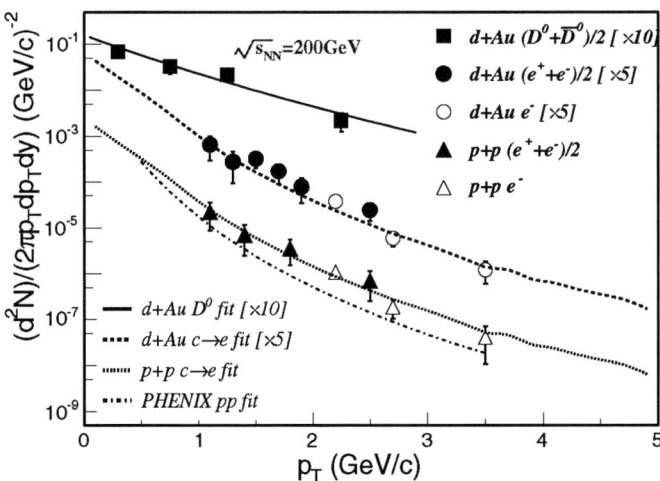

FIGURE 10. $p_\perp$ distribution of D^0 mesons and non-photonic electrons from semi-leptonic decays of charm mesons.

STAR measurements of D meson yields using event-mixing methods from TPC tracks suffer from limited statistics. A future vertex detector upgrade capable of measuring the D decay vertex displacement is essential for both STAR and PHENIX heavy flavor physics programs.

Figure 11 presents the STAR preliminary transverse momentum spectrum of D mesons from dAu collisions normalized by the number of binary collisions, where the $p_\perp$ shapes of D^* and $D^\pm$ are assumed to be the same as that of D^0 [66]. The shape of the $p_\perp$ distribution coincides with the bare charm quark $p_\perp$ distribution from the Fixed-Order-Next-Leading-Log (FONLL) pQCD calculation from M. Cacciari *et al.* [67]. If a fragmentation function such as the Lund fragmentation scheme [68] or the Peterson function [69] is introduced for D meson production, the resulting $p_\perp$ distribution will be significantly below the measurement at the high $p_\perp$ region. This observation raises an outstanding question regarding the $p_\perp$ distribution and the formation mechanism of D mesons in hadro-production. Recently, a $k_\perp$ factorization scheme has been found to significantly change the D meson p_T distribution from nuclear collisions as well [70].

The fact that the D meson $p_\perp$ distribution can be better described by the $p_\perp$ of bare charm quarks from pQCD calculations has been observed in previous fixed target experiments [71]. With a fragmentation function such as the Peterson function the D meson $p_\perp$ distribution is too steep to explain the measured $p_\perp$ distribution. In order to match the calculation with the data one has to introduce a $k_\perp$ kick to the parton distribution. The scale of $\langle k_\perp^2 \rangle$ is on the order of 1 (GeV/c)^2, much larger than the typical Λ_{QCD} scale for strong interaction. Furthermore, the Feynman x_F of the D meson distribution was also found to coincide with the bare charm quark x_F distribution [71]. The fragmentation function will have a large impact on the x_F distribution from bare charm quark to D meson, which cannot be negated by introducing any k_L longitudinal boost of reasonable

223

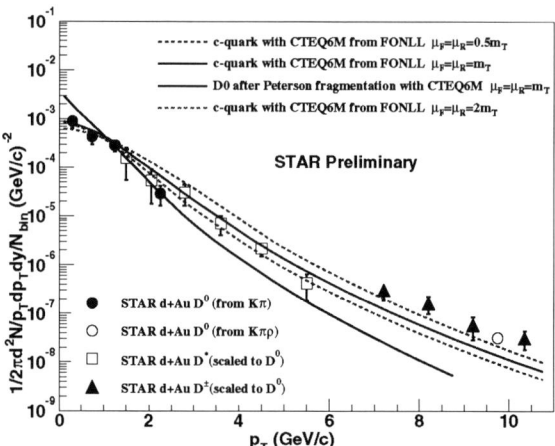

FIGURE 11. STAR preliminary measurement of $p_\perp$ distribution of D mesons from dAu collisions normalized by the number of binary collisions. The shape of the $p_\perp$ distribution is compared to pQCD FONLL calculations.

scale as in the case for $k_\perp$ kick in the transverse momentum direction.

The transverse momentum distribution of particles produced at RHIC energies are considerably flatter than those at lower energies. The $k_\perp$ kick scheme does not change the shape of the $p_\perp$ distribution significantly. The STAR measurement of the D meson $p_\perp$ distribution suggests either that the charm quark fragmentation may be close to a delta function or that a charm formation mechanism such as the recombination model may be important in hadro-production. The recombination model takes a charm quark and combines it with a light quark, presumably of low $p_\perp$, to form a charmed meson. Therefore, the $p_\perp$ of the meson is not significantly different from that of the bare charm quark. In that way, measurements of charmed meson production provide unique probes for the hadron formation dynamics and for the transport dynamics of heavy quarks in the dense nuclear medium produced at RHIC energies.

Observation of heavy quark hydrodynamic flow would indicate that heavy quarks, once created in the initial state, must have participated in the partonic hydrodynamic evolution over a sufficiently long period of time to reach a substantial flow magnitude. This would be a unique probe for the early stage of a partonic phase [72]. Figure 12 shows preliminary STAR [73] and PHENIX [74] measurements of the v_2 for electrons from charm leptonic decays, which have been demonstrated to be closely correlated with charmed meson v_2 [50].

SUMMARY

Among the most important aspect of many results presented is that the gluon degrees of freedom are not explicitly manifested in the hadron formation. Quarks are the dominant

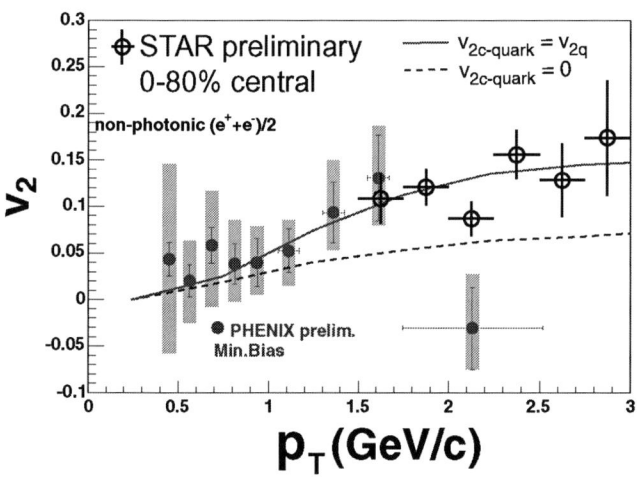

FIGURE 12. Preliminary measurement of non-photonic electron v_2 as a function of p_T from PHENIX and STAR. The curves are from a calculation by V. Greco *et al.* [72].

degrees of freedom at the boundary of quark–hadron transition. It is a critical step to firmly connect this mainly experimental insights on the properties of the quark matter at the boundary of hadronization with the LQCD calculations. The disappearance of the gluon degrees of freedom from the initial state and the emergence of constituent quarks at hadronization are some of the critical conceptual questions to be addressed in the imminent future [75].

Strange (anti) baryon production enhancement at RHIC as function of centrality, energy and $p_\perp$ is significant. For $p_\perp \lesssim 5$ GeV/c, relative baryon to meson yields have shown distinct features that are drastically different from the fragmentation processes in elementary collisions. A novel feature of meson and baryon dependence has been observed in the nuclear modification factor and the angular anisotropy v_2 of π, $K^\pm$, K_S, K^*, p, Λ and Ξ particles at the intermediate $p_\perp$ of 2–5 GeV/c. A constituent quark number scaling has been observed for the v_2 measurement. These experimental measurements suggest that quarks have developed a collective v_2 as a function of $p_\perp$; and the hadron formation at the intermediate $p_\perp$ is likely through multi-parton dynamics such as recombination or coalescence process.

This physical picture emerging from the experimental measurements complements the Lattice QCD results. Spectral function calculations have indicated that hadrons, particularly heavy quarkonium, do not melt completely at critical temperature. It appears plausible that the constituent quark degrees of freedom or even hadron-like quasi-particles play a dominant role at the hadronization of bulk partonic matter though further confirmation of the picture from LQCD is needed. We note the related theoretical models [76], which invoke the notion of quasi-hadrons to describe properties of the dense matter.

We have also described how heavy (charm) quark production and its transport dynamics in dense nuclear medium probe QCD properties of the matter. The charm quark flow measurement will provide a significant insight on recombination or coalescence hadronization mechanism and partonic collectivity of the dense matter. Charmonium enhancement arises from a dense charm rich state. Future detector upgrades from STAR and PHENIX will greatly enhance their heavy quark measurement capabilities at RHIC.

Despite intriguing experimental observations of hadronization from a deconfined bulk partonic matter, a smoking colt signatures for the quark–hadron phase transition remain elusive. On the other hand, the totality of experimental results can be understood invoking deconfinement.

ACKNOWLEDGMENTS

We thank An Tai, Hui Long, Paul Sorensen, Xin Dong, Frank Laue, Zhangbu Xu, Nu Xu, Charles Whitten Jr., Jean Letessier, Giorgio Torrieri, Robert Thews for many stimulating discussions on physics topics in this article. We thank Nora Brambilla, Giovanni Prosperi, and the local organizers of the "VI Quark Confinement and the Hadron Spectrum" conference, for their very kind hospitality in Sardinia.

REFERENCES

1.   J. Adams *et al.*, Phys. Rev. Lett. **91**, 072304 (2003).
2.   S. S. Adler *et al.*, Phys. Rev. Lett. **91**, 027303 (2003).
3.   B.B. Back *et al.*, Phys. Rev. Lett. **91**, 027302 (2003).
4.   I. Arsene *et al.*, Phys. Rev. Lett. **91**, 027305 (2003).
5.   M. Gyulassy and L. McLerran, Preprint nucl-th/0405013 (2004).
6.   K. Adcox *et al.*, Preprint nucl-ex/0410003 (2004).
7.   J. Adams *et al.*, Preprint nucl-ex/0501009 (2005).
8.   B. Muller, Preprint nucl-th/0404015 (2004).
9.   R. D. Field and R. P. Feynman, Nucl. Phys. B**136**, 1 (1978).
10. K. Abe *et al.*, Phys. Rev. D**69**, 072003 (2004).
11. Bo Andersson, 'The Lund Model' Cambridge Monogr. Part. Phys. Nucl. Phys. Cosmol. 7 (1997).
12. Bo Andersson *et al.*, Phys. Scripta **32**, 574 (1985).
13. E. Fermi, Prog. Theor. Phys. **5**, 570 (1950).
14. I. Pomeranchuk, Proc. USSR Academy of Sciences (in Russian) **43**, 889 (1951). Translated in: "Quark–Gluon Plasma: Theoretical Foundations" An annotated reprint collection; Edited by J. Kapusta, B. Muller and J. Rafelski ISBN: 0-444-51110-5 , 836 pages , Elsevier, New York, (2003).
15. R. Hagedorn, Suppl. Nuovo Cimento **2**, 147 (1965).
16. R. Hagedorn and J. Rafelski, Phys. Lett. B **97**, 136 (1980).
17. P. Koch, B. Muller and J. Rafelski, Phys. Rept. **142**, 167 (1986).
18. J. Rafelski and M. Danos, Phys. Lett. B **192**, 432 (1987).
19. D. Molnar and S. A. Voloshin, Phys. Rev. Lett. **91**, 092301 (2003).
20. Z. W. Lin and C. M. Ko, Phys. Rev. Lett. **89**, 202302 (2002).
21. R. J. Fries *et al.*, Phys. Rev. Lett. **90**, 202303 (2003).
22. R. C. Hwa and C. B. Yang, Phys. Rev. C**67**, 034902 (2003).
23. J. Rafelski and B. Muller, Phys. Rev. Lett. **48**, 1066 (1982) [Erratum-ibid. **56**, 2334 (1986)].
24. J. Rafelski, Phys. Rept. **88**, 331 (1982).
25. E. Andersen *et al.* [WA97 Collaboration],Phys. Lett. B **449**, 401 (1999).
26. D. Elia [NA57 Collaboration], Preprint nucl-ex/0410034 (2004).
27. S. S. Adler *et al.*, Phys. Rev. C**69**, 034909 (2004).

28. C. Adler *et al.*, Phys. Rev. Lett. **89**, 092301 (2002).
29. J. Adams *et al.*, Phys. Rev. Lett. **92**, 112301 (2004).
30. J. Adams *et al.*, Phys. Lett. B**595**, 143 (2004).
31. K. Adcox *et al.*, Phys. Rev. Lett. **88**, 242301 (2002).
32. H. Caines, Preprint nucl-ex/0412015 (2004).
33. J. Rafelski and J. Letessier, Phys. Lett. B **469**, 12 (1999).
34. J. Letessier, A. Tounsi and J. Rafelski, Phys. Lett. B **389**, 586 (1996).
35. C. Adler *et al.*, Phys. Rev. Lett. **89**, 202301 (2002).
36. J. Adams *et al.*, Phys. Rev. Lett. bf 91, 172302 (2003)
37. K. Adcox *et al.*, Phys. Rev. Lett. **88**, 022301 (2002)
38. S. S. Adler *et al.*, Phys. Rev. Lett. **91**, 072301 (2003).
39. H. Sorge, Phys. Rev. Lett. **82**, 2048 (1999).
40. S. S. Adler *et al.*, Phys. Rev. Lett. **91**, 182301 (2003).
41. J. Adams *et al.*, Phys. Rev. Lett. **92** 052302 (2003).
42. P. Huovinen *et al.*, Phys. Lett. B**503**, 58 (2001).
43. B. Muller, Phys. Rev. C**67**, 06190 (2003).
44. M. Gyulassy *et al.*, Phys. Lett. B**526**, 301 (2002).
45. J. Adams *et al.*, Preprint nucl-ex/0409033 (2004).
46. E. V. Shuryak, Phys. Rev. C**66**, 027902 (2002).
47. J. Rafelski and J. Letessier, Phys. Rev. Lett. **85**, 4695 (2000).
48. P. R. Sorensen, Ph.D. Thesis UCLA, nucl-ex/0309003 (2003).
49. V. Greco and C. M. Ko, Phys. Rev. C**70**, 024901 (2004).
50. X. Dong *et al.*, Phys. Lett. B**597**, 328 (2004).
51. J. Rafelski and J. Letessier, Acta Phys. Polon. B **34**, 5791 (2003).
52. J. Rafelski, J. Letessier and G. Torrieri, Preprint nucl-th/0412072 (2004).
53. Z. Fodor and S. D. Katz, JHEP **0404**, 050 (2004).
54. G. Torrieri and J. Rafelski, New J. Phys. **3**, 12 (2001).
55. W. Broniowski and W. Florkowski, Phys. Rev. Lett. **87**, 272302 (2001).
56. R. L. Thews, M. Schroedter and J. Rafelski, Phys. Rev. C **63**, 054905 (2001).
57. P. Petreczky, J. Phys. G:Nucl. Part. Phys. **30**, S431 (2004).
58. M. Asakawa and T. Hatsuda, Phys. Rev. Lett. **92**, 012001 (2004).
59. F. Karsch, J. Phys. G**30**, S887 (2004).
60. Y. L. Dokshitzer and D. E. Kharzeev Phys. Lett. B**519**, 199 (2001).
61. M. Djordjevic and M. Gyulassy, Phys Lett B**560**, 37 (2003).
62. R. L. Thews, Preprint hep-ph/0412323 (2004).
63. S. Kelly (The PHENIX Collaboration), J. Phys. G**30**, S1189 (2004).
64. J. Adams *et al.*, Preprint nucl-ex/0407006 (2004).
65. R. Vogt, Preprint hep-ph/0203151 (2002).
66. A. Tai (The STAR Collaboration), J. Phys. G**30**, S809 (2004).
67. M. Cacciari, S. Frixione and P. Nason, JHEP **0103**, 006 (2001).
68. T. Sjostrand *et al.*, PYTHIA 6.3: Physica and Manual hep-ph/0308153 (2003).
69. C. Peterson *et al.*, Phys. Rev. D**27**, 105 (1983).
70. D. Kharzeev and K. Tuchin, Nucl. Phys. A**735**, 248 (2004).
71. S. Frixione *et al.*, Adv. Ser. Direct. High Energy Phys. 15 (1998); Preprint hep-ph/9702287 (1997).
72. V. Greco, C. M. Ko and R. Rapp, Phys. Lett. B**595**, 202 (2004).
73. F. Laue (The STAR Collaboration), Preprint nucl-ex/0411007 (2004).
74. M. Kaneta (The PHENIX Collaboration), J. Phys. G**30**, S1217 (2004).
75. A. Maas, J. Wambach, B. Gruter and R. Alkofer, Preprint hep-ph/0411289 (2004), in this volume.
76. G. E. Brown *et al.*, Preprint nucl-th/0402207 (2004).

QCD, hadrons and beyond

G. Nardulli

Department of Physics, University of Bari and INFN-Bari, Italy

Abstract. I give a summary of Section E of the sixth edition of the Conference *Quark confinement and the hadron spectrum*. Papers were presented on different subjects, from spectroscopy, including pentaquarks and hadron structure, to new physics effects (non commutative field theories, supersymmetry and extra dimensions) and the problem of color confinement, both in ordinary Yang-Mills models and in supersymmetric Yang-Mills.

Keywords: Pentaquark, SUSY, non-commutative field theories, vortex, confinement
PACS: 12.39.Dc,12.60.Jv,11.10.Kk,11.10.Nx,12.38.Aw

PENTAQUARKS, SPECTROSCOPY AND HADRON STRUCTURE

Starting from January 2003 several experiments have produced evidence for a few baryonic resonances whose simplest interpretation in terms of the quark model is that of a bound state of five quarks, more exactly four quarks and an antiquark. The first observed pentaquark has been the baryonic resonance $\Theta^+(1540)$, reported by several experiments: LEPS [1], DIANA [2], CLAS [3, 4], SAPHIR [5], HERMES [6], as well as by analyses of old bubble chamber experiments [7]. Several new experimental results on this state have been presented at this conference, and a lively discussion took place also in Section E. All the experiments giving evidence of $\Theta^+(1540)$ show that this resonance decays into K^+n or $K_s^0 p$ with a width compatible with experimental resolution ($\Gamma \sim$ a few MeV). The former decay, with strangeness $S = +1$ and baryonic number $B = 1$ is exotic. In principle one might model this state as a molecule K^+n; this interpretation would require an interplay of an attractive interaction with a range of ~ 1 fermi and the centrifugal barrier ($\ell \neq 0$); however such a model is disfavored since it would produce too large widths. As remarked above, the simplest quark model interpretation is that of a pentaquark, i.e. an exotic state formed by five quarks: $udud\bar{s}$. Other narrow exotic cascade states, e.g. a Ξ^{--} state with quantum numbers $B = 1, Q = S = -2$, and also a Ξ^- and Ξ^0 state have been reported by the NA49 Collaboration, see [8]. Also these signals can be interpreted as pentaquark states, e.g. for Ξ^{--}, $dsds\bar{u}$.

In the first year there has been an almost continuous flow of experimental results, but, starting from January 2004, new data appeared, many with negative results [9], e.g. BES, OPAL, PHENIX, DELPHI, ALEPH, CDF, BaBar, E690.

Negative results from the Hera-B experiment were presented in Section E by M. Medinnis. The search for $\Theta^+(1540)$ and Ξ^{--} in pA collisions at 920 GeV from this experiment only resulted in upper bounds. More precisely, looking at the pK_s^0 invariant mass, Hera-B finds the upper limit

$$\mathcal{B}d\sigma/dy|_{y=0} = 3.7\mu b/N \tag{1}$$

CP756, *Quark Confinement and the Hadron Spectrum VI*
edited by N. Brambilla, U. D'Alesio, A. Devoto, K. Maung, G.M. Prosperi and S. Serci
© 2005 American Institute of Physics 0-7354-0241-8/05/$22.50

at 1530 MeV/c^2 and

$$\mathscr{B}d\sigma/dy|_{y=0} = 2.4\mu b/N \qquad (2)$$

at 1540 MeV/c^2. The upper limit for Ξ^{--} production is 2.5μb/N at 1862 MeV/c^2, i.e. in the mass region where NA49 finds positive results.

J. Engelfried reported results on the search of strange and charmed pentaquark states at Selex. This collaboration finds no evidence of the strange pentaquark Θ^+, while for the charmed pentaquark Θ_c no conclusion can be drawn yet.

The charmed pentaquark state Θ_c^0 has been searched by the H1 and ZEUS Collaborations at DESY. L. Gladilin reported results from these two experiments. H1 [10] finds a 5σ signal in deep inelastic scattering and a signal in photoproduction at the same mass (~ 3.1 GeV). This narrow anticharmed state is seen through its decays into $D^{*-}p + c.c.$ and its minimal quark content is $udud\bar{c}$. On other hand ZEUS does not find it [11]. As to the strange pentaquark Θ^+, ZEUS [12] observes 221 ± 48 events in the channel pK_s^0 with a mass of the Θ^+ equal to 1521.5 ± 1.5 MeV. On the other hand no signals of the Ξ^{--} pentaquark is found by this collaboration.

Much experimental effort is expected in the near future to clarify these experimental issues. The origin of the discrepancies might be in the difference of production mechanisms, leading to different yields in different experiments. A careful analysis of the different assumptions in the experimental analyses would be certainly welcome, in particular those related to the kinematical cuts. In any event high statistics experiments should provide an answer in the near future. For the time being we can certainly assert that the appearance of exotic states, coming after years of fruitless experimental researches of exotica, has revived theoretical interest in QCD spectroscopy and its low energy models. Pentaquark states were indeed predicted long ago in the framework of the Chiral Soliton Model [13, 14], which is an extension to three flavors [15, 16, 17] of the Skyrme model [18, 19]. In the Chiral Quark Soliton Model [20, 21] all baryonic states are interpreted as arising from quantizing the chiral nucleon soliton and the pentaquark emerges as the third rotational excitation with states belonging to an antidecuplet with spin $s = 1/2$. Other interpretations have been proposed after the first results on $\Theta^+(1540)$, most notably the one of Jaffe and Wilczek [22, 23] who propose that the Θ^+ comprises two highly correlated ud pairs (diquarks: $\mathcal{Q}$) and an $\bar{s}$. Diquarks properties are similar to those of the diquark condensates of QCD in the high density color-flavor-locking (CFL) phase [24]. Both diquarks are in spin 0 state, antisymmetric in color and flavor. Together they produce a $\mathcal{QQ}$ state in the flavor-symmetric $\mathbf{6_f}$ that must be antisymmetric in color and in p−wave to satisfy Bose statistics. When combined with the antiquark the diquarks produce a $\overline{\mathbf{10}}_\mathbf{f}$ with spin 1/2 and positive parity (they can also produce a $\mathbf{8_f}$, and mixing is possible).

The hypothesis that the attractive interaction in the antisymmetric color channel may play a role both at low and high density quark matter is especially interesting in the light of the quark hadron continuity which has been suggested [25] to exist between the CFL and the hypernuclear phase. Due to the formation of the CFL condensate that breaks color, flavor and the electric charge, though preserving a combination of the electric charge and of the color generator T_8, the physical states are obtained by dressing the quarks by diquarks. The result is that in this phase eight quarks have exactly the same quantum numbers of baryons. Also the ninth quark corresponds to a singlet with a gap which is twice the gap of the octet. The same phenomenon takes place for the other states. For instance, the gluons are dressed by a pair $\overline{Q}Q$ giving rise to vector states with the same quantum number of the octet of vector resonances (ρ, etc.).

Quark-hadron-continuity plays a role in relating quark and baryons in the low-lying octet. Apparently it also matters in assigning a role to diquark attraction at zero baryonic densities. In a recent paper [26] it has been suggested that another sign of it is the existence of baryon chiral solitons also at finite density. The mechanism of its formation is based on the existence of condensates giving rise to color superconductivity and Nambu Goldstone Bosons (NGB) associated to the breaking of global symmetries. One starts with the effective lagrangian for the Nambu-Goldstone bosons written in terms of the fields bosonic fields X and Y associated to the right handed and left handed spin 0 condensates [26]:

$$\mathcal{L} = -\frac{F_T^2}{4}\text{Tr}\left[\left(X\partial_0 X^\dagger - Y\partial_0 Y^\dagger)^2\right)\right] - \alpha_T\frac{F_T^2}{4}\text{Tr}\left[\left(X\partial_0 X^\dagger + Y\partial_0 Y^\dagger + 2g_0)^2\right)\right]$$
$$+\frac{F_S^2}{4}\text{Tr}\left[\left|X\nabla X^\dagger - Y\nabla Y^\dagger\right|^2\right] + \alpha_S\frac{F_S^2}{4}\text{Tr}\left[\left|X\nabla X^\dagger + Y\nabla\hat{Y}^\dagger + 2\mathbf{g}\right|^2\right]$$
$$+\frac{1}{2}(\partial_0\phi)^2 - \frac{v_\phi^2}{2}|\nabla\phi|^2 - \frac{1}{g_s^2}Tr[\varepsilon\mathbf{E}^2 - \frac{1}{\lambda}\mathbf{B}^2]. \tag{3}$$

where the gluon strength is

$$F_{\mu\nu} = \partial_\mu g_\nu - \partial_\nu g_\mu - [g_\mu, g_\nu] \tag{4}$$

and

$$E_i = F_{0i}, \quad B_i = \frac{1}{2}\varepsilon_{ijk}F_{jk} . \tag{5}$$

The parameters ε and λ are the dielectric constant and the magnetic permeability of the dense condensed medium. In the CFL vacuum the gluons g_0^a and g_i^a acquire Debye and Meissner masses given by

$$m_D^2 = \alpha_T g_s^2 F_T^2, \quad m_M^2 = \alpha_S g_s^2 F_S^2 = \alpha_S g_s^2 v^2 F_T^2. \tag{6}$$

where

$$v^2 = \frac{F_S^2}{F_T^2} . \tag{7}$$

It should be stressed that these are not the true rest masses of the gluons, since there is a large wave function renormalization effect making the gluon masses of the order of the gap Δ, rather than μ [27]. One can decouple the gluons solving their classical equations of motion neglecting the kinetic term. The result from Eq. (3) is

$$g_\mu = -\frac{1}{2}\left(X\partial_\mu X^\dagger + Y\partial_\mu Y^\dagger\right). \tag{8}$$

By substituting this expression in Eq. (3), and performing a gauge rotation to get $Y = 1$, one gets

$$\mathcal{L} = \frac{F_T^2}{4}\left(\text{Tr}[\dot{\Sigma}\dot{\Sigma}^\dagger] - v^2\text{Tr}[\vec{\nabla}\Sigma\cdot\vec{\nabla}\Sigma^\dagger]\right) + \frac{1}{2}\left(\dot{\phi}^2 - v_\phi^2|\vec{\nabla}\phi|^2\right) - \frac{1}{g_s^2}Tr[\varepsilon\mathbf{E}^2 - \frac{1}{\lambda}\mathbf{B}^2]. \tag{9}$$

with

$$E_i = \frac{1}{4}[\Sigma\partial_0\Sigma^\dagger, \Sigma\partial_i\Sigma^\dagger], \quad B_i = \frac{1}{8}\epsilon_{ijk}[\Sigma\partial_j\Sigma^\dagger, \Sigma\partial_k\Sigma^\dagger]. \tag{10}$$

Apart for the breaking of the Lorentz symmetry, one recognizes in the first term the chiral lagrangian and, in the last one, the Skyrme term [19]. This effective lagrangian enforces the idea of the quark-hadron continuity between the CFL and the hypernuclear matter phase with three flavors. A numerical estimate of the soliton mass based on these assumptions is in Fig. 1 (for $\Delta = 40$ MeV).

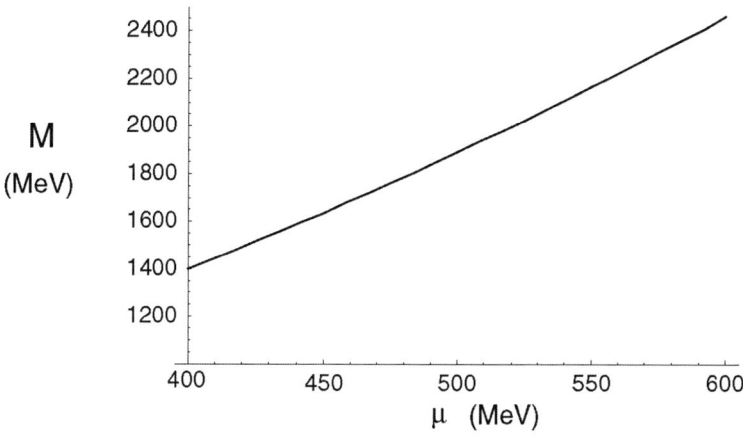

FIGURE 1. The soliton mass M at finite density in the CFL phase as a function of the baryonic chemical potential μ, for the value of the gap $\Delta = 40$ MeV.

Around 400 MeV the soliton mass is in the range of 1200-1400 MeV, which, in the light of the quark-hadron-continuity, is in the right ball-park. If pentaquarks exist at zero density, they should be connected continuously with states comprising two diquarks and an antiquark existing at finite density.

Spectroscopy and the study of the hadron structure may appear as old-fashioned sectors of hadron physics. They are however from time to time also source of surprises in theoretical physics. We have discussed the recent excitement about pentaquarks. Another much debated subject a decade ago concerned the proton spin. On this subject U. D'Alesio gave a talk on the role of the intrinsic partonic k_T. The dependence on intrinsic partonic k_T in parton distribution function (pdf) can be parameterized by a gaussian shape and is born by a twist 2 operator (Sivers) whose effect on asymmetries is competitive with another source of asymmetry, the so called Collins effect. Its presence in the pdf is of interest in single spin asymmetries for inclusive production by high energy transversely polarized hadron hadron scattering. D'Alesio and collaborators [28, 29] give predictions for several transverse single spin asymmetries, e.g. at RHIC. In particular single spin asymmetries in Drell Yan processes can provide a tool to extract quark Sivers distribution functions, while the inclusive production of D by high energy transversely polarized hadron hadron scattering could be a tool to extract gluon distribution function. Related to this study is another talk given in Section E, by I. Vukotic for the HERA-B Collaboration, who presented results for charmonium and open charm production obtained by this collaboration. In particular a new limit

$$\sigma(D^0 \to \mu^+\mu^-) = 2.0 \times 10^{-6} \qquad (90\%CL) \qquad (11)$$

was presented, currently the best published upper limit for this decay.

NEW PHYSICS EFFECTS

The discussion on soliton states at finite density in the CFL is a useful reminder that actually we do not yet know the true ground state of QCD at intermediate densities, i.e. those corresponding to quark chemical potentials $\mu \sim 400$ MeV. It is possible that the QCD ground state is characterized, at these densities, by inhomogeneous color superconductivity. This state is called LOFF state [30], [31], [32]. Its prominent feature is diquark condensation with non vanishing total momentum of the Cooper pair, so that $\Delta = \Delta(\mathbf{r})$. The dynamics of these phases was investigated by P. Castorina in the framework of non-commutative field theories (NCFT) with cut-off [33, 34].

The hypothesis that space coordinates do not commute:

$$[x_\mu, x_\nu] = \Theta_{\mu\nu} , \qquad (12)$$

can be traced back to Snyder [35]. The condition (12) can be realized in quantized motions of particles in strong magnetic field H and the non commuting coordinates can be appropriately chosen on a plane perpendicular to H. Castorina gave a talk on this subject with on overview on NCFT. Several cases of NCFT have been studied so far. They are based on the use of the Moyal product of fields:

$$(f \star g)(x) = \exp\left\{ i\frac{\Theta_{\alpha\beta}}{2} \partial_\alpha \partial'_\beta \right\} f(x)g(x')_{x=x'} \qquad (13)$$

that implements non commutativity by means of the parameters $\Theta_{\alpha\beta}$. Castorina and collaborators study transitions from ordered phases with homogeneous order parameters to phases with inhomogeneous order parameters. NCFT with cutoff Λ are used as an effective approach to describe the dynamical mechanism underlying these transitions. They consider two applications, one for $\lambda\phi^4$, in the context of Bose-Einstein Condensation, and another one for Nambu Jona Lasinio four-fermion coupling, that can be used either in the context of superconductivity or for spontaneous breaking of chiral symmetry. In the latter case one considers:

$$\mathcal{L} = i\bar{\psi}\gamma \cdot \partial \psi + g\bar{\psi}_\alpha \star \psi_\alpha \star \bar{\psi}_\beta \psi_\beta - g\bar{\psi}_\alpha \star \bar{\psi}_\beta \psi_\alpha \star \psi_\beta . \qquad (14)$$

For g larger than a critical value g_c one has chiral symmetry breaking to a phase with inhomogeneous order parameter. The space modulation of the gap is similar to the LOFF case with a Cooper pair momentum $P \propto 1/\Theta\Lambda^2$.

C. Corianò and A. Feo gave talks on supersymmetric models. In the analysis of spectra and hadron multiplicities for collisions induced by Ultra High Energy Cosmic

Rays (UHECR) it is important to model the effects of new physics, to get a quantitative understanding of their role in experimental observables. Corianò presented results on the modifications induced by SUSY models not only on UHECR but also in deep inelastic scattering. The simulations he presented [36, 37] include effects due to low energy gravity scales induced by extra dimensions. Numerical effects can be significant indeed, and might manifest themselves in new generation experiments, be they collider physics experiments or cosmic rays observations.

Feo presented a study of dynamical breaking of supersymmetry by non perturbative lattice techniques, using the hamiltonian formalism in a class of $d = 2, N = 1$ Wess Zumino models [38, 39] (see also [40]). Their study includes an analysis of the phase diagram by analytical strong coupling expansions and numerical simulations. All results with cubic prepotential indicate unbroken SUSY, while for quadratic potential

$$V = \lambda_2 \phi^2 + \lambda_0 \tag{15}$$

they confirm the existence of two phases. At high λ_0 there is a phase characterized by broken SUSY with unbroken Z_2. At low λ_0 SUSY is unbroken SUSY and Z_2 is broken. The critical value of λ_0 they find is

$$\lambda_0^c = -0.48 \pm 0.01 \ . \tag{16}$$

CONFINEMENT IN YANG-MILLS AND SUPER YANG-MILLS

The last two talks I wish to summarize were presented in Section E by K. Konishi and A. Niemi. They have in common the study of topological effects in the discussion of confinement in Yang Mills theories. As is well known, the simplest gauge model with monopole solution is the Georgi-Glashow model, based on the group $G = SU(2)$ and containing a triplet of Higgs fields:

$$\mathcal{L} = -\frac{1}{4}F^2 + \frac{1}{2}(D\phi)^2 + \lambda(\phi^2 - v^2)^2 \tag{17}$$

where

$$D_\mu \phi^a = \partial_\mu \phi^a + g\varepsilon_{abc}A_\mu^b \phi^c \ . \tag{18}$$

After symmetry breaking a symmetry subgroup (e.m.) remains: $H = U(1)_{e.m.}$. The model has soliton solutions with charges

$$q = 0, \qquad g_M = \frac{1}{g} \neq 0 \ . \tag{19}$$

The model can be generalized to other groups G, H usually in the small λ limit (Bogomol'nyi, Prasad, Sommerfield). It is a simplified model but full of interest. Among the other things, it teaches us that microscopic variables in the lagrangian (e.g. in $\mathcal{L}$ the fields ϕ, A) do not necessarily coincide with observed quanta, that are massive gauge bosons and monopoles. In QCD similar differences arise between microscopic (quarks, gluons) and macroscopic (hadrons) degree of freedoms. Which ideas on confinement arise from this analogy? A popular vision of confinement is based on the idea that two color charges at large distances R form a flux tube, i.e. a string like, one-dimensional object with an energy $V(R) \sim \sigma R$. Confinement might be explained by analogy with type II superconductors. Let us consider two magnetic monopoles. Since magnetic flux is conserved and cannot vanish everywhere, it remains confined to vortex lines (Abrikosov vortices) characterized by

$$E/R \simeq \sigma = const. \tag{20}$$

Now in QCD one needs chromoelectric, not chromomagnetic flux tubes, i.e. one needs Cooper condensation of pairs of magnetic charges (dual Meissner effect). In normal QCD magnetic monopoles do not exist as particles, but in supersymmetric QCD (SQCD) they can exist, as first shown by Seiberg and Witten.

Konishi discussed several models [41, 42] of monopole confinement by vortices. After the construction of nonabelian BPS vortices he proves that nonabelian monopoles occur as infrared degree of freedom (e.g. in $N = 2, SU(N_c)$ SQCD). This suggests that softly broken $N = 2$ theories might model QCD confinement. The existence of nonabelian monopoles in these models is essentially a quantum mechanical phenomenon. A peculiar feature of these models is that massless flavor symmetry is important to keep H unbroken and for being part of the dual transformation itself.

Niemi discussed a model with glueballs as closed strings [43]. These topological solutions are stabilized by the existence of twists or knots. Their stability follows from topological considerations, e.g. the existence of twist or knots of the closed string. The model is based on the Yang-Mills theory with two colors. In the infrared regime one separates A_μ in a $U(1)$ e.m. component A_μ and charged $A_\mu^\pm$. By these components a composite gauge field Γ_μ can be constructed for an internal group $U(1)_\Gamma$. The model can be casted in a form similar to a dual superconductor with 2 condensates (as in metallic hydrogen or high T_c superconductors). The vortex lines have a two-sheet structure which allows for twisting and knots. Stability of the closed strings follows from this non trivial topological structure.

These two talks have shown once again that topological solitons (vortex lines) constitute a lively subject of study. Therefore a deeper understanding of the topological structure of QCD, or some SUSY extension of QCD, can shed light on the dynamics of color confinement, perhaps rendering this *millennium problem* (see the site http://www.claymath.org/millennium) eventually solvable.

ACKNOWLEDGMENTS

It is a pleasure to thank N. Brambilla, U. D'Alesio, G. Prosperi and all the organizers for this beautiful conference in a spectacular site.

REFERENCES

1. T. Nakano, et al., *Phys. Rev. Lett.*, **91**, 012002 (2003), hep-ex/0301020.
2. V. V. Barmin, et al., *Phys. Atom. Nucl.*, **66**, 1715–1718 (2003), hep-ex/0304040.
3. S. Stepanyan, et al., *Phys. Rev. Lett.*, **91**, 252001 (2003), hep-ex/0307018.

4. V. Kubarovsky, et al., *Phys. Rev. Lett.*, **92**, 032001 (2004), hep-ex/0311046.
5. J. Barth, et al. (2003), hep-ex/0307083.
6. A. Airapetian, et al., *Phys. Lett.*, **B585**, 213 (2004), hep-ex/0312044.
7. A. E. Asratyan, A. G. Dolgolenko, and M. A. Kubantsev, *Phys. Atom. Nucl.*, **67**, 682–687 (2004), hep-ex/0309042.
8. C. Alt, et al., *Phys. Rev. Lett.*, **92**, 042003 (2004), hep-ex/0310014.
9. J. Pochodzalla (2004), hep-ex/0406077.
10. A. Aktas, et al., *Phys. Lett.*, **B588**, 17 (2004), hep-ex/0403017.
11. S. Chekanov, et al. (2004), hep-ex/0409033.
12. S. Chekanov, et al., *Phys. Lett.*, **B591**, 7–22 (2004), hep-ex/0403051.
13. A. V. Manohar, *Nucl. Phys.*, **B248**, 19 (1984).
14. M. Chemtob, *Nucl. Phys.*, **B256**, 600–608 (1985).
15. E. Witten, *Nucl. Phys.*, **B223**, 433–444 (1983).
16. G. S. Adkins, C. R. Nappi, and E. Witten, *Nucl. Phys.*, **B228**, 552 (1983).
17. E. Guadagnini, *Nucl. Phys.*, **B236**, 35 (1984).
18. T. H. R. Skyrme, *Proc. Roy. Soc. Lond.*, **A260**, 127–138 (1961).
19. T. H. R. Skyrme, *Nucl. Phys.*, **31**, 556–569 (1962).
20. H. Walliser, *Nucl. Phys.*, **A548**, 649–668 (1992).
21. D. Diakonov, V. Petrov, and M. V. Polyakov, *Z. Phys.*, **A359**, 305–314 (1997), hep-ph/9703373.
22. R. L. Jaffe, and F. Wilczek, *Phys. Rev. Lett.*, **91**, 232003 (2003), hep-ph/0307341.
23. R. Jaffe, and F. Wilczek, *Eur. Phys. J.*, **C33**, s38–s42 (2004), hep-ph/0401034.
24. M. G. Alford, K. Rajagopal, and F. Wilczek, *Nucl. Phys.*, **B537**, 443–458 (1999), hep-ph/9804403.
25. T. Schafer, and F. Wilczek, *Phys. Rev. Lett.*, **82**, 3956–3959 (1999), hep-ph/9811473.
26. R. Casalbuoni, and G. Nardulli, *Phys. Lett.*, **B602**, 205–211 (2004), hep-ph/0406030.
27. R. Casalbuoni, R. Gatto, and G. Nardulli, *Phys. Lett.*, **B498**, 179–188 (2001), hep-ph/0010321.
28. M. Anselmino, M. Boglione, U. D'Alesio, E. Leader, and F. Murgia, *Phys. Rev.*, **D 70**, 074025 (2004), hep-ph/0407100.
29. U. D'Alesio, and F. Murgia, *Phys. Rev.*, **D 70**, 072009 (2004), hep-ph/0408092.
30. A. J. Larkin, and Y. N. Ovchinnikov, *Zh. Exsp. teor. Fiz.*, **47**, 1136 (1964).
31. P. Fulde, and R. A. Ferrell, *Phys. Rev.*, **135**, A550 (1964).
32. R. Casalbuoni, and G. Nardulli, *Rev. Mod. Phys.*, **76**, 263 (2004), hep-ph/0305069.
33. P. Castorina, G. Riccobene, and D. Zappala, *Phys. Rev.*, **D69**, 105024 (2004), hep-th/0402188.
34. P. Castorina, G. Riccobene, and D. Zappalá (2004), hep-th/0405093.
35. H. S. Snyder, *Phys. Rev.*, **71**, 38–41 (1947).
36. A. Cafarella, C. Coriano', and T. N. Tomaras (2004), hep-ph/0410190.
37. A. Cafarella, C. Coriano', and T. N. Tomaras (2004), hep-ph/0410358.
38. M. Beccaria, G. F. De Angelis, M. Campostrini, and A. Feo, *Phys. Rev.*, **D 69**, 095010 (2004), hep-lat/0402007.
39. M. Beccaria, G. F. De Angelis, M. Campostrini, and A. Feo, *Phys. Rev.*, **D 70**, 035011 (2004), hep-lat/0405016.
40. A. Feo, *Nucl. Phys. Proc. Suppl.*, **119**, 198–209 (2003), hep-lat/0210015.
41. R. Auzzi, S. Bolognesi, J. Evslin, K. Konishi, and A. Yung, *Nucl. Phys.*, **B 673**, 187 (2003), hep-th/0307287.
42. R. Auzzi, S. Bolognesi, J. Evslin, K. Konishi, and H. Murayama, *Nucl. Phys.*, **B 701**, 207 (2004), hep-th/0405070.
43. A. Niemi (2004), hep-th/0403175.

Chiral transition of $N_f = 2$ and Confinement

M. D'Elia*, A. Di Giacomo† and C. Pica†

*Dipartimento di Fisica dell'Università di Genova & INFN, Sezione di Genova, Italy
†Dipartimento di Fisica dell'Università di Pisa & INFN, Sezione di Pisa, Italy

Abstract. We present a lattice investigation of the order of the chiral transition of $N_f = 2$ QCD. The specific heat and the susceptibility of the chiral condensate are determined for different spatial sizes of the system. A finite size scaling analysis excludes a second order transition in the predicted universality class and hints at a possible first order phase transition. The same indication comes from the analysis of the system by an order parameter related to the condensation of magnetic charges.

Keywords: Confinement; Lattice QCD; Chiral transition
PACS: 11.15.Ha, 12.38.Aw, 14.80.Hv, 64.60.Cn

INTRODUCTION

$N_f = 2$ QCD can provide fundamental insight into the mechanism of confinement. The phase transition is well understood at high masses ($m \geq 2.5 \; GeV$), where the quarks decouple; the transition is first order and the Polyakov line is a good order parameter. At $m \simeq 0$ a chiral transition exists, where chiral symmetry is restored, and the chiral condensate $\langle \bar{\psi}\psi \rangle$ is an order parameter. At some temperature also the axial $U_A(1)$ is expected to be restored: indeed the topological susceptibility drops to zero around T_c [1]. In principle at $m \simeq 0$ there are 3 transitions (chiral, axial $U_A(1)$, deconfinement): it is not clear if they coincide, and the question cannot even be asked if there is no independent definition of deconfinenement. An effective description of the chiral transition can be given in terms of an effective free energy [2]. Assuming that the scalar and pseudoscalar modes are the relevant critical degrees of freedom, there is no infrared stable fixed points for $N_f \geq 3$ and the transition is expected to be first order. For $N_f = 2$ the transition is first order if the anomaly is negligible ($m_{\eta'} \approx 0$) at T_c; it can be second order with symmetry $O(4)$ if the anomaly survives the chiral transition. In the first case the transition surface around $m = 0$ is first order. If the chiral transition is instead second order the surface is a crossover, and a tricritical point is expected in the $\mu - T$ plane (see e.g. [3]), detectable by heavy ion experiments. If the latter is the case, the deconfining transition cannot be order-disorder, there is no order parameter for confinement, and a state of a free quark can continuously be transfered below the "deconfining temperature". Existing literature is admittedly not conclusive [4, 5, 6, 7, 8], with a diffuse tendency to assume a second order $O(4)$ and crossover. The problem is fundamental and deserves additional study.

SCALING ANALYSIS

The order of the chiral transition can be studied by lattice simulations and standard finite size scaling techniques. We have used staggered fermions on $L_t \times L_s^3$ lattices, with $L_t = 4$

and $L_s = 12, 16, 20, 24, 32$. The input parameters of the simulations are $\beta = 6/g^2$ and the quark mass m. The temperature is $T = 1/(L_t a)$ with $a(\beta, m)$ the lattice spacing. The reduced temperature $\tau \equiv (1 - T/T_c)$ is then $\tau = 1 - a(\beta_c, 0)/a(\beta, m)$ or, in a sufficiently small neighborhood of the critical point

$$\tau = \left.\frac{\partial \ln a}{\partial \beta}\right|_{(\beta_c, 0)} [\beta_c - \beta + km] \quad ; \quad k = \left.\frac{\frac{\partial \ln a}{\partial m}}{\frac{\partial \ln a}{\partial \beta}}\right|_{(\beta_c, 0)} \tag{1}$$

For the specific heat C_V and for the susceptibility χ of the order parameter, the following scaling laws are expected

$$C_V - C_0 = L_s^{\alpha/\nu} \Phi_C(\tau L_s^{1/\nu}, am_q L_s^{y_h}) \tag{2}$$

$$\chi - \chi_0 = L_s^{\gamma/\nu} \Phi_\chi(\tau L_s^{1/\nu}, am_q L_s^{y_h}) \tag{3}$$

holding when the correlation length $\xi \gg a$.

In order to test the $O(4)$ option, we have run, among others, at fixed value of the scaling variable $am_q L_s^{y_h}$ with $y_h = 2.49$, which is valid for $O(4)$, and different spatial sizes L_s. The expectation is then, from Eqs. 2-3, that the peak values scale as

$$(C_V - C_0)^{peak} \propto L_s^{\alpha/\nu} \tag{4}$$

$$(\chi - \chi_0)^{peak} \propto L_s^{\gamma/\nu}. \tag{5}$$

For $O(4)$ $\alpha = -0.23$, $\nu = 0.75$, $\gamma = 1.48$: the peak for $C_V - C_0$ is expected to decrease at high L_s. Data show instead a rapid raise. The quantities $(C_V - C_0)^{peak}/L_s^{\alpha/\nu}$ and $(\chi - \chi_0)^{peak}/L_s^{\gamma/\nu}$ should be constant if $O(4)$ were the symmetry. The χ^2/dof for a constant is very high: $O(4)$ symmetry is excluded. Since y_h for $O(2)$ is equal within errors to that of $O(4)$, $O(2)$ symmetry turns out to be equally bad. Our action is not "improved": however we do not expect that an infrared property like Eqs. 4-5 are affected by ultraviolet improvement. We can safely state that $O(4)$ and $O(2)$ are excluded and with them the crossover scenario, leaving the possibility of a first order phase transition open. Other scaling tests can be performed (for instance by studying the equation of state), which hint at the possibility of a first order transition. A full account of our studies is presented elsewhere (see e.g. [9]).

DUAL SUPERCONDUCTIVITY OF THE VACUUM

A disorder parameter $\langle \mu \rangle$ detecting dual superconductivity of the vacuum has been developed by our group during the last years and proved to be a good parameter for the quenched theory. The parameter can be defined equally well in the presence of dynamical quarks and works well independently of the abelian projection chosen to define magnetic charges [10]. $\langle \mu \rangle$ is strictly zero in the deconfined phase and it drops to zero at the transition line, as shown by the fact that the susceptibility $\rho = (\partial/\partial\beta)\ln\langle\mu\rangle$ has a sharp negative peak coincident within errors with the peak of C_V. Around T_c

$$\langle \mu \rangle = L_s^{k/\nu} \Phi_{\langle\mu\rangle}(\tau L_s^{1/\nu}, am_q L_s^{y_h}) \tag{6}$$

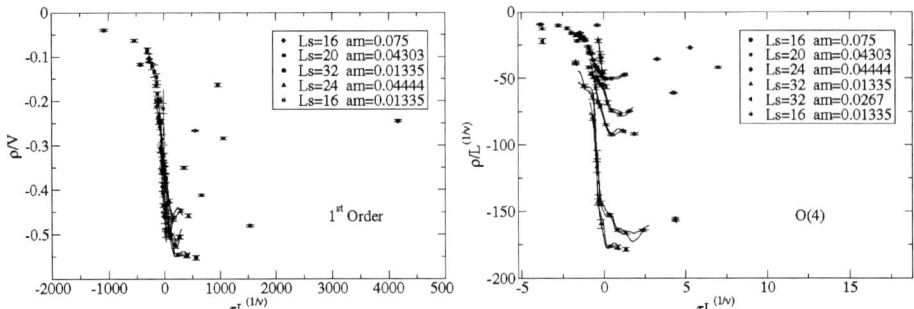

FIGURE 1. $\rho/L_s^{1/\nu}$ scaling according to Eq. 8 for different scaling hypotheses.

continuity arguments require

$$\langle\mu\rangle \simeq (am_q)^{-k/(\nu y_h)}\tilde{\Phi}_{\langle\mu\rangle}(\tau L_s^{1/\nu}) \tag{7}$$

and therefore

$$\rho/L_s^{1/\nu} \simeq f_{\langle\mu\rangle}(\tau L_s^{1/\nu}) \tag{8}$$

Independence on m is expected. Eq. 8 allows a determination of ν. If this agrees with the value obtained from C_ν, a legitimation of $\langle\mu\rangle$ as order parameter follows. Eq. 8 is compatible with first order (Fig. 1).

CONCLUSIONS

Finite size scaling analysis of $N_f = 2$ QCD excludes $O(4)$ –and $O(2)$– symmetry at the chiral critical point, and hints at a first order transition. The investigation is continuing with larger lattices and improved actions. Dual superconductivity of the vacuum is confirmed as a good symmetry for the order parameter.

REFERENCES

1. B. Alles, M. D'Elia and A. Di Giacomo, Phys. Lett. B **483** 139 (2000)
2. R.D. Pisarski and F. Wilczek, Phys. Rev. D **29**, 338 (1984); F. Wilczek, Int. J. Mod. Phys. A **7**, 3911 (1992); K. Rajagopal and F. Wilczek, Nucl. Phys. B **399**, 395 (1993)
3. M. A. Stephanov, K. Rajagopal and E. V. Shuryak, Phys. Rev. Lett. **81** 4816 (1998)
4. M. Fukugita, H. Mino, M. Okawa and A. Ukawa, Phys. Rev. Lett. **65**, 816 (1990); Phys. Rev. D **42**, 2936 (1990)
5. F.R. Brown et al., Phys. Rev. Lett. **65**, 2491 (1990)
6. F. Karsch, Phys. Rev. D **49**, 3791 (1994); F. Karsch and E. Laermann, Phys. Rev. D **50**, 6954 (1994)
7. S. Aoki et al. (JLQCD collaboration), Phys. Rev. D **57**, 3910 (1998).
8. C. Bernard et al. (MILC collaboration), Phys.Rev. D **61**, 054503 (2000)
9. M. D'Elia, A. Di Giacomo and C. Pica, hep-lat/0408008.
10. J.M. Carmona, M. D'Elia, A. Di Giacomo, B. Lucini, G. Paffuti, Phys. Rev. **D64**, 114507 (2001); A. Di Giacomo, hep-lat/0206018.

Lattice Study of Glue-Dynamics
– Gauge Dependent Objects –

A. Nakamura* and T. Saito*

*RIISE, Hiroshima Univ., Higashi-Hiroshima 739-8521, Japan

Abstract. We present our recent lattice QCD simulation results on gluon propagators and Polyakov loop correlations in Coulomb gauge in the confinement and deconfinement regions.

Keywords: Gluon Propagator, Heavy Quark Potential
PACS: 11.10.Wx, 12.38.Aw, 12.38.Gc

GAUGE FIXING ON THE LATTICE

QCD vacuum consists of the quarks and gluons. They are colored objects, but they are confined and hadrons are colorless, i.e., QCD world is colorful, but it looked colorless. Can we see colorful inside to understand the QCD dynamics more deeply ? For this purpose, we need the gauge fixing.

Since the lattice calculation explores non-perturbative regions, we encounter Gribov copies[1], i.e., there can be more than one solution for $\sum_\mu \partial_\mu A_\mu = 0$. Singer proved that[2] one cannot define gauge fixing for SU(N) on S^d $(d \geq 3)$ $(S^d$: d-Sphere), and Killingback showed the same no-go theorem for gauge fixing[3], SU(N) on T^d $(d \geq 3)$ and U(1) on T^d $(d \geq 2)$ $(T^d$: d-Torus). On the lattice the first Gribov copies were reported in Ref.[4].

Gribov proposed that we restrict our path-integral in the first Gribov region in which the Faddeev-Popov operator is positive definite. This phenomena is related to the QCD confinement mechanism, but our understanding is far from the goal. See e.g., Ref.[5]. We would like to stress that the uniqueness of the gauge fixing is a wanted-condition, but does not automatically mean that it is a good gauge to understand the confinement. The QCD confinement is a dynamical issue and we look for a gauge where the confinement mechanism is clarified. Zwanziger's stochastic gauge is a step for this direction [6, 7].

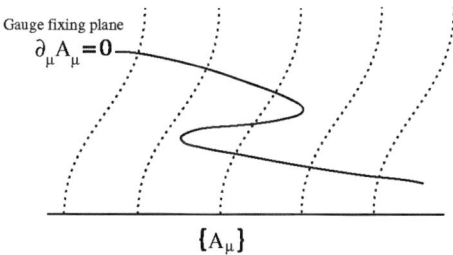

FIGURE 1. Gauge fixing and Gribov copies.

CP756, *Quark Confinement and the Hadron Spectrum VI*
edited by N. Brambilla, U. D'Alesio, A. Devoto, K. Maung, G.M. Prosperi and S. Serci
© 2005 American Institute of Physics 0-7354-0241-8/05/$22.50

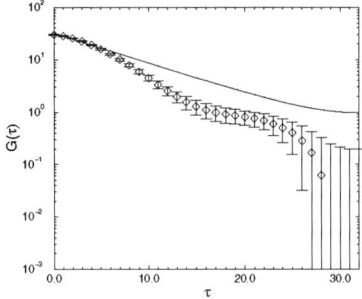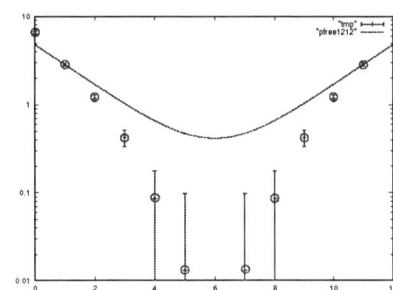

FIGURE 2. Gluon propagator in Landau Gauge on $48 \times 48 \times 48 \times 64$ lattice at $\beta = 6.8$ (left). Gluon propagator in Coulomb Gauge on 12^4 lattice at $\beta = 5.88$ (right).

GLUON PROPAGATORS

Quarks and gluons are two ingredients of QCD. The confinement of quarks is understood in the picture that the inter-quark potential calculated from Wilson loops shows linear rising behavior.

The confinement of gluons, on the contrary, is not well understood. Phenomenologically their mass becomes heavier and heavier as the distance becomes larger. Theoretically their propagator should not have a physical pole. Perturbative calculations do not give us any hint concerning the gluon's confinement which is essentially long distance behavior.

At the finite temperature, we can calculate the magnetic and electric masses of gluons, which are important quantities for the screening effects. Indeed there have been many trials to calculate the gluon propagators at zero temperature [8, 9, 10, 11, 13, 14, 15] and at finite temperature [16, 17, 18, 19, 20].

In Fig.2, we show gluon transverse propagators at zero temperature in Lorentz and Coulomb gauges. Gluon propagators at finite temperature were reported in Ref.[19].

POLYAKOV LOOP CORRELATIONS

McLerran and Svetisky formulated the Polyakov line correlations to obtain heavy quark potentials including color singlet and octet [21] of $q\bar{q}$ system. See Ref.[22] and references therein. This can be extended to qq system[23].

Since the Polyakov line vanishes at zero temperature, this method can not be applied in the confinement phase. However, Greensite, Olejnik and Zwanziger recently reported an interesting work to employ the finite length Polyakov lines,

$$L_n(\vec{x}, t_0) = U_4(\vec{x}, t_0)U_4(\vec{x}, t_0 + 1) \cdots U_4(\vec{x}, t_0 + n), \qquad (1)$$

in SU(2) gauge [24] using an old idea in Ref.[25]. In Fig.3, we show our preliminary results of the qq and $q\bar{q}$ potentials in SU(3). We will report extensive studies of the

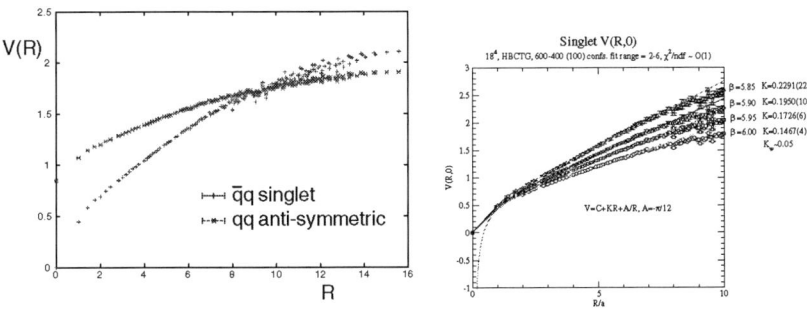

FIGURE 3. qq and $q\bar{q}$ potentials obtained from the finite length Polyakov line. $\beta = 6.0$ on 18^4 lattice. (left). $q\bar{q}$ potentials determined by the one length Polyakov lines and fitting curvs by $V(r) = c + kr + a/r$ with $a = -\pi/12$. k_W is the value of string tension by Wilson loop at $\beta = 6.0$. (right).

color-dependent potentials obtained in this method soon.

ACKNOWLEDGMENTS

We would like to thank Dan Zwanziger for his continuous encouragement and helpful discussion.

REFERENCES

1. V.N.Gribov, Nucl.Phys. **B139** (1978) 1.
2. I.M.Singer, Commun.math.Phys. **60** (1978) 7.
3. T.P.Killingback, Phys.Lett. **B138** (1984) 87.
4. A.Nakamura and M.Plewnia, Phys.Lett. **B255** (1991) 274.
5. C. Feuchter and H. Reinhardt, hep-th/0408236.
6. D.Zwanziger, Nucl.Phys. **B192** (1981) 259.
7. E.Seiler, I.O.Stamatescu and D.Zwanziger, Nucl.Phys. **B239** (1984) 177, 201.
8. G. Mandula and M. Ogilvie, Phys. Lett. B185 (1987) 127.
9. V. G. Bornyakov, V. K. Mitrjushkin, M. Muller- Preussker and F. Pahl, Phys. Lett. B317 (1993) 596.
10. C. Bernard, C. Parrinello and A. Soni, Phys. Rev. D49(1994) 1585.
11. A. Nakamura et al., Nucl. Phys. Proc. Suppl. 42 (1995) 899.
12. P. Marenzoni, G. Martinelli and N. Stella, Nucl. Phys. B455 (1995) 339
13. D. B. Leinweber, J. I. Skullerud, A. G. Williams and C. Parrinello, Phys. Rev. D60 (1999) 094507
14. A. Cucchieri, Phys. Lett. B422 (1998) 233
15. H. Nakajima and S. Furui, Nucl. Phys. Proc. Suppl. 73 (1999) 635.
16. U. M. Heller, F. Karsch and J. Rank, Phys. Lett. B355 (1995) 511.
17. A. Cuccieri, F. Karsch, Nucl.Phys. (PS) 83 (2000) 357
18. A. Cucchieri, F. Karsch and P. Petreczky, Phys. Lett. B497 (2001) 80, Phys. Rev. D64 (2001) 0306001
19. A. Nakamura, I. Pushkina, T. Saito, S.Sakai, Phys. Lett. B549 (2002) 133- 138.
20. A. Nakamura, T. Saito, S. Sakai, PRD69 (2004) 014506
21. L. McLerran and Svetisky, Phys.Rev.D24(1981)450
22. A. Nakamura and T. Saito, Prog.Theor.Phys. 111 (2004) 733, hep-lat/0404002
23. A. Nakamura, and T. Saito, Prog. Theor. Phys. 112 (2004) 183.
24. J. Greensite, S. Olejnik and D. Zwanziger Phys. Rev. D69 (2004) 074506, hep-lat/0401003.
25. E. Marinai, Paciello, G. Parisi and Taglienti, Phys. Lett. B298 (1993) 400.

Finite temperature QCD with two flavors of dynamical quarks on $24^3 \times 10$ lattice

Y. Nakamura[*], V.G. Bornyakov[†], M.N. Chernodub[**], Y. Mori[*], S.M. Morozov[**], M.I. Polikarpov[**], G. Schierholz[‡], A.A. Slavnov[§], H. Stüben[¶] and T. Suzuki[*]

[*]Institute for Theoretical Physics, Kanazawa University, Kanazawa 920-1192, Japan
[†]Institute for High Energy Physics, RU-142284 Protvino, Russia
[**]ITEP, B.Cheremushkinskaya 25, RU-117259 Moscow, Russia
[‡]NIC/DESY Zeuthen, Platanenallee 6, D-15738 Zeuthen, Germany
[§]Steklov Mathematical Institute, Vavilova 42, RU-117333 Moscow, Russia
[¶]Konrad-Zuse-Zentrum für Informationstechnik Berlin, D-14195 Berlin, Germany

Abstract. We present results obtained in QCD with two flavors of non-perturbatively improved Wilson fermions at finite temperature on $16^3 \times 8$ and $24^3 \times 10$ lattices. We determine the transition temperature in the range of quark masses $0.6 < m_\pi/m_\rho < 0.8$ at lattice spacing a≈0.1 fm and extrapolate the transition temperature to the continuum and to the chiral limits.

INTRODUCTION

In order to obtain predictions for the real world from lattice QCD, we have to extrapolate the lattice data to the continuum and to the chiral limits. Recently the Bielefeld group [1] and the CP-PACS collaboration [2] using different fermion actions obtained consistent values for the critical temperature T_c in the chiral limit, albeit on rather coarse lattices at $N_t = 4$ and 6. Edwards and Heller [3] determined T_c for $N_t = 4$, 6 using nonperturbatively improved Wilson fermions. We compute T_c on finer lattices with $N_t = 8$ and 10 with high statistics. Our results for $N_t = 8$ were reported in Ref. [4].

SIMULATION

We use non-perturbatively improved Wilson fermions with c_{sw} was calculated in [5] and Wilson action. Configurations are generated on $16^3 \times 8$ ($\beta = 5.2$ and 5.25) and $24^3 \times 10$ ($\beta = 5.2$) lattices at various κ. The values of κ and the corresponding number of configurations for $16^3 \times 8$ and $24^3 \times 10$ lattices can be found in Ref. [4] and Table 1, respectively. The number of configurations for $24^3 \times 10$ lattice is not large enough and results for this lattice are preliminary. We use results obtained at T=0 to fix the scale. The contour plot of lines of constant r_0/a and m_π/m_ρ is shown in Ref. [6].

κ	0.1348	0.1352	0.1354	0.1355	0.1356	0.1358	0.1360
# conf.	678	679	1,234	799	2,429	480	617

Table 1:Simulation statistics on $24^3 \times 10$.

CP756, *Quark Confinement and the Hadron Spectrum VI*
edited by N. Brambilla, U. D'Alesio, A. Devoto, K. Maung, G.M. Prosperi and S. Serci
© 2005 American Institute of Physics 0-7354-0241-8/05/$22.50

CRITICAL TEMPERATURE

We use the Polyakov loop susceptibility to determine the transition point. The critical value of κ turns out to be $\kappa_t = 0.1354$. Using $r_0 = 0.5$ fm and interpolating r_0/a to the critical point, we obtain for the crtical temperature:

$$T_c = 196(4)\text{MeV}, \quad m_\pi/m_\rho = 0.64(3)$$

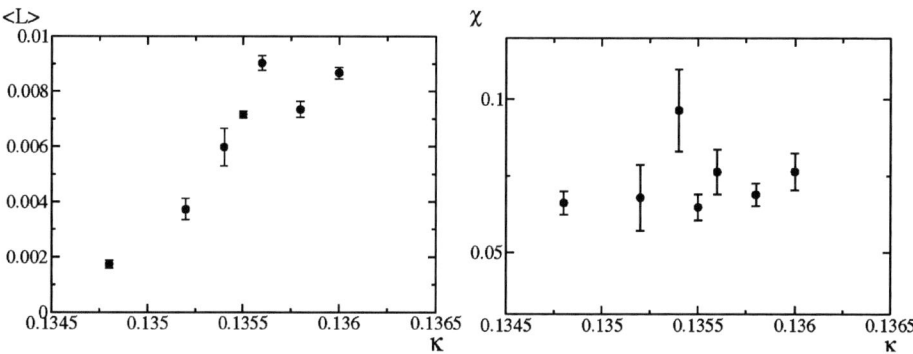

FIGURE 1. Polyakov loop (left) and its susceptibility (right) at $\beta = 5.2$ on $24^3 \times 10$.

CONTINUUM AND CHIRAL LIMITS

At small enough lattice spacing and quark mass one can extrapolate the critical temperature T_c to the continuum and the chiral limits using formula:

$$T_c r_0 = T_c^{m_q, a \to 0} r_0 + C_a \left(\frac{a}{r_0}\right)^2 + C_q \left(\frac{1}{\kappa} - \frac{1}{\kappa_c}\right)^{\frac{1}{\beta\delta}},$$

where $T_c^{m_q, a \to 0}$ corresponds to the extrapolated value of the critical temperature and β and δ are critical indices.

We make an attempt to fit four values for $T_c r_0$ (see Table 3), obtained at rather large quark masses, to estimate the parameters in this extrapolation expression.

$T_c r_0$	a/r_0	β	κ_t	L_t	
0.50(1)	0.201(4)	5.2	0.1354	10	prelim.
0.53(1)	0.234(4)	5.2	0.1345	8	Ref. [4]
0.56(1)	0.225(4)	5.25	0.1341	8	Ref. [4]
0.57(2)	0.29(1)	5.2	0.1330	6	Ref. [3]

Table 3: Available data for $T_c r_0$.

We extrapolate the value of the critical temperature using different values of 0.54 and 1 as $1/\beta\delta$. If the transition in two-flavor QCD is second order, the transition is expected to belong to the universality class of the $3D$ $O(4)$ spin model with $1/\beta\delta \approx 0.54$. If the transition is first order, then $1/\beta\delta = 1$. Table 4 presents fitting results.

$\frac{1}{\beta\delta}$	$T_c r_0$	C_a	C_q	χ^2/dof	order
0.54	0.44(2)	$-0.9(5)$	0.5(1)	0.26	2nd
1	0.51(3)	$-1.3(7)$	0.9(2)	0.13	1st

Table 4:Fitting results.

We get the critical temperature in the continuum and in the chiral limits. In the case of $1/\beta\delta = 0.54$,

$$T_c^{m_q, a \to 0} = 174(8) \text{ MeV}. \tag{1}$$

This value agrees with values obtained in Refs.[1, 2]. In the case of $1/\beta\delta = 1$,

$$T_c^{m_q, a \to 0} = 201(12) \text{ MeV}. \tag{2}$$

Although some lattice studies [1, 2] indicate second order chiral transition in two-flavor QCD, there are also results [7] supporting first order transition. Results of our fits do not allow to discriminate between first and second order transitions because of rather large errors in $T_c r_0$ values. We are continuing simulations on $24^3 \times 10$ lattice to get better precision of T_c value on this lattice.

ACKNOWLEDGMENTS

This work is supported by the SR8000 Supercomputer Project of High Energy Accelerator Research Organization (KEK). A part of numerical measurements has been done using NEC SX-5 at Research Center for Nuclear Physics (RCNP) of Osaka University. The numerical simulations of this work were done using RSCC computer clusters in RIKEN. We wish to acknowledge the support of the computer center at RIKEN. VGB, MNC and MIP are supported by grants RFBR 02-02-17308, 04-02-16079, RFBR-DFG-03-02-04016, DFG-RFBR 436 RUS 113/739/0, INTAS-00-00111, CRDF award RPI-2364-MO-02 and MK-4019.2004.2, A.A.S. is supported by grant Scient.school grant 2052-2003.1. T.S. is supported by JSPS Grant-in-Aid for Scientific Research on Priority Areas 13135210 and (B) 15340073.

REFERENCES

1. F. Karsch, A. Peikert, E. Laermann, Nucl. Phys. **B605** (2001) 579.
2. A. Ali Khan et al., (CP-PACS), Phys. Rev. **D63** (2001) 034502.
3. R. G. Edwards, U. M. Heller, Phys. Lett. **B462** (1999) 132.
4. V. Bornyakov et al., hep-lat/0401014
5. K. Jansen and R. Sommer (ALPHA collaboration), Nucl. Phys. **B530** (1998) 185.
6. S. Booth et al., Phys. Lett. **B519** (2001) 229.
7. M. D'Elia et al., hep-lat/0408008

Evidence for non-trivial holonomy in semiclassical non-Abelian lattice gauge fields

M. Müller-Preussker

Humboldt-Universität zu Berlin, Institut für Physik, D-12489 Berlin, Germany

Abstract. The status of the search for Kraan-van Baal calorons in $SU(2)$ and $SU(3)$ lattice gauge fields at non-zero temperature is reviewed.

Keywords: lattice gauge theory, finite temperature, calorons,non-trivial holonomy
PACS: 11.15.Ha, 12.38.Gc, 11.10.Wx

In this talk a minireview of recent work in collaboration with E.-M. Ilgenfritz, D. Peschka, S. Shcheredin (HU Berlin), B. Martemyanov, A. Veselov (ITEP Moscow) and partially with P. van Baal, F. Bruckmann (Univ. of Leiden), C. Gattringer, A. Schäfer (Univ. Regensburg) and others is given. Details can be found in [1–7]. The investigations finally aim at a reformulation [8] of the semiclassical approximation of the QCD partition function at $T \neq 0$ [9] - originally based on Harrington-Shepard caloron solutions [10] - in terms of more general (anti)selfdual solutions with non-trivial (asymptotic) holonomy. Such solutions have been found for unit topological charge [11, 12] and in some special cases also for higher charge [13]. Non-trivial holonomy means that for the localised solutions the Polyakov loop variable behaves as $P(\vec{x}) \to P_\infty \notin \mathbf{Z}$ for $|\vec{x}| \to \infty$. For $SU(N_c)$ a unit charge caloron consists of N_c monopole constituents which - at large distances - become well-separated lumps of fractional topological charge and represent static BPS monopoles. The positions $\vec{x}_i$, $i = 1, 2, \cdots, N_c$ of the monopole constituents can be determined from the coincidence of eigenvalues of $P(\vec{x})$. For $SU(2)$ this means that $L(\vec{x}_i) \equiv \frac{1}{N_c} \mathrm{tr}\, P(\vec{x}_i) = \pm 1$, $i = 1, 2$, i.e. the constituents lead to a dipole-like behavior for $L(\vec{x})$. This holds independently of the distance between the constituents and therefore, irrespective of their status being dissociated or not. These remarkable solutions are further characterized by a specific behavior of the zero-modes of the Dirac-operator. The modes change their localisation from one constituent to another, when interpolating the boundary conditions for the fermionic fields in the (imaginary) time direction between periodic and antiperiodic ones with a varying phase factor [14]. In comparison with the 'old' Harrington-Shepard calorons one should emphasize that for $N_c > 2$ the 'new' Kraan-van Baal calorons are more general than simply $SU(2)$ embeddings.

In our papers we have collected increasing evidence that for $T \leq T_c$ semiclassical background fields couple dominantly to non-trivial holonomy and really consist of monopole constituents. On asymmetric as well as on symmetric lattices we have found that lumps of integer topological charge commonly interpreted in terms of 'old' calorons or BPST instantons with trivial holonomy have to be treated as objects with non-trivial holonomy and consist of N_c monopole constituents. Iterative minimization of the action

CP756, *Quark Confinement and the Hadron Spectrum VI*
edited by N. Brambilla, U. D'Alesio, A. Devoto, K. Maung, G.M. Prosperi and S. Serci
© 2005 American Institute of Physics 0-7354-0241-8/05/$22.50

starting from Monte Carlo generated lattice gauge fields ('cooling') was our first method chosen [1, 5] (see also [3]). On metastable plateaus for the action, where the equations of motion become optimally satisfied, the cooled fields were investigated with various observables: the topological density, the mean violation of the equations of motion, the (non-)staticity in time, the local behavior of the trace of the Polyakov loop or of its eigenvalues, the localisation of the real (would-be zero-) modes of the (clover-improved) Wilson-Dirac operator, and the Abelian monopole currents in the maximally Abelian gauge. For cooled semiclassical lattice gauge fields with action values of a few instanton

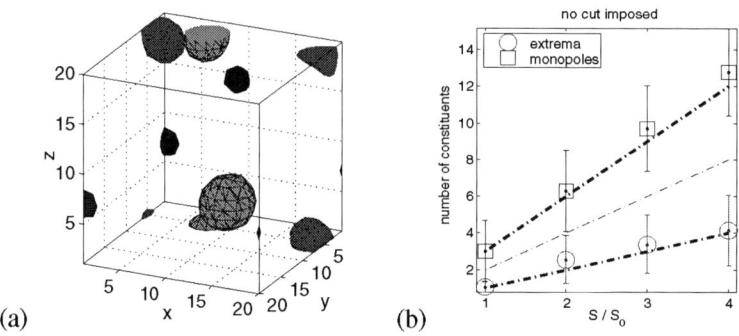

(a)   (b)

FIGURE 1. (a) $Q_t = 2$ caloron with six dissociated constituents found with cooling on a lattice of size $20^3 \times 4$. (b) Average number of monopole constituents and of extrema of the topological density for various action plateaus S/S_0 ($\beta = 5.6$, $12^3 \times 4$; S_0 denotes the instanton action unit).

(or caloron) units our main results can be summarized as follows.

- We determined the 'asymptotic holonomy' L_∞ averaging the Polyakov loop over areas of minimal action density. The distributions of L_∞ are clearly peaking at maximal distance from the group center values [2].
- The smaller the aspect ratio N_t/N_s for the linear lattice sizes is, the larger is the probability to find calorons dissociated into well separated lumps of fractional topological charge [4]. A nice example obtained for $SU(3)$ with the total topological charge $Q_t = 2$ is shown in Fig. 1a [5].
- On symmetric lattices only topological lumps with integer Q_t have been found [2, 3]. This coincides with older lattice searches for instantons.
- The majority of (even undissociated) calorons (and for $N_t/N_s = 1$ the instantons) show N_c intrinsic constituents from the local (for $SU(2)$ dipole-like) Polyakov loop behavior. For the $SU(3)$ case this is illustrated in Fig. 1b, where we show - after employing cooling down to a first visible plateau - the number of maxima of the action density in comparison with number of monopole constituents. The latter is taken from the number of those positions, for which the smallest difference of the Polyakov loop eigenvalues reaches minimal values [5]. The number of constituents is clearly larger than the number of topological lumps and scales nicely with N_c.
- For configurations with dissociated calorons the hopping of the real fermionic modes of the Wilson-Dirac operator is observed between different caloron constituents when changing continuously the fermionic boundary condition [1, 5].

An analogous behavior was established even for equilibrium configurations with charge $|Q_t| = 1$ by Gattringer and coauthors [15].

Our observations clearly show that calorons and instantons found after cooling in the confinement phase have to be interpreted in terms of solutions with non-trivial holonomy, i.e. Kraan-van Baal calorons. For $T = 0$ no analytic solutions with non-trivial holonomy are known so far.

In a further step we have investigated $SU(2)$ lattice gauge fields after applying a moderate four-dimensional smearing technique [16]. The aim was to keep 'closer' to the equilibrium fields. After 50 to 100 smearing steps providing configurations with an action larger by one order of magnitude (compared to cooling as described above) we have selected two classes of clusters of topological charge with the help of Abelian monopole currents [4]. The first one with approximately integer topological charges exhibits the mentioned intrinsic dipole-like Polyakov loop structure within each lump, i.e. they can be interpreted as undissociated calorons. The second class showing equal-sign Polyakov loop values throughout the clusters have approximately half-integer topological charges, i.e. look like separate constituent monopoles.

Thus, we can conclude that a semiclassical approach to the QCD path integral should definitely start from non-trivial holonomy configurations as discussed in [8]. What the physical implications of such a modified model will be, remains to be seen.

REFERENCES

1. E.-M. Ilgenfritz et al., *Phys. Rev.* **D66**, 074503 (2002), hep-lat/0206004.
2. E.-M. Ilgenfritz, B. Martemyanov, M. Müller-Preussker, and A. Veselov, *Phys. Rev.* **D69**, 114505 (2004), hep-lat/0402010.
3. F. Bruckmann, E.-M. Ilgenfritz, B. Martemyanov, and P. van Baal, *Phys. Rev.* **D70**, 105013 (2004), hep-lat/0408004.
4. E.-M. Ilgenfritz, B. Martemyanov, M. Müller-Preussker, and A. Veselov, hep-lat/0412028.
5. D. Peschka, Master Thesis, Humboldt-University Berlin, August 2004; E.-M. Ilgenfritz, M. Müller-Preussker, and D. Peschka, in preparation.
6. C. Gattringer et al., *Nucl. Phys. Proc. Suppl.* **129**, 653 (2004), hep-lat/0309106.
7. F. Bruckmann et al., Contribution to LATTICE 2004, *Nucl. Phys. Proc. Suppl.* (2005) to appear, hep-lat/0408036.
8. D. Diakonov, *Prog. Part. Nucl. Phys.* **51**, 173 (2003); D. Diakonov, N. Gromov, V. Petrov, and S. Slizovskiy, *Phys. Rev.* **D70**, 036003 (2004).
9. D.J. Gross, R.D. Pisarski, and L.G. Yaffe, *Rev. Mod. Phys.* **53**, 43 (1981).
10. B.J. Harrington, H.K. Shepard, *Phys. Rev.* **D17**, 2122 (1978).
11. T.C. Kraan and P. van Baal, *Phys. Lett.* **B428**, 268 (1998); **B435**, 389 (1998); *Nucl. Phys.* **B533**, 627 (1998).
12. K. Lee, *Phys. Lett.* **B426**, 323 (1998); K. Lee and C. Lu, *Phys. Rev.* **D58**, 025011 (1998).
13. F. Bruckmann and P. van Baal, *Nucl. Phys.* **B645**, 105 (2002); F. Bruckmann, D. Nógrádi, and P. van Baal, *Nucl. Phys.* **B698**, 233 (2004).
14. M.García Pérez, A.González-Arroyo, C.Pena, and P. van Baal, *Phys. Rev.* **D60**, 031901 (1999); M.N. Chernodub, T.C. Kraan, and P. van Baal, *Nucl. Phys. Proc. Suppl.* **83**, 556 (2000); F. Bruckmann, D. Nógrádi, and P. van Baal, *Nucl. Phys.* **B666**, 197 (2003).
15. C. Gattringer, *Phys. Rev.* **D67**, 034507 (2003); C. Gattringer and S. Schaefer, *Nucl. Phys.* **B654**, 30 (2003); C. Gattringer and R. Pullirsch, *Phys. Rev.* **D69**, 094510 (2004).
16. T. DeGrand, A. Hasenfratz, and T.G. Kovacs, *Nucl. Phys.* **B520**, 301 (1998).

Lattice Gauge Actions for Fixed Topology

χLF Collaboration: W. Bietenholz*, K. Jansen†, K.-I. Nagai†, S. Necco**,
L. Scorzato* and S. Shcheredin*

*Institut für Physik, Humboldt Universität zu Berlin, Newtonstr. 15, 12489 Berlin, Germany
†NIC/DESY Zeuthen, Platanenallee 6, D-15738 Zeuthen, Germany
**Centre de Physique Théorique, Luminy, Case 907, F-13288 Marseille Cedex 9, France

Abstract. We test a set of lattice gauge actions for QCD that suppress small plaquette values and in this way also suppress transitions between topological sectors. This is well suited for simulations in the ε-regime and it is expected to help in numerical simulations with dynamical quarks.

Our aim is to study the possibility of simulating lattice QCD with a gauge action that strongly reduces the occurrence of small plaquette values. A gauge background with such a feature is expected to improve the locality properties [1] of the Overlap-Dirac operator D_{ov} [2]. By the same argument one also expects to ease the numerical evaluation of D_{ov} itself, and help in general dynamical simulations. It can be proven [1, 3] that as long as all plaquette values in a gauge configuration satisfy: $S_P := 1 - \frac{1}{3}\text{Re Tr}(U_P) < 1/20.5$ then no change of topological sector is possible. Hence a suppression of low plaquette values entails a suppression of Q_{top} changes.

Simulations constrained in a fixed topological sector can be problematic for evaluating physical observables in QCD, where all sectors have to be taken into account with the correct weight. However such a constraint is perfectly suited for studying QCD in the ε-regime [4], where predictions exist for observables defined in fixed topological sectors [5], which turns the limitation mentioned above into an advantage. However there are some constraints. The physical volume should be at least, $L \gtrsim 1.1$ fm [6, 7]. Moreover, in order to reach very small pion masses, the chiral properties of the Dirac operator are crucial. Finally a sound definition of Q_{top} is important to compare with predictions in fixed topological sectors. All of these requirements are provided by Ginsparg-Wilson fermions. They have an exact, lattice modified chiral symmetry [8], and the fermionic index defines Q_{top} [9]. Results in the ε−regime with Ginsparg-Wilson fermions were obtained for the Dirac spectrum [6, 10, 11] and for meson correlation functions [7, 12, 13, 14], which can be compared with quenched Chiral Perturbation Theory [15, 16, 17].

Simple examples of gauge actions that suppress small plaquette values (still expected to be in the same universality class as S_P) are

$$\beta \, S_{\varepsilon,n}^{\text{hyp}}(U_P) \;=\; \beta \, \frac{S_P}{(1-\varepsilon^{-1}S_P)^n} \qquad \text{if } S_P < \varepsilon, \text{ and } +\infty \text{ otherwise} \tag{1}$$

$$\beta \, S_{\varepsilon,n}^{\text{pow}}(U_P) \;=\; \beta \, S_P + \varepsilon^{-1}S_P^n \tag{2}$$

$$\beta \, S_{\varepsilon,n}^{\text{exp}}(U_P) \;=\; \beta \, S_P \exp[\varepsilon^{-1}S_P^n] \tag{3}$$

CP756, *Quark Confinement and the Hadron Spectrum VI*
edited by N. Brambilla, U. D'Alesio, A. Devoto, K. Maung, G.M. Prosperi and S. Serci
© 2005 American Institute of Physics 0-7354-0241-8/05/$22.50

TABLE 1. *Results for the $S^{hyp}_{\varepsilon,1}$ on a 16^4 lattice for various ε^{-1} and β. β_W is the coupling resulting in the same r_0 with the Wilson action.*

ε^{-1}	β	r_0/a	β_W	τ^{plaq}	$\tau^{\text{plaq}}(\beta_W)$	f_J	dt	Acceptance
0	6.18	7.14(3)	6.18	7(1)	7(1)	0.015	0.1	$>99\%$
1.00	1.5	6.6(2)	6.13(2)	2.0(1)	n.a.	0.0027	0.05	$>99\%$
1.18	1.0	7.2(2)	6.18(2)	1.3(1)	7(1)	0.0014	0.02 - 0.01	$>99\%$
1.25	0.8	7.0(1)	6.17(1)	1.1(1)	9(1)	0.0025	0.1	$>99\%$
1.52	0.3	7.3(4)	6.19(4)	0.8(1)	7(1)	0.0008	0.1	$\sim 95\%$
1.64	0.1	6.8(3)	6.15(3)	1.0(1)	n.a.	0.0007	0.1	$\sim 65\%$

The first choice above (for $n = 1$) was introduced by M. Lüscher for conceptual purposes [18], and applied by Fukaya and Onogi in Schwinger model simulations [19, 20]. The question is whether one can conciliate the advantages mentioned above, with reasonable lattice sizes (say $La \sim 1 - 2$ fm), without increasing lattice artifacts and with a correct and reasonably decorrelated sampling of interesting observables. A first report of our ongoing study was presented in [21].

Results. Since gauge actions of the type (1,2,3) are non-linear in the link variables, the heat-bath algorithm cannot be applied. Instead we use a local HMC algorithm [22], which is competitive with heat-bath in the standard case. HMC trajectories have discretization dt quoted in Table 1, and the trajectory length is 1. The volume is chosen to 16^4 in order to allow a reliable determination of $r_0/a \simeq 7$ and in order have mild finite volume effects. We computed r_0/a following a standard procedure [23]. Our preliminary results are summarized in Table 1. We estimated the topological charge with cooling and searching for the first plateau [24]. Since we cannot reliably measure the autocorrelation of the Q_{top}, we quote – as an indicator of stability – the number of jumps of Q_{top} divided by the number of trajectories in the full history (f_J). Since we save a configuration only every 50 trajectories, the measured f_J is only a lower bound on the frequency of jumps, which can be reliable only for $f_J \ll 0.01$ (which is the interesting case for us). In particular for the Wilson action at $\beta \sim 6.17$ we expect $f_J \gg 0.015$, which we measure. The stability of Q_{top} has to be compared with the autocorrelation of a typical observable (we quote the plaquette value under τ^{plaq}). It is interesting to see that the latter is strongly decreased for non-zero ε^{-1} (at fixed r_0/a). We have also studied actions of type (3), which have a smooth bound on the plaquette value, and therefore are better suited for efficient simulations with global HMC and dynamical fermions. Results will be presented elsewhere. We also checked that the results are consistent independently on the starting configuration. This is important because the constraint on the plaquette value could in principle generate more obstructions than the topological ones, and this would not be noticed simply from the autocorrelation.

It has been pointed out [25] that the actions (1) do not allow for the existence of a positive definite transfer matrix. However we checked that all the actions (1,2,3) have *site*-reflection positivity, which at least ensures the existence of a positive squared transfer matrix [26]. Moreover we have checked that the unphysical behaviour of the short distance force – which was observed in some cases, and related to the lack of a positive transfer matrix [27] – does not appear in our cases.

Conclusions. Topology conserving gauge actions could be highly profitable in QCD simulations. The suppression of small plaquette values may speed up the simulations with dynamical quarks. A stable Q_{top} is useful in particular in the ε-regime. We are investigating such actions, in view of the physical scale and the topological stability.

Ackowledgements. We thank M. Creutz, H. Fukaya, S. Hashimoto, T. Onogi and R. Sommer for helpful discussions and correspondence. The Deutsche Forschungsgemeinschaft through SFB-TR 9-03, and the European Union under contract HPRN-CT-2002-00311 (EURIDICE), are acknowledged for financial support.

REFERENCES

1. P. Hernández, K. Jansen and M. Lüscher, Nucl. Phys. **B552**, 363 (1999), [hep-lat/9808010].
2. H. Neuberger, Phys. Lett. **B417**, 141 (1998), [hep-lat/9707022].
3. H. Neuberger, Phys. Rev. **D61**, 085015 (2000), [hep-lat/9911004].
4. J. Gasser and H. Leutwyler, Phys. Lett. **B188**, 477 (1987).
5. H. Leutwyler and A. Smilga, Phys. Rev. **D46**, 5607 (1992).
6. W. Bietenholz, K. Jansen and S. Shcheredin, JHEP **07**, 033 (2003), [hep-lat/0306022].
7. W. Bietenholz, T. Chiarappa, K. Jansen, K. I. Nagai and S. Shcheredin, JHEP **02**, 023 (2004), [hep-lat/0311012].
8. M. Lüscher, Phys. Lett. **B428**, 342 (1998), [hep-lat/9802011].
9. P. Hasenfratz, V. Laliena and F. Niedermayer, Phys. Lett. **B427**, 125 (1998), [hep-lat/9801021].
10. L. Giusti, M. Lüscher, P. Weisz and H. Wittig, JHEP **11**, 023 (2003), [hep-lat/0309189].
11. QCDSF-UKQCD, D. Galletly *et al.*, Nucl. Phys. Proc. Suppl. **129**, 453 (2004), [hep-lat/0310028].
12. L. Giusti, P. Hernández, M. Laine, P. Weisz and H. Wittig, JHEP **01**, 003 (2004), [hep-lat/0312012].
13. L. Giusti, P. Hernández, M. Laine, P. Weisz and H. Wittig, JHEP **04**, 013 (2004), [hep-lat/0402002].
14. L. Giusti *et al.*, hep-lat/0409031.
15. P. H. Damgaard, M. C. Diamantini, P. Hernández and K. Jansen, Nucl. Phys. **B629**, 445 (2002), [hep-lat/0112016].
16. P. H. Damgaard, P. Hernández, K. Jansen, M. Laine and L. Lellouch, Nucl. Phys. **B656**, 226 (2003), [hep-lat/0211020].
17. G. Colangelo, hep-lat/0409111.
18. M. Lüscher, Nucl. Phys. **B549**, 295 (1999), [hep-lat/9811032].
19. H. Fukaya and T. Onogi, Phys. Rev. **D68**, 074503 (2003), [hep-lat/0305004].
20. H. Fukaya and T. Onogi, Phys. Rev. **D70**, 054508 (2004), [hep-lat/0403024].
21. S. Shcheredin *et al.*, hep-lat/0409073.
22. P. Marenzoni, L. Pugnetti and P. Rossi, Phys. Lett. **B315**, 152 (1993), [hep-lat/9306013].
23. S. Necco and R. Sommer, Phys. Lett. **B523**, 135 (2001), [hep-ph/0109093].
24. E.-M. Ilgenfritz, M. L. Laursen, G. Schierholz, M. Müller-Preussker and H. Schiller, Nucl. Phys. **B268**, 693 (1986).
25. M. Creutz, Phys. Rev. **D70**, 091501 (2004), [hep-lat/0409017].
26. I. Montvay and G. Munster, Cambridge, UK: Univ. Pr. (1994) 491 p. (Cambridge monographs on mathematical physics).
27. S. Necco, Nucl. Phys. **B683**, 137 (2004), [hep-lat/0309017].

Spontaneous CP-breaking in the domain vacuum

A.C. Kalloniatis[*] and S.N. Nedelko[†]

[*]Centre for the Subatomic Structure of Matter, University of Adelaide, South Australia 5005, Australia
[†]Bogoliubov Laboratory of Theoretical Physics, JINR, 141980 Dubna, Russia

Abstract. The prediction that CP-symmetry can be spontaneously broken for discrete values of the θ parameter in QCD is a constraint on self-consistent models of the non-perturbative QCD vacuum. We show that the domain model, which successfully exhibits confinement, spontaneous symmetry breaking and resolution of the axial $U(1)$ problem, correctly reproduces this feature.

Keywords: CP violation, theta parameter, chiral symmetry breaking, domains, QCD vacuum
PACS: 12.38.Aw, 12.38.Lg, 14.64.Bt, 11.30.Rd, 11.30.Er

The inclusion of the theta term in the action of quantum chromodynamics, for example in Euclidean space,

$$S_\theta = iq\theta, \quad q = \frac{g^2}{32\pi^2} \int d^4x F_{\mu\nu}\tilde{F}^{\mu\nu}, \tag{1}$$

introduces explicit CP-violation in QCD. However in the presence of spontaneously broken chiral symmetry this explicit CP-violation can itself become *spontaneous* for $\theta = \pi$ due to vacuum doubling, the so-called Dashen phenomenon [1]. Though outside the physical range $|\theta| \leq 10^{-9}$ as determined by [2, 3], this behaviour is intimately related to both the mechanism of spontaneous chiral symmetry breaking and the non-appearance of an axial $U(1)$ pseudoscalar Goldstone boson, as shown in the approach of anomalous Ward identities [5] and of effective chiral Lagrangians [6] for large number of colours N_c (see also [7] in these proceedings). We shall show that the domain model of the QCD vacuum [4] reproduces the Dashen phenomenon. Unlike [5, 6], which do not unveil the actual mechanism of non-perturbative vacuum properties, the domain model [4] explicitly assumes a particular vacuum structure here based on a statistical ensemble of domain-like gluon fields, and out of this both vacuum and mesonic properties are derived. Both confinement [4] and chiral symmetry realisation [8, 9] have been verified for the domain vacuum *ansatz*.

The domain model is defined in terms of the partition function for a statistical ensemble of hyperspheres with a constant mean-field (anti-)self-dual gluon field strength [4]. The model parameters are the average domain gluon field strength B and the average domain size R. From the string tension these are fixed at values $\sqrt{B} = 947$MeV and 0.26fm. On domain boundaries we posit singular pure gauge fields which force constraints on gluon and quark fluctuations in the ensemble. The quark boundary condition for the i-th domain is the four-dimensional version of the bag condition,

$$i\,\hat{n}(x)e^{i\alpha_i\gamma_5}\psi(x) = \psi(x) \tag{2}$$

CP756, *Quark Confinement and the Hadron Spectrum VI*
edited by N. Brambilla, U. D'Alesio, A. Devoto, K. Maung, G.M. Prosperi and S. Serci
© 2005 American Institute of Physics 0-7354-0241-8/05/$22.50

where in the partition function all possible sets of chiral angles $\{\alpha_1, \ldots, \alpha_N\}$ are summed ensuring the chiral invariance of the ensemble of $N \to \infty$ domains. The phase of the determinant of the Dirac operator in the presence of such boundary conditions was computed in [9] with an α-dependent imaginary part $\pm iq\arctan(\tan(\alpha))$ emerging. Here $q = B^2 R^4 / 16 = 0.15$ is the absolute value of topological charge in a domain, and the overall sign $(-)+$ corresponds to an (anti-)self-dual domain. This is basically the chiral anomaly. Taken as a contribution to the free energy of the domain ensemble and averaging over self-dual and anti-self-dual domains leads to a contribution $\ln[\cos(q\arctan(\tan(\alpha)))]$ which has only discrete minima for $\alpha = 0, \pi$. In the thermodynamic limit $N \to \infty$ this suppresses continuous axial $U(1)$ symmetry in the ground state; the surviving discrete symmetry is spontaneously broken generating a chiral condensate and the pseudoscalar meson spectrum contains no corresponding Goldstone boson [9].

We now include the CP-violating parameter in the model by the additional term Eq.(1) in the action. We consider N_f degenerate fermions with infinitesimally small mass m. To guarantee full flavour chiral invariance of the ensemble, the boundary condition Eq.(2) must be generalised to $\alpha_i \to \alpha_i + \beta_i^a T^a / 2$. Integrating over the fermions and averaging over (anti-)self-dual configurations gives for the ensemble free energy density

$$F = \mathsf{v}^{-1} \ln \cos q[W_{N_f} - \theta] - m\aleph \sum_{i=1}^{N_f} \cos \Phi_i, \tag{3}$$

where $\mathsf{v} = \pi^2 R^4 / 2$ is the volume of a single domain, $W_{N_f} = \sum_{i=1}^{N_f} \arctan(\tan \Phi_i)$, with $\Phi_1 = \alpha$ for $N_f = 1$, $\Phi_1 = \alpha + \frac{|\vec{\beta}|}{2}, \Phi_2 = \alpha - \frac{|\vec{\beta}|}{2}$, for $N_f = 2$, and

$$\Phi_1 = \alpha + b^3 + b^8/\sqrt{3}, \quad \Phi_2 = -b^3 + b^8/\sqrt{3}, \Phi_3 = -2b^8/\sqrt{3}, \quad \text{for } N_f = 3. \tag{4}$$

The functions b_3, b_8 depend on the non-singlet angles β^a. The constant $\aleph$ arises from the asymmetry of the spectrum of the Dirac operator in a domain background gluon field, and takes numerical value $\aleph = (237.8 \text{ MeV})^3$ after B and R are fixed as above. The quark condensate will be seen to be proportional to $\aleph$. In the absence of the mass term, the minima of the free energy are determined by the solutions

$$\alpha_{kl}(\theta) = \frac{\theta}{N_f} + \frac{2\pi l}{q N_f} + \frac{\pi k}{N_f} \quad (k \in \mathbb{Z}, l \in \mathbb{Z}), \tag{5}$$

to $\cos(q W_{N_f} - q\theta) = 1$. Evidently there are multiple solutions labelled by k, l arising from the various periodic functions appearing in Eq.(3). Requiring that the free energy be minimised also for the non-singlet angles β^a forces that k be an even integer. Similarly, for rational $q = n_1/n_2$ the number l can take any integer value except if n_1 is even, for which l must be even. The scalar and pseudoscalar quark condensates turn out to be

$$\langle \bar{\psi}(x)\psi(x) \rangle = \aleph \cos(\alpha_{kl}(\theta)), \langle \bar{\psi}(x)\gamma_5 \psi(x) \rangle = -\aleph \sin(\alpha_{kl}(\theta)), \tag{6}$$

for k, l specified as above. For $\theta = 0$, the only solution is $\alpha = 0$ so that $\langle \bar{\psi}\psi \rangle = -(237.8 \text{ MeV})^3$. For $N_f = 3$ and $q = 0.15$ these condensates are plotted in Fig.1.

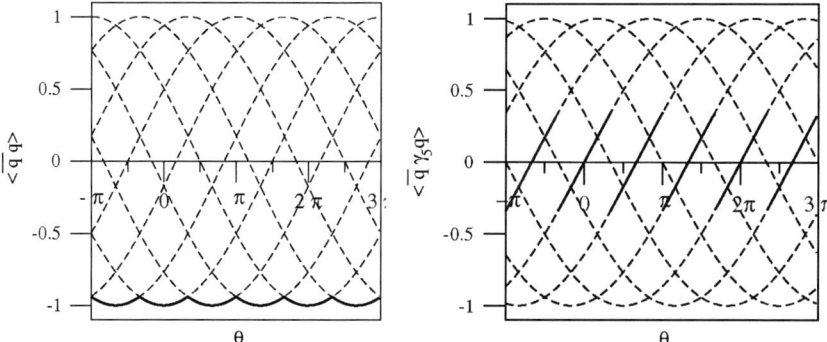

FIGURE 1. The scalar (left) and pseudoscalar (right) quark condensates as a function of θ for $N_f = 3$ and $q = 15/100$ in units of $\aleph$. The dashed lines correspond to discrete minima of the free energy density which are degenerate for $m \equiv 0$. The solid line denotes the minimum which becomes global for a given θ in the presence of infinitesimally small mass term.

We see from the cusps in $\langle \bar{\psi}\psi \rangle$ that $\theta = \pi, \pm\pi/3 \pmod{2\pi}$ are critical θ where vacua become degenerate. These vacua are distinguished by their CP properties, as seen in the corresponding jumps in $\langle \bar{\psi}\gamma_5\psi \rangle$. If we set $q = 1/N_f$ only $\theta = \pi$ survives, consistent with [5]. Periodicity under $\theta \to \theta + 2\pi$ emerges as a consequence of free energy minimisation. It is easy to see that if q is irrational then there is dense set of critical θ and CP is broken spontaneously for any theta. The free energy density in the presence of a mass term is independent of θ, while $\theta = 0$ is the limit of a sequence of critical θ, not belonging to this set. More details are presented in [10].

ACKNOWLEDGMENTS

ACK is supported by the Australian Research Council. SNN was supported by the DFG, contract SM70/1-1 and partially by the grant RFBR 04-02-17370.

REFERENCES

1. R. Dashen, Phys. Rev. D **3**, 1879 (1971).
2. K.F. Smith, *et al.* Phys. Lett. **B 234**, 191 (1990); I.S. Altarev, *et al.*, Phys. Lett. **B 276**, 242 (1992).
3. R.J. Crewther, P. di Vecchia, G. Veneziano, E. Witten, Phys. Lett. **B 88**, 123 (1979); *Erratum, ibid* **B 91**, 487 (1980).
4. A.C. Kalloniatis, S.N. Nedelko, Phys. Rev. D **64**, 114025 (2001).
5. R.J. Crewther, *Chiral Properties of Quantum Chromodynamics*, NATO Adv. Study Inst. Ser. B Phys. **55**, 529 (1980).
6. E. Witten, Ann. Phys. **128**, 363 (1980); P. Di Vecchia, G. Veneziano, Nucl. Phys. **B171**, 253 (1980).
7. M. Creutz, hep-lat/0410043; see also hep-ph/0312018 and hep-ph/0312225.
8. A.C. Kalloniatis, S.N. Nedelko, Phys.Rev. D **66**, 074020 (2002).
9. A.C. Kalloniatis, S.N. Nedelko, *Phys.Rev. D* **69**, 074029 (2004); *Erratum*: Phys.Rev. D, in press.
10. A.C. Kalloniatis, S.N. Nedelko, hep-ph/0412042. Submitted to Phys.Rev.D.

Monopole-Antimonopole pair in an external magnetic field

Yasha Shnir

Institut für Physik, Universität Oldenburg, D-26111, Oldenburg, Germany

Abstract. We consider the interaction of static axially symmetric monopole-antimonopole solution of the $SU(2)$ Yang-Mills-Higgs theory with an external uniform magnetic field. For a non-BPS solution there is a local minimum of the corresponding potential of interaction between the poles.

Keywords: Non-BPS Monopoles, Yang-Mills-Higgs theory
PACS: 14.80.Hv, 11.15.Kc

The structure of the vacuum of the $SU(2)$ Yang-Mills-Higgs (YMH) theory is rather nontrivial, there are spherically symmetric monopoles with unit topological charge [1], axially symmetrical multimonopoles of higner topological charge [2, 3, 4], and solutions with platonic symmetry [5]. There is also a monopole-antimonopole pair (MAP) static solution, which is a deformation of the topologically trivial sector [6, 7]. Recently, we found static axially symmetric solutions representing monopole-antimonopole chains and votrex rings [8]. These solutions are characterized by two integers, the winding number m in polar angle θ and the winding number n in azimuthal angle φ. The structure of the nodes of the Higgs field depends on the values of these integers, there are both chains of zeros and rings. However only winding number n has a meaning of topological charge.

In the Bogomol'nyi-Prasad-Sommerfield (BPS) limit of vanishing Higgs potential spherically symmetric monopole and axially symmetric multimonopole solutions, which satisfy the first order Bogomol'nyi equations [9] as well as the second order field equations, are known analytically [4].

Taubes proved that in the $SU(2)$ YMH theory a smooth, finite energy magnetic dipole solution of the second order field equations, which do not satisfy the Bogomol'nyi equations could exist [10]. The simplest such solution resides in the topologically trivial sector and forms a saddlepoint of the energy functional. It possesses axial symmetry, and there are two zeros of the Higgs field which are located symmetrically on the positive and negative z-axis. Such a configuration corresponds to a magnetic dipole [6, 7].

As shown long ago [11], there is a balance of two long-range interactions between the poles, mediated by the massless photon and the massless scalar particle respectively. Thus, a monopole and an antimonopole can only be in static equilibrium, if they are close enough to experience a repulsive force [7, 8]. In this case we have the complicated pattern of short-range interactions and the structure of the nodes of the Higgs fields strongly depends on the scalar coupling.

Evidently, an external electromagnetic field may be used to as a dipstick to test the structure of the net potential of the interaction between the poles. In this note we discuss

CP756, Quark Confinement and the Hadron Spectrum VI
edited by N. Brambilla, U. D'Alesio, A. Devoto, K. Maung, G.M. Prosperi and S. Serci
© 2005 American Institute of Physics 0-7354-0241-8/05/$22.50

the interaction of the magnetic dipole with an external uniform magnetic field.

The Lagrangian of the Yang-Mills-Higgs model coupled with an external magnetic field B^{ext} is given by

$$-L = \int \left\{ \frac{1}{2} \text{Tr} \left(F_{\mu\nu} F^{\mu\nu} \right) + \frac{1}{4} \text{Tr} \left(D_\mu \Phi D^\mu \Phi \right) + \frac{\lambda}{8} \text{Tr} \left[\left(\Phi^2 - \eta^2 \right)^2 \right] + L_{int} \right\} d^3 r, \quad (1)$$

with $su(2)$ gauge potential $A_\mu = A_\mu^a \tau^a / 2$, field strength tensor $F_{\mu\nu} = \partial_\mu A_\nu - \partial_\nu A_\mu + ie[A_\mu, A_\nu]$, and covariant derivative of the Higgs field $\Phi = \phi^a \tau^a$ in the adjoint representation $D_\mu \Phi = \partial_\mu \Phi + ie[A_\mu, \Phi]$. Here e denotes the gauge coupling constant, η the vacuum expectation value of the Higgs field and λ the strength of the Higgs selfcoupling.

The gauge invariant interaction term $L_{int} = \frac{1}{2} \text{Tr}(\varepsilon_{nmk} F_{mk} \Phi) B_n^{ext}$ describes the direct interaction between the magnetic field of the dipole and the external homogenious magnetic field [12].

To construct solutions numerically we employ the static axially symmetric ansatz [8] for the gauge and the Higgs fields

$$A_\mu dx^\mu = \left(\frac{K_1}{r} dr + (1 - K_2) d\theta \right) \frac{\tau_\varphi^{(n)}}{2e} - n \sin\theta \left(K_3 \frac{\tau_r^{(n,m)}}{2e} + (1 - K_4) \frac{\tau_\theta^{(n,m)}}{2e} \right) d\varphi$$

$$\Phi = \Phi_1 \tau_r^{(n,m)} + \Phi_2 \tau_\theta^{(n,m)}. \quad (2)$$

written in the basis of $su(2)$ matrices $\tau_r^{(n,m)}, \tau_\theta^{(n,m)}$ and $\tau_\varphi^{(n)}$.

The regular solutions with finite energy density and correct asymptotic behaviour are constructed numerically by imposing the boundary conditions as in [12]. The magnitude of the external field is a perturbation parameter of the model. Unperturbed solution is a magnetic dipole with two zeros of the Higgs field. In the BPS limit its mass is $M_0 = 1.69$ and the distance between the nodes is $z_0 = 4.18$ in units of v.e.v. of the scalar field [6, 7]. As λ increases the equilibrium distance is getting smaller and for $\lambda = 0.5$ we have $z_0 = 3.24$ and $M_0 = 2.48$.

Let us briefly recapitulate the results of calculations. If the external magnetic field is directed along negative direction of the z axis there is an additional electromagnetic attraction in the system. We can see that the poles behave according to expectable electrodynamical picture of interaction of two pointlike having opposite charges with an external homogenious field. When the magnitude B_{ext} increases from zero, the nodes of the scalar field, which are accociated with position of the monopoles are approaching to each other and the potential of interaction between the monopoles, which is calculated as difference $E - M_0$, increases linearly (see Fig.1). We observed similar behavior both for a finite and vanishing scalar coupling λ. This case is similar to the situation when gravity is coupled to the magnetic dipole [13]. Thus, the configuration cannot annihilate into trivial sector preserving the axial symmetry of the ansatz, only the modes which are orhogonal to it may contribute to this process.

If the external field is directed along positive ditection of the z axis, the situation depends on λ. We do not find no static solution in the BPS limit. However for a finite value of scalar coupling a branch of solutions emerges from non-interacting solution.

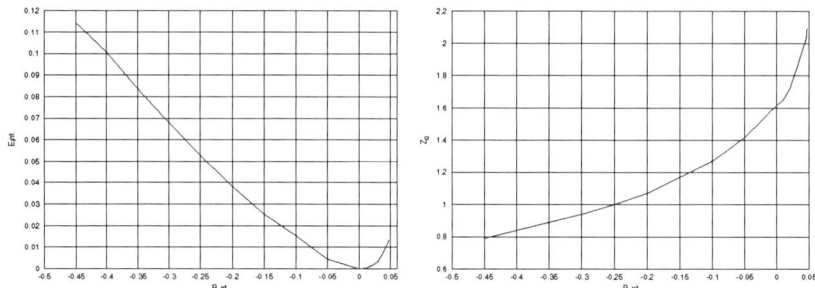

Figure 1. Energy of interaction between the poles of magnetic dipole (left) and the position of the nodes of the Higgs field as function of the external field (right) at $\lambda = 0.5$.

As magnitude B_{ext} increases the distance between the nodes also is increasing. However the potential of interaction is no longer linear (see Fig.1). This branch is extendend up to some critical value of B_{ext} beyond which external field becomes too strong for static dipole to persist. Remarkable at this point the distance between the poles is exactly as for the BPS pair with no external field.

Thus, we conclude that there is a local minimum of the potential of interaction between the poles of the axially symmetric non-BPS dipole. The potential well is getting deeper as λ increases, with a linear shape of the left side and an exponential shape of the right side respectively. Such a potential of monopole-antimonopole interaction allows an intermediate long-living breather state in the annihilation of the configuartion into the trivial vacuum.

This short note is based on a work in colaboration with Burkhard Kleihaus and Jutta Kunz [8]. Further details will be given elsewhere.

REFERENCES

1. G. 't Hooft, *Nucl. Phys.*, **B**79 276 (1974); A.M. Polyakov, *Pis'ma JETP*, **20** 430 (1974).
2. E.J. Weinberg and A.H. Guth, *Phys. Rev.*, **D**14 1660 (1976).
3. C. Rebbi and P. Rossi, *Phys. Rev.*, **D**22 2010 (1980).
4. R.S. Ward, *Comm. Math. Phys.*, 79 317 (1981); P. Forgacs, Z. Horvath and L. Palla, *Phys. Lett.*, 99B 232 (1981); M.K. Prasad, *Comm. Math. Phys.*, 80 137 (1981); M.K. Prasad and P. Rossi, *Phys. Rev.*, **D**24 2182 (1981) 2182.
5. see e.g. P. M. Sutcliffe, *Int. J. Mod. Phys.*, **A** 12 4663 (1997) 4663; C. J. Houghton, N. S. Manton and P. M. Sutcliffe, *Nucl. Phys.*, **B** 510 507 (1998).
6. Bernhard Rüber, Thesis, University of Bonn 1985.
7. B. Kleihaus and J. Kunz, *Phys. Rev.*, **D**61 025003 (2000) .
8. B. Kleihaus, J. Kunz and Ya. Shnir, *Phys. Lett.*, **B**570 237 (2003); B. Kleihaus, J. Kunz and Ya. Shnir, *Phys. Rev.*, **D**68 101701 (2003); B. Kleihaus, J. Kunz and Ya. Shnir, *Phys. Rev.*, **D**70 065010 (2004).
9. E.B. Bogomol'nyi, *Yad. Fiz.*, 24 861 (1976); M.K. Prasad and C.M. Sommerfeld, *Phys. Rev. Lett.*, 760 35 (1975).
10. C. H. Taubes, *Commun. Math. Phys.*, 473 97 (1985); ibid 86 257 (1982); ibid 86 (1982) 299.
11. N.S. Manton, *Nucl. Phys.*, **B** 126 525 (1977).
12. V.G. Kiselev and Ya.M. Shnir, *Phys. Rev.*, 5174 **D**57 (1997); Ya.M. Shnir, *Mod. Phys. Lett.* 287, **A** 19 (2004).
13. B. Kleihaus and J. Kunz, *Phys. Rev. Lett.*, 2430 85 (2000).

Flux tube counting or Casimir scaling

Sedigheh Deldar* and Shahnoosh Rafibakhsh*

*Department of Physics, University of Tehran, Iran

Abstract. QCD confines quarks in all representations. From lattice calculations and fat center vortices model, we discuss that the coefficient of the linear term in the potential is proportional to both casimir scaling and the number of fundamental strings.

Explaining the confinement of quarks in QCD is still a challenging problem. Even though perturbative techniques describe very well the behavior of quarks at short distances, the mechanism of confinement which prevents quarks and gluons to be found free, is still a puzzle. Many lattice calculations confirm the confinement of quarks in the fundamental and higher representations at intermediate distances, but phenomenological models have not been fully successful in describing the infrared behavior of quarks and gluons. One of the features of the confinement is that the potential between quarks increases by distance. Recent numerical calculations [1, 2] show that the string tension, the coefficient of the linear term in the potential, is representation dependent and proportional to the eigenvalue of the quadratic casimir operator of the representation. This proportionality is called "Casimir scaling". The proportionality of the potential with casimir operator is expected for short distances where the force between quarks can be described by one gluon exchange and the coupling is proportional to the quadratic casimir operator. But this behavior is still not understood for intermediate distances. On the other hand, some people believe the string tension between two quarks of a representation can be obtained by multiplying the fundamental string tension times the number of fundamental flux tubes embedded into that representation. The fundamental (elementary) string is a string which connects a fundamental heavy quark with an anti-quark. In this paper, we take a closer look at lattice results and explain that string tensions may also agree with the flux tube counting idea. We also discuss results obtained for potentials between quarks from the fat center vortices model for SU(2), SU(3) and especially our recent calculations for SU(4) and show that string tensions are in agreement with both casimir scaling and flux tube counting but they agree better with flux tube counting especially for larger gauge groups.

Although lattice calculations show a qualitative agreement between string tensions and casimir operators, looking more carefully at lattice results, one can interpret string ratios with the idea of flux tube counting. Cross signs on the plot on the left hand side of figure 1 show ratios of string tensions of SU(3) sources of various representations to that of the fundamental representation obtained from lattice calculations [1]. In this paper, we sometimes call sources in higher representations as "quarks". Circles indicate casimir ratios and diamonds show the number of fundamental strings in each representation. As claimed by lattice people, ratios of string tensions are proportional to casimir operators.

CP756, *Quark Confinement and the Hadron Spectrum VI*
edited by N. Brambilla, U. D'Alesio, A. Devoto, K. Maung, G.M. Prosperi and S. Serci
© 2005 American Institute of Physics 0-7354-0241-8/05/$22.50

On the other hand, the plot shows that there is a rough agreement with the number of fundamental tubes as well. A. Armoni *et al.* [3] have explained why ratios from the lattice are larger than the number of fundamental fluxes. If we define the potential between two quarks as the potential between fundamental strings, then, when strings are very far from each other, they do not have any interaction and the string tension is equal to the string tension of quarks in the fundamental representation times the number of strings. On the other hand, if we put quarks in an appropriate distance, the elementary strings attract each other and this attraction reduces the string tension to some number less than the potential of elementary strings. If quarks get even closer, an overlap between strings happens. Therefore, they repel each other and the string tension between two sources decreases such that it gets larger than the potential between fundamental tubes. As Armoni *et al.* have discussed, the typical length/thickness ratio of the fundamental string of lattice calculations is not large enough. Thus, an overlap between fundamental strings may exist which leads to a repulsion and therefore makes string ratios larger than the number of fundamental fluxes.

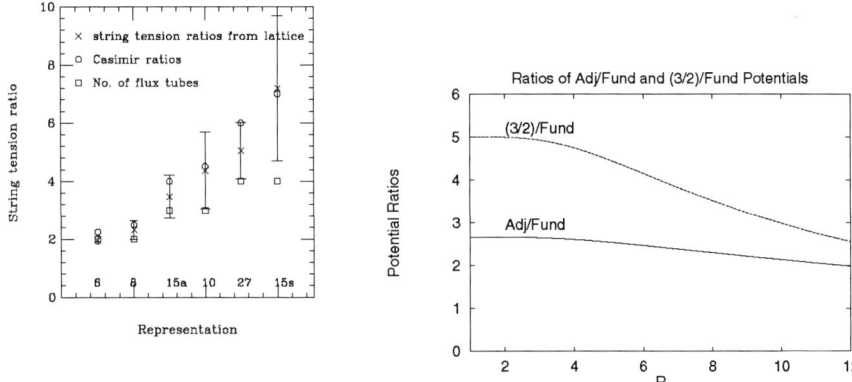

FIGURE 1. The left plot shows ratios of string tensions of SU(3) quarks of different representations to the string tension of quarks in the fundamental representation. Considering the lattice data errors, a good agreement with both casimir scaling and flux tube counting is observed. Potential ratios between quark charges of $j = 1$ (adjoint) and $j = 3/2$ to that of the fundamental quark ($j = 1/2$) are shown in the right. Ratios start up at corresponding casimirs which are $8/3$ and 5 for $j = 1$ and $j = 3/2$, respectively.

Now, we discuss results obtained from the fat center vortices model for SU(2), SU(3) and SU(4) gauge groups. The center vortices model is a phenomenological model which claims that the QCD vacuum is a condensate of color magnetic vortices. These vortices are responsible for confining color charges. For intermediate distances, center vortex model has predicted the confinement of quarks in the fundamental representation. M. Faber *et al.* [4] have observed the confinement of quarks for all representations by making the vortices thick enough. The plot on the right hand side of figure 1 shows ratios of string tensions for SU(2) sources [4]. Potentials for quark charges in the $j = 1/2, 1, 3/2$ are calculated and potentials ratios of sources with the $j = 1$ (adjoint) and $j = 3/2$ to that of the fundamental quark ($j = 1/2$) are plotted. As indicated in the figure, ratios start up at of casimir ratios which are $8/3$ and 5 for $j = 1$ and $j = 3/2$, respectively. Figure

2 shows potential ratios of quarks in various representations to that of the fundamental one, for SU(3) [5] and SU(4) [6] gauge groups. Again, ratios start up roughly at ratios of corresponding casimirs but change so that at some region, which is different for each representation, get close to the number of fundamental strings of that representation. The agreement with flux tube counting in the most linear part of the potential is better for SU(4) than SU(3) [7]. This is in agreement with reference [8] which claims that by increasing the number of gauge groups, the interaction between fundamental strings decreases and the total string tension would be the fundamental string tension times the number of strings.

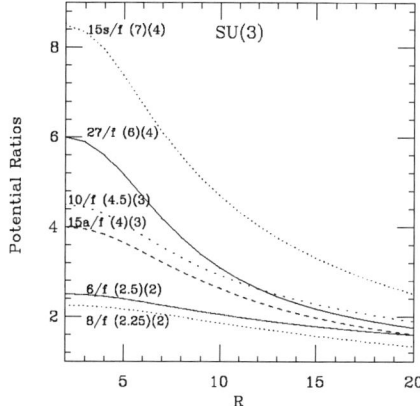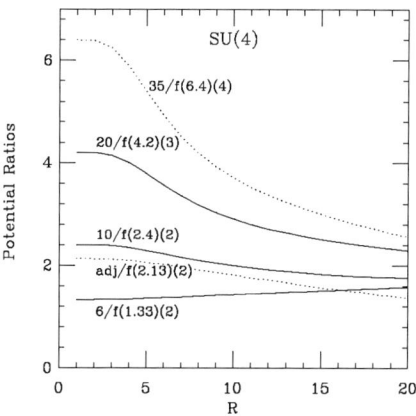

FIGURE 2. Ratios of potentials between quarks of higher representations to that of the fundamental one, for SU(3) and SU(4) gauge groups. Casimir ratios and the number of fundamental strings are shown in the first and second parentheses, respectively. Ratios start at corresponding casimirs but get close to the number of fundamental strings.

We conclude that both lattice calculations and the fat center vortices model predict a linear regime for the potential between quarks of the fundamental and higher representations. The string tension in that region is proportional to both casimir scaling and the number of fundamental tubes. The proportionality with the number of fundamental tubes seems to be better from fat center vortices model especially for SU(4) gauge group.

We would like to thank the research council of the University of Tehran for supporting this work.

REFERENCES

1. S. Deldar, Phys. Rev., D62, p. 034509, 2000.
2. G. Bali, Phys. Rev., D62, p. 114503, 2000.
3. A. Armoni, M. Shifman, Nucl. Phys., B671, p. 67, 2003.
4. M. Faber, J. Greensite, S. Olejník, Phys. Rev., D57, p. 2603, 1998.
5. S. Deldar, JHEP 0101 p. 013, 2001.
6. S. Deldar, S. Rafibakhsh, hep-ph/0411184.
7. S. Deldar, hep-ph/0411269.
8. G. Bali, Phys. Rept. 343, p1 ,2001.

Center Vortex Model for the Infrared Sector of $SU(3)$ Yang-Mills Theory

Markus Quandt*, Michael Engelhardt† and Hugo Reinhardt*

*Institute for Theoretical Physics, University of Tübingen, D-72076 Tübingen, Germany
†Physics Department, New Mexico State University, Las Cruces, NM 88003, U.S.A.

Abstract. In this talk, we review some recent results of the center vortex model for the infrared sector of $SU(3)$ Yang-Mills theory. Particular emphasis is put on the order of the finite-temperature deconfining phase transition and the geometrical structure of vortex branchings. We also present preliminary data for the 't Hooft loop operator and the dual string tension near the phase transition.
Keywords: Yang-Mills theory, lattice, vortex, (de)confinement, finite temperature, $SU(3)$
PACS: 11.15.Ha, 12.38.Aw, 12.38.Gc

Introduction

The vortex picture of the Yang-Mills vacuum, initially proposed as a possible mechanism of colour confinement, has recently attracted a renewed attention. This is mainly due to the advent of new gauge fixing techniques which permit the detection of center vortex structures directly within lattice Yang-Mills configurations. Numerical studies have revealed that the center projection vortices detected in this way do locate true physical objects (rather than lattice artifacts) [1], and there is by now ample evidence that the infrared properties of Yang-Mills theory can be accounted for in terms of vortices [2].

Based on these ideas, a random vortex world-surface model was introduced as an effective low-energy description of $SU(2)$ Yang-Mills theory [3]; it has recently been extended to the gauge group $SU(3)$ [4]. The fundamental assumption is that the long-range structure of Yang-Mills theory is dominated by extended tubes of center flux tracing out closed surfaces in space-time. Consequently, we realise our model on a space-time lattice in which the fixed spacing a represents the transverse thickness of vortices. The random surfaces created on this lattice are weighted by a model action containing a Nambu-Goto and a curvature term, with dimensionless coupling constants ε and c, respectively. Physically, this means that vortices have a certain surface tension and they tend to be *stiff*. For further details on our model and the determination of the parameters ε and c (as well as the vortex extension $a = 0.39\,\text{fm}$), the reader is referred to [4].

Finite Temperature Phase Transition and Vortex Branching

Fig. 1 shows histograms of the action densities measured on $30^3 \times 2$ lattices at the critical points for the two gauge groups $G = SU(3)$ (left panel) and $G = SU(2)$ (right panel). As can be clearly seen, the $SU(3)$ transition exhibits the shallow double-peak

CP756, *Quark Confinement and the Hadron Spectrum VI*
edited by N. Brambilla, U. D'Alesio, A. Devoto, K. Maung, G.M. Prosperi and S. Serci
© 2005 American Institute of Physics 0-7354-0241-8/05/$22.50

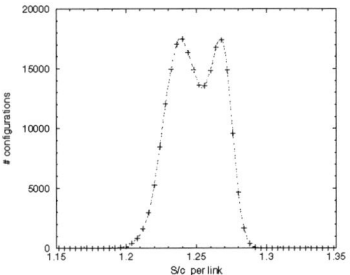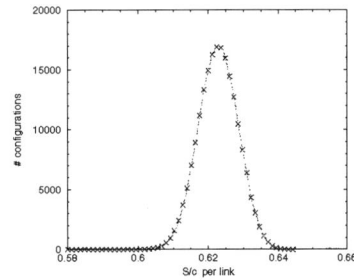

FIGURE 1. Histograms of the average action density at the critical point for $30^3 \times 2$ lattices; the left and right panel show $G = SU(3)$ and $G = SU(2)$, respectively.

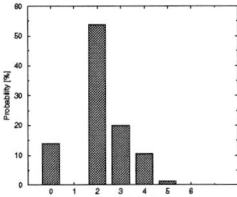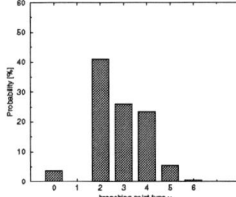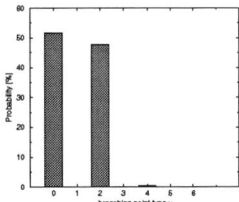

FIGURE 2. Volume fractions occupied by points of a certain branching type v within $3D$ lattice slices. The left panel shows the distribution at zero temperature ($c = 0.21$, confined region). The middle and right panel both correspond to $T > T_c$, with the middle referring to a *time slice* and the right to a *space slice*.

structure characteristic for a weak *first order transition*, while the $SU(2)$ transition is continuous (*second order*). This qualitative behaviour is in agreement with findings from lattice gauge theory.

Since triality is only conserved *mod 3*, an arbitrary number $v = 0, \ldots, 6$ of vortex surfaces can meet at each link. The odd values $v = 3, 5$ are not allowed in $SU(2)$ and represent genuine $SU(3)$ *vortex branchings*. This phenomenon is best studied in $3D$ slices of the lattice, whence possible branching links are projected onto *points* of type v. From fig. 2, we conclude that the largest volume fraction in the confined phase corresponds to non-branching vortex matter ($v = 2$), with a considerable probability of both branchings ($v = 3, 5$) and self-intersections ($v = 4, 6$). Only 15 % of the volume is not occupied by vortices ($v = 0$). In the deconfined phase ($T > T_c$), the situation is qualitatively unchanged for *time-slices*, while *space slices* show virtually zero branchings above T_c. This can be understood if the vortices undergo a *(de)percolation phase transition* above T_c and most vortex clusters wind directly around the short time direction [3],[4].

't Hooft Loop

The 't Hooft loop can be viewed as a vortex creation operator [5] that implements twisted boundary conditions when extended over an entire lattice plane [6]. It has been

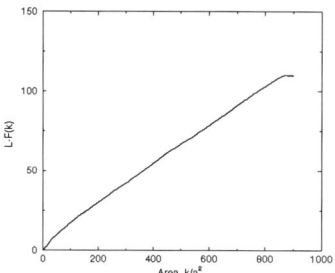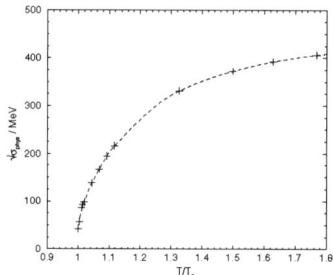

FIGURE 3. Left Panel: Free energy of (incomplete) 't Hooft loops as a function of the loop area within the deconfined phase $(T/T_c = 1.093)$. Right panel: The dual string tension $\tilde{\sigma}$ as a function of the temperature. Measurements were performed on a large $30^3 \times N_0$ lattice with $N_0 = 1, 2$.

shown to be an alternative (dis)-order parameter for the deconfinement phase transition whose behaviour is *dual* to the Wilson loop [6].

This expectation is confirmed in our model: The left panel of fig. 3 exhibits a linear rise of the free energy with the area of the 't Hooft loop, which permits to define a *dual string tension* in the deconfined phase. As we approach the phase transition from above, the dual string tension quickly vanishes (cf. right panel of fig. 3). Precise measurements close to the transition reveal a small discontinuity $\Delta\tilde{\sigma} \simeq (34\,\mathrm{MeV})^2$, which should be compared to the ordinary zero-temperature string tension $\sigma_0 = (440\,\mathrm{MeV})^2$ setting the overall scale. This demonstrates the weakness of the first order transition for $G = SU(3)$.

Conclusions

In this talk, the physical foundation of the center vortex model for the infrared sector of $SU(3)$ Yang-Mills theory has been outlined. Only a selection of the results obtained so far could be presented. Among the effects discussed were the order of the deconfinement phase transition, the structure of branching points and the exact determination of the discontinuity in the free energy of the 't Hooft loop. Interesting questions for future investigations are the study of deconfinement in higher colour groups, in particular the influence of complex branching patterns, as well as the coupling to quarks and the generation of a chiral condensate for $G = SU(3)$.

REFERENCES

1. K. Landfeld, H. Reinhardt, O. Tennert, Phys. Lett. **B419** (1998) 317.
2. see eg. J. Greensite, Prog. Part. Nucl. Phys. **51** (2003) 1; M. Faber, J. Greensite, S. Olejnik, Phys. Lett. **B474** (2000) 177; P. de Forcrand, M. D'Elia, Phys. Rev. Lett. **82** (1999) 4582.
3. M. Engelhardt, H. Reinhardt, Nucl. Phys. **B585** (2000) 591.
4. M. Engelhardt, M. Quandt, H. Reinhardt, Nucl. Phys. **B685** (2004) 227.
5. H. Reinhardt, Phys. Lett. **B557** (2003) 317.
6. P. de Forcrand, M. D'Elia, M. Pepe, Phys. Rev. Lett. **86** (2001) 1438; P. de Forcrand, L. Smekal, Phys. Rev. **D66** (2002) 011504.

Unconstrained Hamiltonian formulation of Yang–Mills theory

Antti Salmela

Theoretical Physics Division, Department of Physical Sciences
P.O. Box 64, 00014 University of Helsinki, Finland

Abstract. A novel method is introduced for incorporating Gauss's law into a Hamiltonian formulation of classical Yang–Mills theory. The method is motivated by Lie's theory of function groups and it produces a local Hamiltonian that decomposes into a finite Laurent series in powers of the coupling constant.

Despite the successes of perturbative and lattice QCD we are still lacking analytical calculation methods in the low-energy regime of strong interactions. In particular, the determination of low-energy states and their energies would become much easier if we knew how to write down a regularised functional Schrödinger equation in QCD. When trying to approach this problem, a natural step in the beginning is to cast classical Yang–Mills theory in the framework of Hamiltonian formalism. Unfortunately it turns out to be difficult to implement Gauss's law in this approach, and one usually ends up with a Hamiltonian that fails to reproduce it. For pure Yang–Mills theory in the temporal gauge ($A_0^a = 0$) such a Hamiltonian is given by

$$H = \int \left(\frac{1}{2} \Pi_{ka} \Pi^{ka} + \frac{1}{4} F_{kl}^a F_a^{kl} \right) d^3\mathbf{x}, \tag{1}$$

$$F_{kl}^a = \partial_l A_k^a - \partial_k A_l^a + g f_{bc}{}^a A_k^b A_l^c.$$

The canonical equations of motion thus reproduce only the dynamical Yang–Mills equations, but Gauss's law is still not completely lost, because the Gauss law generators

$$G_a = \partial^k \Pi_{ka} - g f_b{}^c{}_a A^{kb} \Pi_{kc}$$

are constants of motion in the Hamiltonian dynamics, i.e., $\dot{G}_a = 0$.

At this point there are two ways to proceed. One can quantise the Hamiltonian (1) and then impose Gauss's law on the states as a physicality condition. On the other hand, the order of quantisation and constraining can also be reversed and one can begin by implementing the Gauss law in classical Yang–Mills theory. In this approach it is convenient to parametrise the generators G_a with such variables that make the implementation of Gauss's law easy. The parametrisation then initiates a canonical transformation from the original fields (A_k^a, Π_{ka}) to a new set of variables. With non-Abelian theories the procedure is not straightforward, because the Poisson algebra of the generators

$$\{G_a(\mathbf{x}), G_b(\mathbf{y})\} = -g f_{ab}{}^c G_c(\mathbf{y}) \delta(\mathbf{x} - \mathbf{y}) \tag{2}$$

CP756, *Quark Confinement and the Hadron Spectrum VI*
edited by N. Brambilla, U. D'Alesio, A. Devoto, K. Maung, G.M. Prosperi and S. Serci
© 2005 American Institute of Physics 0-7354-0241-8/05/$22.50

must be compatible with the canonical structure of the variables used in the parametrisation of G.

A novel solution to this problem in the SU(2) case is presented in Ref. [1]. Writing the Gauss law generators in the form

$$
\begin{aligned}
G_1 &= \sqrt{p_1^2 - p_2^2}\,\cos(g\,q_2) \\
G_2 &= -\sqrt{p_1^2 - p_2^2}\,\sin(g\,q_2) \\
G_3 &= p_2,
\end{aligned}
\tag{3}
$$

we see that the algebra (2) with $f_{ab}{}^c = \varepsilon_{ab}{}^c$ then follows from the fundamental Poisson bracket relations of the variables (p_1, q_2, p_2). The idea behind this parametrisation goes back to Lie's work on function groups [2], that is sets of variables equipped with Poisson brackets that close on the set. Lie has proved that every function group can be transformed into a form where the Poisson bracket relations between the variables are canonical. Formula (3) now represents such a transformation as applied to the function group (G_1, G_2, G_3).

Completing the canonical transformation initiated by the parametrisation (3) is a lengthy procedure. If ξ_i denotes any new canonical variable, excluding (q_1, p_1, q_2, p_2), then the Poisson brackets between ξ_i and the previously fixed variables imply that the remaining variables must be gauge-invariant and independent of p_1. Gauge-invariant variables are constructed by transforming the non-gauge-invariant variables (q_1, q_2, p_2) away. The elimination of q_2 and p_2 then leads to an Abelian gauge [3] where the G_a's are rotated into a specific direction in colour space. Before the canonical transformation is complete, one has to deal with the residual U(1) gauge invariance and with an additional internal U(1) symmetry of the gauge-invariant variables.

In the end we have a canonical transformation connecting two sets of variables

$$
(A_k^a, \Pi_{ka}) \quad \Longleftrightarrow \quad
\begin{cases}
(q_i, p_i), & i = 1, 2, 3 \\
(Q_1^1, P_{11}) \\
(Q_2^a, P_{2a}), & a = 1, 3 \\
(Q_3^a, P_{3a}), & a = 1, 2, 3.
\end{cases}
$$

Certain pairs (Q_k^a, P_{ka}) are not regarded as free variables but as functions defined by

$$
\begin{aligned}
Q_1^2 &= -\frac{1}{g\,P_{11}} \sum_{k=2}^{3} \left(\partial_k P_{k3} - g\,\varepsilon_b{}^c{}_3\,Q_k^b\,P_{kc} \right) + \frac{1}{g}\,p_3\,\frac{P_{21}}{P_{11}P_{23}} \\
Q_1^3 &= \frac{1}{g\,P_{11}} \left(\partial_3 P_{32} - \sum_{k=2}^{3} g\,\varepsilon_b{}^c{}_2\,Q_k^b\,P_{kc} \right) \\
Q_2^2 &= \frac{1}{g\,P_{23}} \left(\partial^k P_{k1} - g\,\varepsilon_b{}^c{}_1\,Q_3^b\,P_{3c} \right)
\end{aligned}
\tag{4}
$$

$$
P_{12} = P_{13} = P_{22} = 0.
$$

The transformation equations read

$$
A_k^a = \left(\Omega^T\right)^a{}_b \left(Q_k^b + \frac{1}{g}\delta_{k1}(O_4)^b{}_2 \frac{p_1}{P_{11}} \sqrt{1 - \left(\frac{p_3}{p_1}\right)^2} - \frac{1}{g}\delta_{k2}\,\delta^b{}_2 \frac{p_3}{P_{23}} \right)
$$

$$
- \frac{1}{2g}\varepsilon_{bc}{}^a \left(\Omega^T \partial_k \Omega\right)^{cb}
$$

$$
\Pi_{ka} = \left(\Omega^T\right)_a{}^b P_{kb},
$$

where Ω and O_4 are orthogonal matrices parametrised by the variables (q_i, p_i). The definitions (4) are in force here.

In the Gauss law limit $G_a \to 0$ it turns out that the momenta p_i tend to zero and the coordinates q_i become ambiguous. The physical variables are then the free pairs (Q_k^a, P_{ka}) with their dynamics governed by the Hamiltonian

$$
H_{phys} = \int \left(\frac{1}{2} P_{ka} P^{ka} + \frac{1}{4}\Phi_{kl}^a \Phi_a^{kl} \right) d^3\mathbf{x},
$$

$$
\Phi_{kl}^a = \partial_l Q_k^a - \partial_k Q_l^a + g\,\varepsilon_{bc}{}^a Q_k^b Q_l^c,
$$

where the definitions (4) are applied with $p_i = 0$. This Hamiltonian is local and its integrand contains poles at points where either of the momenta P_{11} or P_{23} vanishes. Expanding the Hamiltonian in powers of the coupling constant, we get a finite series

$$
H_{phys} = \frac{1}{2g^2}H^{(0)} + \frac{1}{g}H^{(1)} + H^{(2)} + g H^{(3)} + \frac{g^2}{2}H^{(4)}.
$$

Note in particular that the Abelian limit $g \to 0$ is singular. The entire Hamiltonian is difficult to quantise, but the weak-coupling term $H^{(0)}$ is easier because it depends on the canonical momenta only.

We have now constructed a classical Yang–Mills Hamiltonian that is unconstrained, local and related to Abelian gauges. Further research is required to find a suitable quantisation and regularisation procedure. Also, in order to apply these results to QCD we must extend the above construction to SU(3) and include fermions in the Hamiltonian.

ACKNOWLEDGMENTS

I would like to thank professor C. Cronström for discussions and comments. This work was partially supported by the Magnus Ehrnrooth foundation.

REFERENCES

1. Salmela, A., hep-th/0409142.
2. Lie, S., *Math. Ann.*, **8**, 215–303 (1875).
3. 't Hooft, G., *Nucl. Phys. B*, **190**, 455–478 (1981).

Gas of monopoles in 3D SU(2) gluodynamics

M. N. Chernodub*, Katsuya Ishiguro† and Tsuneo Suzuki†

*ITEP, B.Cheremushkinskaja 25, Moscow, 117259, Russia
†Institute for Theoretical Physics, Kanazawa University, Kanazawa 920-1192, Japan

Abstract. The Abelian monopoles in the Maximal Abelian projection of the three dimensional pure SU(2) gauge model are studied on the lattice. Using a method of blocking from continuum we find that the behavior of the (squared) monopole lattice density can be described by a Coulomb gas of continuum monopoles. The monopoles treated within our blocking method provide about 75% contribution to the non–Abelian Debye screening length.

The dual superconductor model [1] was invented to explain the confinement of color in non-Abelian gauge theories. The model assumes that the non–Abelian vacuum can be considered as a medium of Abelian monopoles. Infrared properties of the non–Abelian gauge theory in the confinement phase are governed by the monopole condensate while in the deconfinement phase the condensate is absent. The Abelian monopoles are singular configurations of the gluonic field. These configurations can be identified with the help of the Abelian projection method [2]. Numerical simulations show that the dual superconductor is realized in four dimensional non–Abelian gauge theories [3].

We discuss non–perturbative features of three–dimensional SU(2) gauge model which has a relation to high-temperature QCD. The most interesting features are the confinement of color and the mass gap generation (analogues of, respectively, "the spatial confinement" and "the magnetic screening", in the 4D SU(2) gauge model at $T \neq 0$). The Abelian monopole dynamics of this model was previously investigated both by analytical (phenomenological) [4] and numerical [5] approaches. Taking into account the success of the monopole confinement mechanism in 4D [3] it is natural to expect that in 3D SU(2) model the dominant contributions both to the string tension and to the screening mass is given by the Abelian monopoles. We try to describe the action of the Abelian monopoles by a Coulomb gas model. This choice is motivated by the well-known analytical result [6] in the 3D Georgi–Glashow model (which confining as well as the SU(2) gauge model):

$$\mathscr{Z} = \sum_{N=0}^{\infty} \frac{\zeta^N}{N!} \left[\prod_{a=1}^{N} \int d^3x^{(a)} \sum_{q_a=\pm 1} \right] \exp\left\{ -\frac{g_M^2}{2} \sum_{a,b=1}^{N} q_a q_b D(x^{(a)} - x^{(b)}) \right\}. \tag{1}$$

Here x_a and q_a are, respectively, the position and the charge (in units of a fundamental magnetic charge, g_M) of ath continuum monopole. ζ is the fugacity parameter and the Coulomb interaction is represented by the inverse Laplacian D, $-\partial_i^2 D(x) = \delta^{(3)}(x)$.

To match the continuum model (1) with the lattice SU(2) model (and, subsequently, to get values of the couplings of the model (1)) we use the method of blocking from continuum (BFC) [7, 8]. This approach resembles the idea of the blocking of the continuum

CP756, *Quark Confinement and the Hadron Spectrum VI*
edited by N. Brambilla, U. D'Alesio, A. Devoto, K. Maung, G.M. Prosperi and S. Serci
© 2005 American Institute of Physics 0-7354-0241-8/05/$22.50

fields to the lattice fields [9]. The BFC procedure is used to describe topological lattice quantities by a continuum model without going to a deep continuum limit.

The lattice magnetic charge inside the lattice cell C_s is $k_s = \int_{C_s} d^3x\, \rho(x)$, where $\rho(x) = \sum_a q_a \delta^{(a)}(x - x^{(a)})$ is the density of the continuum monopoles. In the low-density approximation the leading order contribution to the (squared) monopole density is [7]:

$$\langle k^2(b) \rangle = \frac{1}{L^3} \langle \sum_{s \in \Lambda} k^2(s) \rangle \equiv \int_{C_s} d^3x \int_{C_s} d^3y\, \langle \rho(x)\rho(y) \rangle = \rho\, b^3\, P(M_D b), \qquad (2)$$

where b is the lattice spacing, and ρ and M_D is the monopole density and the Debye screening mass in the *continuum* limit. The function P is

$$P(\mu) = 1 - \mu^2 \int \frac{d^3q}{(2\pi)^3} \frac{1}{q^2 + \mu^2} \prod_{i=1}^{3} \left[\frac{2\sin(q_i/2)}{q_i} \right]^2. \qquad (3)$$

To compare the analytical predictions with the analytical prediction (2), (3) we have numerically calculated the (squared) monopole density in the pure SU(2) gauge model on the lattice. We have generated 200 configurations of the gauge fields for each chosen value of the coupling constant, $\beta = 2.083, \dots, 14.5$, on the lattice 48^3. We perform Abelian projection in the Maximally Abelian gauge [10] for each $SU(2)$ configuration. To get the lattice density for the monopoles of the large sizes, $b = na$, we applied the blockspin transformation [11]: $k^{(n)}(s) = \sum_{i,j,l=0}^{n-1} k(ns + i\hat{\mu} + j\hat{\nu} + l\hat{\rho})$ with blocking factors $n = 1, \dots, 12$. Here a is the spacing of the fine lattice. We have also removed ultraviolet artifacts in the form of the tightly ($r_{m\bar{n}} < 2a$)) bound monopole–anti-monopole pairs. All dimensional quantities are shown in units of the string tension, σ.

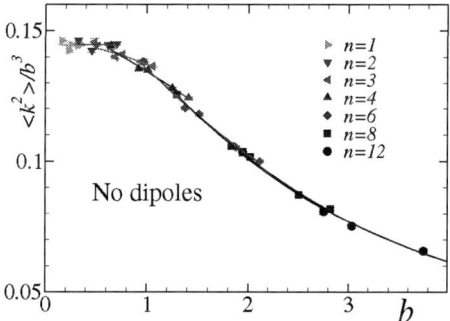

FIGURE 1. The (normalized) squared monopole density, $\langle k_s^2 \rangle / b^3$, and the corresponding fits (2) *vs. b*.

The values of the parameters of the Coulomb gas model in the continuum limit, Eq. (1), can be obtained by fitting the numerical results for $\langle k_s^2 \rangle$ by the analytical prediction (2). The quality of the fit – shown in Figure 1 – is very good, $\chi^2/d.o.f. \sim 1$. We get the following values of the monopole density and the monopole contribution to the Debye screening mass, respectively:

$$\rho/\sigma^{3/2} = 0.174(2), \qquad M_D/\sigma^{1/2} = 1.77(4). \qquad (4)$$

Note that the low-density requirement, $\rho/M_D^3 \gg 1$ (equivalent to $\rho \ll g^6$) is violated according to Eq. (4): $\rho/M_D^3 \approx 0.03 \ll 1$. On the other hand the validity of the low-density approximation can be checked with the help of the quantity $C = M_D\,\sigma/\rho$. In our case this value is $C = 10.1(1)$, and this is about 25% larger than low-density prediction [6], $C_{sp}^{CG} = 8$. The fact that the quantity C is close to its low-density value means that the chosen approximation may still work in our case. Let us check this below.

In the Abelian projection picture, mass of the ground state glueball must be twice bigger than the Debye screening mass, $M_{O^{++}} = 2M_D$. The comparison of our result (4) with the result of Refs. [12], $M_{O^{++}} = 4.72(4)\sqrt{\sigma}$, gives $2M_D/M_{O^{++}} = 0.75(4)$. On the other hand, the direct measurement of the Abelian Debye mass in 3D SU(2) gauge model [13], $m_D^{SU(2)} = 1.39(9)\sqrt{\sigma}$, is consistent with our data: $m_D/m_D^{SU(2)} = 1.27(11)$. Thus the deviations for masses are of the order of 25% similarly to the case of C.

We conclude that the Abelian monopoles in the three–dimensional SU(2) gauge model form the dense Coulomb gas. The Debye screening mass is dominated by contributions from the Abelian monopoles.

ACKNOWLEDGMENTS

This work is supported by JSPS Grant-in-Aid for Scientific Research on Priority Areas 13135210, (B) 15340073, JSPS grant S04045, grants RFBR 01-02-17456, DFG 436 RUS 113/73910, RFBR-DFG 03-02-04016, MK-4019.2004.2. The numerical simulations have been performed on NEC SX-5 at RCNP, Osaka University.

REFERENCES

1. G. 't Hooft, in *High Energy Physics*, ed. A. Zichichi, EPS International Conference, Palermo (1975); S. Mandelstam, *Phys. Rept.* **23**, 245 (1976).
2. G. 't Hooft, *Nucl. Phys.* **B190**, 455 (1981).
3. For a review, see T. Suzuki, Nucl. Phys. Proc. Suppl. **30**, 176 (1993); M. N. Chernodub and M. I. Polikarpov, "Abelian projections and monopoles", in "Confinement, duality, and nonperturbative aspects of QCD", Ed. by P. van Baal, Plenum Press, p. 387, hep-th/9710205; R.W. Haymaker, Phys. Rept. **315**, 153 (1999).
4. S. R. Das and S. R. Wadia, Phys. Rev. D **53**, 5856 (1996).
5. V. Bornyakov and R. Grigorev, Nucl. Phys. Proc. Suppl. **30**, 576 (1993); H. D. Trottier, G. I. Poulis and R. M. Woloshyn, Phys. Rev. D **51**, 2398 (1995).
6. A.M. Polyakov, *Nucl. Phys.* **B120**, 429 (1977).
7. M. N. Chernodub, K. Ishiguro and T. Suzuki, JHEP **0309**, 027 (2003).
8. M. N. Chernodub, K. Ishiguro and T. Suzuki, Phys. Rev. D **69**, 094508 (2004).
9. W. Bietenholz and U.J. Wiese *Nucl. Phys.* **B464**, 319 (1996); Phys. Lett. B **378**, 222 (1996); W. Bietenholz, Int. J. Mod. Phys. A **15**, 3341 (2000)
10. A. S. Kronfeld, M. L. Laursen, G. Schierholz and U. J. Wiese, Phys. Lett. **B198**, 516 (1987); A. S. Kronfeld, G. Schierholz and U. J. Wiese, Nucl. Phys. **B293**, 461 (1987).
11. T.L. Ivanenko, A.V. Pochinsky and M.I. Polikarpov, *Phys. Lett.* **B252**, 631 (1990).
12. M. Teper, Phys. Lett. B **311**, 223 (1993), Phys. Rev. D **59**, 014512 (1999).
13. F. Karsch, M. Oevers and P. Petreczky, Phys. Lett. B **442**, 291 (1998).

The Yang-Mills vacuum in Coulomb gauge[1]

D. Epple, C. Feuchter, H. Reinhardt

Auf der Morgenstelle 14, 72076 Tübingen, Germany

Abstract. The Yang-Mills Schrödinger equation is solved in Coulomb gauge for the vacuum by the variational principle using an ansatz for the wave functional, which is strongly peaked at the Gribov horizon. We find an infrared suppressed gluon propagator, an infrared singular ghost propagator and an almost linearly rising confinement potential. Using these solutions we calculate the eletric field of static color charge distributions relevant for mesons and baryons.

We report on the variational solution of the Yang-Mills Schrödinger equation in Coulomb gauge $\partial_i A_i = 0$ performed in [1]. In this gauge the Yang-Mills Hamiltonian reads

$$H = \frac{1}{2}\int J^{-1}[A]\Pi J[A]\Pi + \frac{1}{2}\int B[A]^2 + \frac{g^2}{2}\int \rho(\hat{D}\partial)^{-1}(-\partial^2)(\hat{D}\partial)^{-1}\rho \ . \tag{1}$$

Here $J[A] = \det(-\hat{D}[A]\partial)$ is the Faddeev-Popov determinant, $B[A]$ is the color magnetic field, $\Pi(x) = \delta/i\delta A(x)$ is the momentum operator representing the color electric field and $\rho(x) = -\hat{A}(x)\Pi(x)$ is the non-Abelian color charge. We use the following ansatz for the Yang-Mills wave functional

$$\Psi = \mathcal{N} J^{-\frac{1}{2}}[A]\exp\left(-\frac{1}{2}\int d^3x d^3x' A(x)\omega(x-x')A(x')\right) , \tag{2}$$

where the kernel $\omega(x-x')$ is determined by minimizing the vacuum energy

$$\langle\Psi|H|\Psi\rangle = \int DAJ[A]\Psi^*[A]H\Psi[A] \ . \tag{3}$$

Thereby we restrict ourselves to 2-loop diagrams. Minimization of the energy gives rise to a set of coupled Schwinger-Dyson equations for the gluon propagator

$$\langle\Psi|A_i^a(x)A_j^b(x')|\Psi\rangle = \frac{1}{2}\delta^{ab}t_{ij}(x)\omega^{-1}(x-x') , \tag{4}$$

the ghost form factor d defined by

$$\langle\Psi|(-\hat{D}\partial)^{-1}|\Psi\rangle = \frac{d}{-\partial^2} , \tag{5}$$

---

[1] Invited talk given by H. Reinhardt at the Confinement conference, Sardinia 2004

CP756, *Quark Confinement and the Hadron Spectrum VI*
edited by N. Brambilla, U. D'Alesio, A. Devoto, K. Maung, G.M. Prosperi and S. Serci
© 2005 American Institute of Physics 0-7354-0241-8/05/$22.50

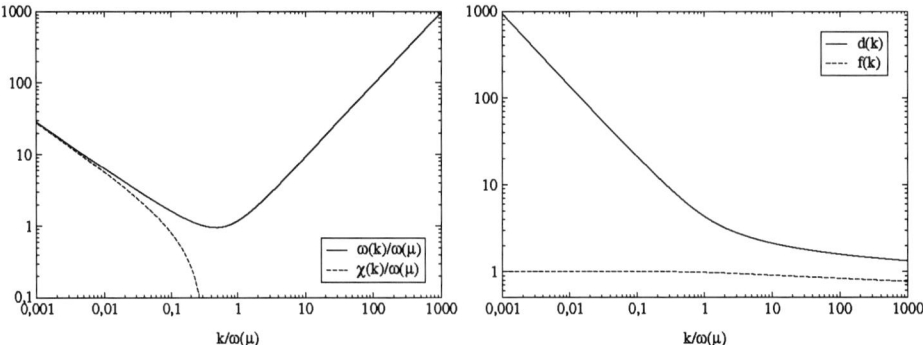

FIGURE 1. Solution for the gap function $\omega(k)$ (left) and Ghost form function $d(k)$ with Coulomb correction $f(k)$ (right).

the Coulomb form factor

$$\langle\Psi|(-\hat{D}\partial)^{-1}(-\partial^2)(-\hat{D}\partial)^{-1}|\Psi\rangle = \frac{d^2 f}{-\partial^2} \tag{6}$$

and the curvature in the space of gauge orbits

$$\chi = -\frac{1}{2}\frac{\delta^2 \ln J[A]}{\delta A \delta A} . \tag{7}$$

Resorting to the angular approximation the Schwinger-Dyson equations can be solved analytically in the infrared $k \to 0$

$$\omega(k) = \chi(k) \sim \frac{1}{k}, \quad d(k) \sim \frac{1}{k}, \quad f(k) \to \text{const} \tag{8}$$

and in the ultraviolet $k \to \infty$

$$\omega(k) \to k, \quad \frac{\chi(k)}{\omega(k)} \to 1/\sqrt{\ln k}, \quad d(k) \sim 1/\sqrt{\ln k}, \quad f(k) \sim 1/\sqrt{\ln k} . \tag{9}$$

Here the ghost form factor has been assumed to fullfil the so-called horizon condition $d(k \to 0) \to \infty$, but otherwise the above aymptotic behaviour is independent of the details of the renormalization. The numerical results are shown in figure 1. The gluon energy $\omega(k)$ is infrared divergent signalling gluon confinement.

The electric field generated by a static charge distribution $\rho(x)$ is given by

$$E(x) = -\partial_x \int d^3x' \langle\Psi|\langle x|(-\hat{D}\partial)^{-1}|x'\rangle|\Psi\rangle\rho(x') . \tag{10}$$

In figure 2 we show the electric field generated by a static quark-antiquark pair. One observes the formation of a color flux tube between the static charges and accordingly we find an (almost) linearly rising confinement potential, see figure 2. Figure 3 shows the

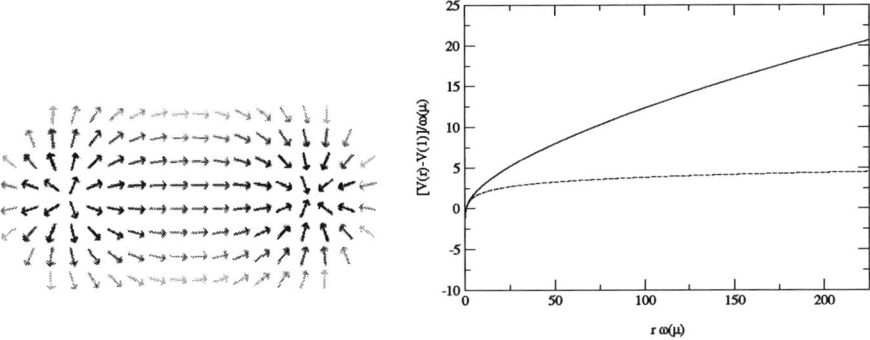

FIGURE 2. (left) Field lines of the longitudinal chromelectric field of a charge-anticharge pair. (right) Coulomb Potential with (full line) and without inclusion of the curvature (dashed line).

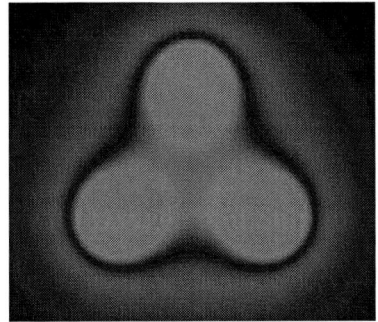

FIGURE 3. The magnitude of the longitudinal chromoelectric field of a three-quark color-singlet state.

module of the static electric field induced by three static color charges in a color singlet state. The flux distribution seems to prefer a so-called Y-shape. Let us also mention that similar calculations, however, with a different ansatz for the wave functional and ignoring the curvature fully [2] or partly [3], have been carried out.

Recently, we have been able to show, that the above presented results do not depend on the detailed ansatz for the wave functional, but does crucially depend on the curvature χ in the space of gauge orbits [4]. In particular, the infrared limit is uniquely determined and to 1-loop order the vacuum wave functional becomes $\Psi[A] = 1$.

The approach presented above is rather encouraging and calls for a more detailed study of the vacuum properties of Yang-Mills theory.

REFERENCES

1. C. Feuchter and H. Reinhardt, Phys. Rev. **D70**, (2004) 105021, hep-th/0408236
2. A.P. Szczepaniak and E.S. Swanson, Phys. Rev. **D65** (2002) 025012
3. A. P. Szczepaniak, Phys. Rev. **D69**, (2004) 074031
4. H. Reinhardt and C. Feuchter, On the Yang-Mills wave functional in Coulomb gauge, hep-th/0408237

Analytic structure of
the Landau gauge gluon propagator

R. Alkofer*, W. Detmold†, C. S. Fischer** and P. Maris‡

**Institute for Theoretical Physics, University of Tübingen, D-72076 Tübingen, Germany*
†*Department of Physics, University of Washington, Box 351560, Seattle WA 98195, USA*
***IPPP, University of Durham, Durham DH1 3LE, U.K.*
‡*U. of Pittsburgh, Dept. of Physics and Astronomy, 100 Allen Hall, Pittsburgh PA 15260, USA*

Abstract. The results of different non-perturbative studies agree on a power law as the infrared behavior of the Landau gauge gluon propagator. This propagator violates positivity and thus indicates
the absence of the transverse gluons from the physical spectrum, *i.e.* gluon confinement. A simple
analytic structure for the gluon propagator is proposed capturing all of its features. We comment
also on related investigations for the Landau gauge quark propagator.

Keywords: Strong QCD, Green's functions, Confinement, Dyson–Schwinger equations
PACS: 12.38.Aw 14.65.Bt 14.70.Dj 12.38.Lg 11.30.Rd 11.15.Tk 02.30.Rz

In this talk a study of the analytic properties of the gluon propagator in Landau
gauge QCD will be presented. Hereby results from non-perturbative calculations of this
propagator are employed. A detailed account of this investigation can be found in ref. [1],
for three-dimensional Yang-Mills theory see also ref. [2].

In the following we will confirm previous results [3, 4] on positivity violation for
the gluon propagator. We will also provide a parameterisation of the gluon propagator
that is analytic everywhere in the complex p^2 plane except on the real timelike axis and
decreases to zero in every direction of the complex p^2 plane[1]. Such a behaviour satisfies
the standard axioms of local quantum field theory except positivity.

In Landau gauge the gluon propagator can be generically written as

$$D_{\mu\nu}(p) \;=\; \left(\delta_{\mu\nu} - \frac{p_\mu p_\nu}{p^2}\right)\frac{Z(p^2)}{p^2}. \tag{1}$$

In Euclidean quantum field theory, positivity of a propagator can be tested by performing
a Fourier transformation with respect to Euclidean time. A violation of the condition

$$\Delta(t) := \int d^3x \int \frac{d^4p}{(2\pi)^4} e^{i(tp_4 + \vec{x}\cdot\vec{p})}\frac{Z(p^2)}{p^2} = \frac{1}{\pi}\int dp_4 \cos(tp_4)\frac{Z(p^2)}{p^2} \geq 0, \tag{2}$$

for the Schwinger function then proves violation of positivity. The Dyson–Schwinger
equations (for recent reviews see *e.g.* [5]) for the ghost, gluon and quark propagators

---

[1] We have also provided parameterisations of the quark propagator [1], some of which are analytic
everywhere in the complex p^2 plane except the timelike real half-axis.

CP756, *Quark Confinement and the Hadron Spectrum VI*
edited by N. Brambilla, U. D'Alesio, A. Devoto, K. Maung, G.M. Prosperi and S. Serci
© 2005 American Institute of Physics 0-7354-0241-8/05/$22.50

in the Landau gauge have been solved recently in a self-consistent truncation scheme [6, 7]. Especially, one analytically obtains

$$Z(p^2) \sim (p^2)^{2\kappa}, \qquad G(p^2) \sim (p^2)^{-\kappa}, \qquad (3)$$

for the gluon and ghost dressing function with exponents related to each other. In this particular truncation κ is an irrational number, $\kappa = (93 - \sqrt{1201})/98 \approx 0.595$ [8, 9]. This result depends only slightly on the employed truncation scheme: Infrared dominance of the gauge fixing part of the QCD action [10] implies infrared dominance of ghosts which in turn can be used to show [8] that the infrared exponents depend only weakly on the dressing of the ghost-gluon vertex [11] and not at all on other vertex functions [12]. Furthermore, investigations based on the Exact Renormalisation Group Equations find the relations (3) with an identical or slightly lower value for κ [13].

The running coupling as it results from numerical solutions for the gluon and ghost propagators can be accurately represented by [6]

$$\alpha_{\text{fit}}(p^2) = \frac{1}{1 + (p^2/\Lambda_{\text{QCD}}^2)} \left(\alpha(0) + \frac{p^2}{\Lambda_{\text{QCD}}^2} \frac{4\pi}{\beta_0} \left(\frac{1}{\ln(p^2/\Lambda_{\text{QCD}}^2)} - \frac{1}{p^2/\Lambda_{\text{QCD}}^2 - 1} \right) \right). \quad (4)$$

with $\beta_0 = (11N_c - 2N_f)/3$. The expression (4) is analytic in the complex p^2 plane except the negative real axis $p^2 < 0$, *i.e.* timelike momenta, where the logarithm produces a cut.

The fact that the exponent κ in eq. (3) is an irrational number has an important consequence: the gluon propagator possesses a cut on the negative real axis. It is possible to fit the non-perturbative solution for the gluon propagator very well without introducing further singularities. The fit to the gluon renormalization function

$$Z_{\text{fit}}(p^2) = w \left(\frac{p^2}{\Lambda_{\text{QCD}}^2 + p^2} \right)^{2\kappa} \left(\alpha_{\text{fit}}(p^2) \right)^{-\gamma} \qquad (5)$$

with $w = 2.65$ and $\Lambda_{\text{QCD}} = 520$ MeV is shown in fig. 1. Similar parametrizations have been explored in ref. [1]. Hereby w is a normalization parameter, and γ is the one-loop value for the anomalous dimension of the gluon propagator.

The Schwinger function $\Delta(t)$ based on the fit (5) is compared to the one of the numerical DSE solution in fig. 1. To enable a logarithmic scale the absolute value is displayed. $\Delta(t)$ has a zero for $t \approx 5/\text{GeV}$ and is negative for larger Euclidean times: One clearly observes positivity violations in the gluon propagator. The overall magnitude w is arbitrary due to renormalization properties. The infrared exponent κ is determined analytically, and for the gluon anomalous dimension γ the one-loop value is used. Thus the parameterization of the gluon propagator has effectively only one parameter, the scale Λ_{QCD}. This and the relatively simple analytic structure gives us confidence that the important features of the Landau gauge gluon propagator are given by (5).

Finally, we want to mention that the regular infrared behaviour of the quark propagator found from the Dyson–Schwinger equations [6] and on the lattice [15] complicates the issue of determining the analytic properties of the quark propagator. Nevertheless there is some evidence that the Schwinger functions related to the quark propagator are positive

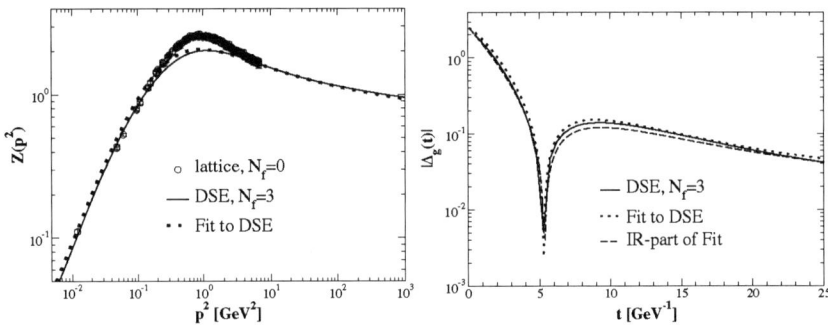

FIGURE 1. Left panel: Dyson–Schwinger [6] and lattice results [14] for the gluon renormalization function $Z(p^2)$ and the fit (5). Right panel: The results for the absolute value of the gluon Schwinger function from the Dyson–Schwinger solution, the fit (5) and the infrared part of this fit.

definite. *E.g.* the quark Schwinger functions can be described accurately by a cut on the negative real axis to the left of a singular point at $p^2 = -m^2$ where $m = 350 \ldots 390$ MeV might be attributed the meaning of an infrared constituent mass.

ACKNOWLEDGMENTS

RA thanks the organizers of *Quark Confinement and the Hadron Spectrum VI* for making this extraordinary conference possible.
This work has been supported by a grant from the Ministry of Science, Research and the Arts of Baden-Württemberg (Az: 24-7532.23-19-18/1 and 24-7532.23-19-18/2) and the Deutsche Forschungsgemeinschaft (DFG) under contract Fi 970/2-1.

REFERENCES

1. R. Alkofer *et al.*, Phys. Rev. D **70**, 014014 (2004).
2. A. Maas *et al.*, Eur. Phys. J. C **37**, 335 (2004); these proceedings [arXiv:hep-ph/0411289].
3. L. von Smekal, R. Alkofer and A. Hauck, Phys. Rev. Lett. **79**, 3591 (1997); Ann. Phys. **267**, 1 (1998).
4. J. E. Mandula, Phys. Rept. **315** (1999) 273.
5. R. Alkofer and L. von Smekal, Phys. Rept. **353**, 281 (2001); C. D. Roberts and S. M. Schmidt, Prog. Part. Nucl. Phys. **45**, S1 (2000); P. Maris and C. D. Roberts, Int. J. Mod. Phys. E **12**, 297 (2003).
6. C. S. Fischer and R. Alkofer, Phys. Rev. D **67**, 094020 (2003); these proceedings.
7. C. S. Fischer and R. Alkofer, Phys. Rev. Lett. **B536**, 177 (2002); C. S. Fischer, R. Alkofer and H. Reinhardt, Phys. Rev. **D65**, 094008 (2002).
8. C. Lerche and L. von Smekal, Phys. Rev. D **65**, 125006 (2002).
9. D. Zwanziger, Phys. Rev. D **65**, 094039 (2002).
10. D. Zwanziger, Phys. Rev. D **69**, 016002 (2004).
11. W. Schleifenbaum *et al.*, arXiv:hep-ph/0411052; arXiv:hep-ph/0411060.
12. F. J. Llanes-Estrada, C. S. Fischer and R. Alkofer, arXiv:hep-ph/0407332; arXiv:hep-ph/0407294.
13. J. M. Pawlowski *et al.*, Phys. Rev. Lett. **93**, 152002 (2004); C. S. Fischer and H. Gies, JHEP **0410**, 048 (2004).
14. F. D. Bonnet *et al.*, Phys. Rev. D **64**, 034501 (2001).
15. J. B. Zhang *et al.*, arXiv: hep-lat/0301018.

Dynamical Chiral Symmetry Breaking in Landau gauge QCD

C. S. Fischer[*] and R. Alkofer[†]

[*]*IPPP, University of Durham, Durham DH1 3LE, U.K.*
[†]*Institute for Theoretical Physics, University of Tübingen, D-72076 Tübingen, Germany*

Abstract. We summarise results for the propagators of Landau gauge QCD from the Green's functions approach and lattice calculations. The nonperturbative solutions for the ghost, gluon and quark propagators from a coupled set of Dyson-Schwinger equations agree almost quantitatively with corresponding lattice results. Similar unquenching effects are found in both approaches. The dynamically generated quark masses are close to 'phenomenological' values. The chiral condensate is found to be large.

Keywords: Confinement, dynamical chiral symmetry breaking, gluon propagator, quark propagator, Dyson-Schwinger equations
PACS: 12.38.Aw 14.65.Bt 14.70.Dj 12.38.Lg 11.30.Rd 11.15.Tk 02.30.Rz

The infrared behaviour of the propagators of Landau gauge QCD has been investigated extensively over the past years in lattice Monte Carlo simulations and the continuum Green's functions approach. Lattice simulations are the only ab initio method known so far and are by now precise enough to pin down these propagators accurately in a large momentum range centered around 1 GeV. In the deep infrared, however, lattice results are inevitably plagued by finite volume effects. In the continuum formulation of QCD the Dyson-Schwinger equations (DSEs) provide a tool complementary to lattice simulations. They can be solved analytically in the infrared. Furthermore numerical solutions over the whole momentum range are available by now. The truncation assumptions necessary to close the DSEs can be checked in the momentum regions where lattice results are available. In general, results from DSEs have the potential to provide a sucessful description of hadrons in terms of quarks and gluons, see [1, 2, 3] and references therein.

The ghost, gluon and quark propagators, $D_G(p)$, $D_{\mu\nu}(p)$ and $S(p)$, in Euclidean momentum space can be generically written as

$$D_G(p) = -\frac{G(p^2)}{p^2}, \tag{1}$$

$$D_{\mu\nu}(p) = \left(\delta_{\mu\nu} - \frac{p_\mu p_\nu}{p^2}\right)\frac{Z(p^2)}{p^2}, \tag{2}$$

$$S(p) = \frac{1}{-i p\!\!\!/ \, A(p^2) + B(p^2)}$$

$$= \frac{Z_Q(p^2)}{-i p\!\!\!/ + M(p^2)}. \tag{3}$$

Here we have chosen Landau gauge which is a fixed point under renormalization [4].

CP756, *Quark Confinement and the Hadron Spectrum VI*
edited by N. Brambilla, U. D'Alesio, A. Devoto, K. Maung, G.M. Prosperi and S. Serci
© 2005 American Institute of Physics 0-7354-0241-8/05/$22.50

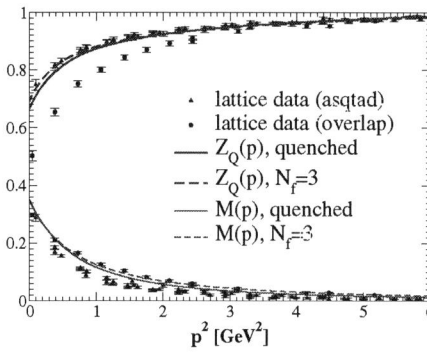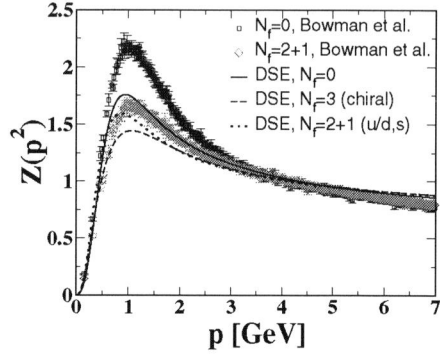

FIGURE 1. Left: The quenched and unquenched quark mass function $M(p^2)$ and the wave function $Z_Q(p^2)$ from the DSE approach [10] compared to results from quenched lattice calculations [11]. Right: The quenched and unquenched gluon dressing function from the DSE approach [10] compared to results from unquenched lattice calculations [13].

The Dyson-Schwinger equations for the ghost and gluon dressing functions, $G(p^2)$ and $Z(p^2)$, have been investigated in refs. [5, 6]. They can be solved analytically in the infrared and one finds simple power laws,

$$
\begin{aligned}
Z(p^2) &\sim (p^2)^{2\kappa}, \\
G(p^2) &\sim (p^2)^{-\kappa},
\end{aligned}
\tag{4}
$$

for the gluon and ghost dressing function with exponents related to each other. The relations (4) can be determined from the ghost-DSE alone and are independent of the truncation scheme. The exponent κ is an irrational number and depends only slightly on the dressing of the ghost-gluon vertex [7, 8]. With a bare vertex one obtains $\kappa = (93 - \sqrt{1201})/98 \approx 0.595$. Recently these results have been confirmed independently in studies of the exact renormalisation group equation [9].

The dynamical generation of quark masses can be studied in the Dyson-Schwinger equation for the quark propagator. It is a genuinely non-perturbative phenomenon and requires a careful treatment of the quark-gluon interaction. In ref. [10] we demonstrated that sizeable nontrivial Dirac-structures in the quark-gluon vertex are necessary to generate dynamical quark masses of the order of 300-400 MeV. Our results for the quenched quark mass function $M(p^2)$ and the wave function $Z_Q(p^2)$ are compared to the quenched lattice results of refs. [11] in fig. 1. The overall qualitative and quantitative agreement between both approaches is very good. The DSE results are within the bounds given by the two different formulations of fermions on the lattice.

Including the backreaction of the quark-propagator on the ghost and gluon system leads to a coupled set of three Dyson-Schwinger equations for the propagators of QCD. These equations have been solved in [10] and allowed a prediction of possible effects of unquenching QCD on the propagators. As can be seen from fig. 1 including $N_f = 3$ chiral quarks in the gluon DSE hardly changes the results for the quark propagator. The chiral condensate is nearly unaffected. It will be interesting to compare these results to unquenched lattice calculations when available.

Unquenched lattice results for the gluon propagator including the effects of two light (up-) and one heavy (strange-) quark have been published recently [13] and are compared to the corresponding results from our DSE-approach in fig. 1. The screening effect from the quark loop is clearly visible in the lattice results for momenta p larger than $p = 0.5$ GeV: the gluonic self interaction becomes less important in this region and the gluon dressing increases. This effect can also be seen in the DSE-approach. In the quenched case there is a discrepancy between the DSE-result and the lattice data, which can be traced back to the fact that not all effects from the gluonic self interaction are accounted for in the DSE truncation. When this part of the gluon interaction becomes less dominant in the unquenched case, both the lattice and the DSE-approach agree very well on a quantitative level, provided similar bare quark masses are taken into account. In the chiral limit the screening effect of the quark loop becomes even stronger as can be seen from the DSE-results in fig. 1. This is expected as the energy needed to create a quark pair out of the vaccuum becomes smaller with decreasing bare quark mass.

Both, the lattice calculations and the Green's functions approach agree in the fact that unquenching does not affect the extreme infrared of the ghost and gluon propagators. Again, this is easily explained from dynamical chiral symmetry breaking: there is not enough energy to generate a quark pair from the vacuum below a certain threshold. Then the quark degrees of freedom decouple from the Yang-Mills sector of the theory.

ACKNOWLEDGMENTS

We thank the organizers of *Quark Confinement and the Hadron Spectrum VI* for their efforts which made this very inspiring conference possible. We are grateful to D. Leinweber, F. Llanes-Estrada, M. Pennington, P. Tandy and A. Williams for helpful discussions. This work has been supported by the Deutsche Forschungsgemeinschaft (DFG) under contract Fi 970/2-1.

REFERENCES

1. P. Maris and C. D. Roberts, Int. J. Mod. Phys. E **12**, 297 (2003).
2. R. Alkofer and L. von Smekal, Phys. Rept. **353**, 281 (2001).
3. C. D. Roberts and S. M. Schmidt, Prog. Part. Nucl. Phys. **45**, S1 (2000).
4. U. Ellwanger, M. Hirsch and A. Weber, Z. Phys. C **69** (1996) 687 [arXiv:hep-th/9506019].
5. L. von Smekal, R. Alkofer and A. Hauck, Phys. Rev. Lett. **79**, 3591 (1997); Annals Phys. **267**, 1 (1998).
6. C. S. Fischer and R. Alkofer, Phys. Lett. B **536**, 177 (2002); C. S. Fischer, R. Alkofer and H. Reinhardt, Phys. Rev. D **65**, 094008 (2002).
7. C. Lerche and L. von Smekal, Phys. Rev. D **65**, 125006 (2002).
8. D. Zwanziger, Phys. Rev. D **65**, 094039 (2002).
9. J. M. Pawlowski, D. F. Litim, S. Nedelko and L. von Smekal, arXiv:hep-th/0312324; C. S. Fischer and H. Gies, JHEP **0410** (2004) 048 [arXiv:hep-ph/0408089].
10. C. S. Fischer and R. Alkofer, Phys. Rev. D **67**, 094020 (2003).
11. P. O. Bowman, U. M. Heller and A. G. Williams, Phys. Rev. D **66**, 014505 (2002); J. B. Zhang *et al.*, arXiv: hep-lat/0301018.
12. F. D. Bonnet *et al.*, Phys. Rev. D **64**, 034501 (2001).
13. P. O. Bowman, U. M. Heller, D. B. Leinweber, M. B. Parappilly and A. G. Williams, Phys. Rev. D **70** (2004) 034509 [arXiv:hep-lat/0402032].

Confinement signatures in Landau gauge QCD

J. M. Pawlowski [*], D. F. Litim[†], S. Nedelko[**] and L. von Smekal[‡]

[*]*Institute for Theoretical Physics, University of Heidelberg, 69120 Heidelberg, Germany*
[†]*TD, CERN, CH-1211 Geneva 23; SPA, U. Southampton, Southampton SO17 1BJ, U.K.*
[**]*BLTP, JINR, 141980 Dubna, Russia*
[‡]*Institute for Theoretical Physics III, Universität Erlangen, D-91058 Erlangen, Germany*

Abstract. We summarise an analysis of the infrared regime of Landau gauge QCD by means of a flow equation approach [1]. The infrared behaviour of gluon and ghost propagators is evaluated. The results provide further evidence for the Kugo-Ojima confinement scenario. We also discuss their relation to results obtained with other functional methods as well as the lattice.

Keywords: confinement, gluon propagator, ghost propagator, flow equation
PACS: 05.10.Cc, 11.15.Tk, 12.38.Aw

In gauge fixed formulations of QCD the infrared behaviour of ghost and gluon propagators provide signatures of confinement: in Landau gauge QCD the confinement scenarios of Kugo-Ojima [2] and Gribov-Zwanziger, e.g. [3], entail an infrared enhancement for the ghost propagation and an infrared suppression for the gluon propagation. This behaviour was first seen within a functional approach using Dyson-Schwinger equations (DSE) [4], giving access to the full momentum regime. It was later confirmed within lattice studies down to scales about 1 GeV. However, in the deep infrared below 1 GeV, lattice studies encounter problems due to finite size effects. This situation calls for an independent confirmation of the infrared behaviour seen in DS-studies. Ideally such a method would still share enough structure with other functional methods such as DSEs in order to benefit from insights and results obtained from these equations. The above features are precisely given for the flow equation, a particular advantage of which is its flexibility when it comes to approximations. So far this approach has been used in Landau gauge QCD for high and intermediate momenta [5].

Here we present results of a flow equation approach to the infrared regime of Landau gauge QCD [1]. Our analysis is based on an integrated flow equation that reads for the scale-dependent effective action Γ_k,

$$\Gamma_0 - \Gamma_k = \frac{1}{2} \int_0^k dk \, \mathrm{Tr} \, \frac{1}{\Gamma_k^{(2)} + R} \, \partial_k R. \tag{1}$$

The scale k is an infrared scale below which Γ_k has no propagating degrees of freedom and $\Gamma^{(n)}$ stands for its nth derivative w.r.t. the fields. Consequently $\Gamma = \Gamma_0$ is the full effective action. Flows for Green functions are obtained by taking field derivatives of (1). The resulting equations share many features with DSEs and stochastic quantisation for Green functions. However, in contradistinction to those approaches built on dressed *and* bare quantities, the flow equation (1) links dressed vertices and propagators exclusively. Moreover, the flow equation and its k-integral (1) are manifestly ultraviolet and infrared finite, no additional renormalisation is required even within truncations. Therefore the flow equation offers an interesting functional method for accessing the infrared regime: its close relation to DSEs makes many of the truncation schemes and results of DSEs accessible to flow equation studies; its qualitative differences and complementary advantages provide additional support for results obtained within both approaches.

We evaluate the integrated flow (1) of ghost and gluon propagators in the deep infrared

$$p^2 \ll \Lambda_{QCD}^2, \tag{2}$$

where Λ_{QCD} is the dynamical mass scale of QCD. For these momenta the integrated flow (1) tends to zero as the flow reaches a (trivial) fixed point at $k = 0$. This enables us to study the leading infrared behaviour of the propagators by means of a fixed point argument developed in [1]. For momenta in the regime (2) and for $k^2 \ll \Lambda_{QCD}^2$ we can expand n-point functions at finite k about that at $k = 0$ with

$$\Gamma_k^{(n)} = \Gamma_0^{(n)}(1 + \delta Z_n), \tag{3}$$

where δZ_n only depend on ratios p_i/k with $i = 1, ..., n-1$. Eq. (3) is valid up to order p_i/Λ_{QCD}. Indeed, (3) can be proven by iterating the integrated flow (1) about $\delta Z_n \equiv 0$. So far we have not relied on any approximation. For the explicit computation we resort to a truncation with dressed vertices with trivial momentum dependence, an approximation which is well in accord with consistency considerations [6] as well as lattice studies [7]. We allow for a general momentum dependence on $x = p^2/k^2$ in the ghost and (transversal) gluon two point functions $\Gamma_{k,C}^{(2)}$ and $\Gamma_{k,A}^{(2)}$

$$\begin{aligned}
\Gamma_{k,C}^{(2)}(x) &\simeq z_C \, p^2 x^{\kappa_C}(1 + \delta Z_C(x)), \\
\Gamma_{k,A}^{(2)}(x) &\simeq z_A \, p^2 x^{-2\kappa_C}(1 + \delta Z_A(x)),
\end{aligned} \tag{4}$$

where we dropped the Lorentz and group structure of the propagators. In (4), $\Gamma_0^{(2)} = z x^\kappa = \hat{z} p^{2\kappa}$ are the leading infrared terms for $k = 0$ with k-independent prefactor $\hat{z}$. The functions δZ entail the transition between the physical infrared regime, $x \gg 1$, and the cutoff regime, $x \ll 1$. In (4) we have also used that non-renormalisation of the ghost-gluon vertex at vanishing k entails $\kappa_A = -2\kappa_C$ and $\alpha_s = g^2/(4\pi z_A z_C^2)$ [8]. Inserting the propagators (4) in (1) leads to two integral equations for δZ_A and δZ_C of the form

$$\delta Z_{A/C}(x) = F_{A/C}[\delta Z_{A/C}, \kappa_C, \alpha_s]. \tag{5}$$

Explicit expressions for the integrals F are given in [1]. The equations (5) are solved iteratively for δZ_A, δZ_C, κ_C and α_s, leading to

$$\kappa_C = 0.59535 \cdots, \qquad \alpha_s = 2.9717 \cdots. \tag{6}$$

The values in (6) are achieved by also invoking an optimisation procedure developed in [1] for eliminating the regulator-dependence.

The results (6) agree with the analytic results obtained within DSEs [8] and stochastic quantisation [3]. For the present truncation within an optimised cut-off scheme one can indeed formally link the integrated flow (1) to a set of DSEs with explicit renormalisation, see [1]. We expect a mild R-dependence for the results as we have resorted to a truncation. Indeed, for general cut-off functions R, the results for κ_C mildly vary in the interval

$$\kappa_C \in [0.539, 0.595].\tag{7}$$

Both bounds have physical interpretations. The upper bound, as we have already discussed, relates to the physical infinite volume result whereas the lower bound can be linked to a finite volume computation. The related regulator is a sharp cut-off that strictly allows no propagation of modes with momenta p^2 smaller than k^2. This is as close to a finite volume (e.g. in lattice studies) as one can get with local momentum cut-off functions. Interestingly, the lower bound $\kappa_C = 0.539$ compares well to very recent lattice results [9]. A further interesting consequence of our analysis is the evaluation of renormalisation procedures for DSEs stemming from the integrated flow (1): for DSEs in the present truncation and subject to a general consistent renormalisation it is impossible to achieve a masslike behaviour for the gluon propagator, $\kappa_C = 1/2$. More generally $\kappa_C \notin [0, 1/2]$, see [1]. With multiplicative renormalisation this was already shown in [8].

The above analysis allows for many interesting extensions. The most important open question in the present truncation concerns the detailed analysis of the transition between ultraviolet and infrared regime. Unfortunately this question has not been resolved completely in subsequent flow studies [10]. Moreover dynamical quarks are investigated which opens the door towards a description of dynamical chiral symmetry breaking.

ACKNOWLEDGMENTS

JMP thanks the organisers of Quark Confinement and the Hadron Spectrum for making this interesting conference possible. This work has been supported by a grant from the Ministry of Science, Research and the Arts of Baden-Württemberg (Az: 24-7532.23-19-18/1 and 24-7532.23-19-18/2), EPSRC, and the DFG under contract SM70/1-1.

REFERENCES

1. J. M. Pawlowski, D. F. Litim, S. Nedelko and L. von Smekal, Phys. Rev. Lett. **93** (2004) 152002 [hep-th/0312324].
2. T. Kugo, I. Ojima, Prog. Theor. Phys. Suppl. **66** (1979) 1.
3. D. Zwanziger, Phys. Rev. D **65** (2002) 094039 [hep-th/0109224].
4. L. von Smekal, A. Hauck, and R. Alkofer, Phys. Rev. Lett. **79** (1997) 3591 [hep-ph/9705242].
5. U. Ellwanger, M. Hirsch and A. Weber, Z. Phys. C **69** (1996) 687 [hep-th/9506019]; Eur. Phys. J. C **1** (1998) 563 [hep-ph/9606468]; B. Bergerhoff and C. Wetterich, Phys. Rev. D **57** (1998) 1591 [hep-ph/9708425].
6. W. Schleifenbaum, A. Maas, J. Wambach and R. Alkofer, hep-ph/0411052.
7. A. Cucchieri, T. Mendes and A. Mihara, hep-lat/0408034.
8. C. Lerche, L. von Smekal, Phys. Rev. D **65** (2002) 125006 [hep-ph/0202194].
9. O. Oliveira and P. J. Silva, these proceedings, hep-lat/0410048.
10. J. Kato, hep-th/0401068; C. S. Fischer and H. Gies, JHEP **0410** (2004) 048 [hep-ph/0408089].

Confinement in the lattice Landau Gauge QCD simulation

Sadataka Furui* and Hideo Nakajima†

*School of Science and Engineering, Teikyo University, Utsunomiya 320-8551, Japan
†Department of Information science, Utsunomiya University, Utsunomiya 320-8585, Japan

Abstract. The running coupling and the Kugo-Ojima parameter of the confinement criterion are measured for the quenched SU(3) $\beta = 6.4, 6.45, 56^4$ lattice and the unquenched $\beta = 5.2, 20^3 \times 48$ lattice of JLQCD, $\beta = 2.1, \kappa = 0.1357, 0.1382, 24^3 \times 48$ lattice of CP-PACS and $\beta_{imp} = 6.76, am_{u,d} = 0.007, 6.83, am_{u,d} = 0.040, 20^3 \times 64$ lattice of MILC collaboration.

The quenched SU(3) 56^4 lattice data suggest presence of infrared fixed point of $\alpha_s = 2.5(5)$ and the approach of the ensemble of the 1st copy to the Gribov boundary. The running coupling of $q > 2$GeV can be fitted by the perturbative QCD(pQCD) $+ c/q^2$ correction. We find the Kugo-Ojima parameter $u(0) = -0.83(3)$.

The rotational symmetry of the gluon propagator of the unquenched SU(3) is partially recovered, but its magnitude depends on whether the Wilson fermion or the Kogut-Susskind(KS) fermion are coupled to the gauge field. The gluon propagator coupled to KS fermion is more suppressed than that coupled to Wilson fermion. The running coupling inherits the same trend. The unquenched running coupling of the of $q > 3$GeV can be fitted by the pQCD without c/q^2 term. The Kugo-Ojima parameter of unquenched configurations with light fermion masses is consistent with $u(0) = -1.0$.

INTRODUCTION

In Landau gauge $\widetilde{MOM}$ schme, we measure the QCD running coupling in terms of gluon dressing functiuon $Z_A(q^2)$ and ghost dressing function $G(q^2)$,

$\alpha_s(q^2) = \dfrac{g^2}{4\pi} \dfrac{G(q^2)^2 Z_A(q^2)}{\tilde{Z}_1^2}$. It is a renormalization group invariant quatity, but in the finite lattice, the vertex renormalization factor $\tilde{Z}_1$ is not necessarily equal to 1 as in pQCD. We fix this value by the fit of the numerical result to the pQCD.

Colour confinement in infrared QCD is characterized by the Kugo and Ojima parameter $u(0) = -c$. The parameter c is related to the renormalization factor as $1 - c = \dfrac{Z_1}{Z_3} = \dfrac{\tilde{Z}_1}{\tilde{Z}_3}$. If the finiteness of $\tilde{Z}_1$ is proved, divergence of $\tilde{Z}_3$ is a sufficient condition. If Z_3 vanishes in the infrared, Z_1 should have higher order 0.

THE GHOST PROPAGATOR AND THE GLUON PROPAGATOR

The ghost propagator is the Fourier transform of an expectation value of the inverse Faddeev-Popov operator $\mathcal{M} = -\partial D = -\partial^2(1 - M)$

$$D_G^{ab}(x,y) = \langle \text{tr}\langle \Lambda^a x | (\mathcal{M}[U])^{-1} | \Lambda^b y \rangle \rangle \tag{1}$$

CP756, *Quark Confinement and the Hadron Spectrum VI*
edited by N. Brambilla, U. D'Alesio, A. Devoto, K. Maung, G.M. Prosperi and S. Serci
© 2005 American Institute of Physics 0-7354-0241-8/05/$22.50

where the outmost $\langle\rangle$ denotes average over samples U.

The ghost propagators $D_G(q^2)$ of quenched and unquenched SU(3) in $\widetilde{MOM}$ scheme can be fitted by the pQCD in $q > 0.4\text{GeV}$ region, and there is no β dependence. It means that there are strong correlation between the string tension and the ghost propagator. In this presentation the gauge field of log-U type[2] is adopted. The ghost propagators are 14% larger when the U-linear definition is adoped.

The gluon propagator of 56^4 lattice is finite at 0 momentum and $Z_3(0)$ is not compatible to 0 in the present lattice size. The gluon propagator of the unquenched SU(3)[4, 5, 6] is measured by adopting the cylindrical cut. Extraction of infrared physical quantities becomes difficult due to lack of symmetry of the four coordinate axes.

We observe reflection positivity violation in the unquenched gluon propagator, and in some polarization components of sample-wise quenched gluon propagator. The reflection positivity violation and the closeness of the Kugo-Ojima parameter to -1 are correlated.

THE QCD RUNNING COUPLING AND THE KUGO-OJIMA PARAMETER

The QCD running coupling $\alpha_s(q)$ of the quenched SU(3) normalized at high momentum region by the 3-loop pQCD is close to the prediction of Dyson-Schwinger calculation of $\kappa = 0.5$. The running coupling of unquenched Wilson fermion (JLQCD, CP-PACS) and that of KS fermion(MILC) are qualitatively different due to different behavior of the gluon propagator in the infrared. The left-most momentum point in the Figure 1 is to be excluded in the cone-cut due to the finite size effect. The running coupling of the MILC in the infrared region is about 2/3 of the CP-PACS.

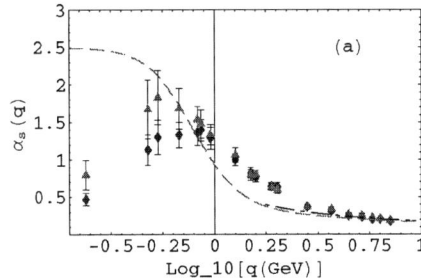 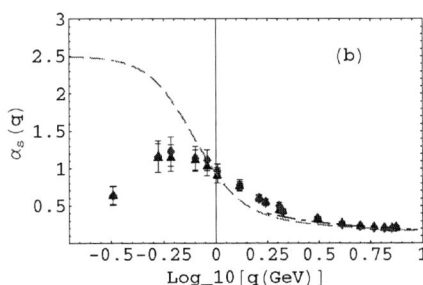

FIGURE 1. (a)The running coupling $\alpha_s(q)$ of the CP-PACS of $K_{sea} = 0.1357$ (diamonds) and that of 0.1382 (triangles). (25 samples) The DSE approach with $\alpha_0 = 2.5$(long dashed line), the fit of the Orsay group $N_f = 2$ perturbative (dash dotted line) are also shown. (b) The running coupling $\alpha_s(q)$ of the MILC $\beta_{imp} = 6.83, am_{u,d} = 0.040$(diamonds) and $6.76, am_{u,d} = 0.007$(triangles). (50 samples)

The Kugo-Ojima parameters are summarized in Table 1. The parameter c becomes larger as the lattice size becomes large. It is about 0.83(3) in 56^4, while in the unquenched simulation it is consistent to 1, when the fermion mass is small. In the case of U-linear definition of the gauge field, c is about 10% smaller than that of the log-U definition.

TABLE 1. The Kugo-Ojima parameter along the spacial directions c_x and that along the time axis c_t and the average c, trace divided by the dimension e/d, horizon function deviation h of quenched Wilson action (The first two rows), unquenched Wilson action (The second two rows), unquenched Wilson improved action (The third two rows), and unquenched KS improved action (The last two rows).

K_{sea} or β	c_x	c_t	c	e/d	h
$\beta = 6.4$	0.827(27)	0.827(27)	0.827(27)	0.954(1)	-0.12(3)
$\beta = 6.45$	0.809(81)	0.809(81)	0.809(81)	0.954(1)	-0.15(8)
$K_{sea} = 0.1340$	0.887(87)	0.723(38)	0.846(106)	0.930(1)	-0.084(106)
$K_{sea} = 0.1355$	1.005(217)	0.670(47)	0.921(238)	0.934(1)	-0.013(238)
$K_{sea} = 0.1357$	0.859(58)	0.763(36)	0.835(68)	0.9388(1)	-0.104(58)
$K_{sea} = 0.1382$	0.887(87)	0.723(38)	0.846(106)	0.9409(1)	-0.051(87)
$\beta_{imp} = 6.76$	1.040(111)	0.741(28)	0.965(162)	0.9325(1)	-0.032(162)
$\beta_{imp} = 6.83$	1.012(153)	0.754(28)	0.947(174)	0.9339(1)	-0.013(174)

SUMMARY

We observed that the running coupling of unquenched Landau gauge QCD in the infrared is enhanced in the case of Wilson fermion and the c/q^2 correction which appeared in the quenched simulation is absent. The presence of the extra taste degrees of freedom in the KS fermion is expected to be the cause of the suppression of the gluon propagator and the running coupling.

This work is supported by the KEK supercomputing project No.04-106. H.N.is supported by a JSPS Grant-in-Aid for Scientific Research on Priority Area No.13135210.

REFERENCES

1. S. Furui and H.Nakajima, Phys. Rev. D**69**,074505(2004), hep-lat/0305010 and references therein.
2. S. Furui and H.Nakajima, Phys. Rev. D**70**,094504(2004), hep-lat/0403021 and references therein.
3. S. Furui and H. Nakajima, in *Confinement IV*, Ed. W. Lucha et.al., World Scientific, Singapore, p.275(2002), hep-lat/0012017.
4. S.Aoki et al., (JLQCD collaboration),Phys. Rev. D**65**,094507(2002).
5. A. AliKhan et al., (CP-PACS collaboration),Phys. Rev. D**65**,054505(2002).
6. C.W. Bernard et al., (MILC collaboration), Phys. Rev. D**64**,054506(2001).

The influence of Gribov copies on the gluon and ghost propagator [1]

A. Sternbeck[*], E.-M. Ilgenfritz[*], M. Müller-Preussker[*] and A. Schiller[†]

[*]Institut für Physik, Humboldt-Universität zu Berlin, D-12489 Berlin, Germany
[†]Universität Leipzig, Institut für Theoretische Physik, D-04109 Leipzig, Germany

Abstract. The dependence of the gluon and ghost propagator in pure $SU(3)$ gauge theory on the choice of Gribov copies in Landau gauge is studied. Simulations were performed on several lattice sizes at $\beta = 5.8$, 6.0 and 6.2. In the infrared region the ghost propagator turns out to depend on the choice, while the impact on the gluon propagator is not resolvable. Also the eigenvalue distribution of the Faddeev-Popov operator is sensitive to Gribov copies.

Keywords: ghost and gluon propagator, Gribov problem, Faddeev-Popov operator eigenvalues
PACS: 11.15.Ha, 12.38.Gc, 12.38.Aw

Studying non-pertubative features of QCD such as confinement, there are two common approaches: lattice gauge theory and Dyson-Schwinger equations. From the latter approach there are promising results in recent years [1] about the infrared behavior of the gluon D and the ghost propagator G. Denoting by Z the dressing functions of the corresponding propagator, in Landau gauge they can be written as

$$D_{\mu\nu}(q^2) = \left(\delta_{\mu\nu} - \frac{q_\mu q_\nu}{q^2}\right)\frac{Z_{gl}(q^2)}{q^2} \quad \text{and} \quad G(q^2) = \frac{Z_{gh}(q^2)}{q^2}. \tag{1}$$

According to [1] in the low-momentum region the dressing functions are proposed to behave as $Z_{gl} \propto (q^2)^{2\kappa}$ and $Z_{gh} \propto (q^2)^{-\kappa}$ with a common value $\kappa \in (0.5, 1)$. The infrared suppression of the gluon propagator and the enhancement of the ghost propagator at low-momentum is in agreement with the Zwanziger-Gribov horizon condition [2–4] as well as with the Kugo-Ojima confinement criterion [5].

Zwanziger [2] has suggested that in the continuum the behavior of both propagators in Landau gauge results from restricting the gauge fields to the Gribov region Ω, where the Faddeev-Popov operator is non-negative. Generically, one gauge orbit has more than one intersection (Gribov copies) within the Gribov region Ω, but expectation values taken over this region are proposed to be equal to those over the fundamental modular region Λ. On a finite lattice, however, this is not expected [2]. In this contribution we assess the importance of the Gribov ambiguity on a finite lattice for the $SU(3)$ ghost and gluon propagators as well as for the lowest eigenvalues of the Faddeev-Popov operator.

To study these propagators in Landau gauge using lattice simulation, all thermalized gauge field configurations $\{U_{x,\mu}\}$ have to be fixed to this gauge. On the lattice the Landau gauge condition is implemented by searching for a gauge transformation

---

[1] Talk presented by A. Sternbeck.

CP756, *Quark Confinement and the Hadron Spectrum VI*
edited by N. Brambilla, U. D'Alesio, A. Devoto, K. Maung, G.M. Prosperi and S. Serci
© 2005 American Institute of Physics 0-7354-0241-8/05/$22.50

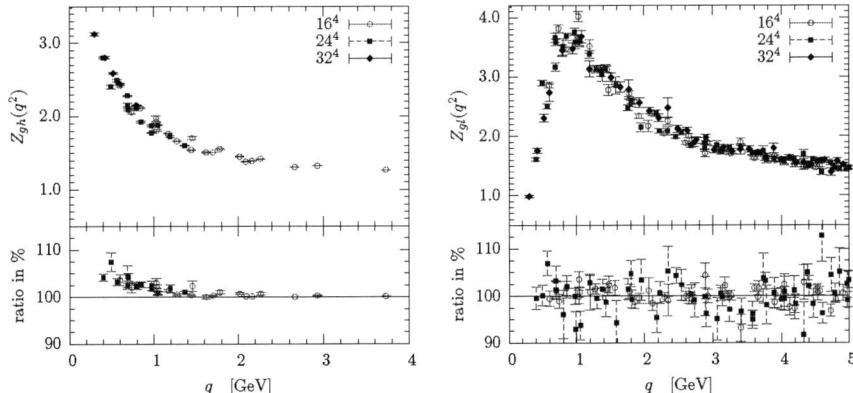

FIGURE 1. The upper parts show the dressing functions of the ghost Z_{gh} and gluon propagator Z_{gl} measured on best gauge copies as functions of the momentum q (scaled to physical units at $\beta = 5.8, 6.0$ and 6.2) using various lattice sizes. The lower parts show the ratio $\langle Z^{(fc)} \rangle / \langle Z^{(bc)} \rangle$ determined from the first (fc) and best (bc) gauge copies.

$^g U_{x,\mu} = g_x U_{x,\mu} g^\dagger_{x+\hat{\mu}}$, while keeping $U_{x,\mu}$ fixed, which maximizes the functional

$$F_U[g] \propto \sum_{x,\mu} \operatorname{Re} \operatorname{Tr} {}^g U_{x,\mu} . \tag{2}$$

This functional has many different local maxima whose number increases as the lattice size increases or the inverse coupling β decreases. The different gauge copies corresponding to those maxima are called Gribov copies, due to its relation to the Gribov ambiguity in the continuum [4]. All Gribov copies $\{^g U\}$ belong to the gauge orbit created by U and satisfy the lattice Landau gauge condition $\partial_\mu{}^g A_{x,\mu} = 0$ with

$$^g A_{x+\hat{\mu}/2,\mu} = \frac{1}{2i} \left({}^g U_{x,\mu} - {}^g U^\dagger_{x,\mu} \right) \Big|_{\text{traceless}} . \tag{3}$$

In the literature it is widely accepted that the gluon propagator does not depend on the choice of Gribov copy, while an impact on the $SU(2)$ ghost propagator has been observed [6–8]. However, in a more recent investigation [9] an influence of Gribov copies on the $SU(3)$ gluon propagator has been demonstrated, too.

Here we report on a combined study of the $SU(3)$ gluon and ghost propagator in Landau gauge on the same gauge field configurations generated at $\beta = 5.8, 6.0$ and 6.2. For each configuration we have taken $N_{cp} = 30, 40$ and 10 random gauge copies for the lattice sizes 16^4, 24^4 and 32^4, respectively. A subsequent gauge-fixing was carried out using standard over-relaxation until $\max_x(\partial_\mu{}^g A_{x,\mu})^2 < 10^{-14}$ was reached.

On each first (fc) and each best (bc) gauge copy — that with largest functional value among N_{cp} copies — both the ghost and the gluon propagator have been measured. The results are shown in Fig. 1. The upper parts show the dressing functions of both propagators measured on the best gauge copies as a function of the momentum q scaled to energy units. In order to compare to other studies [9, 10] we have used $a^{-1} = 1.53$, 1.885 and 2.637 GeV for $\beta = 5.8, 6.0$ and 6.2, respectively. Looking at the lower parts

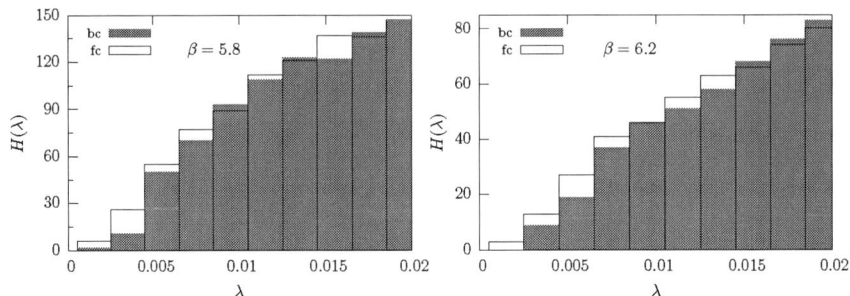

FIGURE 2. The frequency $H(\lambda)$ of the lowest eigenvalues λ of the Faddeev-Popov operator is shown. Full boxes represent the distribution obtained on the best gauge copies, while empty boxes represent those on the first gauge copies.

of this figure it becomes clear that the ghost propagator is affected by the choice of the Gribov copy the more the momentum is decreased. The impact on the gluon propagator stays inside the statistical error. For further details we refer to [11].

In trying to fit to the proposed power laws of the dressing function at lowest momenta, mentioned at the beginning, it turns out the lattice sizes used are to small to confirm such a behavior.

We also calculated the eigenvalue distribution of the Faddeev-Popov operator on the first and best gauge-fixed configurations as shown in Fig. 2. Looking at this figure it is obvious that the distribution $H(\lambda)$ of the lowest lying eigenvalues λ on the best gauge copies is slightly shifted towards larger eigenvalues compared to that determined on arbitrary first gauge copies. Thus better gauge-fixing seems to increase the gap between the lowest eigenvalues and the Gribov horizon.

All simulations were done on the IBM pSeries 690 at HLRN. We thank R. Alkofer for discussions and H. Stüben for contributing parts of the program code. This work has been supported by the DFG under contract FOR 465. A. Sternbeck acknowledges support of the DFG-funded graduate school GK 271.

REFERENCES

1. R. Alkofer and L. von Smekal, *Phys. Rept.*, **353**, 281 (2001) and references therein.
2. D. Zwanziger, *Phys. Rev.*, **D69**, 016002 (2004).
3. D. Zwanziger, *Nucl. Phys.*, **B412**, 657–730 (1994).
4. V. N. Gribov, *Nucl. Phys.*, **B139**, 1 (1978).
5. T. Kugo and I. Ojima, *Prog. Theor. Phys. Suppl.*, **66**, 1 (1979); T. Kugo (1995), hep-th/9511033.
6. A. Cucchieri, *Nucl. Phys.*, **B508**, 353–370 (1997).
7. T. D. Bakeev, E.-M. Ilgenfritz, V. K. Mitrjushkin, and M. Müller-Preussker, *Phys. Rev.*, **D69**, 074507 (2004).
8. H. Nakajima and S. Furui, *Nucl. Phys. Proc. Suppl.*, **129**, 730–732 (2004); hep-lat/0408001.
9. P. J. Silva and O. Oliveira, *Nucl. Phys.*, **B690**, 177–198 (2004).
10. D. B. Leinweber, J. I. Skullerud, A. G. Williams, and C. Parrinello, *Phys. Rev.*, **D60**, 094507 (1999).
11. A. Sternbeck, E.-M. Ilgenfritz, M. Müller-Preussker, and A. Schiller (2004), hep-lat/0409125.

Determination of $\alpha_s(p)$ from gluon and from ghost propagators

A. Cucchieri and T. Mendes

Instituto de Física de São Carlos, Universidade de São Paulo,
C.P. 369, 13560-970, São Carlos, SP, Brazil

Abstract. We present a numerical study of the running coupling constant in $4d$ pure-$SU(2)$ lattice gauge theory. The running coupling is evaluated by fitting data for gluon and ghost propagators in minimal Landau gauge. The fitting formulae are obtained by a simultaneous integration of the β function and of a function coinciding (in the momentum-subtraction scheme) with the anomalous dimension of the propagator. We use these formulae at two, three and four loops. In the ghost case, a careful analysis of finite-size and hypercubic effects reveals a smooth behavior of the data over the whole range of momenta considered (i.e. up to about 25 GeV). This allows a precise determination of $\Lambda_{\overline{MS}}$ and an investigation of possible anomalous running (power corrections) in the strength of the coupling.

Keywords: running coupling constant, gluon and ghost propagators, lattice gauge theory
PACS: 12.38.Aw, 12.38.Gc

INTRODUCTION

The strong coupling constant $\alpha_s(\overline{\mu})$, taken at a fixed reference scale $\overline{\mu}$, has been numerically evaluated in lattice QCD using several different methods [1]. In particular, results for the Λ parameter with an accuracy of the order of 5% in the quenched case and 12% in the full-QCD case have been obtained using an iterative finite-size-scaling method [2] and a numerical determination of the plaquette [3]. Several lattice groups also study the running of the QCD coupling (in some scheme) over a range of energies and fit the resulting data using a two- or three-loop expression for $\alpha_s(\mu)$, considering the scale Λ as a fitting parameter.

Here we evaluate the running coupling constant $\alpha_s(p)$ using the (bare) gluon and ghost propagators (in Landau gauge), following the analysis introduced for the gluon propagator $D_B(\mu^2)$ in [4]. Let us recall that, after defining $Z_D(\mu^2) = \mu^2 D_B(\mu^2)$, the gluon data can be fitted by considering the two coupled differential equations

$$\frac{d\log Z_D(\mu^2)}{d\log\mu^2} = \Gamma_D(\alpha_s)\,, \qquad \frac{d\alpha_s}{d\log\mu} = \beta(\alpha_s)\,, \qquad (1)$$

in any renormalization scheme. At the lowest order these equations imply the well-known result $D_B(\mu^2) \sim 1/\left[\mu^2 \log^c(\mu^2/\Lambda^2)\right]$ with $c = \gamma_0/\beta_0 = 13/22$. A solution for these two equations depends on the values of $Z_D(\mu_0^2)$ and $\alpha_s(\mu_0)$ at some scale μ_0, and a direct fit of the data provides a determination of these two constants. The value $\alpha_s(\mu_0)$ obtained from the fit can then be evolved, using renormalization group, to the $\overline{\mu}$ scale and related to the $\overline{MS}$ scheme, using the perturbative relation between the two schemes.

CP756, *Quark Confinement and the Hadron Spectrum VI*
edited by N. Brambilla, U. D'Alesio, A. Devoto, K. Maung, G.M. Prosperi and S. Serci
© 2005 American Institute of Physics 0-7354-0241-8/05/$22.50

NUMERICAL SIMULATIONS AND RESULTS

In this work we apply the analysis described above to the gluon and ghost propagators in the $4d$ pure-$SU(2)$ case (details of the simulations can be found in [5]). In the ghost case, a careful analysis of finite-size and hypercubic effects (using an extrapolation similar to that introduced in [4]) reveals a smooth behavior of the data over the whole range of momenta considered, i.e. up to about 25 GeV (see Figure 1).

In order to minimize discretization effects we take the propagators as a function of the lattice momentum p with components $p_\mu = 2 \sin(k_\mu/2)$. For both propagators we have considered two different types of momenta, namely momenta with components $(0, 0, 0, k)$ and with components (k, k, k, k). For the fitting formulae we use the two-, three- and four-loop cases (see again [5] for details). We also employ several different cuts of the data at low momenta and four different renormalization schemes studied in [6], which are indicated as $\widetilde{\text{MOM}}$, $\widetilde{\text{MOM}}^q$, $\widetilde{\text{MOM}}^g$, $\widetilde{\text{MOM}}^{gg}$. They correspond to subtracting, respectively, the ghost-gluon, quark-gluon, and three-gluon (in two different ways) vertices. Let us note that any pair of such schemes satisfies the condition

$$\left| \beta_3^{(1)} - \beta_3^{(2)} \right| \ll 4\pi \beta_2^{(1,2)}, \tag{2}$$

which was used in [4] to select a reasonable domain of "good schemes". Let us also stress that the same relation is not verified if one considers the $\overline{MS}$ scheme.

In the ghost case we obtain a good fit to the data (see Figure 1), giving $\Lambda_{\overline{MS}}$ with a very small statistical error. Also, the dependence on the scale μ_0, on the considered renormalization scheme and on the lattice side is very small. Moreover, the variation in the determination of $\Lambda_{\overline{MS}}$ when considering three- or four-loop formulae is of the order of 40 MeV. On the other hand, the value of $\Lambda_{\overline{MS}}$ is strongly affected (see again Figure 1) by varying the set of data considered for the fit, i.e. by excluding different sets of data in the infrared (IR) region.

In the gluon case the above method cannot be implemented directly, since the data clearly show large systematic effects due to the breaking of rotational invariance. In particular, the data corresponding to momenta with components (k, k, k, k) cannot be fitted without a careful analysis. Considering only data corresponding to momenta with components $(0, 0, 0, k)$, we find that there is a systematic tendency to produce smaller values of $\Lambda_{\overline{MS}}$ when compared to the ghost case. We are currently investigating the hyper-cubic effects for the gluon propagator. Let us stress that when fitting the gluon propagator the statistical error is not negligible, being usually of the order of 5% in the evaluation of $\Lambda_{\overline{MS}}$.

CONCLUSIONS

We believe that the use of the ghost propagator and of the method considered here will allow a good determination of $\Lambda_{\overline{MS}}$. Indeed, we have shown that statistical errors and most systematic errors are very small, yielding a combined total error of order of 1%. The only remaining large systematic effect is related to the choice of the IR cut used for the fits. We are now verifying if this effect can be reduced at the percent level considering

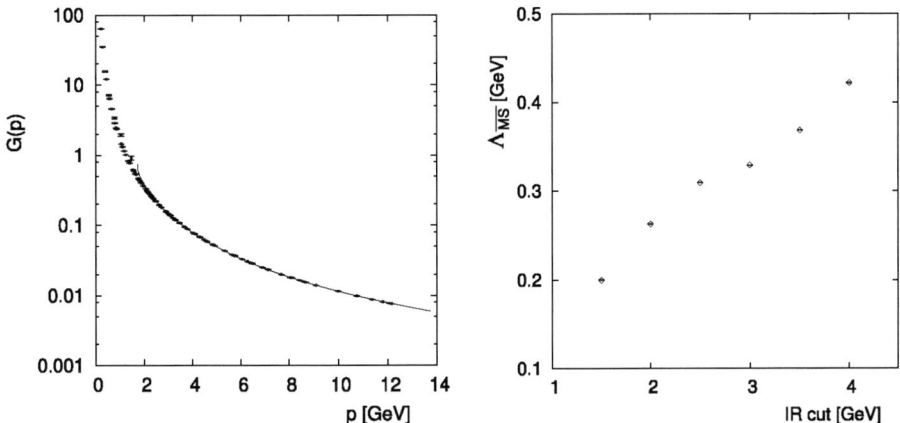

FIGURE 1. Left: fit of the ghost propagator using an IR cut at 2 GeV. Right: value of $\Lambda_{\overline{MS}}$ as a function of the IR cut considered for the fit. In both cases we considered the four-loop fitting formula and $\mu_0 = 24$ GeV.

(estimated) 5-loop coefficients and/or possible (non-perturbative) power corrections. Once this is done, this method could become the method of choice for handling the full-QCD case.

ACKNOWLEDGMENTS

This work was supported by FAPESP (Project No. 00/05047-5). Partial support from CNPq is also acknowledged.

REFERENCES

1. P. Weisz, *Nucl. Phys. Proc. Suppl.*, **47**, 71–83 (1996); A. Cucchieri, in *Hadron Physics 2002*, edited by C. A. Z. Vasconcellos et al., World Scientific, Singapore, 2003, pp. 137–146, *hep-lat/0209076*.
2. M. Luscher, P. Weisz, and U. Wolff, *Nucl. Phys.*, **B359**, 221–243 (1991); M. Luscher, R. Sommer, U. Wolff, and P. Weisz, *Nucl. Phys.*, **B389**, 247–264 (1993); M. Luscher, R. Sommer, P. Weisz, and U. Wolff, *Nucl. Phys.*, **B413**, 481–502 (1994); G. de Divitiis, et al., *Nucl. Phys.*, **B437**, 447–470 (1995); A. Bode, et al., *Phys. Lett.*, **B515**, 49–56 (2001).
3. G. P. Lepage, and P. B. Mackenzie, *Phys. Rev.*, **D48**, 2250 (1993); C. T. H. Davies, et al., *Phys. Rev.*, **D56**, 2755 (1997); S. Booth, et al. *Nucl. Phys. Proc. Suppl.*, **106**, 308 (2002); S. Booth, et al., *Phys. Lett.*, **B519**, 229 (2001).
4. D. Becirevic, et al., *Phys. Rev.*, **D60**, 094509 (1999); D. Becirevic, et al., *Phys. Rev.*, **D61**, 114508 (2000).
5. A. Cucchieri, and T. Mendes, *Running coupling constant from lattice studies of gluon and ghost propagators*, presented at the *IX Hadron Physics and VII Relativistic Aspects of Nuclear Physics Workshops*, Angra dos Reis, Rio de Janeiro, Brazil (March 28–April 3, 2004).
6. K. G. Chetyrkin, and A. Retey, *hep-ph/0007088*; K. G. Chetyrkin, *hep-ph/0405193*.

The Infrared Landau Gauge Gluon Propagator from Lattice QCD

O. Oliveira* and P. J. Silva*

*Centro de Física Computacional, Departamento de Física, Universidade de Coimbra, 3004-516 Coimbra, Portugal

Abstract. The quenched Landau gauge gluon propagator is investigated in lattice QCD with large assimetric lattices, accessing momenta as low as $q \sim 100$ MeV or smaller. Our investigation focus on the IR limit of the gluon dressing function, testing the compatibility with recent solutions of the Dyson-Schwinger equations. In particular, the low energy parameters κ and $\alpha(0)$ are measured.

The confinement of quarks in hadrons and the chiral symmetry breaking mechanism are believed to be linked to the low energy properties of QCD. In particular, the gluon and ghost propagators can provide us information on the mechanism of confinement.

Two first principles approaches to non-perturbative problems are Dyson-Schwinger equations (DSE) and lattice methods. In lattice QCD, the gluon propagator has been revisited a number of times. However, due to the lattice sizes used in previous studies, the access to the IR region was limited. For the Dyson-Schwinger solution for the gluon propagator, recently in [4, 5] the low energy behaviour of the propagator was solved analytically. Moreover, the authors found a parametrization for the gluon dressing function that fitted the numerical solution of the DSE. The lattice being complementary to DSE, allows a check of compatibility between the two methods. In particular, we would like to check for the behaviour of small and zero momenta gluon propagator.

Our investigation uses pure gauge, Wilson action, SU(3) configurations, $\beta = 6.0$ ($a^{-1} = 1.885$ GeV), on large assimetric lattices[1]: $16^3 \times 128$ and $16^3 \times 256$. They were generated using combinations of over-relaxation (OVR) and Cabibbo-Mariani (HB) updates. For the smaller (larger) lattice, a combined sweep of 7 OVR and 2 HB (7 OVR and 4 HB) was used and 3000 combined sweeps for thermalization. The 160 (70) configurations were saved with a separation of 3000 (1500) combined sweeps. Note that, due to the large extension on the time direction, the lowest momenta considered is about 93 MeV for the smallest lattice and 46 MeV for the largest lattice.

For notation and details see [1]. The data for the Landau gauge gluon propagator[2]

$$\langle A_\mu^a(p) \, A_\mu^a(p') \rangle = V \delta \left(p + p'\right) \delta^{ab} \left(\delta_{\mu\nu} - \frac{p_\mu p_\nu}{p^2} \right) \frac{Z(p^2)}{p^2} \tag{1}$$

shows finite volume effects. To disentangle the finite volume effects, our data was

---

[1] All configurations were generated with MILC code http://physics.indiana.edu/~sg/milc.html.

[2] For gauge fixing we used the steepest descent Fourier accelerated method described in [2].

CP756, *Quark Confinement and the Hadron Spectrum VI*
edited by N. Brambilla, U. D'Alesio, A. Devoto, K. Maung, G.M. Prosperi and S. Serci
© 2005 American Institute of Physics 0-7354-0241-8/05/$22.50

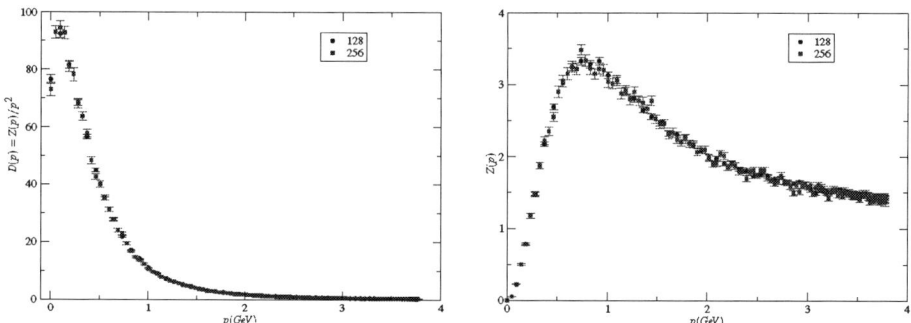

FIGURE 1. Gluon propagator. Statistical errors were computed using the jackknife procedure.

compared with the results reported in [3]. No essential differences were seen between our pure temporal data and the Leinweber *et al* data. From now on, we refer only to pure temporal data – see figure 1. Note that the data seems to support a vanishing zero momenta gluon propagator $D(p^2)$.

To test the compatibility between lattice and the DSE results [5], our data for the gluon dressing function was fitted to

$$Z(p^2) = \omega \left(\frac{p^2}{\Lambda_{QCD}^2 + p^2} \right)^{2\kappa} \left(\alpha(p^2) \right)^{-\gamma}, \tag{2}$$

where $\gamma = -13/22$, using two ansatz for the running coupling [5, 6]:

$$\alpha_1(p^2) = \frac{1}{1 + \frac{p^2}{\Lambda_{QCD}^2}} \left[\alpha(0) + \frac{p^2}{\Lambda_{QCD}^2} \frac{4\pi}{\beta_0} \left(\frac{1}{ln(p^2/\Lambda_{QCD}^2)} - \frac{1}{p^2/\Lambda_{QCD}^2 - 1} \right) \right], \tag{3}$$

$$\alpha_2(p^2) = \frac{\alpha(0)}{ln \left[e + a_1(p^2/\Lambda_{QCD}^2)^{a_2} \right]}, \tag{4}$$

with $\beta_0 = 11$.

The IR propagator was investigated fitting the lattice data to

$$Z_{IR1}(p^2) = \omega \left(p^2 \right)^{2\kappa} \qquad Z_{IR2}(p^2) = \omega \left(\frac{p^2}{\Lambda_{QCD}^2 + p^2} \right)^{2\kappa} \tag{5}$$

Only when the first three lowest momenta of the larger lattice ($|q| \leq 139 MeV$) were considered, we were able to fit the IR analytical solution of DSE Z_{IR1} with an acceptable $\chi^2/d.o.f. = 0.39$, meaning that the solution $Z(q^2) \sim (q^2)^{2\kappa}$ seems to be valid only for momenta below 150 MeV. The measured[3] $\kappa = 0.5003^{+32}_{-29}$ does not provide a clear

---

[3] Statistical errors for fit parameters were computed with 2000 bootstrap samples.

answer about the zero momenta gluon propagator. The results for the IR2 approximation are shown in table 1. For both lattices, the fit to the largest range of momenta supports an infrared finite gluon propagator.

TABLE 1. Z_{IR2} fits.

Lattice	Range	κ	Λ_{QCD} (MeV)	$\chi^2/d.o.f.$
$16^3 \times 128$	$\|q\| \leq 461 MeV$	0.5020^{+46}_{-49}	429^{+10}_{-9}	0.50
$16^3 \times 128$	$\|q\| \leq 553 MeV$	0.5122^{+42}_{-46}	403^{+8}_{-7}	1.41
$16^3 \times 256$	$\|q\| \leq 644 MeV$	0.5199^{+26}_{-31}	377^{+5}_{-5}	1.19

The fits to all data distinguishes the functional forms $\alpha_1(p^2)$ and $\alpha_2(p^2)$ for the running coupling constant. For $\alpha_1(p^2)$, the $\chi^2/d.o.f. > 2$. Moreover, $\alpha(0) \sim 10$ is much larger than the DSE result: $\alpha(0) = 2.972$ [5, 4]. However, if one uses $\alpha_2(p^2)$, the lattice data is well described by (2) - see table 2. $\alpha(0)$ can be computed from the asymptotic behaviour of the QCD β function and $\alpha_2(p^2)$. Then, $\alpha(0) = (4\pi/\beta_0)a_2$ giving $\alpha(0) = 2.78^{+2}_{-2}$ and 2.74^{+1}_{-2} for the smaller and larger lattices, respectively. The analysis of all momenta favours a finite zero momenta gluon propagator. Our figures for κ are smaller than those reported in the DSE studies, $\kappa \sim 0.595$ [4]. However, our κ values reported in table 2 are within the figures discussed in [7, 8].

TABLE 2. Fits to all lattice data.

Lattice	κ	Λ_{QCD} (MeV)	a_1	a_2	$\chi^2/d.o.f.$
$16^3 \times 128$	0.5439^{+36}_{-41}	352^{+4}_{-4}	0.0063^{+4}_{-3}	2.43^{+2}_{-1}	1.74
$16^3 \times 256$	0.5314^{+25}_{-24}	354^{+3}_{-3}	0.0065^{+4}_{-3}	2.40^{+1}_{-2}	1.56

We are currently involved in improving the statistics of our analysis and, simultaneously, trying to understand the role of Gribov copies[4] relying on the method discussed in [9]. We plan also to compute the ghost propagator and running coupling constant directly from the lattice.

P.J.Silva acknowledges financial support from FCT via grant SFRH/BD/10740/2002.

REFERENCES

1. P.J.Silva, O.Oliveira, Nucl. Phys. B**690**(2004) 177 [hep-lat/0403026].
2. C.T.Davies *et al*, Phys. Rev. **D37**(1988)1581.
3. D. B. Leinweber *et al* Phys. Rev. **D60** (1999) 094507; erratum-ibid Phys. Rev. **D61** (2000) 079901 [hep-lat/9811027].
4. Ch. Lerche, L. von Smekal, Phys. Rev. **D65**(2002), 125006 [hep-ph/0202194].
5. R. Alkofer *et al*, hep-ph/0309078, references therein and these proceedings.
6. C. S. Fischer, R. Alkofer, Phys. Rev. **D67**(2003)094020 [hep-ph/0301094].
7. J. M. Pawlowski *et al*, Phys. Rev. Lett. **93**(2004)152002 [hep-th/0312324] and these proceedings.
8. C. S. Fischer, H. Gies, hep-ph/0408089.
9. O.Oliveira, P.J.Silva, Comp. Phys. Comm. **158** (2004) 73 [hep-lat/0309184].

---

[4] In [1], the effect of the Gribov copies was estimated as a 2 to 3σ effect. This does not change our predictions for the zero momenta gluon propagator.

292

Approximated Analytical Solution of the Faddeev-Niemi Model

Andrzej Wereszczyński

Institute of Physics, Jagiellonian University, Reymonta 4, Kraków, Poland

Abstract. A systematic way of construction of multi-knotted solutions with non-trivial Hopf index is presented. Such knotted configurations emerge from the eikonal equation which appears to be an integrability condition for the Faddeev-Niemi model. It is shown that these knots give good approximation to the Faddeev-Niemi hopfions.

The Faddeev-Niemi model [1] is considered as an effective model for the low energy regime of the quantum gluodynamics. In particular, it can describe glueballs as knotted solitons. Indeed, existence of such knotted solutions has been confirmed in some numerical work [2], [3]. However, the analytic investigation of the model is still in its infancy. The Faddeev-Niemi model is given by the following action

$$S = \int d^4x \frac{1}{2} m^2 (\partial_\mu \vec{n})^2 - \frac{1}{4e^2} [\vec{n} \cdot (\partial_\mu \vec{n} \times \partial_\nu \vec{n})]^2, \tag{1}$$

where $\vec{n}$ is an unit, three component vector field. In order to have finite energy solutions the unit field must tend to a constant value at the spatial infinity. Then, such configurations are maps from S^3 onto S^2 and can be divided into disconnected homotopy classes $\pi_3(S^2)$ and classified by the Hopf index.
It has been recently observed [4] that this model possesses an integrable submodel if the following condition is fulfilled

$$L_\mu \partial^\mu u = 0, \tag{2}$$

$$L_\mu = m^2 \partial_\mu u - \frac{4}{e^2} \frac{K_\mu}{(1+|u|^2)^2}, \quad K_\mu = (\partial^\nu u \partial_\nu u^*)\partial_\mu u - (\partial^\nu u)^2 \partial_\mu u^*$$

where the stereographic projection has been taken into account: $\vec{n} = 1/(1+|u|^2)(u + u^*, -i(u - u^*), |u|^2 - 1)$, u is a complex field. Then, infinite family of local conserved currents can be constructed. It is common wisdom known form the standard soliton theory in $(1+1)$ and $(2+1)$ dimensions that the existence of such currents can lead to topologically non-trivial solutions. Thus, one can ask whether also in the case of the model (1) the integrability condition (2) can teach us something about knotted solitons. The condition can be rewritten in the form $L_\mu \partial^\mu u = m^2 (\partial_\mu u)^2$. This formula vanishes if $m = 0$ or the eikonal equation is satisfied. The case with vanishing mass parameter is trivial. It means that the kinetic term in the action is absent and no stable solitons can be constructed. Thus, the eikonal equation is unique non-trivial integrability condition for the Faddeev-Niemi model

$$(\partial_\nu u)^2 = 0. \tag{3}$$

CP756, Quark Confinement and the Hadron Spectrum VI
edited by N. Brambilla, U. D'Alesio, A. Devoto, K. Maung, G.M. Prosperi and S. Serci
© 2005 American Institute of Physics 0-7354-0241-8/05/$22.50

Two questions immediately arise: what kind of solutions is admitted by (3), what is their topological structure? And, if they can have anything to do with the original Faddeev-Niemi knots?

One has to be aware that full integrable submodel consists of the dynamical equation [4]

$$\partial^\mu \left[m^2 \partial_\mu u - \frac{4}{e^2} \frac{\partial^\nu u \partial_\nu u^*}{(1+|u|^2)^2} \partial_\mu u \right] = 0. \tag{4}$$

The correct solutions of the Faddeev-Niemi model in the integrable regime should fulfill both the dynamical equation (4) as well as the constrain (3).

Let us now solve the eikonal equation. We introduce the toroidal coordinates $x = \tilde{a} \sinh \eta \cos \phi / q$, $y = \tilde{a} \sinh \eta \sin \phi / q$, $z = \tilde{a} \sin \xi / q$, where $q = \cosh \eta - \cos \xi$ and $\tilde{a} > 0$ fixes the scale. The eikonal equation takes the form

$$\frac{q^2}{\tilde{a}^2} \left[(\partial_\eta u)^2 + (\partial_\xi u)^2 + \frac{1}{\sinh^2 \eta} (\partial_\phi u)^2 \right] = 0. \tag{5}$$

The generalized Adam solution [5] is given by the following expression [6]

$$u = \sum_{j=1}^{N} f_j^\pm(\eta) e^{i(m_j \xi + n_j \phi)} + c \tag{6}$$

where

$$f_j^\pm = A_j \sinh^{\pm|n_j|} \eta \, \frac{\left(|m_j| \cosh \eta + \sqrt{n_j^2 + m_j^2 \sinh^2 \eta} \right)^{\pm|m_j|}}{\left(|n_j| \cosh \eta + \sqrt{n_j^2 + m_j^2 \sinh^2 \eta} \right)^{\pm|n_j|}}, \tag{7}$$

and integer, positive parameters m, n are not independent but satisfy relation $m_j n_k = m_k n_j$. A_j and $c = c_0 e^{i\psi}$ are complex constants. The most important fact is that these configurations possess non-vanishing Hopf index [6]

$$Q_H = -\max\{m_j n_j, \; j = 1...N\}$$

and can form, identical as the Faddeev-Niemi hopfions, really knotted configurations. In the simplest $N = 1$ case, we obtain K elementary knots, with $Q_e = -pq$ Hopf charges, located on a torus with a constant radius. Here $m/n = (K \cdot p)/(K \cdot q)$ and p, q are relative prime number. The position of the core of the eikonal knots for various m, n parameters is presented in Fig. 1. For two component Anzatz, there is an additional central (un)knot located at $\eta = \infty$ which carries $Q_c = -\min\{m_i n_i, \; i = 1, 2\}$ and has shape of circle.

The crucial feature which allows us to believe that eikonal knots can approximate Faddeev-Niemi solitons follows from the observation that they lead to a finite value of the total energy integral for Faddeev-Niemi model. The situation is even better since the minimal energy configurations (given by unknots with $c_0 = 0$ and $E = E_{min}$) are only approximately 20% heavier than numerically found Faddeev-Niemi hopfions $E = E_{num}$ [2] (see Tab. 1). However, really knotted eikonal solutions can be obtained already for a

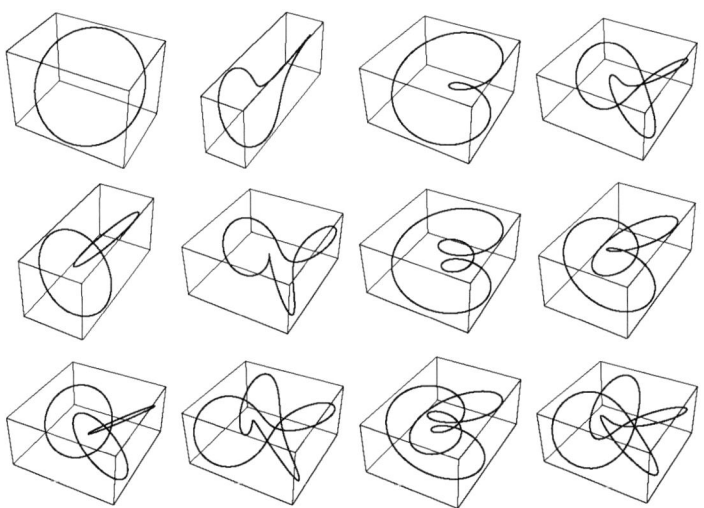

FIGURE 1. Position of the core of the eikonal knot for various (m, n) parameters: $(1,1), (2,2), (3,3)$; $(1,2), (1,3), (2,4)$; $(2,1), (3,1), (4,2)$; $(2,3), (3,2), (3,4)$.

TABLE 1. Energy of the eikonal knots and Faddeev-Niemi hopfions

type of hopfion	$|A|$	c_0	E_{min}	c_0	E	E_{num}
m=1,n=1	1.252	0	304.3	0.2	311.2	272.0
m=1,n=2	0.357	0	467.9	0.1	471.9	417.5
m=1,n=3	0.065	0	658.1	0.05	659.5	578.5
m=2,n=1	5.23	0	602.7	0.2	622.3	417.5
m=2,n=3	0.3	0	1257.0	0.1	1269.0	990.5

little bit larger energy E [6].

I am grateful to prof. Antti Niemi for stimulating discussion. This work is partially supported by Foundation for Polish Science FNP and ESF "COSLAB" programme.

REFERENCES

1. L. Faddeev and A. Niemi, Nature **387**, 58 (1997); Phys. Rev. Lett. **82**, 1624 (1999).
2. R. A. Battye and P. M. Sutcliffe, Phys. Rev. Lett. **81**, 4798 (1998); Proc.Roy.Soc.Lond. A **455**, 4305 (1999).
3. J. Hietarinta and P. Salo, Phys. Lett. B **451**, 60 (1999); Phys. Rev. D **62**, 81701 (2000).
4. H. Aratyn, L. A. Ferreira and A. H. Zimerman, Phys. Lett. B **456**, 162 (1999).
5. C. Adam, math-ph/0312031.
6. A. Wereszczyński, hep-th/0410148 .

QCD Green functions in a gluon field

Stéphane Peigné

LAPTH, B.P. 110, F-74941 Annecy-le-Vieux Cedex, France

Abstract. We formulate a dressed perturbative expansion of QCD, where Green functions are evaluated in the presence of an external zero-momentum gluon field. The approach preserves gauge and Poincaré symmetries. Quarks and gluons decay at asymptotic times because of the appearance of a branch cut in their propagators, and dressed color singlet states seem to be confined.

Keywords: Perturbative QCD, Infrared Singularities, $1/N$ Expansion
PACS: 12.38.Aw, 11.15.Pg.

MOTIVATION

Some of the long-distance properties of the strong interaction might be a direct consequence of the presence of quark and gluon condensates in the QCD vacuum. In particular, such condensates might prevent the propagation of coloured particles to large distances, though acting as a superfluid for color singlet states.

In order to explore this question, we will assume that the effects of the gluon condensate can be modelled by the presence of an external gluon field, denoted as Φ_μ in the following, which couples to standard quark and gluon fields $\psi(x)$ and $A_\mu(x)$.

This talk summarizes the results of Ref. [1].

MODEL FOR VACUUM EFFECTS

The couplings of quarks and gluons to Φ can be inferred from gauge invariance by considering a shift in the gluon field, $A^\mu \to A^\mu + \Phi^\mu$. Under this shift the quark part of the QCD lagrangian generates the term

$$-g\,\bar{\psi}\Phi_\mu\gamma^\mu\psi \ . \tag{1}$$

This is the form of the coupling we will use between Φ and the quark field. Gauge invariance thus requires Φ^μ to transform under an SU(N) (we consider N colors) gauge transformation as:

$$\Phi^\mu \to U(x)\Phi^\mu U(x)^\dagger \ . \tag{2}$$

Under the shift the field strength tensor $F_{\mu\nu} = \partial_\mu A_\nu - \partial_\nu A_\mu + ig\,[A_\mu, A_\nu]$ becomes

$$F_{\mu\nu} \ \to \ F_{\mu\nu} + F_{\mu\nu}^\Phi + \Phi_{\mu\nu} \tag{3}$$

$$F_{\mu\nu}^\Phi \ = \ \partial_\mu\Phi_\nu - \partial_\nu\Phi_\mu + ig([\Phi_\mu, A_\nu] - [\Phi_\nu, A_\mu]) \tag{4}$$

$$\Phi_{\mu\nu} \ = \ ig\,[\Phi_\mu, \Phi_\nu] \tag{5}$$

CP756, *Quark Confinement and the Hadron Spectrum VI*
edited by N. Brambilla, U. D'Alesio, A. Devoto, K. Maung, G.M. Prosperi and S. Serci
© 2005 American Institute of Physics 0-7354-0241-8/05/$22.50

The $F_{\mu\nu}^{\Phi}$ and $\Phi_{\mu\nu}$ tensors transform similarly to $F_{\mu\nu}$ under a gauge transformation, $F_{\mu\nu}^{\Phi} \to U F_{\mu\nu}^{\Phi} U^{\dagger}$ and $\Phi_{\mu\nu} \to U \Phi_{\mu\nu} U^{\dagger}$. The simplest gauge invariant interaction term we can build between the gluon and Φ fields is of the form

$$-\text{Tr}\left[F^{\mu\nu} F_{\mu\nu}^{\Phi} \right] \; . \tag{6}$$

Since the Φ field is meant to describe some long-wavelength vacuum effects, we will approximate Φ to be independent of the space-time coordinate x, which guarantees translation invariance. Of course, in general a gauge transformation (2) induces some x-dependence of Φ, and this is actually needed to fix the form of the coupling (6). We can however choose Φ_{μ}^{a} to be constant in a given gauge, for instance in covariant gauge. We thus work with the modified (massless) QCD lagrangian

$$\mathcal{L} = \bar{\psi} i \slashed{D} \, \psi - \frac{1}{2}\text{Tr}\left[F_{\mu\nu} F^{\mu\nu} \right] - \frac{1}{\xi}\text{Tr}\left[(\partial_{\mu} A^{\mu})^2 \right] - \bar{c}\partial^{\mu} D_{\mu} c - g\bar{\psi}\slashed{\Phi}\,\psi - \text{Tr}\left[F^{\mu\nu} F_{\mu\nu}^{\Phi} \right] \tag{7}$$

which includes gauge fixing and ghost terms, and where $F_{\mu\nu}^{\Phi} = ig([\Phi_{\mu}, A_{\nu}] - [\Phi_{\nu}, A_{\mu}])$.

The lagrangian (7) involves only linear terms in Φ and might be viewed as the simplest model to include the effects of a (gluon) vacuum field. There are infinitely many ways to build gauge invariant couplings between Φ and quarks and gluons, for instance $\bar{\psi}\Phi_{\mu}\Phi^{\mu}\psi$, $\text{Tr}[\Phi^{\mu\nu} F_{\mu\nu}^{\Phi}]$, etc... Such terms may be added to (7) as a refined model for the vacuum effects. In the following we stress that the simple model (7) is already rich enough to imply some features of confinement.

In order to maintain Lorentz and gauge invariance, we average over Φ_{μ}^{a} using the gaussian weight $\exp[\Phi_{\mu}^{a}\Phi_{a}^{\mu}/(2\Lambda^2)]$, which introduces the mass scale Λ and implies the following 'propagator' for the Φ field:

$$iD_{\Phi,\mu\nu}^{ab}(p) = -\Lambda^2 g_{\mu\nu}\delta^{ab}(2\pi)^4\delta^4(p) \; . \tag{8}$$

Being a constant in space-time the external Φ field carries no momentum.

'DRESSED TREE' RESULTS

In the presence of the constant vacuum field Φ coupled to quarks and gluons as in (7), any QCD Green function gets modified. It is now expressed as a double perturbative expansion, in terms of the standard coupling g^2, and of a new parameter Λ^2/P^2, where P is a generic momentum argument of the Green function. The higher-twist corrections of order $(\Lambda^2/P^2)^n$ have no effect on the standard high-energy perturbative QCD predictions. They however drastically affect the infrared domain of the theory. In order to single out the effects of the zero-momentum field Φ, at each order in g^2 the Green function is dressed to all orders in Λ^2/P^2.

We now present the results we obtained in Ref. [1] for the dressed quark and gluon propagators and photon polarization tensor, at the zeroth (tree) order in g^2. The calculations were performed in the 't Hooft large N limit [2].

The dressed tree quark propagator is given by a simple Dyson-Schwinger type equation, which admits both chiral symmetry conserving and breaking solutions. Those solutions were previously obtained in a different framework [3], and we give here the chiral symmetry conserving solution,

$$S(p) = \frac{2\not{p}}{p^2 + \sqrt{p^2(p^2 - 4\mu^2)}} \xrightarrow[p^2 \to \pm\infty]{} \frac{1}{\not{p}} \ , \tag{9}$$

where $\mu^2 = g^2 N \Lambda^2$. The quark propagator has a branch cut $\sim 1/\sqrt{p^2}$ instead of a pole at $p^2 \to 0$, and thus decays at large times as $|S(t, \vec{p})| \sim 1/\sqrt{|t|}$. The quark is thus removed from the set of asymptotic states.

The calculation of the dressed tree gluon propagator in covariant gauge yields [1]:

$$iD_{\mu\nu}^{ab}(p) = \frac{-i}{p^2} \left[\left(g_{\mu\nu} - \frac{p_\mu p_\nu}{p^2} \right) d(-2\mu^2/p^2) + \xi \frac{p_\mu p_\nu}{p^2} \right] \delta^{ab} \tag{10}$$

$$d(x) = \frac{1}{2x} \left[1 - {}_2F_1\left(-\tfrac{1}{2}, \tfrac{1}{2}, 2, 16x\right) \right] \xrightarrow[x \to 0]{} 1 \ . \tag{11}$$

Similarly to the quark case, the pole at $p^2 = 0$ of the bare gluon propagator is removed and shifted to a $1/\sqrt{p^2}$ branch cut, removing also the gluon from asymptotic states.

We also studied the effect of the presence of the Φ field on the propagation of color singlet states such as the quark (loop) fluctuation of a virtual photon. When the number of couplings between the *massless* quark loop and the external Φ field increases, infrared singularities appear. The typical size of the quark-antiquark fluctuation can be arbitrarily large, suggesting that zero-momentum external gluons might not decouple from the color singlet quark loop.

Indeed, although in this case fixed order perturbative QCD is ill-defined, we have shown that the resummation to all orders in Λ^2 of the couplings between Φ and the quark loop leads to a dressed photon polarization tensor which is *infrared finite and non-zero*. The color singlet quark pair is effectively 'confined' at a distance of order μ^{-1}. This result is consistent with the finite propagation time $\sim \mu^{-1}$ found for the dressed partons.

Our study suggests that expanding around a non-empty vacuum might help understanding some long-distance properties of QCD.

Indeed, modelling vacuum effects by the presence of an external zero-momentum gluon field implies several features of confinement. Partons in a color singlet state seem to be confined, which is consistent with the analytic structure found for the parton propagators. The latter have $1/\sqrt{p^2}$ branch cuts at $p^2 = 0$, removing coloured partons from asymptotic states.

REFERENCES

1. P. Hoyer and S. Peigné, hep-ph/0304010 and hep-ph/0410235.
2. G. 't Hooft, *Nucl. Phys. B*, **72**, 461 (1974); *Nucl. Phys. B*, **75**, 461 (1974).
3. H. J. Munczek and A. M. Nemirovsky, *Phys. Rev. D*, **28**, 181 (1983);

A Model for the Generalized Parton Distribution of the Pion

P.Stassart*, F.Bissey†, J.R.Cudell*, J.Cugnon* and J.P. Lansberg*

*Physique Théorique Fondamentale, Bt. B5a, Univ. de Liège, Sart Tilman, B-4000 Liège 1
†Institute of Fund. Sciences, Massey Univ., Private Bag 11 222, Palmerston North, NZ

Abstract. We calculate the off-forward structure function of the pion within a simple model where the size of the pion is accounted for using a momentum cut-off. Twist-two and twist three generalized parton distributions are extracted. Relations between twist-three and twist-two contributions are obtained, the origin of which is not kinematical as they differ from those arising from the Wandzura-Wilczek approximation.

INTRODUCTION

Recent interest has focused on off-forward parton distributions [1] as they carry information on correlations between parton inside hadrons. Based on the model we built to calculate the diagonal structure functions of the pion in a gauge-invariant, regularization-independent way [2], we have performed the calculations of off-forward structure functions and link them to generalized parton distributions.

In this simple model, the pion field is related to the constituent quarks through a γ^5 vertex and pion size effects are introduced through a gauge-invariant cut-off procedure, by requiring that the relative momentum squared of the quarks inside the pion be smaller than the cut-off value.

In the following, we shall calculate the imaginary part of the off-forward scattering amplitude and link it to the structure functions that are the coefficients of the five independent tensors in this amplitude. We shall write these structure functions in terms of vector and axial-vector form factors [3] and link them to the generalized parton distributions H, H^3 and $\tilde{H}^3$.

THE MODEL

The diagrams contributing to the imaginary part of the amplitude are displayed on Fig. 1. The Lorentz invariants are $t = \Delta^2$, $Q^2 = -q^2$, $x = Q^2/2p \cdot q$, $\xi = \Delta \cdot q/2p \cdot q$. The hadronic tensor reads

$$T_{\mu\nu} = -\mathcal{P}_{\mu\sigma}g^{\sigma\tau}\mathcal{P}_{\tau\nu}F_1 + \frac{\mathcal{P}_{\mu\sigma}p^\sigma p^\tau \mathcal{P}_{\tau\nu}}{p \cdot q}F_2 + \frac{\mathcal{P}_{\mu\sigma}(p^\sigma(\Delta^\tau - 2\xi p^\tau) + (\Delta^\sigma - 2\xi p^\sigma))\mathcal{P}_{\tau\nu}}{2p \cdot q}F_3$$
$$+ \frac{\mathcal{P}_{\mu\sigma}(p^\sigma(Delta^\tau - 2\xi p^t au) - (\Delta^\sigma - 2\xi p^\sigma)p^\tau)\mathcal{P}_{\tau\nu}}{pp \cdot q}F_4 + \mathcal{P}_{\mu\sigma}(\Delta^\sigma - 2\xi p^\sigma)(\Delta^\tau - 2\xi p^\tau)\mathcal{P}_{\tau\nu}F_5. \quad (1)$$

CP756, Quark Confinement and the Hadron Spectrum VI
edited by N. Brambilla, U. D'Alesio, A. Devoto, K. Maung, G.M. Prosperi and S. Serci
© 2005 American Institute of Physics 0-7354-0241-8/05/$22.50

where $\mathcal{P}$ is the projector built on the metric tensor together with the momenta of the ingoing and outgoing photons. The Pion Quark coupling is described by the Lagrangian density $\mathcal{L}_{int} = ig(\overline{\psi}\vec{\tau}\gamma_5\psi).\vec{\pi}$, where ψ stands for the quark field and $\vec{\pi}$ for the pion field while $\vec{\tau}$ is the isospin operator. The cut-off, which accounts for the finite size of the pion, is imposed by requiring that the relative four-momentum squared of the quarks inside the pion be smaller than Λ^2 at one of the quark-pion vertices of each diagram [4]. One should note that imposing these conditions leads to a suppression of the crossed diagrams by a factor Λ^2/Q^2 compared with the box diagrams [2]. In the following we keep the value of the coupling constant g to its diagonal case value, obtained by imposing the sum rule on F_1, that is imposing that there be two valence quarks inside the pion.

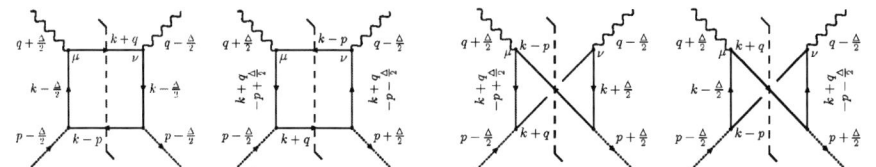

FIGURE 1. Lowest order diagrams contributing to the imaginary part of the $\gamma^\star\pi \to \gamma^\star\pi$ scattering amplitude

RESULTS FOR THE STRUCTURE FUNCTIONS

Having obtained the imaginary parts of the five structure functions F_i by projection of the amplitude on the corresponding tensors, we look for their asymptotic behaviour at large Q^2, and find relations between them at leading order. These relations read

$$F_2 = 2xF_1 + \mathcal{O}(1/Q^2), \quad F_3 = \frac{2x\xi}{\xi^2 - 1}F_1 + \mathcal{O}(1/Q^2),$$
$$F_4 = \frac{2x}{\xi^2 - 1}F_1 + \mathcal{O}(1/Q^2), \quad F_5 = \mathcal{O}(1/Q^2). \tag{2}$$

Getting such simple relations, similar to the Callan-Gross relation between the diagonal form factors, constitutes a remarkable result of our model. We shall see that these *Generalised Callan-Gross Relations* will lead to relations between the generalized parton distributions.

GENERALIZED PARTON DISTRIBUTIONS

Following the twist-three tensor analysis of Ref. [3] that links the twist-two $\mathcal{H}$ and twist-three $\mathcal{H}^3, \tilde{\mathcal{H}}^3$ form factors to the tensorial content of $T_{\mu\nu}$, we can relate the F_i structure functions to the form factors [4]. The imaginary parts of the form factors yield the corresponding Generalized Parton Distributions up to a constant factor.

Normalizing the structure functions by use of the same factor, we can relate them to the GPD's, getting:

$$F_{1n} = H, \ F_{2n} = 2xH, \ F_{3n} = \frac{2x}{x^2 - \xi^2} \left(H^3 x^2 + \tilde{H}^3 \xi x - H\xi \right),$$

$$F_{4n} = \frac{2x}{x^2 - \xi^2} \left(H^3 \xi x + \tilde{H}^3 x^2 - Hx \right), \ F_{5n} = \mathcal{O}(1/Q^2). \tag{3}$$

So that, with the help of our *Generalised Callan-Gross Relations* we can write

$$\tilde{H}^3 = \frac{(x-1)}{x(\xi^2 - 1)} H \text{ and } H^3 = \frac{(x-1)\xi}{x(\xi^2 - 1)} H = \xi \tilde{H}^3. \tag{4}$$

We compared relations (4) with the results of the Wandzura-Wilczek approximation [5] applied to our results for H, and found these to be very different.

Hence we conclude that our relations (4) are new, and do not correspond to kinematical twist corrections but come from the dynamics of the model we use, including its finite-size content.

ACKNOWLEDGMENTS

The authors thank M.V.Polyakov for useful comments. This work was performed with the help of the ESOP collaboration (European Union contract HPRN-CT-2000-00130)

REFERENCES

1. W. Broniowski it et al., *Phys. Lett. B*, **574**, 57 (2003) ; S. Dalley, *Phys. Lett. B* **570**, 191 (2003) [arXiv:hep-ph/0306121]; L. Theussl *et al.*, *Eur. Phys. J. A*, **20**, 483 (2004) .
2. F. Bissey, J.R. Cudell, J. Cugnon, M. Jaminon, J.P. Lansberg and P. Stassart, *Phys. Lett. B*, **547**, 210 (2002) [arXiv:hep-ph/0207107]; J.P. Lansberg, F. Bissey, J.R. Cudell, J. Cugnon, M. Jaminon and P. Stassart, *AIP Conf. Proc.* **660** 339 (2003) [arXiv:hep-ph/0211450].
3. A. V. Belitsky *et al.*, *Phys. Rev. D*, **64**, 116002 (2001) [arXiv:hep-ph/0011314].
4. F. Bissey, J.R. Cudell, J. Cugnon, J.P. Lansberg and P. Stassart, *Phys. Lett. B*, **587**, 189 (2004).
5. S. Wandzura and F. Wilczek, *Phys. Lett. B*, **72**, 195 (1977) .

Stringification of Chiral Dynamics: Wess-Zumino interaction

A.A.Andrianov*, D. Espriu† and A. Prats†

*V.A.Fock Institute of Physics, SPbGU, Russia
†Departament d'ECM, Universitat de Barcelona, Spain

Abstract. The QCD hadronic string is supplemented with the boundary interaction to background chiral fields associated with pions in a way compatible with the conformal symmetry. The modification of boundary interaction necessary to induce the parity-odd Chiral Dynamics (WZW action) is discussed.

Keywords: Quantum Chromodynamics, Hadron String,Chiral Lagrangian
PACS: 12.38.-t,12.39.Fe,12.40.-y

Introduction: pion coupling to QCD string. String description of QCD in the hadronization regime is a long-standing problem with a number of theoretical arguments [1] and phenomenological evidences [2, 3] as well as with the recent lattice simulations [4] in favor to its viability at intermediate energies (hadron masses).

The crucial low-energy phenomenon in QCD which makes influence on Hadron String building is Chiral Symmetry Breaking. It determines the QCD vacuum and results in the formation of light (massless in the chiral limit) pseudoscalar mesons. For the string dynamics the background chiral fields $U(x)$ add new couplings [5] involving the string variable $x_\mu(\tau,\sigma)$, on the boundary of the string where flavor is attached. A consistent string propagation in this non-perturbative background has been realized in [5] where it was based on the essential property of string theory - conformal invariance.

The boundary quark fields $\psi_L(\tau)$, $\psi_R(\tau)$ transform in the fundamental representation of the light-flavor group $SU(N)$. The subscripts L,R are related to the *chiral* spinors. A local hermitian action $S_b = \int d\tau L^{(f)}$ is introduced on the boundary $\sigma = 0$, $-\infty < \tau < \infty$ to describe the interaction with background chiral fields $U(x(\tau)) = \exp(i\pi(x)/f_\pi)$ where $f_\pi \simeq 90 MeV$, the weak pion decay constant, relates the field $\pi(x)$ to a π-meson one.

The boundary Lagrangian is chosen to be reparametrization invariant and in its bare minimal form reads

$$L^{(f)}_{min} = \frac{1}{2}i\left(\bar{\psi}_L U (1-z)\dot{\psi}_R - \dot{\bar{\psi}}_L U (1+z)\psi_R\right) + \text{h.c.} \quad , \qquad (1)$$

where a dot implies a τ derivative. It has been proved [5] to provide the E.o.M. of Chiral dynamics and thereby the Chiral Lagrangian for the parity-even sector from the conformal symmetry of this boundary QFT.

However this Lagrangian does not contain any vertices which can finally entail the anomalous P-odd part of the Chiral Dynamics. In our talk we outline the modification of the boundary interaction which might bring the Wess-Zumino-Witten Chiral action and other parity-odd vertices.

CP756, *Quark Confinement and the Hadron Spectrum VI*
edited by N. Brambilla, U. D'Alesio, A. Devoto, K. Maung, G.M. Prosperi and S. Serci
© 2005 American Institute of Physics 0-7354-0241-8/05/$22.50

Chiral dynamics on the line. To approach the required modification we guess on what might be the form of boundary Lagrangian if one derives it from the essential part of the Chiral Quark Model projecting it on the string boundary. The constituent quark fields control properly the chiral symmetry during the "ein-bein" projection, $Q_L \equiv \xi^\dagger \psi_L$, $\qquad Q_R \equiv \xi \psi_R$, $\qquad \xi^2 \equiv U$. In these variables and in the chiral limit the CQM Lagrangian density and the pertinent E.o.M. read

$$\mathscr{L}_{CQM} = i\bar{Q}\left(\overleftrightarrow{\partial} + \slashed{v} + g_A \slashed{a}\gamma_5\right)Q; \quad i\left(\overleftrightarrow{\partial} \mid \slashed{v} + g_A \slashed{a}\gamma_5\right)Q = 0, \tag{2}$$

where

$$v_\mu \equiv \frac{1}{2}(\xi^\dagger(\partial_\mu\xi) - (\partial_\mu\xi)\xi^\dagger), \qquad a_\mu \equiv -\frac{1}{2}(\xi^\dagger(\partial_\mu\xi) + (\partial_\mu\xi)\xi^\dagger), \tag{3}$$

and $g_A \equiv 1 - \delta g_A$ is an axial coupling constant of quarks to pions. We relegate the effects of constituent quark mass to the gluodynamics encoded in the string interaction. Then one can decouple the left and right components of boundary fields in the process of dim-1 projection.

Let's assume the quark fields to be located on the dim-1 boundary with coordinates $x_\mu \equiv x_\mu(\tau)$. The first step in projection of the E.o.M. (2) can be performed by their multiplication on $\gamma^\mu \dot{x}_\mu$ which leads to the following boundary equations,

$$\left\{i\left(\partial_\tau + \dot{x}_\mu v^\mu + g_A \gamma_5 \dot{x}_\mu a^\mu\right) + \sigma^{\mu\nu}\dot{x}_\mu\left(\partial_\nu + v_\nu + g_A \gamma_5 a_\nu\right)\right\}Q = 0; \ \sigma^{\mu\nu} \equiv \frac{1}{2}i[\gamma^\mu\gamma^\nu]. \tag{4}$$

We notice that this projected Dirac-type equation seems to be associated to the boundary action with a Lagrangian of type (1).

Let us restore the current quark basis of fields ψ_L thereby going back to the original chiral fields U,

$$\frac{1}{2}\left\{i\left(\{\partial_\tau, U^\dagger\} + z\dot{U}^\dagger\right) + \sigma^{\mu\nu}\dot{x}_\mu\left(\{\partial_\nu, U^\dagger\} + g_A \partial_\nu U^\dagger\right)\right\}\psi_L = 0;$$

$$\frac{1}{2}\left\{i\left(\{\partial_\tau, U\} + z\dot{U}\right) + \sigma^{\mu\nu}\dot{x}_\mu\left(\{\partial_\nu, U\} + g_A \partial_\nu U\right)\right\}\psi_R = 0. \tag{5}$$

Now the culminating point of the "ein-bein" projection consists of making the quark fields ψ truly one-dimensional. Namely we define their gradient in terms of the tangent vector $\dot{x}_\mu$:

$$\{\partial_\mu, U^\dagger\}\psi_L \Rightarrow \frac{\dot{x}_\mu}{\dot{x}_\nu\dot{x}^\nu}\{\partial_\tau, U^\dagger\}\psi_L; \quad \{\partial_\mu, U\}\psi_R \Rightarrow \frac{\dot{x}_\mu}{\dot{x}_\nu\dot{x}^\nu}\{\partial_\tau, U\}\psi_R. \tag{6}$$

Finally, the projected equations are originated from the boundary Lagrangian,

$$L^{(f)} \equiv \frac{1}{2}i\left\{\bar{\psi}_L\left[\{\partial_\tau, U\} + \widehat{F}^{\mu\nu}\dot{x}_\mu\partial_\nu U\right]\psi_R + \bar{\psi}_R\left[\{\partial_\tau, U^\dagger\} - \widehat{F}^{\mu\nu}_\sharp\dot{x}_\mu\partial_\nu U^\dagger\right]\psi_L\right\};$$

$$\widehat{F}^{\mu\nu} \equiv zg^{\mu\nu} + g_\sigma\sigma^{\mu\nu}; \quad \widehat{F}^{\mu\nu}_\sharp \equiv \gamma_0\left(\widehat{F}^{\mu\nu}\right)^\dagger\gamma_0, \tag{7}$$

303

where, keeping in mind a certain ambiguity in the projection procedure, we consider both constants z and g_σ as arbitrary ones and search for their values from the consistency of the Hadron string with chiral fields on its boundary.

Two-dimensional QCD and beyond. The above constructed projection is unambiguously verified in the two-dimensional version of QCD where the bosonization fixes basic coupling constants in the Chiral lagrangian. As in two dimensions $\gamma_0 = \sigma_1$; $\gamma_1 = -i\sigma_2$, γ_2 (*i.e.* "γ_5") $= \sigma_3$ the Lorentz algebra is generated by $\sigma_{\mu\nu} = i\varepsilon_{\mu\nu}\gamma_2$ which must be used in the boundary action (7).

To develop the string perturbation theory we expand the function $U(x)$ in powers of the string coordinate field $x_\mu(\tau) = x_{0\mu} + \tilde{x}_\mu(\tau)$, expand the boundary action in powers of $\tilde{x}_\mu(\tau)$ and look for divergences. At one loop one obtains the following condition to remove the divergences (β-function$= 0$ to preserve conformal symmetry),

$$-\partial_\mu^2 U + \frac{1}{2}(3 + z^2 - g_A^2)\partial_\mu U U^\dagger \partial^\mu U - ig_A \varepsilon_{\mu\nu} \partial^\mu U U^\dagger \partial^\nu U = 0. \tag{8}$$

Unitarity of chiral fields (= local integrability of Eqs. of Motion (8)) constrains the coupling constants to $g_A^2 - z^2 = 1$. The choice in accordance with the QCD bosonization is $z = 0, g_A = 1$. It corresponds to the correct value of the dim-2 anomaly (last term in (8)). Thus in QCD$_2$ the hadron string induces the WZW action from the vanishing the boundary β function already at one-loop level.

In dim-4 QCD the anomaly and the WZW action have dimension 4 and therefore they are generated by cancelation of two-loop divergences. The antisymmetric tensor $\varepsilon_{\mu\nu\rho\lambda}$ in anomalies arises from the well-known algebra of $\sigma_{\mu\nu}$ matrices. At one-loop level the interplay between coupling constants z and g_σ takes place as well with the unitarity condition, $3g_A^2 - z^2 = 1$. But now their values are determined from the consistency (local integrability) of the two-loop equations providing β-function$= 0$.

Acknowledgments. One of us (A.A.) is grateful to the Organizing Committee of the 6th Conference on Quark Confinement and the Hadron Spectrum for financial support. A.A. was also supported by Grant RFBR and the Program "Universities of Russia: Basic Research". The work of D.E was supported by the EURIDICE Network, grant FPA-2001-3598 and grant 2001SGR-00065.

REFERENCES

1. A. M. Polyakov, *"Gauge Fields and Strings"* (Harwood, Chur, Switzerland, 1987)
2. P. Frampton, *"Dual Resonance Models"* (Benjamin, New York, 1974); A. V. Anisovich, V. V. Anisovich and A. V. Sarantsev, *Phys. Rev.* **D62**, 051502 (2000).
3. A. A. Andrianov, V. A. Andrianov and S. S. Afonin, *5th Int. Conf. on Quark Conf. and HS - Gargnano 2002*, World Sci. 2003, p.361, hep-ph/0212171; A. A. Andrianov, V. A. Andrianov, S. S. Afonin and D. Espriu, *JHEP* **0404** (2004) 039.
4. K. J. Juge. J. Kuti and C. Morningstar, *Phys. Rev. Lett.* **90**, 161601 (2003).
5. J. Alfaro, A. Dobado and D. Espriu, *Phys. Lett.* **B460** (1999) 447; A. A. Andrianov and D. Espriu, *Theor. Math. Phys.* **135/3** (2003) 745; *AIP Conf.Proc.* **660** (2003) 17; *5th Int. Conf. Quark Conf. and HS - Gargnano 2002*, World Sci. 2003, p.291; J. Alfaro, A. A. Andrianov, L. Balart and D. Espriu, *Int. J. Mod. Phys.* **A18** (2003) 2501.

Matrix product variational formulation for lattice gauge theory

Takanori Sugihara

RIKEN BNL Research Center, Brookhaven National Laboratory, Upton, New York 11973, USA

Abstract. For hamiltonian lattice gauge theory, we introduce the matrix product anzats inspired from density matrix renormalization group. In this method, wavefunction of the target state is assumed to be a product of finite matrices. As a result, the energy becomes a simple function of the matrices, which can be evaluated using a computer. The minimum of the energy function corresponds to the vacuum state. We show that the $S = 1/2$ Heisenberg chain model are well described with the ansatz. The method is also applied to the two-dimensional $S = 1/2$ Heisenberg and U(1) plaquette chain models.

RHIC experiments have started to test the fundamental properties of quantum chromodynamics (QCD). The importance of first-principle analysis of QCD has increased largely in the context of color confinement. There is the expectation that quarks and gluons deconfine in extreme conditions such as heavy ion collision. However, lattice gauge theory at finite density has been stuck for a long time. The most ideal treatment of non-equilibrium quantum physics is to trace time-evolution of quantum states based on the Schrödinger equation.

Density Matrix Renormalization Group (DMRG) is the variational method that gives the most accurate results in one-dimensional quantum systems [1]. In these days, DMRG has been used as a standard method to complement quantum Monte Carlo and successful in solving one-dimensional (zero and finite temperature) and two-dimensional (zero temperature) quantum systems. Application of DMRG to elemanrary particle and molecular physics has started some years ago. Recent interesting progress of DMRG is its application to non-equilibrium quantum physics and quantum information theory [2].

The first application of DMRG to particle physics was the massive Schwinger model with the θ term [3]. There is an old prediction by S. Coleman that quarks deconfine at $\theta = \pi$ [4]. The model has not been analyzed accurately with Monte Carlo because of the sign problem. On the other hand, DMRG has been successful in describing the details of the phase transition with large lattices because DMRG is free from the sign problem.

The second application has been given by the author [5]. It is a preliminary work for study of gauge theory. In bosonic lattice systems, each site has infinite degrees of freedom and therefore hamiltonian is infinite dimensional differently from spin and fermion systems. It is not evident whether DMRG truncation works for bosonic degrees of freedom. DMRG needs to be tested in a simpler bosonic model before going to gauge theory. In Ref. [5], DMRG has been applied to a (1+1)-dimensional $\lambda\phi^4$ model. The DMRG result for the critical exponent $\beta = 0.1264 \pm 0.0073$ is consistent with the exact one $\beta = 1/8 = 0.125$.

CP756, *Quark Confinement and the Hadron Spectrum VI*
edited by N. Brambilla, U. D'Alesio, A. Devoto, K. Maung, G.M. Prosperi and S. Serci
© 2005 American Institute of Physics 0-7354-0241-8/05/$22.50

TABLE 1. Numerical results for ground-state energy per site in the $S = 1/2$ Heisenberg chain model. The exact values have been obtained with the Bethe ansatz.

$M \backslash L$	10	100	1000	10000
6	−0.4092	−0.4372	−0.4371	−0.4368
12	−0.4092	−0.4427	−0.4425	−0.4425
Exact	−0.4515	−0.4438	−0.4431	−0.4431

Matrix product variational method is a result of large simplification of DMRG [6]. In this method, the energy function has a simple form and easy to evaluate. However, the advantage is lost if a constraint is imposed to variational space directly. The Gauss law in gauge theory is one of the most important examples of constraints. To avoid this difficulty, the author has developed a method to introduce constraints with undetermined multipliers in Hamiltonian [7]. Also, the author has generalized the matrix product variational method so that it can be applied to higher-dimensional general systems [8].

We introduce matrix product states according to Ref. [6]. With the knowledge from DMRG, wavefunction is represented as a product of matrices.

$$|\Psi\rangle = \sum_{s_1,\ldots,s_L=1}^{K} \text{tr}[A[s_1]\ldots A[s_L]]|s_1\rangle \ldots |s_L\rangle, \tag{1}$$

where periodicity is assumed. $A[s]$ has the following normalization condition.

$$\sum_{\beta=1}^{M} \sum_{s=1}^{K} A^*[s]_{\alpha\beta} A[s]_{\alpha'\beta} = \delta_{\alpha\alpha'}. \tag{2}$$

Each of the matrices $A[s]$ can be parameterized with an appropriate number of independent variables [7]. It is expected that better results is obtained for larger M. Let us apply the ansatz to the $S = 1/2$ Heisenberg chain model with periodic boundary conditions

$$H = \sum_{i=1}^{L} (S_i^x S_{i+1}^x + S_i^y S_{i+1}^y + S_i^z S_{i+1}^z). \tag{3}$$

Periodicity simplifies calculation of the energy function

$$E[A] = \frac{\langle\Psi|H|\Psi\rangle}{\langle\Psi|\Psi\rangle} = \sum_a \frac{\text{tr}(\hat{S}^a \hat{S}^a \hat{1}^{L-2})}{\text{tr}(\hat{1}^L)}, \tag{4}$$

where $\hat{S}^a = \sum_{s,s'}\langle s|S_1^a|s'\rangle A^*[s] \otimes A[s']$ and $\hat{1} = \sum_s A^*[s] \otimes A[s]$. The minimum of the energy function corresponds to the ground state. In actual numerical calculation, the matrix $\hat{1}$ is diagonalized to simplify calculation of the powers of $\hat{1}$.

Table 1 shows numerical results for ground-state energy per site, which are compared with the exact ones. M is the size of the matrices $A[s]$. When the lattice size L is small, convergence is poor even for large M. On the other hand, when L is large, the numerical

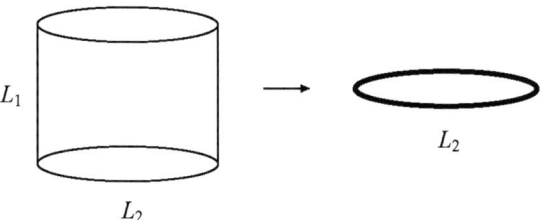

L_1

L_2

L_2

FIGURE 1. Reduction of two dimensional lattice to one dimensional

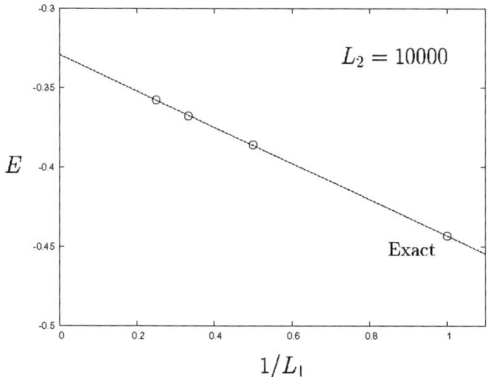

$L_2 = 10000$

E

Exact

$1/L_1$

FIGURE 2. Energy per bond in the two-dimensional $S = 1/2$ Heisenberg model.

results approaches the exact one as M becomes larger. In the best case, error is less than 1%. The lattice size dependence of energy is small when $L \geq 100$.

In the same way, the two-dimensional $S = 1/2$ Heisenberg model is analyzed. To use the matrix product ansatz, one-dimensional structure needs to be found on the two-dimensional lattice (see Fig. 1). The tube and ring sizes are denoted as L_1 and L_2, respectively. Figure 2 plots energy per bond as a function of $1/L_1$ for a very large ring with $L_2 = 10000$. The circle for $L_1 = 1$ is the exact result from the Bethe ansatz. The other three circles have been calculated using the matrix product ansatz. We are going to make a comparison in the thermodynamic limit $L_1 \rightarrow \infty$ by extrapolating the obtained points to the limit. In Table 2, energy per bond is compared among various results in the thermodynamic result. "Lattice size" means the largest lattice size used for calculation.

The method is applied to a U(1) plaquette chain model. The hamiltonian is

$$H = \sum_l E_l^2 - x \sum_p (U_p + U_p^\dagger),$$

where $x \equiv 1/g^4$ [13]. Figure 3 plots energy as a function of the parameter x and compares it with the results of Ref. [14]. In this calculation, the Gauss law $\nabla \cdot \mathbf{E} = 0$ has not been imposed on the variational space. Since vacuum wavefunction is available, we can check gauge invariance of the obtained vacuum state by calculating the vacuum expectation value of the electric field. The obtained result $\sim 10^{-3}$ is small compared to energy,

TABLE 2. Energy per bond of the two-dimensional $S = 1/2$ Heisenberg model in the thermodynamic limit.

Method	E	Lattice size	Year	Reference
Monte Carlo	-0.3347	16^2	1999	[9]
DMRG	-0.3347	12^2	2001	[10]
DMRG	-0.3321	20^2	2003	[11]
TPVA	-0.3272	Very large	2004	[12]
Matrix product	-0.3292 ± 0.0005	10000×4	2004	This work

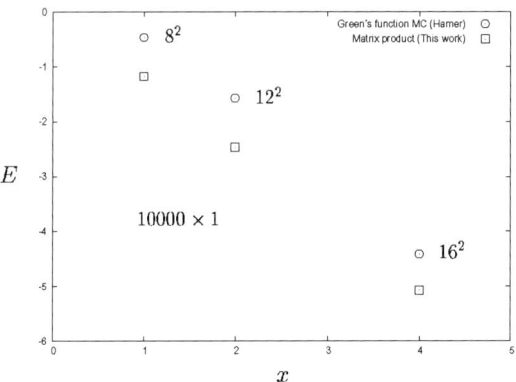

FIGURE 3. Energy per site vs x in the $U(1)$ hamiltonian lattice gauge model.

which shows approximate gauge invariance of the vacuum state. Currently, calculation with the Gauss law constraint is being conducted. Further precise analysis with larger lattices will be given elsewhere to refine the results shown in this presentation.

The numerical calculations were carried on the RIKEN RSCC system. This work has been partially supported by RIKEN BNL.

REFERENCES

1. S. R. White, Phys. Rev. Lett. **69**, 2863 (1992); Phys. Rev. B **48**, 10345 (1993).
2. U. Schöllwoeck, cond-mat/0409292.
3. T. Byrnes, P. Sriganesh, R. J. Bursill, and C. J. Hamer, Phys. Rev. D **66**, 013002 (2002).
4. S. R. Coleman, Annals Phys. **101**, 239 (1976).
5. T. Sugihara, JHEP **0405**, 007 (2004).
6. S. Östlund and S. Rommer, Phys. Rev. Lett. **75**, 3537 (1995).
7. T. Sugihara, in preparation.
8. T. Sugihara, in preparation.
9. A. W. Sandvik, Phys. Rev. B **56**, 678 (1997).
10. T. Xiang, J. Lou, and Z. Su, Phys. Rev. B **64**, 104414 (2001).
11. D. J. J. Farnell, Phys. Rev. B **68**, 134419 (2003).
12. Y. Nishio, N. Maeshima, A. Gendiar, and T. Nishino, cond-mat/0401115.
13. J. B. Kogut and L. Susskind, Phys. Rev. D **11**, 395 (1975).
14. C. J. Hamer, R. J. Bursill and M. Samaras, Phys. Rev. D **62**, 054511 (2000).

Nonperturbative calculations in SU(3) gauge theory

Vladimir Dzhunushalicv[*], Douglas Singleton[†] and Tatyana Nikulicheva[*]

[*]Dept. Phys. and Microel. Eng., KRSU, Kievskaya Str. 44, 720021, Bishkek, Kyrgyz Republic.
[†]Physics Dept., CSU Fresno, M/S MH37 Fresno, CA 93740-8031, USA.

Keywords: confinement, nonperturbative quantization
PACS: 12.38.Aw, 12.38.Lg

Modern perturbative quantum field theories (QFT) are based on the concept of a particle or field quanta. This perturbative paradigm gives the most accurate description for some aspects of the physical world (*e.g.* the *g*-factor of the electron). Mathematically the elementary particles that arise in perturbative quantum field theory are quantized harmonic excitations of fundamental fields. The quanta are defined through the creation and annihilation operators, $a^\dagger$ and a, of the "second-quantized" theory.

The physical reason of the appearance of quanta is that in linear theories, such as electrodynamics, a general solution can be obtained by making the Fourier expansion:

$$A_\mu^{linear} = \int d^3k \sum_{\lambda=0}^{3} [a_k(\lambda)\varepsilon_\mu(k,\lambda)e^{-ik\cdot x} + a_k^\dagger(\lambda)\varepsilon_\mu^*(k,\lambda)e^{ik\cdot x}], \tag{1}$$

where ε_μ is the polarization vector. The key point is that the general solution can be obtained by a linear superposition of the plane wave solutions of the theory.

For a theory based on a non-Abelian group, like quantum chromodynamics (QCD), this can no longer be done in the strong coupling limit, due to the nonlinear nature of Yang-Mills equations. There are certain nonperturbative field configurations (for example, the hypothesized color electric flux tube that is thought to be important in the dual superconducting picture of confinement) in which the field distribution can not be explained as a cloud of quanta. One way of looking at this situation is that the fields of these nonperturbative configurations are split into ordered fields (the fields inside flux tube stretched between quark and antiquark) and disordered fields (the fields outside the flux tube). These fields can not be interpreted as a (perturbative) cloud of quanta. In such situations the fields play a primary role over the particles.

The need for nonperturbative techniques in strongly interacting, nonlinear quantum field theories is an old problem, and much effort has gone into trying to find an appropriate frame in which to carry on calculations for these theoreies. Despite this the problem is not yet fully resolved.

In [1] a possible approach was suggested based on a version of a quantization method originally due to Heisenberg. Starting with the classical SU(3) Yang-Mills equations

$$\partial_\nu \mathscr{F}^{B\mu\nu} = 0 \tag{2}$$

CP756, *Quark Confinement and the Hadron Spectrum VI*
edited by N. Brambilla, U. D'Alesio, A. Devoto, K. Maung, G.M. Prosperi and S. Serci
© 2005 American Institute of Physics 0-7354-0241-8/05/$22.50

($\mathscr{F}_{\mu\nu}^{B} = \partial_{\mu}\mathscr{A}_{\nu}^{B} - \partial_{\nu}\mathscr{A}_{\mu}^{B} + gf^{BCD}\mathscr{A}_{\mu}^{C}\mathscr{A}_{\nu}^{D}$ is the SU(3) field strength) one replaces the classical fields by field operators $\mathscr{A}_{\mu}^{B} \to \widehat{\mathscr{A}}_{\mu}^{B}$. This yields the following differential equations for the operators

$$\partial_{\nu}\widehat{\mathscr{F}}^{B\mu\nu} = 0. \tag{3}$$

These nonlinear equations for the field operators of the nonlinear quantum fields can be used to determine expectation values for the field operators $\widehat{\mathscr{A}}_{\mu}^{B}$. The simple gauge field expectation values, $\langle\mathscr{A}_{\mu}^{B}(x)\rangle$, are obtained by averaging Eq. (3) over some quantum state $|Q\rangle$

$$\left\langle Q\left|\partial_{\nu}\widehat{\mathscr{F}}^{B\mu\nu}\right|Q\right\rangle = 0. \tag{4}$$

One problem in using these equations to obtain expectation values like $\langle\mathscr{A}_{\mu}^{B}\rangle$, is that these equations involve not only powers or derivatives of $\langle\mathscr{A}_{\mu}^{B}\rangle$ (i.e. terms like $\partial_{\alpha}\langle\mathscr{A}_{\mu}^{B}\rangle$ or $\partial_{\alpha}\partial_{\beta}\langle\mathscr{A}_{\mu}^{B}\rangle$) but also contain terms like $\mathscr{G}_{\mu\nu}^{BC} = \langle\mathscr{A}_{\mu}^{B}\mathscr{A}_{\nu}^{C}\rangle$. Starting with Eq. (4) one can generate an operator differential equation for this product $\widehat{\mathscr{A}}_{\mu}^{B}\widehat{\mathscr{A}}_{\nu}^{C}$ allowing the determination of the Green's function $\mathscr{G}_{\mu\nu}^{BC}$

$$\left\langle Q\left|\widehat{\mathscr{A}}^{B}(x)\partial_{\gamma\nu}\widehat{\mathscr{F}}^{B\mu\nu}(y)\right|Q\right\rangle = 0. \tag{5}$$

However this equation will in turn contain other, higher order Green's functions. Repeating these steps leads to an infinite set of equations connecting Green's functions of ever increasing order. This construction does not have an exact, analytical solution and so must be handled using some approximation.

In order to do some calculations we give an approximate method which leads to the 2 and 4-points Green's functions only [1]. We will consider two cases: in the first one the fields are in completely disordered phase. Starting with the pure SU(3) Lagrangian

$$\widehat{\mathscr{L}}_{SU(3)} = \frac{1}{4}\widehat{\mathscr{F}}_{\mu\nu}^{B}\widehat{\mathscr{F}}^{B\mu\nu} \tag{6}$$

one can arrive at following effective, pure scalar Lagrangian (see [1] for details)

$$\frac{g^{2}}{4}\mathscr{L}_{eff} = -\frac{1}{2}\left(\partial_{\mu}\phi^{A}\right)^{2} + \frac{\lambda_{1}}{4}[\phi^{a}\phi^{a} - \phi_{0}^{a}\phi_{0}^{a}]^{2} + \frac{\lambda_{2}}{4}[\phi^{m}\phi^{m} - \phi_{0}^{m}\phi_{0}^{m}]^{2} + (\phi^{a}\phi^{a})(\phi^{m}\phi^{m}) \tag{7}$$

where the vector gauge fields have been replaced by effective scalar field, ϕ^{A}, through the ansatz

$$\mathscr{G}_{\alpha\beta}^{BC}(x,y) \approx -\eta_{\alpha\beta}f^{BDE}f^{CDF}\phi^{E}(x)\phi^{F}(y) \tag{8}$$

The a index in eq. (7) refers to the SU(2) subgroup of SU(3), and the m index refers to the coset SU(3)/SU(2), and two separate couplings (λ_{1} and λ_{2}) have been introduced between the SU(2) subgroup and the coset. A numerical investigation of eq. (7) yields a regular solution with finite energy. The profile of the energy density for this solution is given in the Fig. 1A. We interpret this solution as a glueball.

310

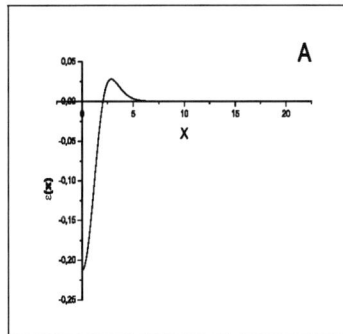 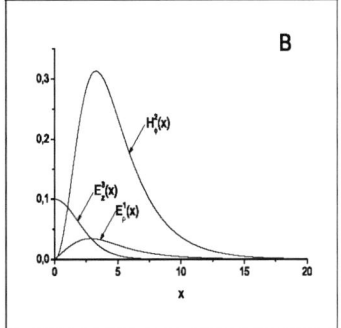

FIGURE 1. **A.** The energy density of glueball/soliton solution. **B.** The color electric $E_z^3(x)$, $E_\rho^1(x)$ and magnetic $H_\varphi^2(x)$ fields

A second example involves both an ordered (described by a vector gauge field) and disordered (described by an effective scalar field) phase. Again after some assumptions and simplifications [1] we have find regular solutions with finite energy density. The effective Lagrangian in this case has the form

$$\mathscr{L}_{eff} = -\frac{1}{4}h_{\mu\nu}^a h^{a\mu\nu} + \frac{1}{2}\left(D_\mu\phi^a\right)^2 - \frac{\lambda}{4}\left(\phi^a(x)\phi^a(x)\right)^2 + \frac{g^2}{2}a_\mu^b \phi^b a^{c\mu}\phi^c \qquad (9)$$

where $h_{\mu\nu}^b = \partial_\mu a_\nu^b - \partial_\nu a_\mu^b + \varepsilon^{bcd}a_\mu^c a_\nu^d$ is the SU(2) gauge field (ordered phase) and ϕ^a is a scalar field which describes 2 and 4-points Green's functions of SU(3)/SU(2) coset fields (disordered phase) via a relationship similar to eq. (8) *i.e.* $\mathscr{G}^{mn}(x,y) = \frac{1}{3}f^{mpb}f^{npc}\phi^b(x)\phi^c(y)$. A numerical investigation of the effective Lagrangian of eq. (9) shows that there are finite energy solutions whose profiles for the longitudinal color electric field and transversal electric and magnetic fields are given on Fig. 1B.

In conclusion by applying a Heisenberg-like quantization method to pure Yang-Mills theory we are able to construct effective Lagrangians that have only scalar fields or a mixture of scalar plus gauge fields. Both of these systems have finite energy solutions which are phenomenologically interesting. The completely disordered phase Lagrangian, containing only effective scalar fields, had finite energy solutions which could be interpreted as glueballs. The second case had a Lagrangian with both ordered (the SU(2) gauge field) and disordered (the effective scalar field) phases. This system had finite energy flux tube-like solutions.

I am very grateful to the Alexander von Humboldt Foundation for the financial support.

REFERENCES

1. V. Dzhunushaliev, D. Singleton and T. Nikulicheva, "Nonperturbative calcutional method in quantum field theory", hep-ph/0402205, to be published in "Focus on quantum field theory", Nova Science Publishers, Inc.

Chiral symmetry, scalar di-quark and pentaquark

H. Toki*, E. Hiyama†, M. Kamimura** and A. Hosaka*

*Research Center for Nuclear Physics (RCNP), Osaka University, Ibaraki, Osaka 567-0047, Japan
†Department of Physics, Nara Women's University, Nara 630-8506, Japan
**Department of Physics, Kyushu University, Fukuoka 812-8581, Japan

Abstract. We model the chiral symmetry feature in terms of a simplest possible non-relativistic quark model for the use of the full many body numerical method developed for nuclear few body problems. We present the numerical results on the properties of the flavor octet baryons, in particular the non-leptonic weak decay of baryons. We apply the model Hamiltonian to the structure of pentaquark system by solving the 5 body system. We find that the $1/2^+$ state appears clearly as the lowest state and have extremely a small width. We see that $1/2^-$ state does not have a clearly distinguished state and is buried in the continuum above the K^+ neutron threshold.

Keywords: Chiral symmetry, scalar diquark, pentaquark
PACS: 11.30.Rd, 12.40.Yx

INTRODUCTION AND NON-RELATIVISTIC QUARK MODEL

It was reported that a signature of pentaquark was observed by the LEPS group in the experiment performed at SPring8.[1] The GeV energy photons were bombarded on ^{12}C and the out-coming K^+ and K^- were detected in the forward direction. The event by event analysis provided a clear signature of a peak at the mass of E~ 2.4GeV and the width of $\Gamma \leq 20$MeV. This energy and the width were the features of the state predicted by the chiral quark soliton model as the lightest member of the flavor anti-decuplet state.[2] After the announcement of the pentaquark there have been many experiments performed at various laboratories with both positive and negative results.[3] At the present, in addition we do not know the spin and parity, although there are now many proposals to identify these properties of the state. There are also some signature with anti-charm quark. We are standing at the door of this fascinating physics of multi-quark particle, which we shall call the quark nuclear physics (QNP).

We model the feature of the chiral symmetry as the NJL model for the use of the numerical technique developed for Nuclear Physics.[4] We make a simplest possible non-relativistic quark model, which is written in the form of the Hamiltonian as,

$$H = \sum_i \frac{\mathbf{p}_i^2}{2m_i} - T_G + V_C + V_S + V_0,$$ (1)

where

$$V_C = \sum_{i<j} \frac{1}{2} K \left(\mathbf{x}_i - \mathbf{x}_j \right)^2,$$ (2)

CP756, Quark Confinement and the Hadron Spectrum VI
edited by N. Brambilla, U. D'Alesio, A. Devoto, K. Maung, G.M. Prosperi and S. Serci
© 2005 American Institute of Physics 0-7354-0241-8/05/$22.50

TABLE 1. *P*-wave non-leptonic weak transition amplitude (in 10^{-7} unit). In the second and forth columns, values in brackets show the results without V_s.

Decay	Pole	others	Total	Exp.
$\Sigma^+ \to p\pi^0$	23.23(13.45)	2.05	25.28(15.50)	26.6
$\Sigma^+ \to n\pi^+$	40.9(8.21)	0.00	40.9(8.21)	42.2
$\Lambda \to n\pi^0$	$-5.29(\;3.54)$	-5.02	$-10.31(-8.56)$	-15.8

$$V_S = \sum_{i<j} \frac{C_{SS}}{m_i m_j} \exp\left[-\left(\mathbf{x}_i - \mathbf{x}_j\right)^2 / \beta^2 \right],\qquad(3)$$

for spin=0 pair and zero for spin=1 pair. In the above Hamiltonian, m_i denotes the constituent quark mass and T_G is the center-of-mass energy. V_0 is a constant to normalize the nucleon mass. As for the quark-anti-quark interaction, we have the interaction in the pseudo-scalar channel to provide the pion mass and the kaon mass. The three parameters, K, C_{SS} and β are fixed from the delta-nucleon mass difference, the baryon sizes and β from the non-leptonic transitions. We have to improve the confinement potential together with the constant term, V_0, as we shall discuss the meson, baryon and pentaquark states, where the confinement will act differently in these different states. We solve the Schroedinger equation, $H\Psi = E\Psi$, using the Gaussian basis quark rearrangement (GBQR) model.[5]

We find good $SU(3)_f$ baryon mass spectrum.[5] The agreement with the baryon masses encourages us to proceed with our approach. The non-leptonic weak transition parity-conserving amplitudes are tabulated in Table I.

We show the pole contributions in the second column and the additional factorization and penguin contributions in the third column. We show the total decay amplitudes in the forth column to be compared with the experiments. It is worth noting that the $\Sigma^+ \to n\pi^+$ parity conserving decay process is completely dominated by the baryon pole diagrams without any factorization or penguin contributions. This fact indicates that the $\Sigma^+ \to n\pi^+$ *P*-wave amplitude is the most appropriate observable to probe the quark-

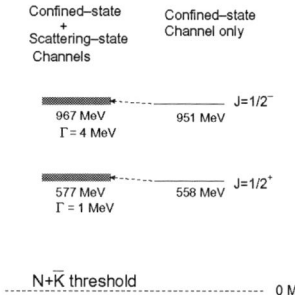

FIGURE 1. The energy level with the isospin, T=0, for the pentaquark state. The two spin states with $J^\pi = 1/2^+$ and $1/2^-$ are shown to have very narrow widths.

quark correlation in the baryons. We find a good agreement for $\Sigma \to N\pi$ decays, while the pole contribution of the $\Lambda \to N\pi$ is not enough. We have evaluated the non-leptonic weak transition amplitudes as well as the mass spectrum.

PENTAQUARK

We apply the non-relativistic quark model to the pentaquark system.[1] We confine ourselves to the case of the total isospin zero state, T=0, since the experiment seems to indicate the zero isospin. We have naturally two coordinate systems to describe the 5 quark system. Anticipating a strong pair correlation among the S=T=0 pair in the H diagram, we restrict the color wave function of the diquark to be $\bar{3}$.[6] The K diagram contains the scattering states of the baryon and meson system and the color of the 3 quark system and of the 2 quark system to have the singlet wave functions. The wave functions of the pentaquark system can be constructed in the manner similar to the 3 quark baryon system in the GEQR model, although it is far more complicated. We calculate the energy levels in the GEQR framework. We show the calculated results in Fig.1.

We calculate the 5 quark states. We find a $J^\pi = 1/2^+$ state much lower than a $J^\pi = 1/2^-$ state, although the $J^\pi = 1/2^+$ state comes out to be about 500MeV higher than the experimental value. There exists a strong pair correlation among the S=0 pair for the $1/2^+$ state, but the relative wave function between the diquarks has to be in the p-state due to the anti-symmetrization among the ud-quarks and the quark kinetic energy prevents the mass from coming down as the experiment demands. As for the $1/2^-$ state, one of the diquark pairs has mainly L=2 component and hence the energy comes out to be very large. The problem of having a large mass should be connected with the treatment of the confining mechanism. We point out that the simplest configuration with $(0s)^5$ quarks with the spin $1/2^-$ is buried in the scattering state and it should never show up as a distinct resonant state in the N-K spectrum. The absolute value for the resonance energy comes out to be about 500MeV higher than the experimental value in the present framework. We should model the confining potential differently between the three quark system and the pentaquark system. We should at the same time improve the treatment of the kinetic energy, since the momentum and the mass of the system are of the same order.

This is a talk presented by H. Toki in International Symposium on Quark confinement, held in Italy through Sep. 28-30, 2004.

REFERENCES

1. T. Nakano *et al.*, Phys. Rev. Lett. **91**, 012002 (2003).
2. D. Diakonov, V. Petrov and M. Polyakov, Z. Physil **A359** (1997) 305.
3. A. Hosaka, Proceedings of Pentaquark04, World Scientific (2004).
4. E. Hiyama, Y. Kino and M. Kamimura, Prog. in Nucl. Part. Phys. **51** (2003) 223.
5. E. Hiyama, K. Suzuki, H. Toki and M. Kamimura, Prog. Theor. Phys. **112** (2004) 99.
6. R. Jaffe and F. Wilczek, Phys. Rev. Lev. **91** (2003) 232003.

Search for pentaquarks in WA97 experiment at CERN

R. A. Fini[a] for the WA97 Collaboration: F. Antinori[h], W. Beusch[e],
I. J. Bloodworth[d], G. E. Bruno[a], R. Caliandro[a], N. Carrer[e], D. Di Bari[a],
S. Di Liberto[j], D. Elia[a], D. Evans[d], K. Fanebust[b], F. Fayazzadeh[g],
R. A. Fini[a], B. Ghidini[a], G. Grella[k], H. Helstrup[c], M. Henriquez[g],
A. K. Holme[g], A. Jacholkowski[e], G. T. Jones[d], J. B. Kinson[d],
K. Knudson[e], I. Králik[f], V. Lenti[a], R. Lietava[d], R. A. Loconsole[a],
G. Løvhøiden[g], V. Manzari[a], M. A. Mazzoni[j], F. Meddi[j], A. Michalon[l],
M. E. Michalon-Mentzer[l], M. Morando[h], P. I. Norman[d], B. Pastirčák[g],
E. Quercigh[e,h], D. Röhrich[b], G. Romano[k], K. Šafařík[e], L. Šándor[e,f],
G. Segato[h], P. Staroba[i], M. Thompson[d], J. Urbán[f], T. Vik[g], O. Villalobos
Baillie[d], T. Virgili[k], M. F. Votruba[d], and P. Závada[i]

[a] Dipartimento I.A. di Fisica dell'Università e del Politecnico di Bari and Sezione INFN, Bari, Italy
[b] Fysisk institutt, Universitetet i Bergen, Bergen, Norway
[c] Høgskolen i Bergen, Bergen, Norway
[d] School of Physics and Astronomy, University of Birmingham, Birmingham, UK
[e] CERN, European Laboratory for Particle Physics, Geneva, Switzerland
[f] Institute of Experimental Physics, Slovak Academy of Sciences, Košice, Slovakia
[g] Fysisk institutt, Universitetet i Oslo, Oslo, Norway
[h] Dipartimento di Fisica dell'Università and Sezione INFN, Padua, Italy
[i] Institute of Physics, Academy of Sciences of the Czech Republic, Prague, Czech Republic
[j] Dipartimento di Fisica dell'Università "La Sapienza" and Sezione INFN, Rome, Italy
[k] Dipartimento di Scienze Fisiche "E.R. Caianiello" dell'Università and INFN, Salerno, Italy
[l] Institut de Recherches Subatomiques, IN2P3/ULP, Strasbourg, France

Abstract. The WA97 experiment at CERN has collected data aiming to study strangeness production in heavy ion collisions at 158 A GeV/c; p-Be data were also taken as reference data. Here we show some results of an analysis of these p-Be data in order to check for evidence of a recently claimed pentaquark state.

Keywords: pentaquark state, exotic baryon resonance, strangeness
PACS: 25.75.Dw, 14.20.Gk

INTRODUCTION

The WA97 experiment is dedicated to study the enhanced production of strange and multistrange baryons and antibaryons (Λ, Ξ, Ω) in Pb-Pb collisions at 158 A GeV/c, as a signature of the formation of Quark Gluon Plasma (QGP) [1]. In order to match such goal the WA97 experimental apparatus was designed to identify and reconstruct hyperon decays with good resolution, in a high multiplicity environment. The main tracking device is a silicon telescope made of pixel and microstrip planes ($5cm \times 5cm$

CP756, *Quark Confinement and the Hadron Spectrum VI*
edited by N. Brambilla, U. D'Alesio, A. Devoto, K. Maung, G.M. Prosperi and S. Serci
© 2005 American Institute of Physics 0-7354-0241-8/05/$22.50

cross dimension, about 0.5 M channels), placed in 1.8 T uniform magnetic field [2].

In order to compare the heavy ion results with those from proton induced interactions, where the onset of QGP is not foreseen, the WA97 experiment has taken data in p-Be and p-Pb interactions at the same beam momentum [3].

The p-Be sample employed for this analysis consists of about 200 M events collected with at least two charged tracks in the tracking telescope. The sample of Ξ found here, is used to study the $\Xi\pi$ system, where the NA49 experiment has reported evidence for the existence of a narrow resonance with mass 1862 ± 2 MeV [4], which could be a pentaquark state [5].

DATA ANALYSIS AND SEARCH FOR PENTAQUARKS

The accepted kinematic window covers about one unit of rapidity at midrapidity and p_t from a few hundreds MeV/c, depending on the decay topology. The acceptance is the same for both negative and positive particles, hyperons and antihyperons. The data were taken with both orientations of the magnetic field. No charged particle identification being available, the charged tracks coming from the main vertex are assumed to be pions. Strange and multistrange baryons are identified through their weak decay, i.e. for Ξ:

$$\Xi^- \rightarrow \Lambda + \pi^-$$
$$\hookrightarrow p + \pi^- \quad (+c.c.)$$

where all the decay particles are reconstructed in the telescope and then are traced backwards to the main vertex, requiring that the Λ and Ξ decay vertices are well separated from each other and from the main vertex at the target. The Λ and Ξ signals in the respective mass spectra are very clean (background below 5%), with ≤ 15 MeV FWHM.

The total p-Be sample consists of about 800 clean Ξ's in the mass peak. The track multiplicity of the events with a reconstructed Ξ is shown in figure 1a, without counting the tracks of the Ξ itself. Assuming that the additional tracks are all due to pions, about 200 $\Xi\pi$ mass combinations are obtained; these are shown in figure 1b. An additional cut on the pion track impact parameter, leaves 111 $\Xi\pi$ mass combinations (figure 1c), without changing the shape of the background significantly. In the mass region where the new narrow resonance at 1862 MeV is expected, indicated by the arrow, there are only 3 events (two $\Xi^-\pi^+$, one $\overline{\Xi}^+\pi^-$, no $\Xi^-\pi^-$). They are compatible with what is expected for the background.

The acceptance for the new particle decaying with these topologies has been computed assuming that its production has the same p_t dependence as the one we found for Ξ (inverse slope of the exponential distribution T=200 MeV/c). The acceptance is found to be $\sim 2\%$ and $\sim 0.8\%$ that of the Ξ, for $(\Xi\pi)^0$ state and $(\Xi\pi)^{--}$ state, respectively.

From the measured Ξ yields, we get an upper limit for the production cross section of the new state: $\sigma=186$ μb for $(\Xi\pi)^0$ system, $\sigma=201$ μb for $(\Xi\pi)^{--}$ system. Note that this estimate depends on the assumed p_t distribution. As low p_t particles have lower acceptance in our apparatus, a different assumption on the p_t distribution will result in a change of the cross section estimates given above. In particular, recent theoretical

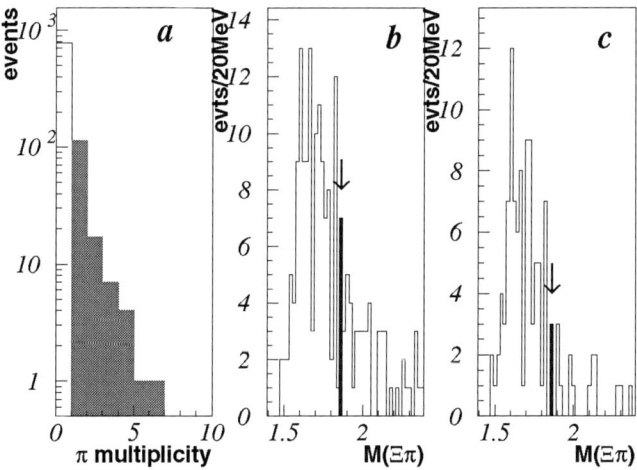

FIGURE 1. a) Multiplicity of charged particles in Ξ events. b) Ξπ invariant mass for all combinations and c) the combinations surviving the π impact parameter cuts. Masses are in GeV.

calculations [6] suggest a steeper p_t dependence: in such a case, the observation of Ξ^{--} would be well below the sensitivity of our experiment.

CONCLUSIONS

In conclusion, we have looked at the $\Xi\pi$ system in p-Be data at 158 GeV/c. We do not find evidence of a narrow resonance at 1862 MeV and we establish the upper limit at 95% CL for its production cross section in the central rapidity region, to be 150–200 μb.

REFERENCES

1. There are dedicated talks at this Conference that explain the role of strangeness as a signature of QGP; see, for example: J. Rafelski, these Proceedings.
2. For the description of the WA97 apparatus, see, for example: E. Andersen *et al.* (WA97 Collaboration), *Phys. Lett.*, **B449**, 401–406 (1999).
3. For a review of the main results of WA97 experiment, see, for example: R. A. Fini *et al.* (WA97 Collaboration), *J. Phys.*, **G27**, 375–381 (2001).
4. C. Alt *et al.* (NA49 Collaboration), *Phys. Rev. Lett.*, **92**, 042003/1–5 (2004).
5. See M. Karliner, these Proceedings.
6. F. M. Liu, H. Stocker, and K. Werner, hep-ph/0404156 (2004).

The jigsaw puzzle of scalar mesons

M. Boglione

Dipartimento di Fisica Teorica, Universita' di Torino, via P. Giuria 1, 10125 Italy

Abstract. This is a brief overview of light scalar meson spectroscopy, addressing longstanding problems, recent developments and future perspectives. In particular, a new comprehensive data analysis is introduced which will help to unravel the structure of the $f_0(980)$.

Keywords: Scalar mesons, hadronic amplitudes, comprehensive data analysis
PACS: 11.55.-m, 13.75.Lb, 14.40.Cs

We all know that the quark model works well for most mesons: nice nonet structures arise when all possible combinations of $q\bar{q}$ pairs are ordered according to their isospin and strangeness. Then, by exploiting the mass and decay properties of the physical mesons delivered by experiments, we can find a slot for each candidate in the nonet. This game can be safely played for vector and tensor mesons and, to some extent, for pseudoscalar mesons, if the appropriate mixing angles are taken into account. Consider, for instance, the $\omega(782)$ and the $\phi(1020)$: experiments tell us that the $\omega(782)$ decays mostly into pions and is lighter than the $\phi(1020)$ which, on the contrary, decays into $K\bar{K}$ 85% of the time. It's mass being close to that of the $\rho(770)$ provides clear indication that the $\omega(782)$ is the $I = 0$ non-strange candidate, whereas $\phi(1020)$ is undoubtedly the $I = 0$ $s\bar{s}$ member of the vector nonet. Similarly for the tensors f_2 and f_2'.

For scalar mesons this does not work. The quark model fails inexorably: first of all, experiments detect many more physical scalar resonances than can fit in a nonet. Secondly, their decay properties are mostly unknown, so there is little guide to their classification, thirdly their spectra cannot be approximated by Breit-Wigner shapes, because they overlap and interfere with each other, some of them being very broad. Therefore, the classical methods of analysing data cannot be applied.

How can we try and disentagle such a complicated picture? *Unitarity* comes to our rescue. Indeed, this property, which follows from conservation of probability, has to be fulfilled whatever the quantum numbers of the $q\bar{q}$ pair, and give very useful constraints for our analyses. Unitarity requires the T matrix for each partial wave to satisfy $\text{Im}T = \rho|T|^2$, where ρ is the phase space matrix. This relation constrains the imaginary part of $(1/T)$ to be $\text{Im}(1/T) = -\rho$, in the simplest case, leaving $\text{Re}(1/T)$ unconstrained. By parametrizing $\text{Re}(1/T)$ by a real matrix $1/K$, one obtains $T = \frac{K}{1-i\rho K}$, which is the usual K-matrix representation. If there is only one channel, like in $\pi\pi \to \pi\pi$ scattering below $K\bar{K}$ threshold, and only one narrow resonance, this resonance will appear like a single pole in the K amplitude, $K = \frac{g^2}{M^2-s}$, and the T amplitude can be approximated by $T = \frac{g^2}{M^2-s-i\rho g}$. The pole of K gives the "bare state" and T has a Breit-Wigner form. This simple picture works only for narrow and well separated resonances, where coupling to hadronic loops has little effect. For the scalar sector, where resonances

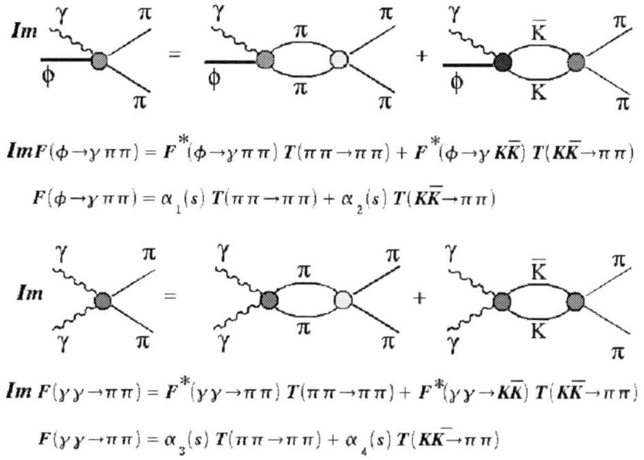

$$Im\,F(\phi \to \gamma\pi\pi) = F^*(\phi \to \gamma\pi\pi)\,T(\pi\pi \to \pi\pi) + F^*(\phi \to \gamma K\bar{K})\,T(K\bar{K} \to \pi\pi)$$

$$F(\phi \to \gamma\pi\pi) = \alpha_1(s)\,T(\pi\pi \to \pi\pi) + \alpha_2(s)\,T(K\bar{K} \to \pi\pi)$$

$$Im\,F(\gamma\gamma \to \pi\pi) = F^*(\gamma\gamma \to \pi\pi)\,T(\pi\pi \to \pi\pi) + F^*(\gamma\gamma \to K\bar{K})\,T(K\bar{K} \to \pi\pi)$$

$$F(\gamma\gamma \to \pi\pi) = \alpha_3(s)\,T(\pi\pi \to \pi\pi) + \alpha_4(s)\,T(K\bar{K} \to \pi\pi)$$

FIGURE 1. Coupled channel unitarity constrains the amplitudes $F(\phi \to \gamma\pi\pi)$ and $F(\gamma\gamma \to \pi\pi)$ in terms of hadronic amplitudes corresponding to final state interactions, $\pi\pi \to \pi\pi$ and $K\bar{K} \to \pi\pi$ for a given I, J. For ϕ-decay, the photon is assumed to be a spectator.

are broad (i.e. their poles are located very far from the real s-axis, where experiments happen), interfeering and overlapping (i.e. their spectra are not made of nicely separated peaks), this simple interpretation breaks down.

Fig. 1 shows how similarly coupled-channel unitarity constrains the partial wave amplitudes F corresponding to two different processes $\phi \to \gamma\pi\pi$ and $\gamma\gamma \to \pi\pi$; scalar meson resonances are produced in the final state interactions $\pi\pi \to \pi\pi$ and $K\bar{K} \to \pi\pi$ and are embodied as poles in the $I = J = 0$ hadronic amplitudes, T. The general solution of the unitarity requirement for the F's is given by a linear combination of the T's, where the coefficients $\alpha_i(s)$ are real functions of s, simple polynomials apart from some factors as explained in [1]. Notice that unitarity requires *consistency* between reactions, in that the same strong interaction amplitudes T, combined and weighted using appropriate α_i coefficients, form the amplitudes corresponding to different reactions. The α-vector formulation embodies *universality*, demanding that poles of the S matrix transmit to all processes with the same quantum numbers in exactly the same position. This indeed makes the determination of the F amplitudes very sensitive to the details of the T's.

Recently, M.R. Pennington and I made an analysis [1] of $\phi \to \gamma\pi\pi$ experimental data [2] based on the coupled channel unitarity constrains of Fig. 1 and showed that huge differences arise in the determination of the relevant couplings and the $\phi \to \gamma f_0(980)$ branching ratio due to different choices of underlying amplitudes T. We chose an old set of hadronic amplitudes called ReVAMP, determined as in [3] and a recent one, obtained by Anisovich and Sarantsev in [4] fitting a much larger amount of data. In the first set of amplitudes, the $f_0(980)$ appears as a narrow resonance, lighter than the $\phi(1020)$. In the second case the $f_0(980)$ is a much broader object, heavier than the $\phi(1020)$. Since the decay rate distribution depends crucially on the *cube* of the photon momentum, i.e. $(m_\phi^2 - s)^3$, and since the $f_0(980)$ is so close to the end of phase space, it turns out that the determination of the couplings and branching ratio is extremely sensitive to the exact position of the $f_0(980)$ pole in the T's. The fit clearly favours the ReVAMP set of amplitudes, which give an excellent quality of results with constant $\alpha_i(s)$ (3 parameter fit), confirming that the $\pi\pi$ final state interactions in this particular process are consistent with those of the processes exploited to determine the ReVAMP amplitudes. Indeed, when the new, high statistics, KLOE data will be released, we will have the chance to test this consistency further.

While for decays like $\phi \to \pi\pi X$ we have to assume X is a spectator to apply unitarity as in Fig.1, for $\gamma\gamma \to \pi\pi$ scattering unitarity and universality apply with no assumptions. A few years ago, M.R. Pennington and I analysed $\gamma\gamma \to \pi\pi$ world data [5] to determine the radiative widths of scalar mesons. The underlying hadronic amplitudes we used were the same ReVAMP set described above. We found two classes of solutions, delivered by fits equally good in quality and giving comparable scalar widths: one where the $f_0(980)$ showed up as a peak, and the other where the $f_0(980)$ showed up as a dip. Shortly, new very high statistics data from BELLE and BaBar will be available: they will allow us a global reanalysis to discern between the two solutions and to test the T underlying hadronic amplitudes.

For these re-analysis, we are considering a different parametrization for the T's. In fact, the simplest solution to the unitarity requirement, as shown above, violates left hand cut analyticity: each ρ matrix element is singular at $s \to \infty$, which constrains the T's in an artificial and unnecessary way. To avoid this, we perform new fits [6] that include recent experimental data in addition to those used for the original ReVAMP analysis, in which $\text{Im}(1/T)$ is given by the Chew-Mandelstam function, which is not affected by that flaw.

Concluding, the main message of this talk is the following: unitarity and analyticity give powerful constraints and must be at the very basis of any data analysis. Unitarity requires *consistency* among different reactions, so that analysing data where final state interactions are important only makes sense if it is done in a global and comprehensive way. It's like a big jigsaw puzzle game: you have to take care of combining appropriately all the single pieces before the total picture is revealed.

ACKNOWLEDGMENTS

It is a pleasure to thank Umberto D'Alesio for inviting me to this conference, and to all the organizers for their warmest hospitality. I am infinitely grateful to M.R. Pennington for many invaluable discussions on this subject and for years of fruitful and enjoyable collaboration.

REFERENCES

1. M. Boglione, and M.R. Pennington, *Eur. Phys. J.* **C30**, 503 (2003).
2. A. Aloisio *et al.*, *Phys. Lett.* **B537**, 21 (2002).
3. D. Morgan, and M.R. Pennington, *Phys. Rev.* **D48**, 1185 (1993).
4. V.V. Anisovich, and A.V. Sarantsev, *Eur. Phys. J.* **A16**, 229 (2003).
5. M. Boglione, and M.R. Pennington, *Eur. Phys. J.* **C9**, 11 (1999).
6. M. Boglione, and M.R. Pennington, in preparation.

Four-Quark Mesons

L. Maiani[*], F. Piccinini[†], A.D. Polosa[**] and V. Riquer[‡]

[*]*Università di Roma 'La Sapienza' and I.N.F.N., Roma, Italy*
[†]*I.N.F.N. Sezione di Pavia and Dipartimento di Fisica Nucleare e Teorica, via A. Bassi, 6, I-27100, Pavia, Italy*
[**]*Dipartimento di Fisica and I.N.F.N., Bari, Italy*
[‡]*CERN Theory Department, CH-1211, Switzerland*

Abstract. The features of a model interpreting the light scalar mesons as diquark-antidiquark bound states and the consequences of its natural extension to include heavy quarks are briefly reviewed.

Keywords: scalar mesons, diquarks, isospin symmetry
PACS: 12.40.Yx,12.39.-x, 14.40.Lb

The $q\bar{q}$ assignment has never really worked for the scalar mesons below 1 GeV. Alternative identifications have been proposed in the past [1], notably the f as a bound $K\bar{K}$ molecule [2] or as a $(q)^2(\bar{q})^2$ state [3]. We illustrate in this contribution the hypothesis, examined in Ref. [4], that the lowest lying scalar mesons are $S-$wave bound states of a diquark-antidiquark pair. Following Ref. [5], the diquark is more likely bound in the $\bar{3}_c$, 0_s (color antitriplet, spin zero) channel. If strange quarks are included, Fermi statistics favors the $\bar{3}_f$ combination. Therefore $(q)^2(\bar{q})^2$ states form a flavor $SU(3)$ nonet. We propose to put the $\sigma(450)$ [6] in the $I = S = 0$ state, and to assign to the $S = \pm 1$ states the $\kappa(800)$, a $K\pi$ resonance seen by several experiments, most recently in the $K\pi\pi$ spectrum from D decays [7]. A simple hypothesis on the way the $(q)^2(\bar{q})^2$ states may transform into a pair of pseudo-scalar mesons is found to give a rather good one-parameter description of the decays allowed by the OZI rule [8]. The extension of the picture to states including one ore more heavy quarks gives quite interesting predictions, accommodating recently discovered narrow states.

Quantum numbers and spectrum. We denote by $[q_1 q_2]$ the fully antisymmetric state of the two quarks q_1 and q_2. The composition of the members of the nonet is as follows:

$$a^+(I = 1, I_3 = +1, S = 0) = [su][\bar{s}\bar{d}]$$

$$a^0(I = 1, I_3 = 0, S = 0) = \frac{1}{\sqrt{2}}\left([su][\bar{s}\bar{u}] - [sd][\bar{s}\bar{d}]\right)$$

$$a^-(I = 1, I_3 = -1, S = 0) = [sd][\bar{s}\bar{u}]$$

$$f_\circ(I = 0, S = 0) = \frac{1}{\sqrt{2}}\left([su][\bar{s}\bar{u}] + [sd][\bar{s}\bar{d}]\right)$$

$$\sigma_\circ(I = 0, S = 0) = [ud][\bar{u}\bar{d}]$$

$$\kappa(I = 1/2, I_3 = +1/2, S = +1) = [ud][\bar{s}\bar{d}]$$

$$\kappa(I = 1/2, I_3 = -1/2, S = +1) = [ud][\bar{s}\bar{u}]$$

CP756, *Quark Confinement and the Hadron Spectrum VI*
edited by N. Brambilla, U. D'Alesio, A. Devoto, K. Maung, G.M. Prosperi and S. Serci
© 2005 American Institute of Physics 0-7354-0241-8/05/$22.50

$$\kappa(I = 1/2, I_3 = +1/2, S = -1) = [us][\bar{d}\bar{u}]$$
$$\kappa(I = 1/2, I_3 = -1/2, S = -1) = [ds][\bar{d}\bar{u}]$$

where the neutral states $f(980)$ and $\sigma(450)$ are superpositions of the isoscalar states $f_\circ$ and $\sigma_\circ$. The mixing angle results to be small because the OZI rule is respected in the physical mass spectrum.

In the limit of exact octet symmetry, the states given above are mass eigenstates, the mass matrix parameterized by α and β, the diquark masses squared with strange and non-strange content. In the most general case of octet symmetry breaking, two more parameters are required to account for symmetry breaking terms [4].

The mass spectrum obtained is inverted with respect to what one would get for a $q\bar{q}$ nonet: the isolated $I = 0$ state is the lightest one and strange particles come next. The same pattern is shown by data and this is a most evident indication in favor of the four-quark nature of the scalar nonet.

Strong decays. Diquarks, being colored objects, cannot be separated by their anti-particles. As soon as the distance between two diquarks in a four-quark state gets large enough, a $q - \bar{q}$ pair is created out of the vacuum and the state should dissociate into a baryon-antibaryon pair. This process is obviously kinematically forbidden as long as four-quark light scalars are considered.

An alternative decay mechanism is the switching of a quark-antiquark pair between the two diquarks to form a pair of color neutral $q\bar{q}$ states (pseudoscalar mesons), which can indefinitely separate from each other. In the exact $SU(3)$ limit there is only one amplitude, $\mathscr{A}$, to describe this process. The amplitude $\mathscr{A}$ describes the tunneling from the bound diquark pair configuration to the meson-meson pair, made by the unbound final state particles. As seen in Ref. [4], the value $\mathscr{A} = 2.6$ GeV gives a good description of the rates, compared to the available experimental information. The large value of $\mathscr{A}$ seems indicative of a short distance effect, making perhaps more justifiable the use of flavor $SU(3)$ symmetry.

Our picture has some connection with baryonium states [9] and with the $K\bar{K}$ molecule picture [2]. In the latter case, however, the analogy is only superficial.

Adding the other three $SU(3)$ allowed (annihilation-)couplings (neglecting a fourth coupling related to a pure singlet-to-singlets amplitude) improves the description of the OZI allowed channels, except for the κ width, which seems to be sensibly smaller than the observed one. Also the OZI forbidden decay $f \to \pi\pi$, turns out to be too small with respect to the experimental rate, even allowing for the full $SU(3)$ effective strong decay Lagrangian. It is quite possible that this mode proceeds via a different mechanism.

However the overall picture is encouraging and reinforces considerably the case of the scalar mesons as $(q)^2(\bar{q})^2$ states.

Open and hidden charm mesons. A natural extension of the present scheme is the existence of analogous states where one or more quarks are replaced by charm or beauty. We consider the case of charm, extension to beauty is obvious. Open charm scalar mesons of the form $S = [cq][\bar{q}\bar{q}]$, fall into characteristic $\mathbf{6} \oplus \bar{\mathbf{3}}$ multiplets of $SU(3)_f$. The $\bar{\mathbf{3}}$ has the same conserved quantum numbers of $c\bar{q}$ states ('cryptoexotic'), but the $\mathbf{6}$ has a pure exotic content. Hidden charm states of the form $[cq][\bar{c}\bar{q}]$ fall into $\mathbf{8} \oplus \mathbf{1}$ multiplets of $SU(3)$. In Ref. [4] a list of possible decays and related thresholds has been given.

Two issues are crucial to the description of open or hidden charm four-quark mesons: *isospin breaking* and *heavy-quark spin symmetry*. These aspects are briefly summarized in the next two paragraphs.

The mesons $a(980)$ and $f(980)$ are degenerate within about 10 MeV [10]. This reflects the smallness of the OZI violating contributions (annihilation graphs) to the mass matrix, which would align the mass eigenstates to pure $SU(3)$ representations. Thus sizeable deviations from the isospin basis are expected. Due to asymptotic freedom, suppressing quark pair annihilation into gluons, we expect annihilation contributions to be even smaller in systems containing heavy quarks. The mass eigenvalues will be aligned with states diagonal with respect to quark masses, even for the light, up and down, quarks [9]. The $D_{sJ}(2632)$ [11], if confirmed, could be interpreted as a $[cd][\bar{d}\bar{s}]$ state, not an isospin eigenstate [12], whose decay into $D^0 K^+$ is OZI forbidden [8].

The approximate spin-independence of heavy quark interactions, which is exact in the limit of infinite charm mass, implies both spin zero and spin one diquarks to form bound states. This implies a rich spectrum of states with $J = 0, 1, 2$. The states with $J^{PC} = 1^{++}$ and 2^{++} could be identified [13] with the $X(3872)$ and $X(3940)$ seen in BELLE data [14].

Also for these states isospin breaking would apply. An indication of the latter phenomenon comes from the observation of the relative decay rate of $X \to J/\psi + \rho$ and $X \to J/\psi + \omega$.

Heavy-light diquarks can be the building blocks of a rich spectrum of states which can accommodate some of the newly observed charmonium-like resonances not fitting a pure $c\bar{c}$ assignment. A new charm spectroscopy could be behind the corner.

FP wishes to thank G.M. Prosperi for his kind invitation.

REFERENCES

1. F.E. Close and N.A. Tornqvist, J. Phys. **G28** (2002) R249 and references therein.
2. J. Weinstein and N. Isgur, Phys. Rev. Lett. **48** (1982) 659.
3. R.L. Jaffe, Phys. Rev. **D15** (1977) 281; hep-ph/0001123.
4. L. Maiani, F. Piccinini, A.D. Polosa and V. Riquer, Phys. Rev. Lett. **93**, 212002 (2004).
5. R.L. Jaffe and F. Wilczek, Phys. Rev. Lett. **91** (2003) 232003.
6. KLOE Collaboration (A. Aloisio et al.), Phys. Lett. **B537** (2002) 21; E.M. Aitala et al., Phys. Rev. Lett. **86** (2001) 770.
7. E.M. Aitala et al., Phys. Rev. Lett. **89** (2002) 121801.
8. G. Zweig, CERN report S419/TH412 (1964), unpublished; S. Okubo, Phys. Lett, **5**(1963) 165; I. Iizuka, K. Okuda and O. Shito, Prog. Theor. Phys. **35** (1966) 1061.
9. G.C. Rossi and G. Veneziano, Nucl. Phys. **B123** (1977) 507; for a recent update see [arXiv:hep-th/0404262].
10. S. Eidelman et al., Phys. Lett. **B592** (2004) 1.
11. A.V. Evdokimov et al., [arXiv:hep-ex/0406045].
12. L. Maiani, F. Piccinini, A.D. Polosa and V. Riquer, Phys. Rev. **D70** (2004) 054009.
13. L. Maiani, F. Piccinini, A.D. Polosa and V. Riquer, [arXiv:hep-ph/0412098].
14. K. Abe et al., [arXiv:hep-ex/0408116].

On the nature of the lightest scalar resonances[1]

Z. X. Sun, L. Y. Xiao, Z. G. Xiao, H. Q. Zheng and Z. Y. Zhou

Department of Physics, Peking University, Beijing 100871, P. R. China

Abstract. We briefly review the recent progresses in the new unitarization approach being developed by us. Especially we discuss the large N_c $\pi\pi$ scatterings by making use of the partial wave S matrix parametrization form. We find that the σ pole may move to the negative real axis on the second sheet of the complex s plane, therefore it raises the interesting question that this 'σ' pole may be related to the σ in the linear σ model.

Keywords: Scalar meson, chiral symmetry, dispersion relations
PACS: 14.40.Cs, 13.85.Dz, 11.55.Bq, 11.30.Rd

The problem of how to restore unitarity and meanwhile respecting chiral perturbation amplitudes at low energies is very interesting and also difficult. A simple solution one has when dealing with such a difficult problem is the Padé approximation and its variations, which achieved some phenomenological success. Nevertheless, the Padé approximation encounters serious problems [1] which can hardly be resolved within the method itself. For this reason, it is worthwhile to make further efforts to study the problem from a more rigorous and different point of view.

In Refs. [2, 3, 4], a new parametrization form – which we call as the 'PKU' parametrization form – for partial wave S matrices in the elastic channel is developed, which, when combined with chiral symmetry, has been proven useful in probing the resonance structure of low energy strong interaction dynamics. For example, it reveals that the existence of the σ [5, 4] meson is fully consistent with chiral symmetry. Combining with crossing symmetry, it further predicts the σ pole mass and width to be $M_\sigma = 470 \pm 50 \text{MeV}$, $\Gamma_\sigma = 570 \pm 50 \text{MeV}$. [4] Also it is shown that there should exist the κ resonance if the πK scattering length in the I,J=1/2,0 channel does not deviate much from the value predicted by chiral perturbation theory. [3] The PKU parametrization form is the following,

$$S^{phy.} = \prod_i S_i^{poles} \cdot S^{cut} , \tag{1}$$

where S_i^{poles} denote various kinds of poles: resonance, bound state and virtual bound state. For resonance poles we have

$$S^R(s) = \frac{M^2(z_0) - s + i\rho(s)sG[z_0]}{M^2(z_0) - s - i\rho(s)sG[z_0]} , \tag{2}$$

---

[1] Talk presented by Zheng at "Quark Confinement and Hadron Spectroscopy VI", 21–25 Sept. 2004, Cagliari, Italy

where

$$M^2(z_0) = \mathrm{Re}[z_0] + \frac{\mathrm{Im}[z_0]\,\mathrm{Im}[z_0\,\rho(z_0)]}{\mathrm{Re}[z_0\,\rho(z_0)]} \; , \quad G[z_0] = \frac{\mathrm{Im}[z_0]}{\mathrm{Re}[z_0\,\rho(z_0)]} \; , \tag{3}$$

where z_0 denotes the resonance pole location on the complex s plane. The Eq. (2) is very interesting as it reveals the remarkable difference between a narrow resonance located far above the threshold and a light and broad resonance. In fact, $s = M^2(z_0)$ is the place where the resonance's contribution to the phase shift passes $\pi/2$. However, a light and broad resonance may correspond to a very large $M^2(z_0)$. The Eq. (2) for a light and broad pole actually nicely summarizes the major contribution to IJ=00 channel $\pi\pi$ scattering phase shift at low energies. The S^{cut} in Eq. (1) no longer contains any pole and for $\pi\pi$ scatterings it can be parameterized as:

$$S^{cut} = e^{2i\rho f(s)} \; , \quad f(s) = \frac{s}{\pi}\int_L \frac{\mathrm{Im}_L f(s')}{s'(s'-s)} + \frac{s}{\pi}\int_R \frac{\mathrm{Im}_R f(s')}{s'(s'-s)} \; . \tag{4}$$

where $L = (-\infty, 0]$ and R denotes physical cuts higher than the 2π cut, and $\mathrm{Im}_{L,R} f = -\frac{1}{2\rho}\log|S^{phy}|$. Before proceeding it should be emphasized that the above parametrization form is only obtainable by assuming analyticity on the whole cut plane, which can be derived from Mandelstam representation but nevertheless not proven rigorously from field theory. However the Lehman–Martin domain of analyticity is large enough for phenomenological applications. Therefore the parametrization form described above may afford a good approximation to the real situation.

At low energies one may approximate S^{phy} appeared in the dispersion integral by $S^{\chi PT}$ on L, to estimate the background contributions from the left after introducing a proper cutoff parameter to truncate the dispersion integral. One may then get more information from the parametrization form discussed above. Rewrite Eq. (1) as

$$\prod_i S_i^{poles} = S^{phy\cdot}(S^{cut})^{-1} \; , \tag{5}$$

as stated before the r.h.s. of the above equation can be expressed by low energy quantities appeared in, for example, the $O(p^4)$ low energy chiral Lagrangian. Expanding both sides of Eq. (5) at threshold, one relates the pole parameters to the low energy constants of the effective Lagrangian. Making use of the N_c counting rule of low energy constants [6] one can thus trace how the pole moves on the complex s plane when N_c varies. Nevertheless, in order to get the pole trajectory we need some further assumptions. Precisely we assume one pole dominates the l.h.s. of Eq. (5) for arbitrarily value of N_c. Such an assumption is of course only a speculation and may be subjected to criticism, though in the case of $N_c = 3$ one pole dominance at low energies is a good approximation. [4] Nevertheless we will proceed with this working assumption to see what happens. Here we keep all the N_c dependence including chiral logs. We make use of $O(p^4)$ χPT results to approximate $S^{phy\cdot}$ and to calculate $f(s)$, and neglect the right hand cut integral and truncated the left hand integral at certain value. In this way we get for $N_c = 3$ the pole mass of σ in rather good agreement with more realistic calculations. The N_c dependence of the pole mass can be traced numerically. The result is shown in Fig. 1

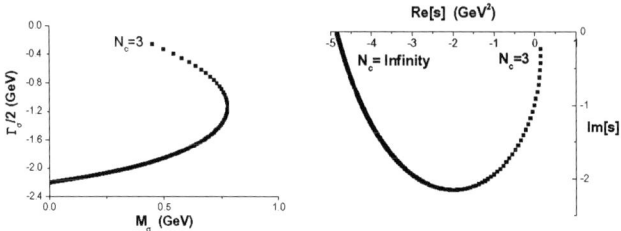

FIGURE 1. The trajectory of the σ pole: left) the complex $\sqrt{s}$ plane; right) the complex s plane.

and it is amazing to see that the pole moves to the negative real axis when N_c approaches infinity. If the pole location is plotted for large but finite value of N_c on the E plane, it will stay somewhere on the complex plane. The latter is the observation made in Ref. [7] However, the pole trajectory on the s plane clearly indicates what is the correct interpretation: the mass square becomes negative when the pole moves towards the real negative axis, rather than that the σ resonance has a large width of $O(1)$. Many phenomenological studies predicted that the σ pole moves towards left, though some finally touch the negative real axis, some do not. [8]. However it should be pointed out that the σ pole trajectory is very flexible. In our scheme, it is actually easy to tune the L_i parameters within 1σ error bar to convert the pole position to move towards the real positive s axis above the threshold. We have checked that such a flexibility also exists in the Padé amplitudes. Therefore our present scheme has a similar prediction on the pole trajectory comparing with the Padé amplitude. Therefore only one definite conclusion can be made by us: the σ pole moves to the real axis in the large N_c limit, and $M^2 \sim O(1)$, $G \sim O(1/N_c)$ (more details will appear soon). Finally we remark that a resonance with negative mass square, though seems odd, does not seem to spoil any fundamental principles because it locates on the second sheet. If the 'σ' discussed in this paper really gets a negative M^2, one wonders wether it has anything to do with the σ in the linear σ model.

Acknowledgment: We would like to thank Frieder Kleefeld and George Rupp for helpful discussions.

REFERENCES

1. Q. Ang et al, Commun. Theor. Phys. **36**(2001)563; G. Y. Qin eta al, Phys.Lett.B542:89-99,2002
2. H. Q. Zheng, hep-ph/0304173; J. Y. He, Z. G. Xiao and H. Q. Zheng, Phys. Lett. **B526**(2002)59 (Erratum: *ibid*. **B549**(2002)362).
3. H. Q. Zheng et al., Nucl. Phys. **A733**(2004)235.
4. Z. Y. Zhou et al, hep-ph/0406271 v2.
5. Z. G. Xiao and H. Q. Zheng, Nucl. Phys. **A695**(2001)273.
6. J. Gasser and H. Leutwyler, Nucl. Phys. B250(1985)465.
7. J. R. Pelaez, Phys. Rev. Lett. **92**(2004)1020001.
8. V. E. Markushin and M. P. Locher, in *Frascati 1999, Hadron spectroscopy* 229-236 (hep-ph/9906246); E. Van Beveren and G. Rupp, Eur. Phys. J. **C22** (2001) 493.

Schwinger-Dyson Equations and Dynamical gluon mass generation

A. C. Aguilar* and A. A. Natale[†]

*Departamento de Física Teórica, Universidad de Valencia–CSIC
E-46100, Burjassot, Valencia, Spain
[†]Instituto de Física Teórica, Universidade Estadual Paulista
Rua Pamplona 145, 01405-900, São Paulo, SP, Brazil

Abstract.
We discuss the solutions obtained for the gluon propagador in Landau gauge within two distinct approximations for the Schwinger-Dyson equations (SDE). The first, named Mandelstam's approximation, consist in neglecting all contributions that come from fermions and ghosts fields while in the second, the ghosts fields are taken into account leading to a coupled system of integral equations. In both cases we show that a dynamical mass for the gluon propagator can arise as a solution.

Keywords: Nonperturbative QCD; Gluon Schwinger-Dyson Equation; Infrared Gluon Propagator
PACS: 12.38-t, 11.15.Tk

The Schwinger-Dyson equations (SDE) are compound by an infinite set of the coupled equations that embody the full structure of the theory and provide in the continuum the appropriate framework to describe the behavior of the Green functions which are beyond of the scope of perturbative theory. In particular, the SDE can shed some light on the infrared (IR) properties of Quantum Chromodynamics (QCD) which can only be accessed through non-perturbative methods.

The structure of this tower of equations is such that it relates the n-point Green function to the (n+1)-point function which naturally must satisfy, in its turn, its own SDE and so on. Since we were not able to deal with this infinite system of equations, some approximations must be imposed in order to obtain a tractable system.

Many attempts have been made to understand the IR gluon propagator behavior through the SDE. In the late seventies Mandelstam initiated the study of the gluon SDE in Landau gauge [1]. Neglecting the ghost fields contribution and imposing cancellations of certain terms in the gluon polarization tensor, he found a highly singular gluon propagator in the infrared. However, these results are discarded by simulations of QCD on the lattice at 95% confidence level [2], where it is shown that the gluon propagator is probably infrared finite.

Infrared finite solutions are also found in the Schwinger-Dyson approach, as result of different procedures. A gluon propagator endowed with a dynamical mass was obtained by Cornwall utilizing the "pinch technique" [3] while solving the gluon-ghost coupled system an infrared vanishing gluon propagator behavior was found by the authors of Ref. [4].

Here we show that in the Mandelstam and in the gluon-ghost coupled system approximations a dynamical gluon mass arise as another possible solution if we adopt a three gluon vertex with a massless pole and follow a specific renormalization procedure [See

CP756, *Quark Confinement and the Hadron Spectrum VI*
edited by N. Brambilla, U. D'Alesio, A. Devoto, K. Maung, G.M. Prosperi and S. Serci
© 2005 American Institute of Physics 0-7354-0241-8/05/$22.50

[5]].

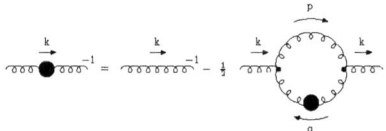

FIGURE 1. The Schwinger-Dyson equation in Mandelstam's approximation.

The full gluon and ghost propagators in Landau gauge are given by

$$D^{\mu\nu}(q^2) = \left(\delta^{\mu\nu} - \frac{q^\mu q^\nu}{q^2} \right) \frac{\mathscr{Z}(q^2)}{q^2} \quad \text{and} \quad D_G(k) = -\frac{\mathscr{F}(k^2)}{k^2}. \tag{1}$$

where $\mathscr{Z}(k^2)$ and $\mathscr{F}(k^2)$ are the dressing of the gluon and ghost propagators respectively. When we set $\mathscr{Z}(k^2) = \mathscr{F}(k^2) = 1$ we recover the perturbative expressions for the propagators at tree level. Therefore these functions measure the transition from the nonperturbative to the perturbative regimes as their values change with the scale.

The truncation scheme known as Mandelstam approximation consists in neglecting the ghosts fields and all terms with four-gluon interactions, and is pictorially represented in Fig.(1). The infinite contribution to the gluon SDE in this approximation is shown in the left-hand side of Eq.(2) together with the gluon propagator renormalization constant Z_3. On the right-hand side we show the behavior of the finite contribution $(1 + \frac{\kappa}{k^2})$ after the infinite contribution is subtracted with the help of the constant Z_3

$$Z_3 - \frac{\alpha(\mu^2)}{4\pi} \int_{k^2}^{\Lambda^2} dq^2\, 7q^2 D(q^2) = 1 + \frac{\kappa}{k^2}. \tag{2}$$

The details of this procedure are explained in the Ref.[5], and it allows for massive solutions of the SDE. We obtain the curves for the gluon propagator, $D(q^2)$, as a function of the momentum q^2 which are plotted in the left panel of Fig.(2). Theses curves were produced utilizing different values to the renormalization point μ^2. The external curves delimit the lower and the higher values of the Λ_{QCD} in the range $[182, 557]$ MeV which is fixed by the choice of μ^2. These numerical solutions for the gluon propagator can be fitted by the following equations

$$D(q^2) = \frac{1}{q^2 + \mathscr{M}^2(q^2)}, \quad \text{where} \quad \mathscr{M}^2(q^2) = \frac{m_0^4}{q^2 + m_0^2}. \tag{3}$$

where $\mathscr{M}^2(q^2)$ is the gluon dynamical mass.

Analising the ratio m_0/Λ_{QCD} which give to us a better idea about dependency on the renormalization point of our solution, we have found an average value $m_0/\Lambda_{QCD} \approx 3$ that is consistent with the previous estimates for this mass [3].

On the other hand, if we take into account the ghost fields, following the same steps of Ref.[4] and applying the renormalization procedure explained in Ref.[5], we obtain, for the gluon fields, the same qualitative behavior shown in the left panel of Fig.(2), however the average value for the ratio m_0/Λ_{QCD} decreases to 2.5.

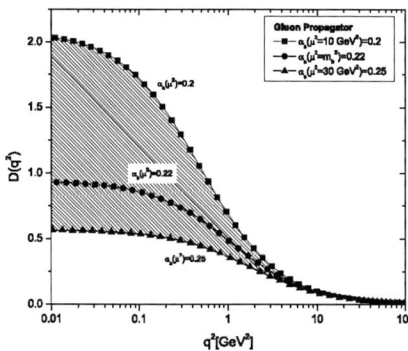 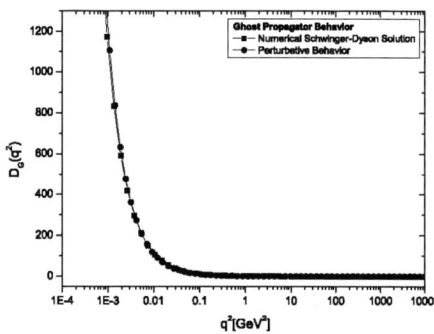

FIGURE 2. Left panel: Gluon propagator, $D(q^2)$, as function of momentum q^2 for different scales. The central curve (line + circle) was obtained when we fix the renormalization point, μ^2, at bottom quark mass, $m_b^2 = (4.5)^2 \, \text{GeV}^2$ with the central value of $\alpha(m_b^2) = 0.22$. Right panel: The behavior of the ghost propagator, $D_G(q^2)$, obtained through the numerical solution of the Schwinger-Dyson equation, when $\alpha_s(\mu^2) = 0.22$ at $\mu^2 = (m_b)^2 = (4.5)^2 \, \text{GeV}^2$, together with its ultraviolet behavior.

Within this approximation, we can see that the ghost propagator develops practically the same perturbative behavior for all spectrum of momenta, as is shown in the right panel of Fig.(2).

CONCLUSIONS

We computed the SDE for the gluon propagator in the Landau gauge within two different approximations where the fermions and quartic vertices are neglected. The full triple gluon vertex is also extended to include the possibility of dynamical mass generation. We show, in both cases, that following a specific renormalization procedure, the gluon propagator develops a dynamical mass whose ratio with the Λ_{QCD} scale has an average value around two or three depending on the approximation utilized, such values are consistent with the previous estimates [3].

ACKNOWLEDGMENTS

This research was supported by the Conselho Nacional de Desenvolvimento Científico e Tecnológico (CNPq) (AAN), Fundação de Amparo à Pesquisa do Estado de São Paulo (FAPESP) and Coordenação de Aperfeiçoamento de Pessoal de Nível Superior (CAPES) (ACA).

REFERENCES

1. S. Mandelstam, Phys. Rev. D20 (1979) 3223.
2. P. Marenzoni, G. Martinelli, N. Stella, e M. Testa, Phys. Lett. B318 (1993) 511; C. Alexandrou, Ph. de Forcrand and E. Follana, Phys. Rev. D65 (2002) 114508; D65 (2002) 117502.
3. J. M. Cornwall, Phys. Rev. D26 (1982) 1453.
4. R. Alkofer and L. von Smekal, Phys. Rept. 353 (2001) 281; L. von Smekal, A. Hauck and R. Alkofer, Ann. Phys. 267 (1998) 1; L. vonSmekal, A. Hauck and R. Alkofer, Phys. Rev. Lett. 79 (1997) 3591.
5. A. C. Aguilar and A. A. Natale, hep-ph/0405024; JHEP 0408 (2004) 57.

Recent Developments in the Bethe-Salpeter Description of Light Mesons

Peter Watson

Institute for Theoretical Physics, University of Giessen, Heinrich-Buff-Ring 16, D-35392 Giessen, Germany

Abstract. Results for the light meson mass spectrum from a Bethe-Salpeter approach are presented. The results obtained in the standard framework are Poincaré covariant and compare favourably with lattice results. Using a more sophisticated scheme, the pseudoscalar, vector and 1^{++} (a_1/f_1) axialvector charge eigenstate masses are unaltered whereas the 1^{+-} (b_1/h_1) axialvector meson mass is raised.

The Bethe-Salpeter equation [BSE] is the fully relativistic description of the two-body bound state problem. It has been found in the last decade that the ladder truncation of the BSE, using as input the quark propagators derived from the rainbow truncation of the Schwinger-Dyson equation [DSE], gives rise to a good description of the light flavor non-singlet pseudoscalar and vector mesons [1]. The underlying mechanism for this is chiral symmetry manifested through the flavor non-singlet axialvector Ward-Takahashi identity [AXWTI]. By ensuring that the kernels of both equations respect the AXWTI, it is shown that the pion emerges as both a bound state of massive constituents and as an almost massless Goldstone boson of the broken chiral symmetry [2].

The ladder truncation of the homogeneous BSE for quark-antiquark mesons is written (working in Euclidean space with Hermitian Dirac matrices obeying $\{\gamma_\mu, \gamma_\nu\} = 2\delta_{\mu\nu}$):

$$\Gamma(p;P) = -\frac{4}{3} \int \frac{d^4k}{(2\pi)^4} g^2 \Delta_{\mu\nu}(p-k)\gamma_\mu S(k_+)\Gamma(k;P)S(k_-)\gamma_\nu \quad , \qquad (1)$$

where Γ is the Bethe-Salpeter amplitude, $k_+ = k + \xi P$ and $k_- = k + (\xi - 1)P$ with $\xi = [0,1]$ the momentum sharing parameter between the two quarks. Invariance of the resulting observables with respect to ξ is a reflection of Poincaré covariance. The total momentum $P = p_+ - p_-$ is such that the equation is solved for $P^2 = -M^2$ where M is the mass of the meson. In Eq. (1), the dressed quark propagators are the solution of the rainbow quark DSE

$$S^{-1}(p) = \iota\not{p} + m + \frac{4}{3} \int \frac{d^4k}{(2\pi)^4} g^2 \Delta_{\mu\nu}(p-k)\gamma_\mu S(k)\gamma_\nu \quad , \qquad (2)$$

where m is the current mass parameter of the quark; the two truncations being consistent with the AXWTI. The effective interaction $g^2\Delta_{\mu\nu}(q)$ has the following form [3]:

$$g^2\Delta_{\mu\nu}(q) = t_{\mu\nu}(q)4\pi^2 D\frac{q^2}{\omega^2}\exp\left(-\frac{q^2}{\omega^2}\right) \qquad (3)$$

CP756, *Quark Confinement and the Hadron Spectrum VI*
edited by N. Brambilla, U. D'Alesio, A. Devoto, K. Maung, G.M. Prosperi and S. Serci
© 2005 American Institute of Physics 0-7354-0241-8/05/$22.50

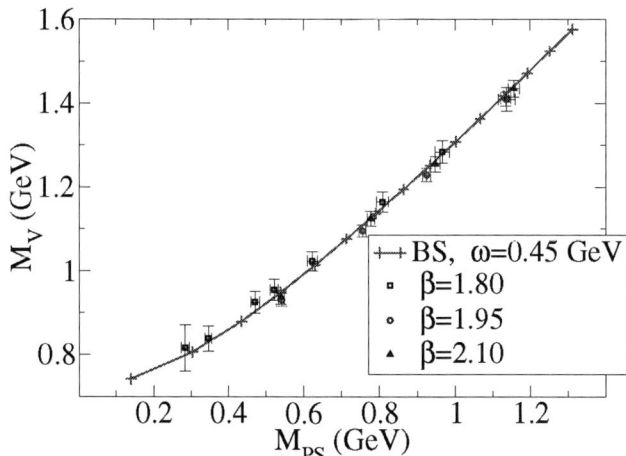

FIGURE 1. Comparison of BSE and lattice results for pseudoscalar and vector meson masses. The lattice data are from CP-PACS (unquenched) [4].

($t_{\mu\nu}$ is the transverse projector). The two parameters ω and D set the length scale and magnitude of the interaction. The integrals are UV convergent and all renormalisation constants are unity. The BSE is solved by writing the amplitude Γ in its most general form consistent with the desired parity and charge conjugation properties and expanding the resulting scalar functions as a Chebyshev series in the angular variable. Numerical results are shown in Table 1 using the parameters $\omega = 0.5 GeV$, $D = 16 GeV^{-2}$, $m_u = m_d = 5 MeV$, $m_s = 115 MeV$. One can see that the pseudoscalar and vector meson results are in good agreement, but the axialvector mesons are $\sim 300 MeV$ too light. Varying the momentum sharing parameter (ξ) one finds that for sufficient numbers of Chebyshev moments ($N \geq 6$), the mass results are stable, demonstrating the Poincaré covariance [3].

TABLE 1. Mass results (in GeV) for the light meson masses using the rainbow/ladder truncation.

J^{PC}	M_{BS}			M_{exp}		
	$\bar{u}u$	$\bar{u}s$	$\bar{s}s$	$\bar{u}u$	$\bar{u}s$	$\bar{s}s$
0^{-+}	0.137	0.492	–	0.135	0.498	–
1^{--}	0.758	0.946	1.078	0.770	0.892	1.020
1^{+-}	0.915	1.075	1.233	1.230	1.270	1.170?
1^{++}	0.936	1.075	1.291	1.230	1.270	1.282

It is interesting to compare the BSE results for the pseudoscalar and vector masses with lattice results at different quark masses. Figure 1 shows a comparison of M_V vs. M_{PS} using unquenched CP-PACS data [4] and the BSE results with $\omega = 0.45 GeV$, $D = 24.4 GeV^{-2}$. There is good agreement between the lattice and the BSE results over a wide range of M_{PS}.

To improve the description of the axialvector mesons one must augment the truncation scheme to include nontrivial vertex corrections whilst maintaining the AXWTI. As an initial attempt we consider an abelian, one-loop correction to the quark-gluon vertex with the following form [5]:

$$\Gamma_\mu(k,p) = \gamma_\mu + \frac{1}{6} \int \frac{d^4q}{(2\pi)^4} \gamma_\rho S(k+q)\gamma_\mu S(p+q)\gamma_\lambda g^2\overline{\Delta}_{\rho\lambda}(q) \quad . \tag{4}$$

To make the system tractable, the interaction in the vertex correction is taken as $g^2\overline{\Delta}_{\rho\lambda}(q) = (2\pi)^4 G\delta^4(q)t_{\rho\lambda}(q)$ which reduces the integral to an algebraic expression. The parameter G is tuned such that the integrated strengths of Δ and $\overline{\Delta}$ are equal. At spacelike momenta the quark propagator is only slightly modified from the rainbow case but this is not true at general complex momenta due to the different analytic structures introduced by the δ-function. As before, one can construct an appropriate kernel for the BSE which preserves the AXWTI and the charge conjugation symmetry. The results show that the charge eigenstate pseudoscalar, vector and 1^{++} (a_1/f_1) axialvector meson masses are largely unaltered but the 1^{+-} (b_1/h_1) axialvector mass increases $\sim 300MeV$, comparing reasonably with the experimental observation [5].

To summarize, the BSE – when using input from the quark DSE whilst respecting the AXWTI – provides a powerful framework to study the meson mass spectrum. There are two main areas with which to proceed: the first being the inclusion of more sophisticated truncations of the kernels; an exploratory attempt being described here. The second area to investigate is to include a mechanism for dynamical meson decay and multiquark states. Such an effort has been started in ref. [6].

It is a pleasure to thank N. Brambilla and the organizing committee for an enjoyable and productive conference. This work was supported by FZ Jülich.

REFERENCES

1. P. Jain and H. J. Munczek, Phys. Rev. D **48**, 5403 (1993) [arXiv:hep-ph/9307221]. P. Maris and C. D. Roberts, Int. J. Mod. Phys. E **12**, 297 (2003) [arXiv:nucl-th/0301049].
2. P. Maris, C. D. Roberts and P. C. Tandy, Phys. Lett. B **420**, 267 (1998) [arXiv:nucl-th/9707003].
3. R. Alkofer, P. Watson and H. Weigel, Phys. Rev. D **65**, 094026 (2002) [arXiv:hep-ph/0202053].
4. A. Ali Khan et al. [CP-PACS Collaboration], Phys. Rev. D **65**, 054505 (2002) [Erratum-ibid. D **67**, 059901 (2003)] [arXiv:hep-lat/0105015]. Y. Namekawa et al. [CP-PACS Collaboration], arXiv:hep-lat/0404014.
5. P. Watson, W. Cassing and P. C. Tandy, Few-Body Systems, in print, [arXiv:hep-ph/0406340].
6. P. Watson and W. Cassing, Few-Body Systems, in print, [arXiv:hep-ph/0405287].

Mesonic states in the generalised Nambu-Jona-Lasinio theories

A. V. Nefediev* and J. E. F. T. Ribeiro†

*Institute of Theoretical and Experimental Physics, 117218, B.Cheremushkinskaya 25, Moscow,
Russia
†Centro de Física das Interacções Fundamentais (CFIF), Departamento de Física, Instituto
Superior Técnico, Av. Rovisco Pais, P-1049-001 Lisboa, Portugal

Abstract. For any Nambu-Jona-Lasinio model of QCD with arbitrary nonlocal, instantaneous, quark current-current confining kernels, we use a generalised Bogoliubov technique to go beyond BCS level (in the large-N_C limit) so as to explicitly build quark-antiquark compound operators for creating/annihilating mesons. In the Hamiltonian approach, the mesonic bound-state equations appear (from the generalised Bogoliubov transformation) as mass-gap-like equations which, in turn, ensure the absence, in the Hamiltonian, of mesonic Bogoliubov anomalous terms. We go further to demonstrate the one-to-one correspondence between Hamiltonian and Bethe-Salpeter approaches to non-local NJL-type models for QCD and give the corresponding "dictionary" necessary to "translate" the amplitudes built using the graphical Feynman rules to the terms of the Hamiltonian, and vice versa. We comment on the problem of multiple vacua existence in such type of models and argue that mesonic states in the theory should be prescribed to have an extra index — the index of the replica in which they are created. Then the completely diagonalised Hamiltonian should contain a sum over this new index. The method is proved to be general and valid for any instantaneous quark kernel.

We study generalised Nambu-Jona-Lasinio models which are expected to mimic the most important low-energy properties of QCD [1, 2]. The Hamiltonian reads:

$$\hat{H} = \int d^3x \bar{\psi}(\vec{x},t)\left(-i\vec{\gamma}\cdot\vec{\bigtriangledown}+m\right)\psi(\vec{x},t) + \frac{1}{2}\int d^3xd^3y\, J_\mu^a(\vec{x},t)K_{\mu\nu}^{ab}(\vec{x}-\vec{y})J_\nu^b(\vec{y},t), \quad (1)$$

and contains the interaction of the quark currents $J_\mu^a(\vec{x},t) = \bar{\psi}(\vec{x},t)\gamma_\mu\frac{\lambda^a}{2}\psi(\vec{x},t)$, parameterised through the instantaneous quark kernel, $K_{\mu\nu}^{ab}(\vec{x}-\vec{y}) = g_{\mu0}g_{\nu0}\delta^{ab}V_0(|\vec{x}-\vec{y}|)$, with a power-like confining potential $V_0(|\vec{x}|) = K_0^{\alpha+1}|\vec{x}|^\alpha$. The standard approach to the theories (1) is the Bogoliubov-Valatin transformation defined by the chiral angle φ_p [2],

$$\begin{cases} u(\vec{p}) &= \frac{1}{\sqrt{2}}\left[\sqrt{1+\sin\varphi_p}+\sqrt{1-\sin\varphi_p}\,(\vec{\alpha}\hat{\vec{p}})\right]u(0), \\ v(-\vec{p}) &= \frac{1}{\sqrt{2}}\left[\sqrt{1+\sin\varphi_p}-\sqrt{1-\sin\varphi_p}\,(\vec{\alpha}\hat{\vec{p}})\right]v(0). \end{cases} \quad (2)$$

Anomalous terms in the Hamiltonian vanish if φ_p obeys the mass-gap equation,

$$m\cos\varphi_p - p\sin\varphi_p = \frac{C_F}{2}\int\frac{d^3k}{(2\pi)^3}V_0(\vec{p}-\vec{k})\left[\sin\varphi_k\cos\varphi_p-(\hat{\vec{p}}\hat{\vec{k}})\cos\varphi_k\sin\varphi_p\right]. \quad (3)$$

CP756, *Quark Confinement and the Hadron Spectrum VI*
edited by N. Brambilla, U. D'Alesio, A. Devoto, K. Maung, G.M. Prosperi and S. Serci
© 2005 American Institute of Physics 0-7354-0241-8/05/$22.50

Eq. (3) is subject to numerical studies. As soon as a nontrivial solution $\varphi_0(p)$ to the mass-gap equation is built, it defines the vacuum of the theory with spontaneously broken chiral symmetry, which is energetically preferable as compared to the unbroken phase. The terms in (1) quartic in quark operators contain the suppressing factor of $1/\sqrt{N_C}$ and thus the Hamiltonian of the theory is diagonalised in the quark sector (BCS level).

To proceed beyond BCS level and reformulate the theory in terms of colourless mesonic states, we notice that only operators creating/annihilating $q\bar{q}$ pairs are allowed:

$$\hat{M}_{ss'}(\vec{p},\vec{p}) = \frac{1}{\sqrt{N_C}}\hat{d}_{\alpha s}(-\vec{p})\hat{b}_{\alpha s'}(\vec{p}) = \sum_{v}[\kappa_v(\hat{\vec{p}})]_{ss'}\sum_{n}[\hat{m}_{nv}\varphi_{nv}^+(p) + \hat{m}_{nv}^\dagger\varphi_{nv}^-(p)], \quad (4)$$

all other operators, like $\hat{b}^\dagger\hat{b}$ and $\hat{d}^\dagger\hat{d}$, being suppressed by N_C [3, 4]. In (4) we consider the $q\bar{q}$ pair at rest and separate the spin-angular and the radial parts of the operator $\hat{M}$. The complete set $\{\kappa_v\}$ is chosen to be the J^{PC} set of states, which are known to diagonalise the Hamiltonian of strong interactions. We also perform a second, generalised, Bogoliubov-like transformation and introduced the mesonic creation/annihilation operators $\hat{m}_{nv}^\dagger/\hat{m}_{nv}$. The Bogoliubov amplitudes $\varphi_{nv}^\pm$ obey the normalisation condition which follows immediately from the commutation relation for the mesonic operators:

$$[\hat{m}_{nv},\hat{m}_{mv}^\dagger] = \int \frac{p^2dp}{(2\pi)^3}\left[\varphi_{nv}^+(p)\varphi_{mv}^+(p) - \varphi_{nv}^-(p)\varphi_{mv}^-(p)\right] = \delta_{nm}. \quad (5)$$

Meanwhile, $\varphi_{nv}^\pm$ also play the role of the two components of the mesonic wave function responsible for the forward and backward in time motion of the $q\bar{q}$ pair in the meson. They are required to be solutions of an eigenvalue problem — the bound-state equation; M_{nv} being the mass of the corresponding meson:

$$\begin{cases} [2E_p - M_{nv}]\varphi_{nv}^+(p) = \int \frac{q^2dq}{(2\pi)^3}[T_v^{++}(p,q)\varphi_{nv}^+(q) + T_v^{+-}(p,q)\varphi_{nv}^-(q)] \\ [2E_p + M_{nv}]\varphi_{nv}^-(p) = \int \frac{q^2dq}{(2\pi)^3}[T_v^{-+}(p,q)\varphi_{nv}^+(q) + T_v^{--}(p,q)\varphi_{nv}^-(q)]. \end{cases} \quad (6)$$

Eq. (6) can be alternatively derived using the Bethe–Salpeter approach to mesonic states [2], and thus a close connection between the Bethe–Salpeter and Hamiltonian approaches to the theory can be established, including graphical rules, which allow one to build the T-amplitudes in Eq. (6) using Feynman-like diagrams [4]. The Hamiltonian (1) takes now a diagonal form in terms of mesonic operators,

$$\hat{\mathcal{H}} = \sum_{n,v}M_{nv}m_{nv}^\dagger m_{nv}, \quad (7)$$

with corrections to the leading regime (7) suppressed by N_C. The first correction, of order $1/\sqrt{N_C}$, is responsible for mesonic decays [3]. Notice that, as soon as the mass-gap equation is solved, no new information, at least in the leading order in N_C, is needed to proceed beyond the BCS level and to introduce mesonic states. Numerical analysis of the mass-gap equation (3) for various confining potentials $V(r)$ demonstrated existence of extra, "excited", solutions — the vacuum replicas [2, 5]. The existence of an infinite tower of such replicas for power-like confining potentials $V(r) \propto r^\alpha$ was proved analytically and verified numerically for all α's from the allowed region $0 \le \alpha \le 2$, as well as for D=2 and D=4 (D is the dimensionality of the space-time) [6]. It was argued in [6, 4] that the appearance of such replicas is a consequence of a very peculiar behaviour of the dressed quark dispersive law E_p in the infrared region and, therefore, is closely related to chiral symmetry breaking. A similar conclusion was also made in a different approach in [7]. It was demonstrated in [4] that, with the proper definition of the chiral angle, one encounters no problem with the imaginary mass of the pion in the replica vacua — all mesons build in replicas being normal hadronic states, but with a heavier mass as compared to mesons created in the ground-state vacuum. It is quite natural then to require that the information about vacuum replicas is transfered beyond the BCS level, the full Hamiltonian should contain the sum over the index $\mathcal{N}$ numerating the replicas,

$$\hat{\mathscr{H}} = \sum_{n,\nu,\mathcal{N}} M_{n\nu\mathcal{N}} m_{n\nu\mathcal{N}}^\dagger m_{n\nu\mathcal{N}}. \tag{8}$$

In conclusion let us notice that the analysis of the generalised Nambu-Jona-Lasinio theories performed above is insensitive to the dimensionality of the space-time, the Lorentz nature of confinement, and its explicit form. We argue that the existence of replicas is a rule rather than an exception for any confining quark kernel and we believe it is not an artifact of the instantaneous interquark interaction used in our analysis in order to bypass the problem of the relative time. An approach to replicas, as to local excitations, was suggested in [8] and an effective diagrammatic technique was derived in order to take into account the effect of replicas in hadronic reactions. Another important ingredient of the theory of vacuum replicas is a mechanism of excitation of replicas as global objects. This work is in progress now and will be reported elsewhere.

One of the authors (A.V.N.) would like to acknowledge discussions with P. Bicudo and Yu. S. Kalashnikova as well as the financial support of the grant NS-1774.2003.2, as well as of the Federal Programme of the Russian Ministry of Industry, Science, and Technology No 40.052.1.1.1112.

REFERENCES

1. Nambu Y., Jona-Lasinio G., *Phys. Rev.*, **122**, 345 (1961).
2. Amer A., Le Yaouanc A., Oliver L., Pene O., and Raynal J.-C., *Phys. Rev. Lett.*, **50**, 87 (1983); Le Yaouanc A., Oliver L., Pene O., and Raynal J.-C., *Phys. Lett.*, **134B**, 249 (1984); *Phys. Rev.*, **D29**, 1233 (1984); Bicudo P. and Ribeiro J. E., *Phys. Rev.*, **D42**, 1611 (1990); 1625 (1990); 1635 (1990); Bicudo P., *Phys. Rev. Lett.*, **72**, 1600 (1994); Bicudo P., *Phys. Rev.*, **C60**, 035209 (1999).
3. Kalashnikova Yu. S. and Nefediev A. V., *Phys. Atom. Nucl.*, **62**, 323 (1999) *Phys. Usp.*, **45**, 347 (2002); Kalashnikova Yu. S., Nefediev A. V., and Volodin A. V., *Phys. Atom. Nucl.*, **63**, 1623 (2000).
4. Nefediev A. V. and Ribeiro J. E. F. T., *Phys. Rev. D*, in press.
5. Bicudo P. J. A., Nefediev A. V., and Ribeiro J. E. F. T., *Phys. Rev.*, **D65**, 085026 (2002).
6. Bicudo P. J. A. and Nefediev A. V., *Phys. Rev.*, **D68**, 065021 (2003); *Phys. Lett.* **573B**, 131 (2003).
7. Osipov A. A. and Hiller B., *Phys. Lett.*, **539B**, 76 (2002).
8. Nefediev A. V. and Ribeiro J. E. F. T., *Phys. Rev.*, **D67**, 034028 (2003).

Pion Corrections in Gribov's Approach to the Dyson-Schwinger Equation

Carlo Ewerz

Università di Milano and INFN, Via Celoria 16 , I-20133 Milano, Italy

Abstract.
Chiral symmetry breaking in QCD leads to the emergence of pions as Goldstone bosons. Their existence in turn affects the Green functions of the theory. Here we study the effect of pion corrections on the light quark's Green function in Feynman gauge using the framework of Gribov's approach to the Dyson-Schwinger equation.

GRIBOV'S APPROACH

The system of Dyson-Schwinger equations (DSEs) for the Green functions is one of the few tools we have for investigating the nonperturbative structure of QCD. In particular, one would like to use the DSEs to understand the mechanism giving rise to chiral symmetry breaking (χSB) and confinement. The difficulty in doing so lies in the fact that the DSEs form a tower of coupled integral equations. Their treatment hence requires truncations or approximations which are in general difficult to control.

Some time ago Gribov suggested a new approach to the DSE for light quarks [1]. His approach is designed to systematically collect the most important contributions to this integral equation originating from the infrared (IR) region in which the dynamics leading to χSB and confinement is expected to take place. The largest contributions to the integrals in the DSE from the IR can be most conveniently identified in Feynman gauge. In that gauge the gluon propagator has the form $D_{\mu\nu}(k) = -\alpha_s(k)g_{\mu\nu}/k^2$, which can be understood as a definition of a nonperturbative coupling $\alpha_s(k)$ at small momenta. The applicability of Gribov's approach requires only mild assumptions on the behavior of $\alpha_s(k)$, namely that is does not diverge at $k = 0$ and does not vary too rapidly with k. One then takes the second derivative of the inverse Green function, $\partial^\mu \partial_\mu G^{-1}(q)$ with $\partial_\mu = \partial/\partial q^\mu$. Using the DSE one finds that the most singular contribution from the IR comes from differentiating twice the gluon Green function because $\partial^2(1/q^2) \sim \delta(q^2)$. With the help of Ward identities one then obtains a differential equation for the light quark's Green function, see eq. (1) below without the last term. It can also be shown to reproduce the correct renormalization group equation in the ultraviolet (UV) region. Less IR-singular terms can be computed systematically in this approach as subleading corrections. In this sense Gribov's approach is an approximation rather than a mere truncation scheme for the quark's DSE.

One can then use Gribov's equation for a detailed investigation of the light quark's Green function in Feynman gauge. According to [1] it should in particular be possible

CP756, *Quark Confinement and the Hadron Spectrum VI*
edited by N. Brambilla, U. D'Alesio, A. Devoto, K. Maung, G.M. Prosperi and S. Serci

to use this approach in order to study the analytic properties of the Green function and to relate them to a picture in which confinement is caused by the phenomenon of supercritical charges. For a recent review of the ideas underlying Gribov's picture of confinement we refer the reader to [2].

PIONS FROM CHIRAL SYMMETRY BREAKING

As was discussed at a previous edition of this conference [3], in Gribov's approach it is found that χSB takes place if the strong coupling α_s exceeds a critical value of $\alpha_c = 0.43$ in some region of momenta in the IR. In this situation the dynamical mass function $M(q^2)$ exhibits oscillations around zero, and as a consequence the relation between the perturbative (or current) quark mass and the renormalized mass $m_R = M(0)$ is no longer one-to-one. Instead one finds that a nonvanishing renormalized mass can be generated even for vanishing perturbative quark mass. It was found though that in the approximation discussed above the analytic structure of the Green function does not correspond to confined quarks [4].

When χSB takes place massless pions are created as Goldstone bosons and appear in the physical spectrum. Their Bethe-Salpeter amplitude can be obtained as $\varphi \sim \{i\gamma_5, G^{-1}\}$ from an equation for $q\bar{q}$ bound states derived in the same approximation as discussed above for the quark [1].

In the phase of χSB in which pions exist as (massless) physical particles it is natural to consider their backreaction on the quark's Green function. In the first approximation described above their effect is not properly taken into account. Instead, they have to be included explicitly. It was argued in [5] that pion corrections can have a crucial effect in particular on the analytic structure of the Green function, possibly giving rise to the confinement of quarks.

PION EFFECTS ON THE QUARK'S GREEN FUNCTION

It turns out that the emission and reabsorption of pions can be included very easily in Gribov's approximation scheme for the quark DSE. The pion propagator $\sim 1/k^2$ in connection with the differential operator ∂^2 makes it again possible to isolate the most important IR contribution in diagrams with pion loops on the quark. Further, we express the Bethe-Salpeter amplitude of the pion in terms of G, as a result of which we are again left with a differential equation for G only. Finally, the pion-quark coupling can be fixed with the help of the Goldberger-Treiman relation. The resulting equation for $G(q)$ reads

$$\partial^\mu \partial_\mu G^{-1} = \frac{C_F \alpha_s}{\pi} \left(\partial^\mu G^{-1}\right) G \partial_\mu G^{-1} - \frac{3}{16\pi^2 f_\pi^2} \left\{i\gamma_5, G^{-1}\right\} G \left\{i\gamma_5, G^{-1}\right\}, \quad (1)$$

where the last term constitutes the modification of the original equation due to pions.

We have performed a numerical study of this improved equation for the Green function of light quarks. Here we concentrate an the dynamical mass function M in the euclidean region, $q^2 < 0$. More detailed results will be presented elsewhere.

Recall that the Green function can be written as $G^{-1} = Z^{-1}(\not{q} - M)$ with the wave function renormalization factor Z^{-1}. From equation (1) one finds analytically that the pion corrections become negligible in the UV region. Hence also the modified equation reproduces the correct RG behavior there. Figure 1 shows our numerical results for the mass function. They have been obtained with an IR frozen coupling $\alpha_s(q)$, see [4]. The

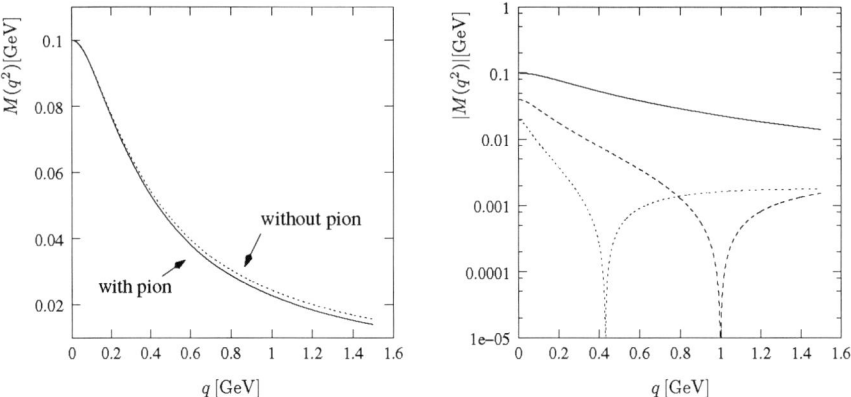

FIGURE 1. Change in the mass function due to pion corrections (left), and the mass function for three different values of the renormalized mass (right)

mass function is affected only very little due to the pion corrections (see example on the left). The figure on the right shows $M(q^2)$ for different values of the renormalized mass. For sufficiently small m_R the mass function oscillates, giving rise to χSB.

In summary, we find only small effects of the pion corrections for spacelike momenta. The pattern of chiral symmetry breaking is the same as in the approximation without pions. Preliminary results indicate that the analytic structure of the quark's Green function in the complex q^2-plane, on the other hand, is considerably changed due to the pion corrections, as was anticipated in [5]. This interesting aspect and its potential consequences for confinement clearly deserve further study.

ACKNOWLEDGMENTS

This work was supported by a Feodor Lynen fellowship of the Alexander von Humboldt Foundation.

REFERENCES

1. Gribov, V. N., *Eur. Phys. J.*, **C10**, 71–90 (1999).
2. Dokshitzer, Y. L., and Kharzeev, D. E., *hep-ph/0404216* (2004).
3. Ewerz, C., "Gribov's Light Quark Confinement Scenario," in *Quark Confinement and the Hadron Spectrum V*, edited by N. Brambilla and G. Prosperi, World Scientific, 2003, pp. 309–311.
4. Ewerz, C., *Eur. Phys. J.*, **C13**, 503–518 (2000).
5. Gribov, V. N., *Eur. Phys. J.*, **C10**, 91–105 (1999).

Covariant Model for Relativistic Three-Body Systems

Philippe Droz-Vincent

LUTH, Observatoire de Paris-Meudon, Place Jules Janssen, F-92195 Meudon, France

Abstract. The system is described by three mass-shell constraints. When at least two masses are equal, this picture has a reasonable nonrelativistic limit. At first post-Galilean order and provided the interaction is not too much energy-dependent, the relativistic correction is tractable like a conventional perturbation problem. A covariant version of harmonic oscillator is given as a toy model.

A system of three particles can be covariantly described by three mass-shell constraints, involving an interaction term referred to as *potential*. These constraints must reduce to three independent Klein-Gordon (or Dirac) equations in the absence of potential. In any case, they determine the evolution of a wave function which depends on three four-dimensional arguments, say p_a with $a, b = 1, 2, 3$, if we chose the momentum representation of quantum mechanics.

Naturally, the potential depends on both configuration and momentum variables, q_a, p_b, and must allow for mutual compatibility of the constraints. Moreover it happens that, just like in the Bethe-Salpeter approach, manifest covariance is paid by the presence of redundant degrees of freedom of which the elimination is by no means straightforward (in contrast to the two-body case). These two important issues have been considered earlier by H. Sazdjian [1] who aimed at solving the general $n-$body case and proposed an approximate solution.

Specially dealing with the *three-boson* case, we have recently exhibited in closed analytic form a new set of variables q'_a, p'_b. In terms of these new variables, admissible expressions for the potential are explicitly available, and two superfluous degrees of freedom can be eliminated [2]. Setting $P = \Sigma p$ we linearly introduce relative variables

$$ z_A = q_1 - q_A, \qquad y_A = \frac{P}{3} - p_A, \qquad A = 2, 3 $$

and similar formulas for z'_A, y'_B in terms of q'_a, p'_b.

The mass-shell constraints can be equivalently replaced by their sum and differences; it is convenient to set $v_A = \frac{1}{2}(m_1^2 - m_A^2)$.

The difference equations, in their original form, yield no simplification. But we perform a *quadratic* change among the momenta, say $p_a \to p'_a$, or equivalently $P, y_A \to P', y'_A$. in order to ensure the elimination of two redundant degrees of freedom; this change is characterized by

$$ (p_1 - p_A)(p_1 + p_A) = (p'_1 - p'_A) \cdot P $$

CP756, *Quark Confinement and the Hadron Spectrum VI*
edited by N. Brambilla, U. D'Alesio, A. Devoto, K. Maung, G.M. Prosperi and S. Serci
© 2005 American Institute of Physics 0-7354-0241-8/05/$22.50

whereas $P' = P$ and the transverse parts of the momenta remain unaffected, say $\tilde{y}' = \tilde{y}$, where the *tilda* on any four-vector refers to its transverse part with respect to P.

Of course, this procedure generates a change of canonical variables [2], in particular we obtain new configuration variables, z'_A.

Three-dimensional Reduction

We impose a sharp value of the total linear momentum, it is a timelike vector k, and we define $k \cdot k = M^2$.

Notations: The *hat* on any vector refers to its transverse part with respect to k.

Underlining any dynamical variable indicates that, in its expression, we substitute k for P and take into account the difference equations

$$3y'_A \cdot k \, \Psi = (4v_A - 2v_B)c^2 \, \Psi \tag{1}$$

We factorize out the relative energies; as a result the sum equation becomes

$$(3\sum m^2 - M^2)c^2 \, \psi = 6(\hat{y}_2^2 + \hat{y}_3^2 + \hat{y}_2 \cdot \hat{y}_3)\psi + (6M^2c^2\underline{\Xi} + 18\underline{V}) \, \psi \tag{2}$$

for a *reduced* wave function ψ which depends only on the transverse relative momenta $\hat{y}'_A = \hat{y}_A$.

The meaning of Ξ is purely kinematic; this term depends only on the momenta and can be expressed in terms of their transverse part and P. Here V denotes the relativistic potential; it may be phenomenological or motivated by considerations of field theory. In particular it may be *formally* constructed as a sum of two-body terms, like in equation (5) below; so doing one uses *the shape* of two-body potentials but (for the sake of compatibility) with the *new* three-body variables as arguments. Not only the total momentum P but also the new configuration variables z'_A mix the two-body clusters, which amounts to automatically incorporate three-body forces. The form of any admissible potential entails that $\underline{V}$ is a function of the *new* variables $\hat{z}_2^2, \hat{z}_3^2$ and M^2c^2.

The reduced equation (2) is actually a nonconventional eigenvalue probem, where the operator to be diagonalized *explicitly depends* on its eigenvalue. This situation is by no means a special drawback of our model, in fact it is common in relativistic quantum mechanics [3], but it renders the general treatment rather involved.

On the other hand, it is natural to expand the formulas in powers of $1/c^2$ and to look for the nonrelativistic limit. For arbitrary masses, the term $M^2c^2\underline{\Xi}$ generally blows up, which leads to consider, instead of (2) an alternative combination of the mass-shell constraints.

Two equal masses.

Drastic simplifications arise when two masses are equal, say $m_2 = m_3 = m$, equivalently $v_2 = v_3 = v$. We find that the Galilean limit of our eigenvalue problem is a Schroedinger equation with effective (or *Galilean*) masses that are generally distinct from the constituent masses m_a. However they still coincide with the constituent masses, at first order in the "mass-dispersion index" v/m^2.

Three Equal Masses.

When $m_a = m$ for all particles, equation (2) can be written as follows, using the rest frame

$$\lambda \psi = (\mathbf{y}_2^2 + \mathbf{y}_3^2 + \mathbf{y}_2 \cdot \mathbf{y}_3)\psi - 3\underline{V}\psi - M^2c^2\underline{\Xi}\psi \tag{3}$$

with $6\lambda = (M^2 - 9m^2)c^2$. Now the last term in (3) remains finite in the nonrelativistic limit. Indeed we can write $Mc^2\underline{\Xi} = \dfrac{1}{M^2c^2}\Gamma_{(0)} + O(1/c^4)$ where

$$\Gamma_{(0)} = \frac{3}{4}\left\{(\hat{y}_2^2)^2 + (\hat{y}_3^2)^2 + 4(\hat{y}_2 \cdot \hat{y}_3)^2 + 2(\hat{y}_2^2 + \hat{y}_3^2)\,(\hat{y}_2 \cdot \hat{y}_3) - \hat{y}_2^2\hat{y}_3^2\right\} \qquad (4)$$

At first order in $1/c^2$ we can, in $\underline{\Xi}$, replace M^2 by $9m^2$, which is independent from λ. Thus we replace $M^2c^2\underline{\Xi}$ by $\Gamma_{(0)}/9m^2c^2$. If the relativistic "potential" V doesnot depend on P^2, or if this dependence is of higher order, equation (3) becomes a conventional eigenvalue problem, tractable by perturbation theory. The last term in(3) brings a negative correction to the value λ_{NR} furnished by the nonrelativistic approximation, say

$$\lambda = \lambda_{NR} - <\Gamma_{(0)}> /9m^2c^2$$

if λ_{NR} corresponds to a nondegenerate level. One has to calculate $<\Gamma_{(0)}>$ in the eigenstate solution of the nonrelativistic problem.

Harmonic Oscillator
A covariant version of the harmonic potential is given by

$$V = 2\kappa\left\{(\vec{z}_2)^2 + (\vec{z}_3)^2 - \vec{z}_2 \cdot \vec{z}_3\right\} \qquad (5)$$

hence $\underline{V}$ in terms of $\vec{z}_A \cdot \vec{z}_B = -\mathbf{z}_A'^2 \cdot \mathbf{z}_B'^2$. In the nonrelativistic limit we recover the naive SU_3 invariant Schroedinger equation. At the first post-Galilean approximation, M^2 can be replaced by $9m^2$, neglecting the dependence on total energy in the reduced equation. At this stage, the eigenvalue problem amounts to diagonalize a nonrelativistic harmonic oscillator, with potential $V_{NR} = -3\underline{V}/m$, submitted to a momentum-dependent perturbation. Expressed in terms of Jacobi-like coordinates, namely $\mathbf{R}_2 = -\mathbf{z}_2' + \mathbf{z}_3'$, $\mathbf{R}_3 = (\mathbf{z}_2' + \mathbf{z}_3')/\sqrt{3}$ and their conjugate momenta, the unperturbed ground state is a Gaussian. If the unit of lenght is choosen such that $\kappa = \dfrac{2}{9}$, one finds $<\Gamma_{(0)}> = 11 + 1/4$.

This approach is intented for applications to confining interactions; future work should implement spin and investigate a possible contact with recents developments [4] of the BS approach.

REFERENCES

1. H. Sazdjian, *Physics Lett.* **B 208**, 470(1988); Annals of Phys. **191**, 52(1989).
2. Ph. Droz-Vincent, *Int. Jour. of Theor. Phys.* **42**, 1809 (2003).
3. V.A. Rizov, H. Sazdjian, I.T.Todorov *Ann. of Phys.* **165**, 59, (1985)
4. J. Bijtebier, *Jour. of Phys. G: Nucl. Part. Phys.* **26**, 871(2000)

Pion distribution amplitude – from theory to data

A. P. Bakulev

Bogoliubov Laboratory of Theoretical Physics, JINR, 141980 Dubna, Russia

Abstract. We describe the present status of the pion distribution amplitude as it originated from two sources: (i) a nonperturbative approach, based on QCD sum rules with nonlocal condensates and (ii) a NLO QCD analysis of the CLEO data on $F^{\gamma \gamma^* \pi}(Q^2)$, supplemented by the E791 data on diffractive dijet production, and the JLab F(pi) data on the pion electromagnetic form factor.

PION DISTRIBUTION AMPLITUDE FROM QCD SUM RULES

The (light cone) pion distribution amplitude (DA)

$$\langle 0 \mid \bar{d}(z)\gamma_\mu \gamma_5 \mathscr{W}(z,0)u(0) \mid \pi(P)\rangle\Big|_{z^2=0} \;=\; if_\pi P_\mu \int_0^1 dx\, e^{ix(zP)}\, \varphi_\pi(x,\mu^2) \tag{1}$$

describes the transition of a pion $\pi(P)$ to a pair of valence quarks u and d, separated by the (straight) Wilson line $\mathscr{W}$, with corresponding momentum fractions xP and $\bar{x}P$, ($\bar{x} \equiv 1-x$).

In order to obtain the pion DA we use a QCD sum rule approach with non-local condensates (NLC) [1], employing the scalar condensate $\langle \bar{q}(0)q(z)\rangle = \langle \bar{q}(0)q(0)\rangle\, e^{-|z^2|\lambda_q^2/8}$. The nonlocality parameter $\lambda_q^2 = \langle k^2 \rangle$ characterizes the average momentum of quarks in the QCD vacuum and has been estimated in QCD SRs [2, 3] and on the lattice [4, 5]: $\lambda_q^2 = 0.45 \pm 0.1$ GeV^2. The NLC sum rules for the pion DA produce [6] a *"bunch"* of 2-parameter models at $\mu^2 \simeq 1$ GeV^2

$$\varphi_\pi(x) = \varphi^{\text{as}}(x)\left[1 + a_2 C_2^{3/2}(2x-1) + a_4 C_4^{3/2}(2x-1)\right] \tag{2}$$

corresponding to $\langle x^{-1}\rangle_\pi^{\text{bunch}} = 3.17 \pm 0.10$, which is in agreement with the result of an independent sum rule, viz., $\langle x^{-1}\rangle_\pi^{\text{SR}} = 3.30 \pm 0.30$ and shown in Fig. 1a for the value $\lambda_q^2 = 0.4$ GeV^2.

NLO LIGHT-CONE SUM RULES (LCSR) AND THE CLEO DATA

The CLEO experimental data on $F^{\gamma \gamma^* \pi}(Q^2)$ allow one to obtain direct constraints on $\varphi_\pi(x)$. Applying the LCSR approach [7], one can effectively account for the long-distance effects of a real photon by using quark-hadron duality in the vector channel and a dispersion relation in q^2.

In our CLEO data analysis [10], we also took into account the relation between λ_q^2 and the twist-4 magnitude $\delta_{\text{Tw-4}}^2 \approx \lambda_q^2/2$ and estimated $\delta_{\text{Tw-4}}^2 = 0.19 \pm 0.02$ at $\lambda_q^2 = 0.4$ GeV^2. We found that even with a 20% uncertainty in $\delta_{\text{Tw-4}}^2$, the Chernyak–Zhitnitsky (CZ) DA [11] was excluded *at least* at the 4σ-level, whereas the asymptotic DA was off the 3σ-level, while our "bunch" was inside the 1σ-region and other nonperturbative models were near the 3σ-boundary [12]. We show in Fig. 1b the plot of $Q^2 F_{\gamma^*\gamma\to\pi}(Q^2)$ for our "bunch" (shaded strip), the CZ DA (upper dashed line), the asymptotic DA (lower dashed line), and two instanton-based models (dotted [13] and dash-dotted [14] lines) in comparison with the CELLO and CLEO data. We see that the BMS "bunch" describes rather well all data for $Q^2 \gtrsim 1.5$ GeV^2.

CP756, *Quark Confinement and the Hadron Spectrum VI*
edited by N. Brambilla, U. D'Alesio, A. Devoto, K. Maung, G.M. Prosperi and S. Serci
© 2005 American Institute of Physics 0-7354-0241-8/05/$22.50

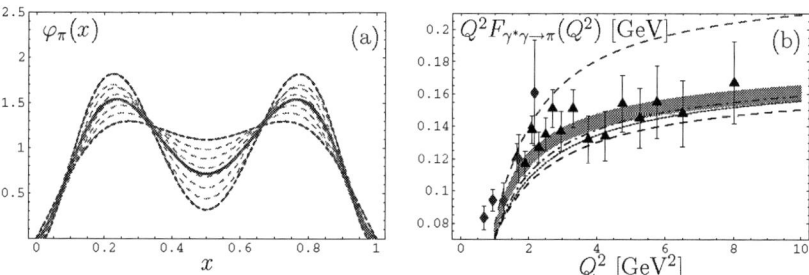

FIGURE 1. (a) "Bunch" of pion DAs extracted from NLC QCD sum rules [6]. (b) $\gamma^*\gamma \to \pi$ transition form factor in comparison with the CELLO ($\blacklozenge$) [8] and the CLEO ($\blacktriangle$) [9] data.

DIFFRACTIVE DIJET PRODUCTION

The diffractive dijet production in $\pi + A$ collisions has been suggested as a tool to extract the profile of the pion DA in [15]. It was argued that the jet distribution with respect to the longitudinal momentum fraction has to follow the quark momentum distribution in the pion and hence provides a direct measurement of the pion DA. But it was shown in [16] (see also [17]) that this proportionality does not hold beyond the leading logarithms in energy and the distribution in the longitudinal momentum fraction of the jets for the non-factorizable contribution is the same as for the factorizable contribution with the asymptotic pion DA. Using the convolution approach of [16], we estimated [12] the distribution of jets in this experiment for our "bunch" of pion DAs and show the results in comparison with φ^{as} and φ^{CZ} in Fig. 2a. The main conclusion from this comparison: *all three DAs are compatible with the E791 data.* Hence, this experiment cannot serve as a safe profile indicator.

PION ELECTROMAGNETIC FORM FACTOR

We have calculated the pion form factor in analytic NLO pQCD [18]

$$F_\pi(Q^2;\mu_R^2) = F_\pi^{LD}(Q^2) + Q^4 \left(2s_0^{2\text{-loop}} + Q^2\right)^{-2} F_\pi^{Fact}(Q^2;\mu_R^2), \qquad (3)$$

taking into account the soft part $F_\pi^{LD}(Q^2)$ via local duality (LD) and correcting the factorized contribution F_π^{Fact} by a power-behaved pre-factor (with $s_0^{2\text{-loop}} \approx 0.6 \text{ GeV}^2$) in order to respect the Ward identity at $Q^2 = 0$ while preserving its high-Q^2 asymptotics. In our analysis $F_\pi^{Fact}(Q^2;\mu_R^2)$ was computed to NLO [19] using Analytic Perturbation Theory [20, 21, 22] and trading the running coupling and its powers for analytic expressions in a non-power series expansion [18]. This procedure provides results practically independent of the scheme/scale setting and in good agreement with the experimental data [23] and φ^{as} (dashed lines)—Fig. 2b.

CONCLUSIONS

The QCD sum rule method with nonlocal condensates produces a "bunch" of admissible pion DAs for each λ_q^2 value [6]. Comparing these results with the CLEO constraints, obtained in the LCSR analysis of the $\gamma^*\gamma \to \pi$-transition form factor, clearly fixes the value of QCD vacuum nonlocality to $\lambda_q^2 = 0.4 \text{ GeV}^2$ [10, 12]. The corresponding "bunch" of pion DAs agrees well

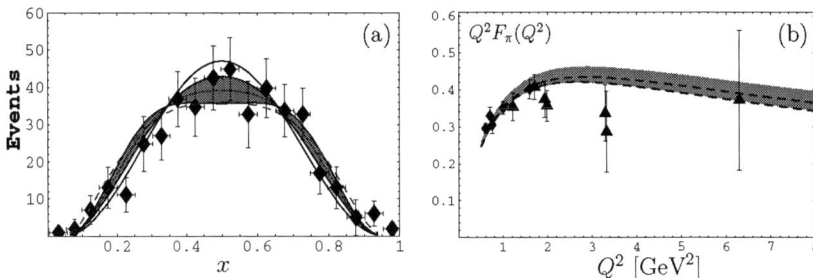

FIGURE 2. (a) Comparison with the E791 data (♦) [24] of φ^{as} (solid line), φ^{CZ} (dashed line), and BMS pion DAs "bunch" (strip). (b) Pion electromagnetic form factor in comparison with the JLab (♦) and Bebek et al. (▲) [23] data. The shaded strip corresponds to the "bunch" and includes the NLC QCD sum-rule uncertainties and scale-setting ambiguities at NLO. The region between the dashed lines denotes the area accessible to φ^{as}.

with both the E791 data on diffractive dijet production and with the JLab F(pi) data on the pion electromagnetic form factor. Analytic perturbation theory with a non-power NLO contribution for the pion form factor diminishes scale-setting ambiguities already at the NLO level [18].

ACKNOWLEDGMENTS

This work was supported in part by the Deutsche Forschungsgemeinschaft (projects 436 KRO 113/6/0-1 and 436RUS113/752/0-1), the Heisenberg-Landau Programme, and the Russian Foundation for Fundamental Research (grants 03-02-16816 and 03-02-04022). I am indebted to the organizers of the Conference for financial support.

REFERENCES

1. S. V. Mikhailov and A. V. Radyushkin, *JETP Lett.* **43**, 712 (1986); *Phys. Rev.* **D45**, 1754 (1992).
2. V. M. Belyaev and B. L. Ioffe, *Sov. Phys. JETP* **57**, 716 (1983);
3. A. A. Ovchinnikov and A. A. Pivovarov, *Sov. J. Nucl. Phys.* **48**, 721 (1988).
4. M. D'Elia, A. Di Giacomo, and E. Meggiolaro, *Phys. Rev.* **D59**, 054503 (1999).
5. A. P. Bakulev and S. V. Mikhailov, *Phys. Rev.* **D65**, 114511 (2002).
6. A. P. Bakulev, S. V. Mikhailov, and N. G. Stefanis, *Phys. Lett.* **B508**, 279 (2001); hep-ph/0104290.
7. A. Khodjamirian, *Eur. Phys. J.* **C6**, 477 (1999).
8. H. J. Behrend et al., *Z. Phys.* **C49**, 401 (1991).
9. J. Gronberg et al., *Phys. Rev.* **D57**, 33 (1998).
10. A. P. Bakulev, S. V. Mikhailov, and N. G. Stefanis, *Phys. Rev.* **D67**, 074012 (2003); *Fizika* **B13**, 423 (2004).
11. V. L. Chernyak and A. R. Zhitnitsky, *Nucl. Phys.* **B201**, 492 (1982).
12. A. P. Bakulev, S. V. Mikhailov, and N. G. Stefanis, *Phys. Lett.* **B578**, 91 (2004); *Annalen Phys.* **13**, 629 (2004).
13. V. Y. Petrov, M. V. Polyakov, R. Ruskov, C. Weiss, and K. Goeke, *Phys. Rev.* **D59**, 114018 (1999).
14. M. Praszalowicz and A. Rostworowski, *Phys. Rev.* **D64**, 074003 (2001).
15. L. Frankfurt, G. A. Miller, and M. Strikman, *Phys. Lett.* **B304**, 1 (1993).
16. V. M. Braun, D. Y. Ivanov, A. Schäfer, and L. Szymanowski, *Nucl. Phys.* **B638**, 111 (2002).
17. N. N. Nikolaev, W. Schafer, and G. Schwiete, *Phys. Rev.* **D63**, 014020 (2001).
18. A. P. Bakulev, K. Passek-Kumerički, W. Schroers, and N. G. Stefanis, *Phys. Rev.* **D70**, 033014 (2004).
19. B. Melić, B. Nižić, and K. Passek, *Phys. Rev.* **D60**, 074004 (1999).
20. D. V. Shirkov and I. L. Solovtsov, *Phys. Rev. Lett.* **79**, 1209 (1997).
21. D. V. Shirkov, *Theor. Math. Phys.* **127**, 409 (2001).
22. N. G. Stefanis, W. Schroers, and H.-C. Kim, *Phys. Lett.* **B449**, 299 (1999); *Eur. Phys. J.* **C18**, 137 (2000).
23. J. Volmer et al., *Phys. Rev. Lett.* **86**, 1713 (2001); C. N. Brown et al., *Phys. Rev.* **D8**, 92 (1973).
24. E. M. Aitala et al., *Phys. Rev. Lett.* **86**, 4768 (2001).

Structure functions of free and in-medium nucleons

I. C. Cloet[*,†], W. Bentz[**] and A. W. Thomas[†]

[*]Special Research Centre for the Subatomic Structure of Matter and
Department of Physics and Mathematical Physics, University of Adelaide, SA 5005, Australia
[†]Jefferson Lab, 12000 Jefferson Avenue, Newport News, VA 23606, U.S.A.
[**]Department of Physics, School of Science, Tokai University, Hiratsuka-shi, Kanagawa 259-1292,
Japan

Abstract. Spin-dependent quark light-cone momentum distributions are calculated for a nucleon in the nuclear medium. We utilize a modified NJL model where the nucleon is described as a composite quark-diquark state. Scalar and vector mean fields are incorporated in the nuclear medium and these fields couple to the confined quarks in the nucleon. The effect of these fields on the spin-dependent distributions is investigated. Our results for the "spin-dependent EMC effect" are also discussed.

Keywords: Spin-dependence, medium modifications, structure functions
PACS: 12.39Fe, 14.20Dh, 25.30Fj

INTRODUCTION

In this paper we determine the valence spin-dependent quark light-cone momentum distributions in a nuclear medium. The theoretical investigation of medium modifications to spin-dependent parton distributions (see *e.g.* [1, 2, 3, 4]) has not experienced the same level of activity as their spin-independent counterparts. However, it is crucial to investigate these effects as they go to the very heart of our understanding of nuclear structure. From a purely practical point of view we need to know how to correctly extract neutron structure functions from nuclear data.

FINITE DENSITY QUARK DISTRIBUTIONS

The spin-dependent light-cone quark distribution per nucleon in a nucleus of mass number A is defined as

$$\Delta f_{q/A}(x_A) = \frac{P_-}{A^2} \int \frac{d\omega^-}{2\pi} e^{iP_- x_A \omega^-/A} \langle A, P | \overline{\psi}_q(0) \gamma^+ \gamma_5 \psi_q(\omega^-) | A, P \rangle, \qquad (1)$$

where ψ_q is the quark field (flavor q) and P^μ the 4-momentum of the nucleus. We evaluate this distribution using the convolution formalism.

In our model, we describe the single nucleon as a bound state of a quark and a scalar diquark in the Nambu-Jona-Lasinio model, and then calculate the nuclear matter equation of state in the mean field approximation. As a result we obtain the mean scalar and vector fields as functions of the baryon density. It is demonstrated in Ref. [5] that given a quark distribution in a free nucleon, the in-medium effect of the scalar

CP756, *Quark Confinement and the Hadron Spectrum VI*
edited by N. Brambilla, U. D'Alesio, A. Devoto, K. Maung, G.M. Prosperi and S. Serci
© 2005 American Institute of Physics 0-7354-0241-8/05/$22.50

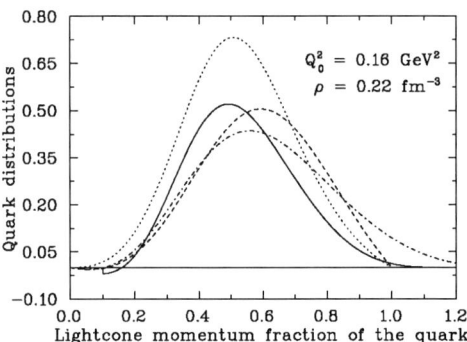

FIGURE 1. All results presented here for the polarized up-quark distribution are at the model scale, $Q_0^2 = 0.16\,\text{GeV}^2$ and each distribution is multiplied by and plotted with respect to the appropriate Bjorken scaling variable. For the meaning of the lines, see text.

field can be included via the effective masses, and after the inclusion of fermi motion via convolution, the effects of the mean vector field can be included via the scale transformation

$$\Delta f_{q/A}(x_A) = \frac{\varepsilon_F}{E_F} \Delta f_{q/A0}\left(\tilde{x}_A = \frac{\varepsilon_F}{E_F} x_A - \frac{V_0}{E_F}\right). \tag{2}$$

Here $\varepsilon_F = \sqrt{p_F^2 + M_N^2} + 3V_0 \equiv E_F + 3V_0$, p_F is the Fermi momentum and V_0 is the zeroth component of the vector field felt by a quark. (The index 0 in Eq.(2) refers to the distribution without the vector field.)

The evaluation of this distribution can be associated with Feynman diagram calculations, see Ref. [6]. The zero density, longitudinally polarized, spin-dependent valence quark distributions, have a term from the "quark diagram", $\Delta f_{q/N0}^{(Q)}$, and the "diquark diagram", $\Delta f_{q/N0}^{(D)}$, and are given by

$$\Delta u_N(x) = \Delta f_{q/N0}^{(Q)} + \frac{1}{2}\Delta f_{q/N0}^{(D)}, \qquad \Delta d_N(x) = \frac{1}{2}\Delta f_{q/N0}^{(D)}, \tag{3}$$

where all quantities involve the free (zero density) masses. Because we include only the scalar diquark channel at this stage we find that $\Delta d_N(x) = 0$. For further details of our calculation and results for the free longitudinally polarized distributions, we refer to Refs. [6, 5].

RESULTS

The results for the finite density spin-dependent up-quark distribution are presented in Fig. 1. The dotted line shows the distribution in a free proton, the dashed line includes the effect of the scalar field, and incorporating Fermi motion via convolution results in the dot-dashed distribution. The effect of the vector field is now simply determined from the scale transformation of Eq. (2), and is indicated by the solid line in Fig. 1.

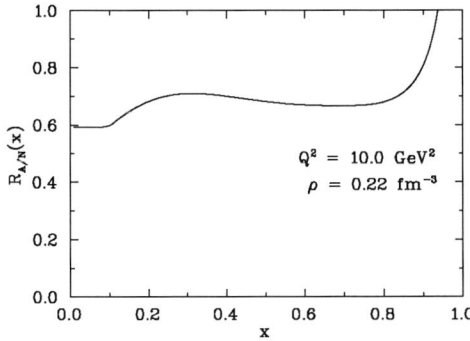

FIGURE 2. Ratio of $\Delta u_A(x_A)$ in the nuclear medium to $\Delta u_N(x)$ in the free proton, at the model saturation density and $Q^2 = 10$ GeV^2.

The curve in Fig. 2 is the ratio of the nuclear and nucleon spin-dependent u-quark distributions

$$R_{A/N}(x) = \frac{\Delta u_A(x_A)}{\Delta u_N(x)} \qquad (4)$$

where the relation $x_A = \frac{M_{N0}}{\varepsilon_F} x = 1.02x$ is used to obtain $R_{A/N}(x)$ as a function of x only. (Here $M_{N0} = 940$ MeV is the free nucleon mass.) The ratio exhibits a plateau between $x \simeq 0.4$ to $x \simeq 0.8$ with an average value of about 0.7, and if we compare this to the corresponding result for the spin-independent case (that is, the EMC ratio), we expect that the medium modifications are more significant for the spin-dependent structure functions. For more quantitative conclusions, however, we have to include in addition also the axial vector diquark channel.

CONCLUSION

Nuclear medium modifications to the spin-dependent quark light-cone momentum distributions have been discussed. We find that the medium effects are significant, more so than the spin-independent case, discussed in Ref. [5], where the same formalism was used.

This work was supported by the Australian Research Council and DOE contract DE-AC05-84ER40150, under which SURA operates Jefferson Lab, and by the Grant in Aid for Scientific Research of the Japanese Ministry of Education, Culture, Sports, Science and Technology, Project No. C2-16540267.

REFERENCES

1. A. W. Thomas, arXiv:hep-ph/9410335.
2. F. C. Khanna and A. Y. Umnikov, arXiv:hep-ph/9609356.
3. F. M. Steffens, K. Tsushima, A. W. Thomas and K. Saito, Phys. Lett. B **447**, 233 (1999).
4. A. Sobczyk and J. Szwed, Acta Phys. Polon. B **32**, 2947 (2001).
5. H. Mineo, W. Bentz, N. Ishii, A. W. Thomas and K. Yazaki, Nucl. Phys. A **735**, 482 (2004).
6. I. C. Cloet, W. Bentz and A.W. Thomas, to be published.

Phase transitions and behavior of kaons in hot and dense matter

C. A. de Sousa*, P. Costa*, M. C. Ruivo* and Yu. L. Kalinovsky†

*Centro de Física Teórica, Departamento de Física, Universidade, P3004-516 Coimbra, Portugal
†Université de Liège, Départment de Physique B5, Sart Tilman, B-4000, LIEGE 1, Belgium

Abstract. We study phase transitions and the behavior of kaons in hot and dense matter, giving special attention to the role of strange quarks. At $T = 0$, it is found that the fraction of the strange valence quarks affects the energy per particle of the system, without changing the nature of the phase transition, and, on the other side, it has meaningful effects on the behavior of kaons and antikaons masses. The phase behavior of kaons in the $T - \rho$ plane is analyzed in connection with the chiral phase transition.

Experimental and theoretical efforts have been done in order to explore the $\mu - T$ phase boundary. There are indications from Lattice QCD that the transition from the hadronic phase to the quark gluon plasma occurs at temperatures of about 150 MeV. While the phase transition with finite chemical potential and zero temperature is expected to be first order, at zero chemical potential and finite temperature there will be a smooth crossover.

Strange quark matter (SQM) has attracted a lot of interest since the suggestion that it could be the absolute ground state of matter. Stable SQM in β–equilibrium is expected to exist in the interior of neutron stars or might also be formed in the earlier stages of heavy-ion collisions (in this case β–equilibrium may not be achieved).

In this paper, we start with the study of the stability condition and order of phase transition at $T = 0$ and $T \neq 0$. Several strange quark matter scenarios are considered, with and without β-equilibrium.

Our research was carried out in the framework of the SU(3) Nambu-Jona-Lasinio (NJL) model with 't Hooft interaction. The Lagrangian is thus given by [1]:

$$\mathscr{L} = \bar{q}(i\partial \cdot \gamma - \hat{m})q + \frac{g_S}{2}\sum_{a=0}^{8}\left[(\bar{q}\lambda^a q)^2 + (\bar{q}(i\gamma_5)\lambda^a q)^2\right]$$
$$+ \ g_D\left[\det\left[\bar{q}(1+\gamma_5)q\right] + \det\left[\bar{q}(1-\gamma_5)q\right]\right]. \tag{1}$$

Following a standard hadronization procedure and the well known Matsubara technique, an effective action is obtained, allowing to calculate several physical quantities at finite temperature and density [2].

Chiral phase transition at zero temperature. We start by analyzing the behavior of quark matter at zero temperature and, in order to discuss the role of the strangeness degree of freedom, we will consider "neutron" matter in chemical equilibrium and

CP756, *Quark Confinement and the Hadron Spectrum VI*
edited by N. Brambilla, U. D'Alesio, A. Devoto, K. Maung, G.M. Prosperi and S. Serci
© 2005 American Institute of Physics 0-7354-0241-8/05/$22.50

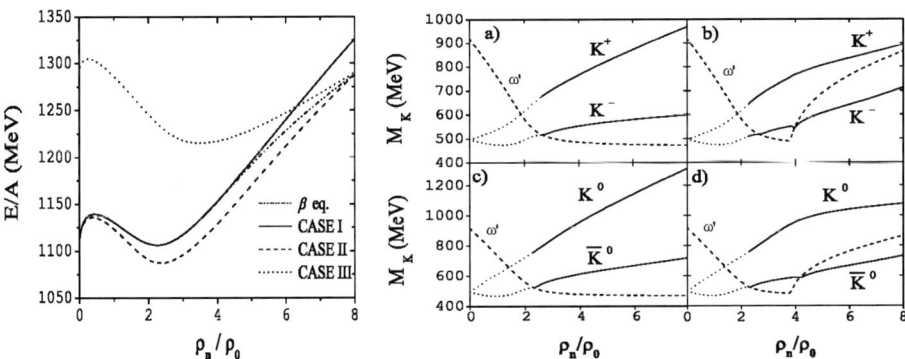

FIGURE 1. Energy per baryon number for all types of quark matter considered at $T = 0$ (left side). Kaon and anti-kaon masses as function of density at $T = 0$ (right side): Case I (a) and c)) and with β–equilibrium (b) and d)). ω' -lower limit of the Dirac sea continuum.

with charge neutrality, and matter without β–equilibrium [2, 4]. For matter without β–equilibrium we consider three cases: Case I – "neutron" matter without strangeness, $(\rho_d = 2\rho_u , \rho_s = 0)$; Case II – matter with equal chemical potentials ($\mu = \mu_d = \mu_u = \mu_s$) with isospin symmetry, $\rho_u = \rho_d, \rho_s = \frac{1}{\pi^2}(\mu^2 - M_s^2)^{3/2}\theta(\mu^2 - M_s^2)$; Case III – matter entirely flavor symmetric ($\rho_d = \rho_u = \rho_s$). We observe that the pressure has a zero at $\rho_n = 0$, the energy per baryon being $3M_u$ for non strange quark matter. If there is another zero of the pressure, at $\rho_n \neq 0$, that corresponds to a minimum of the energy, the criterion for stability of the system at that point is $\mu_u + 2\mu_d < 3M_u$ [3, 4].

We notice that in all cases, whether valence strange quarks are present or not, there is an absolute minimum of the energy per particle at non zero density and zero pressure, lower than the vacuum constituent quark masses; so we have a first order phase transition with the formation of quark droplets (see Fig. 1, left side).

Chiral phase transition at finite temperature and density. Now we discuss the phase transition in hot and dense matter and we consider only matter in β–equilibrium (the other cases are qualitatively similar). We observe that for very low temperatures the absolute minimum of the energy turns to be at zero density: the phase transition is still first order but the system is unstable against expansion. With increasing temperature, we will have a crossover above $T > T_{cl} = 56$ MeV. The critical end point, that connects the first order phase transition and the crossover regions, is found at $T = 56$ MeV and $\rho_n = 1.53\rho_0$. The numerical results indicate a clear manifestation of the restoration of chiral symmetry for the light quarks with increasing temperature and density. A more smooth behavior of the strange quark is observed, with the chiral symmetry showing a slow tendency to get restored in the strange sector. **Kaons in cold quark matter**. As it was already shown in other works [2], two kinds of solutions may be found in asymmetric matter for kaonic modes in the NJL model, corresponding respectively to excitations of the Dirac sea and excitations of the Fermi sea; here, only excitations of the Dirac sea will be discussed. In order to appreciate the role of the strangeness degree of freedom, we start by comparing the results obtained in matter with β–equilibrium and without, Case I. We observe the expected splitting between charge multiplets: the increase of the masses of K^+, K^0 with respect to those of $K^-, \bar{K}^0$, respectively, is due to Pauli blocking, and at the critical density the antikaons enter in the continuum. By comparing the kaonic behavior in matter with and without strange quarks (See Fig. 1, right side) one concludes that the presence of strange quarks has two effects: it is responsible for the antikaons becoming again bound states above a certain density ($\approx 4\rho_0$) and contributes to reduce the splitting between kaon and antikaon masses.

Kaons in hot and dense matter. In order to illustrate the combined effect of temperature and density on the behavior of kaons, we discuss the case of hot matter in weak equilibrium. It is found that, since the threshold of the $\bar{q}q$ continuum is now at the sum of the constituent quark masses, the mesons dissociate at densities and temperatures close to the critical ones. The numerical results indicate a line (a kind of Mott circle) separating the region where the meson are bound states from the region where they are in the continuum.

To conclude and summarize, in the present paper we have investigated phase transitions in hot and dense matter, and the in–medium behavior of kaonic mesons. For a suitable choice of the parameters at zero temperature we have a mixed phase. We notice that, in flavor asymmetric matter, the minimum of the energy is in a region where strange valence quarks are still absent. However, for higher densities the energy density is reduced by having three Fermi seas instead of just two. Only for the case of equal number of quarks u, d, s we found stable SQM, but with a higher energy per particle then atomic nuclei. Concerning the masses of the kaons, there is a splitting between the flavor multiplets in flavor asymmetric matter. In the high density region the splitting is reduced in matter with strangeness. In hot and dense matter the phase transition becomes a crossover above the critical end point, $T = 56 \, \text{MeV}$ and $\rho = 1.53\rho_0$, the system having a mixed phase before that point. The main feature is the dissociation of mesons at the Mott transition point that occurs when the meson masses equals the sum of the masses of their constituents. After that point the mesons cease to be bound states and become resonances.

Work supported by grant SFRH/BD/3296/2000 (P. Costa), CFT and by FEDER/FCT under project POCTI/FIS/451/94.

REFERENCES

1. P. Rehberg, S. P. Klevansky and J. Hüfner, *Phys. Rev.* **C 53**, 410 (1996).
2. P. Costa and M. C. Ruivo, *Europhys. Lett.* **60**(3), 356 (2002); P. Costa, M. C. Ruivo and Yu. L. Kalinovsky, *Phys. Lett.* **B 560**, 171 (2003); *Phys. Lett.* **B 577**, 129 (2003).
3. M. Buballa and M. Oertel, *Nucl. Phys.* **A 642**, 39 (1998); *Phys. Lett.* **B 457**, 261 (1999).
4. P. Costa, M. C. Ruivo, C. A de Sousa and Yu. L. Kalinovsky, *Phys. Rev.* **C 70**, 025204 (2004).

Is the $U_A(1)$ symmetry restored at finite temperature or density?

P. Costa*, M. C. Ruivo*, C. A. de Sousa* and Yu. L. Kalinovsky†

*Centro de Física Teórica, Departamento de Física, Universidade, P3004-516 Coimbra, Portugal
†Université de Liège, Départment de Physique B5, Sart Tilman, B-4000, LIEGE 1, Belgium

Abstract. We investigate the full U(3)⊗U(3) chiral symmetry restoration, at finite temperature and density, on the basis of the three flavor Nambu-Jona-Lasinio model with the anomaly term given by the 't Hooft interaction. We implement a temperature (density) dependence of the anomaly coefficient motivated by lattice results for the topological susceptibility. The results suggest that the axial part of the symmetry is restored before the possible restoration of the full U(3)⊗U(3) chiral symmetry can occur.

The important role of Quantum Chromodynamics (QCD) at finite temperature and density to describe relevant features of particle physics in the early universe, in neutron stars and in heavy-ion collisions, is nowadays more and more recognized, bringing together researches in lattice QCD, effective models like the Nambu-Jona-Lasinio (NJL) model, and compact star physics calculations. In fact, restoration of symmetries and deconfinement are expected to occur under extreme conditions that may be achieved in ultra relativistic heavy-ion collisions or in the interior of neutron stars. An interesting question is whether both chiral $SU(N_f)\otimes SU(N_f)$ and axial $U_A(1)$ symmetries are restored and which observables could carry information about the possible restorations.

Several studies have been done linking the decrease with temperature of the topological susceptibility, χ, with the restoration of the $U_A(1)$ symmetry [1]. There are also preliminary lattice results which indicate the existence of a drop in the behavior of χ with increasing baryonic density [2].

We perform our calculations in the framework of an extended SU(3) Nambu–Jona-Lasinio model Lagrangian density that includes the 't Hooft determinant:

$$\mathcal{L} = \bar{q}(i\gamma.\partial - \hat{m})q + \frac{g_S}{2}\sum_{a=0}^{8}[(\bar{q}\lambda^a q)^2 + (\bar{q}i\gamma_5\lambda^a q)^2]$$
$$+ g_D\{\det[\bar{q}(1+\gamma_5)q] + \det[\bar{q}(1-\gamma_5)q]\}. \tag{1}$$

By using a standard hadronization procedure, an effective meson action is obtained, leading to gap equations for the constituent quark masses and to meson propagators from which several observables are calculated [3].

In the present work we follow the methodology of [4, 5], and extract the temperature dependence of the anomaly coeficient g_D from the lattice results for the topological susceptibility [1]. Other dependences for g_D were studied in [6]. At temperatures around $T \approx 200$ MeV the mass of the light quarks drops to the current quark mass, indicating a

CP756, *Quark Confinement and the Hadron Spectrum VI*
edited by N. Brambilla, U. D'Alesio, A. Devoto, K. Maung, G.M. Prosperi and S. Serci

washed-out crossover. The strange quark mass also starts to decrease significantly in this temperature range, however even at $T = 400$ MeV it is still 2 times the strange current quark mass. So, chiral symmetry shows a slow tendency to get restored in the s sector. In fact, as $m_u = m_d < m_s$, the (sub)group SU(2)⊗SU(2) is a much better symmetry of the NJL Lagrangian. So, the effective restoration of the above symmetry implies the degeneracy between the chiral partners (π^0, σ) and (a_0, η) which occurs around $T \approx 250$ MeV. At $T \approx 350$ MeV both a_0 and σ mesons become degenerate with the π^0 and η mesons, showing an effective restoration of both chiral and axial symmetries. So, we recover the SU(3) chiral partners (π^0, a_0) and (η, σ) which are now all degenerated. However, the η' and f_0 masses do not yet show a clear tendency to converge in the region of temperatures studied [7].

Recent calculations on lattice QCD at finite chemical potential motivates also the study of the restoration of the $U_A(1)$ symmetry at finite density. Since there are no firmly lattice results for the density dependence of χ, to be used as input, we have to extrapolate from our previous results for the finite temperature case and proceed by analogy. Here we present an example (see Fig. 1, left panel) where we consider quark matter simulating "neutron" matter. This "neutron" matter is in β–equilibrium with charge neutrality, and undergoes a first order phase transition [3]. To begin with, we calculate the mixing angles for scalar and pseudoscalar mesons, θ_S and θ_P, respectively. We observe that θ_S starts at $16°$ and increases up to the ideal mixing angle $35.264°$. A different behavior is found for the angle θ_P that changes sign at $\rho_B \approx 4\rho_0$. In fact, it starts at $-5.8°$ and goes to the ideal mixing angle $35.264°$, leading to a change of identity between η and η'. We think this result might be a useful contribution for the understanding of the somewhat controversial question: under extreme conditions will the pion degenerate with η or η'? We found that the change of sign and the corresponding change of identity between η and η', effects that we do not observe in the finite temperature case, is related to the small fraction of the strange quarks that only appear in the medium for $\rho_B \approx 4\rho_0$ [3].

The meson masses, as function of the density, are plotted in Fig. 1, right panel. The results for constant g_D are also presented (middle panel) for comparison purposes. The SU(2) chiral partners (π^0, σ) are bound states and become degenerated at $\rho_B = 3\rho_0$. With respect to the SU(2) chiral partners (η, a_0), the a_0 meson is always a purely non strange quark system. For $\rho_B < 0.8\rho_0$ a_0 is above the continuum and, when $\rho_B \geq 0.8\rho_0$, a_0 becomes a bound state. At $\rho_B = 0$, the η has a strange component and, as the density increases, η becomes degenerated with a_0 at $4.0\rho_0 \leq \rho_B \leq 4.8\rho_0$ as expected. In this range of densities (η, a_0) and (π^0, σ) are all degenerated. Suddenly the η mass separates from the others becoming a purely strange state. This is due to the behavior of θ_P that changes the sign and goes to $35.264°$ at $\rho_B \approx 4.8\rho_0$. On the other hand, the η', that starts as an unbounded state and becomes bounded at $\rho_B > 3.0\rho_0$, turns into a purely light quark system and degenerates with π^0, σ and a_0 mesons. Taking into account the presented arguments, we conclude that the $U_A(1)$ symmetry is effectively restored at $\rho_B > 4\rho_0$ [7]. In fact, the $U_A(1)$ violating quantities show a tendency to vanish, which means that the four meson masses are degenerated and the topological susceptibility goes to zero. Without the restoration of the axial symmetry, the a_0 (σ) mass was moved upwards and never met the π^0 (η') mass as can be seen in Fig. 1, middle panel. We remember that the determinant term acts in an opposite way for the scalar and pseudoscalar mesons.

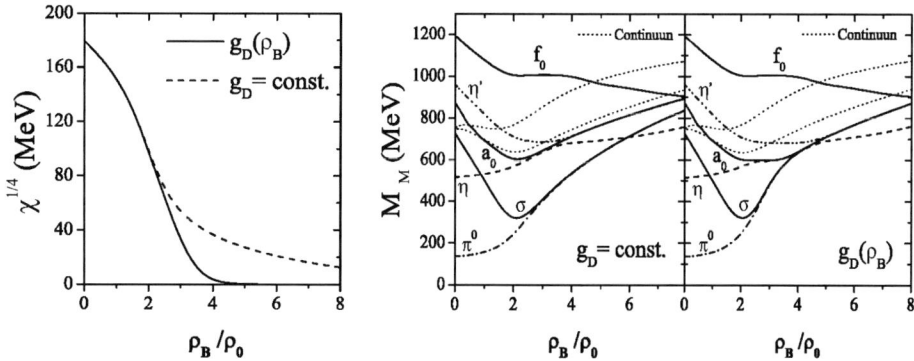

FIGURE 1. Topological susceptibility (left panel): the solid (dashed) line represents our fitting with constant (density dependent) g_D. Meson masses, as functions of density, with g_D constant (middle panel) and $g_D(\rho_B)$ (rigth panel). The dotted lines indicate the density dependence of the limits of the Dirac sea continua, defining $q\bar{q}$ thresholds for a_0 and η' mesons.

In summary, we have implemented a criterion which combines a lattice-inspired behavior of the topological susceptibility with the convergence of appropriate chiral partners to explore effective restoration of symmetries. However, the role of $U_A(1)$ symmetry for finite temperature, and mainly for finite density media, has not been so far investigated and this question is still controversial and not settled yet. We hope that new studies, especially lattice based and experimental ones, can finally clarify it.

Work supported by grant SFRH/BD/3296/2000 (P. Costa), CFT and by FEDER/FCT under project POCTI/FIS/451/94.

REFERENCES

1. B. Allés, M. D'Elia, and A. Di Giacomo, *Nucl. Phys.* **B 494**, 281 (1997).
2. B. Allés, M. D'Elia, M. P. Lombardo and M. Pepe, *Nucl. Phys.* (Proc. Suppl.) **94**, 441 (2001).
3. P. Costa and M. C. Ruivo, *Europhys. Lett.* **60** (3), 356 (2002); P. Costa, M. C. Ruivo, C. A de Sousa and Yu. L. Kalinovsky, *Phys. Rev.* **C 70**, 025204 (2004).
4. K. Fukushima, K. Ohnishi, and K. Ohta, *Phys. Rev.* **C 63**, 045203 (2001); *Phys. Lett.* **B 514**, 200 (2001).
5. J. Schaffner-Bielich, *Phys. Rev. Lett.* **84**, 3261 (2000).
6. R.Alkofer, P. A. Amundsen and H. Reinhardt, *Phys. Lett.* **B 218**, 75 (1989); T. Kunihiro, *Phys. Lett.* **B 219**, 363 (1989).
7. P. Costa, M. C. Ruivo, C. A de Sousa and Yu. L. Kalinovsky, hep-ph/0408177.

Octet contributions in radiative Υ decays

Xavier Garcia i Tormo

Departament d'Estructura i Constituents de la Matèria. Universitat de Barcelona. Diagonal 647.
E-08028 Barcelona. Catalonia. Spain

Abstract. We analyze the end-point region of the photon spectrum in semi-inclusive radiative decays of very heavy quarkonium ($m\alpha_s^2 \gg \Lambda_{QCD}$). The S- and P-wave octet shape functions are calculated. When they are included in the analysis of the photon spectrum of the $\Upsilon(1S)$ system the agreement with data becomes excellent.

The standard Non-Relativistic QCD (NRQCD) factorization [1] (operator product expansion) breaks down at the end-point region of the photon spectrum in semi-inclusive radiative decays of heavy quarkonium; this is due to the fact that collinear degrees of freedom, that are relevant in this kinematic situation, are not included in NRQCD. A proper effective field theory treatment of the end-point region of the spectrum requires, thus, the combination of NRQCD with the Soft-Collinear Effective Theory (SCET) [2]. With this approach factorization formulas has been derived for this process and the resummation of Sudakov logarithms has been performed [3, 4].

If one considers the initial heavy quarkonium state as a Coulombic state, the octet shape functions (that appear in the factorization formulas) can be calculated with a combination of potential NRQCD (pNRQCD) [5] and SCET. The relevant contributions that need to be calculated are depicted in figure 1. From the calculation of the two diagrams in that figure we obtain the S-wave and P-wave octet shape functions in the weak coupling regime, the result is the following:

$$S_S(l_+) := \frac{4\alpha_s(\mu_u)}{3\pi N_c} \left(\frac{C_F}{2m}\right)^2 \int_0^\infty dx \left(2\psi_{10}(0)I_S(\frac{l_+}{2}+x) - I_S^2(\frac{l_+}{2}+x)\right) \quad (1)$$

$$S_{P1}(l_+) := \frac{\alpha_s(\mu_u)}{6\pi N_c} \int_0^\infty dx \left(2\psi_{10}(0)I_P(\frac{l_+}{2}+x) - I_P^2(\frac{l_+}{2}+x)\right)$$

$$S_{P2}(l_+) := \frac{\alpha_s(\mu_u)}{6\pi N_c} \int_0^\infty dx \frac{8l_+x}{(l_++2x)^2} \left(\psi_{10}^2(0) - 2\psi_{10}(0)I_P(\frac{l_+}{2}+x) + I_P^2(\frac{l_+}{2}+x)\right) \quad (2)$$

$$I_S(\frac{k_+}{2} \mid x) := m\sqrt{\frac{\gamma}{\pi}} \frac{\alpha_s N_c}{2} \frac{1}{1-z'} \left(1 - \frac{2z'}{1+z'} \, {}_2F_1\left(-\frac{\lambda}{z'}, 1, 1 - \frac{\lambda}{z'}, \frac{1-z'}{1+z'}\right)\right) \quad (3)$$

$$I_P(\frac{k_+}{2}+x) := \sqrt{\frac{\gamma^3}{\pi}} \frac{8}{3}(2-\lambda) \frac{1}{4(1+z')^3} \left(2(1+z')(2+z') + (5+3z')(-1+\lambda)+\right.$$

$$\left. +2(-1+\lambda)^2 + \frac{1}{(1-z')^2}\left(4z'(1+z')(z'^2-\lambda^2)\left(-1 + \frac{\lambda(1-z')}{(1+z')(z'-\lambda)}+\right.\right.\right.$$

CP756, *Quark Confinement and the Hadron Spectrum VI*
edited by N. Brambilla, U. D'Alesio, A. Devoto, K. Maung, G.M. Prosperi and S. Serci
© 2005 American Institute of Physics 0-7354-0241-8/05/$22.50

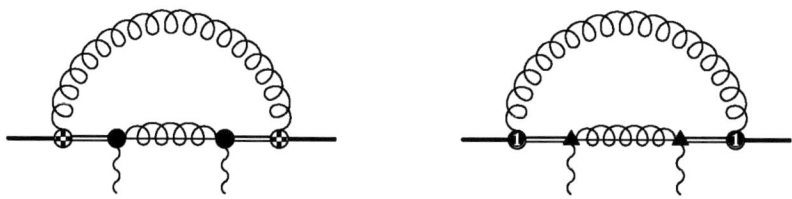

FIGURE 1. Color octet contributions. • represents the color octet S-wave current, ▲ represents the color octet P-wave current. The notation for the other vertices is $\clubsuit := \frac{igc_F}{\sqrt{N_c T_F}} \frac{(\sigma_1 - \sigma_2)}{2m} \text{Tr} \left[T^b \mathbf{B} \right]$ and $\mathbf{0} := \frac{ig}{\sqrt{N_c T_F}} \mathbf{x} \text{Tr} \left[T^b \mathbf{E} \right]$. The solid line represents the singlet field, the double line represents the octet field and the gluon with a line inside represents a collinear gluon.

$$+ {}_2F_1 \left(-\frac{\lambda}{z'}, 1, 1 - \frac{\lambda}{z'}, \frac{1-z'}{1+z'} \right) \Big) \Big) \Big) \tag{4}$$

where

$$\gamma = \frac{mC_f \alpha_s}{2} \quad z' = \frac{\kappa}{\gamma} \quad -\frac{\kappa^2}{m} = E_1 - \frac{k_+}{2} - x \quad \lambda = -\frac{1}{2N_c C_f} \quad E_1 = -\frac{\gamma^2}{m} \tag{5}$$

These shape function are ultraviolet divergent and need to be regularized and renormalized (see [6] for the details of the procedure); some finite parts coming from a linear divergence must also be subtracted to have a smooth connection with the leading order NRQCD results.

Since there is good evidence that the $\Upsilon(1S)$ can be understood as a weak coupling bound state, we compare the results with the experimental data for the $\Upsilon \to X\gamma$ decay [7]. The calculation in pNRQCD+SCET is reliable in the region $z \in [0.7, 0.95]$ (where $z = 2E_\gamma/M$, M being the mass of the heavy quarkonium state), in order to have a reliable prediction for the whole spectrum we adapt the interpolation formula used in [3] for the inclusion of the color octet contributions, that is

$$\frac{1}{\Gamma_0} \frac{d\Gamma_{int}}{dz} = \frac{1}{\Gamma_0} \frac{d\Gamma_{LO}^{dir}}{dz} + \left(\frac{1}{\Gamma_0} \frac{d\Gamma_{resum}^{sing}}{dz} - z \right) + \left(\frac{1}{\Gamma_0} \frac{d\Gamma_{resum}^{oct}}{dz} - z(4 + 2\log(1-z))(1-z) \right) \tag{6}$$

Furthermore the low photon energy region of the spectrum $z \lesssim 0.4$ is dominated by the so called fragmentation contributions; these contributions must be added to the previous ones to obtain the total decay rate; we will use the formulas presented in section II.B of the first reference in [3] and the evaluation of the NRQCD color octet matrix elements derived in the appendix of [6] to obtain these fragmentation contributions.

The comparison with data is depicted in figure 2, the solid curve consists of the color octet contribution (with the Sudakov resummation performed in [4] included) and the color singlet contribution [3], interpolated according to (6) with the leading order NRQCD result (which is reliable away from the end-point), plus the fragmentation contributions. No error bars are presented in this plot (neither for the experimental data points nor for the theoretical curve) since the aim is just to illustrate that a good description of the spectrum is possible (a detailed description of this analysis will be

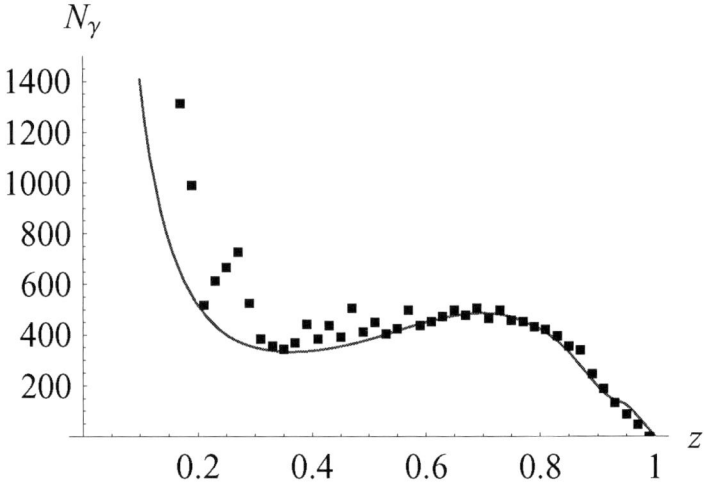

FIGURE 2. Photon spectrum in semi-inclusive Υ decay. The points are the CLEO data [7], the solid line is the theoretical prediction described in the text.

presented in [8]). Taking into account that the data points below $z = 0.4$ have large errors we can affirm that the agreement of the theoretical curve with data is excellent.

ACKNOWLEDGMENTS

It is a pleasure to thank J. Soto for past and present collaboration on this subject. Financial support from a CICYT-INFN 2003 collaboration contract, the MCyT and Feder (Spain) grant FPA2001-3598, the CIRIT (Catalonia) grant 2001SGR-00065, the Departament d'Universitats, Recerca i Societat de la Informació of the Generalitat de Catalunya and the Acciones Integradas España-Italia, project HI2003-0362 is acknowledged.

REFERENCES

1. G. T. Bodwin, E. Braaten and G. P. Lepage, Phys. Rev. D **51** (1995) 1125 [Erratum-ibid. D **55** (1997) 5853].
2. C. W. Bauer, S. Fleming, D. Pirjol and I. W. Stewart, Phys. Rev. D **63** (2001) 114020.
3. S. Fleming and A. K. Leibovich, Phys. Rev. D **67** (2003) 074035. S. Fleming and A. K. Leibovich, Phys. Rev. Lett. **90** (2003) 032001. S. Fleming and A. K. Leibovich, arXiv:hep-ph/0407259.
4. C. W. Bauer, C. W. Chiang, S. Fleming, A. K. Leibovich and I. Low, Phys. Rev. D **64** (2001) 114014.
5. A. Pineda and J. Soto, Nucl. Phys. Proc. Suppl. **64** (1998) 428.
6. X. Garcia i Tormo and J. Soto, Phys. Rev. D **69** (2004) 114006.
7. B. Nemati *et al.* [CLEO Collaboration], Phys. Rev. D **55** (1997) 5273.
8. X. Garcia i Tormo and J. Soto, in preparation.

Recent Developments in the Treatment of Heavy Quarks

Matthias Steinhauser

Institut für Theoretische Teilchenphysik, Universität Karlsruhe, D-76128 Karlsruhe

Abstract. We discuss new results obtained for the description of a system of two heavy quarks close to the production threshold. Special emphasis is put on the resummation of logarithms in the quark velocity. As physical applications we discuss the determination of the mass of the η_b meson and the ratio of the photon mediated production or annihilation rates of spin triplet and spin singlet heavy quarkonium states.

Keywords: Quarkonium, non-relativisic QCD, renormalization group
PACS: 12.38.Bx, 13.25.Gv

INTRODUCTION

One of the most important tasks of a future linear collider is the precise measurement of the cross section for the production of top quark pairs close to their production threshold. Next to the mass and the width of the top quark also the strong coupling and — in case the Higgs boson is not too heavy — also the top quark Yukawa coupling can be determined with a quite high accuracy. In order to match the expected experimental precision [1] it is important to compute higher-order quantum corrections to this process. In this context we refer to the Refs. [2, 3, 4, 5].

In these proceedings we want to concentrate on the system of two bottom quarks. In particular we report about results based on the renormalization group equation within the framework of non-relativistic QCD [6, 7, 8]. In particular, in the next Section we discuss the determination of the mass of the pseudoscalar η_b meson, $M(\eta_b)$, and the last Section deals with the ratio of the photon mediated production or annihilation rates of spin triplet and spin singlet heavy quarkonium states.

PREDICTION OF $M(\eta_B)$

In order to determine $M(\eta_b)$ we exploit the relation $M(\eta_b) = M(\Upsilon(1S)) - E_{\text{hfs}}$ where $M(\Upsilon(1S)) = 9.46030(26)$ GeV and E_{hfs} is the hyperfine splitting. The latter can be determined from the spin-dependent part of the effective Lagrangian. The next-to-leading order (NLO) approximation of E_{hfs} is easily obtained from the N^3LO corrections to the energy level [3]. In order to improve the approximation one has to resum the logarithms in the velocity of the heavy quarks contained in the corresponding matching coefficient. This leads to E_{hfs} in next-to-leading logarithmic (NLL) accuracy. Its computation can be divided into three steps: First, the renormalization group (RG) equations have to be established within potential non-relativistic QCD (pNRQCD) [6]. One has to make sure to include the running of all relevant operators and to consider the soft, potential and

CP756, *Quark Confinement and the Hadron Spectrum VI*
edited by N. Brambilla, U. D'Alesio, A. Devoto, K. Maung, G.M. Prosperi and S. Serci
© 2005 American Institute of Physics 0-7354-0241-8/05/$22.50

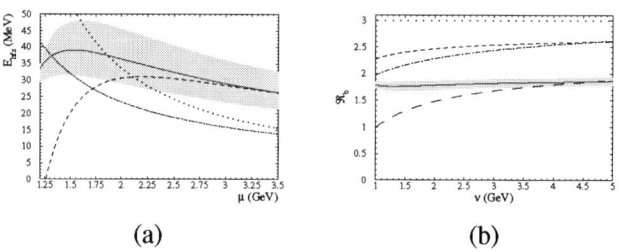

(a)   (b)

FIGURE 1. (a) HFS of 1S bottomonium as a function of the renormalization scale μ in the LO (dotted line), NLO (dashed line), LL (dot-dashed line), and NLL (solid line) approximations. (b) $\mathscr{R}_q$ as the function of the renormalization scale v in LO$\equiv$LL (dotted line), NLO (short-dashed line), NNLO (long-dashed line), NLL (dot-dashed line), and NNLL (solid line) approximation for the bottomonium ground state. The (yellow) band reflects the errors due to $\alpha_s(M_Z) = 0.118 \pm 0.003$.

ultra-soft regions. In a second step the RG equations have to be solved. In the case at hand this could be done analytically. As a result one obtains the matching coefficient of the spin-dependent operator, $D^{(2)}_{s^2,s}$, to the NLL accuracy. This expression is used in the third step to evaluate within perturbation theory the corrections to the energy level. These steps have been performed in Refs. [9, 10].

A compact analytical expression for the HFS to NLL, $E^{\text{NLL}}_{\text{hfs}}$, which is a function of $\alpha_s(\mu)/\alpha_s(m_b)$, is given in Eq. (1) of Ref. [9]. In Fig. 1(a), the HFS for the bottomonium ground state is plotted as a function of μ in the LO, NLO, LL, and NLL approximations. As can be seen, the LL curve shows a weaker scale dependence compared to the LO one. The scale dependence of the NLO and NLL expressions is further reduced, and, moreover, the NLL approximation remains stable up to smaller scales than the fixed-order calculation. At the scale $\mu' \approx 1.3$ GeV. The NLL correction vanishes. Furthermore, at $\mu'' \approx 1.5$ GeV, the result becomes independent of μ; *i.e.*, the NLL curve shows a local maximum. This suggests a nice convergence of the logarithmic expansion despite the presence of the ultrasoft contribution with α_s normalized at the rather low scale $\bar{\mu}^2/m_b \sim 0.8$ GeV. From these observations the following prediction of the mass of the as-yet undiscovered η_b meson has been obtained [9]

$$M(\eta_b) = 9421 \pm 11 \, (\text{th})^{+9}_{-8} \, (\delta\alpha_s) \, \text{MeV} \,, \tag{1}$$

where the errors due to the high-order perturbative corrections and the nonperturbative effects are added up in quadrature in "th", whereas "$\delta\alpha_s$" stands for the uncertainty in $\alpha_s(M_Z) = 0.118 \pm 0.003$. If the experimental error in future measurements of $M(\eta_b)$ will not exceed a few MeV, the bottomonium HFS will become a competitive source of $\alpha_s(M_Z)$ with an estimated accuracy of ± 0.003, as can be seen from Fig. 1(a).

SPIN DEPENDENCE OF HEAVY QUARKONIUM PRODUCTION AND ANNIHILATION

In Ref. [11] a further step was undertaken and the ratio of the photon mediated production or annihilation rates of spin triplet and spin singlet heavy quarkonium states has been considered. In particular, we define

$$\mathscr{R}_q = \frac{\sigma(e^+e^- \to \mathscr{Q}(n^3S_1))}{\sigma(\gamma\gamma \to \mathscr{Q}(n^1S_0))} = \frac{\Gamma(\mathscr{Q}(n^3S_1) \to e^+e^-)}{\Gamma(\mathscr{Q}(n^1S_0) \to \gamma\gamma)}, \tag{2}$$

which can be written as

$$\mathscr{R}_q = \frac{c_s^2(v)}{3Q_q^2} \frac{|\psi_n^v(0)|^2}{|\psi_n^p(0)|^2} + \mathscr{O}(\alpha_s v^2). \tag{3}$$

$\psi_n^{(v,p)}(\vec{r})$ are the spin triplet (vector) and spin singlet (pseudoscalar) quarkonium wave functions of the principal quantum number n. The expression for $|\psi_n^v(0)|^2/|\psi_n^p(0)|^2$ to NNLL can be found in Ref. [11]. c_s is the ratio of the corresponding matching coefficients. With the help of the NLL approximation of $D_{S^2,s}^{(2)}$ it was possible to obtain the NNLL corrections of c_s and thus for the quantity $\mathscr{R}_q$ [11], i.e. for the first time for a physical quantity.

It is interesting to consider the various approximations for $\mathscr{R}_q$ as a function of the renormalization scale, v. One observes a very good convergence for the top quark system [11]. In the case of the bottom quark the result is shown in Fig. 1(b). A nice convergence of the logarithmic expansion is observed despite the presence of ultrasoft contributions with α_s normalized at a rather low scale v^2/m_b. At the same time, the perturbative corrections are important and reduce the leading order result by approximately 41%. More details can be found in Ref. [11].

ACKNOWLEDGMENTS

I would like to thank B. Kniehl, A. Penin, A. Pineda and V. Smirnov for a very pleasant and fruitful collaboration. This work was supported by HGF Grant No. VH-NG-008.

REFERENCES

1. M. Martinez and R. Miquel, Eur. Phys. J. C **27** (2003) 49.
2. A.H. Hoang et al., Eur. Phys. J.direct C **3**, 1 (2000).
3. B.A. Kniehl, A.A. Penin, V.A. Smirnov, and M. Steinhauser, Nucl. Phys. **B635**, 357 (2002); A.A. Penin and M. Steinhauser, Phys. Lett. B 538 (2002) 335.
4. B.A. Kniehl, A.A. Penin, V.A. Smirnov, and M. Steinhauser, Phys. Rev. Lett. 90 (2003) 212001; Erratum *ibid.* 91 (2003) 139903.
5. A.H. Hoang, Phys. Rev. D 69 (2004) 034009.
6. A. Pineda and J. Soto, Nucl. Phys. Proc. Suppl. 64 (1998) 428; N. Brambilla, A. Pineda, J. Soto, and A. Vairo, Nucl. Phys. B 566 (2000) 275.
7. M.E. Luke, A.V. Manohar, and I.Z. Rothstein, Phys. Rev. D **61** (2000) 074025.
8. A. Pineda, Phys. Rev. D 65 (2002) 074007; Phys. Rev. D 66 (2002) 054022.
9. B.A. Kniehl, A.A. Penin, A. Pineda, V.A. Smirnov, and M. Steinhauser, Phys. Rev. Lett. **92** (2004) 242001.
10. A.A. Penin, A. Pineda, V.A. Smirnov, and M. Steinhauser, Phys. Lett. B **593** (2004) 124
11. A.A. Penin, A. Pineda, V.A. Smirnov, and M. Steinhauser, Nucl. Phys. B **699** (2004) 183.

Scalar mesons and Adler zeros

George Rupp*, Frieder Kleefeld* and Eef van Beveren†

*Centro de Física das Interacções Fundamentais, Instituto Superior Técnico,
P-1049-001 Lisboa, Portugal
†Centro de Física Teórica, Departamento de Física, Universidade de Coimbra,
P-3004-516 Coimbra, Portugal

Abstract. A simple unitarized quark-meson model, recently applied with success to light and charmed scalar mesons, is shown to encompass Adler-type zeros in the amplitude, due to the use of relativistic kinematics in the scattering sector. These zeros turn out to be crucial for the description of the $K_0^*(800)$ resonance, as well as the new charmed scalar mesons $D_{s0}^*(2317)$ and $D_0^*(2300)$.

Keywords: Scalar mesons, Adler zeros, relativistic kinematics, $K_0^*(800)$, $D_{s0}^*(2317)$, $D_0^*(2300)$
PACS: 14.40.-n, 11.80.Cr, 12.40.Yx, 13.75.Lb

The light scalar mesons are still causing lots of headaches to many theorists as well as experimentalists, despite the growing, yet far from general, consensus that unitarization in some form should represent the necessary setting. For the purpose of the present short note, we limit ourselves to refer to a recent brief review [1], which, albeit voicing mainly our personal view-points, contains many references to work in this field.

An important contribution to unraveling the conundrum of the light scalar mesons from a data-analysis perspective was recently made by D. V. Bugg [2], in successfully describing elastic $\pi\pi$ and $K\pi$ scattering as well as the corresponding σ ($f_0(600)$) and κ ($K_0^*(800)$) resonances. The clue to this achievement was the inclusion of the respective Adler zeros for these processes directly into the energy-dependent widths contained in the relativistic Breit–Wigner amplitudes used in the phase-shift fits.

Such T-matrix zeros below threshold were determined by S. Weinberg [3] for elastic $\pi\pi$ scattering, elaborating upon consistency conditions among strong-interaction amplitudes derived by S. L. Adler [4] for $\pi\pi$, πN, and $\pi\Lambda$ processes, on the basis of PCAC. In the case of elastic $\pi\pi$ and $K\pi$ scattering, the Adler zeros lie at $s_A^{\pi\pi} \approx m_\pi^2/2$ and $s_A^{K\pi} \approx m_K^2 - m_\pi^2/2$, respectively, which were the values used by D. V. Bugg.

On the other hand, as early as in 1986 two of us co-authored a paper [5] in which, for the first time, a complete light nonet of scalar-meson resonances [6] was predicted with a simultaneous reasonable description of the corresponding elastic S-wave meson-meson phase shifts. Although the latter coupled-channel model results were obtained without any fit in the scalar sector, its predictions for the resonance pole positions lied very close to the present-day world averages. Nevertheless, no attempt was made in this work to account for dynamical Adler zeros from PCAC or (approximate) chiral symmetry, besides the use of a physical input pion mass. Moreover, in a more recent paper [7], a simplified yet less model-dependent version of the mentioned coupled-channel approach was employed to fit the S-wave $K\pi$ phase shifts up to 1.6 GeV, thereby extracting both the established $K_0^*(1430)$ and the now also listed $K_0^*(800)$ resonances [8]. Finally, we used the very same modified Breit–Wigner formula, with adjusted quark and meson

CP756, *Quark Confinement and the Hadron Spectrum VI*
edited by N. Brambilla, U. D'Alesio, A. Devoto, K. Maung, G.M. Prosperi and S. Serci
© 2005 American Institute of Physics 0-7354-0241-8/05/$22.50

masses, to successfully describe [9] the couple of just discovered charmed scalar mesons consisting of the very narrow $D_{s0}^*(2317)$ [10] and the broad $D_0^*(2300)$ [11].

In view of these surprising results, we analyze here in more detail the behavior of the latter model amplitude below threshold and in the complex-momentum plane (see also Ref. [12]). The corresponding K^{-1} matrix is, for the 1×1 scalar case, simply given by

$$\cot g\left(\delta(s)\right) = \frac{n_0(pa)}{j_0(pa)} - \left\{ 2\lambda^2 \mu(s) \, pa j_0^2(pa) \left[\frac{1.0}{\sqrt{s} - E_1} + \frac{0.2}{\sqrt{s} - E_2} - 1 \right] \right\}^{-1}, \quad (1)$$

where j_0 and n_0 are spherical Bessel and Neumann functions, respectively, E_1 and E_2 are the energies of the lowest bare $J^{PC} = 0^{++}$ $q\bar{q}$ states, λ is the coupling for $^3P_0$ quark-pair creation, a is the corresponding interaction radius, p is the relativistic relative momentum of the two-meson system, and $\mu(s)$ is the associated relativistic reduced mass

$$\mu(s) \equiv \frac{1}{2} \frac{dp^2}{d\sqrt{s}} = \frac{\sqrt{s}}{4} \left[1 - \left(\frac{m_1^2 - m_2^2}{s} \right)^2 \right]. \quad (2)$$

We immediately see [12] from the latter expression that $\mu(s)$ vanishes at $s = \pm(m_1^2 - m_2^2)$. Therefore, at this point below threshold, the factor with λ in Eq. (1) is squashed to zero, so that the amplitude also vanishes. This property was shared by the model of Ref. [5].

Nice illustrations of the effect of these kinematical Adler-type zeros are the S-matrix pole trajectories in the complex-p plane for S-wave DK and $D\pi$ scattering (see the Figure). Note that the parametrization here is slightly different from the one in Ref. [9], as we now scale λ and a with the reduced mass of the $q\bar{q}$ system, so as to guarantee rigorous flavor invariance (see Refs. [1], [12] for details). The result is that, in the DK case, the two pole trajectories interchange their roles, so that now it is the bare state which gives rise to the $D_{s0}^*(2317)$ instead of the continuum state. However, despite the

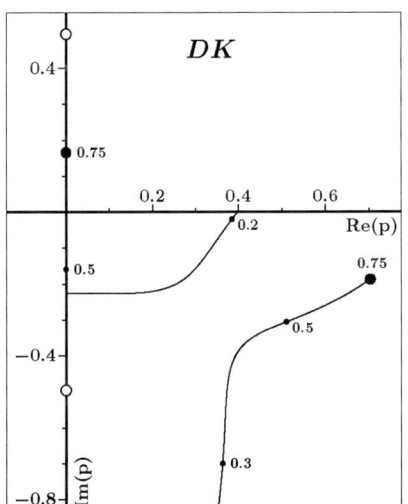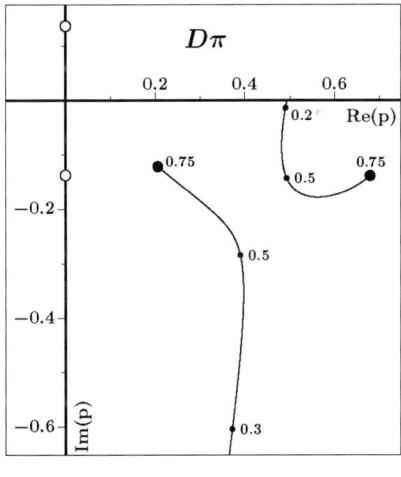

FIGURE. Pole trajectories as a function of coupling λ_0, and kinematical Adler-type zeros (○), in complex-momentum plane (GeV units), for S-wave DK and $D\pi$ scattering.

dramatic change in the trajectories themselves, their end points, corresponding to the value 0.75 $(\text{GeV}^{-3/2})$ of the universal coupling λ_0, only suffer a modest shift, which even produces an improved value for the $D^*_{s0}(2317)$ mass, viz. 2.327 GeV. In the $D\pi$ case, the trajectories remain qualitatively unaltered, but the predicted $D^*_0(2300)$ pole position also improves, i.e., to $2.114 - i0.118$ GeV. However, the crucial message from the Figure appears to be the large effect of the Adler-type zeros, indicated by the open circles. Namely, while in the DK case the relatively distant zeros, located at $p_A = \pm im_K$, allow the lower pole to travel all the way to the upper imaginary-p axis, for $D\pi$ scattering the nearby zeros at $p_A = \pm im_\pi$ slow down this pole so as to settle above threshold. Note that these kinematical zeros deviate less than 1% from the theoretical Adler zeros. Thus, we get a (quasi-)bound state in the former case, and a quite broad resonance in the latter, similarly to what happens with the κ meson (see Ref. [12] for the κ-pole trajectory).

As a final test, we check what happens to the κ by removing the Adler-type zero. So we substitute the relativistic reduced mass (2) by a nonrelativistic one, and then refit λ, a to the $K\pi$ phase shifts. The result is a somewhat unphysical κ pole at $E = 483 - i274$ MeV. Defining also the relative momentum nonrelativistically then completely kills the κ, while the $K^*_0(1430)$ still survives. This may provide a clue to the absence of a light κ in the analysis of Ref. [13], in which a distant, negative Adler zero at $s_A = -0.41$ GeV^2 was used, in contrast with e.g. the unitarized chiral approaches of Ref. [14].

ACKNOWLEDGMENTS

We wish to thank D. V. Bugg for drawing our attention to the importance of Adler zeros in scalar-meson data analysis. One of us (G.R.) also acknowledges useful discussions with N. A. Törnqvist and H. Q. Zheng.

This work was supported by the *Fundação para a Ciência e a Tecnologia* of the *Ministério da Ciência, Inovação e Ensino Superior* of Portugal, under contract no. POCTI/FNU/49555/2002 and grant no. SFRH/BPD/9480/2002.

REFERENCES

1. E. van Beveren and G. Rupp, *Mod. Phys. Lett. A* **19**, 1949 (2004) [hep-ph/0406242].
2. D. V. Bugg, *Phys. Lett. B* **572**, 1 (2003) [*Erratum-ibid.* **595**, 556 (2004)]; *Phys. Rept.* **397**, 257 (2004).
3. S. Weinberg, *Phys. Rev. Lett.* **17**, 616 (1966).
4. S. L. Adler, *Phys. Rev.* **137**, B1022 (1965); **139**, B1638 (1965).
5. E. van Beveren, T. A. Rijken, C. Metzger, C. Dullemond, G. Rupp, and J. E. Ribeiro, *Z. Phys. C* **30**, 615 (1986).
6. M. D. Scadron, *Phys. Rev. D* **26**, 239 (1982).
7. Eef van Beveren and George Rupp, *Eur. Phys. J. C*, **22**, 493 (2001) [hep-ex/0106077].
8. S. Eidelman *et al.* [Particle Data Group Collaboration], *Phys. Lett. B* **592**, 1 (2004).
9. Eef van Beveren and George Rupp, *Phys. Rev. Lett.* **91**, 012003 (2003) [hep-ph/0305035].
10. B. Aubert *et al.* [BABAR Collab.], *Phys. Rev. Lett.* **90**, 242001 (2003) [hep-ex/0304021].
11. K. Abe *et al.* [Belle Collab.], *Phys. Rev. D* **69**, 112002 (2004) [hep-ex/0307021].
12. F. Kleefeld, *AIP Conf. Proc.* **717**, 332 (2004) [hep-ph/0310320].
13. Nils A. Törnqvist, *Z. Phys. C* **68**, 647 (1995) [hep-ph/9504372].
14. H. Q. Zheng *et al.*, *Nucl. Phys. A* **733**, 235 (2004) [hep-ph/0310293]; J. A. Oller, E. Oset, and J. R. Pelaez, *Phys. Rev. D* **59**, 074001 (1999) [*Erratum-ibid.* **60**, 099906 (1999)] [hep-ph/9804209].

Open and Hidden Beauty Production at HERA-B

M.Mevius for the HERA-B collaboration

DESY, Hamburg

Abstract. HERA-B is a fixed target experiment that used the 920 GeV proton beam of HERA at DESY on targets of different materials. The results of the analysis of 150M dilepton triggered events on b-quark production are presented. The preliminary results for the $b\bar{b}$ cross section yields: $\sigma(b\bar{b}) = 9.8 \pm 1.4_{stat} \pm 2.0_{sys}$ nb/nucleon. For the Υ production cross section a preliminary result of: $\mathrm{Br}(\Upsilon \rightarrow l^+ l^-) \times \frac{d\sigma(\Upsilon)}{dy}\big|_{y=0} = 3.4 \pm 0.8_{stat}$ pb/nucleon is reported.

Keywords: hadroproduction, bottom mesons
PACS: 13.85.Ni,14.65.Fy

The Experiment

The HERA-B detector is a multiparticle spectrometer located at the 920 GeV proton beam of HERA at DESY. Interactions are produced by inserting target wires of different materials into the halo of the beam. A vertex detector which consists of 8 layers of double-sided silicon strip detectors and which is located directly behind the targets, provides adequate resolution to separate b-decays from primary interactions. Downstream of the vertex detector a magnet followed by several layers of tracking stations are placed. Behind the magnet there are also 3 detectors for particle identification. First, the Cherenkov counter (RICH) for π/K separation, second the electromagnetic calorimeter (ECAL) for electron identification and finally the muon tracking chambers (MUON) for muon identification. The latter two detectors also serve to provide an input for the trigger, which is designed to select dilepton events.

During the physics run of 2002/2003 about 150M dilepton triggered events were stored on tape. The preliminary results presented in this paper concern the analysis of these events.

Open Beauty Production

The open beauty production cross section ($\sigma(b\bar{b})$) is measured in the inclusive channel $b \rightarrow J/\psi X$, where the J/ψ decays into two leptons. The x_F acceptance for J/ψ ranges from -0.35 to 0.15. To minimize the systematic error resulting from uncertainties in trigger and reconstruction efficiencies and to avoid the measurement of absolute luminosity, $\sigma(b\bar{b})$ is measured relative to the direct (*prompt*) J/ψ cross section ($\sigma(J/\psi)$):

$$\sigma(b\bar{b}) = \sigma(J/\psi) \cdot \frac{n_B}{n_P} \cdot \frac{1}{\varepsilon_R \cdot \varepsilon_B^{\Delta z} \cdot \mathrm{Br}(b\bar{b} \rightarrow J/\psi X)}, \quad (1)$$

CP756, *Quark Confinement and the Hadron Spectrum VI*
edited by N. Brambilla, U. D'Alesio, A. Devoto, K. Maung, G.M. Prosperi and S. Serci
© 2005 American Institute of Physics 0-7354-0241-8/05/$22.50

where n_B, n_P are the number of observed $b \to J/\psi X$ and prompt J/ψ events, respectively. Most of the uncertainties on trigger and reconstruction efficiency cancel in the ratio of efficiencies for J/ψ selection (ε_R) in $b \to J/\psi X$ and prompt J/ψ events. ε_R is determined from MC to be close to 1. $\varepsilon_B^{\Delta z}$ is the efficiency of the b-selection criteria. The branching ratio $Br(b\bar{b} \to J/\psi X)$ is the probability of having at least one J/ψ in a $b\bar{b}$ event and is taken as two times the LEP1 measurement of $Br(b \to J/\psi X) = (1.16 \pm 0.10)\%$ [1].

The $b \to J/\psi X$ events are selected by making use of the long lifetime of the B-meson. A J/ψ that is inconsistent with being produced in the target must be the product of a b-decay. Therefore we select dilepton candidates which have a common vertex which is displaced from the target (8 mm on average). Furthermore, also both leptons are required to be inconsistent with being produced in the primary interaction, by demanding large impact parameters with the target. The efficiency of the selection ranges between 30 and 40 %, depending on the target configuration. In Fig. 1 the mass of the remaining dilepton candidates with a common vertex downstream of the target, is shown, both in the $\mu^+\mu^-$ and in the e^+e^- channel. A clear J/ψ signal is visible in both channels, where no signal is visible in the unphysical upstream region. The background can be described by combinatorics and double-semileptonic $b\bar{b}$ and $c\bar{c}$ events.

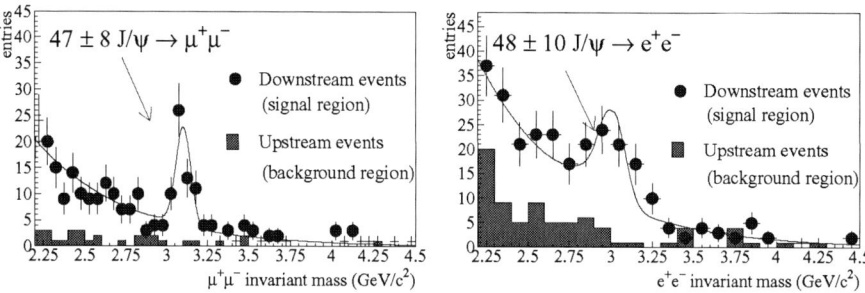

FIGURE 1. Dilepton mass for detached events: left: $\mu^+\mu^-$, right: e^+e^-. A clear J/ψ signal of about 50 events in each channel is visible. In each plot the filled histogram shows the spectrum for the unphysical upstream region, where no signal can be observed.

Combining the results in the electron and muon channel, we measured a preliminary value of the cross section ratio:

$$\frac{\sigma(b\bar{b})}{\sigma(J/\psi)} = 0.027 \pm 0.004_{stat} \pm 0.005_{sys} \qquad (2)$$

The J/ψ cross section has been measured by two Fermilab experiments at energies close to the HERA-B energy [2] [3]. Their combined measurement scaled to the HERA-B cms energy yield: $\sigma(J/\psi) = 357 \pm 2_{stat} \pm 36_{sys}$ nb/nucleon. Inserting this value into equation 2, the preliminary value for the $b\bar{b}$ cross section becomes: $\sigma(b\bar{b}) = 9.8 \pm 1.4_{stat} \pm 2.0_{sys}$ nb/nucleon.

The statistics of $b\bar{b}$ events of the HERA-B 2002/2003 measurement surpasses that of all earlier fixed target experiments by at least a factor 5. The result is within 2 sigma comparable with the HERA-B 2000 measurement [4].

Hidden Beauty Production

At the high end of the dilepton mass spectra (Fig. 2) a total of about 60 $\Upsilon \to l^+l^-$ decays (both channels) is observed. Since the contributing states could not be resolved, the ratio of the different Υ states is fixed to the results of the E605 experiment [5]. At high masses the background is dominated by Drell-Yan production, the shape of which is estimated from MC, while at low masses it is dominated by the combinatorial background, estimated from either like-sign (*muons*) or mixed (*electrons*) events.

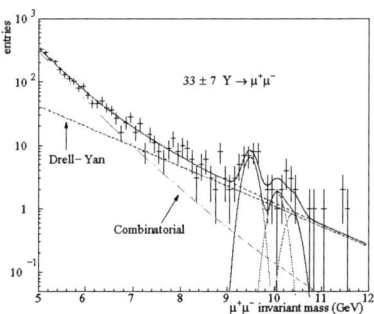 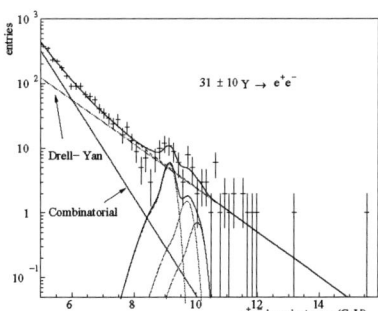

FIGURE 2. Dilepton mass: left: $\mu^+\mu^-$, right: e^+e^-. The ratio for the different Υ states is fixed from the results of the E605 experiment [5]. The different background contributions (Drell-Yan and combinatorics) are shown separately.

These signals are used to measure the Υ cross section ($\sigma(\Upsilon)$) times the branching ratio ($\mathrm{Br}(\Upsilon \to l^+l^-)$) at midrapidity. $\mathrm{Br}(\Upsilon \to l^+l^-)$ is the weighted average over the 3 Υ states of the Υ branching fraction into either e^+e^- or $\mu^+\mu^-$ (assumed equal). As in the case of open b production, the measurement is done relative to the J/ψ cross section:

$$\mathrm{Br}(\Upsilon \to l^+l^-) \times \frac{d\sigma(\Upsilon)}{dy}\Big|_{y=0} = \mathrm{Br}(J/\psi \to l^+l^-) \cdot \sigma(J/\psi) \cdot \frac{N_\Upsilon}{N_{J/\psi}} \frac{\varepsilon_{J/\psi}}{\varepsilon_\Upsilon} \frac{1}{\Delta y_{eff}}, \quad (3)$$

with N_Υ, $N_{J/\psi}$ the number of observed Υ and J/ψ events, respectively, and $\varepsilon_\Upsilon, \varepsilon_{J/\psi}$ the respective efficiencies. The Υ production model gives $\Delta y_{eff} = 1.068 \pm 0.002$. The preliminary combined muon and electron result yields:

$$\mathrm{Br}(\Upsilon \to l^+l^-) \times \frac{d\sigma(\Upsilon)}{dy}\Big|_{y=0} = 3.4 \pm 0.8_{stat} \text{ pb/nucleon} \quad (4)$$

The uncertainty on this measurement is comparable to or better than that of earlier measurements at energies close to the HERA-B energy.

REFERENCES

1. S. Eidelman *et al.*, Phys. Lett. **B592** (1) 2004.
2. T. Alexopoulos *et al.*, Phys. Rev. **D55** (1997) 3927.
3. M. H. Schub *et al.*, Phys. Rev. **D52** (1995) 1307.
4. I. Abt *et al.*, Eur. Phys. J. **C26** (2003) 345.
5. G. Moreno *et al.*, Phys. Rev. **D43** (1991) 2815.

Effective Field Theories for QQQ and QQq baryons

Nora Brambilla

Dipartimento di Fisica dell'Università di Milano and INFN via Celoria 16, 20133 Milan, Italy

Abstract. Effective field theories describing baryons with three heavy quarks (QQQ) and baryons with two heavy quarks (QQq) are discussed.

Keywords: Heavy Baryons, QCD, NRQCD, HQET
PACS: 12.38.-t,14.20.-c

INTRODUCTION

Baryons made of more than one heavy quark are interesting systems for several reasons.

Baryons made by two heavy quarks and a light quark QQq combine the slow motion of the two heavy quarks with the fast motion of the light quark. Thus, a treatment combining in two steps an EFT for the QQ interaction and an EFT for the QQ degrees of freedom with the light quark is the most appropriate one (at least when the momentum transfer between the two heavy quarks is larger than Λ_{QCD}). While recent lattice results from several groups on three quark static potentials exist [1], such potentials have not yet been calculated for the situation containing two static sources at distance r, accompanied by a light quark. Direct evaluation of the masses have been performed using NRQCD or anisotropic lattices [2]. The interest of these states is also related to the fact that the SELEX experiment recently announced the evidence of four doubly charmed baryon states [3, 4]. In the Section *Baryons with two heavy quarks* we briefly summarizes the results of the EFT that describes QQq systems.

Baryons made of three heavy quarks QQQ have not yet been observed. On the other hand, the three-quark static Wilson loop is studied [1] as a source of information about the baryon heavy quark potential and about the type of nonperturbative configurations that bind together the three quarks. It is therefore interesting to elucidate also for the case of three quarks, the relation between the interaction potential and the three-quark Wilson loop inside the effective field theory framework, as has been done for the case of two heavy quarks [5]. In the Section *Baryons with three heavy quarks* we briefly summarize the results of the effective field theory (EFT) that describes the QQQ systems in all the dynamical regimes. The matching procedure gives the form of the three-quark potentials.

BARYONS WITH THREE HEAVY QUARKS

In the case of a bound state formed by three heavy quarks, a hierarchy of physical scales similar to the quarkonium case exists [4, 6]. Consequently, starting from NRQCD and integrating out the scale of the momentum transfer $\simeq mv$, it is possible to write the pNRQCD Lagrangian for heavy baryons [7, 8]. In this case, of course, we have a center of mass coordinate and two Jacobi coordinates, two reduced masses and two relative momenta. What matters is that the two relative momenta have the same counting in the velocity v. Then two different dynamical situations may occur: the momentum transfer is much larger than Λ_{QCD}, or it is of order Λ_{QCD}.

In the first case the matching is perturbative and the Lagrangian is similar to the pNRQCD Lagrangian for the two quark case [6] with more degrees of freedom for the quark part: two octets, one singlet and one decuplet (as it comes from the color decomposition of $3 \times 3 \times 3$) [7].

In the second case the matching is nonperturbative and the Lagrangian reduces to the usual Schrödinger form with only the three-quark singlet as degree of freedom. This result relies on lattice measurements of the first three-quark hybrid static potential [9] that shows a gap of order Λ_{QCD} with respect to three-quark potential, as in the quark-antiquark case [5]. The (matching) potentials that appear in the EFT Lagrangian are nonperturbative objects [7]. They appear factorized in a part containing the high energy dynamics (and calculable in perturbation theory) which is inherited from the NRQCD matching coefficients, and a part containing the low energy dynamics given in terms of Wilson loops and chromo-electric and chromo-magnetic insertions in the Wilson loop. Such low energy contributions can be calculated on the lattice or evaluated in QCD vacuum models. The Wilson loop approach calculation of the three-quark potential [12] was missing all the matching coefficients.

These results should in principle be used in the phenomenological description [10, 11] of the heavy baryon spectrum with potentials and, as explained above, the potentials will have a different form if the physical system is such that the soft scale is comparable to Λ_{QCD} or such that it is bigger than Λ_{QCD}.

BARYONS WITH TWO HEAVY QUARKS

In this case the non relativistic motion of the two heavy quarks is similar to quarkonium while the light quark is moving relativistically around the slowly moving QQ pair. This is used in several phenomenological approaches [10, 11] to simplify the treatment, using e.g. a diquark approximation or a two step procedure in the potential calculation. More precisely, for $mv \gg \Lambda_{QCD}$, the QQ state may be treated as a point-like spin color antitriplet [13]. Then the system is similar to a $\bar{Q}q$ system and HQET and spin symmetry relations can be used to put in relation the hyperfine separation of these baryons with the hyperfine separation of B and D mesons [13]. However, the situation is much more interesting because more scales are active and m is finite. For instance the QQ has internal excitations. If one constructs first the EFT for the two heavy quarks more degreees of freedom enter and depending on the dynamical situation of the physical system, these degrees of freedom may or may not have a role. A rich spectrum of excitations is thus expected, which combines both the excitations of the QQ system and those of the light degrees of freedom.

Working under the condition that the momentum transfer between the two heavy quarks is larger than Λ_{QCD}, it is possible to construct a pNRQCD Lagrangian of the type given in [6] for the $Q\bar{Q}$ case, with a triplet and a sextet as QQ degrees of freedom (for all the details see [7]). Such degrees of freedom, besides giving origin to more energy levels in the spectrum, could also be relevant for the study of double charmonia production [14].

The assumption that the two heavy quarks behave like a nonrelativistic perturbative system has to be confirmed by the data. It may well be that the scales of the physical system are such that the system is completely nonperturbative. In this case the two step matching procedure which factorizes the QQ dynamics from the light quark one is no longer applicable and we should rely on a lattice calculation of the QQ potential in presence of the light quark. Such a calculation resembles the lattice evaluation of the static hybrid potential [8].

REFERENCES

1. G. S. Bali, Phys. Rept. **343**, 1 (2001) [arXiv:hep-ph/0001312]; T. T. Takahashi, H. Suganuma and H. Ichie, arXiv:hep-lat/0401001; H. Suganuma, T. T. Takahashi and H. Ichie, arXiv:hep-lat/0312031; C. Alexandrou, P. De Forcrand and A. Tsapalis, Phys. Rev. D **65**, 054503 (2002) [arXiv:hep-lat/0107006]; V. G. Bornyakov *et al.* [DIK Collaboration], arXiv:hep-lat/0401026; F. Okiharu and R. M. Woloshyn, Nucl. Phys. Proc. Suppl. **129-130**, 745 (2004) [arXiv:hep-lat/0310007].
2. R. Lewis, N. Mathur and R. M. Woloshyn, Phys. Rev. D **64**, 094509 (2001) [arXiv:hep-ph/0107037]; N. Mathur, R. Lewis and R. M. Woloshyn, Phys. Rev. D **66**, 014502 (2002) [arXiv:hep-ph/0203253]; J. M. Flynn, F. Mescia and A. S. B. Tariq [UKQCD Collaboration], JHEP **0307**, 066 (2003) [arXiv:hep-lat/0307025].
3. M. Mattson *et al.* [SELEX Collaboration], Phys. Rev. Lett. **89**, 112001 (2002) [arXiv:hep-ex/0208014].
4. N. Brambilla *et al.*, arXiv:hep-ph/0412158; http://www.qwg.to.infn.it
5. N. Brambilla, A. Pineda, J. Soto and A. Vairo, Phys. Rev. D **63**, 014023 (2001) [arXiv:hep-ph/0002250]; A. Pineda and A. Vairo, Phys. Rev. D **63**, 054007 (2001) [Erratum-ibid. D **64**, 039902 (2001)] [arXiv:hep-ph/0009145].
6. N. Brambilla, A. Pineda, J. Soto and A. Vairo, arXiv:hep-ph/0410047. A. Pineda and J. Soto, Nucl. Phys. Proc. Suppl. **64**, 428 (1998) [arXiv:hep-ph/9707481]; N. Brambilla, A. Pineda, J. Soto and A. Vairo, Nucl. Phys. B **566**, 275 (2000) [arXiv:hep-ph/9907240]; N. Brambilla, A. Pineda, J. Soto and A. Vairo, Phys. Rev. D **60**, 091502 (1999) [arXiv:hep-ph/9903355].
7. N. Brambilla, T. Rösch, A. Vairo "QCD Effective Lagrangians for Heavy Baryons" IFUM-808-FT; T. Rösch, Diploma thesis, Heidelberg 2003.
8. J. Soto, arXiv:hep-ph/0301138.
9. T. T. Takahashi and H. Suganuma, arXiv:hep-lat/0210024;
10. J. M. Richard, arXiv:hep-ph/0212224. J. M. Richard, Phys. Rept. **212** (1992) 1; J. G. Korner, M. Kramer and D. Pirjol, Prog. Part. Nucl. Phys. **33**, 787 (1994) [arXiv:hep-ph/9406359].
11. V. V. Kiselev and A. K. Likhoded, Phys. Usp. **45**, 455 (2002) [Usp. Fiz. Nauk **172**, 497 (2002)] [arXiv:hep-ph/0103169]; D. Ebert, R. N. Faustov, V. O. Galkin and A. P. Martynenko, Phys. Rev. D **66**, 014008 (2002) [arXiv:hep-ph/0201217].
12. N. Brambilla, P. Consoli and G. M. Prosperi, Phys. Rev. D **50**, 5878 (1994) [arXiv:hep-th/9401051]; N. Brambilla, G. M. Prosperi and A. Vairo, Phys. Lett. B **362**, 113 (1995) [arXiv:hep-ph/9507300].
13. M. J. Savage and M. B. Wise, Phys. Lett. B **248**, 177 (1990); M. J. Savage and R. P. Springer, Int. J. Mod. Phys. A **6**, 1701 (1991).
14. J. P. Ma and Z. G. Si, Phys. Lett. B **568**, 135 (2003) [arXiv:hep-ph/0305079].

P-wave Radial distributions of a Heavy-light meson on a lattice

UKQCD Collaboration, A.M. Green*, J. Koponen† and C. Michael**

*Helsinki Institute of Physics
P.O. Box 64, FIN–00014 University of Helsinki, Finland
†Department of Physical Sciences and Helsinki Institute of Physics
P.O. Box 64, FIN–00014 University of Helsinki, Finland
**Department of Mathematical Sciences, University of Liverpool, L69 3BX, UK

Abstract. This is a follow-up to our earlier work for the charge(vector) and matter(scalar) distributions for S-wave states in a heavy-light meson, where the heavy quark is static and the light quark has a mass about that of the strange quark. The calculation is again carried out with dynamical fermions on a $16^3 24$ lattice with a lattice spacing of about 0.14 fm. It is shown that several features of the S- and P-wave distributions are in qualitative agreement with what one expects from a simple one-body Dirac equation interpretation.

INTRODUCTION

Experimentalists are often able to tell us several properties of a given meson, such as its energy, width and angular momentum. However, usually they can not tell us the structure of the meson. For example, with B_s states — the topic of interest here — they can not say whether these states are $b\bar{s}$, $b\bar{s}u\bar{u}$ or BK. Unfortunately, when theoreticians try to describe B_s states, they have to decide beforehand the state structure to be used in some model, which often has sufficient freedom to fit the data with any of the possible structures. In an attempt to clarify the Experiment $\longleftrightarrow$ Theory comparison, we suggest the use of lattice QCD. In principle, lattice QCD should give us all we need to know about B_s states. However, in practice, the results need to be corrected for the lattice spacing (a), finite lattice size (L) and quark mass (m_q) effects — but these are usually under control and with the advent of more computer resources they will decrease in importance.

The strategy followed here is to concentrate on the simplest of quark states, namely, the $Q\bar{q}$ system, where Q is an infinitely heavy quark (i.e. static) and q is a quark with about the strange quark mass. This system is sufficiently simple to enable state-of-the-art lattice calculations to generate much more "data" than can be achieved by direct experiment. This data consists of the ground state energies of S-, P- and D-wave states and also the spin average F-wave energy[1]. Not only are the ground state energies extracted, but also those of the corresponding excited states containing at least one radial node. In addition, the vector (charge) and scalar (matter) radial distributions of these states can be measured. This is an abundance of data, far beyond what has been done experimentally in the B_s meson. Given all this data the challenge is now for theorists to make models to explain it. In this quest there are two simplifications. Firstly, since Q is static, the system

CP756, Quark Confinement and the Hadron Spectrum VI
edited by N. Brambilla, U. D'Alesio, A. Devoto, K. Maung, G.M. Prosperi and S. Serci
© 2005 American Institute of Physics 0-7354-0241-8/05/$22.50

is essentially reduced to a one-body problem involving only the light quark q_s. Secondly, in the lattice calculations, since the energies and radial distributions are extracted from $Q\bar{q}$ correlations propagating in Euclidean time, it is expected that the resultant states are indeed $Q\bar{q}$ states with little contamination from other possible multiquark components. Support for this expectation(hope) is seen in Fig. 2 of Ref. [2]. There a $Q\bar{Q}$ correlation generates a linearly rising potential for interquark distances far larger than expected i.e. way beyond where $(Q\bar{q})(\bar{Q}q)$ configurations should appear through string breaking. To see this effect needs the the explicit introduction of $(Q\bar{q})(\bar{Q}q)$ correlations [3].

The "data base" for properties of $Q\bar{q}$ states measured on lattices is so far incomplete. In Ref. [4] we concentrated on the S-wave energies and radial distributions. The latter showed two distinct features. Firstly, the excited state distributions exhibited nodes as expected. Secondly, the charge distribution $(x_C(R))$ was of longer range than the matter distribution $(x_M(R))$ but with $x_C(0) \approx x_M(0)$. This is readily explained by the one-body Dirac equation, since there $x_C(R) = G(R)^2 + F(R)^2$ and $x_M(R) = G(R)^2 - F(R)^2$, where for S-waves $G(0) \gg F(0)$ with $G(R)$ and $F(R)$ becoming comparable for large R. In Ref. [1] the data base was extended to include the energies of the excited states $P_{+/-}$ and $D_{+/-}$ and also the spin averaged combinations D_{+-} and F_{+-}. Here P_- is the P-wave state with $j_q = 1/2$, since the spin of the Q does not play a dynamical role. In this note we return to measuring the charge and matter distributions as in Ref. [4], but concentrate on the P_--state and its excitations. This state is of particular interest, since — as shown in Fig. 4 of Ref. [1] — the indications are that the predicted $B_s(0^+)$ is below the BK-threshold and so should be very narrow as was found for the $c\bar{s}$ counterpart. The outcome is seen in Figs. 1 for the ground state distributions x^{11} and the off-diagonal distributions between the ground state and the first excited state x^{12}. The latter show single nodes as

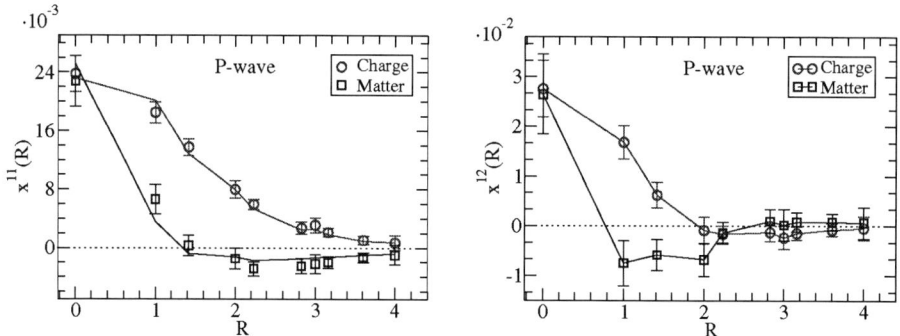

FIGURE 1. The P_- charge and matter distributions for $x^{11}(R)$ and $x^{12}(R)$.

expected from a first excited state. However, x^{11} shows two features which at first sight seem surprising:

1) The distributions are *finite* at $R = 0$, even though they are P-waves.

2) The matter distribution has a *node*, even though it involves only the ground state.

But again this is precisely what one expects from solutions of the Dirac equation, where for the P_--state both G and F are *non-zero* at $R = 0$ and, furthermore, the "small" component F can be *larger* than G at small R. In Figs. 2 the above data, now with a factor of R^2 included, are compared with the Dirac distributions using a quark mass of

100 MeV and an interquark potential $V = -a/R + bR$, where $a = 0.6$ and $b=1.3$ GeV/fm. These three parameters are very sensible and were not tuned to get an optimal fit.

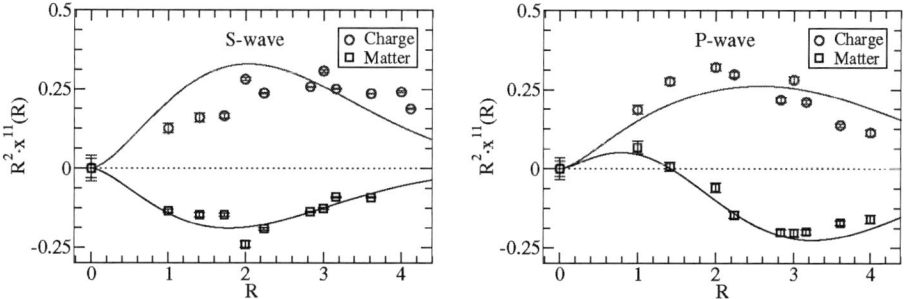

FIGURE 2. The S and P_- radial distributions from Fig. 1 compared with Dirac equation fits.

CONCLUSIONS

The S-and P-wave charge and matter distributions in Figs. 2 suggest that they can be understood qualitatively in terms of the one-body Dirac equation. The challenge is now to see to what extent there is an analogous *quantitative* description of the energies and distributions of all the S-, P-, D- and F-wave states — both ground and excited, when they become available. It is possible that the strategy used in studying the NN-potential is appropriate, namely, to first concentrate on the higher partial waves and so avoid or reduce complications, such as the effect of form factors needed to regulate the one-gluon-exchange potential and also instanton-induced interactions, that enter at small values of R.

ACKNOWLEDGMENTS

The authors wish to thank the Center for Scientific Computing in Espoo, Finland and the ULGrid project of the University of Liverpool for making available ample computer resources. Two of us, AMG and JK, acknowledge support by the Academy of Finland contract 54038 and the EU grant HPRN-CT-2002-00311 Euridice. JK thanks the Finnish Cultural Foundation and the Magnus Ehrnrooth Foundation for financial support.

REFERENCES

1. Green, A., Koponen, J., McNeile, C., Michael, C., Thompson, G., and UKQCD, *Phys. Rev. D*, **69**, 094505 (2004).
2. Bolder, B. e. a., *Phys. Rev. D*, **63**, 074504 (2001).
3. Michael, C., Pennanen, P., and UKQCD, *hep-lat/000105* (2000).
4. Green, A., Koponen, J., Michael, C., Pennanen, P., and UKQCD, *Eur. Phys. J. C*, **28**, 79 (2003).

Polarization in charmless $B \to VV$ decays

Fulvia De Fazio

Istituto Nazionale di Fisica Nucleare, Sezione di Bari, Italy

Abstract. Recent data for B decays to two light vector mesons show that the longitudinal amplitude dominates in $B^0 \to \rho^+\rho^-$, $B^+ \to \rho^+\rho^0$, $\rho^0 K^{*+}$ decays and not in $B^0 \to \phi K^{*0}$, $B^+ \to \phi K^{*+}$. We consider rescattering mediated by charmed resonances, finding that in $B \to \phi K^*$ it can be responsible of the suppression of the longitudinal amplitude. A similar result is found for $B \to \rho K^*$.

Recently, decay widths and polarization fractions of several B decays to two light vector mesons were measured [1, 2]. The branching ratios are of $\mathcal{O}(10^{-5})$. The measured polarization fractions, collected in Table 1, show that in penguin induced $B \to \phi K^*$ transitions the longitudinal amplitude does not dominate. However, using factorization and the heavy quark limit one can show that the light VV final state should be mainly longitudinally polarized. Actually, the decay $B^0 \to \phi K^{*0}$ is described by the amplitude

$$\mathcal{A}(B^0(p) \to \phi(q,\varepsilon)K^{*0}(p',\eta)) = \mathcal{A}_0\,\varepsilon^* \cdot \eta^* + \mathcal{A}_2(\varepsilon^* \cdot p)(\eta^* \cdot q) + i\mathcal{A}_1\,\varepsilon^{\alpha\beta\gamma\delta}\varepsilon^*_\alpha \eta^*_\beta p_\gamma p'_\delta,$$

with $\varepsilon(q,\lambda)$, $\eta(p',\lambda)$ the ϕ, K^* polarization vectors and $\lambda = 0,\pm 1$ the helicities. Since the B meson is spinless, the final mesons have the same helicity. In terms of $\mathcal{A}_{0,1,2}$ (describing S, P, D wave decays, respectively) the helicity amplitudes $\mathcal{A}_L$, $\mathcal{A}_\pm$ read: $\mathcal{A}_L = -[(p \cdot p' - M_{K^*}^2)\mathcal{A}_0 + M_B^2|\vec{p}'|^2\mathcal{A}_2]/(M_\phi M_{K^*})$, $\mathcal{A}_\pm = -\mathcal{A}_0 \mp M_B|\vec{p}'|\mathcal{A}_1$. The transverse amplitudes are defined as $\mathcal{A}_{\parallel,\perp} = (\mathcal{A}_+ \pm \mathcal{A}_-)/\sqrt{2}$, while the polarization fractions are $f_i = |\mathcal{A}_i|^2/|\mathcal{A}|^2$, $i = L, \parallel, \perp$.

Considering the effective weak Hamiltonian inducing the $\bar{b} \to \bar{s}s\bar{s}$ transitions [3] the amplitude $\mathcal{A}(B^0 \to \phi K^{*0})$ admits a factorized form $\mathcal{A}_{fact}(B^0 \to \phi K^{*0}) = -(G_F/\sqrt{2})V_{tb}^*V_{ts}a_W\langle K^{*0}(p',\eta)|(\bar{b}s)_{V-A}|B^0(p)\rangle\langle\phi(q,\varepsilon)|(\bar{s}s)_V|0\rangle$, with a_W a combination of Wilson coefficients. Using $\langle\phi(q,\varepsilon)|\bar{s}\gamma^\mu s|0\rangle = f_\phi M_\phi \varepsilon^{*\mu}$ and

$$\langle K^*(p',\eta)|(\bar{b}s)_{V-A}|B(p)\rangle = -i\varepsilon_{\mu\nu\rho\sigma}\eta^{*\nu}p^\rho p'^\sigma \frac{2V(q^2)}{M_B + M_{K^*}} - [(M_B + M_{K^*})A_1(q^2)\eta^*_\mu$$

$$-\frac{A_2(q^2)}{M_B + M_{K^*}}(\eta^* \cdot p)(p+p')_\mu - 2M_{K^*}\frac{(A_3(q^2) - A_0(q^2))}{q^2}(\eta^* \cdot p)q_\mu],\ (1)$$

$(q = p - p')$, one can write the polarization fractions and check that, for large M_B: $\mathcal{A}_L \propto M_B^3[(A_1(M_\phi^2) - A_2(M_\phi^2)) + \frac{M_{K^*}}{M_B}(A_1(M_\phi^2) + A_2(M_\phi^2))]$, $\mathcal{A}_\parallel \propto M_B A_1(M_\phi^2)$, $\mathcal{A}_\perp \propto M_B V(M_\phi^2)$. For $M_B \to \infty$, $q^2 = 0$ it was found that: $A_2/A_1 = V/A_1 = 1$ [4], giving $f_L \simeq 1 + \mathcal{O}(M_B^{-2})$, $f_\parallel/f_\perp \simeq 1$. Using generalized factorization, considering the a_i as effective parameters, one may reproduce the experimental branching ratio, but not the polarization fractions, since the dependence on the a_i cancels in the ratios. Hence, to explain the small f_L one has to look either at finite mass corrections and effects beyond factorization, or at new

TABLE 1. Polarization fractions in charmless $B \to VV$ transitions.

Mode	Pol. fraction	Belle [1]	BaBar [2]	Average
$B^+ \to \phi K^{*+}$	f_L	$0.49 \pm 0.13 \pm 0.05$	$0.46 \pm 0.12 \pm 0.03$	0.47 ± 0.09
	$f_\perp$	$0.12^{+0.11}_{-0.08} \pm 0.03$		
$B^0 \to \phi K^{*0}$	f_L	$0.52 \pm 0.07 \pm 0.05$	$0.52 \pm 0.05 \pm 0.02$	0.52 ± 0.04
	$f_\perp$	$0.30 \pm 0.07 \pm 0.03$	$0.22 \pm 0.05 \pm 0.02$	0.24 ± 0.04
$B^+ \to \rho^0 K^{*+}$	f_L		$0.96^{+0.04}_{-0.15} \pm 0.04$	
$B^+ \to \rho^0 \rho^+$	f_L	$0.95 \pm 0.11 \pm 0.02$	$0.97^{+0.03}_{-0.07} \pm 0.04$	0.96 ± 0.07
$B^0 \to \rho^+ \rho^-$	f_L		$0.98^{+0.02}_{-0.08} \pm 0.03$	

physics [5]. Here we present the study in [6]. Since, using form factors computed for finite heavy quark mass, the polarization fractions in $B^0 \to \phi K^{*0}$ do not fit data, in [6] we considered rescattering of intermediate charm states, already studied in [7]-[9] and more recently in [10], showing that they can invalidate the dominance of the longitudinal configuration in $B \to \phi K^*$ without affecting $B \to \rho\rho$ modes.

The decay $B^0 \to \phi K^{*0}$ can also be induced by rescattering through $B \to D_s^{(*)} D^{(*)} \to \phi K^*$. Such effects could be sizeable since they involve Wilson coefficients of current-current operators ($\mathcal{O}(1)$), while the coefficients of penguin operators in $\bar b \to \bar s s \bar s$ are $\mathcal{O}(10^{-2})$. Besides, there is no CKM suppression ($|V_{tb}^* V_{ts}| \simeq |V_{cb}^* V_{cs}|$). The vertices $D_s^{(*)} D^{(*)} K^*$, $D_s^{(*)} D_s^{(*)} \phi$ can be estimated using an effective Lagrangian describing the interactions of heavy hadrons with light vector mesons [11]. We write
$$\langle D_s^{(*)+} D^{(*)-} | H_W | B^- \rangle = (G_F/\sqrt{2}) V_{cb} V_{cs}^* a_1 \langle D^{(*)-} | (V-A)^\mu | B^0 \rangle \langle D_s^{(*)+} | (V-A)_\mu | 0 \rangle,$$
with $a_1 \simeq 1$, since there is empirical evidence that factorization works in these modes. In the heavy quark limit the above matrix elements involve the Isgur-Wise function ξ and a single quantity $f_{D_s} = f_{D_s^*}$; we use $\xi(y) = (2/(1+y))^2$ and $f_{D_s} = 240$ MeV.

Since the exchanged mesons are off-shell, we write the couplings $g_i(t) = g_{i0} F(t)$, g_{i0} being on-shell couplings and $F(t) = (\Lambda^2 - M_{D_s^*}^2)/(\Lambda^2 - t)$ to satisfy QCD counting rules. The relative sign of rescattering and factorized amplitude is unknown, as well as the role of diagrams involving excitations, hence we fix $\Lambda = 2.3$ GeV analyzing the sum $\mathcal{A} = \mathcal{A}_{fact} + r \mathcal{A}_{resc}$ in terms of the parameter r. In $\mathcal{A}_{fact}$ we use the $B \to K^*$ form factors in [12, 13], with Wilson coefficients from [3]. The result is shown in fig.1 [6]. For the model [12], $r \simeq 0.08$ gives the experimental branching ratio and $f_L \simeq 0.55$. Using the form factors in [13], we reproduce the branching ratio for $r \simeq -0.05$, though increasing f_L. However, this conclusion depends on the value of the Wilson coefficients. For smaller a_W, in both cases a similar long-distance contribution is required, reducing f_L.

Our conclusion is that rescattering can modify the helicity amplitudes in penguin dominated modes. On the other hand, such effects are too small to affect $B \to \rho\rho$ decays. Actually, while in the tree diagram in $B^0 \to \rho^+ \rho^-$ the CKM factor ($V_{ub}^* V_{ud}$) has similar size to that in the rescattering diagrams ($V_{cb}^* V_{cd}$), the Wilson coefficient in current-current transition is $\mathcal{O}(1)$. We expect to observe FSI effects in colour-suppressed and other penguin induced $B \to VV$ decays. For $B^+ \to \rho^0 K^{*+}$, including the rescattering term, we get $f_L \simeq 0.7$, i.e. smaller (though compatible within 2-σ) than the datum in Table 1. Hence, our approach can give a small $f_L(B \to \phi K^*)$ at the price of a smaller

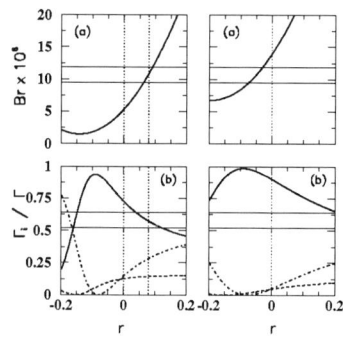

FIGURE 1. Dependence of branching ratio and polarization fractions of $B^0 \to \phi K^{*0}$ on the long distance term. $B \to K^*$ form factors in [12] (left) and [13] (right) are used in the factorized amplitude. $r = 0$ corresponds to absence of rescattering. The three curves in (b) refer to f_L (continuous), $f_\perp$ (dashed) and $f_\parallel$ (dot-dashed). The horizontal lines represent the data for the branching ratio (a) and for f_L (b).

$f_L(B \to \rho K^*)$. For $B^+ \to K^{*0}\rho^+$ very recent measurements reported lower longitudinal fractions: $f_L = 0.79 \pm 0.08 \pm 0.04 \pm 0.02$ [14], $f_L = 0.50 \pm 0.19^{+0.05}_{-0.07}$ [15].

Other analyses of non factorizable effects have been proposed [10, 16], but they will not be discussed here. It is only worth mentioning the common conclusion that there seems to be no need of non standard explanations to understand the polarization fractions in $B \to VV$ transitions, even though more refined analyses are required.

I thank P. Colangelo and T.N. Pham for collaboration. Partial support from the EC Contract No. HPRN-CT-2002-00311 (EURIDICE) is acknowledged.

REFERENCES

1. K.-F. Chen *et al.* [BELLE Collab.], Phys. Rev. Lett. **91** (2003) 201801; J. Zhang *et al.* [BELLE Collab.], Phys. Rev. Lett. **91** (2003) 221801; K.Abe et al., [BELLE Collab.], hep-ex/0408141.
2. B. Aubert *et al.* [BABAR Collab.], Phys. Rev. Lett. **91** (2003) 171802; Phys. Rev. D **69** (2004) 031102; hep-ex/0408017.
3. A. Ali, G. Kramer and C. D. Lu, Phys. Rev. D **58** (1998) 094009.
4. J. Charles *et al.*, Phys. Rev. D **60** (1999) 014001.
5. See Y. Grossman, Int. J. Mod. Phys. A **19** (2004) 907 and references therein.
6. P. Colangelo, F. De Fazio and T. N. Pham, Phys. Lett. B **597**, 291 (2004).
7. P. Colangelo *et al.*, Z. Phys. C **45** (1990) 575.
8. C. Isola *et al.*, Phys. Rev. D **64** (2001) 014029; Phys. Rev. D **68** (2003) 114001.
9. P. Colangelo *et al.*, Phys. Lett. B **542** (2002) 71; Phys. Rev. D **69** (2004) 054023.
10. M. Ladisa *et al.*, hep-ph/0409286; H. Y. Cheng *et al.*, hep-ph/0409317.
11. R. Casalbuoni *et al.*, Phys. Lett. B **292** (1992) 371.
12. P. Colangelo *et al.*, Phys. Rev. D **53** (1996) 3672 [Erratum-ibid. D **57** (1998) 3186].
13. P. Ball, eConf **C0304052** (2003) WG101 [arXiv:hep-ph/0306251].
14. B. Aubert [BABAR Collab.], hep-ex/0408093.
15. K. Abe et al. [BELLE Collab.], hep-ex/0408102.
16. A. L. Kagan, Phys. Lett. B **601** (2004) 151; W. S. Hou and M. Nagashima, hep-ph/0408007; H. n. Li and S. Mishima, hep-ph/0411146; H. n. Li, hep-ph/0411305.

Classical Solutions of SU(3) Yang-Mills Theory and Heavy Quark Phenomenology

O. Oliveira

Centro de Física Computacional, Departamento de Física, Universidade de Coimbra, 3004-516 Coimbra, Portugal

Abstract. It is showed that potentials derived from classical solutions of the SU(3) Yang-Mills theory can provide confining potentials that reproduce the heavy quarkonium spectrum within the same level of precision as the Cornell potential.

In order to solve the classical Yang-Mills equations of motion, usually one writes an ansatz that simplifies the Euler-Lagrange equations and, hopefully, includes the relevant dynamical degrees of freedom. In [1] it was proposed a generalized Cho-Faddeev-Niemi-Shabanov ansatz for the gluon field, where the gluon is given in terms of two vector fields, $\hat{A}_\mu$ and Y^a_μ, and a covariant constant real scalar field n^a,

$$A^a_\mu = n^a \hat{A}_\mu + \frac{3}{2g} f_{abc} n^b \partial_\mu n^c + Y^a_\mu \qquad (1)$$

with the constraints

$$D_\mu n^a = 0, \quad n^a Y^a_\mu = 0. \qquad (2)$$

In [1] it was showed that the above decomposition of the gluon field is gauge invariant but not necessarily complete. In the weak coupling limit, $g \rightarrow 0$, a finite gluon field requires either $n = 0$ or $\partial_\mu n = 0$. If $n = 0$, the gluon field is reduced to a vector field in the adjoint representation of SU(3) gauge group. For the other case, $\partial_\mu n = 0$, the gluon is writen in terms of the vector fields $\hat{A}_\mu$ and Y^a_μ and includes the previous solution as a particular case. Accordingly, a field such that $n \neq 0$ or $\partial_\mu n \neq 0$ does not produce a finite gluon field in the weak coupling limit and, in this sense, can be viewed as a nonperturbative field. Among this class of fields, the simplest parametrisation for the covariant scalar field[1] is $n^a = \delta^{a1}(-\sin\theta) + \delta^{a2}(\cos\theta)$. Then

$$A^a_\mu = n^a \hat{A}_\mu + \delta^{a3} \frac{1}{g} \partial_\mu \theta + \delta^{a8} C_\mu, \qquad (3)$$

where $C_\mu = Y^8_\mu$. The classical Lagrangian and equations of motion are independent of θ and are abelian like in $\hat{A}_\mu$ and C_μ. Among the possible nonperturbative gluons given

---

[1] From the constraint equation $Dn = 0$ it follows that n^2 is constant. Our choice was $n^2 = 1$. A different value for the norm of n is equivalent to a rescaling of $\hat{A}$.

CP756, *Quark Confinement and the Hadron Spectrum VI*
edited by N. Brambilla, U. D'Alesio, A. Devoto, K. Maung, G.M. Prosperi and S. Serci
© 2005 American Institute of Physics 0-7354-0241-8/05/$22.50

by (1), the simplest configuration has $\hat{A} = C = 0$. The coupling to the fermionic fields requires only the Gell-Mann matrix λ^3, decoupling the different colour components. This suggests, naively, that such a field is able to produce either confining, non-confining or free particle solutions for the quarks.

The classical equations of motion are independent of θ. However, a choice of a gauge condition, provides an equation for this field. For the Landau gauge, θ verifies a Klein-Gordon equation for a massless scalar field. Note that there is no boundary condition for θ, i.e. the usual free particle solutions of the Klein-Gordon equation are not the only possible ones. Indeed, writing $\theta(t,\vec{r}) = T(t)V(\vec{r})$, then

$$\frac{T''(t)}{T(t)} = \frac{\nabla^2 V(\vec{r})}{V(\vec{r})} = \Lambda^2 > 0 , \tag{4}$$

$$T(t) = ae^{\Lambda t} + be^{-\Lambda t} , \tag{5}$$

$$V(\vec{r}) = \sum_{l,m} V_l(r) Y_{lm}(\Omega) , \tag{6}$$

$$V_l(r) = \frac{\alpha_l}{\sqrt{z}} I_{l+1/2}(z) + \frac{\beta_l}{\sqrt{z}} K_{l+1/2}(z) , \tag{7}$$

where $z = \Lambda r$ and $I_{l+1/2}(z)$ and $K_{l+1/2}(z)$ are modified spherical Bessel functions of the 1st and 1rd kind[2]. The lowest multipole solution is

$$V_0(r) = A\frac{\sinh(\Lambda r)}{r} + B\frac{e^{-\Lambda r}}{r} \tag{8}$$

and the associated gluon field is given by

$$A_0^3 = \Lambda \left(e^{\Lambda t} - be^{\Lambda t} \right) V_0(r) , \tag{9}$$

$$\vec{A}_0^3 = - \left(e^{\Lambda t} - be^{\Lambda t} \right) \nabla V_0(r) . \tag{10}$$

From the lowest multipole solution one can derive a potential, which maybe suitable to describe heavy quarkonium. Indeed, assuming that quarks do not exchange energy, in the nonrelativistic approximation and leading order in $1/m$, the spatial function in A_0^3, $V_0(r)$, can be viewed as a nonrelativistic potential[3] and one can try to solve the associated Schrödinger equation. For the potential (8), the wave function goes to zero faster than an exponential for large quark distances,

$$\psi(\vec{r}) = \exp\left\{ \frac{-2}{\Lambda} \sqrt{\frac{2A}{m}} \exp\left(\frac{\Lambda r}{2}\right) \right\} . \tag{11}$$

As a first try to compute the heavy quarkonium spectra, we fixed A, B and Λ minimising the square of the difference between $V_0(r) + Constant$ and the Cornell potential [2] $V_{Cornell} = e/r + \sigma r$ ($e = -0.25$, $\sqrt{\sigma} = 427$ MeV) integrated between 0.2 fm and 1 fm.

---

[2] Note that, by definition, the mass scale Λ is independent of a rescaling of the gluon field.
[3] The potential is $\sim 1/r$ for short distances and goes to infinity for large quark distances.

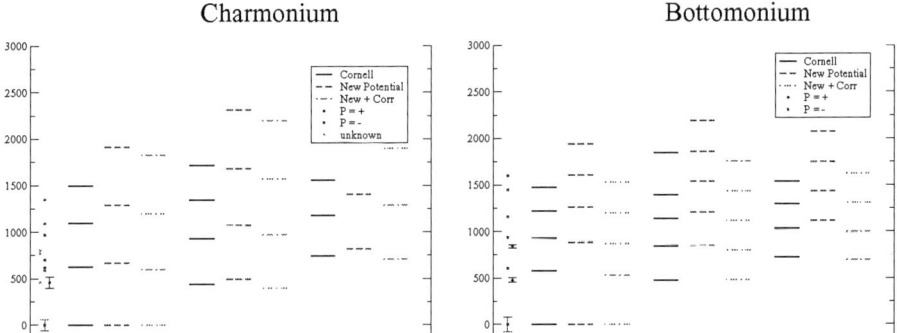

FIGURE 1. Heavy quarkonia spectra in MeV. The plots include the spin averaged experimental values.

This optimisation provides the following parameters $A = 5.4$, $B = -1.0$, $\Lambda = 281$ MeV, $Constant = -1190$ MeV; for these values $-24 MeV \leq V_{Cornell} - (V_0 + Constant) \leq 64$ MeV in the integration range considered. Then, we can compare the Schrödinger equation spectrum for the charmonium ($m_c = 1.25$ GeV) and for the bottomonium ($m_b = 4.25$ GeV) for the two potentials. The spectrum for the new potential shows an equal level spacing for both the charmonium and bottomonium spectra. If the V_0 charmonium spectrum is quite close to the Cornell spectrum, the botomonium shows clear deviations; see figure 1. The differences are the result of overestimating the strength of $V_0(r)$ for smaller distances. Indeed, one can improve our potential linearising the full QCD equations around the above configuration. To lowest order, this is equivalent to add a term like k/r to V_0. Computing k perturbatively[4] adjusting the $M[(1P)] - M[(1S)]$ bottomonium mass difference, gives $k = 0.2448251$. The heavy quarkonia spectra, including this correction, is given in figure 1.

In conclusion, classical configurations seem to be able to produce a spectra close to the Cornell potential. Hopefully, this is an indication that these configurations can be of help to understand strong interaction physics. Of course, there are a number of issues that need to be further investigated (definition of the potential parameters, inclusion of time dependence, decay rates). We are currently working on these topics and will provide a report soon.

REFERENCES

1. O.Oliveira, R. A. Coimbra, hep-ph/0305305
2. See G. S. Bali, Phys. Rep.343(2001)1-136 [hep-ph/0001312] and references therein.

---

[4] For each stationary, the shift in energy due to this term is compatible with a perturbative treatment. Corrections are clearly below 10-20%.

Theory and phenomenology of heavy quarkonia and new Higgs physics

G.A. Kozlov

Joint Institute for Nuclear Research, Joliot-Curie st.,6, 141980 Dubna, Russia

Abstract.
We study the effect of Higgs-boson exchange on the heavy quarkonium $\bar{Q}Q$ which induces a strong attractive force between a heavy quark Q and an antiquark $\bar{Q}$. The criterion for making the $\bar{Q}Q$ bound state is studied. The nonperturbative effects due to gluonic field fluctuations are rather small in such a heavy quark sector.

A study on quarkonia $T(\bar{Q}Q)$ composed of a heavy quark Q and an antiquark $\bar{Q}$ (a possible and interesting candidate for Q is the up(U)- and/or down(D)-quarks in the fourth generation family) is required in current particle physics for testing the Standard Model (SM) and/or searching for signals for physics beyond the SM. In particular, no theoretical arguments are seen to rule out the heavy quarks and the heavy quarkonium states with the masses around a hundred GeV or even a few TeV (see, e.g. [1] and the references therein).

Once the Higgs boson H is discovered, one needs to measure its couplings to other particles, including H production coming from the decay of very heavy quarkonia, e.g., $T(\bar{Q}Q) \to HZ$, $T(\bar{Q}_1 Q_1) \to HT(\bar{Q}_2 Q_2)$, $T(\bar{Q}Q) \to \gamma H$ and $T(\bar{Q}Q) \to ggH, \gamma\gamma H$, where $T(\bar{Q}Q)$ carries quantum numbers of $J^{PC} = 1^{--}, 0^{-+}$.

Relevant to the vector states it was a complement to the Wilczek radiative decay of the spin-1 particle with an emission of a Higgs boson [2].

The criterion for existence of a heavy quarkonium is that the binding energy ε_B should be larger at least than the total decay width Γ_{tot}, namely $c = (\Gamma_{tot}/\varepsilon_B) < 1$.

Creation of $\bar{Q}Q$ out of vacuum may be resulting also in a screening of quark color charges at large distances. Hence, one cannot exclude the possibility of the Higgs-boson interaction which dominates significantly over the one-gluon exchange $\sim -(4/3)\alpha_s(m_Q)/r$ for the very heavy quarkonium, where α_s is the strong coupling constant depending on the quark mass m_Q and r is a distance between Q and $\bar{Q}$.

It is well known that the exact QCD vacuum should contain fluctuations of gluonic fields at large scales [3]. These nonperturbative fluctuations cause the distortion of interactions between Q and $\bar{Q}$ allocated in the gluonic vacuum. We find that the nonperturbative fluctuation effect of the gluonic field gives a negligible result.

In the lowest bound state $\bar{Q}Q$, the quark Q and the antiquark $\bar{Q}$ are assumed to be located at a distance $r \sim [m_Q \lambda(m_Q)]^{-1}$ which is small compared to the scale of strong interactions; $\lambda(m_Q)$ is the strength of the interaction between Q and $\bar{Q}$. The wave function of the lowest bound state is proportional to $\exp(-\mu r)$ with $\mu \sim m_Q \lambda(m_Q)$. Let us consider a simple model where the dominant effective potential for a $\bar{Q}$ and Q system

CP756, *Quark Confinement and the Hadron Spectrum VI*
edited by N. Brambilla, U. D'Alesio, A. Devoto, K. Maung, G.M. Prosperi and S. Serci
© 2005 American Institute of Physics 0-7354-0241-8/05/$22.50

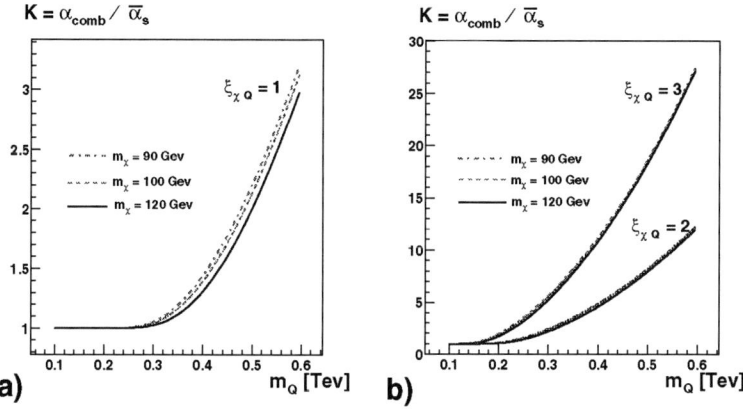

FIGURE 1.

at a small distance looks like [4], [5]

$$V_{eff}(r) \sim -\frac{C_F}{r}\alpha_s(m_Q) - \frac{\lambda(m_Q,\xi_{\chi Q})}{r}\exp(-m_\chi r), \quad \lambda(m_Q,\xi_{\chi Q}) = \frac{m_Q^2}{4\pi v^2}\xi_{\chi Q}^2, \quad (1)$$

and $\xi_{\chi Q}$ reflects the model "flavor" in the strength of the interaction between the scalar χ-boson and a heavy quark Q ($\xi_{\chi Q} = 1$ in the minimal SM, otherwise $\xi_{\chi Q} > 1$); $v = 246$ GeV ($v = \sqrt{v_1^2 + v_2^2}$, v_1 and v_2 are two neutral Higgs field vacuum expectation values) and $C_F = 4/3$ is the color factor for the color group $SU(3)$. In Fig.1, we show the ratio of the combined coupling $\alpha_{comb} = (4/3)\alpha_s(m_Q) + \lambda(m_Q,\xi_{\chi Q}) \cdot \exp(-m_\chi r)$ (see (1)) to the pure QCD coupling $\bar{\alpha}_s = (4/3)\alpha_s$, as a function of a heavy quark mass m_Q for different values of $\xi_{\chi Q}$ and m_χ. The second term in α_{comb} is appropriate for $r \sim [m_Q \lambda(m_Q)]^{-1}$. The ratio becomes somewhat bigger for smaller Higgs-boson masses. Because of our demand $\lambda(m_Q,\xi_{Q\chi}) > C_F\alpha_s(m_Q)$ for a relevance of the χ-boson interaction, the lower bound on m_Q is given as $m_Q > (v/\xi_{\chi Q})(4\pi C_F \alpha_s)^{1/2}$ which leads to $m_Q \geq m_t$ even if $\xi_{\chi Q} = 2$. The ratio $c = (\Gamma_{tot}/\varepsilon_B) < 1$ could be guaranteed for, e.g., $T(\bar{U}U)$ quarkonia to be formed by a strong attractive force via scalar Higgs-boson exchange with a sufficiently "hard" Yukawa coupling $\lambda(m_Q,\xi_{\chi Q})$. The total decay width Γ_{tot} is given by a sum of two terms $\Gamma_{tot} = \Gamma_T + \Gamma_U$, where the width Γ_T is defined by the following decay channels: $T(\bar{U}U) \to hZ, \gamma Z, \gamma h, W^+W^-, \bar{b}b, \bar{t}t, \tau^+\tau^-, \mu^+\mu^-, ggg$, and the single quark decay width Γ_U is a sum of the following contributions: $U \to DW^+, bW^+, bH^+$ (H^+ is the charged Higgs-boson). For the case of U quarks, one can expect that Γ_{tot} is given by $\Gamma_{tot} = \Gamma_T(T(\bar{U}U) \to hZ, W^+W^-) + \Gamma_U(U \to DW^+, bW^+)$. The main contribution to Γ_T arises from the channel $T(\bar{U}U) \to hZ$.

The peak of the cross-section at a $\bar{Q}Q$ resonance due to the one-gluon exchange is given by $\sigma_c \sim \alpha_s^3 \left(\frac{m_Q}{\Gamma_Q}\right) \frac{1}{s}$ for a given center-of-mass energy s. This cross-section has a strong sensitivity to α_s and decreases sharply with increasing m_Q because of the rapid growth of Γ_Q. To leading order, the scalar χ-boson exchange effect is taken

379

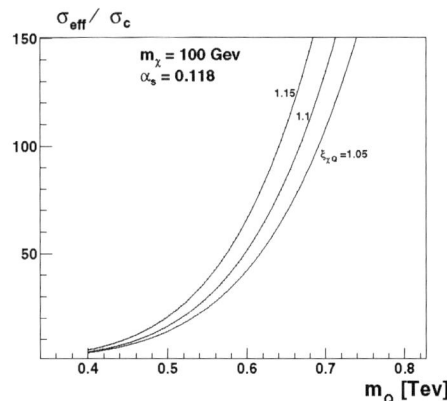

FIGURE 2.

into account simply by making the replacement $\alpha_s \to \alpha_s + \tilde{\alpha}(m_Q, m_\chi, \xi_{\chi Q})$, where $\tilde{\alpha}(m_Q, m_\chi, \xi_{\chi Q}) = \frac{3m_Q^2}{16\pi v^2} \xi_{\chi Q}^2 \exp(-\varepsilon)$ with $\varepsilon = m_\chi/(m_Q\lambda) \ll 1$. Here, we see a simple increase of the coupling strength between $\bar{Q}$ and Q. We find a relative enhancement of the effective cross-section $\sigma_{eff} = \sigma_c \left(1 + 3\tilde{\alpha}\,\alpha_s^{-1} + 3\tilde{\alpha}^2\,\alpha_s^{-2} + \tilde{\alpha}^3\,\alpha_s^{-3}\right)$ due to the χ-boson exchange effect. In Fig.2 we show the ratio of the effective cross-section σ_{eff} to the one-gluon exchange cross-section as a function of m_Q.

We have estimated the corrections due to non-perturbative fluctuations of the gluonic field in the decay width $\Gamma(T(\bar{U}U) \to hZ) = \Gamma_0(T(\bar{U}U) \to hZ)(1 + \delta_{NP})$, where $\Gamma_0(T(\bar{U}U) \to hZ)$ is the decay width thanks to the potential model, while the nonperturbative correction factor δ_{NP} is defined by $\delta_{NP} = \Gamma_{NP}/\Gamma_0$ with the non-perturbative amplitude A_{NP}, $\Gamma_{NP}(T(\bar{U}U) \to hZ) = \frac{1}{16\pi m_T} |A_{NP}|^2 \left(1 - \frac{4m_h^2}{m_T^2}\right)^{1/2}$. The δ_{NP} correction is obtained at the order of magnitude of 10^{-8}. Finally, some comments are order in the following: (a) the (super)heavy quarks are considered to be nonrelativistic; (b) if the χ Higgs-boson mass is $m_\chi \sim \mathcal{O}(100\ GeV)$, then the decoupling limit ($\xi_{\chi Q} \sim \mathcal{O}(1)$) is appropriate only for $m_Q \geq 2m_t$; (c) for the $(\bar{t}t)$-bound state with $\xi_{\chi t} \sim O(1)$, the mass m_χ should be smaller than the lower bound given by the LEP 2 experiments; (d) if the heavy quarkonium mass is of the order $2m_t$, the χ-boson contribution to the combined potential (1) becomes appreciable only for large values of $\xi_{\chi t} \geq 6$.

REFERENCES

1. G. A. Kozlov et al., *J. Phys. G: Nucl. Part. Phys.*, **30**, 1201–1218 (2004).
2. F. Wilczek, *Phys. Rev. Lett.*, **39**, 1304–1306 (1977).
3. M. B. Voloshin, *Nucl. Phys. B*, **154**, 365–387 (1979).
4. H. Inazawa and T. Morii, *Phys. Lett. B*, **247**, 107–112 (1990).
5. K. Hagiwara et al., *Nucl. Phys. B*, **344**, 1–32 (1990).

Singular behavior of $^1P_1^{+-}$ quarkonium and positronium annihilation decays and relevance of relative energy

Dieter Gromes

Institut für Theoretische Physik, Philosophenweg 16, D-69120 Heidelberg
E - mail: d.gromes@thphys.uni-heidelberg.de

Abstract. Using a four dimensional approach, we show that the singularities for small gluon momenta, which arise in the usual three dimensional treatment, disappear if all poles in the relative energy are taken into account correctly in the integration. We obtain an explicit formula for the decay width, not only for the familiar logarithmic dependence on the binding energy, but also for the constant to be added to the logarithm. It turns out to be large and negative.

The usual non relativistic approach for calculating the decay width of the $^1P_1^{+-}$-state into three gluons has infrared divergences originating from soft gluon momenta. In the original paper [1] the problem was overcome by introducing a fictituous width for the decay into a real and a virtual gluon. The result involved a singular logarithmic dependence on the binding energy, the constant which has to be added to the logarithm could, however, not be calculated.

In a series of papers Bodwin, Braaten, and Lepage [2] clarified important aspects of the problem. The divergences found previously are canceled by considering the contributions from the quark-antiquark-glue component of the wave function.

In a further paper on the subject the present author [3] calculated the glue content in heavy quarkonia in order to obtain more definite statements. Although this calculation was succesful and gave reasonable results, it turned out that it was, at least up to now, not helpful for calculating the decay width.

Here we go back to a more fundamental four dimensional formalism. It involves the T-matrix for the annihilation of the quark-antiquark pair into gluons, as well as the four dimensional Bethe Salpeter wave function, taken in the leading approximation coming from the static Salpeter equation. Both quantities depend on the relative energy $p^0 = (p_1 - p_2)^0/2$ of the quark-antiquark pair (Fig. 1).

$$< \mathbf{k}_1^a \mathbf{k}_2^b \mathbf{k}_3^c | K >_{tree} = -\frac{(2\pi)^4 \delta^{(4)}(k_1 + k_2 + k_3 - K)}{\sqrt{(2\pi)^5 m}} \times \tag{1}$$

$$\int Tr[T(\mathbf{p}, p^0)\Gamma(\mathbf{p})] \frac{(|E| + \mathbf{p}^2/m)\ \tilde{\psi}(\mathbf{p})\ dp^0\ d^3p}{(p^0 + |E|/2 + \mathbf{p}^2/2m - i\varepsilon)\ (p^0 - |E|/2 - \mathbf{p}^2/2m + i\varepsilon)}.$$

Here $\Gamma(\mathbf{p})$ is the spin wave function of the bound state. For a singlet state it reads $\Gamma(\mathbf{p}) = \Lambda(\mathbf{p}_1)(\gamma^5/\sqrt{2})\Lambda^c(\mathbf{p}_2)$, where $\Lambda(\mathbf{p}_k)$ are the projectors to positive energy states.

CP756, *Quark Confinement and the Hadron Spectrum VI*
edited by N. Brambilla, U. D'Alesio, A. Devoto, K. Maung, G.M. Prosperi and S. Serci
© 2005 American Institute of Physics 0-7354-0241-8/05/$22.50

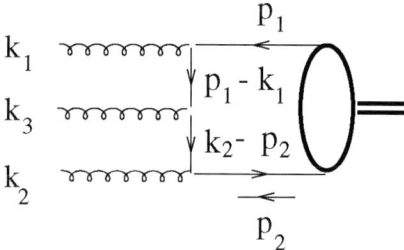

FIGURE 1. *The tree graph contribution to the three gluon decay.*

Finally $\tilde{\psi}(\mathbf{p})$ is the Schrödinger wave function, while E is the binding energy.

The BS wave function has two poles at $p^0 = \varepsilon_w^{\pm} \equiv \mp\, (|E|/2 + \mathbf{p}^2/(2m) - i\varepsilon\,)$. The usual non relativistic approach is equivalent to considering only the pole ε_w^{+} in the upper half plane, say, when integrating over p^0. The p^0-dependence of the T-matrix, which has four poles $\varepsilon_1^{\pm}$ and $\varepsilon_2^{\pm}$ from the two quark propagators, is, however, ignored by fixing the quarks on mass shell there. This procedure is usually legitimate, but fails in the present case if one of the gluons is ultrasoft. In this situation some poles of the quark propagators move close to poles of the wave function or to each other. A careful discussion shows that spurious poles in k_j cancel if all residues are considered. The poles $\sim 1/k_j$ for $j = 1, 2$, which arise in the usual three dimensional approach, disappear, and are replaced by the non singular but non local expressions $1/(k_j + |E| + \mathbf{p}^2/m)$. After integrating over p^0, eq. (1) becomes (we dropped ultrasoft contributions wherever legitimate):

$$
< \mathbf{k}_1^a \mathbf{k}_2^b \mathbf{k}_3^c | K >_{tree} = \left(\frac{d^{abc}}{4\sqrt{N_c}} \right) i(2\pi)^4 \delta^{(4)}(k_1 + k_2 + k_3 - K)\frac{g^3}{\pi^{3/2}m^{3/2}} \times
$$

$$
\int \left\{ \frac{[\varepsilon_1 \times \varepsilon_3] \cdot \mathbf{k}_1(\varepsilon_2 \cdot \mathbf{p})}{k_1(k_2 + |E| + \mathbf{p}^2/m)} + \frac{1}{4mk_1k_2}\left(\frac{(\mathbf{k}_1 \cdot \mathbf{p})}{k_1} - \frac{(\mathbf{k}_2 \cdot \mathbf{p})}{k_2} \right) \times \right. \tag{2}
$$

$$
\left. \left[(k_1k_2 + (\mathbf{k}_1 \cdot \mathbf{k}_2)[\varepsilon_1 \times \varepsilon_2] \cdot \varepsilon_3 + (\varepsilon_1 \cdot \varepsilon_2)[\mathbf{k}_1 \times \mathbf{k}_2] \cdot \varepsilon_3 + 2(\mathbf{k}_2 \cdot \varepsilon_1)[\varepsilon_3 \times \varepsilon_2] \cdot \mathbf{k}_1 \right] \right\} \times
$$

$\tilde{\psi}(\mathbf{p})\, d^3p +$ permutations of gluons.

The first term in the integrand arises from the product of the terms $\sim p$ in the trace, times the terms ~ 1 of the denominator (type 1), the remaining three from the terms ~ 1 of the trace, times the terms $\sim p$ of the denominator (type 2). Note that the P-state wave function also contains a factor of $\mathbf{p}$ from the spherical harmonics.

In QCD, contrary to QED, we encounter a phenomenon which we found already in [3]. If $\mathbf{k}_2$, say, is ultrasoft, a non abelian loop graph, where the gluon $\mathbf{k}_2$ couples to two Coulomb gluons, while the latter couple to the two quark lines, contributes to the same order as the tree graph in Fig. 1 (although it does not lead to a singular logarithmic dependence). The reason is that the tree graph is suppressed by the coupling of a transverse gluon to a non relativistic quark, while the loop graph has unsuppressed couplings to the quarks via Coulomb gluons.

Due to the non local dependence on the momentum **p** in (2), one cannot simply express the result by the derivative of the wave function at the origin. Instead one has to perform the p-integration explicitly. For the lowest Coulomb P-state wave function the width becomes

$$\Gamma(^1P_1^{+-} \to 3 \text{ gluons}) = \frac{5}{18} \frac{8}{\pi} \frac{\alpha_s^3}{m^4} |R'_{21}(0)|^2 \times \tag{3}$$

$$\left[\ln \frac{m}{|E|} - (\frac{25}{9} + \frac{\pi^2}{64} + 2\ln 2) - (\frac{3\pi^2}{32} - \frac{3}{4}) + (\frac{49\pi^2}{192} - \frac{5}{2}) - \frac{3\sqrt{3}\pi}{2\sqrt{2}} + \frac{3\pi}{8} \right].$$

The logarithmic term coincides with [1]. The five constants, which we were able to calculate explicitly, arise as follows: Square of type 1, mixed term between type 1 and type 2, square of type 2, mixed term between type 1 and loop graph, square of loop graph. (Only if terms which are singular for $k_2 \to 0$ come together, one obtains a relevant contribution from the loop graph, therefore mixed terms between type 2 and the loop graph don't show up.) Numerically the five contributions are:
$-4.318 - 0.175 + 0.019 - 3.256 + 1.178 = -6.553$.

This is a rather striking result. The constant is negative and large, to be compared, e.g. with $\ln(m/|E|) = 5.416$ for a Coulomb potential with $\alpha_s = 0.2$. This would imply, that the whole approach only makes sense for much smaller α_s. Clearly the result depends drastically on the correctness of relative signs (there was an error and an omission in [4]), an independent check would be highly welcome!

We made use of the $Q\bar{Q}$- wave function only. To make contact with the three dimensional approach, the contributions from the poles in the T-matrix can, alternatively, be interpreted as contributions from $Q\bar{Q}g$ components (with a soft gluon) in the wave function. In our approach we can get this contribution immediately in an explicit form.

At present there are no established candidates for the $^1P_1^{+-}$ states h_c and h_b. The decay of the spin triplet state $^3P_1^{++}$ into a real gluon plus a light quark-antiquark pair shows the same infrared problems, and can be treated along the same lines. The corresponding charmonium state $\chi_{c1}(1P)$ is experimentally well established and will be investigated in forthcoming work.

REFERENCES

1. R. Barbieri, R. Gatto, E. Remiddi, Phys. Lett. **61** B, 465 (1976).
2. G. T. Bodwin, E. Braaten, G. P. Lepage, Phys. Rev. D **46**, R1914 (1992), Phys. Rev. D **51**, 1125 (1995); Phys. Rev. D **55**, 5853(E) (1997).
3. D. Gromes, Eur. Phys. J. C **30**, 233 (2003).
4. D. Gromes, Eur. Phys. J. C **36**, 169 (2004), hep-ph/0403290, and errata to appear.

Spin Correlations and Velocity-Scaling in NRQCD Matrix Elements [1]

Geoffrey T. Bodwin*, Jungil Lee† and D. K. Sinclair*

*HEP Division, Argonne National Laboratory, 9700 South Cass Avenue, Argonne, IL 60439
†Department of Physics, Korea University, Seoul 136-701, Korea

Abstract. We compute spin-dependent matrix elements for decays of S-wave quarkonia in lattice NRQCD. They appear to be in approximate agreement with the velocity-scaling rules of NRQCD.

PACS: 12.38Ge,13.25Gv

The effective field theory Nonrelativistic QCD (NRQCD) [1, 2, 3] is an elegant formalism within which to describe heavy-quarkonium physics. It is believed that, within NRQCD, inclusive cross sections for the production of a quarkonium at large transverse momentum p_T can be written in a factorized form [3]:

$$\sigma(H) = \sum_n \sigma_n(\Lambda)\langle 0|\mathcal{O}_n^H(\Lambda)|0\rangle, \tag{1}$$

where the σ_n are perturbatively calculable short-distance coefficients, and the $\langle 0|\mathcal{O}_n^H(\Lambda)|0\rangle$ are matrix elements in the vacuum state of four-fermion operators of the form $\mathcal{O}_n^H = \chi^\dagger \kappa_n \psi \left(\sum_X |H+X\rangle\langle H+X|\right) \psi^\dagger \kappa_n' \chi$. Here, H is the quarkonium state, ψ is the Pauli spinor field that annihilates a heavy quark, χ is the Pauli spinor field that creates a heavy antiquark, and Λ is the ultraviolet cutoff of NRQCD. κ contains Pauli matrices, color matrices, and the covariant derivatives. NRQCD predicts the leading scaling behavior of the matrix elements with v, the heavy-quark velocity in the quarkonium rest frame [3]. As a consequence of these v-scaling rules, the sum over operator matrix elements can be regarded as an expansion in powers of v, where $v^2 \approx 0.3$ for charmonium and $v^2 \approx 0.1$ for bottomonium. A similar factorization formula applies to inclusive quarkonium decays [3], except that the matrix elements are now between quarkonium states, rather than vacuum states, and the four-fermion operators have the form $\mathcal{O}_n = \psi^\dagger \kappa_n \chi \chi^\dagger \kappa_n' \psi$.

At large p_T at the Tevatron, the dominant mechanism for J/ψ production is gluon fragmentation into a $Q\bar{Q}$ pair, which then evolves nonperturbatively into the J/ψ. The nonperturbative evolution is described by the NRQCD matrix elements. The NRQCD v-scaling rules predict that the spin-flip matrix elements are suppressed by v^2 in comparison with the non-spin-flip matrix elements. Therefore, in existing calculations, the spin-flip effects are neglected, and the J/ψ is assumed to take on the polarization of the fragmenting gluon. Consequently, the J/ψ is expected to have a significant transverse polarization at large p_T (Ref. [4]). Surprisingly, the CDF data for the J/ψ polarization

---

[1] Talk presented by G. T. Bodwin.

CP756, *Quark Confinement and the Hadron Spectrum VI*
edited by N. Brambilla, U. D'Alesio, A. Devoto, K. Maung, G.M. Prosperi and S. Serci
© 2005 American Institute of Physics 0-7354-0241-8/05/$22.50

[5] show decreasing transverse polarization with increasing p_T and disagree with the NRQCD prediction [6] in the largest p_T bin.

One can question whether the neglect of spin-flip processes is justified in the case of the J/ψ. The v-scaling rules predict the leading power of v in a matrix element, but not its coefficient. It could happen that the spin-flip matrix elements, although suppressed by v^2 relative to the non-spin-flip matrix elements, are not actually smaller numerically. It has also been suggested that the v-scaling rules themselves may need to be modified for charmonium [7, 8, 9, 10, 11].

Ideally, we would settle these issues by computing the J/ψ production matrix elements in a lattice QCD simulation. Unfortunately, it is not known how to formulate the computation of production matrix elements in lattice simulations. However, we can instead use lattice calculations of *decay* matrix elements to test the validity of using the v-scaling rules to estimate the sizes of matrix elements.

We take for our lattice action the discretized version of the NRQCD action through next-to-leading order in v that is given in Ref. [12]. We also include improvements of relative-order a to the terms in the lattice action that are of leading order in v^2, and we implement tadpole improvement [13] as described in Ref. [12]. Our initial computations were carried out on quenched gauge-field configurations, which should reproduce the qualitative features of QCD. Our preliminary results are based on 400 gauge-field configurations on $12^3 \times 24$ lattices at $\beta = 5.7$. We use the values $am_c = 0.8$ and $am_b = 3.15$ for the heavy-quark masses in lattice units. We take the parameter n that appears in the action of Ref. [12] to be 1 for bottomonium and 4 for charmonium.

Since, at $\beta = 5.7$, $a = 0.81$ GeV^{-1} for charmonium [12] and $a = 0.73$ GeV^{-1} for bottomonium [14], the quarkonium is well contained in the lattice volume. However, since the quarkonium radius is approximately $1/(mv)$, which is about 1.2 GeV^{-1} for charmonium and about 0.6 GeV^{-1} bottomonium, the lattice spacing is fairly coarse and warrants the implementation of order-a improvements.

Our preliminary results for S-wave quarkonium states are shown in Table 1. As can be seen, the hierarchy of matrix elements that is predicted by v-scaling is observed. However, the sizes of the matrix elements decrease faster with increasing powers of v than one would expect from the powers of v alone. The NRQCD heavy-quark spin symmetry [3] predicts that the singlet-to-triplet transition and the triplet-to-singlet transition should be in a ratio of 3:1, up to corrections of order v^2. The lattice results agree with this prediction. Furthermore, the ratios of bottomonium matrix elements to charmonium matrix elements for the larger, well-measured matrix elements agree within about a factor of two with expectations from v scaling. Taken together, these results suggest that v scaling is fairly well obeyed, but that additional factors, beyond the powers v, must be taken into account in order to estimate the sizes of the matrix elements accurately.

As can be seen from Table 1, the triplet-to-singlet transition rate is comparable to the triplet-to-triplet transition rate. This suggests that, at large p_T at the Tevatron, where the color-octet, spin-triplet process dominates S-wave quarkonium production, the η_c production rate may be comparable to the J/ψ production rate.

The spin-triplet up-to-longitudinal transition rate is small compared with the spin-triplet up-to-up transition rate. That is, the spin-flip matrix elements are suppressed, as is expected from v-scaling. Hence, our preliminary results support the prediction of large transverse polarization of J/ψ's produced at large p_T at the Tevatron.

TABLE 1. Quarkonium decay matrix elements. A spin transition labeled "$S_i \rightarrow S_f$" refers to the matrix element of the color-octet S-wave operator of spin S_f in an S-wave color-singlet state of spin S_i. The matrix elements are normalized to the matrix element of the $^3S_1$ color-singlet operator in a $^3S_1$ color-singlet state. We average over unspecified spins in the quarkonium state and sum over unspecified spins in the operators. The column labeled "scaling" gives the v scaling of the normalized matrix element. Here we include the color factor $1/(2N_c)$ that arises in the free $Q\overline{Q}$ matrix elements, as is suggested in Ref. [15]. The columns labeled "bottom." and "charm." give the lattice results for the normalized matrix elements in the lowest-lying S-wave bottomonium and charmonium states. The columns labeled "bottom. est." and "charm. est." give the numerical values of the velocity-scaling estimates in the "scaling" column, taking $v^2 = 0.1$ for bottomonium and $v^2 = 0.3$ for charmonium. The quoted uncertainties are statistical only. Systematic uncertainties could be considerably larger.

spin Transition	scaling	bottom.	bottom. est.	charm.	charm. est.
singlet → triplet	$v^3/(2N_c)$	$2.72(4) \times 10^{-4}$	5.3×10^{-3}	$2.90(3) \times 10^{-3}$	2.7×10^{-2}
triplet → singlet	$v^3/(2N_c)$	$9.0(1) \times 10^{-5}$	5.3×10^{-3}	$1.13(2) \times 10^{-3}$	2.7×10^{-2}
singlet → singlet	$v^4/(2N_c)$	$6.5(5) \times 10^{-5}$	1.7×10^{-3}	$9.7(2) \times 10^{-4}$	1.5×10^{-2}
triplet → triplet	$v^4/(2N_c)$	$6.9(6) \times 10^{-5}$	1.7×10^{-3}	$1.02(1) \times 10^{-3}$	1.5×10^{-2}
triplet up → triplet up	$v^4/(2N_c)$	$6.9(6) \times 10^{-5}$	1.7×10^{-3}	$1.02(1) \times 10^{-3}$	1.5×10^{-2}
triplet up → triplet long.	$v^6/(2N_c)$	1–2×10^{-6}	1.7×10^{-4}	$2.8(7) \times 10^{-6}$	4.5×10^{-3}
triplet up → triplet down	$v^6/(2N_c)$	$< 5 \times 10^{-8}$	1.7×10^{-4}	$1.4(2) \times 10^{-6}$	4.5×10^{-3}

ACKNOWLEDGMENTS

Work in the High Energy Physics Division at Argonne National Laboratory is supported by the U. S. Department of Energy, Division of High Energy Physics, under Contract No. W-31-109-ENG-38.

REFERENCES

1. W. E. Caswell and G. P. Lepage, Phys. Lett. B **167**, 437 (1986).
2. B. A. Thacker and G. P. Lepage, Phys. Rev. D **43**, 196 (1991).
3. G. T. Bodwin, E. Braaten, and G. P. Lepage, Phys. Rev. D **51**, 1125 (1995) [Erratum-ibid. D **55**, 5853 (1997)] [hep-ph/9407339].
4. P. L. Cho and M. B. Wise, Phys. Lett. B **346**, 129 (1995) [hep-ph/9411303].
5. T. Affolder *et al.* [CDF Collaboration], Phys. Rev. Lett. **85**, 2886 (2000) [hep-ex/0004027].
6. E. Braaten, B. A. Kniehl, and J. Lee, Phys. Rev. D **62**, 094005 (2000) [hep-ph/9911436].
7. M. Beneke, [hep-ph/9703429].
8. N. Brambilla, A. Pineda, J. Soto, and A. Vairo, Nucl. Phys. B **566**, 275 (2000) [hep-ph/9907240].
9. S. Fleming, I. Z. Rothstein, and A. K. Leibovich, Phys. Rev. D **64**, 036002 (2001) [hep-ph/0012062].
10. M. A. Sanchis-Lozano, Int. J. Mod. Phys. A **16**, 4189 (2001) [hep-ph/0103140].
11. N. Brambilla, D. Eiras, A. Pineda, J. Soto, and A. Vairo, Phys. Rev. D **67**, 034018 (2003) [hep-ph/0208019].
12. C. T. H. Davies, K. Hornbostel, G. P. Lepage, A. J. Lidsey, J. Shigemitsu, and J. H. Sloan, Phys. Rev. D **52**, 6519 (1995) [hep-lat/9506026].
13. G. P. Lepage and P. B. Mackenzie, Phys. Rev. D **48**, 2250 (1993) [hep-lat/9209022].
14. C. T. H. Davies, K. Hornbostel, A. Langnau, G. P. Lepage, A. Lidsey, J. Shigemitsu and J. H. Sloan, Phys. Rev. D **50**, 6963 (1994) [hep-lat/9406017].
15. A. Petrelli, M. Cacciari, M. Greco, F. Maltoni, and M. L. Mangano, Nucl. Phys. B **514**, 245 (1998) [hep-ph/9707223].

Renormalization group analysis in cross section of b-quark production

Ali N. Khorramian [*†], A. Mirjalili[1][**†] and S. Atashbar Tehrani[‡†]

*Physics Department, Semnan University, Semnan, Iran
†Institute for Studies in Theoretical Physics and Mathematics (IPM),
P.O.Box 19395-5531, Tehran, Iran
**Physics Department, Yazd University, Yazd, Iran
‡Physics Department, Persian Gulf University 75168, Boushehr, Iran

Abstract. There is a sizable discrepancy between experimental data and theoretical results on the b-quark production in proton and anti-proton collisions which is coming from the renormalization and factorization scale dependence of finite order perturbation calculations. The "Complete RG-improvement " approach is used to separate the perturbation series into infinite subsets of terms which when summed are renormalization scheme(RS)- invariants. We use from this approach in moment space and obtain parton distribution functions (PDFs) which is based on using the valon model. Total cross section for producing b-quarks is calculated which is coming from the multiplicity of partonic cross section and PDFs. There is an expectation to get a better consistency between theoretical results and available experimental date for b-quark production.

INTRODUCTION

In studying the event shape in DIS at HERA, it is found that recent data on b-quark production in proton and anti-proton collisions lie systematically above the median of current theoretical calculations. The QCD calculations for above cases depend on a number of inputs and finally the choice of renonrmalization and factorization scales μ and M. The idea of "Complete RG-improvement (CORGI)" is introduced [1]. In this approach it was proposed that one should perform a resummation to all-orders of *all* renormalization group-predictable terms at each order of perturbation theory. This procedure automatically organizes the series into infinite subsets of terms which are separately renormalization scheme-invariant, and crucially also, by completely resumming ultraviolet logarithms, generates the correct asymptotic dependence on the single dimensionful parameter 'Q' on which the observable depends.

We extract PDF's for quark singlet, non-singlet and gluon distribution function which is based on using the phenomenological valon model. In this model, proton consists of two "up" and one "down" valons. These valons thus bear the quantum numbers of respective valence quarks.

We use from the convolution of moments and employ inverse Mellein technique to get PDFs. In this stage CORGI approach is employed and we convolute the PDF's

---

[1] E-mail address: mirjalili@ ipm.ir

CP756, *Quark Confinement and the Hadron Spectrum VI*
edited by N. Brambilla, U. D'Alesio, A. Devoto, K. Maung, G.M. Prosperi and S. Serci
© 2005 American Institute of Physics 0-7354-0241-8/05/$22.50

with partonic cross section to get the analytic result for total cross section of b-quark production. we expect to get better theoretical results which are in more consistence with available experimental data.

EVALUATION OF PARTON DISTRIBUTION FUNCTIONS

For hadrons the factorization scale dependence of PDF (M) is determined by the system of evolution equations for quark singlet, non-singlet and gluon distribution functions [2]. This evaluation for instance for quark non-singlet is given by

$$\frac{dq_{NS}(M)}{d\ln M^2} = P_{NS}(M) \otimes q_{NS}(M) + P_{qG}(M) \otimes G(M) ,$$ (1)

The cross symbol in above equation indicates the convolution

$$f(x) \otimes g(x) = \int_x^1 \frac{dy}{y} f(\frac{x}{y}) g(y)$$ (2)

The spiliting functions admit expansion in powers of $\alpha_s(M)$

$$P_{ij}(x,M) = \frac{\alpha_s(M)}{2\pi} P_{ij}^{(0)}(x) + \left(\frac{\alpha_s(M)}{2\pi}\right)^2 P_{ij}^{(1)}(x) + \ldots$$ (3)

where $P_{ij}^{(0)}(0)$ are unique, where as all higher splitting functions $P_{kl}^{(j)}$, $j > 0$ depend on the choice of the factorization scheme (FS). Conversely, they can be taken as defining the FS.

Consider the dimensionless QCD observable $R(Q)$, dependent on the single energy scale Q (we assume massless quarks). Without loss of generality, by raising to a power and scaling, we can arrange that $R(Q)$ has a perturbation series of the form,

$$R(Q) = a + r_1 a^2 + r_2 a^3 + \ldots + r_n a^{n+1} + \ldots ,$$ (4)

where $a \equiv \alpha_s(\mu)/\pi$ is the RG-improved coupling. In CORGI approach this QCD perturbatice series can be written as [1]

$$R(Q) = a_0 + X_2 a_0^3 + X_3 a_0^4 + \ldots + X_n a_0^{n+1} + \ldots ,$$ (5)

where for instance $X_i(n)$ is a factorization and renormalization scheme-invariant. It can be computed in any scheme, for instance, in $\overline{MS}$ scheme. Here $a_0 = a(0,0,0,)$ is the coupling in this scheme and satisfies

$$\frac{1}{a_0} + c\ln\left(\frac{ca_0}{1+ca_0}\right) = b\ln\left(\frac{Q}{\Lambda_R}\right) .$$ (6)

and can be written in closed form in terms of the Lambert W-function [3], defined implicitly by $W(z)\exp(W(z)) = z$,

$$a_0 = -\frac{1}{c[1 + W(z(Q))]} ,$$ (7)

where $z(Q) \equiv -\frac{1}{e}\left(\frac{Q}{\Lambda_R}\right)^{-b/c}$.

GENERAL FROM OF $\sigma_{TOT}(Q\bar{Q})$

According to the factorization theorem the total cross section of b-quark production in $p\bar{p}$ collision at the center of mass energy $\sqrt{s}$ has the form

$$\sigma_{tot}(p\overline{p} \to Q\overline{Q}, s) = \int\int dx dy \sum_{i,j} D_i^{\bar{p}}(x,M) D_j^{p}(y,M) \sigma_{ij}(s=xyS,M) \tag{8}$$

where partonic cross section σ_{ij} as well as PDF of the beam particles depend on the factorization scale M. Fixed order perturbation theory enters if we we insert in above equation the solution of evolution equations with the splitting function P_{ij} and calculate σ_{ij} as power expansion in the coupling $\alpha_s(\mu)$,

$$\sigma_{ij}(S,M) = \alpha_s^2(\mu)\sigma_{ij}^2(s) + \alpha_s^3(\mu)\sigma_{ij}^3(s,M,\mu) + ... \tag{9}$$

The latter dependence is cancelled by that of PDF provided the splitting function P_{ij} in the evolution equation retaken to all orders. At NLO, i.e. taking into account the first two terms in related perturbative series of above equation, we get

$$\sigma_{tot}^{NLO}(M,\mu) = \alpha_s^2(\mu)\left\{ \int\int dx dy \sum_{i=1}^{2n_f} q_i(x,M)q_i(y,M) \left[\sigma_{q\bar{q}}^{(2)}(xy) + \alpha_s^2(\mu)\sigma_{q\bar{q}}^{(3)}(xy,M,\mu)\right] + \right.$$

$$2\int\int dx dy \Sigma(x,M)G(y,M)\alpha_s(\mu)\sigma_{qG}^{(3)}(xy,M) + \int\int dx dy G(x,M)G(y,M)[\sigma_{GG}^{(2)}(xy)+$$

$$\left. \alpha_s(\mu)\sigma_{GG}^{(3)}(xy,M,\mu)]\right\} \; .$$

The kinematic relationship between daughter production (P_T^{μ} spectrum) and parent b quark (P_T^b spectrum) is used to extract b quark production cross section [4], integrated form P_T^{min} to infinity, over the rapidity range $| y^b |< 1$. In this stage we can employ CORGI approach where we need to use it first in moment apace to extract parton distributions and in other stage we have to reformulate the partonic cross section in this approach and consequently we expect to get better consistency between our theoretical model and experimental data for cross section of $b\bar{b}$ production.

REFERENCES

1. C.J.Maxwell and A.Mirjalili, *Nucl . Phys.* **B 577**, 209 (2000).
2. R.K.Ellis, W.J.Stirling, B.R. Webber;QCD and Collider Physics, Cambridge University Press (1996).
3. Einan Gardi,Georges Grunberg and Marek Karliner, JHEP **07** 007 (1998).
4. D0 Collaboration (B. Abbott et al.), *Phys.Rev.Lett.***85**: *5068 (2000).*

Elliptic Flow and the Strongly Coupled QGP

Raimond Snellings[1]

NIKHEF, Kruislaan 409, 1098 SJ Amsterdam, The Netherlands

Abstract. These proceedings give a very brief description of the strong collective flow, in particular the elliptic flow, observed at RHIC and how it is interpreted.

INTRODUCTION

In a QCD system at very high temperatures the quarks and gluons are expected to become quasi free, after which the bulk properties of this QGP can than be described by an ideal gas Equation of State (EOS). For such an EOS, bulk properties like energy density and pressure, are sensitive to the basic degrees of freedom.

The properties of the QCD system at lower temperatures can be calculated using lattice QCD. Lattice QCD provides from the QCD Lagrangian quantitative information on e.g. the QCD phase transition between confined and deconfined matter and its EOS. Figure 1a shows the from lattice QCD calculated [1] energy density. This energy density changes rapidly at $T_c \approx 170$ MeV due to the rapid increase in effective degrees of freedom at the phase transition between confined and deconfined matter. The corresponding pressure, shown in Fig. 1b, changes much more slowly at T_c, which leads to a rapid drop in the velocity of sound at the phase transition, as shown in Fig. 1c. The asymptotic ideal Stefan-Boltzmann gas behavior is given by the arrows in Figs. 1a and b. From these lattice calculations it is clear that even at $4 \times T_c$ the bulk properties of the system still differ from an ideal Stefan-Boltzmann gas approximation.

ELLIPTIC FLOW

Experimentally heavy-ion collisions are a unique tool to study deconfinement and the EOS of hot QCD matter in the laboratory. The hot and dense system created in a heavy-ion collision will expand and cool down. In this time evolution the system probes a range of energy densities and temperatures, and possibly various phases. Experimental information on the properties of this system and its EOS above T_c can be obtained by early probes which are not significantly altered by the late hadronic phase.

The particle yield in a non-central heavy-ion collision has an azimuthal dependence (due to interactions in the created almond shaped system, see Fig. 2a) which can be characterized in a Fourier expansion. Due to the almond shaped geometry of the collision zone the second order coefficient, v_2, is expected to be dominant. This coefficient, v_2, is

---

[1] E-mail: Raimond.Snellings@nikhef.nl

CP756, *Quark Confinement and the Hadron Spectrum VI*
edited by N. Brambilla, U. D'Alesio, A. Devoto, K. Maung, G.M. Prosperi and S. Serci
© 2005 American Institute of Physics 0-7354-0241-8/05/$22.50

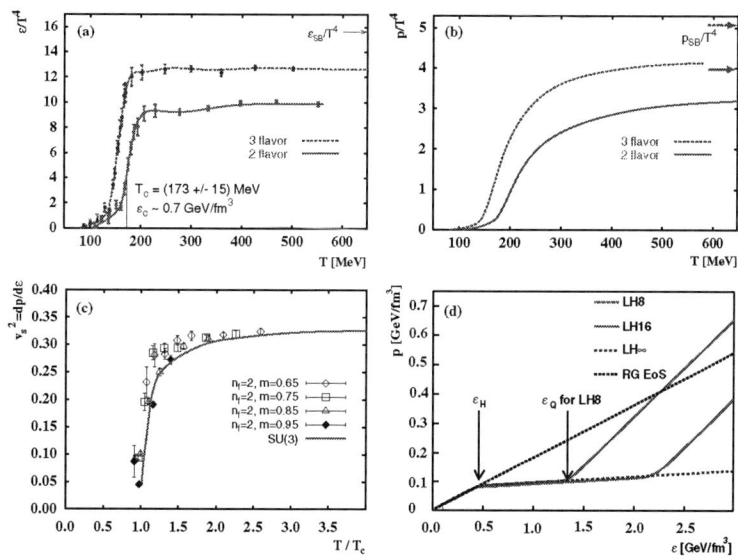

FIGURE 1. a-c) Lattice calculations from [1] d) Pressure versus energy density for various equations of state with different amounts of latent heat, from [2]

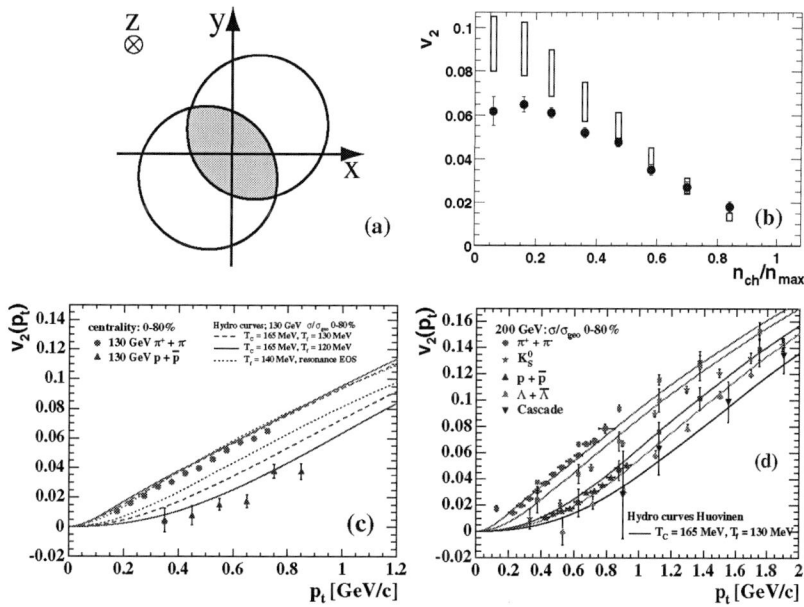

FIGURE 2. a) Transverse geometry of a non-central heavy-ion collision b) Integrated charged particle elliptic flow versus centrality [3] c) Elliptic flow and its sensitivity to the EOS [4, 9] d) Identified particle elliptic flow [6, 7, 8, 9]

commonly known as elliptic flow and is an early probe of the system. Figure 2b shows the elliptic flow versus the centrality of the collision. The boxes show the range of hydrodynamical predictions for the different EOS. The large magnitude of the elliptic flow shows that the system is strongly interacting. The agreement with ideal hydrodynamical calculations (for mid-central to central collisions) is consistent with early thermalization and ideal hydrodynamic expansion of the created system.

The predicted drop in the velocity of sound (see Fig. 1c) during the phase transition will affect the build-up of elliptic flow. This can be calculated in hydrodynamical calculations using a lattice inspired EOS (see Fig. 1d). Figure 2c shows hydrodynamical calculations of $v_2(p_t)$ for a resonance gas EOS (no phase transition) and an EOS with a phase transition to a QGP at 165 MeV. Clearly the heavier particles are more sensitive to the difference in EOS and the measurements are better described using an EOS which includes a phase transition. Detailed elliptic flow measurements ranging from pions to Ξs are shown in Fig. 2d and compared to hydrodynamical calculations with a QGP EOS. It shows that the mass dependence from pions to Ξs can, to first order, be understood with a single EOS and a common freeze-out. While the QGP EOS gives the best description there are still significant uncertainties in the choice of EOS. This is illustrated in Fig. 2c which shows that the same QGP EOS using a different freeze-out temperature can lead to a difference of the same order of magnitude as the difference between a resonance EOS and QGP EOS.

CONCLUSIONS

The large elliptic flow measured at RHIC indicates that a strongly interacting system during the early hot and dense phase is created. Its agreement with ideal hydrodynamical calculations is evidence that the system is, to good approximation, in early thermal equilibrium and, perhaps as surprising, behaves as an ideal fluid. The observed elliptic flow is consistent with hydrodynamical calculations with an EOS which incorporates a QGP phase transition. However due to the simplification of the freeze-out in the hydrodynamical calculations the EOS is not well constrained yet.

High precision measurements of low-p_t multi-strange and charm elliptic flow at RHIC and the first flow results from the heavy-ion program at the LHC will provide additional important constraints or perhaps even new insights to our current understanding.

REFERENCES

1. F. Karsch, and E. Laermann (2003), hep-lat/0305025.
2. D. Teaney, J. Lauret, and E. V. Shuryak (2001), nucl-th/0110037.
3. K. H. Ackermann, et al., *Phys. Rev. Lett.*, **86**, 402–407 (2001), nucl-ex/0009011.
4. C. Adler, et al., *Phys. Rev. Lett.*, **87**, 182301 (2001), nucl-ex/0107003.
5. P. Huovinen (2001), nucl-th/0108033.
6. S. S. Adler, et al., *Phys. Rev. Lett.*, **91**, 182301 (2003), nucl-ex/0305013.
7. J. Castillo, *J. Phys.*, **G30**, S1207–S1212 (2004), nucl-ex/0403027.
8. J. Adams, et al. (2004), nucl-ex/0409033.
9. P. Huovinen (private communication, 2004).

Overview of Charm Production – J/ψ's and Open Charm

Michael J. Leitch

Los Alamos National Laboratory
P-25 MS H846, Los Alamos, NM 87544 USA

Abstract. Heavy quark production provides a sensitive window for the study of the gluon structure of nucleons and its modification in nuclei. It also provides a very important means of studying the hot-dense conditions created in high-energy collisions of heavy nuclei and a critical probe to look for de-confinement in this hot-dense matter. I will review, from an experimental point of view, the physics issues as seen in current experimental results for charm production and point out the remaining puzzles.

Keywords: Heavy quarks, J/ψ, ψ' and Υ, nuclear effects, shadowing.
PACS: 13.85.Ni, 14.65.Dw, 24.85.+p, 25.75.-q, 25.75.Dw

J/ψ, ψ' and Υ's are produced primarily from gluons in the projectile and target. Open charm or beauty production shares this sensitivity to the gluons as well as to other initial state effects in nuclei such as initial-state gluon energy loss and multiple scattering causing p_T broadening. However, only the bound heavy-quark states suffer absorption in the final state. A longstanding problem in J/ψ production is that models that produced singlet $c\bar{c}$ states predicted cross sections that were several orders of magnitude smaller than those observed in CDF[1]. Although color-octet production (COM) is able to reproduce these cross sections, the matrix elements determined are not universal and do not work for photo production of J/ψ's. A serious problem is that the COM predicts transverse polarization at high p_T, but all measurements (CDF, E866/NuSea[2]) so far see no substantial polarization. One exception to this is in the Υ sector (Fig 1a), where the Υ_{1S}, like the J/ψ, has no polarization whereas the Υ_{2S+3S} has maximal polarization[3]. It is possible this is because the higher Υ states have little feed-down from higher mass states, while both the J/ψ and Υ_{1S} have substantial feed-down (~30% for the J/ψ[4]) which would tend to destroy any polarization. Clearly a measurement of the ψ' polarization would be of interest since it does not suffer from feed-down. For single charm (D meson) production, recent results from RHIC at √S = 200 GeV are presently confused by a factor of two (but only 3-sigma) disagreement between results from STAR and PHENIX[5].

Effects of the nuclear medium on production of heavy quarks include shadowing (depletion at small x) of the gluon distributions in a nucleus, energy loss and multiple scattering of the gluons before the hard interaction, and, for onia, final-state absorption. Shadowing is thought to involved coherence effects that effectively "shadow" the partons inside a nucleus or can also be thought of as a saturation effect

CP756, *Quark Confinement and the Hadron Spectrum VI*
edited by N. Brambilla, U. D'Alesio, A. Devoto, K. Maung, G.M. Prosperi and S. Serci
© 2005 American Institute of Physics 0-7354-0241-8/05/$22.50

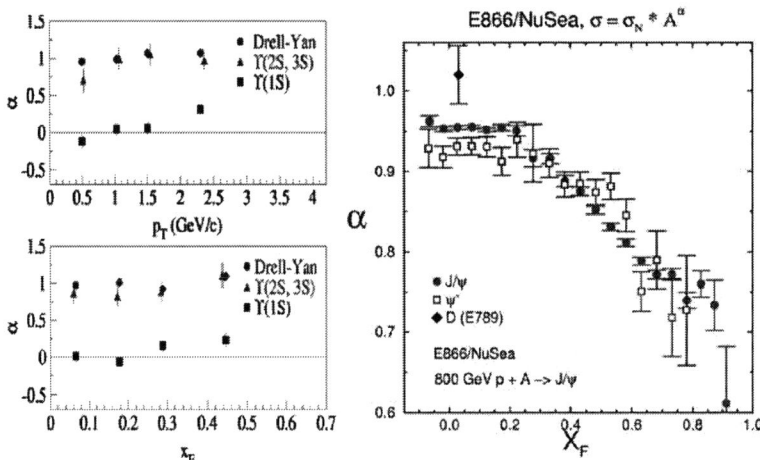

FIGURE 1. a) Polarization of Υ production at 800 GeV showing none for the Υ_{1S} and maximal for the Υ_{2S+3S}. b) Nuclear dependence versus x_F for J/ψ and ψ' production at 800 GeV.

in a nucleus, where at small enough x the gluons from different nucleons overlap and interact with each other – causing a promotion through gg→g interactions to higher x and a corresponding reduction in their populations at small x, e.g. as in a recent model called the color-glass condensate (CGC)[6].

The E866 nuclear dependence versus x_F is shown in Fig. 1b, where $\alpha = 1$ would correspond to no change in the per nucleon cross section. Near $x_F = 0$ the suppression of the resonances and lack of suppression of open-charm is thought to be due to absorption of the resonances. At these small x_F values the $c\bar{c}$'s are moving slowly out of the nucleus after their creation should be starting to hadronize in the nucleus and the larger suppression of the ψ' relative to the J/ψ would be due to the larger size and looser binding of the ψ' compared to the J/ψ. While at larger x_F they traverse the nucleus as a $c\bar{c}$ pair, only hadronizing way outside, and therefore experience the same effects. The strong increase of the suppression as x_F increases is thought to be due to a combination of shadowing (large x_F is small x in the nuclear target) and also energy loss of the gluon in the initial state[7]. As shown in Fig. 2a, new data from PHENIX at $\sqrt{s} = 200$ GeV also shows similar features, but so far with much worse statistical precision. However, if one looks at the suppression seen at three different energies ($\sqrt{s} = 19$ GeV (NA3), 39 GeV (E866) and 200 GeV (PHENIX)) the suppression does not scale with x_2 but does, as shown in Fig. 2b, with x_F. This appears to indicate that shadowing (which should scale with x_2), is not the dominant physics in the large x_F behavior of the J/ψ. The reason for the apparent scaling with x_F remains a puzzle, although I suppose it conceivable that energy loss provides the bulk of the high-x_F suppression at the lower energies but that it vanishes at higher energies (RHIC) where only a weak shadowing is seen. Early studies at RHIC also indicate, that open-charm is not modified in either d-Au or Au-Au collisions.

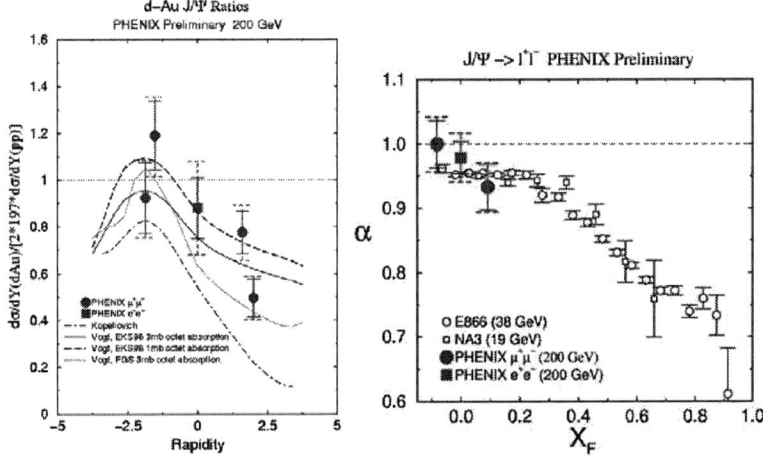

FIGURE 2. a) Nuclear dependence of J/ψ production versus rapidity at √s = 200 GeV. b) Same versus x_F.compared to lower energy measurements.

In high-energy nucleus-nucleus collisions heavy quarks and the onia are thought to provide an important tool for the detection and study of the strongly interacting de-confined matter that may be made. In particular the color-screening in a de-confined media would heavily suppress J/ψ production[8]. On the other hand, recent models driven by the large amount of charm created in these collisions also predict an enhancement from recombination of that charm[9]. At CERN when the observed suppression in Pb-Pb collisions was compared to simple estimates using a fixed nuclear cross section for the normal suppression (as determined from p-nucleus studies)[10]; an anomalous suppression at high energy density (small impact parameter collisions) was observed and, by some, is cited as evidence of creation of a quark-gluon plasma. However, the estimate of the normal nuclear suppression seems inadequate to clearly determine if the new de-confined state of matter was actually seen. So far similar studies for the higher-energy collisions at RHIC have been inconclusive due to the low luminosity at RHIC so far and the corresponding low statistics for J/ψ in Au-Au collisions[11]. We are looking forward to more substantial results in the near future as the analysis is completed of the much higher RHIC Au-Au run just completed.

[1] Beneke, Kramer, Phys. Rev. D55, 5269 (1997).
[2] T. Affolder et al. (CDF) Phys. Rev. Lett. 85, 2886 (2000) & T. Chang et al. (E866), Phys. Rev. Lett. 91, 211801 (2003).
[3] C.N. Brown et al. (E866/NuSea), Phys. Rev. Lett. 86, 2529 (2001).
[4] L. Antoniazzi et al. (E705), Phys. Rev. Lett. 70, 383 (1993) & preliminary results from HERA-B.
[5] J. Adams, et al. (STAR) nucl-ex/0407006; PHENIX preliminary, Quark Matter 2004.
[6] L.D. McLerran & R. Venugopalan, Phys. Rev. D49, 2233 (1994) ; Phys. Rev. D49, 3352 (1994).
[7] B. Kopeliovich et al., Nucl. Phys. A696, 669 (2001); R. Vogt, hep-ph/9907317.
[8] T. Matsui, H. Satz, Phys. Lett. B 178, 416 (1986).
[9] R.L. Thews et al., Phys. Rev. C63, 054905 (2001); L. Grandchamp et al. hep-ph/0403204.
[10] M.C. Abreu et al. (NA50), Phys. Lett. B477, 28 (2000) & Phys. Lett. B 521, 195 (2001).
[11] S.S. Adler et al. (PHENIX), Phys. Rev. C69, 014901 (2004).

Expected Further Evidence for Hadron Deconfinement at LHC

Laszlo Gutay

E-735 Collaboration
Purdue University, Department of Physics
West Lafayette, IN 47907

Abstract: Deconfinement results from $\sqrt{s}$=1.8 TeV are extrapolated to $\sqrt{s}$=14 TeV . Expected deconfinement signatures in Alice and CMS detectors are discussed.

Keywords: Deconfinement

PACS: 25.75

DECONFINEMENT SINGATURES AT $\sqrt{s}$=1.8 TeV

Experimental signature of first order phase transition for classical gases in the T-S plane for fixed pressure have been known for a century[1]. Hagedorn, Van Hove and Wieskopf suggested[2] to carry out an experiment at the FNAL $\bar{p}$-p collider to search for evidence for hadron deconfinement in centrally produced hadronic matter in minimum biased events in the -3<η<+3, n_c>20 kinematic region. They identified the entropy S and temperature T of an event as:

$$S = \frac{dn_c}{d\eta} \; ; \; T = \text{inverse logarithmic slope of the transverse momentum distribution}$$

According to Bjorken[3], the deconfined matter expands, its energy density and temperature drop, and at a critical temperature T_c and energy density ε_c, it decays into hadronic matter. The Duke-Fermilab-Iova-Notre Dame-Purdue-Wisconsin collaboration was formed and the search for the deconfinement was approved[4] (FNAL-E735) at the CO Collider region at the new 1.8 TeV $\bar{p}$-p collider. The event multiplicity was measured by a 4π scintillator hadoscope and the transverse momenta of some of the particles were measured by a 90° spectrometer.

As expected for a first order phase transition signal, we found[5] a sudden rise and subsequent flattening of <p_t> as function of $dn_c/d\eta$ for pions and kaons. The temperature remained constant from $dn_c/d\eta$ = 9 until there was n_c data. It's value was T_c = 179.5±5 MeV and ε_c=1.1±0.26 GeV/f^3. Using the above temperature, energy density and the measured volume at deconfinement, we determined the number of degrees of freedom as G(T) = 24.8±6 which is only 3 for a pion gas. Putting this in

CP756, *Quark Confinement and the Hadron Spectrum VI*
edited by N. Brambilla, U. D'Alesio, A. Devoto, K. Maung, G.M. Prosperi and S. Serci
© 2005 American Institute of Physics 0-7354-0241-8/05/$22.50

the form $\varepsilon_c/T^4 = \pi^2/30\ G(T) = 8.15 \pm 2$ is in agreement with the lattice gauge calculation of Karsch et al.[6] See Fig.1.

Extrapolations to $\sqrt{s} = 14$ TeV

The expected relevant experimental quantities at LHC are n_c, σ_{NSD} and $<n_c>_{nsd}$ according to:

1. Walker from FNAL-E735
   $\sigma_{NSD} = 64.2 \pm 6$ mb, $< n_c >_{NSD} \approx 100$

2. Safarik using Polikov – Dokshitzer distribution
   $\sigma_{NSD} = 70$ mb, $<n_c>_{NSD} \approx 100$

3. Sjostrand and Van Zijl using the Lund Model
   $\sigma_{NSD} \approx 70$ mb, $<n_c>_{NSD} \approx 100$

Assuming a startup luminosity of $10^{28} cm^{-2} s^{-1}$ and $\sigma_{NSD} = 70$mb we will get 7×10^9 events / year. Assuming an exponential fall off of the multiplicity distribution, with this high statistics, we can explore the multiplicity region
$$n_{cmax} < n_c < 7 n_{cmax}$$
$$100 < n_c < 700$$
This corresponds to an $25 < dn_c/d\eta < 70$. This particle rapidity density is only a factor of 3 larger then our FNAL results at $\sqrt{s} = 1.8$ TeV.

This can be understood by noting that
$$\Delta\eta \approx \ln s \text{ and } < n_c >_{NSD} \approx A + B \ln (s)$$
With this large $dn_c/d\eta$ range and high statistics at $\sqrt{s} = 14$ TeV we can

(a) Check the linear dependence of deconfinement length on $dn_c/d\eta$
   See Fig.2.
(b) Look for electromagnetic signal emerging from the high energy events for which the initial energy density may be as high as:
   $$\varepsilon = \varepsilon_c (dn_c/d\eta)/8 = 1 \text{ GeV} \times 70/8 \approx 9 \text{GeV}/f^3$$
(c) Study the $dn_c/d\eta$ dependence of $\bar{d}/\bar{p}$ ratio
(d) If deconfinement takes place at the critical point, cluster size fluctuation is expected as observed[7] at $\sqrt{s} = 1.8$ TeV. It can be verified with good statistics
(e) Search for baryon non conservation

The low magnetic field and particle identification of ALICE is ideal for hadron deconfinement experiments.

The E.M. and Hadron calorimeters of CMS make it ideal detector to study the reaction: $g + g \rightarrow \gamma + jet$, which might take place at high $dn_c/d\eta$ and study J-Ψ extinction.

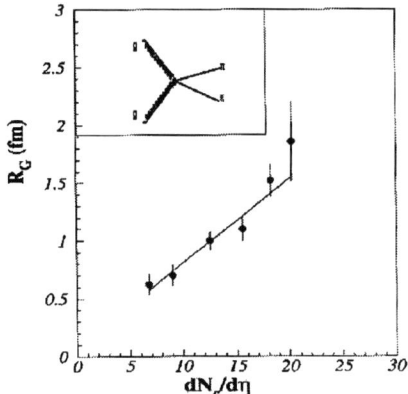

FIGURE 1. Dependence of the longitudinal Gaussian radius on $dN_c/d\eta$, from reference 5.

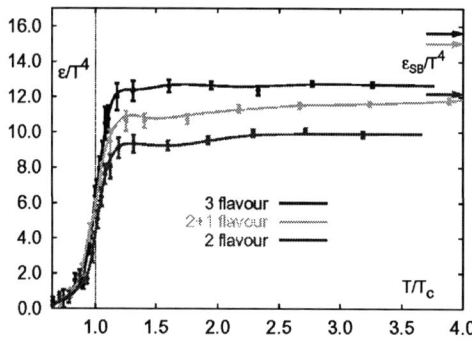

FIGURE 2. Lattice gauge calculation of Karsch, from Reference 6.

ACKNOWLEDGMENTS

Help from R.P. Scharenberg, W.D. Walker, K. Safarik and Bolek Wyslouch is greatly appreciated.

REFERENCES

1. R. Hagedorn, L. Van Hove, V.F. Weisskopf, private communication (1981)
2. Siers and M.W. Zemansky, Heat and Thermodynamics, Third Edition p213.
3. J.D. Bjorken, Phys. Rev. D27, 140 (1983)
4. Search for Hadron Deconfinement, Accepted at Fermilab in 1983 as FNAL E-735.
5. T. Alexopoulos et al. Phys. Lett. B478, 447 (2002)
6. F. Karsch et al. Phys.Lett. B478, 447 (2000)
7. T. Alexopoulos et al. Phys. Lett. B353, 155 (1995)

Physics potential of ALICE experiment at LHC

A. Sandoval and P. Foka for the ALICE collaboration

CERN, CH-1211 Geneva 23, Switzerland

Abstract. The ALICE experiment under construction at the CERN LHC is going to study heavy-ion collisions at a novel energy regime. The physics potential and performance of the experiment is discussed on the basis of a few selected observables.

Keywords: Heavy-ion physics, ALICE, flow, heavy quarks

The ultimate goal of heavy-ion physics is to understand the QCD phase diagram mapping different regimes of temperature and baryochemical potential with experimental measurements from different accelerators. The existence of a new phase of strongly interacting matter, the Quark-Gluon Plasma (QGP), characterised by deconfinement and chiral-symmetry restoration is a well established prediction of the theory, and is expected at temperatures higher than a critical temperature. Many of the predicted QGP signatures have been already confirmed by the CERN SPS fixed target experiments. The results from BNL RHIC Collider, at twelve times higher energy, not only confirm the SPS results but add new observables accessible at the higher energy regime. ALICE at LHC is expected to study strongly interacting matter at conditions of vanishing baryochemical potential and access a novel energy regime, reached so far only in the interactions of the highest energy cosmic rays.

Heavy-ion collisions at LHC, at about 30 times higher energy than that at RHIC and 300 times higher than at SPS will access a quantitatively different regime of much higher energy density and temperature providing ideal conditions to study QCD Thermodynamics and the QGP equation of state. Figure 1 (left) shows the temperature dependence of the energy density of strongly interacting matter obtained by Lattice QCD calculations [1]. This increase in energy will be reflected in many parameters characterizing the QGP phase. Estimates of some of those parameters [2] are shown in Table 1.

In addition, a qualitatively new regime of parton kinematics becomes available. The energy increase at LHC will make accessible the small Bjorken-x region where x is substantially smaller than at RHIC (by a factor 30, same factor as the energy increase). The x-values of partons involved in particle production at mass-square scale M^2 for the SPS, RHIC and LHC energies are shown in Fig. 1. The cross section for hard processes will increase dramatically. It is expected that the production of 20 GeV transverse-momentum hadrons will be enhanced by three orders of magnitude compared to RHIC. Thus, at LHC, we will be able to really study, for first time, hard QCD in heavy-ion collisions.

ALICE will study Pb–Pb collisons at $\sqrt{s_{NN}} = 5.5$ TeV. Runs of lighter ions, pp and pA are foreseen for system-size dependence systematic studies and baseline measurements. The aim of ALICE is to measure all relevant observables in the same experiment. The challenge for the optimisation of the experimental apparatus is the expected high par-

CP756, Quark Confinement and the Hadron Spectrum VI
edited by N. Brambilla, U. D'Alesio, A. Devoto, K. Maung, G.M. Prosperi and S. Serci

TABLE 1. Comparison of the most relevant—model-dependent—parameters characterizing central nucleus–nucleus collisions at different energy scales [2].

Parameter		SPS	RHIC	LHC
$\sqrt{s_{NN}}$	[GeV]	17	200	5500
dN_{gluons}/dy		450	1200	4700
dN_{ch}/dy		350	800	3000
$Q_{sat,A}$	[GeV]	0.71	1.13	2.13
Initial temperature	[GeV]	0.38	0.6	>1
Initial energy density	[GeV/fm^3]	~5	~25	~250
Freeze-out volume	[fm^3]	~10^3	~10^4	~10^5
Life time	[fm]	<2	2-4	>10

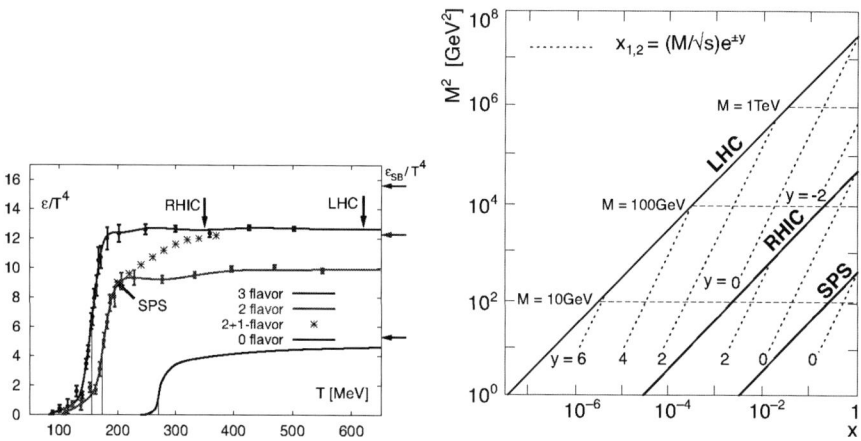

FIGURE 1. Left: Energy density ε normalized to T^4 as a function of temperature T for two or three light quark flavours and for two light and one heavier quarks. Right: The range of Bjorken-x and M^2 relevant for particle production in AA collisions at the top SPS, RHIC and LHC energies. Lines of constant rapidity are indicated.

ticle multiplicity and the corresponding large uncertainties. Extrapolations from lower energies give estimates of the charged particle rapidity density at mid-rapidity for central Pb–Pb collisions at LHC in the range 1500 - 3000. The ALICE detectors are optimized for charged particle density $dN_{ch}/dy = 4000$. However, its performance is also checked for multiplicities up to $dN_{ch}/dy = 8000$, showing that there is a safety margin of two for most observables.

The expected luminosity for Pb–Pb collisions is about $10^{27} \mathrm{cm}^{-2}\mathrm{s}^{-1}$ which results in a rather low minimum-bias interaction rate of 8 kHz. In the heavy-ion mode the detectors will have to operate at more than three orders of magnitude lower rates compared to pp running; however, with up to three orders of magnitute higher particle densities.

ALICE is optimised for a comprehensive measurement of identified particles in a wide momentum range. Low-p_T coverage down to ~ 100 MeV/c gives access to all soft observables directly related to the bulk thermodynamic properties of the system. Momentum resolution better than 10% at p_T of 100 GeV/c gives access to high-p_T observables whose medium dependence reflects specific properties of this bulk. The wide p_T coverage allows in addition, to cross correlate all relevant observables.

Special emphasis has been put the last few years on the detailed study of high-p_T observables and in particular the study of jet quanching and heavy flavour production, reported in the proceeding of this conference by C. Loizides [3] and R. Turrisi [4], respectively. As an example of the completeness of the experiment and the possibility to cross correlate observables we refere further to the ALICE capabilities for reaction plane reconstruction and elliptic flow measurement, as well as heavy flavour studies.

Experience from SPS and RHIC shows that among the first physics results that become available are measurements of azimuthal correlations relative to the reaction plane. The reaction plane can be reconstructed in the ALICE central barrel, using charged particles measured in the Time Projection Chamber (TPC), see below, or the Inner Tracking System (ITS). In addition, the Photon Multiplicity Detector (PMD) will measure the anisotropy in the emitted photons in the forward region $2.3 < \eta < 3.5$ [5]. Hence we will be able to study particle yields and spectra as a function of their angle of emission with respect to the reaction plane; for example two-particle correlations, production of different particle like J/ψ and Υ [6] or jet quenching. All produced high-p_T particles will have different in-medium path lengths within the reaction plane or orthogonal to it. A quantitative study of the properties of the medium becomes then possible via a systematic study of the medium depence of hard probes.

Elliptic flow and reaction plane reconstruction. One of the most important hadronic observables is elliptic flow; its measurement provides direct information about the collective transverse dynamics due to pressure build-up in the collision zone. Based on elliptic flow measurements and their p_T and centrality dependence one can probe the degree to which the medium reach the thermodynamic limit and get the time scale of thermalisation, as well as estimate the initial energy density.

A few fully simulated and reconstructed ALICE events were used to test the flow analysis procedure, showing no problems. The p_T dependence of v_2 measured by STAR was used to generate the flow signal and was successfully reconstructed. Here we present results from high-statistics fast simulation studies. Figure 2 summarizes the resolution of the reaction plane angle ψ and the Fourier parameter v_2 as a function of the particle multiplicity (top) and intensity of the effect (bottom). To estimate the resolution of ψ and v_2 reconstruction as a function of multiplicity, the v_2 parameter was fixed to the value measured by STAR, $v_2 = 0.05$ and the multiplicity in the range 1000–20000 tracks in the TPC. The resolution of ψ, varies from 11° for multiplicity 1000 to 2° for multiplicity 20000. The standard deviation of v_2, varies in the range $dv_2 = 0.015$ ($dv_2/v_2 \sim 30\%$) for the low multiplicity to $dv_2 = 0.004$ ($dv_2/v_2 \sim 8\%$) for multiplicity 20000.

To estimate the resolution of ψ and v_2 reconstruction as a function of the parameter v_2, the multiplicity was fixed to 5000. The simulation was performed for four values of v_2 in the range 0.02–0.08. The resolution of the reaction plane angle ψ, for the v_2 value measured by STAR, is 5° and shows $1/v_2$ behaviour approximately. The resolution of v_2, only slightly depends on v_2 itself. For $v_2 = 0.05$ the standard deviation is 20% of the effect.

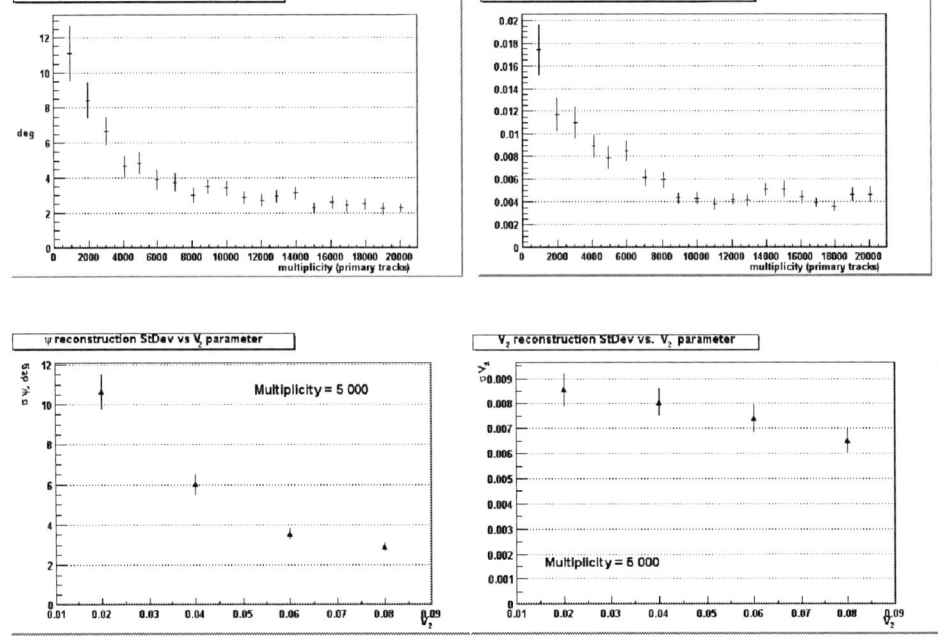

FIGURE 2. The resolution of the reaction plane angle ψ and the Fourier parameter υ_2 as a function of the particle multiplicity (top) and intensity of the effect (bottom).

Heavy flavour production. It is expected that the suppression pattern of quarkonium states will give us an indispensable information about the matter produced in heavy-ion collisions. At the LHC energies, the quarkonium production rates are high enough for accurate measurements of the yields and p_T spectra of most of the interesting states, as J/ψ, ψ', Υ, Υ', Υ''. ALICE will provide such measurements in both, dielectron and dimuon decay channels in the central barrel region and the forward muon spectrometer, respectively, and comprehensive measurements in pp, pA and AA collisions [5].

However, quarkonium suppression may result not only from colour screening, but also from nuclear effects like shadowing and absorption. In order to isolate pure QGP effects, it was proposed to study the p_T dependence of quarkonium ratios instead of single quarkonium p_T distributions. By doing so, nuclear effects are washed out, at least in the p_T variation of the ratio. An extensive study of the properties of the Υ/Υ'ratio in heavy-ion collisions at the LHC is presented in [7]. It was shown that the p_T dependence of such ratios is a direct probe of the QGP physics.

Details on the capability of the ALICE muon spectrometer to measure the p_T dependance of the Υ/Υ'ration are presented in [8]. The results show that with the statistics collected in one month of data taking, the measured Υ/Υ'ratios exhibit a strong sensitivity to the characteristics of the QGP. The properties of the deconfined medium can be further pin down by systematic measurements of the centrality dependence.

Because of uncertainties in the heavy-quark production rates in heavy-ion collisions, the best normalisation of quarkonium yields will be the production rates of other particles containing the same heavy-flavour, i.e. D or B mesons. ALICE has demonstrated the capability for such measurements [4] and [5]. In particular, the D^0 production in central Pb–Pb collisions using the $K^- \pi^+$ decay mode can be measured in a wide transverse momentum range, from 0.5 to 15 GeV/c [4].

The measurement of open heavy-flavour production is interesting not only for quarkonium normalisation but also in itself. High-p_T heavy quarks, similarly to other partons, will suffer from medium-induced energy losses. However, due to their large mass, at intermediate p_T a part of the cone around the heavy-quark direction is forbidden for gluon radiation [9]. This phenomenon, called dead-cone effect, will effectively suppress the jet quenching for heavy quarks at intermediate p_T, comparable with the heavy-quark mass, until it becomes ultra-relativistic. The dead-cone effect may be clearly observed in the p_T dependence of the ratio of the two nuclear modification factors (production yield in AA collisions normalised to that in elementary pp collisions, properly scaled with the number of binary collisions), the one for D mesons and that for light-flavour hadrons [10]. This observable is very sensitive to the medium properties and in particular to the transport coefficient (defined as the mean value of the transverse momentum squared transfere between the parton and the medium per unit length), which can then be quantitatively inferred from such measurement.

Summary

In heavy-ion collisions at the LHC, a completely new energy domain for the study of hot and dense strongly interacting matter becomes accessible. With the detectors well in the construction phase and the first volume of the Physics Performance Report (PPR Vol. I) published [11], the collaboration is now evaluating in detail the physics performance of the experiment, to be published in the PPR Vol. II. First results show that ALICE will succesfully combine the traditional heavy-ion observables with a new list of high-p_T observables that become available at LHC; such studies are expected to shed additional light on the structure of the phase diagram of nuclear matter and more generally on strong interaction physics in a very dense, longlived partonic environment.

REFERENCES

1. F. Karsch *Lect. Notes Phys.*, **583**, 209-249 (2002).
2. K.J. Eskola, K. Kajantie, P.V. Ruuskanen, and K. Tuominen *Nucl. Phys. B*, **570**, 379-389 (2000).
3. C. Loizides, proceedings of this conference
4. A. Dainese and R. Turrisi, Proceedings of Quark Confinement 10-24 Sep. 2002, World Scientific R. Turrisi, proceedings of this conference.
5. ALICE PPR Vol. II http://alice.web.cern.ch/Alice/ppr/web/PPRVIICurrentVersion.html
6. A. Morsch, R. Raniwala, S. Raniwala, Y.P. Viyogi, *ALICE Note*, **ALICE-INT-2001-22**.
7. J. F. Gunion, and R. Vogt, *Nucl. Phys. B*, **492**, 301 (1997).
8. E. Dumonteil and P. Crochet, *ALICE Note*, **ALICE-INT-2005-002**.
9. Yu. L. Dokshitzer and D. E. Kharzeev, *Phys. Lett. B*, **519**, 1999 (2001).
10. A. Dainese, *Eur. Phys. J.*, **C33**, 495 (2004).
11. ALICE Collaboration, *J. Phys. G.*, **30**, 1517-1763 (2004).

Jet physics in heavy-ion collisions with ALICE

C. Loizides (for the ALICE collaboration)

University of Frankfurt, Max-von-Laue-Straße 1, D-60438 Frankfurt

Abstract. This contribution summarizes the performance of the ALICE detector for the exclusive measurement of high-energy jets at mid-pseudo-rapidity in ultra-relativistic nucleus–nucleus collisions at LHC and their potential for the characterization of the matter created in these collisions.

Keywords: cone finder, jet fragmentation, jet quenching
PACS: 25.75.-q

In the first stage of ultra-relativistic nucleus–nucleus collisions, interactions between initial, high-p_T partons and the dense, partonic matter produced in the collision lead to parton energy loss via medium-induced gluon radiation [1].

Evidence of jet quenching has been found in Au–Au collisions at RHIC, at $\sqrt{s_{NN}} = 200$ GeV (and at lower energies), in the suppression of single-particle yields at high p_T, the disappearance, in central collisions, of jet-like correlations in the azimuthally-opposite side (away-side) of a high-p_T leading particle, together with the absence of these (final-state) effects in d–Au collisions at the same centre-of-mass energy. However, the centre-of-mass energy at RHIC is too low to produce outstanding, high-energy jets to be identified, event-by-event, on top of the heavy-ion background.

At LHC, lead ions are foreseen to collide at centre-of-mass energies of $\sqrt{s_{NN}} = 5.5$ TeV, and at these energies, jets with $E_T \gtrsim 20$ GeV will be copiously produced. ALICE, one of the four experiments preparing for LHC, is dedicated to the study of heavy-ion collisions [2]. As such, the detector system has been designed to measure properties of the bulk (soft hadronic, large cross section, physics) and of rare probes (hard, small cross section, physics), and to operate in an environment of extreme charged-particle density. Recently, it has been proposed to add a large electromagnetic calorimeter at central acceptance. Combined with the barrel tracking detectors, it improves the jet energy resolution, and can be used for jet triggering.

The detector system will allow the study of jets over a broad energy range, from about 5 GeV to about 250 GeV, and, therefore, to discriminate between various theoretical models of jet modification in heavy-ion collisions. The jet production spectrum steeply falls with increasing jet energy; only about $3 \cdot 10^5$ jets with $E_T > 100$ GeV are produced within the EMCAL acceptance in one year of ALICE min. bias Pb–Pb running. These high-energy jets may be reconstructed on event-by-event basis, and their analysis, challenged by the large background present in heavy-ion collisions, are the topic of this contribution.

In order to simulate jets in central Pb–Pb collisions, jets generated by PYTHIA [3], representing a realistic jet spectrum, are embedded into background events generated by HIJING [4]. The response of the detectors is not included in the simulation; rather a grid of calorimeter towers is constructed in $\eta \times \phi$ with a granularity of 0.05, in which the transverse energy of all detectable (e.g. charged) particles in the central acceptance is summed. In our approach [5], jets are then reconstructed with a cone

CP756, Quark Confinement and the Hadron Spectrum VI
edited by N. Brambilla, U. D'Alesio, A. Devoto, K. Maung, G.M. Prosperi and S. Serci
© 2005 American Institute of Physics 0-7354-0241-8/05/$22.50

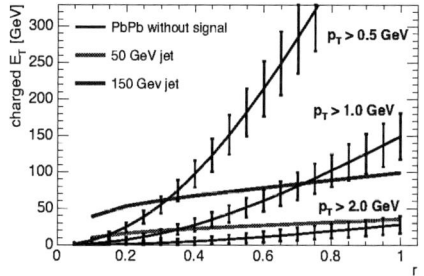 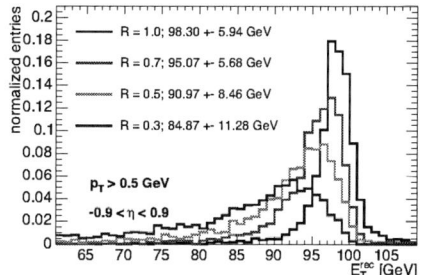

FIGURE 1. Left: Average, charged, energy content of 50 and 150 GeV jets compared to the average (and rms) content of randomly chosen cones in 0–10% central Pb–Pb events as a function of the cone size, r. The background is suppressed by p_T-cuts of 0.5, 1 and 2 GeV. Right: Out-of-cone fluctuations for fixed energy jets of 100 GeV (defined for $R = 1$), visible in the reconstructed, total, jet energy spectrum for smaller cones of $R = 0.7, 0.5$ and 0.3.

jet finder, as typically done for jet measurements in hadronic collisions. However, as we show in Fig. 1, the presence of numerous mini-jets in the heavy-ion environment makes it necessary to reduce the cone size, the radius in the plane spanned by pseudo-rapidity and azimuth, from its nominal value of $R \sim 0.7$ used at hadron colliders to $R = 0.3$. In addition, the high-particle multiplicity density of the soft bulk at mid-pseudo-rapidity demands to introduce a cut in transverse momentum for charged hadrons with $p_T < 2$ GeV. In this way, jets of about 50 GeV and higher will be measurable with ALICE, but intrinsic fluctuations in the jet fragmentation, out-of-cone fluctuations and the remaining underlying mini-jet background limit the energy resolution. Without the EMCAL, the resolution is mainly dominated from intrinsic fluctuations in ratio of charged-to-neutral particles in the jet fragmentation. The mean reconstructed fraction amounts to 50% and a width of about 50%. Including the EMCAL, the mean fraction increases to about 60% and a width of 30%. In both cases, however, the spatial direction of the jet axis can be reconstructed with a resolution of better than 10% at 50 GeV, improving with increasing energy.

Quenched jets are simulated by a modified version of PYTHIA, which introduces parton energy loss via final-state gluon radiation in a rather crude way: Before the partons originating from the hard 2-to-2 process (and the gluons originating from ISR) are subject to fragment, they are replaced according

$$\text{parton}_i(E) \rightarrow \text{parton}_i(E - \Delta E) + n(\Delta E)\, \text{gluon}(\Delta E / n(\Delta E))$$

conserving energy and momentum. We distinguish two toy models:

- *fixed* energy loss: $\Delta E / E = 0.2$ and $n = 1$ independent of parton type and parton production point in the system (geometry);
- *variable* energy loss: ΔE given by PQM [6] dependent on medium density, parton type and parton production point in the system (geometry).

The physics signature of medium-induced gluon radiation generally is believed to be visible in the modification of the jet fragmentation function as measured through the longitudinal and transverse momentum distributions of associated hadrons within the jet [7]. The momenta parallel to the jet axis, $p_L = p_{\text{hadron}} \cos(\theta_{\text{jet}}, \theta_{\text{hadron}})$, are

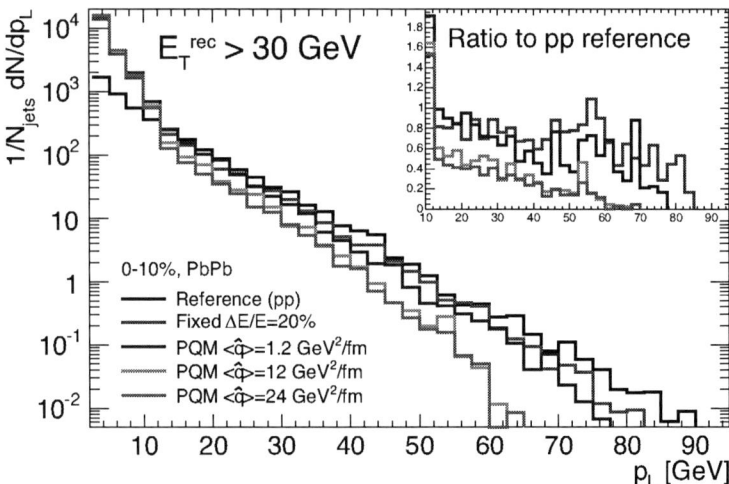

FIGURE 2. Longitudinal momentum distribution, with respect to the jet axis, for charged particles in jets with $E_T^{rec} > 30$ GeV for different quenching conditions in 0–10% central Pb–Pb compared to pp collisions. The inset shows the ratio of the distribution for the different Pb–Pb cases to the distribution obtained in pp. The jets are identified with cone finder using $R = 0.3$ and a cut of $p_T > 2$ GeV.

expected to be reduced (jet quenching), while the momenta in the transverse direction to be increased (transverse heating). Figure 2 shows the longitudinal distribution for different quenching scenarios in 0–10% central Pb–Pb compared to pp collisions taking into account all reconstructed jets where $E_T^{rec} > 30$ GeV. Qualitatively, the expected behavior is clearly visible: higher medium density leads to stronger suppression of the momenta along the jet axis and enhancing of smaller momenta. At $p_L \gtrsim 25$ GeV, the effect of the additional gluon radiation becomes apparent for both of the dense cases. However, at low $p_L \lesssim 10$ GeV, the strong modification of the shape of the distribution can be predominantly attributed to the remaining hadrons of the underlying background, and in this region all models agree. The influence of the background can be reduced by increasing the cut on the jet energy, but, at the same time, the demands on the statistics reach the limits achievable with ALICE. Jet quenching will be measurable in the ratio of the longitudinal distribution obtained for jets in Pb–Pb with respect to pp. Within the limited scope of the introduced quenching method, however,the quantitative distinction between, for example, the two dense PQM scenarios seems to be impossible.

REFERENCES

1. M. Gyulassy, and M. Plümer, *Phys. Lett.*, **B243**, 432–438 (1990).
2. F. C. and others (ALICE collaboration), *J. Phys. G: Nucl. Part. Phys.*, **30**, 1517–1763 (2004).
3. T. S. and others, *Comput. Phys. Commun.*, **135**, 238 (2001), [arXiv:hep-ph/0010017].
4. M. Gyulassy, and X. Wang, *Comput. Phys. Commun.*, **83**, 307–331 (1994).
5. C. Loizides, *Jet physics in ALICE*, Ph.D. thesis, Uni. of Frankfurt (2005), arXiv:nucl-ex/0501017.
6. A. Dainese, C. Loizides, and G. Paic, *Eur. Phys. J.*, **C38**, 461–474 (2005), [arXiv:hep-ph/0406201].
7. C. Salgado, and U. Wiedemann, *Phys. Rev. Lett.*, **93**, 042301 (2004), [arXiv:hep-ph/0310079].

Charm and beauty with ALICE at LHC

Rosario Turrisi

Dipartimento di Fisica "Galileo Galilei", via Marzolo, 8, I35131, Padova, Italy
for the ALICE Collaboration

Abstract. Main results are presented on the ALICE performance for the measurement of heavy quark production via the detection of D^0 and B mesons and for charm production quenching.

PACS: 25.75.-q, 25.75.DW, 14.65.Dw, 13.85.Ni, 13.85.Qk

Heavy-ion physics at the LHC, where Pb–Pb collisions at 5.5A TeV will be produced, is focussed on the study of the behavior of strongly interacting matter at extreme energy density over a large volume, where the formation of the Quark Gluon Plasma (QGP) is expected. Heavy flavors are, in this context, an especially interesting observable. The energy loss of b and c quarks, which can be studied by comparing the p_t spectra from AA and pp collisions, can provide information on the composition of the medium. The measurement of the heavy quark cross sections will provide the natural normalization for the analysis of quarkonia suppression, one of the main signals of QGP formation, and the amount of $b \to J/\psi$ contamination to the direct J/ψ yield. In ALICE[1], the LHC experiment dedicated to heavy-ion physics, the performance for the detection of beauty hadrons in the semi-electronic decay channel and of charm mesons in the exclusive hadronic channel $D^0(\overline{D}^0) \to K^{\mp}\pi^{\pm}$ has been studied.

The ALICE detector is designed to handle a multiplicity currently estimated as 4000-8000 charged particles per rapidity unit at midrapidity per event. Among the many detectors which make up ALICE, three are functional for this study. The Time Projection Chamber[2] (TPC) and the Inner Tracking System[3] (ITS) provide tracking and vertexing at p_t as low as 200 MeV with a magnetic field $B = 0.4$ T. The Transition Radiation Detector[4] (TRD) allows electron/hadron separation. The TPC can also be used for particle identification purposes by the dE/dx signal.

In these paper we outline the basic points of the strategy and the results, for more details the reader is referred to other papers [5, 6, 7]. We assumed as reference cross sections of charm and beauty 45 and 1.79 barn respectively. The corresponding yields are ~ 0.53 Kπ pairs per rapidity unit (from the D^0 decays) around zero rapidity and ~ 0.9 electrons (from beauty decays) per event (see [8] for details about the cross section extrapolation to LHC energy and from pp to Pb–Pb), while the background multiplicity is fixed at 6000 charged particles per rapidity unit at midrapidity per event; about 2 rapidity units are covered by the central barrel, i.e. $\sim 10^4$ charged particles are expected in the acceptance.

D^0 meson detection is based mainly on the selection of two tracks with large impact parameter d_0 [1] and small angle between the D^0 momentum, as reconstructed from the daughters, and its flight-line. Particularly effective is the combination of the two variables which, along with kaon identification in the TOF detector, allows us to reach a 11% signal/background ratio with a significance of 37 in the $M_{D^0} \pm 1\sigma$ mass window. The accessible p_t range is 1–14 GeV/c with a statistical error better than 15–20% and a systematic error better than 20% .

We studied [6] the sensitivity for a comparison of the energy loss of charm quarks and of massless partons by considering the *nuclear modification factor* of D mesons and the ratio of the nuclear modification factors of D mesons and of charged hadrons, as a function of p_t (NN=nucleon-nucleon, AA=nucleus-nucleus):

$$R^D_{AA}(p_t) \equiv \frac{dN^D_{AA}/dp_t/\text{binary NN collisions}}{dN^D_{pp}/dp_t}, R_{D/h}(p_t) \equiv R^D_{AA}(p_t)\Big/R^h_{AA}(p_t) \qquad (1)$$

$R^D_{AA}(p_t)$ would be equal to 1 if the AA collision were a mere superposition of independent NN collisions, without nuclear or medium effects. The results for R^D_{AA} are presented in Fig. 1, left panel. The experimental sensitivity expected from simulation studies is indicated around the curve corresponding to no energy-loss (the uncertainties are discussed in Refs. [6, 5]). The effect of nuclear shadowing, introduced via the EKS98 parameterization [9], is clearly visible in R_{AA} without energy loss for $p_t \lesssim 7$ GeV/c. Above this p_t, only parton energy loss is expected to affect the nuclear modification factor of D mesons. Upper and lower pair of curves reflect different assumptions and can be assumed as an estimation of sensitivity of the calculation. The relative importance of the energy-loss and dead-cone effects can be disentangled using the $R_{D/h}$ ratio (not shown here), which can be measured with good precision since it is a double ratio (AA/pp)/(AA/pp), many systematic uncertainties canceling out. We find that this ratio is enhanced with respect to 1 only by the dead cone effect and, consequently, it appears as a clean tool to investigate and quantify this prediction of QCD (see for details [6]).

The strategy to separate a sample of beauty-originated electrons relies on the peculiar properties of B-mesons decay products, namely the hard transverse momentum spectrum and the high impact parameter, due to the mean decay length of B's (of the order of 500μm). Also identification plays a fundamental role. The main sources of background electrons to beauty signal are: (a) pions misidentified as electrons, (b) Dalitz decays of light mesons (mainly pions), (c) semi-electronic decays of D mesons, (d) photon conversions in the detector materials, (e) strange particle decays.

Because of the overwhelming number of pions produced in nucleus-nucleus reactions at ultra-relativistic energies, source (a) is by far the most abundant, at the level of 10^2 times the signal in the simulation, even at high d_0 , where the pion contribution is still high because of scattering in the detector materials or because originated in strangeness decays. A combined TPC-TRD identification technique is able to reduce this background of a 10^{-4} factor. Dalitz decays of light mesons and photon conversions are the main source of electrons that can contaminate the signal. Such electrons have originally d_0 =0 (they

---

[1] The impact parameter d_0 of a track is defined here as the distance of closest approach of the particle trajectory to the interaction vertex in the plane orthogonal to the beam.

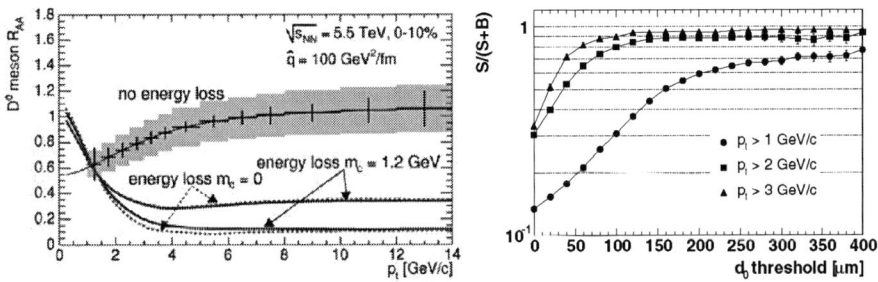

FIGURE 1. Left: R_{AA} of D^0 mesons, without energy loss ('data' points), with energy loss for massless quarks ($m_c = 0$; dashed line) and with energy loss for massive quarks ($m_c = 1.2$ GeV; solid line). The reported errors are: bars = statistical, shaded area = systematic contributions combined. Right:S/(S+B) ratio of reconstructed electron tracks. Background includes all sources other than beauty.

come from very short-lived particles), so a threshold in d_0 will reject most of them. An upper limit of 500 μm on d_0 is imposed in order to reject the electrons coming from decays of strange particles. Electrons originating from the decays of D^0 mesons have lower d_0 and a softer p_t spectrum than beauty-decay electrons and they can be efficiently removed by cuts on both variables.

In figure 1, right panel, we report the S/(S+B) (S=signal, B=background) ratio as a function of the threshold on d_0 , where S=electrons from beauty and B=sources a-e. Three p_t thresholds are shown: 1, 2 and 3 GeV/c. For instance, for $d_0 > 180$ μm and $p_t > 2$ GeV/c, the expected contamination of about 10%, mainly given by the decays $D \rightarrow e + X$. Given these conditions, the estimated statistics of electrons from B decays is 5×10^4 for 10^7 central Pb–Pb events (one month run). The present analyses confirm the good capability of ALICE for the study of heavy flavor production in the high track-density environment of heavy–ion collisions at the LHC and constitutes the basis for future comparison of quark energy loss between heavy (both charm and beauty) and light probes.

REFERENCES

1. ALICE Physics Performance Report, Volume I, CERN/LHCC 2003-049 (2003);
2. ALICE Time Projection Chamber Technical Design Report, CERN/LHCC 2000-001.
3. ALICE Inner Tracking System Technical Design Report, CERN/LHCC 99-12.
4. ALICE Transition Radiation Detector Technical Design Report, CERN/LHCC 2001-021.
5. A. Dainese, Ph.D. Thesis, arXiv:nucl-ex/0311004.
6. A. Dainese, Eur. Phys. J. C **33**, 495 (2004).
7. R. Turrisi, M. Lunardon, ALICE Internal Note, to be submitted.
8. N. Carrer and A. Dainese, ALICE-INT-2003-019 (2003), arXiv:hep-ph/0311225.
9. K.J. Eskola, V.J. Kolhinen, C.A. Salgado, Eur. Phys. J. C **9**, 61 (1999).

Color superconductivity in high density QCD

Roberto Casalbuoni

Dipartimento di Fisica dell' Universita' di Firenze and Sezione INFN, Via G. Sansone 1, 50019 Sesto Fiorentino (Firenze), Italy. E-mail: casalbuoni@fi.infn.it

Abstract. We describe the effects of a strange quark mass and color and electric neutrality on the superconducing phases of QCD.

It is now a well established fact that at zero temperature and sufficiently high densities quark matter is a color superconductor [1, 2]. The study starting from first principles was done in [3, 4, 5]. At densities much higher that the masses of the quarks u, d and s, the favored state is the so-called Color-Flavor-Locking (CFL) state, whereas at lower densities the strange quark decouples and the relevant phase is called two-flavor color superconducting (2SC).

An interesting possibility is that in the interior of compact stellar objects (CSO) some color superconducting phase may exist. In fact we recall that the central densities for these stars could be up to $1 \div 1.5$ fm^{-3}, whereas the temperature is of the order of tens of keV. However the usual assumptions leading to show, for instance, that with three flavors the favored state is CFL, should now be reviewed. Matter inside a CSO should be electrically neutral and should not carry any color. Also conditions for β-equilibrium should be fulfilled. As far as color is concerned, it is possible to impose a simpler condition, that is color neutrality, since in [6] it has been shown that there is no free energy cost in projecting color singlet states out of color neutral states. With the further observation that using β-equilibrium one has $\mu_e = -\mu_Q$, where μ_e and μ_Q are the chemical potentials associated to the electrons and to the electric charge respectively, we see that all the previous conditions can be satisfied by requiring that the electric-charge, T_3 and T_8 densities vanish. This is equivalent to require that the derivatives of the free energy with respect to the corresponding chemical potentials are zero:

$$\frac{\partial \Omega}{\partial \mu_e} = \frac{\partial \Omega}{\partial \mu_3} = \frac{\partial \Omega}{\partial \mu_8} = 0 \tag{1}$$

A color superconducting state is also characterized by a non-zero expectation value of a diquark operator which depends, in the homogeneous case, on several constants, the gaps. The free energy is evaluated by starting from a microscopic description of the quark interaction, which is usually assumed to be a four-fermi interaction. Of course this is an approximation to the real case, but it turns out to be quite effective. In this way one gets the free-energy as a function of the chemical potentials and of the energy gaps, Δ_i. Therefore, besides requiring the conditions (1), we have to minimize the free energy

CP756, Quark Confinement and the Hadron Spectrum VI
edited by N. Brambilla, U. D'Alesio, A. Devoto, K. Maung, G.M. Prosperi and S. Serci
© 2005 American Institute of Physics 0-7354-0241-8/05/$22.50

also with respect to Δ_i

$$\frac{\partial \Omega}{\partial \Delta_i} = 0 \qquad (2)$$

Another important point to be considered in these applications is the mass of the strange quark, since the relevant chemical potentials are of order 400-500 MeV. Both the strange quark mass and the β-equilibrium in conjunction with electrical neutrality have the effect of producing a mismatch in the Fermi momenta of the pairing fermions. For increasing mismatch the BCS pairing mechanism is lost and the system can undergo a phase transition either to the normal state or to a different phase.

Let us review how the mismatch in the Fermi momenta is originated. Consider a massive fermion and a massless one at the same chemical potential μ. The corresponding Fermi momenta are $p_{F_1} = \sqrt{\mu^2 - M^2} \approx \mu - \frac{M^2}{2\mu}$ and $p_{F_2} = \mu$. Therefore the mismatch is of order M^2/μ. For the second instance, consider that β-equilibrium requires the chemical potential for the electrons to be the opposite of the chemical potential associated to the electric charge. Then, the neutrality condition determines uniquely μ_e in terms of the other chemical potentials. Since μ_e is not arbitrary one is forced to consider the free energy along the neutrality line rather than long a line of fixed μ_e. The result of this analysis, at first sight, is a little bit surprising, since where one has unstable phases moving along lines of constant μ_e, now one gets stable phases along neutrality lines [7].

If we characterize the mismatch in terms of the difference among effective chemical potentials, call it $2\delta\mu$, the typical spectrum of quasi-fermions is $E = \pm\delta\mu + \sqrt{(p-\mu)^2 + \Delta^2}$. We see that if $\delta\mu \geq \Delta$ gapless modes are present in correspondence of momenta $p = \mu \pm \sqrt{\delta\mu^2 - \Delta^2}$. The point $\delta\mu = \Delta$ plays a special role, since the energy cost for pairing the fermions belonging to Fermi spheres with a mismatch is $2\delta\mu$ whereas the gain for pairing is 2Δ. Therefore when $\delta\mu > \Delta$ fermions start to lose their BCS pairing.

Examples of the previous situation are the gapless phases g2SC [7] and gCFL [8, 9]. In the g2SC phase the fermion condensate has the same structure as in the 2SC phase, but there is a mismatch $\delta\mu = \mu_e/2$, and correspondingly a phase transition from 2SC to g2SC at $\mu_e = 2\Delta$. The analysis shows that the g2SC phase is stable along the neutrality line. In this particular example the quarks involved are the quark up and the quark down, so there is no effect from the strange quark mass, but only from the interplay of β-equilibrium and electrical neutrality.

The gCFL has a fermionic condensate given by

$$\langle 0|\psi_{aL}^\alpha \psi_{bL}^\beta|0\rangle = \Delta_1 \varepsilon^{\alpha\beta 1}\varepsilon_{ab1} + \Delta_2 \varepsilon^{\alpha\beta 2}\varepsilon_{ab2} + \Delta_3 \varepsilon^{\alpha\beta 3}\varepsilon_{ab3} \qquad (3)$$

whereas in CFL, $\Delta_1 = \Delta_2 = \Delta_3$. In this case the mass strange quark mechanism for the mismatch plays the driving role. In fact, the mismatch among the blue-down and green-strange quarks is given by $M_s^2/(2\mu)$ and a phase transition between the CFL and the gCFL (where $\Delta_3 > \Delta_2 > \Delta_1$) occurs exactly at $M_s/\mu = 2\Delta$. From the analysis made in [9] it turns out that gCFL is the favored phase over 2SC and g2SC up to about $M_s^2/\mu \approx 130$ MeV, for a choice of the gap Δ in CFL given by 25 MeV.

In presence of a mismatch among the Fermi momenta of the pairing fermions there is another interesting possibility which has been considered in the literature, that is the so called LOFF phase [10] (see also the reviews [11]). In this case the mechanism proposed is a different way of pairing in which each of the pairing fermions stays close to its own Fermi surface. As a consequence the pair has a non zero total momentum leading to a breaking of translational and rotational invariance [12]. This phase is particularly interesting since it may give rise to a crystalline structure. This possibility was widely explored in [13] within the context of a Ginzburg-Landau expansion. Since the validity of this expansion in the present context is not completely justified the result of the analysis was a conjecture about the most favored structure, a face centered cube. In [14] a different approximation to the problem was proposed. This approximation holds in a region far from a second order transition, and in this sense is complementary to the Ginzburg Landau expansion. It was found that when the BCS phase is lost a LOFF phase with a crystalline structure corresponding to an octahedron (or a body centered cube) takes place. This happens at $\delta\mu = \Delta/\sqrt{2}$. This phase remains the favored one up to $\delta\mu \approx 0.95\Delta$ where the face-centerd cube crystal becomes energetically favored. Then this phase persists up to 1.32Δ, when the system goes back to the normal phase. According to the authors of ref. [9] this would imply (extrapolating the results of ref. [14] obtained in the 2SC case), that the LOFF phase should take place at about $M_s^2/\mu \approx 120$ MeV up to 225 MeV. If this would be the case the phases 2SC and g2SC would play no role since increasing M_s one would go from CFL to gCFL through a second order transition and then via a first order one to the LOFF phase.

REFERENCES

1. B. Barrois, *Nuclear Physics* **B129**, 390 (1977); S. Frautschi, *Proceedings of workshop on hadronic matter at extreme density*, Erice 1978; D. Bailin and A. Love, *Physics Report* **107**, 325 (1984).
2. M. Alford, K. Rajagopal, and F. Wilczek, Phys. Lett. B **422**, 247 (1998); R. Rapp, T. Schafer, E. V. Shuryak and M. Velkovsky, Phys. Rev. Lett. **81**, 53 (1998).
3. D.T. Son, Phys. Rev. D **59**, 094019 (1999); T. Schäfer and F. Wilczek, Phys. Rev. D **60**, 114033 (1999); D.K. Hong, V.A. Miransky, I.A. Shovkovy, and L.C.R. Wijewardhana, Phys. Rev. D **61**, 056001 (2000); S.D.H. Hsu and M. Schwetz, Nucl. Phys. **B572**, 211 (2000); W.E. Brown, J.T. Liu, and H.-C. Ren, Phys. Rev. D **61**, 114012 (2000).
4. R.D. Pisarski and D.H. Rischke, Phys. Rev. D **61**, 051501 (2000).
5. I.A. Shovkovy and L.C.R. Wijewardhana, Phys. Lett. B **470**, 189 (1999); T. Schäfer, Nucl. Phys. **B575**, 269 (2000).
6. P. Amore, M. C. Birse, J. A. McGovern and N. R. Walet, Phys. Rev. D **65**, 074005 (2002) [arXiv:hep-ph/0110267].
7. I. Shovkovy and M. Huang, Phys. Lett. **B564**, 205 (2003) [arXiv:hep-ph/0302142].
8. M. Alford, C. Kouvaris and K. Rajagopal, Phys. Rev. Lett. **92**, 222001 (2004) [arXiv:hep-ph/0311286].
9. M. Alford, C. Kouvaris and K. Rajagopal, [arXiv:hep-ph/0406137].
10. A. I. Larkin and Yu. N. Ovchinnikov, Sov. Phys. JETP **20** (1965), 762; P. Fulde and R. A. Ferrell, Phys. Rev. **135** (1964), A550.
11. R. Casalbuoni and G. Nardulli, Rev. of Mod. Phys. **76**, 263 (2004) [arXiv:hep-ph/0305069]; J. A. Bowers, [arXiv:hep-ph/0305301].
12. M. G. Alford, J. A. Bowers and K. Rajagopal, *Phys. Rev.* **D63** (2001) 074016, [arXiv:hep-ph/0008208].
13. J. A. Bowers and K. Rajagopal, *Phys. Rev.* **D66** (2002) 065002, [arXiv:hep-ph/0204079].
14. R. Casalbuoni, M. Ciminale, M. Mannarelli, G. Nardulli, M. Ruggieri and R. Gatto, [arXiv:hep-ph/0404090].

Search for universal scaling at the chiral phase transition in 2-flavor lattice QCD

T. Mendes

Instituto de Física de São Carlos, Universidade de São Paulo,
C.P. 369, 13560-970, São Carlos, SP, Brazil

Abstract. We consider the problem of universal scaling at the QCD transition in the case of two degenerate light-quark flavors, discretized with the staggered formulation. A comparison to the predicted $O(4)$ scaling function can be made in an unambiguous way by a suitable normalization of the QCD data, using the observed scaling along the so-called pseudocritical line. We show that reasonably good scaling is obtained for the larger quark masses (thus supporting a second-order phase transition) and discuss possible reasons for the deviations from scaling at the smaller masses.

Keywords: finite-temperature QCD, lattice simulations, chiral phase transition, scaling, universality
PACS: 12.38.Aw, 12.38.Gc, 25.75.Nq, 05.70.Jk, 05.70.Ce

INTRODUCTION

Lattice-QCD simulations allow a nonperturbative description of the phase transition in hadronic matter at high temperatures. For the quenched case one studies the deconfining transition itself, while for the full-QCD case one must consider the chiral phase transition. This transition occurs when the chiral symmetry — exact in the limit of zero quark masses and spontaneously broken at low temperatures — is restored at high temperature. In the case of two dynamic quarks (i.e. considering dynamic effects of only two degenerate light-quark flavors) one would expect to observe a second-order phase transition with critical behavior in the universality class of the $3d$ $O(4)$ spin model [1]. This prediction has been investigated numerically by lattice simulations for over ten years, yet there is still no agreement about the order of the transition or about its scaling properties [2]. More precisely, the predicted $O(4)$ scaling has been observed in the Wilson-fermion case [3], but not in the staggered-fermion case, believed to be the appropriate formulation for studies of the chiral region. In this case, extensive numerical studies and scaling tests have been done by the Bielefeld [4], JLQCD [5] and MILC [6] groups. It was found that the chiral-susceptibility peaks scale reasonably well with the predicted exponents, but no agreement is seen in a comparison with the $O(4)$ scaling function. At the same time, some recent numerical studies with staggered fermions suggest that the deconfining transition may be of first order [7, 8].

Here we present plots of staggered-fermion data — including data points from Refs. [4, 5, 6] above and some new points at a higher mass (partially presented in [9]) — suitably normalized for a direct comparison to the universal $O(4)$ scaling function. Our normalization procedure [10, 11] consists in using the observed $O(4)$ scaling along the *pseudocritical line* — defined by the points where the chiral susceptibility shows a

CP756, *Quark Confinement and the Hadron Spectrum VI*
edited by N. Brambilla, U. D'Alesio, A. Devoto, K. Maung, G.M. Prosperi and S. Serci
© 2005 American Institute of Physics 0-7354-0241-8/05/$22.50

(finite) peak — to find the two non-universal normalization constants that are necessary for relating systems in a given universality class to their common universal scaling form. This provides an unambiguous way to normalize the QCD data, without privileging data at lower or higher quark masses. [We do note however that the data points at very small mass are known to show poor $O(4)$ scaling even on the pseudocritical line.] Using this normalization we find reasonably good support for the predicted universal scaling, especially at higher quark masses (see Figs. 1 left and right). We conclude with a discussion of possible sources of systematic errors at small values of the quark mass.

SCALING TESTS

The scaling Ansatz for a second-order transition implies that the order parameter — e.g. the magnetization M for a spin system — be described by a universal function

$$M/h^{1/\delta} = f_M(t/h^{1/\beta\delta}) \, ,$$

where β, δ are critical exponents defining a universality class and

$$t = (T - T_c)/T_0, \quad h = H/H_0$$

are the reduced temperature and magnetic field, respectively. The function f_M is *universal*, i.e. once the non-universal normalization constants T_0 and H_0 are determined for a given system in the universality class, its order parameter M scales according to the scaling function f_M for this class. The comparison of (normalized) scaling functions between two systems is a good test of universality, especially in cases where the critical exponents cannot be determined with great accuracy.

For QCD the order parameter is given by the chiral condensate $< \overline{\psi} \psi >$, the analogue of the magnetic field is the quark mass m_q, and (on the lattice) the reduced temperature t is usually taken to be proportional to $6/g^2 - 6/g_c^2(0)$, where g is the lattice bare coupling. As described in [10, 11], we use the observed scaling along the pseudocritical line and universal quantities from the $O(4)$ model [12, 13, 14] to determine the normalization constants H_0, T_0. This allows an unambiguous comparison of the QCD data to the scaling function f_M. This comparison is shown in Fig. 1 (left) below. We have also redone the comparison using the form proposed in [8] for the reduced temperature t, which includes a contribution from the quark mass. The comparison in this case is shown in Fig. 1 (right). We note that in the determination of the critical coupling $6/g_c^2(0)$ using the first definition of t we have dropped the data points with the lowest mass ($m_q = 0.01$ in lattice units), in order to obtain a good χ^2/dof for the fit. This was not necessary when using the second definition, i.e. including a term in m_q (and thus adding an extra fit parameter).

CONCLUSIONS

Our plots show reasonably good $O(4)$ scaling for data points at relatively large quark masses. One interpretation of this result is that data at smaller masses (closer to the physical values) may suffer from systematic errors in the simulations, possibly related

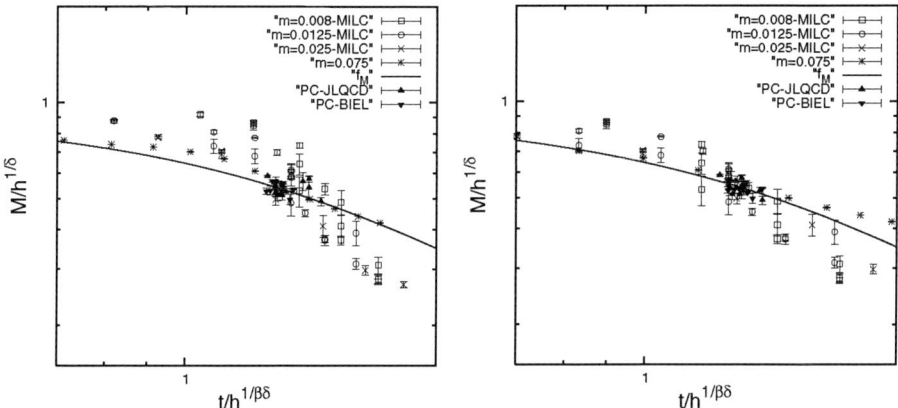

FIGURE 1. Comparison between the normalized QCD data and the universal $O(4)$ scaling function f_M. Data points represented by triangles correspond to the pseudocritical line. Left: usual definition for t. Right: definition for t introduced in [8].

to the use of the so-called R algorithm [15]. We are currently studying the behavior of the algorithm as a function of the quark mass and several run parameters, using the MILC code. Note that data at smaller masses are also more likely to show finite-size effects, although no such effects were found in the finite-size-scaling analyses done in [10, 13].

The author thanks A. Cucchieri and A. Di Giacomo for discussions. This work was supported by FAPESP (Project No. 00/05047-5). Partial support from CNPq is also acknowledged.

REFERENCES

1. R. Pisarski, and F. Wilczek, *Phys. Rev.*, **D29**, 338 (1984).
2. S.D. Katz, *Nucl. Phys. Proc. Suppl.*, **B129**, 60 (2004).
3. Y. Iwasaki, K. Kanaya, S. Kaya, and T. Yoshié, *Phys. Rev. Lett.*, **78**, 179 (1997).
4. F. Karsch, and E. Laermann, *Phys. Rev.*, **D50**, 6954 (1994).
5. S. Aoki et al. (JLQCD Collaboration), *Phys. Rev.*, **D57**, 3910 (1998).
6. C. Bernard et al. (MILC Collaboration), *Phys. Rev.*, **D61**, 054503 (2000).
7. P. Cea, L. Cosmai, and M. D'Elia, *JHEP*, **0402**, 018 (2004).
8. M. D'Elia, A. Di Giacomo, and C. Pica, hep-lat/0408008.
9. A. Cucchieri, and T. Mendes, to appear in the proceedings of *Hadron Physics 2004*.
10. T. Mendes, in *Statistical QCD 2001*, edited by F. Karsch and H. Satz, Elsevier, Amsterdam, 2002, pp. P29–P30, hep-lat/0111044.
11. T. Mendes, in *Hadron Physics 2002*, edited by C.A.Z. Vasconcellos et al., World Scientific, Singapore, 2003, pp. 227–230.
12. J. Engels, and T. Mendes, *Nucl. Phys.*, **B572**, 289 (2000).
13. J. Engels, S. Holtmann, T. Mendes, and T. Schulze, *Phys. Lett.*, **B514**, 299 (2001).
14. A. Cucchieri, and T. Mendes, hep-lat/0406005.
15. M.A. Clark, B. Joó, and A.D. Kennedy, *Nucl. Phys. Proc. Suppl.*, **B119**, 1015 (2003).

Finite Density Lattice QCD
– How to Fight against the Complex Fermion Determinant

Atsushi Nakamura*, Yuji Sasai[1][†] and Tetsuya Takaishi**

*RIISE, Hiroshima University, Higashi-Hiroshima 739-8421, Japan
†Oshima National College of Maritime Technology, Yamaguchi 742-2193, Japan
**Hiroshima University of Economics, Hiroshima 731-0192, Japan

Abstract. In finite density lattice QCD, once the quark chemical potential μ is introduced, the fermion determinant appearing in the path integral measure becomes complex. Therefore the numerical simulation becomes difficult due to the sign problem. To examine phase effects of the $SU(3)$ lattice QCD, we study the range of μa from 0.1 to 1.4, where a is the lattice spacing. We confirmed that the conjugate gradient calculation is possible in the higher density region though it was difficult in the intermediate density region (approximately $\mu a = 0.3 \sim 1.0$). We calculated the quark number density, the chiral condensate, Polyakov line and the phase effect as a function of μa at $6/g^2 = 5.30$.

Keywords: Lattice QCD, Finite density, Phase effect
PACS: 11.10.Wx, 12.38.Aw, 12.38.Gc

INTRODUCTION

Lattice QCD is a powerful method to obtain the information of nonperturbative regions. Recently the numerical simulation at finite density has come to be performed actively. In finite temperature and finite density system, it has been expected that the investigation about the phase transition between the confinement phase and the deconfinement phase gives the information of aspects of nuclear matter such as neutron stars, heavy nucleus and the early universe. There are several recent experimental results such as RHIC at BNL and SPS at CERN. Baryon number chemical potential is introduced at finite density. In $SU(3)$ lattice QCD at finite density, the fermion determinant appearing in the path integral measure becomes complex. Therefore the numerical simulation becomes difficult due to the sign problem. In this case the ordinary Monte Carlo method by which gauge configurations are generated with Boltzman weight does not work well [1, 2].

The lattice partition function for Kogut-Susskind (KS) fermions is given by

$$Z = \int DU (\det \Delta(\mu))^{N_f/4} e^{-\beta S_g}, \tag{1}$$

where N_f, μ and $\Delta(\mu)$ denote a flavor number, a baryon chemical potential and a fermion matrix respectively. In the continuum field theory, the chemical potential is introduced through the substitution $p_4 \rightarrow p_4 - i\mu$. On the other hand, on lattice, the chemical potential is introduced as multiplying $\exp(\mu)$ by the temporal link variables.

---

[1] Presenter

CP756, *Quark Confinement and the Hadron Spectrum VI*
edited by N. Brambilla, U. D'Alesio, A. Devoto, K. Maung, G.M. Prosperi and S. Serci

For $N_f = 2$, the expectation value of an operator O is given by

$$\langle O \rangle = \frac{1}{Z} \int DU (\det\Delta(\mu))^{1/2} O e^{-\beta S_g} = \frac{\int DU |\det\Delta|^{1/2} e^{i\theta/2} O e^{-\beta S_g}}{\int DU |\det\Delta|^{1/2} e^{i\theta/2} e^{-\beta S_g}}$$

$$= \frac{\int DU |\det\Delta|^{1/2} e^{i\theta/2} O e^{-\beta S_g}}{\int DU |\det\Delta|^{1/2} e^{-\beta S_g}} \Big/ \frac{\int DU |\det\Delta|^{1/2} e^{i\theta/2} e^{-\beta S_g}}{\int DU |\det\Delta|^{1/2} e^{-\beta S_g}} = \frac{\langle O e^{i\theta/2}\rangle_0}{\langle e^{i\theta/2}\rangle_0}, \quad (2)$$

where $\langle \ldots \rangle_0$ is the expectation value with the phase-quenching measure. We use the phase-quenching measure $DU |\det\Delta(\mu)|^{N_f/4} e^{-\beta S_g}$ in our Monte Carlo simulations.

NUMERICAL INVESTIGATION

We perform calculations on a $8^3 \times 4$ lattice at $m = 0.05$ for $N_f = 2$ using the R-algorithm, where m is a quark mass in Kogut-Suskind fermions. The trajectory length is set to be 0.5 with a step size of $\Delta t = 0.01$. The first 1000 trajectories were discarded for thermalization. To examine phase effects of the $SU(3)$ lattice QCD, we calculated $\det\Delta(\mu)$ at $\mu a = 0.1, 0.2$ and 0.25 in the range of $\beta = 6/g^2 = 5.20 \sim 5.40$. Fig.1 shows the fluctuations of $\langle \cos(\theta/2)\rangle_0$[3]. The fluctuations increase with increasing μa and also decrease for $\beta > \beta_c$, where critical coupling β_c is around 5.3. Using the phase θ of $\det\Delta(\mu)$, we estimate the reweighted values of the chiral condensate $\langle \bar{\psi}\psi \rangle$ and Polyakov line $\langle L \rangle$ according to the reweighting formula Eq.(2). There are almost no differences with the phase-quenched (no phase) cases and the reweighted cases. Fig.2 shows the results of the chiral condensate. Values for the reweighted case are very similar to those for the phase-quenched case. In the low density region, it seems that there is no phase effect.

We are interested whether the phase effect appears at more large μa. Conjugate gradient (CG) calculation to obtain quark propagator does not work well beyond $\mu a = 0.28$ at $\beta = 5.20$. We have faced such a difficulty in running our simulations, so we tried to perform CG calculation in the range of $\mu a = 0.1 \sim 1.4$. As a result, we confirmed that

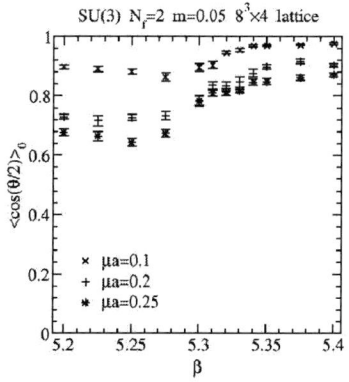

Fig. 1. Phase fluctuation as a function of β

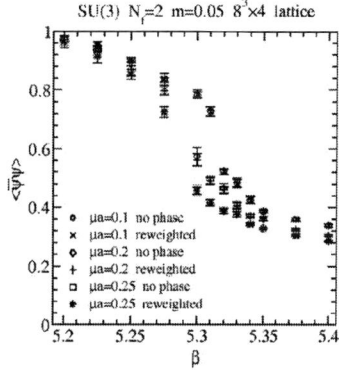

Fig. 2. Chiral condensate as a function of β

CG calculation is also possible in the higher density region though it was difficult in the intermediate density region. Because pion mass is $m = 457\text{MeV}$ ($ma = 0.57$) in our calculation, it seems that the calculation becomes difficult at pion mass. By examining the eigenvalue distributions for KS fermions, we confirmed that zero eigenvalues occur in the intermediate density region. In Fig.3 we plot typical eigenvalue distributions. Fig.4 shows that chiral symmetries restore in the high density region. Fig.5 shows the results of Polykov line. It seems that the phase becomes deconfinement phase as μa increases. At a glance there is confinement phase in the high density region. In this region the quark free energy is infinity due to the lattice artifact.

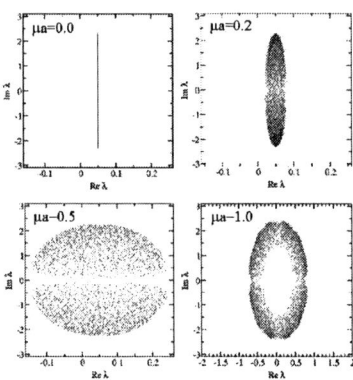

Fig. 3. Eigenvalue distributions

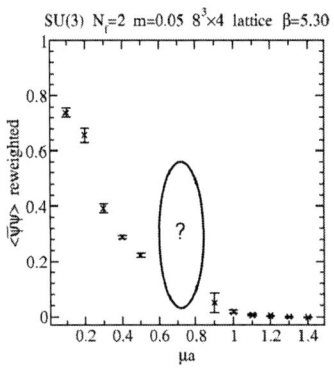

Fig. 4. Chiral condensate as a function of μa

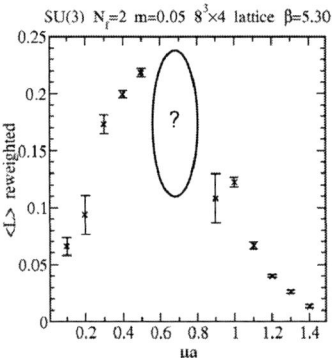

Fig. 5. Polyakov line as a function of μa

SUMMARY

We examined the phase effect of $SU(3)$ lattice QCD (2 flavors) on a $8^3 \times 4$ lattice. We calculated chiral condensates and Polyakov lines in both cases of phase-quenched and reweighted. We cannot find those differences until now. CG calculation cannot converge due to zero eigenvalues in the intermediate density region. However we confirmed that calculation is possible in the higher density region. Our simulations were performed on a super computer system SX5 at RCNP, Osaka University.

REFERENCES

1. Z.Fodor and S.D.Katz, *JHEP*, **0203** (2002) 014.
2. Ph.de Forcrand, S. Kim and T. Takaishi, *Nucl.Phys. (Proc.Suppl.)*, **119** (2003) 541.
3. Y. Sasai, A. Nakamura and T. Takaishi, *Nucl.Phys. (Proc.Suppl.)*, **129-130C** (2004) 539.

Spinodal decomposition in pure-gauge QCD

G. Krein

Instituto de Física Teórica, Universidade Estadual Paulista
Rua Pamplona, 145, 01405-900, São Paulo, SP - Brazil

Abstract. Spinodal decomposition in a model of pure-gauge $SU(2)$ theory that incorporates a deconfinement phase transition is investigated by means of real-time lattice simulations of the fully nonlinear Ginzburg-Landau equation. Results are compared with a Glauber dynamical evolution using Monte Carlo simulations of pure-gauge lattice QCD.

Keywords: Spinodal decomposition, color confinement, lattice QCD, Ginzburg-Landau equation
PACS: 12.38.Mh, 12.38.Gc, 74.20.De, 64.75.+g

Relativistic heavy-ion experiments produce highly excited hadronic matter. On the basis of quantum chromodynamics (QCD) one expects that at sufficiently high excitation energies this matter is composed by quarks and gluons that are not confined in the interior of hadrons. One important question in the study of the properties of the produced excited matter is the understanding of the dynamics of the quark-gluon deconfining process. One possible scenario [1] is that it proceeds via a process similar to spinodal decomposition of condensed matter physics [2]. Spinodal decomposition is an instability driven by infinitesimal amplitude, long wavelength fluctuations of an order parameter after a sudden temperature quench that brings the system into an unstable state. In pure gauge QCD, the equilibrium order parameter is the thermal average of the trace of the Polyakov loop in the fundamental representation, $\langle L_F \rangle$. For the $SU(2)$ gauge theory one can write [1]

$$ L_F = \frac{1}{2}\mathrm{Tr}_F L, \qquad \text{where} \qquad L = \begin{pmatrix} e^{i\pi q} & 0 \\ 0 & e^{-i\pi q} \end{pmatrix}, \qquad 0 \le q \le 1. \qquad (1) $$

For temperatures T below a critical value T_d, one has $\langle L_F \rangle = 0$, and for $T > T_d$, $\langle L_F \rangle \ne 0$. Like the magnetization of the Ising model, $\langle L_F \rangle$ is a nonconserved order parameter, in the sense that there is no associated conservation law.

In condensed matter problems the time evolution towards equilibrium of an nonconserved order parameter is characterized by a relaxation process and is commonly described by means of phenomenological macroscopic Ginzburg-Landau (GL) field equations with Langevin dynamics [2]

$$ \frac{\partial \psi}{\partial t} = -\Gamma \frac{\delta S_{eff}[\psi]}{\delta \psi} + \varsigma, \qquad (2) $$

where $S_{eff}[\psi]$ is the coarse grained effective action for the order parameter $\psi(\mathbf{x},t)$, Γ is a dissipation coefficient and ς is a noise term that mimics the thermal fluctuations of the system. The long-time equilibrium distribution $\rho[\psi_{eq}]$ at temperature T is given by

CP756, Quark Confinement and the Hadron Spectrum VI
edited by N. Brambilla, U. D'Alesio, A. Devoto, K. Maung, G.M. Prosperi and S. Serci

$\rho[\psi] = \exp(-F[\psi]/T)$ where $F[\psi] = T S_{eff}[\psi]$ is the coarse-grained free-energy, or effective potential. The real time description of the Polyakov loop relaxation process directly from QCD is an impossible task at the moment. But one can use local Monte Carlo updating algorithms of lattice QCD simulations to imitate real time thermal fluctuations of the Polyakov loop, very much like the Glauber dynamical evolution [3] commonly used for spin systems [4]. This has been pursued some time ago by Miller and Ogilvie [1].

In this paper we present results of a real-time lattice simulation of the fully nonlinear GL equation of Eq. (2), and compare results with lattice Monte Carlo simulations of pure gauge $SU(2)$ theory. For the simulation of the GL equation we use the one-loop effective action $S_{eff}[\psi]$ [5], augmented with a phenomenological term to model the second order phase transition of $SU(2)$ pure gauge QCD [6]

$$S_{eff} = \int d^3x \left[\frac{\pi^2 T}{2g^2} (\nabla \psi)^2 + \frac{\pi^2 T^3}{12} (1 - \psi^2)^2 - \frac{M^2 T}{4} (1 - \psi^2) \right], \qquad (3)$$

where $\psi(\mathbf{x},t)$ is related to q that parameterizes L in Eq. (1) by $q = (1 - \psi)/2$. This action presents a second order deconfining phase transition at temperature $T_d = \sqrt{3/2}\,M/\pi$. At $T \ll T_d$, the minimum of the action density is at $\psi(\mathbf{x},t) = 0$. Now, if at $t = 0$ the temperature is rapidly increased to $T \gg T_d$, the system is brought to an unstable state and therefore will start "rolling down" to the two new minima of the effective action density. Therefore, at short times, when $\psi(\mathbf{x},t) \approx 0$, the ψ^4 term in Eq. (3) can be neglected and one can write the (noiseless) solution of Eq. (2) as

$$\psi(\mathbf{x}, t \approx 0) = \int d^3k \, e^{i\mathbf{k}\cdot\mathbf{x} - \Gamma \frac{\pi^2 T}{g^2}\left[\mathbf{k}^2 - \frac{g^2 T^2}{3}\varepsilon(T)\right]t} \, \widetilde{\psi}(\mathbf{k},0), \qquad (4)$$

where $\varepsilon(T) = 1 - T_d^2/T^2$, and $\widetilde{\psi}(\mathbf{k},0)$ is the Fourier transform of the $\psi(\mathbf{x}, t = 0)$. For $T \gg T_d$, one has $\varepsilon(T) \simeq 1$, and there will be an explosive exponential growth of long wavelengths modes, signalling the spinodal decomposition with a nonconserved order parameter. For a given temperature, there will exponential growth for modes with $k < k_c$, where $k_c^2 = \varepsilon(T)g^2 T^2/3$. Obviously there is a qualitative change with respect to the model Ref. [1] where $T_d = 0$, because spinodal decomposition will happen only for $T > T_d$. For $T \gg T_d$, the model of Ref. [1] and give the same short-time exponential growth of the order parameter.

As time increases, the order parameter increases and the linear approximation is no longer valid and the use of the complete effective action is necessary. From the point of view of the heavy-ions experiments, understanding the short-time behavior of $\langle L_F \rangle$ is important for knowing the time scales of the explosive growth of the order parameter compared to the expansion of the system. Since there is no hope that in a foreseeable future one will able to simulate an spectacular event like an heavy-ion simulation with lattice QCD, the use of phenomenological equations will be the only alternative available. The comparison between results of lattice QCD simulations and of real-time phenomenological equations like Eq. (2) is however essential for extracting parameters entering the phenomenological equations.

We solve the full nonlinear GL equation for $T = 2T_d$ and compare with a lattice simulation. Although we do not present results for dynamical critical exponents here, we are interested in the short-time growth of the order parameter to investigate the role of the nonlinearities. We use $T_d = 240$ MeV and $g = 3$, which gives $m_D(T) \simeq 210$ MeV for the Debye screening mass. We prepare several 64^3 lattices with random distributions around $\psi(\mathbf{x}, 0) \approx 0$ and then average the time-evolved solutions over all lattices. For the QCD simulation we thermalize several lattices with the heath-bath algorithm for $64^3 \times 4$ lattice sites with $\beta = 4/g^2 = 2.0$. From these thermalized lattices we evolve in "time" with $\beta = 3.0$ (which corresponds roughly to $T = 2T_d$) and then record the value of $\langle L_F \rangle$ for each heath-bath sweep, in the same lines of Ref. [1]. In Fig. 1 we present our results. On the left panel, are plotted the GL results on the right panel are the lattice QCD results.

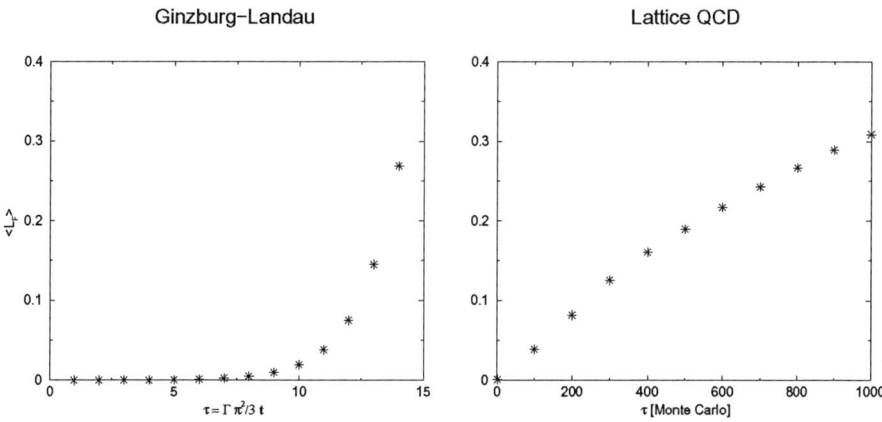

FIGURE 1. Left: GL equation. Right: lattice QCD.

Based on these results, one can make the point that in GL evolution, it seems that the nonlinearities start operating quite late, as compared to the MC time evolution. The MC early-time behavior seems nonlinear from the very beginning. While the early time evolution is exponential for the GL equation, in the MC evolution it seems to be power-law. The early-time nonlinearity might indicate that the dissipation constant Γ in a GL equation type of equation is not really a constant, but rather field dependent [7].

ACKNOWLEDGMENTS

Work partially supported by CNPq and FAPESP (Brazilian agencies).

REFERENCES

1. T. Miller and M. Ogilvie, *Phys. Lett. B*, **488**, 313-318 (2000).
2. A. J. Bray, *Adv. Phys.*, **43**, 357-459 (1994).
3. R. Glauber, *J. Math. Phys.*, **4**, 294-307 (1963).
4. B. A. Berg, U. M. Heller, H. Meyer-Ortmanns, and A. Velytsky, *Phys. Rev. D*, **69**, 034501 (2004).
5. D. Gross, R. Pisarski and L. Yaffe, *Rev. Mod. Phys.*, **53**, 43-80 (1981).
6. P. N. Meisinger, T. R. Miller and M. C. Ogilvie, *Phys. Rev. D*, **65**, 034009 (2002).
7. G. Krein, work in progress.

Phase transition from nuclear to quark matter

S. Lawley*, W. Bentz†, T. Horikawa† and A. W. Thomas**

*Special Research Centre for the Subatomic Structure of Matter,
University of Adelaide, Adelaide SA 5005, Australia
†Department of Physics, School of Science, Tokai University,
1117 Kita-Kaname, Hiratsuka 259-1207, Japan
**Jefferson Lab, 12000 Jefferson Avenue, Newport News, VA 23606, U.S.A.

Abstract. We use the flavor SU(2) NJL model to describe both nuclear and quark matter, and construct phase diagrams to illustrate the phase transitions to normal quark matter (NQM) and color superconducting quark matter (SQM). We calculate the corresponding charge neutral equations of state using the Gibbs conditions to generate the mixed phases.

Keywords: Nambu–Jona-Lasinio Model, Phase Diagrams, Phase Transitions
PACS: 12.39.Fe, 12.39.Ki, 64.70.Fx

Many ideas in the area of finite density phase transitions have been developed in the context of neutron star calculations, in particular the phase transitions from nuclear matter (NM) to quark matter (QM) [1, 2]. In this work we calculate both the NM and QM phases in β-equilibrium using the flavor SU(2) Nambu–Jona-Lasinio (NJL) model. Recently, the problem of NM saturation has been addressed with a method to simulate confinement in the description of the nucleon as a quark-diquark bound state[3]. In the QM phase, we allow for the possibility of scalar diquark condensation (color superconductivity)[4], and the mixed phases are calculated using the method of Glendenning [1].

The effective potential for NM is $V^{NM} = V_{vac} + V_N + V_\omega + V_\rho + V_e$, where the vacuum term V_{vac} accounts for the polarization of the quark Dirac sea due to the valence nucleons [5], and the Fermi motion of the nucleons contributes

$$V_N = -\sum_{N=n,p} \gamma_N \int \frac{d^3k}{(2\pi)^3} \Theta(k_{F_N} - k) \left[\sqrt{k_{F_N}^2 + M_N^2} - \sqrt{k^2 + M_N^2} \right], \tag{1}$$

where the nucleon mass $M_N(M)$ is calculated from the quark-diquark bound state equation. The omega meson, rho meson and electron terms are $V_\omega = -9G_\omega(\rho_n + \rho_p)^2$, $V_\rho = -9G_\rho(\rho_p - \rho_n)^2$ and $V_e = -\frac{k_{F_e}^4}{12\pi^2}$, where G_ω and G_ρ are coupling constants, fixed by the empirical values for the saturation point and the asymmetry coefficient respectively. The nucleon chemical potentials have the form $\mu_N = \sqrt{k_{F_N}^2 + M_N^2} + 18G_\omega(\rho_p + \rho_n) \pm 18G_\rho(\rho_p - \rho_n)$, from which one can define the chemical potentials for baryon number and isospin as $\mu_B = (\mu_p + \mu_n)/2$ and $\mu_I = (\mu_p - \mu_n)/2$. The electron chemical potential is fixed by β-equilibrium: $\mu_e = k_{F_e} = -2\mu_I$.

CP756, *Quark Confinement and the Hadron Spectrum VI*
edited by N. Brambilla, U. D'Alesio, A. Devoto, K. Maung, G.M. Prosperi and S. Serci
© 2005 American Institute of Physics 0-7354-0241-8/05/$22.50

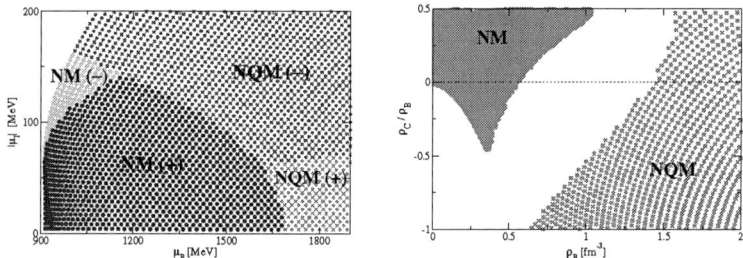

FIGURE 1. Phase diagrams for nuclear matter and normal quark matter phases.

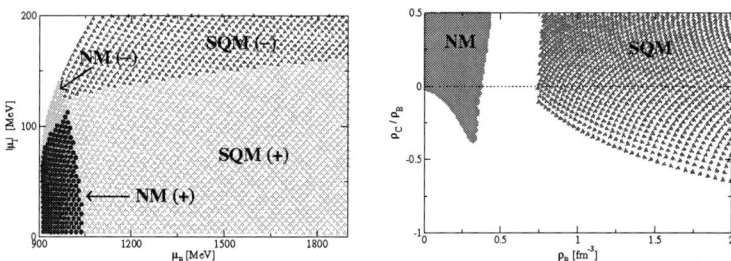

FIGURE 2. Phase diagrams for nuclear matter and color superconducting quark matter phases.

The effective potential for QM is $V^{QM} = V_{vac} + V_Q + V_e + V_\Delta$, where the vacuum and electron terms are as above and the term for the Fermi motion of the quarks V_Q is analogous to V_N, with the sum being over up and down quarks instead of neutrons and protons. Color superconducting quark matter (SQM) includes a contribution from quark pairing arising at the Fermi surface. This is expressed as follows,

$$V_\Delta = \sum_{\alpha = \pm} 2i \int \frac{d^4 k}{(2\pi)^4} \left[\ln \frac{k_0^2 - (\varepsilon_\alpha + \mu_I)^2}{k_0^2 - (E_\alpha + \mu_I)^2} + \ln \frac{k_0^2 - (\varepsilon_\alpha - \mu_I)^2}{k_0^2 - (E_\alpha - \mu_I)^2} \right] + \frac{\Delta^2}{6 G_s}, \qquad (2)$$

where G_s is the coupling constant in the scalar diquark channel, and we have defined $\varepsilon_\pm = \sqrt{(E_k \pm \mu_q)^2 + \Delta^2}$, $E_\pm = |E_k \pm \mu_q|$ and $E_k = \sqrt{k^2 + M_Q^2}$. Here $\mu_q = \mu_B/3$. We set the quark mass M_Q to zero in these calculations, as it is known to decrease rapidly with density [5]. To construct the phase diagram we compare the NM and QM effective potentials for each baryon and isospin chemical potential.

Examples for the resulting phase diagrams are shown in Fig. 1 for the case $G_s = 0$ (normal quark matter, NQM), and in Fig. 2 for the case $G_s/G_\pi = 0.25$, where G_π is the coupling constant in the pionic channel which is fixed by adjusting the pion mass.

Consider first the diagrams on the left hand side, which show the phases in the plane of the chemical potentials μ_B and μ_I. The charge neutral equations of state trace the lines between the positively and negatively charged regions. For example, in Fig. 2 the equation of state begins in the NM region, proceeds along the boundary between negatively charged NM and positively charged SQM, and ends in the SQM phase. The boundary between the NM(−) phase and the SQM(+) phase, which is rather short in this

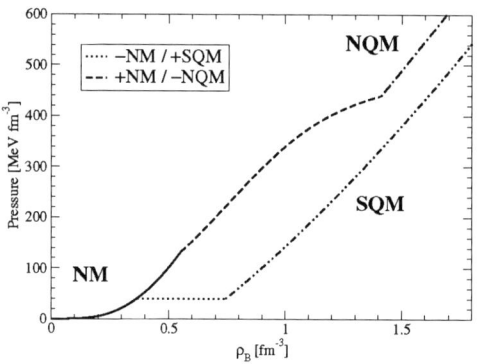

FIGURE 3.   Phase transitions to NQM and SQM

case, corresponds to the NM/SQM mixed phase. The volume fractions of NM and SQM in the mixed phase are obtained by the requirement of charge neutrality. It is important to observe that the coupling constant $G_s = 0.25\,G_\pi$ is sufficiently strong to support a superconducting gap which is stable for isospin asymmetric matter. The figures on the right hand side show the phases in the plane of the charge density ρ_C and the baryon density. We see that charge neutral matter will have a pure NM phase in the lower density region and a pure QM phase at high densities. In each case the space in between allows for a mixed phase at intermediate densities.

The pressure of the charge neutral equation of state is shown as a function of baryon density in Fig. 3 for the two cases corresponding to Fig.1 and 2. For finite pairing interaction $(G_s/G_\pi = 0.25)$, the phase transition from NM to SQM resembles very much the result of a naive Maxwell construction. In order to further investigate this point, however, the effects of finite quark mass should be taken into account.

This work was supported by the Australian Research Council, the DOE contract DE-AC05-84ER40150, under which SURA operates Jefferson Lab, and the Grant in Aid for Scientific research of the Japanese Ministry of Education, Culture, Sports, Science and Technology, C2-16540267.

REFERENCES

1. Glendenning, N. K., *Phys. Rev. D,* **46**, 1992, pp. 1274.
2. Glendenning, N. K., *Compact Stars*, Springer, New York, 2000.
3. Bentz, W., and Thomas, A. W., *Nucl. Phys. A* **696** 2001, pp. 138.
4. Alford, M., Rajagopal, K., and Wilczek, F., *Phys. Lett. B*, **422**, 1998, pp. 247.
5. Bentz, W., Horikawa, T., Ishii, N., and Thomas, A. W., *Nucl. Phys. A* **720** 2003, pp. 95.

Finite-Temperature Yang-Mills Theory in Landau Gauge

A. Maas[*], J. Wambach[*,†], B. Grüter[**] and R. Alkofer[**]

[*]Institute for Nuclear Physics, Darmstadt University of Technology, Schloßgartenstraße 9,
D-64289 Darmstadt, Germany
[†]Gesellschaft für Schwerionenforschung mbH, Planckstr. 1, D-64291 Darmstadt, Germany
[**]Institute of Theoretical Physics, Tübingen University, Auf der Morgenstelle 14,
D-72076 Tübingen, Germany

Abstract. The gluon and ghost propagators in Landau Gauge Yang-Mills Theory are investigated. Self-consistent solutions are obtained from their equations of motion above and below the presumed phase transition. Gluon confinement is manifest in these solutions and can be read off the infrared behavior of the gluon and ghost propagators. Confinement prevails below the presumed phase transition. Above and in the infinite temperature limit, a qualitative change is observed: The chromoelectric sector exhibits screening, while long-range chromomagnetic interactions, mediated by soft modes, are still observed. At least part of the gluon spectrum is still confined. These findings agree with corresponding lattice results.

Keywords: Finite-temperature field theory; Quantum chromodynamics; Other nonperturbative calculations; Quark-gluon plasma; Gluons
PACS: 11-10.Wx 12.38.-t 12.38.Lg 12.38.Mh 14.70.Dj

It is by now well established that QCD undergoes a phase transition, or at least a rapid crossover, at some critical temperature. There is significant evidence that the equilibrium system in the high-temperature phase is a strongly interacting one. This was already anticipated by the infinite-temperature limit of the equilibrium state which is described by a confining three-dimensional (3d) theory. This has been confirmed by lattice calculations, see e.g. [1], and also by the present work. The latter investigates the equilibrium properties of gluons at different temperatures using Dyson-Schwinger equations (DSEs) [2].

The following investigations are restricted to pure Yang-Mills theory, as substantial evidence exists that the non-perturbative features of QCD are generated in the gauge sector. The equilibrium theory is governed by the Euclidean Lagrangian

$$\mathscr{L} = \frac{1}{4}F_{\mu\nu}^a F_{\mu\nu}^a + \bar{c}^a \partial_\mu D_\mu^{ab} c^b$$

$$F_{\mu\nu}^a = \partial_\mu A_\nu^a - \partial_\nu A_\mu^a - g f^{abc} A_\mu^b A_\nu^c \qquad D_\mu^{ab} = \delta^{ab}\partial_\mu + g f^{abc} A_\mu^c \,,$$

with the field strength tensors $F_{\mu\nu}^a$ and the covariant derivative D_μ^{ab}. A_μ^a denotes the gluon field and $\bar{c}^a$ and c^a are the ghost fields describing part of the quantum fluctuations of the gluon field. The Landau gauge is used as it is best suited for the purpose at hand [2].

Important properties of the gluon and ghost fields are characterized by their respective two-point Green's functions, i.e. the propagators. Their infrared properties are linked to the presence of confinement. Especially, a particle is absent from the physical spectrum

CP756, *Quark Confinement and the Hadron Spectrum VI*
edited by N. Brambilla, U. D'Alesio, A. Devoto, K. Maung, G.M. Prosperi and S. Serci
© 2005 American Institute of Physics 0-7354-0241-8/05/$22.50

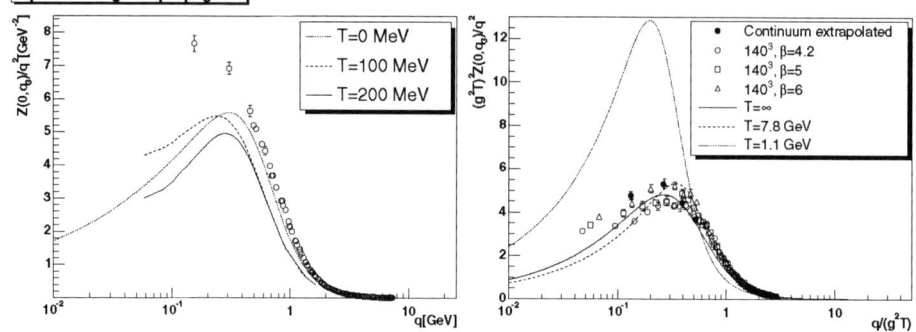

FIGURE 1. The transverse gluon propagator: The left panel displays $T < T_c$ results compared to $T = 0$ lattice data [9] while the right panel shows the $T > T_c$ case compared to corresponding $T \to \infty$ lattice data [1, 10].

if its propagator $D(q^2)$ vanishes in the infrared [2]

$$\lim_{q \to 0} D(q^2) = 0. \qquad (1)$$

The propagators can be used to define dressing functions as

$$D_G(q) = -\frac{G(q_0^2, q_3^2)}{q^2},$$

$$D_{\mu\nu}(q) = P_{\mu\nu}^T(q)\frac{Z(q_0^2, q_3^2)}{q^2} + P_{\mu\nu}^L(q)\frac{H(q_0^2, q_3^2)}{q^2}.$$

Here G denotes the ghost dressing function, and Z and H the ones for gluons being transverse or longitudinal w.r.t. the heat bath. To obtain these functions, a truncated set of DSEs for the propagators has been solved [3, 4, 5, 6]. The necessity to truncate the infinite set of coupled Dyson-Schwinger equations generates several problems concerning gauge invariance. These have been dealt with and the errors are, at least qualitatively, under control [3, 5, 6]. Recently, some of the assumptions have been confirmed by lattice calculations [7] and semi-perturbative methods [8].

Calculations have been performed at $T = 0$ [3], $T < T_c$ [4], $T > T_c$ [5], and $T \to \infty$ [6]. Those at $T < T_c$ have used a toroidal discretized space-time while the others were done in the continuum. Note that, for $T \to \infty$, the theory becomes dimensionally reduced to a 3d Yang-Mills theory plus an adjoint Higgs field, where Z corresponds to the (chromomagnetic) 3d gluon and H to the (chromoelectric) Higgs.

The ghost dressing function is found to diverge at zero momentum. This behavior is not affected qualitatively by temperature [4, 5, 6]. The divergence indicates the mediation of long-range forces which are still present even at $T \to \infty$. In addition, this divergence is connected to the confinement mechanism [2]. This is an indication for some residual gluon confinement in the high-temperature phase.

The results for the transverse gluon are shown in Fig. 1. Below the phase transition, the propagator exhibits confinement by virtue of condition (1). It becomes steeper in

426

the infrared with increasing temperature. In the high-temperature phase there is no qualitative difference. Hence, the transverse gluon is always confined. This can be interpreted as over-screening. In fact, the temperature dependence is surprisingly small. The lattice data at $T = 0$ show strong finite-volume effects in the infrared, but have very recently been shown to bend over when large volumes are employed, see e.g. [11], as is the case in the 3d theory, where significantly larger lattices can be used.

At $T = 0$, $H = Z$ and thus gluons longitudinal w.r.t. the heat bath are confined as well. At $T < T_c$, H becomes shallower with temperature in the infrared [4]. It, however, still exhibits confinement. In the high-temperature phase, the situation is much different [5, 6]. In contrast to the transverse sector, it shows screening and behaves similar to a massive particle. Nevertheless, a comparison to lattice results and perturbation theory indicates that the Higgs propagator contains sizeable non-trivial effects [6].

In conclusion, the infrared behavior of gluon and ghost propagators, employing DSEs, has been analyzed. At vanishing and small temperatures, the results show manifest gluon confinement. It is found that, even in the $T \rightarrow \infty$ limit, strong long-range correlations are present, leading to a non-perturbative behavior of the soft modes. Part of the gluons are still confined: The Gribov-Zwanziger scenario (see e.g. [12]) applies at *all* temperatures. These results, together with the ones of lattice calculations, demonstrate that the high-temperature phase is far from trivial. Furthermore, this may have significant impact on the thermodynamic properties near the phase transition [13] and thus on experimentally accessible observables.

ACKNOWLEDGMENTS

A.M. thanks the organizers of Quark Confinement and the Hadron Spectrum VI for the very inspiring conference and the opportunity to present this work. This work is supported by the BMBF under grant number 06DA116, by the European Graduate School Basel-Tübingen (DFG contract GRK683) and by the Helmholtz association (Virtual Theory Institute VH-VI-041).

REFERENCES

1. A. Cucchieri, F. Karsch and P. Petreczky, *Phys. Rev.* **D64**, 036001 (2001).
2. R. Alkofer and L. von Smekal, *Phys. Rept.* **353**, 281 (2001).
3. C. S. Fischer and R. Alkofer, *Phys. Lett.* **B536**, 177 (2002); *Phys. Rev.* **D67**, 094020 (2003).
4. B. Grüter, R. Alkofer, A. Maas and J. Wambach, arXiv:hep-ph/0408282.
5. A. Maas, J. Wambach, and R. Alkofer, in preparation. A. Maas, PhD thesis, TU Darmstadt, 2004.
6. A. Maas, J. Wambach, B. Grüter and R. Alkofer, Eur. Phys. J. C **37** (2004) 335 [arXiv:hep-ph/0408074].
7. A. Cucchieri, T. Mendes and A. Mihara, arXiv:hep-lat/0408034.
8. W. Schleifenbaum, A. Maas, J. Wambach and R. Alkofer, arXiv:hep-ph/0411052.
9. P. O. Bowman, U. M. Heller, D. B. Leinweber, M. B. Parappilly and A. G. Williams, arXiv:hep-lat/0402032.
10. A. Cucchieri, T. Mendes and A. R. Taurines, *Phys. Rev.* **D67**, 091502 (2003).
11. O. Oliveira and P. J. Silva, arXiv:hep-lat/0410048 and references therein.
12. D. Zwanziger, *Phys. Rev.* **D69**, 016002 (2004) and references therein.
13. D. Zwanziger, arXiv:hep-ph/0407103.

Λ and Ω⁻ Polarization in Pb+Pb Collisions at 160 A GeV/c

The NA57 Collaboration

F. Antinori[k], P.A. Bacon[e], A. Badalà[f], R. Barbera[f], A. Belogianni[a], I.J. Bloodworth[e], M. Bombara[h], G.E. Bruno[b], S.A. Bull[e], R. Caliandro[b], M. Campbell[g], W. Carena[g], N. Carrer[g], R.F. Clarke[e], A. Dainese[k], D. Di Bari[b], S. Di Liberto[n], R. Divià[g], D. Elia[b], D. Evans[e], G.A. Feofilov[p], R.A. Fini[b], P. Ganoti[a], B. Ghidini[b], G. Grella[o], H. Helstrup[d], K.F. Hetland[d], A.K. Holme[j], A. Jacholkowski[f], G.T. Jones[e], P. Jovanovic[e], A. Jusko[e], R. Kamermans[r], J.B. Kinson[e], K. Knudson[g], V. Kondratiev[p], I. Králik[h], A. Kravčáková[i], P. Kuijer[r], V. Lenti[b], R. Lietava[e], G. Løvhøiden[j], V. Manzari[b], M.A. Mazzoni[n], F. Meddi[n], A. Michalon[q], M. Morando[k], P.I. Norman[e], A. Palmeri[f], G.S. Pappalardo[f], B. Pastirčák[h], R.J. Platt[e], E. Quercigh[k], F. Riggi[f], D. Röhrich[c], G. Romano[o], K. Šafařík[g], L. Šándor[h], E. Schillings[r], G. Segato[k], M. Sené[l], R. Sené[l], W. Snoeys[g], F. Soramel[k*], M. Spyropoulou-Stassinaki[a], P. Staroba[m], R. Turrisi[k], T.S. Tveter[j], J. Urbán[i], P. van de Ven[r], P. Vande Vyvre[g], A. Vascotto[g], T. Vik[j], O. Villalobos Baillie[e], L. Vinogradov[p], T. Virgili[o], M.F. Votruba[e], J. Vrlakova[i] and P. Závada[m].

a. Physics Department, University of Athens, Greece
b. Dipartimento IA di Fisica dell'Università e del Politecnico di Bari and INFN, Bari, Italy
c. Fysisk Institutt , Universitetet i Bergen, Bergen, Norway
d. Høgskolen i Bergen, Bergen, Norway
e. University of Birmingham, Birmingham, UK
f. University of Catania and INFN, Catania, Italy
g. CERN, European Laboratory for Particle Physics, Geneva, Switzerland
h. Institute of Experimental Physics Slovak Academy of Science, Kosice, Slovakia
i. P.J. Safárik University, Kosice, Slovakia
j. Fysisk institutt, Universitetet i Oslo, Oslo, Norway
k. University of Padua and INFN, Padua, Italy
l. Collège de France, Paris, France
m. Institute of Physics, Prague, Czech Republic
n. University ``La Sapienza" and INFN, Rome, Italy
o. Dipartimento di Scienze Fisiche ``E.R. Caianiello" dell'Università and INFN, Salerno, Italy
p. State University of St. Petersburg, St. Petersburg, Russia
q. IReS/ULP, Strasbourg, France
r. Utrecht University and NIKHEF, Utrecht, The Netherlands

*   Permanent address: University of Udine, Udine, Italy

presented by Emanuele Quercigh

Abstract. We study the transverse polarization of Λ's and Ω⁻'s produced in Pb + Pb interaction at 160 A GeV/c. We also present a study of the longitudinal polarization of Λ's produced in the weak decays of Ξ's

Keywords: Hyperon; Polarization; Nuclear Interactions; Quark-Gluon Plasma.
PACS: 27.75.Gz; 25.75.Nq

INTRODUCTION

After eighteen years of activity, the field of high energy heavy ions physics has now reached maturity. The remarkable progress achieved by the experimental techniques has allowed experimenters to unravel a variety of interesting signals, from the thousands of particles produced in these collisions. At the CERN-SPS, the experiments have provided strong evidence [1] of a new state of matter which behaves like the long-sought plasma of quarks and gluons is predicted to behave.

At the BNL-RHIC collider, a wealth of new results on such a state have been obtained [2] at center of mass energies up to about 10 times larger than those available at the SPS.

Up to now however, little effort has been devoted to looking for the spin polarization of final state particles and for spin correlations between them; one reason for this delay being that the study of spin effects requires a thorough understanding of experimental biasses, which is rather uncommon for experiments of the first generations. An important result has been the observation of a significant polarization of Λ's produced at forward rapidity in Au + Au collisions at 11.6 A GeV/c beam momentum [3].

In this paper we present a study of transverse polarization for Λ [4] and Ω^- produced in Pb + Pb interaction at 160 A GeV/c.

This is intended as a first step towards the more difficult study of spin correlations between hyperon pairs e.g. $\Lambda\Lambda$, $\Lambda\overline{\Lambda}$: a study aimed at a better understanding of the mechanism of strangeness production.

As a check of the reconstruction software and of the various correction for acceptance and efficiency losses used in the analysis, we also present a study of the longitudinal polarization of Λ's produced in the weak decay of Ξ^-: $\Xi^- \rightarrow \Lambda + \pi^-$. For these Λ's in fact, the longitudinal polarization is expected to be equal to the weak decay parameter of the Ξ: $\alpha_\Xi = -0.458 \pm 0.012$ [5].

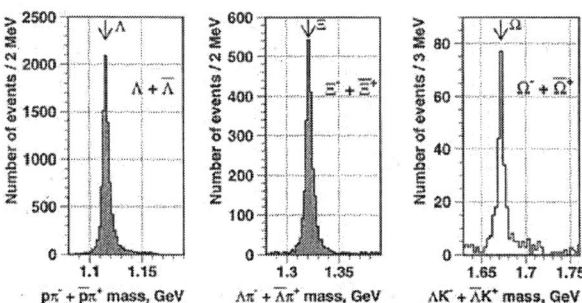

FIGURE 1. Proton π^-, $\Lambda \pi^-$ and ΛK^- (+ charge conjugate) mass spectra showing the Λ, Ξ^- and Ω^- signals.

The data come from the NA57 experiment at the CERN SPS [6]. The Λ, Ξ^- and Ω^- signals (Fig. 1) were identified above a negligible background, by a kinematic analysis of their decay:

$$\Lambda \rightarrow p\,\pi^- \;;\; \Xi^- \rightarrow \Lambda\,\pi^- \;;\; \Omega^- \rightarrow \Lambda\,K^-$$

The trigger selected the most central 56% of the inelastic cross section.

The kinematic window for this study was limited to transverse momenta ≥ 1 Gev/c and to the central unit of the rapidity interval i.e. to values of the Feynman variable $x_F < 0.15$. At such low values of x_F, hyperon polarizations have been found to decrease to zero in pp and pA collisions over a large energy range. From simple symmetry considerations, polarization is expected to vanish at $x_F = 0$ in the collision between identical particles.

Λ AND Ω TRANSVERSE POLARIZATION

Due to parity conservation, particles produced in strong interaction can only be polarized along the normal to the production plane defined as:
$$N = (p_{beam} \times p_{particle})/ \mid p_{beam} \times p_{particle} \mid$$
where p_{beam} and $p_{particle}$ are the momentum vectors of the beam and the particle, respectively.

For a polarized Λ (Ω^-) sample, parity violation in the weak decay leads to a center of mass angular distribution for the decay particles which is asymmetric with respect to the production plane; for the proton (Λ) one will then have:
$$\text{(1)} \qquad dN/d\cos\theta = N_0 \cdot (1 + P \cdot \alpha \cdot \cos\theta)$$
Where N_0 is a normalization constant, α is the decay asymmetry parameter ($\alpha_\Lambda = 0.642 \pm 0.013$, $\alpha_\Omega = -0.026 \pm 0.023$ [5]), P is the Λ (Ω^-) polarization and θ the angle between the momentum vector of the decay proton (Λ) and the normal to the production plane. Thus from the decay angular distribution one can measure the polarization. NA57 has performed such an analysis for two bins centered around the x_F values: 0.09 and 0.14. The results [4], from a sample of 55000 Λ's, are shown as a function of x_F in Fig. 2, together with previous results from pp, pA and Au Au collisions; the predictions of three theoretical models are also shown. The new data confirm the vanishing of polarization at low x_F; a behaviour which appears to be very similar between pp, pA and AA collisions.

The same method cannot be applied to Ω^- since its decay asymmetry parameter α_Ω is consistent with zero, thus preventing us from determining the Ω^- polarization using its decay asymmetry. The Ω^- polarization could instead be deduced by measuring the polarization of the daughter Λ [7]. However, to find out whether the Ω^- polarization is significantly different from zero, it is sufficient to test whether the Ω^- decay shows the anysotropic distribution expected from a polarized spin 3/2 object (e.g. $W(\cos\theta \propto 1 + 3\cos^2\theta$, for the assumption of spin 3/2 and helicity 1/2 [8]). The NA57 Ω^- statistics of 275 event, compares favourably with the statistic needed to establish that the Ω^- spin was $\geq 3/2$ [8]. Fig. 3 shows the angular distribution of the decay Λ in the Ω^- c.m.s., a) as measured for 275 Ω^- and b) as expected for a sample of unpolarized Ω^-. A Kolmogoroff test gives a probability of agreement of more than 90%, indicating that we do not observe a significant polarization for our Ω^- sample.

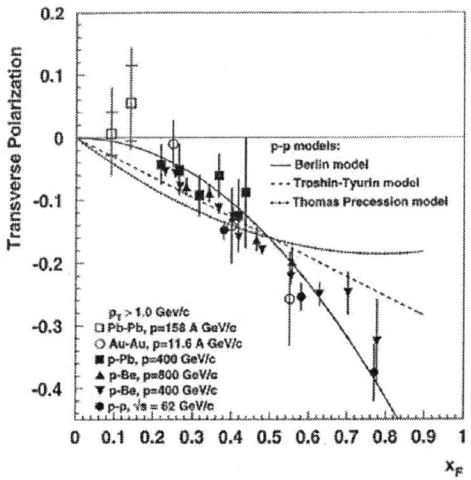

FIGURE 2. The transverse Λ polarization measured by NA57 [4] compared with the results from other experiments.The error given is the combined error, where the horizontal lines indicates the statistical erors only.

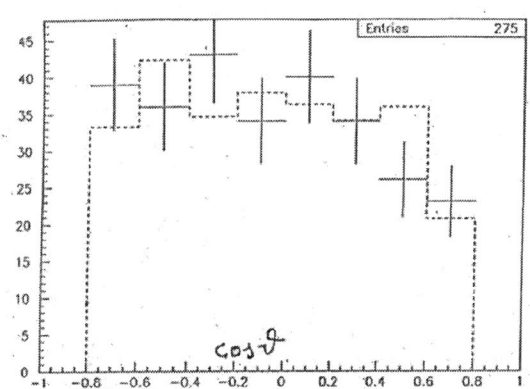

FIGURE 3. Cosθ distribution for the decay $\Omega^- \rightarrow \Lambda + K^-$, θ is the c.m.s. angle between the Λ and Ω^- line of flight. The crosses correspond to the observed distribution for 275 decays, while the hystogram represents the distribution expected in the NA57 apparatus in absence of Ω^- polarization.

$\Xi^- \rightarrow \Lambda + \pi^-$: Λ LONGITUDINAL POLARIZATION

Parity violation in the Ξ decay will induce a non-zero longitudinal polarization for the Λ : $p_{long} = \alpha_\Xi$. In 1987 it was suggested [9] that the relatively large abundance of $\overline{\Xi}$ expected from a quark-gluon plasma formed in relativistic nuclear collisions, could be observed by measuring the longitudinal polarization of $\overline{\Lambda}$ from $\overline{\Xi}$ decay. This measurement was never tried since experiments indeed succeeded in reconstructing significant samples of Ξ and $\overline{\Xi}$ decays, notwithstanding the high event multiplicities, as first shown by the WA85 experiment for S+W [10] and by WA97 for Pb+Pb collisions [11].

Today however, such a study offers the opportunity of checking the completeness of our understanding of the experimental setup and it can thus be considered as a necessary preliminary to a search for spin effects in heavy ion collisions. The decay angular distribution of the Λ from an unpolarized sample of Ξ^- decays is given by (1) where P= -0.458 ± 0.012 and θ is the c.m.s. decay angle of the proton with respect to the Λ line of flight. This is valid even if Ξ's are produced with a transverse polarization provided that the apparatus acceptance is left-right symmetric with respect to the incident beam [12], which is the case for the NA57 setup.

The experimental $\cos\theta$ distribution from a sample of 830 Λ's produced in Ξ^- decays, has been compared using the Kolmogoroff test with the distributions expected a) for Λ longitudinally polarized b) for unpolarized Λ's. The probability of agreement were found to be 35% and 0.6% respectively, as expected, thus giving us confidence in our understanding of the experiment.

CONCLUSIONS AND OUTLOOK

The study of spin effects in heavy ion physics, is now still in its infancy.

Polarizations appear to vanish at low x_F values and thus may not be of interest for the ALICE experiment at LHC, where the rapidity coverage corresponds to values of x_F smaller than 10^{-3}. However a recent paper [13] suggests, for non-central collisions, the possibility of hyperon polarization normal to the reaction plane as determined from the analysis of azimuthal asymmetries in the final state.

The study of spin correlations instead, may well provide a new window on the mechanism of strangeness production, even at LHC. Such a study could already start at SPS by looking for decay correlations between hyperon pairs, as $\Lambda\overline{\Lambda}$, for which statistics of the order of thousands of events have been collected e.g. by the NA57 experiment.

REFERENCES

1. www.cern.ch/CERN/Announcements/2000/NewStateMatter See also Heinz, U., and Jacob, M., -nucl-th/00022042 and references therein.
2 See *Proceedings of the Quark Matter 2004 Conference* – Oakland, USA, 11–17 January 2004/*Journ. Phys. G*; 30–8–2004 and references therein.
3 Bellwied, R., *Nucl. Phys.* A698 (2002) 499c–502c.
4 Eelco Schillings. Thesis University of Utrecht. Subatomic Physics Department, May 2003.
5 Review of Particle Physics; *Phys. Lett. B* Issues 1–4 (2004).
6 Manzari, V., et al., *Journ. Phys.* G 27 (2001) 383 and references therein.
7 Lee, T.D., and Yang, C.N., *Phys. Rev.* 108, 1615 (1957) see also *Review of Particle Physics Phys. Lett.* 170B, 154 (1986).
8 See for instance Baubillier, M.. et al. *Phys. Lett.* 78B, 342 (1978).
9 Jacob, M., and Rafelski, J., *Phys. Lett.* 190B, 173 (1987).
10 Abatzis, S., et al. *Phys. Lett.*. 259B, 508 (1991).
11 Holme, A.K., et al. *Journ. Phys.* G 23 1851 (1997).
12 Jacob, M., Z. Phys. *C-Particle and Fields* 38, 273 (1988).
13 Liang, Z. T., and Wang, X. N., *Nucl-th*/0410079 3.

Indications for the onset of deconfinement in nucleus nucleus collisions

D. Flierl for the NA49 collaboration

Institut für Kernphysik, Johann Wolfgang Goethe Universität, Frankfurt am Main, Germany
E-mail: flierl@ikf.uni-frankfurt.de

Abstract. The hadronic final state of central Pb+Pb collisions at 20, 30, 40, 80, and 158 AGeV has been measured by the CERN NA49 collaboration. The mean transverse mass of pions and kaons at midrapidity stays nearly constant in this energy range, whereas at lower energies, at the AGS, a steep increase with beam energy was measured. Compared to p+p collisions as well as to model calculations, anomalies in the energy dependence of pion and kaon production at lower SPS energies are observed. These findings can be explained, assuming that the energy density reached in central A+A collisions at lower SPS energies is sufficient to force the hot and dense nuclear matter into a deconfined phase.

INTRODUCTION

At top CERN SPS and at RHIC energies the initial energy density reached in heavy ion collisions is apparently large enough to create a deconfined phase at the early stage of the collision: the Quark Gluon Plasma [1][2]. On the other hand, at lower energies, at the AGS, the maximum energy density is probably not sufficient to produce such a state of matter. Therefore, an energy scan progam in the intermediate energy range, covered by the CERN SPS, was initiated. The goal was to find anomalies in the energy dependence which could be related to the change of the number of degrees of freedom connected to a phase transition of the hot and dense matter created in heavy ion collisions.

In the course of the SPS energy scan program, the NA49 collaboration measured the hadronic final state of central Pb+Pb collisions at 20, 30, 40, 80, and 158 AGeV beam energy. A detailed description of the NA49 setup is given in [3]. In the following, the energy dependence of selected hadronic observables will be discussed, with the main focus on where we start to see indications of a deconfined phase of matter.

ENERGY DEPENDENCE OF TRANSVERSE MASS SPECTRA

A model independent way to compare transverse momentum spectra is to consider the mean transverse mass $\langle m_t \rangle - m_0$. In Figure 1 the energy dependence of this quantity is shown for pions, kaons and protons. If energy density and hence pressure increase with beam energy, a stronger transverse expansion is expected at higher energies. Assuming that the strength of the transverse expansion is reflected in the mean transverse mass, the observable $\langle m_t \rangle - m_0$ will rise with $\sqrt{s_{NN}}$. At AGS energies we indeed observe a strong increase of the mean transverse mass of pions, kaons and protons with energy. However,

CP756, *Quark Confinement and the Hadron Spectrum VI*
edited by N. Brambilla, U. D'Alesio, A. Devoto, K. Maung, G.M. Prosperi and S. Serci
© 2005 American Institute of Physics 0-7354-0241-8/05/$22.50

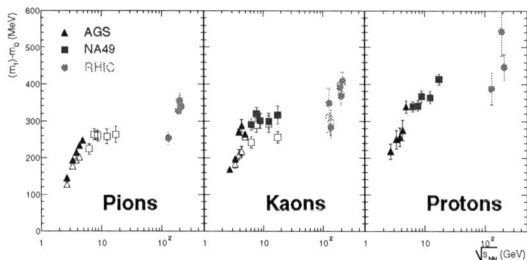

FIGURE 1. Energy dependence of the mean transverse mass of pions, kaons and protons. Open symbols indicate negatively charged particles.

at lower SPS energies this behaviour changes: the mean transverse mass increases only weakly with beam energy.

This charateristic change in the energy dependence has been interpreted as a signature for a phase transition [4]. Because the equation of state - the relation between energy density and pressure - changes at the phase boundary and with it the strength of the transverse expansion.

This observation supports the hypothesis, that the initial energy density reached in central A+A collisions at lower SPS energies is already large enough, to force a phase transition of the strongly interacting nuclear matter.

ENERGY DEPENDENCE OF PARTICLE YIELDS

The total pion multiplicity per wounded nucleon as a function of Fermi's measure $F \approx \sqrt{\sqrt{s_{NN}}}$ is shown in Figure 2 left. The relative pion production in central Pb+Pb collisions increases with collision energy, but starting from lower SPS energies the rate of increase grows. This feature is not visible in p+p interactions. The steepening of the energy dependence can be explained by an increase of the effective number of degrees of freedom due to the onset of deconfinement at about 30 AGeV beam energy [4]. Figure 3 shows the energy dependence of relative strangeness production. The ratios of

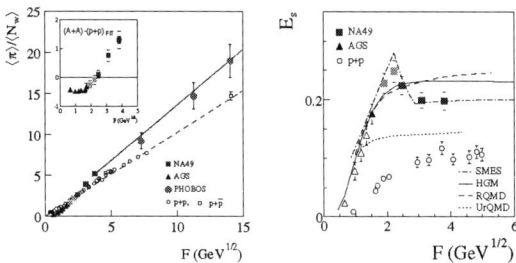

FIGURE 2. Energy dependence of the mean number of pions per wounded nucleon in A+A and p+p collisions. Lines are only drawn to guide the eye (left panel).
Energy dependence of the strangeness to pion ratio and model calculations (right panel).

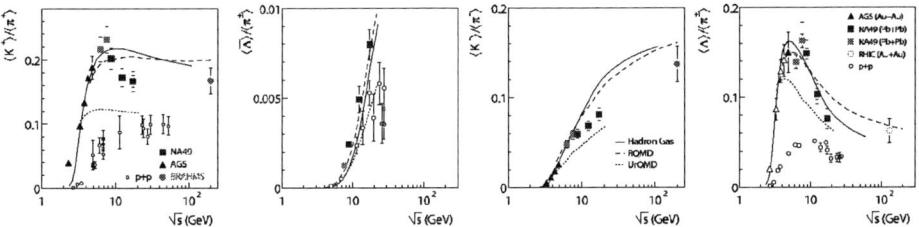

FIGURE 3. Energy dependence of particle ratios $\langle K^+\rangle/\langle\pi^+\rangle$, $\langle\overline{\Lambda}\rangle/\langle\pi\rangle$, $\langle K^-\rangle/\langle\pi^-\rangle$, and $\langle\Lambda\rangle/\langle\pi\rangle$ in central A+A collisions and p+p interactions. Lines indicate model calculations.

the total multiplicities $\langle K^+\rangle/\langle\pi^+\rangle$, $\langle K^-\rangle/\langle\pi^-\rangle$, $\langle\Lambda\rangle/\langle\pi\rangle$ and $\langle\overline{\Lambda}\rangle/\langle\pi\rangle$ in central Pb+Pb collisions are compared to model calculations and when available to p+p interactions. While $\langle K^-\rangle/\langle\pi^-\rangle$ and $\langle\overline{\Lambda}\rangle/\langle\pi\rangle$ ratios rise continuously with energy, a distinct maximum is visible in the energy dependence of the ratios $\langle K^+\rangle/\langle\pi^+\rangle$ and $\langle\Lambda\rangle/\langle\pi\rangle$. Hadron gas [5] (HGM) and microscopic models [6] (RQMD, UrQMD) roughly describe the trend of the energy dependence, but especially the pronounced peak in the $\langle K^+\rangle/\langle\pi^+\rangle$ ratio is not reproduced. As demonstrated in Figure 3, this feature is also absent in p+p interactions.

The total strangeness production to pion rate can be approximated by $E_S = (2(\langle K^-\rangle + \langle K^-\rangle) + \langle\Lambda\rangle)/\langle\pi\rangle$. As seen from Figure 2 (right), we find that the hadron gas and microscopic models describe the trend of the data, but they do not show the distinct maximum at lower SPS energies. Only a model [4] (SMES) including a phase transition around 30 AGeV exhibits a peak as seen in the data.

SUMMARY AND OUTLOOK

The energy dependence of several observables shows anomalies at lower SPS energies. These signatures indicate, that the energy density reached in A+A collisions at beam energies of about 30 AGeV are already sufficient to produce a deconfined phase.

The next question to address is, how the system behaves when the volume is varied. At which intermediate system size appear indications for a deconfined phase? Future experiments at the SPS could deliver the answer to this question if the proposed light ion program will be approved [7].

REFERENCES

1. U. W. Heinz and M. Jacob, arXiv:nucl-th/0002042.
2. M. Gyulassy and L. McLerran, arXiv:nucl-th/0405013.
3. S. Afanasev *et al.* [NA49 Collaboration], Nucl. Instrum. Meth. A **430**, 210 (1999).
4. M. Gazdzicki and M. I. Gorenstein, Acta Phys. Polon. B **30**, 2705 (1999),
   M. I. Gorenstein, M. Gazdzicki and K. A. Bugaev, Phys. Lett. B **567** (2003) 175.
5. J. Cleymans and K. Redlich, Phys. Rev. C **60** (1999) 054908.
6. H. Weber, E. L. Bratkovskaya, W. Cassing and H. Stöcker, Phys. Rev. C **67** (2003) 014904.
7. J. Bartke *et al.*, CERN-SPSC-2003-038, SPSC-EOI-01.

Chiral Lagrangians at finite temperature and the Polyakov Loop [1]

E. Megías*, E.Ruiz Arriola* and L.L. Salcedo*

*Departamento de Física Moderna, Universidad de Granada. 18071-Granada (Spain)

Abstract. Heat kernel expansions at finite temperature of massless QCD and chiral quark models generate effective actions relevant for both low and high temperature QCD. The key relevance of the Polyakov Loop to maintain the large and non-perturbative gauge invariance at finite temperature is stressed.

In the imaginary time formulation of quantum field theory [1], finite temperature is introduced by imposing periodic or anti-periodic boundary conditions for bosons and fermions respectively. This approach necessarily breaks the Lorentz invariance of the starting Lagrangian since the heat bath is a static infinitely heavy system which is assumed to be at rest and communicates energy at no expense with the fundamental degrees of freedom. This fact generates an unpleasant plethora of possible Lorentz contributions to Feynman diagrams which fortunately are ultraviolet finite due to the presence of suppressing kinetic Boltzmann factors. As a gratifying consequence, the finite temperature renormalizability of a theory relies indeed on its renormalizability at zero temperature and provides any fundamental theory with definite predictive power, if the parameters entering the Lagrangian are assumed to be temperature independent. This standard procedure might be called *minimal thermal coupling* of the heat bath since, at least in perturbation theory, finite temperature (and hence Lorentz breaking) effects are indeed quantum effects. For an effective field theory of composite particles the situation may be not that simple since the corresponding Low Energy Constants (LEC's) encode the microscopic information of the underlying substructure. Actually, the constituents are also heated up and it is clear that LEC's inherit a finite temperature dependence. Moreover, genuinely Lorentz breaking and temperature dependent terms might be further added to the effective Lagrangian.

The previous discussion is actually relevant to QCD at finite temperature [2, 3]. At low energies and temperatures, one uses Chiral Perturbation Theory (ChPT) [4] and assumes that in the effective theory the LEC's are temperature independent; the T dependence is generated through thermal pion loops [5]. This assumption relies strongly on the existence of a mass gap in the physical hadronic spectrum due to the presence of would-be massless pseudo-scalar Goldstone bosons, so one expects the leading temperature effects, well below the phase transition, to be $\mathscr{O}(e^{-M_\pi/T})$, whereas the next corrections

---

[1] Presented by E.R. Arriola at the VI Conference on "Quark Confinement 2004", Villasimius, Sardinia, Italy, 21-25 September 2004.

CP756, *Quark Confinement and the Hadron Spectrum VI*
edited by N. Brambilla, U. D'Alesio, A. Devoto, K. Maung, G.M. Prosperi and S. Serci
© 2005 American Institute of Physics 0-7354-0241-8/05/$22.50

might be $\mathcal{O}(e^{-M_\rho/T})$. At present, one cannot compute the LEC's from QCD itself, but some insight on the base of chiral quark models can be gained. Actually, the way how these Boltzmann factors actually arise in chiral quark models is not at all trivial.

To illustrate the occurrence of non-minimal thermal couplings in effective Lagrangeans, we have used based on previos work [6, 7] the heat kernel method [8] (see also Ref. [9, 10]) for an application to high temperature QCD) and computed recently [11, 12] the chiral Lagrangian using chiral quark models where pseudo-scalar Goldstone bosons arise as bound $q\bar{q}$ states. At finite temperature it can be written in the standard Gasser and Leutwyler [4] form [2]

$$\mathcal{L}_q^{*(2)} = \frac{f^{*2}}{4}\mathrm{tr}_f\left(\mathbf{D}_\mu U^\dagger \mathbf{D}_\mu U + (\overline{\chi}^\dagger U + \overline{\chi} U^\dagger)\right),$$

$$\mathcal{L}_q^{*(4)} = \sum_{i=1}^{10} L_i^* \mathcal{L}_i + \sum_i L_i' \mathcal{L}_i' \qquad (1)$$

Here, f^* is the pion weak decay constant at finite T in the chiral limit, and is given by

$$f^{*2} = 4M^2 T \mathrm{Tr}_c \sum_{\omega_n} \int \frac{d^3k}{(2\pi)^3} \frac{1}{[\omega_n^2 + k^2 + M^2]^2}. \qquad (2)$$

with M the constituent quark mass and $\omega_n = 2\pi T(n+1/2)$ the fermionic Matsubara frequency and Tr_c is the colour trace. The constants $L_{1,...,10}^*$ appearing in the higher order terms are displayed in Ref. [12] are the temperature dependent LEC's of the Lorentz preserving terms whereas there are appear new ones which break explicitly the symmetry and obviously vanish at zero temperature. Note that at lowest order in the chiral counting there is no Lorentz breaking terms. These expressions illustrate our point, that the LEC's may depend themselves on the temperature. On the other hand, it also raises a puzzle, because the leading thermal effects in the large N_c limit are $\mathcal{O}(N_c e^{-M/T})$ (see also Ref. [13]), but in ChPT there are no leading large N_c corrections.

The solution to the puzzle has to do with the fact that for gauge theories like QCD at finite temperatures [1, 2, 3] the non Abelian gauge manifest non perturbatively. In the Polyakov gauge, where $\partial_4 A_4 = 0$ and A_4 is a diagonal and traceless $N_c \times N_c$ matrix, and N_c is the number of colors, there is still some freedom in choosing the gluon field. Under periodic gauge transformations [6, 7] the requirement of gauge invariance really implies identifying all Gribov replicas differing by a multiple of $2\pi/\beta$, which means periodicity in the diagonal amplitudes of A_4 of period $2\pi/\beta$. Perturbation theory, which corresponds to expanding in powers of small A_4 fields manifestly breaks gauge invariance at finite temperature. A way of avoiding this explicit breaking is by considering the Polyakov loop Ω as an independent variable,

$$\Omega = e^{i\beta A_4(\vec{x})} \qquad (3)$$

We have recently developed an expansion keeping these symmetries in general theories and applied it to QCD at the one quark+gluon loop level [8, 9]. When these ideas are

---

[2] We use an asterisk as upper-script for finite temperature quantities, i. e. $\mathcal{O}^* = \mathcal{O}_T$.

incorporated into the chiral quark model framework at the leading one quark loop approximation there appears an accidental center symmetry similar to the one of QCD in the quenched approximation, under aperiodic gauge transformations [3] of the center $Z(N_c)$ of the group $SU(N_c)$. The Polyakov loop transforms as the fundamental representation of the $Z(N_c)$ group, and hence $\langle \Omega \rangle = 0$ in the unbroken center symmetric and confining phase. More generally, $\langle \Omega^n \rangle = 0$ for $n \neq mN_c$ with m an arbitrary integer. The anti-periodic quark fields at the end of the Euclidean imaginary interval transforms also as the fundamental representation of the center group, so that the center symmetry is explicitly broken by the presence of dynamical quarks, and hence in the quenched approximation non-local condensates fulfill a selection rule of the form, $\langle \bar{q}(n\beta)q(0) \rangle = 0$ for $n \neq mN_c$. This selection rule has some impact on chiral quark models. In practice, the Polyakov loop can be coupled to the chiral quark model using the modified fermionic Matsubara frequencies [6, 7]

$$\hat{\omega}_n = 2\pi T (n + 1/2 + v), \quad v = (2\pi i)^{-1} \log \Omega \qquad (4)$$

which are shifted by the logarithm of the Polyakov loop which we assume for simplicity to be $\vec{x}$ independent, as suggested also in Refs. [14, 15, 16]. The Tr_c in Eq. (2) does not simply give N_c because this coupling introduces a colour source into the problem for a fixed A_4 field. After projection onto the colour neutral states by gauge averaging, at the one quark loop level, there is an accidental $Z(N_c)$ symmetry in the model which generates a similar selection rule as in pure gluodynamics, from which a strong thermal suppression, $\mathcal{O}(e^{-N_c M/T})$ follows in agreement with a previous observation [17]. In this way compliance with ChPT at finite temperature can be achieved.

This work is supported in part by funds provided by the Spanish DGI with grant no. BMF2002-03218, Junta de Andalucía grant no. FM-225 and EURIDICE grant number HPRN-CT-2003-00311.

REFERENCES

1. N. P. Landsman and C. G. van Weert, Phys. Rept. **145** (1987) 141.
2. D. J. Gross, R. D. Pisarski and L. G. Yaffe, Rev. Mod. Phys. **53**, 43 (1981).
3. B. Svetitsky, Phys. Rept. **132** (1986) 1.
4. J. Gasser and H. Leutwyler, Nucl. Phys. B **250**, 465 (1985).
5. J. Gasser and H. Leutwyler, Phys. Lett. B **184**, 83 (1987).
6. L. L. Salcedo, Nucl. Phys. B **549**, 98 (1999)
7. C. Garcia-Recio and L. L. Salcedo, Phys. Rev. D **63**, 045016 (2001)
8. E. Megías, E. Ruiz Arriola and L. L. Salcedo, Phys. Lett. B **563**, 173 (2003)
9. E. Megías, E. Ruiz Arriola and L. L. Salcedo, Phys. Rev. D **69** (2004) 116003
10. E. Megias, arXiv:hep-ph/0407052.
11. E. Megias, E. Ruiz Arriola and L. L. Salcedo, arXiv:hep-ph/0410053.
12. E. Megías, E. Ruiz Arriola and L. L. Salcedo (in preparation)
13. W. Florkowski and W. Broniowski, Phys. Lett. B **386**, 62 (1996)
14. A. Gocksch and M. Ogilvie, Phys. Rev. D **31**, 877 (1985).
15. P. N. Meisinger, T. R. Miller and M. C. Ogilvie, Nucl. Phys. Proc. Suppl. **119**, 511 (2003)
16. K. Fukushima, Phys. Lett. B **591** (2004) 277
17. M. Oleszczuk and J. Polonyi, Annals Phys. 227 (1993) 76,

Spectroscopy and pentaquarks at HERA

Leonid Gladilin

(on behalf of the H1 and ZEUS Collaborations)

DESY, ZEUS experiment, Notkestr. 85, 22607 Hamburg, Germany
E-mail: gladilin@mail.desy.de
On leave from Moscow State University, supported by the U.S.-Israel BSF

Abstract.
Results of the H1 and ZEUS Collaborations on spectroscopy of light and charmed mesons and on pentaquark searches, obtained using the HERA I data, are summarised.

Keywords: spectroscopy, pentaquarks
PACS: 13.60.Le, 13.60.Rj, 14.20.Lq, 14.40.Lb

INTRODUCTION

Light and charmed hadrons are produced copiously in ep collisions with a centre-of-mass energy of 318 GeV at HERA. During first phase of HERA operation (1992-2000), the H1 and ZEUS Collaborations accumulated data samples corresponding to $\sim 120\,\mathrm{pb}^{-1}$ each. The H1 and ZEUS results on hadron spectroscopy and pentaquark searches are summarised in this note.

SPECTROSCOPY OF LIGHT AND CHARMED MESONS

Inclusive photoproduction of η, ρ^0, $f_0(980)$ and $f_2(1270)$ mesons was measured at an average photon-proton centre-of-mass energy $W = 210\,\mathrm{GeV}$ [1]. The differential cross sections for those mesons and for charged pions as a function of $p_T + m$, where m is the meson's nominal mass, show similar power-law behaviour. The results suggest a similar mechanism of the mesons production in fragmentation processes.

Measurement of inclusive $K_s^0 K_s^0$ production in deep inelastic scattering (DIS) revealed a state at 1537 MeV, consistent with $f_2'(1525)$, and another at 1726 MeV [2]. The state at 1726 MeV has a mass consistent with $f_0(1710)$, and is found in a gluon-rich region of phase space. This observation indicates that $f_0(1710)$ has a sizeable gluonic component.

The production of excited charmed and charmed-strange mesons was studied using their decays to final states involving $D^{*\pm}$ [3]. The measured rates of c quarks hadronising as D_1^0, D_2^{*0} and $D_{s1}^{\pm}$ mesons agree with those obtained in e^+e^- annihilations. The measured value of the helicity parameter for $D_{s1}^{\pm}$ mesons is consistent with the observation of the CLEO Collaboration that the spin-parity of the $D_{s1}^{\pm}$ is 1^+. A search for the radially excited $D^{*\prime\pm}$ meson revealed no signal. The upper limit on the product of the fraction of c quarks hadronising as a $D^{*\prime+}$ meson and the branching ratio of the $D^{*\prime+}$ decay to $D^{*+}\pi^+\pi^-$ was estimated to be 0.7% (95% C.L.).

CP756, *Quark Confinement and the Hadron Spectrum VI*
edited by N. Brambilla, U. D'Alesio, A. Devoto, K. Maung, G.M. Prosperi and S. Serci
© 2005 American Institute of Physics 0-7354-0241-8/05/$22.50

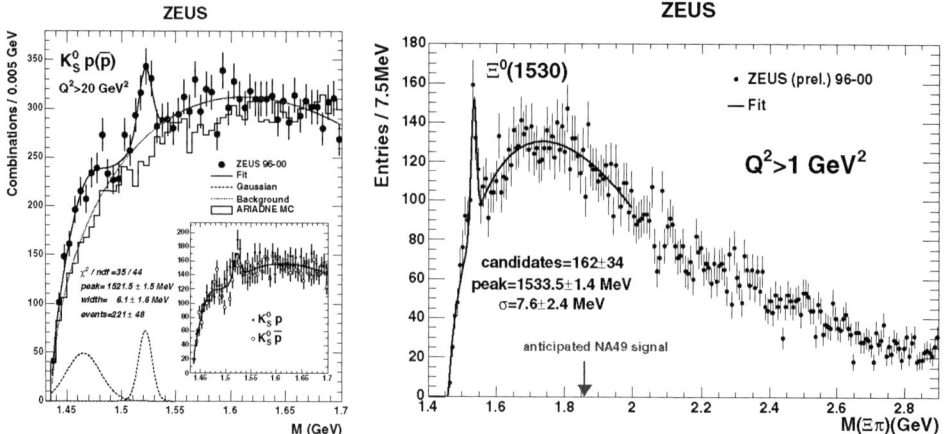

FIGURE 1. Invariant mass spectrum for the $K_s^0 p(\bar{p})$ combinations for $Q^2 > 20\,\mathrm{GeV}^2$ (left) and for $\Xi\pi$ combinations for $Q^2 > 1\,\mathrm{GeV}^2$ (right)

STRANGE PENTAQUARKS

A peak in the $K_s^0 p(\bar{p})$ invariant mass spectrum around $1520\,\mathrm{MeV}$ was observed in DIS by the ZEUS Collaboration [4]. In Fig. 1(left), the spectrum is shown for exchanged photon virtuality $Q^2 > 20\,\mathrm{GeV}^2$. The statistical significance of the signal varies between 3.9σ and 4.6σ depending upon the treatment of the background. The signal is seen in both $K_s^0 p$ and $K_s^0 \bar{p}$ samples. If the signal corresponds to the pentaquark Θ^+, this provides the first evidence for an anti-pentaquark with a quark content $\bar{u}\bar{u}d\bar{d}s$. A ratio of the Θ^+ and Λ^0 production cross sections was measured to be $4.2 \pm 0.9(\mathrm{stat.})^{+1.2}_{-0.9}(\mathrm{syst.})\%$ [5].

The ZEUS Collaboration performed also a search for two other pentaquarks, reported by the NA49 Collaboration, and observed no signal in the $\Xi\pi$ invariant mass spectrum [6]. In Fig. 1(right), the spectrum is shown for $Q^2 > 1\,\mathrm{GeV}^2$. A clear peak with more than 160 $\Xi^0(1530)$ baryons indicates that the statistical sensitivity of the search is similar to that of the NA49 Collaboration.

CHARMED PENTAQUARKS

An observation of a candidate for the charmed pentaquark state, $\Theta_c^0 = uudd\bar{c}$, decaying to $D^{*\pm}p^{\mp}$ was reported by the H1 Collaboration [7]. Fig. 2(left) shows the $D^{*\pm}p^{\mp}$ invariant-mass distributions in DIS with $Q^2 > 1\,\mathrm{GeV}^2$ and in photoproduction with smaller Q^2 values. A fit of the signal in DIS yielded 50.6 ± 11.2 signal events and the mass of $3099 \pm 3(\mathrm{stat.}) \pm 5(\mathrm{syst.})\,\mathrm{MeV}$. The observed resonance was reported to contribute roughly 1% of the $D^{*\pm}$ production rate in the kinematic range studied in DIS.

The observation of the H1 Collaboration was challenged by the ZEUS collaboration [8]. Using a larger sample of $D^{*\pm}$ mesons, ZEUS observed no signature of the narrow resonance in the $M(D^{*\pm}p^{\mp})$ spectra shown in Fig. 2(right). The Monte Carlo

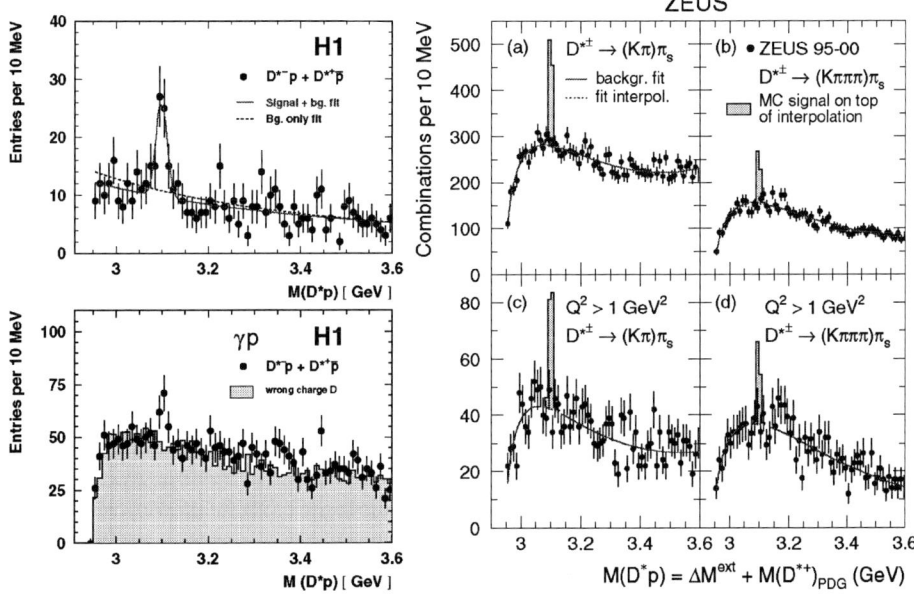

FIGURE 2. The distributions of $M(D^{*\pm}p^{\mp})$ obtained by the H1 Collaboration (left) and by the ZEUS Collaborations (right)

Θ_c^0 signals normalised to 1% of the number of reconstructed $D^{*\pm}$ mesons are shown on top of the fitted backgrounds. The upper limit on the fraction of $D^{*\pm}$ mesons originating from Θ_c^0 decays was evaluated to be 0.23% (95% C.L.). The upper limit for DIS with $Q^2 > 1\,\mathrm{GeV}^2$ is 0.35% (95% C.L.).

REFERENCES

1. H1 Collab., Abstract 6-0184, International Conference on High Energy Physics, Beijing, China (ICHEP 2004), August 2004. Available from http://www-h1.desy.de/.
2. ZEUS Collab., S. Chekanov et al., *Phys. Lett.* **B 578** (2004) 33.
3. ZEUS Collab., Abstract 854, International Conference on High Energy Physics, Osaka, Japan (ICHEP 2000), July-August 2000.
   ZEUS Collab., Abstract 497, International Europhysics Conference on High Energy Physics, Budapest, Hungary (EPS 2001), July 2001. Available from http://www-zeus.desy.de/.
4. ZEUS Collab., S. Chekanov et al., *Phys. Lett.* **B 591** (2004) 7.
5. ZEUS Collab, Abstract 10-0273, International Conference on High Energy Physics, Beijing, China (ICHEP 2004), August 2004. Available from http://www-zeus.desy.de/.
6. ZEUS Collab, Abstract 10-0293, International Conference on High Energy Physics, Beijing, China (ICHEP 2004), August 2004. Available from http://www-zeus.desy.de/.
7. H1 Collab., C. Atkas et al., *Phys. Lett.* **B 591** (2004) 7.
8. ZEUS Collab., S. Chekanov et al., DESY 04-164 (2004), accepted by *Eur. Phys. J.* **C**.

Search for Pentaquarks in 920 GeV Proton-Nucleus Collisions at HERA-B

M. Medinnis, for the HERA-B Collaboration

DESY, Notkestraße, 85, Hamburg, D-22603 Germany

Abstract. HERA-B has searched in vain for evidence of the production of two recently reported states which have been identified as possible pentaquarks: the $\Theta^+(1540)$, decaying into $p\bar{K}^0_s$, and the $\Xi^{--}_{3/2}$, decaying into $\Xi^-\pi^-$. Upper limits on production cross sections at mid-rapidity and on ratios of the production cross sections to those of well-known resonances in 920 GeV proton-nucleus interactions are reported.

Keywords: pentaquark, hadron spectroscopy
PACS: 13.30.Eg, 13.85.Rm, 14.80.-j

INTRODUCTION

Motivated by recent reports of observations of possible pentaquark candidates in pK^0_s [1] and $\Xi\pi$ [2] channels, we have searched a large data set of proton-nucleus collisions at $\sqrt{s} = 41.6$ GeV for supporting evidence. The final results of that search are briefly summarized. A more detailed account [3] is also available.

HERA-B is a fixed target experiment operating at the 920 GeV proton ring of HERA at DESY. Thanks to its large aperture, excellent particle identification system, vertexing capabilities and good momentum and mass resolution, it is well suited for studies of the production of narrow resonances at mid-rapidity. The results presented here are based on the analysis of a sample of 200 million interactions on carbon, titanium and tungsten targets taken with a minimum bias trigger. A summary of the statistics of relevant signals and their resolutions is given in Table 1.

TABLE 1. Statistics and experimental resolutions σ of the relevant signals (charge-conjugate modes indicated by c.c.).

Signal	C target	All targets	σ/MeV
K^0_s	2.2M	4.9M	4.9
Λ [c.c.]	440k [210k]	1.1M [520k]	1.6
$\Lambda(1520)$ [c.c.]	1.3k [760]	3.5k [2.1k]	2.3
Ξ^- [c.c.]	4.7k [3.4k]	12k [8.2k]	2.6
$\Xi(1530)^0$ [c.c.]	610 [380]	1.4k [940]	2.9

SEARCH FOR $\Theta^+ \rightarrow PK^0_S$

Protons are identified with tight cuts on RICH proton likelihood and combined with K^0_s candidates to produce the mass spectrum shown in Fig. 1(a) for the carbon target sample.

CP756, *Quark Confinement and the Hadron Spectrum VI*
edited by N. Brambilla, U. D'Alesio, A. Devoto, K. Maung, G.M. Prosperi and S. Serci
© 2005 American Institute of Physics 0-7354-0241-8/05/$22.50

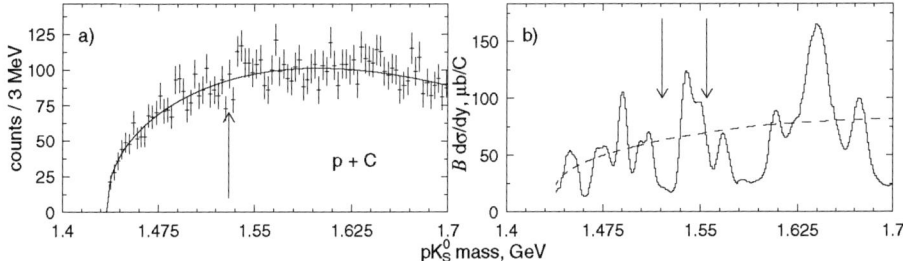

FIGURE 1. a) The pK_s^0 invariant mass distribution for the carbon target data set. The solid line is the background estimate from event mixing. b) The 95% confidence level upper limit (solid curve) for mid-rapidity production of a narrow resonance in p+C interactions. The dashed line indicates the experimental sensitivity.

The solid line shows the background estimate from event mixing. No significant narrow structures are visible over the entire mass range. Mass spectra from the titanium and tungsten target samples as well as the combined sample similarly show no significant structure. To estimate the upper limit on the number of events which can be attributed to a narrow resonance, we fit the spectrum to a Gaussian shape with resolution consistent with the experimental resolution plus a fixed background shape. The fixed mean of the Gaussian is stepped across the mass plot in increments of 1 MeV. The resolution ranges from 2.6 to 6.1 MeV over the considered mass range and is 3.9 MeV at 1530 MeV (approximately the mean value of reported Θ^+ masses in the pK_s^0 channel). Using the prescription of Feldman and Cousins [4], and combining with the experimental sensitivity we arrive at the upper limit (95%) curve for $\mathscr{B} \cdot d\sigma/dy|_{y\approx 0}$ per carbon nucleus shown as the solid curve in Fig. 1(b). The dashed line indicates the experimental sensitivity. The upper limit curve derived from the full sample is within ±30% of the carbon sample limit given in Fig 1(b).

The range of reported mass values of Θ^+ candidates is indicated by arrows in Fig. 1. Over this range the upper limit on $\mathscr{B} \cdot d\sigma/dy|_{y\approx 0}$ per nucleon varies from 4 – 16 μbarns per nucleon, assuming an atomic mass dependence of $A^{0.7}$.

To provide another basis for comparison with other experiments and models, we have evaluated the ratios of the upper limit to the $\Lambda(1116)$ and $\Lambda(1520)$ cross sections. Assuming $\mathscr{B}(\Theta^+ \to pK_s^0) = 1/4$, the upper limit (95%) for the production ratios are: $\frac{\Theta^+}{\Lambda(1116)} = 0.92\%$ and $\frac{\Theta^+}{\Lambda(1520)} = 2.7\%$ at a pK_s^0 mass of 1530 MeV, using the full data sample. For a pK_s^0 mass of 1540 MeV, these values must be multiplied by ≈ 4.

SEARCH FOR $\Xi_{3/2} \to \Xi\pi$

We have searched for the four charge states of the $\Xi_{3/2}$ by combining the Ξ^- and $\overline{\Xi}^+$ candidates with pions of both charges. Only loose particle identification requirements (RICH pion likelihood > 0.01 and calorimeter electron likelihood < 0.9) are applied to the pion candidates. The resulting mass spectra are shown in Fig. 2. The smooth

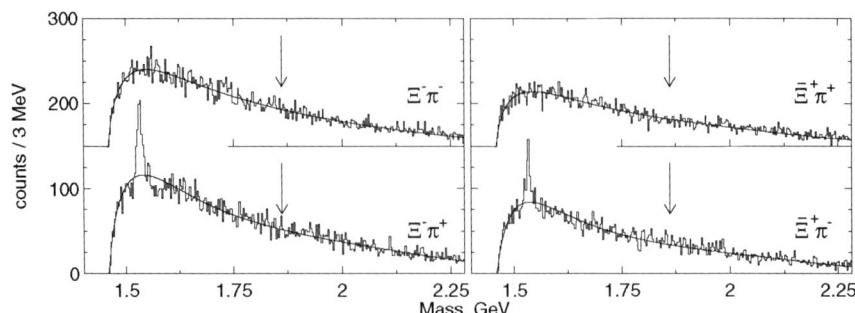

FIGURE 2. The $\Xi\pi$ invariant mass distributions in the indicated charge combinations. The arrow marks the location (1862 MeV) of the narrow peak observed by NA49 [2] in all four charge combinations.

curves are background estimates from event mixing, normalized to the data. Apart from a prominent signal for $\Xi(1530)$ in the neutral combinations, no evidence is seen for a narrow resonance in any of the spectra. The experimental resolution at 1862 MeV is 6.6 MeV. Using an analogous technique to that described in the previous section, we derive an upper limit (95%) for $\mathscr{B} \cdot d\sigma/dy|_{y\approx0}$ of a narrow resonance at a mass of 1862 MeV of 2.5, 2.3, 0.85, and 3.1 μb/nucleon for $\Xi^-\pi^-$, $\Xi^-\pi^+$, $\overline{\Xi}^+\pi^+$, and $\overline{\Xi}^+\pi^-$ respectively for the carbon target data set and 2.7, 3.2, 0.94, and 3.1 μb/nucleon for the full data set.

The upper limit (95%) of production cross section ratios at mid-rapidity of $\frac{\Xi^{--}}{\Xi^-}$ and $\frac{\Xi^{--}}{\Xi(1530)}$ are $3\%/\mathscr{B}(\Xi^{--} \to \Xi^-\pi^-)$ and $4\%/\mathscr{B}(\Xi^{--} \to \Xi^-\pi^-)$, respectively. For comparison, we take the estimated number of Ξ^{--} candidates from the NA49 publication [2] and the estimated number of $\Xi(1530)$ events (150) from the same data set [5] and, assuming the relative efficiencies for $\Xi(1530)$ and Ξ^{--} of NA49 are similar to those of HERA-B, we arrive at a production cross section ratio of $\approx 18\%/\mathscr{B}$ for the NA49 signal, significantly larger than the upper limit of $4\%/\mathscr{B}$ obtained here.

REFERENCES

1. T. Nakano *et al.* (LEPS Coll.), Phys. Rev. Lett. **91**, 012002 (2003);
   V. V. Barmin *et al.* (DIANA Coll.), Yad. Fiz. **66**, 1763 (2003);
   A. E. Asratyan *et al.* (ITEP Coll.), Yad. Fiz. **67**, 704 (2004);
   A. Airapetian *et al.* (HERMES Coll.), Phys. Lett. **B585**, 213 (2004);
   M. Abdel-Bary *et al.* (COSY-TOF Coll.), hep-ex/0403011;
   P.Zh. Aslanyan, V.N. Emelyanenko, G.G. Rikhkvitzkaya, hep-ex/0403044;
   A. Aleev *et al.* (SVD Coll.), hep-ex/0401024;
   S. Chekanov *et al.* (ZEUS Coll.), Phys. Lett. **B591**, 7 (2004).
2. C. Alt *et al.* (NA49 Coll.), Phys. Rev. Lett. **92**, 042003 (2004).
3. I. Abt *et al.* (HERA-B Coll.), Phys. Rev. Lett. **93**, 212003 (2004).
4. G.J. Feldman, R.D. Cousins, Phys. Rev. **D57**, 3873 (1998).
5. H. G. Fischer, S. Wenig, hep-ex/0401014.

Open and Hidden Charm Production in 920 GeV Proton-Nucleus Collisions

Ilija Vukotić for the HERA-B Collaboration

DESY, D-15738 Zeuthen, Germany

Abstract. The production of the charmonium states J/ψ, $\psi(2S)$ and χ_c in proton-nucleus collisions at a center-of-mass energy of 42 GeV is studied using the HERA-B detector. We present J/ψ and ψ' differential distributions in the ranges: $-0.3 < x_F < 0.1$ and $0 < p_T < 4.8$ GeV/c. The dependence of the production cross sections on the target nucleus is studied on a data sample obtained by simultaneous use of two different target materials. Preliminary results on production cross sections and cross section ratios for open charm mesons are presented. This paper includes the final result on a search for the FCNC decay $D^0 \rightarrow \mu^+\mu^-$.

Keywords: Charmonium, J/ψ, x_F, p_T, A-Dependence
PACS: 14.40.Lb, 25.40.Ep, 29.85.+c

Introduction

There are several competing theoretical models of charmonium and open charm production in nuclear media [1, 2, 3]. These QCD based models have to make assumptions concerning non-perturbative effects, such as colour neutralization by soft-gluons, formation of bound states, etc. Although the models qualitatively agree with the results of the previous experiments, additional high precision measurements in a broad kinematical range are needed. HERA-B [4] is unique in its ability to measure J/ψ and ψ' production and nuclear dependence in the negative x_F range. Preliminary results on J/ψ, ψ' and χ_c production are presented and, in addtion, preliminary results on D-meson production cross sections.

The HERA-B Detector and 2002/2003 data

The HERA-B detector is a fixed-target large aperture particle spectrometer, operating at the 920 GeV proton beam of HERA at DESY. The detector features a highly selective di-lepton trigger, good momentum resolution for charged tracks, precise secondary vertex reconstruction, and excellent particle identification capabilities.

The analyses presented here are based on the HERA-B data acquired in 2002/2003. From the 150 million dilepton triggered data, approximately 170,000 $J/\psi \rightarrow \mu^+\mu^-$, 5,000 $\psi(2S) \rightarrow \mu^+\mu^-$ and 130,000 $J/\psi \rightarrow e^+e^-$ decays have been reconstructed. In addition around 200 million minimum bias events produced on three target materials (C, Ti, W) were recorded.

CP756, *Quark Confinement and the Hadron Spectrum VI*
edited by N. Brambilla, U. D'Alesio, A. Devoto, K. Maung, G.M. Prosperi and S. Serci
© 2005 American Institute of Physics 0-7354-0241-8/05/$22.50

Hidden Charm

From the dilepton triggered data, J/ψ differential x_F and p_T distributions are obtained. The differential p_T distributions of the J/ψ are well fitted by the function $\frac{d\sigma}{dp_T^2} = A\left[1+(p_T/p_0)^2\right]^{-6}$, leading to the average p_T, $\langle p_T \rangle = \frac{35\pi}{256}p_0$ given in Table 1. The differential x_F distributions of the J/ψ are well described by the NRQCD model [5] (Fig. 1a).

TABLE 1. Average transverse momentum of J/ψ produced in pC and pW interactions (preliminary).

Target	max p_T	$\langle p_T \rangle (e^+e^-) [GeV/c]$	$\langle p_T \rangle (\mu^+\mu^-) [GeV/c]$
Carbon	4.8 GeV/c	1.24 ± 0.01(stat)	1.24 ± 0.03(stat)
Tungsten	4.5 GeV/c	1.29 ± 0.01(stat)	1.30 ± 0.04(stat)

Nuclear effects in heavy quark production are commonly parameterized by the power law: $\sigma_{pA} = \sigma_{pN} \cdot A^{\alpha(x_F, p_T)}$. For the results presented here, only data with simultaneously operating carbon and tungsten targets have been included in order to reduce systematic uncertainties. A preliminary measurement of the A-dependence as a function of x_F, obtained from 10% of the total data, is consistent with the E866 result [6], and indicates no change of the α parameter in the negative x_F region (Fig. 1b.)

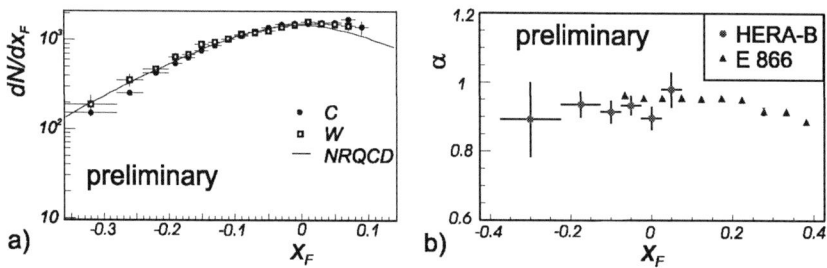

FIGURE 1. a) $J/\psi \to \mu^+\mu^-$ x_F distributions. b) α vs. x_F for two target configurations of the carbon and tungsten wires, compared to the result of E866 [6].

In order to reduce systematic uncertainties, the $\psi(2S)$ production cross section is measured relative to the J/ψ cross section. The acceptance ratio for the two states is determined from detailed Monte Carlo simulation. The preliminary results listed in Table 2, are compared to results of other experiments as a function of the center-of-mass energy and target atomic mass number (see Fig. 2).

TABLE 2. $\psi(2S)$ to J/ψ production cross section ratios (preliminary).

Target	$\frac{B(\psi' \to e^+e^-)\sigma(\psi')}{B(J/\psi \to e^+e^-)\sigma(J/\psi)}[\%]$	$\frac{B(\psi' \to \mu^+\mu^-)\sigma(\psi')}{B(J/\psi \to \mu^+\mu^-)\sigma(J/\psi)}[\%]$
Carbon	1.6 ± 0.2	1.6 ± 0.1
Tungsten	1.8 ± 0.4	1.5 ± 0.2

An additional check of the hadronic charmonium production models is the fraction of J/ψ originating from χ_c meson decays. A preliminary result from the analyzed part of

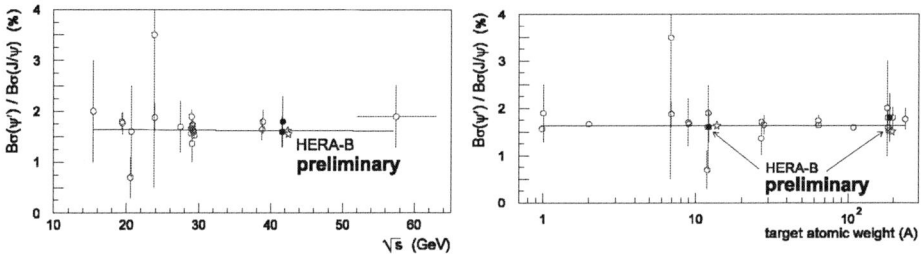

FIGURE 2. $\psi(2S)$ to J/ψ production ratio as a function of the center-of-mass energy and the atomic mass number A. The HERA-B preliminary results: solid circles (electron channel) and open stars (muon channel).

the data gives $R(\chi_c) = 0.21 \pm 0.05$(stat.) The published HERA-B result from the data acquired in 2000 is $R(\chi_c) = 0.32 \pm 0.06$(stat.)$\pm 0.04$(syst.) [7].

Open Charm

Clean signals of $D^0 \to K^- \pi^+$, $D^+ \to K^- \pi^+ \pi^+$ and $D^{*+} \to D^0 \pi^+$ (and charge conjugate decays) are obtained from the minimum bias data after imposing cuts on the separation of the reconstructed vertex from the primary vertex for the D decays. Preliminary results of these measurements are summarized in Tab. 3.

TABLE 3. Production cross sections and cross section ratios for D^0, D^+ and D^{*+} decays.

Preliminary [μb/nucl]	$-0.1 < x_F < 0.05$	full x_F range
σ_{D^0}	$21.4 \pm 3.2 \pm 3.6$	$56.3 \pm 8.5 \pm 9.5$
σ_{D^+}	$11.5 \pm 1.7 \pm 2.2$	$30.2 \pm 4.5 \pm 5.8$
$\sigma_{D^{*+}}$	$10.0 \pm 1.9 \pm 1.4$	$27.8 \pm 5.2 \pm 3.9$
$\sigma_{D^+}/\sigma_{D^0}$		$0.54 \pm 0.11 \pm 0.14$
$\sigma_{D^{*+}}/\sigma_{D^0}$		$0.49 \pm 0.12 \pm 0.10$

Dimuon triggered data are used to derive an upper limit on the flavour changing neutral current decay $D^0 \to \mu^+ \mu^-$. We have measured: $B(D^0 \to \mu^+ \mu^-) < 2.0 \times 10^{-6}$ (90 % confidence level).

REFERENCES

1. R. Baier, and R. Ruckl, *Z. Phys.*, **C19**, 251 (1983).
2. H. Fritzsch, *Phys. Lett.*, **B67**, 217 (1977).
3. G. T. Bodwin, E. Braaten, and G. P. Lepage, *Phys. Rev.*, **D51**, 1125–1171 (1995), hep-ph/9407339.
4. E. Hartouni, et al., *DESY-PRC 95/01* (1995).
5. R. Vogt, *HERA-B Note 01-04-02* (2001).
6. M. J. Leitch, et al., *Phys. Rev. Lett.*, **84**, 3256–3260 (2000), nucl-ex/9909007.
7. I. Abt, et al., *Phys. Lett.*, **B561**, 61–72 (2003), hep-ex/0211033.
8. I. Abt, et al., *Phys. Lett.*, **B596**, 173–183 (2004), hep-ex/0405059.

Hadron Spin Structure: Novel Effects from Transverse Single Spin Asymmetries

U. D'Alesio

Dipartimento di Fisica, Università di Cagliari, and INFN, Sezione di Cagliari, C.P. 170, I-09042 Monserrato (CA), Italy

Abstract. Transverse single spin asymmetries can be a challenging tool in our understanding of the internal structure of hadrons. Some aspects and recent results are discussed.

Keywords: perturbative QCD, inclusive hadron production, spin effects
PACS: 12.38.Bx, 13.85.Ni, 13.88.+e

Several experimental results clearly show that transverse single spin asymmetries (SSA) in high-energy hadronic collisions can be, in particular kinematics regions, very large. Two relevant examples are the transverse polarization, P_T, of Λ produced in unpolarized hadron collisions and the left-right asymmetry, A_N, observed in $p^\uparrow p \to \pi X$:

$$P_T^\Lambda = \frac{d\sigma^{AB \to \Lambda^\uparrow X} - d\sigma^{AB \to \Lambda^\downarrow X}}{d\sigma^{AB \to \Lambda^\uparrow X} + d\sigma^{AB \to \Lambda^\downarrow X}} \qquad A_N = \frac{d\sigma^{A^\uparrow B \to CX} - d\sigma^{A^\downarrow B \to CX}}{d\sigma^{A^\uparrow B \to CX} + d\sigma^{A^\downarrow B \to CX}} \qquad (1)$$

where $d\sigma$ stands for the corresponding invariant differential cross section and $\uparrow, \downarrow$ denote transverse polarization with respect to the hadron production plane. While these observables can reach in size values up to 30%-40%, it is easy to see that in the usual collinear partonic kinematics, perturbative QCD (pQCD) predicts almost vanishing SSA. In fact at the partonic level single spin asymmetries are related to helicity flip amplitudes and to relative phases, both of which are absent in the perturbative, chirality conserving, leading order interactions of quarks and gluons. SSA are then sensitive to higher twist contributions, or non perturbative effects in the long distance physics, and are expected to vanish in the truly asymptotic, high-energy, large Q^2 (or p_T) regions.

Among the attempted explanations of A_N and P_T^Λ we consider here the approach based on pQCD dynamics, through a generalization of the factorization scheme: one starts from the leading twist, collinear configuration scheme and generalizes it with the inclusion of transverse motion of partons in distribution functions (PDF) and hadrons in fragmentation functions (FF). This leads, for the inclusive cross section for $AB \to CX$, to

$$d\sigma^{AB \to CX} = \sum_{a,b,c,d} \hat{f}_{a/A}(x_a, \boldsymbol{k}_{\perp a}) \otimes \hat{f}_{b/B}(x_b, \boldsymbol{k}_{\perp b}) \otimes d\hat{\sigma}^{ab \to cd}(\hat{s}, \hat{t}) \otimes \hat{D}_{C/c}(z, \boldsymbol{k}_{\perp C}), \quad (2)$$

where $\otimes$ stands for convolution both on $x_i(z)$ and $\boldsymbol{k}_{\perp i}$.

As discussed in [1] the inclusion of intrinsic $\boldsymbol{k}_\perp$ effects could also play a relevant role in reducing the gap (in some cases very large) between the theoretical pQCD estimates and the experimental data for unpolarized inclusive particle production.

CP756, *Quark Confinement and the Hadron Spectrum VI*
edited by N. Brambilla, U. D'Alesio, A. Devoto, K. Maung, G.M. Prosperi and S. Serci
© 2005 American Institute of Physics 0-7354-0241-8/05/$22.50

For polarized processes the introduction of $\boldsymbol{k}_\perp$ and spin dependences opens up the way to many possible spin effects; these can be summarized by the new functions:

$$\Delta^N f_{q/p^\uparrow} \equiv \hat{f}_{q/p^\uparrow}(x,\boldsymbol{k}_\perp) - \hat{f}_{q/p^\downarrow}(x,\boldsymbol{k}_\perp) = \hat{f}_{q/p^\uparrow}(x,\boldsymbol{k}_\perp) - \hat{f}_{q/p^\uparrow}(x,-\boldsymbol{k}_\perp) \quad (3)$$

$$\Delta^N f_{q^\uparrow/p} \equiv \hat{f}_{q^\uparrow/p}(x,\boldsymbol{k}_\perp) - \hat{f}_{q^\downarrow/p}(x,\boldsymbol{k}_\perp) = \hat{f}_{q^\uparrow/p}(x,\boldsymbol{k}_\perp) - \hat{f}_{q^\uparrow/p}(x,-\boldsymbol{k}_\perp) \quad (4)$$

$$\Delta^N D_{h/q^\uparrow} \equiv \hat{D}_{h/q^\uparrow}(z,\boldsymbol{k}_\perp) - \hat{D}_{h/q^\downarrow}(z,\boldsymbol{k}_\perp) = \hat{D}_{h/q^\uparrow}(z,\boldsymbol{k}_\perp) - \hat{D}_{h/q^\uparrow}(z,-\boldsymbol{k}_\perp) \quad (5)$$

$$\Delta^N D_{h^\uparrow/q} \equiv \hat{D}_{h^\uparrow/q}(z,\boldsymbol{k}_\perp) - \hat{D}_{h^\downarrow/q}(z,\boldsymbol{k}_\perp) = \hat{D}_{h^\uparrow/q}(z,\boldsymbol{k}_\perp) - \hat{D}_{h^\uparrow/q}(z,-\boldsymbol{k}_\perp). \quad (6)$$

The functions in Eq.s (3) and (5) are respectively the so-called Sivers [2] and Collins [3] functions; in Eq.s (4) and (6) we have the functions introduced by Boer and Mulders [4] and the so-called "polarizing" FF [4, 5]. Moreover the ones in Eq.s (4) and (5) are chiral-odd, while the other two are chiral-even. All the above functions vanish when $k_\perp = 0$, are naïvely T-odd and have a clear partonic interpretation. For instance, the Sivers mechanism corresponds to the azimuthal dependence (around the light-cone direction of the parent nucleon) of the number density of unpolarized partons inside a transversely polarized nucleon; the Collins mechanism corresponds to the azimuthal dependence (around the light-cone direction of the fragmenting parton) of the number density of unpolarized hadrons resulting from the fragmentation of a transversely polarized quark. Similar functions can be found in the literature with different notations [4, 6].

In principle both the Sivers and the Collins mechanisms could be responsible for the observed A_N at E704 [7] (see [8]), while the polarizing FF in Eq. (6) could explain the measured transverse Λ polarization in unpolarized hadron collisions [5].

Recent phenomenological studies [9] have shown how the detailed microscopic dynamics, with all the correct azimuthal angular dependences, produces a strong suppression of the transverse SSA arising from the Collins mechanism. The Sivers effect is not suppressed [1]. In Fig. 1 we show our latest results for A_N both in terms of the Sivers effect alone (including only valence contributions) and with the Collins effect alone (maximizing and including all contributions). A complete study of single (and double) spin asymmetries within the helicity formalism including $\boldsymbol{k}_\perp$ effects is underway [10].

A few words on some theoretical developments are mandatory. In fact only in the last years the role played by the gauge link (Wilson line) entering the operator definition of these functions has been exploited. As a result we expect that whereas deep inelastic scattering (DIS) and Drell-Yan processes probe the same unpolarized PDF, they select Sivers PDF with opposite sign [11]. This poses obviously a question on universality.

As pointed out above, usually, more than one mechanism might in principle contribute to the same SSA. Therefore it is crucial to find proper ways to isolate each of them.

To this aim a combined experimental analysis of transverse SSA in Drell-Yan processes and semi-inclusive deep inelastic scattering (SIDIS) would be extremely useful. First data on azimuthal SSA arising from Sivers effect in SIDIS are now available [12] and more are coming from HERMES and COMPASS collaborations. On the theoretical side it has been shown how by proper suitable integration over the angular dependence of the lepton pair, a measurement of A_N in polarized Drell-Yan processes would give a direct access to the (quark) Sivers function [13]. A tool to learn on possible Sivers effect from gluons could be through the process $p^\uparrow p \to DX$ at RHIC energies (heavy meson production), being dominated by the partonic subprocess $gg \to c\bar{c}$ [14].

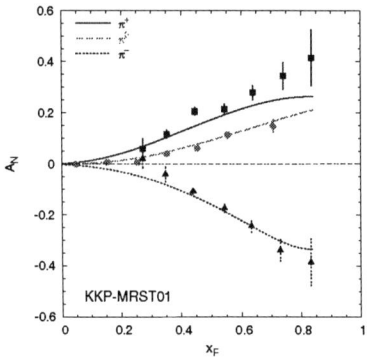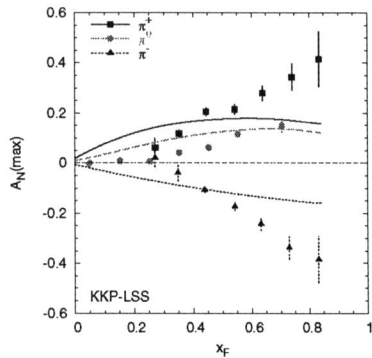

FIGURE 1. $A_N(p^\uparrow p \to \pi X)$ vs. x_F at $p_{lab} = 200$ GeV/c and fixed $p_T = 1.5$ GeV/c. Curves are obtained with Sivers effect (left plot), and with Collins effect (right plot). See [1, 9] for details. Data are from [7].

Finally, Collins effect could play a crucial role in the extraction of the the still unknown, leading twist, transversity PDF, h_1, describing the quark transverse polarization inside a transversely polarized proton. Indeed due to its chiral oddness, the Collins function can be a partner of h_1 in the azimuthal SSA observed in $\ell p^\uparrow \to \ell' \pi X$.

SSA offer a unique access to new information on hadron structure. This new class of spin and $k_\perp$ dependent functions can give a much deeper insight of non perturbative and long-range physics. New data, soon available, will help in their interpretation. On the theoretical side, a better understanding of their fundamental properties, like universality, QCD evolution, factorizability, and classification would be extremely useful.

This brief overview is based on a series of papers in collaboration with M. Anselmino, M. Boglione, E. Leader, S. Melis and F. Murgia.

REFERENCES

1. U. D'Alesio and F. Murgia, *Phys. Rev.* **D70**, 074009 (2004).
2. D. Sivers, *Phys. Rev.* **D41**, 83 (1990); *Phys. Rev.* **D43**, 261 (1991).
3. J.C. Collins, *Nucl. Phys.* **B396**, 161 (1993).
4. P.J. Mulders, and R.D. Tangerman, *Nucl. Phys.* **B461**, 197 (1996), erratum ibid. **B484**, 538 (1997); D. Boer, and P.J. Mulders, *Phys. Rev.* **D57**, 5780 (1998); D. Boer, *Phys. Rev.* **D60**, 014012 (1999).
5. M. Anselmino, D. Boer, U. D'Alesio and F. Murgia, *Phys. Rev.* **D63**, 054029 (2001).
6. A. Bacchetta, U. D'Alesio, M. Diehl, C. Andy Miller, *Phys. Rev.* **D70**, 117504 (2004).
7. D.L. Adams *et al.*, *Phys. Lett.* **B264** (1991) 462; **B261** 201 (1991).
8. M. Anselmino, M. Boglione and F. Murgia, *Phys. Lett.* **B362**, 164 (1995); *Phys. Rev.* **D60**, 054027 (1999); M. Boglione and E. Leader, *Phys. Rev.* **D61** 114001 (2000).
9. M. Anselmino, M. Boglione, U. D'Alesio, E. Leader and F. Murgia, *Phys. Rev.* **D71** 014002 (2005).
10. M. Anselmino, M. Boglione, U. D'Alesio, E. Leader, S. Melis and F. Murgia, in preparation.
11. S.J. Brodsky, D.S. Hwang, and I. Schmidt, *Phys. Lett.* **B530**, 99 (2002); *Nucl. Phys.* **B642**, 344 (2002); J.C. Collins, *Phys. Lett.* **B536**, 43 (2002).
12. A. Airapetian *et al.* (HERMES Coll.) e-Print Archive: hep-ex/0408013.
13. M. Anselmino, U. D'Alesio, and F. Murgia, *Phys. Rev.* **D67**, 074010 (2003).
14. M. Anselmino, M. Boglione, U. D'Alesio, E. Leader, and F. Murgia, *Phys. Rev.* **D70**, 074025 (2004).

Study of dynamical supersymmetry breaking for the two dimensional lattice Wess-Zumino model [1]

Matteo Beccaria*, Gian Fabrizio De Angelis*, Massimo Campostrini[†] and
Alessandra Feo**

*INFN, Sezione di Lecce, and Dipartimento di Fisica dell'Università di Lecce, Via Arnesano, ex
Collegio Fiorini, I-73100 Lecce, Italy
[†]INFN, Sezione di Pisa, and Dipartimento di Fisica "Enrico Fermi" dell'Università di Pisa, Via
Buonarroti 2, I-56125 Pisa, Italy
**Dipartimento di Fisica, Università di Parma and INFN Gruppo Collegato di Parma, Parco Area
delle Scienze, 7/A, 43100 Parma, Italy

Abstract. A new approach to the study of the transition point in a class of two dimensional Wess-Zumino models is presented. The method is based on the calculation of rigorous lower bounds on the ground state energy density in the infinite lattice limit. Such bounds are useful in the discussion of supersymmetry phase transition. The transition point is then determined and compared with recent results based on large-scale Green Function Monte Carlo simulations with good agreement.

The simplest theoretical laboratory to study non perturbative dynamical supersymmetry breaking is the two dimensional $N = 1$ Wess-Zumino model that involves chiral superfields with no vector multiplets. Let us remind the (continuum) $N = 1$ supersymmetry algebra, $\{Q_\alpha, Q_\beta\} = 2(PC)_{\alpha\beta}$. Since P_i are not conserved on the lattice, a lattice formulation of a supersymmetric model must break this algebra explicitly. A very important advantage of the Hamiltonian formulation is the possibility to conserve exactly a key subalgebra of this relation; specializing to $1 + 1$ dimensions, in a Majorana basis, $\gamma_0 = C = \sigma_2$, $\gamma_1 = i\sigma_3$, the algebra becomes $Q_1^2 = Q_2^2 = P^0 \equiv H$, $\{Q_1, Q_2\} = 2P^1 \equiv 2P$. On the lattice, since H is conserved but P is not, we can pick up one of the supercharges, say, $Q_1^2 = H$, build a discretized version Q_L and define the lattice Hamiltonian $H = Q_L^2$.

The explicit lattice model is built by considering a spatial lattice with L sites. On each site we place a real scalar field φ_n together with its conjugate momentum p_n such that $[p_n, \varphi_m] = -i\delta_{n,m}$. The associated fermion is a Majorana fermion $\psi_{a,n}$ with $a = 1, 2$ and $\{\psi_{a,n}, \psi_{b,m}\} = \delta_{a,b}\delta_{n,m}$, $\psi_{a,n}^\dagger = \psi_{a,n}$. The discretized supercharge $Q_L = \sum_{n=1}^{L}[p_n\psi_{1,n} - (\frac{\varphi_{n+1} - \varphi_{n-1}}{2} + V(\varphi_n))\psi_{2,n}]$, with arbitrary $V(\varphi)$ (called prepotential) can be used to define a semipositive definite lattice Hamiltonian $H = Q_L^2$. Notice that $Q_1^2 = H$ is enough to guarantee that $E_0 \equiv \langle\Psi_0|H|\Psi_0\rangle \geq 0$, that all eigenstates of H with $E > 0$

---

[1] Talk presented by A. Feo.

CP756, *Quark Confinement and the Hadron Spectrum VI*
edited by N. Brambilla, U. D'Alesio, A. Devoto, K. Maung, G.M. Prosperi and S. Serci
© 2005 American Institute of Physics 0-7354-0241-8/05/$22.50

are paired in doublets and that $E_0 = 0$ if and only if supersymmetry is unbroken, i.e., the ground state is anihilated by Q_1.

Rigorous results from the continuum can be found in [1]. On the lattice, accurate numerical results are available [2, 3], although a clean determination of the supersymmetry breaking transition remains rather elusive. All predictions and results indicate: for the model with cubic prepotential, $V = \varphi^3$, unbroken supersymmetry; for the model with a quadratic prepotential, $V = \lambda_2 \varphi^2 + \lambda_0$, dynamical supersymmetry breaking. Along a line of constant λ_2 the results for a quadratic potential indicate the existence of two phases: a phase of broken supersymmetry with unbroken discrete Z_2 at high λ_0 and a phase of unbroken supersymmetry with broken Z_2 at low λ_0, separated by a single phase transition. On the other hand, strong coupling expansion demonstrate that for a polynomial $V(\varphi)$, the relevant parameter is just its degree q. For odd q, strong coupling expansion and tree-level results agree and supersymmetry is expected to be unbroken. This conclusion gains further support from the nonvanishing value of the Witten index [4]. For even q in strong coupling expansion, the ground state has a positive energy density also for $L \to \infty$ and supersymmetry appears to be broken for all λ_0. On the other hand, weak coupling expansion predicts unbroken supersymmetry when $\lambda_0 < 0$ [2].

We used two different approaches to investigate the pattern of dynamical supersymmetry breaking in a class of two dimensional Wess-Zumino models. In the first one, [2, 3], the numerical simulations were performed using the Green Function Monte Carlo (GFMC) algorithm and strong coupling expansion. The GFMC is a method that computes a numerical representation of the ground state energy density on a finite lattice with L sites in terms of the states carried by an ensemble of K walkers. By performing numerical simulations along a line of constant $\lambda_2 = 0.5$, we determined the numerical value of λ_0 separating a phase of broken supersymmetry from a phase of unbroken supersymmetry. The usual technique for the study of a phase transition is the crossing method applied to the Binder cumulant [2]. The crossing method consists in plotting B vs. λ_0 for several values of L. The crossing point $\lambda_0^{\text{cr}}(L_1, L_2)$, determined by the condition $B(\lambda_0^{\text{cr}}, L_1) = B(\lambda_0^{\text{cr}}, L_2)$ is an estimator of $\lambda_0^{(c)}$. The value obtained is $\lambda_0^{(c)} = -0.48 \pm 0.01$. The main source of systematic errors in this method is the need to extrapolate to infinity both K and L. For this reason, an indepentend method to test the numerical results of [2] is welcome.

The second method is based on a new approach to the study of the supersymmetry phase diagram introduced in Ref. [5] and is based on the calculation of rigorous lower bounds on the ground state energy density in the infinite lattice limit. The Hamiltonian lattice version of the Wess-Zumino model conserves enough supersymmetry to prove that the ground state has a non negative energy density $\rho \geq 0$, as its continuum limit. Moreover, the ground state is supersymmetric if and only if $\rho = 0$, whereas it breaks (dynamically) supersymmetry if $\rho > 0$. Therefore, if an exact positive lower bound ρ_{LB} is found with $0 < \rho_{\text{LB}} \leq \rho$, we can claim that supersymmetry is broken.

The relevant quantity for our analysis is the ground state energy density ρ evaluated on the infinite lattice limit, $\rho = \lim_{L \to \infty} \frac{E_0(L)}{L} = \lim_{L \to \infty} \rho(L)$. It can be used to tell between the two phases of the model: supersymmetric with $\rho = 0$ or broken with $\rho > 0$. We now explain how to exploit the sequence of bounds $\rho^{(L)}$ (indexed by the cluster size L) to determine the phase at a particular point in the coupling constant space. In order to do

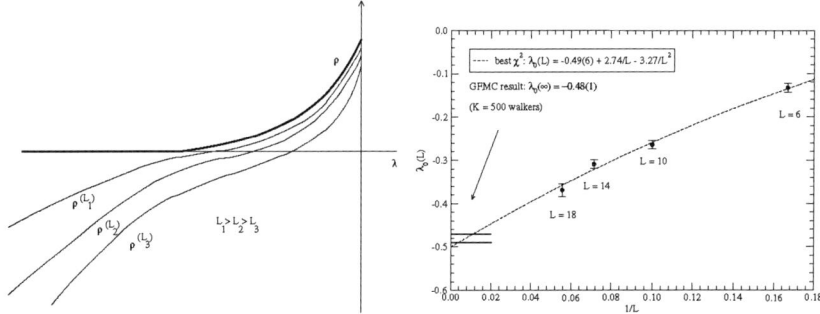

FIGURE 1. Left: Qualitative plot of the function $\rho(\lambda_0)$ and $\rho^{(L)}(\lambda_0)$. Right: Best fit with a quadratic polynomial in $1/L$ together with the best GFMC result [2] obtained with $K = 500$ walkers.

so, we compute numerically $\rho^{(L)}$ at various values of the cluster size L. If $\rho^{(L)} > 0$ for some L, we can immediately conclude that we are in the broken phase. On the other hand, if we find a negative bound we cannot conclude in which phase we are. However, we know that $\rho^{(L)} \to \rho$ for $L \to \infty$ and the study of $\rho^{(L)}$ as a function of both L and the coupling constants permit the identification of the phase in all cases. To test the method we studied in detail the case of the quadratic prepotential and discussed the dependence of ρ on λ_0 at a fixed value of $\lambda_2 = 0.5$ (as in Ref. [2]).

On the left side of Fig. 1 a reasonable qualitative pattern of the curves representing $\rho^{(L)}(\lambda_0)$ is shown. Notice that a single zero is expected in $\rho^{(L)}(\lambda_0)$ at some $\lambda_0 = \lambda_0(L)$. Since $\lim_{L\to\infty} \rho^{(L)} = \rho$, we expect that $\lambda_0(L) \to \lambda_0^*$ for $L \to \infty$, allowing for a determination of the critical coupling λ_0^*. The results of the energy for the lower bound $\rho^{(L)}(\lambda_0)$ for all cluster sizes behaves as expected: it is positive around $\lambda_0 = 0$ and decreases as λ_0 moves to the left. At a certain unique point $\lambda_0^*(L)$, the bound vanishes and remains negative for $\lambda_0 < \lambda_0^*(L)$. This means that supersymmetry breaking can be excluded for $\lambda_0 > \min_L \lambda_0^*(L)$. Also, consistency of the bound means that $\lambda_0^*(L)$ must converge to the infinite-volume critical point as $L \to \infty$. Since the difference between the exact Hamiltonian and the one used to derive the bound is $\mathcal{O}(1/L)$, we can fit $\lambda_0^*(L)$ with a polynomial in $1/L$. This is shown on the right side of Fig. 1 [5], where also the GFMC result is quoted. The best fit with a parabolic function gives $\lambda_0^* = -0.49 \pm 0.06$, quite in agreement with the previous $\lambda_{0,\mathrm{GFMC}}^* = -0.48 \pm 0.01$ using the GFMC algorithm.

REFERENCES

1. A. Jaffe and A. Lesniewski, *Supersymmetric quantum fields and infinite dimensional analysis*, lectures published in Cargese Summer Inst. (1987) and references therein.
2. M. Beccaria, M. Campostrini and A. Feo, Phys. Rev. **D69** 095010 (2004); M. Beccaria, M. Campostrini and A. Feo, Nucl. Phys. Proc. Suppl. **129** 874 (2004); M. Beccaria, M. Campostrini and A. Feo, Nucl. Phys. Proc. Suppl. **119** 891 (2003).
3. M. Beccaria and C. Rampino, Phys. Rev. **D67**, 127701 (2003).
4. E. Witten, Nucl. Phys. **B188**, 513 (1981); E. Witten, Nucl. Phys. **B202**, 253 (1982).
5. M. Beccaria, G. De Angelis, M. Campostrini and A. Feo, Phys. Rev. **D70** 035011 (2004).

Nonabelian Monopoles, Vortices and Confinement

Kenichi Konishi

Dipartimento di Fisica, "E. Fermi"
Università di Pisa,
Largo Pontecorvo, 3, Ed. C
56127 Pisa, Italy
E-mail: konishi@df.unipi.it

Abstract. We discuss possible relevance of quantum mechanical and topological aspects of nonabelian monopoles to the confinement.

Keywords: Supersymmetry, Gauge Theories, Confinement, Monopoles, Vortices
PACS: 11.10.-z,11.15.-q,11.15.Tk

CONFINEMENT IN $SU(N)$ YM THEORY

The test charges in $SU(N)$ YM theory take values in $(Z_N^{(M)}, Z_N^{(E)})$ where Z_N is the center of $SU(N)$ and $Z_N^{(M)}, Z_N^{(E)}$ refer to the magnetic and electric center charges. $(Z_N^{(M)}, Z_N^{(E)})$ classification of phases follows [1, 2]. Namely, (i) If a field with $x = (a,b)$ condenses, particles $X = (A,B)$ with $\langle x, X \rangle \equiv aB - bA \neq 0 \pmod N$ are confined. (e.g. $\langle \phi_{(0,1)} \rangle \neq 0$ → Higgs phase); (ii) Quarks are confined if some magnetically charged particle χ condenses, $\langle \chi_{(1,0)} \rangle \neq 0$; (iii) In softly broken $N = 2$ gauge theories, in a typical vacuum in the confinement phase, nonabelian monopoles condense and give rise to a dual superconductivity of nonabelian variety. What is χ in QCD? How do they interact? Is chiral symmetry breaking related to confinement?

A familiar idea is that the ground state of QCD is a dual superconductor [2]. Although there exist no elementary nor soliton monopoles in QCD, monopoles can be detected as topological singularities (lines in 4D) of Abelian gauge fixing, $SU(3) \rightarrow U(1)^2$, as suggested by 't Hooft. Although there is some evidence in lattice QCD [3] for "Abelian dominance", there remain several questions. Most significantly, does dynamical $SU(N) \rightarrow U(1)^{N-1}$ breaking occur? That would imply a richer spectrum of mesons ($T_1 \neq T_2$, etc) than expected. Both in Nature and presumably in QCD there is only one "meson" state, $\sum_{i=1}^{N} |q_i \bar{q}_i\rangle$, *i.e.*, 1 state vs $\left[\frac{N}{2}\right]$ states. It does not suffice to assume that $SU(N) \rightarrow U(1)^{N-1} \times Weyl$ symmetry, with an extra discrete symmetry, to solve the problem: the multiplicity would be wrong. If nonabelian degrees of freedom are important, after all, how do they manifest themselves?

NA MONOPOLES ARE QUANTUM MECHANICAL

Nonabelian monopoles turn out to be essentially quantum mechanical. In fact, finding semiclassical degenerate monopoles, as studied earlier, is not sufficient to conclude that they form a multiplet of $\tilde{H}$, as H can break itself dynamically at lower energies and break the degeneracy among the monopoles. We must ensure that this does not take place. Nonabelian monopoles are in this sense never really semi-classical, even if $\langle \phi \rangle \gg \Lambda_H$: (*e.g.*, Pure $N = 2$, $SU(3)$).

In this connection, there is a famous "no go theorem" which states that there are no "colored dyons"[4]. Actually, those results does not exclude existence of sets of monopoles transforming as members of a dual multiplet (even if at present the explicit form of such nonlocal transformations are not known).

PHASES OF SOFTLY BROKEN $N = 2$ GAUGE THEORIES

Fully quantum mechanical results about the phases of $SU(n_c)$, $USp(2n_c)$ and $SO(n_c)$ theories with n_f hypermultiplets (quarks), perturbed by the superpotential

$$W(\phi, Q, \tilde{Q}) = \mu \operatorname{Tr}\Phi^2 + m_i \tilde{Q}_i Q^i, \qquad m_i \to 0$$

are known [5, 6]. The confining vauca of $SU(n_c)$ theories are classified as follows. From these results we learn, in particular, that the spectrum of the "dual quarks" in

TABLE 1. Phases of $SU(n_c)$ gauge theory with n_f flavors. $\tilde{n}_c \equiv n_f - n_c$.

Deg.Freed.	Eff. Gauge Group	Phase	Global Symmetry
monopoles	$U(1)^{n_c-1}$	Confinement	$U(n_f)$
monopoles	$U(1)^{n_c-1}$	Confinement	$U(n_f-1) \times U(1)$
NA monopoles	$SU(r) \times U(1)^{n_c-r}$	Confinement	$U(n_f-r) \times U(r)$
rel. nonloc.	-	Confinement	$U(n_f/2) \times U(n_f/2)$
NA monopoles	$SU(\tilde{n}_c) \times U(1)^{n_c-\tilde{n}_c}$	Free Magnetic	$U(n_f)$

the infrared theory (charges, multiplicity, flavor) is identical to what is expected from the semiclassical abelian or nonabelian monopoles. We note in particular that the $r-$ vacua (*i.e.* vacua with a low-energy effective $SU(r)$ gauge group) exist only for $r < \frac{n_f}{2}$, namely as long as the sign-flip of the beta function occurs:

$$b_0^{(dual)} \propto -2r + n_f > 0, \qquad b_0 \propto -2n_c + n_f < 0.$$

Indeed, analogous r vacua exist semiclassically for all values up to $min(n_f, n_c)$, but quantum mechanical vacua exist only with $r \le n_f/2$. Also, when the sign flip is not possible (e.g. $N = 2$ YM or on a generic point of the quantum moduli space) dynamical abelianization is expected and does take place!

These observations lead us to conclude that the "dual quarks" belonging to the fundamental representation of the infrared $SU(r)$ gauge group, actually *are* the Goddard-Nuyts-Olive-Weinberg monopoles, which have become massless by quantum effects [8].

Most importantly, we are led to the general criterion for nonabelian monopoles to survive quantum effects: the system must produce, upon symmetry breaking, a sufficient number of massless flavors to protect H from becoming too strongly-coupled. Natural embedding in $N = 2$ systems for various cases in the Table has been discussed in Ref.[7]

A very subtle hint about the nature of the nonabelian monopoles come from the recent discovery of nonabelian vortices.[9, 10, 11]

NONABELIAN MONOPOLES AND CONFINEMENT

Let us summarize:

(1) Nonabelian monopoles in $G \to H$ are essentially quantum mechanical;

(2) Massless flavor symmetry is important for (i) keeping H unbroken; and for (ii) being part of the dual transformation among monopoles;

(3) Nonabelian BPS *vortices* do exist quantum mechanically; again their existence requires appropriate massless flavors in the theory;

(4) Nice "models" of monopole confinement by vortices can be constructed;

(5) Nonabelian monopoles do occur as infrared degrees of freedom (examples: r vacua in $N = 2$, $SU(N_c)$ SQCD); actually they can be further classified in two types of vacua: (i) nonabelian superconductor in terms of weakly coupled nonabelian monopoles; (ii) nonabelian superconductor in terms of strongly coupled nonabelian monopoles (almost superconformal vacua);

(6) Softly broken $N = 2$ theories thus give deep hints about confinement in QCD.

REFERENCES

1. G. 't Hooft, **Nucl. Phys. B138** (1978) 1; *ibid B153* (1979) 141, *B190* (1981) 455.
2. G. 't Hooft, **Nucl. Phys. B190** (1981) 455; S. Mandelstam, **Phys. Lett. 53B** (1975) 476.
3. L. Del Debbio, A. Di Giacomo and G. Paffuti, Nucl. Phys. B (Proc. Suppl.) 42 (1995) 231; A. Di Giacomo, hep-lat/0206018.
4. A. Abouelsaood, **Nucl. Phys. B226** (1983) 309; P. Nelson and A. Manohar, **Phys. Rev. Lett. 50** (1983) 943; A. Balachandran, G. Marmo, M. Mukunda, J. Nilsson, E. Sudarshan and F. Zaccaria, **Phys. Rev. Lett. 50** (1983) 1553; P. Nelson and S. Coleman, **Nucl. Phys. B227** (1984) 1.
5. A. Hanany and A. Oz, **Nucl. Phys. B452** (1995) 283, hep-th/9505075, P. Argyres, M. Plesser and N. Seiberg, **Nucl. Phys. B471** (1996) 159, hep-th/9603042,
6. G. Carlino, K. Konishi and H. Murayama, **JHEP 0002** (2000) 004, hep-th/0001036; **Nucl. Phys. B590** (2000) 37, hep-th/0005076; G. Carlino, K. Konishi, Prem Kumar and H. Murayama, **Nucl. Phys. B608** (2001) 51, hep-th/0104064.
7. R. Auzzi, S. Bolognesi, Jarah Evslin, K. Konishi, H. Murayama, **Nucl. Phys. B701** (2004) 207, hep-th/0405070.
8. S. Bolognesi and K. Konishi, **Nucl. Phys. B645** (2002) 337, hep-th/0207161.
9. R. Auzzi, S. Bolognesi, Jarah Evslin, K. Konishi and A. Yung, **Nucl. Phys. B 673** (2003) 187, hep-th/0307287.
10. R. Auzzi, S. Bolognesi, J. Evslin and K. Konishi, **Nucl. Phys. B686** (2004) 119, hep-th/0312233.
11. A. Hanany and D. Tong, **JHEP 0404** (2004) 066, hep-th/0403158; M. Shifman and A. Yung, **Phys.Rev.D70** (2004) 045004, hep-th/0403149.

QCD, Supersymmetry and Low Scale Gravity

Alessandro Cafarella* Claudio Corianò* and T. N. Tomaras†

*Dipartimento di Fisica dell' Universita' di Lecce and INFN-Lecce, Via Arnesano, 73100 Lecce,
Italy. E-mail: alessandro.cafarella@le.infn.it, claudio.coriano@le.infn.it
†Department of Physics and Institute for Plasma Physics, University of Crete and FORTH,
Heraklion, Crete, Greece E-mail: tomaras@physics.uoc.gr

Theories with large extra dimensions [1] invoke a brane picture of the universe, with matter confined on a brane embedded in a higher D-dimensional space $(D = 4+N)$, and only gravity free to propagate in the extra dimensions. A certain number, say n, of the N extra dimensions may be large, with size of the order of a millimeter. These scenarios are characterized by a low fundamental scale for gravity, M_*, related to the Planck scale M_{Pl} by $M_{Pl}^2 = M_*^{n+2} V_{(n)}$, with $V_{(n)}$ the volume of the extra dimensions. For $n = 2$, M_* can be of the order of a TeV if the typical size of the large extra dimensions is a millimeter. At the LHC, for QCD factorization scales above M_*, gravity becomes strong and hadronic collisions should be characterized by a rich new phenomenology. In particular, mini black holes of mass $M_{BH} \approx M_*$ are expected to be produced copiously [2] (see [3] for a discussion of some quantum aspects).

Mini black holes are hot, characterized by a temperature which is inversely proportional to their mass M_{BH}. Their formation takes place in (parton-parton) collisions for impact parameters of the order of the size of the horizon (r_H)

$$r_H = \frac{1}{\sqrt{\pi} M_*} \left(\frac{M_{BH}}{M_*} \right)^{\frac{1}{n+1}} \left(\frac{8\Gamma \left(\frac{n+3}{2} \right)}{n+2} \right)^{\frac{1}{n+1}}. \tag{1}$$

corresponding to the collision energy $E \sim M_{BH}$ in the center of mass frame. For $n > 0$ the relation between r_H and M_{BH} becomes nonlinear and the presence of M_* in the denominator of Eq. (1) in place of M_{Pl} increases the size of the horizon for a given M_{BH}. For $M_{BH}/M_* \sim 5$ and $M_* = 1$ TeV the size of the horizon is around 10^{-4} fm and decreases with increasing n. A good approximation to the partonic cross section for producing a mini black hole is $\sigma_{BH} \approx \pi r_H^2$, the geometrical one. It can be folded with parton distributions $(f(x, Q^2))$ to give predictions, for instance, for total cross sections

$$\sigma(pp \to BH + X) = \frac{1}{s} \sum_{a,b} \int_{M_{BH,min}^2}^{s} dM_{BH}^2 \times \int_{x_{1,min}}^{1} \frac{dx_1}{x_1} f_a(x_1, Q^2) \sigma_{BH} f_b(x_2, Q^2), \tag{2}$$

---

[1] Presented by C. Corianò at the VI Conference on Quark Confinement and the Hadron Spectrum, Villasimius, Sardinia, Italy, 21-25 September 2004

CP756, Quark Confinement and the Hadron Spectrum VI
edited by N. Brambilla, U. D'Alesio, A. Devoto, K. Maung, G.M. Prosperi and S. Serci
© 2005 American Institute of Physics 0-7354-0241-8/05/$22.50

where x_1 and $x_2 = M_{BH}^2/(x_1 s)$ are the momentum fractions of the initial partons and $x_{1,min} = M_{BH}^2/s$. The factorization scale Q is of the order of $1/r_H$. The absorption/emission cross section depends sensitively on the greybody factors of the black hole, which are energy dependent. The choice of either constant or full energy-dependent greybody factors, which are known for static (Schwarzschild) mini black hole solutions [4], gives widely different results [5]. The known analytical expressions of the greybody factors at low frequencies are of limited help in the prediction of the event rates at the LHC, but these can be computed numerically [6]. Of particular relevance would be the numerical study of the greybody factors for Kerr solutions, since black holes, in general, will be produced with non-vanishing angular momentum.

Studies of the p_T distributions show a much larger signal compared to the fast falling QCD background [5], even for M_* as high as 5 TeV, starting at $P_T \sim 50 - 200$ GeV and up. The dependence on the number of extra dimensions n is also significant. A second sensitivity in the prediction of event rates comes from the integration over the invariant mass M_{BH} for M_{BH} close to M_{Pl}, since the semiclassical picture of the formation and decay of the black hole is not valid any longer. In all the studies presented so far larger multiplicites of the final states and broader p_T distributions appear to be a striking signature of mini black hole formation in hadron collisions. In the most optimistic scenario in which both low energy gravity and supersymmetry will be discovered at the LHC, then the multiplicities of the final state in the decay of the black hole should grow even faster from what inferred from these studies. However, it is important to keep in mind that a part of the energy available in the collision is loss into gravitational emission, and only a fraction of it remains available for the hadronization, which would imply reduced multiplicities.

The time scales for the black hole decay into partons and the QCD hadronization scale are largely separated and the decay of the black hole is, essentially, instantaneous. Hadronization takes place soon after the partons, which are emitted in an approximate s-wave, cross the horizon. The emissions of single partons are assumed to be uncorrelated, and can be described by a multinomial distribution, while the hadronization is studied either using Monte Carlo [7] or renormalization group equations [8].

The computation of the cumulative probabilities to produce any number (K) of hadrons of type h by the decay of the black hole are obtained from the multinomial distribution multiplied by the fragmentation probabilities of each elementary state to h and summing over all possible emissions [8]

$$\mathrm{Pr}_{\mathrm{cum\,h}}(K,Q) \equiv \sum_{n_f,n_i} \frac{K!}{\prod_f n_f! \prod_i n_i!} \prod_f \left(p_f < D_f^h(Q_F) > \right)^{n_f} \prod_i \left(p_i < D_i^h(Q_F) > \right)^{n_i},$$
(3)

where i is summed over gluons, photons and a set or remainder states, f runs over the quark flavours, while $K = n_f + n_i$. In (3) the $< D_{i,f}^h(Q_F) >$ are the first moments of the fragmentation functions of a parton/photon k to a hadron h at a scale Q_F. The sum is over all the main hadronic states. The fragmentation scale Q_F is related to the number of fundamental decaying states N_m to which the black hole couples in a democratic way and to its mass M_{BH} by $Q_f = M_{BH}/N_m$, where our knowledge of the multiplicity N_m is clearly approximate. Obtaining a good estimate of N_m is important for studies of the multi-jet structure of the events at the LHC, but is less relevant for cosmic ray studies. In this latter case the evolution of the air shower after the decay of the black hole washes out the information on small variations in the original multiplicities in the decay. At this time, the only known formulas available for N_m come from a semiclassical analysis. We recall that in cosmic ray physics mini black hole events can be triggered by neutrinos scattering off nucleons in the atmosphere. An analysis of the lateral distributions of showers and of the corresponding multiplicities shows that intermediate mini black hole resonances are respectively much wider and larger compared to ordinary air showers [8], in agreement with the fireball picture of the decay which has emerged from LHC studies.

Proposals for the best approximation to N_m are several. In [2] was suggested to use

$$N_m = \frac{2\pi}{n+1} \left(\frac{M_{BH}}{M_*}\right)^{\frac{n+2}{n+1}} \left(\frac{8\Gamma\left(\frac{n+3}{2}\right)}{n+2}\right)^{\frac{1}{n+1}} \frac{1}{\sqrt{\pi}}, \tag{4}$$

but there are variants of it. Other expressions include a correction factor ρ coming from a more detailed analysis of the Hawking formula for the semiclassical decay which takes into account the corresponding greybody (Γ_s) factors more accurately [9]. Then $N_m = \rho S_0$ with S_0 being the entropy of the black hole and

$$\rho = \frac{\sum_s c_s f_s \Gamma_s \Gamma(3)\,\zeta(3)}{\sum_s c_s f'_s \Gamma_s \Gamma(4)\,\zeta(4)}, \tag{5}$$

which is expressed in terms of the greybody factors and certain numerical coefficients (c_s, f_s, f'_s) dependent on the spin s of the fields propagating over the black hole background. As we have already mentioned, the issue of gravitational energy emission during the formation of the black hole and during its decay remains open. Work in this direction can follow closely some of the recent results on the study of quasi-normal modes for ordinary black holes in 4 dimensions aimed at the detection of gravitational waves [10].

REFERENCES

1. N. Arkani-Hamed, S. Dimopoulos and G. Dvali, Phys. Lett. **B429**, 263 (1998); I. Antoniadis, N. Arkani-Hamed, S. Dimopoulos and G. Dvali, Phys. Lett. **B436**, 257 (1998); L. Randall and R. Sundrum, Phys. Rev. Lett. **83**, 3370 (1999); L. Randall and R. Sundrum, Phys. Rev. Lett. **83**, 4690 (1999).
2. S. Dimopoulos and G. Landsberg, Phys. Rev. Lett. **87**, 161602 (2001).
3. S.D.H. Hsu, Phys.Lett.**B555** 92, (2003); V. P. Frolov, D. Stojkovic, Phys.Rev.**D 67** 084004, (2003).
4. P. Kanti and J. March-Russell, Phys. Rev.**D 67**, 104019 (2003).
5. I. Mocioiu, Y. Nara and I. Sarcevic, Phys. Lett. B **557**, 87 (2003).
6. P. Kanti and C.M. Harris, JHEP **0310** 014,(2003).
7. C.M. Harris, P. Richardson and B.R. Webber, JHEP 0308,033, (2003).
8. A. Cafarella, C. Corianò and T.N. Tomaras, [arXiv:hep-ph/0410358], [arXiv:hep-ph/0410190].
9. M. Cavaglià Phys.Lett. **B569**, 7, (2003), M. Cavaglià and S. Das, Class.Quant.Grav. **21**, 4511, (2004);
10. V. Cardoso, O. J.C. Dias, J.P.S. Lemos, Phys. Rev. **D 67**, 064026 (2003); K.D. Kokkotas, Living Rev.Rel.**2**, 2, (1999); see also H. Kodama and A. Ishibashi, Prog.Theor.Phys.**110** 701, (2003).

Symmetry breaking in non-commutative cut-off field theories

Paolo Castorina and Dario Zappalà

Dept. of Physics, University of Catania and INFN, Sezione di Catania- Italy

Abstract. The relation between symmetry breaking in non-commutative cut-off field theories and transitions to inhomogeneous phases in condensed matter and in finite density QCD is discussed. The non-commutative dynamics, with its peculiar infrared-ultraviolet mixing, can be regarded as an effective description of the mechanisms which lead to inhomogeneous phase transitions and a roton-like excitation spectrum.

Keywords: Symmetry breaking, non-commutative field theories
PACS: 11.10.Nx 11.30.Qc

The relation between symmetry breaking in quantum field theory and phase transitions in condensed matter is well known. The previous correspondence is usually restricted to the case of constant order parameter because this guarantees the translational and rotational invariance of the field theory. On the other hand, in condensed matter and, more recently, in high density QCD one considers transitions from homogeneous to inhomogeneous phases with non-constant order parameters. For a bosonic system, according to [1], this kind of transition is associated with an order parameter which for large distances is an oscillating function and with a roton-like behavior of the excitation spectrum. For the fermionic systems it is possible to build inhomogeneous superconducting states, with energy lower than the BCS state, where the particle-particle (p-p) Cooper pairs or the particle-hole (p-h) pairs have non-zero total momentum and then the corresponding fermionic condensates are not uniform. The former phase is described in condensed matter systems by the LOFF state [2] and in QCD is related with a diquark condensation with interesting theoretical and phenomenological implications [3, 4]. The latter has been originally proposed in condensed matter by Overhauser [5] (the (p-h) instability is called spin density wave) and in QCD the corresponding phase, called chiral density wave, can compete with the QCD-BCS phase in the strong coupling regime and in an intermediate region of the chemical potential [6].

In [7, 8, 9, 10, 11] it has been shown that the previous condensed matter and QCD inhomogeneous transitions are also typical features of quantum field theory with non-commutative coordinates, $[x_\mu, x_\nu] = i\theta_{\mu\nu}$. As discussed in this talk, in the non-commutative generalization of $\lambda\phi^4$ theory [7, 8, 11] the spontaneous symmetry breaking occurs for a non-uniform stripe phase and in the non-commutative Gross-Neveu (GN) model there is an inhomogeneous chiral symmetry breaking corresponding to spin density waves [9]. These results are mainly due to the infrared/ultraviolet (IR/UV) connection which characterizes the non commutative field theories (for a review see [12]).

We shall first discuss the relation between the symmetry breaking in non-commutative self-interacting scalar field theory and the roton excitation in BEC and then we consider

CP756, Quark Confinement and the Hadron Spectrum VI
edited by N. Brambilla, U. D'Alesio, A. Devoto, K. Maung, G.M. Prosperi and S. Serci
© 2005 American Institute of Physics 0-7354-0241-8/05/$22.50

the fermionic systems. On general grounds, in a bosonic condensate the roton spectrum is due to a non-local interatomic potential $V(\vec{r}-\vec{r}')$, with a momentum dependent Fourier transform. Since the BEC with the local (pseudo)potential $\delta(\vec{r}-\vec{r}')$ is analogous to the spontaneous symmetry breaking in $\lambda\phi^4$ theory, one can assume that some relevant physical effects due to the non-local repulsive interaction can be described by generalizing the self-interacting field theory in such a way to introduce an effective non-local coupling.

A simple approach is to consider the non-commutative $\lambda\phi^4$ theory with action

$$S(\phi) = \int d^4x \left(\frac{1}{2}\partial_\mu\phi \, \partial^\mu\phi - \frac{1}{2}m^2\phi^2 - \frac{\lambda}{4!}\phi^{4*} \right) \tag{1}$$

where the star (Moyal)product is defined by $(i, j = 1,..,4)$

$$\phi^{4*}(x) = \phi(x) * \phi(x) * \phi(x) * \phi(x) =$$

$$\exp\left[\frac{i}{2}\sum_{ij}\theta_{\mu\nu}\partial_{x_i}^\mu\partial_{x_j}^\nu \right] \left(\phi(x_1)\phi(x_2)\phi(x_3)\phi(x_4) \right)\Big|_{x_i=x} \tag{2}$$

The "deformation" of the self-interaction term by the Moyal product gives a momentum dependent repulsive effect which is responsible, as we shall see below, for the roton spectrum and for the phase transition to an inhomogeneous background. In [8], for $\theta_{ij} = \varepsilon_{ijk}\theta^k$ with $\vec{\theta} = (0,0,\theta)$, the spontaneous symmetry breaking for the theory in Eqs. (1) and (2) has been analyzed with the following results: **1)** the transition occurs to a stripe phase where the order parameter is $\phi(\vec{x}) = A\cos\vec{Q}\cdot\vec{x}$; **2)** A, Q and the energy excitation $\omega(p)$ are fixed by minimizing the energy; **3)** $\vec{Q}$ is orthogonal to $\vec{\theta}$ and Q is small for large θ; **4)** the excitation spectrum can be approximated by $\omega^2(\vec{p}) = p^2 + M^2(\vec{p})$ where the function $M(\vec{p})$ will be discussed later. As discussed in detail in [8], since Q is small, the inhomogeneous background is a smooth function of x and then, the breaking of translational (smooth) and rotational invariance is approximated by a translational invariant propagator with a momentum dependent mass term. In the particular case $\vec{Q} = (Q/\sqrt{2}, Q/\sqrt{2}, 0)$ and large values of $\theta\Lambda^2$ (Λ is the UV cut-off), it turns out that $Q^2/\Lambda^2 = (\lambda/24\pi^2)^{1/2}(1/\theta\Lambda^2)$ and that $M(\vec{p})$, for small p, is given by

$$M^2(\vec{p})\big|_{p\to 0} \sim \alpha + \frac{\lambda}{6\pi^2}\frac{1}{|\vec{p}\times\vec{\theta}|^2} \tag{3}$$

where α is a constant and $\times$ indicates the usual vector product. The peculiar behavior for small p of the last term in the previous equation is due to the IR/UV connection of the non-commutative field theory and gives a divergent mass term in the IR region and a minimum in the irreducible two-point function. However the effective theory has a natural self-generated IR cut-off Q where it is more correct to cut the small momenta and then the excitation spectrum should be correctly identified for $p \geq Q$ by the previous expressions for $\omega(p)$ and $M^2(p)$. It has a roton-like dip at a typical scale of order Q.

The Moyal-deformed term in Eq. (1) can mimic effective interactions which are non-local and globally repulsive. Then, the previous results of the non-commutative theory can describe some interesting physical effects of the Bose -Einstein trapped condensates where the non-contact repulsive interaction is dominant. This analysis has been performed in detail in ref. [10] where the correlation between roton-like spectrum and non-uniform background of the non commutative theory has been compared with the results obtained for BE trapped condensates, in analogous dynamical conditions.

Let us now consider the informations coming from non-commutative effective field theories for fermionic systems. In [9] the transition from homogeneous to inhomogeneous phase has been obtained by generalizing the GN model, in four dimensions, to the non-commutative case with lagrangian

$$L(x) = i\bar{\psi}_\alpha \partial\!\!\!/ \psi_\alpha + g\bar{\psi}_\alpha * \psi_\alpha * \bar{\psi}_\beta * \psi_\beta - g\bar{\psi}_\alpha * \bar{\psi}_\beta * \psi_\alpha * \psi_\beta . \qquad (4)$$

For g larger than some critical value, one finds, as in tha commutative case, chiral symmetry breaking but, this time, in an inhomogeneous phase where the pair correlation function has a dependence on a total momentum, $\vec{P}$ of the ("Cooper") pair, with $P/\Lambda \simeq (1/\theta\Lambda^2)$. The order parameter turns out to be an oscillating function of $\vec{x}$ and one has the breaking of translational, rotational and chiral invariance : $< \bar{\psi}(x)\psi(x) >= [1 + cP^2\cos(Px)] < \bar{\psi}\psi >_0$, where $< \bar{\psi}\psi >_0$ is the constant order parameter of the commutative case and c is a numerical constant. Also in this case [9] the spectrum has roton-like dip in the plane orthogonal to $\vec{\theta}$ and one recovers the dynamical relation with the non-uniform ground state [1]. The previous field theoretical model is then analogous to a system with a non-local, strong four-fermion interaction, with an inhomogeneous phase where the particle-hole (p-h) pairs have non-zero total momentum.As discussed, this phase in QCD has an energy close to the BCS phase only for strong coupling , corresponding to intermediate value of the density, that is before entering the perturbative regime where the QCD- LOFF phase is realized.

Finally, the non commutative field theoretical results indicate that the phase transitions to inhomogeneous condensates are first order and one expects similar behavior for the corresponding condensed matter and QCD systems.

REFERENCES

1. S.A. Brazovskii, Zh. Eksp. Teor. Fiz. **68** (1975)175.
2. A. J. Larkin, Y. N. Ovchinnikov, Zh. Exsp. Theor. Fiz. **47** (1964) 1136; P. Fulde and R.A. Ferrell, Phys. Rev. **135** (1964) A550.
3. K. Rajagopal and F. Wilczek, "Handbook of QCD", Edited by M. Shifman, World Scientific 2001.
4. R. Casalbuoni and G. Nardulli, Rev. Mod. Phys. **76** (2004) 263.
5. A.W.Overhauser Phys. Rev. **128** (1962) 1437.
6. R. Rapp, E. Shuryak and I. Zahed, Phys. Rev. **D63** (2001) 034008.
7. S.S. Gubser and S.L. Sondhi, Nucl. Phys. **B605** ;(2001) 395.
8. P.Castorina, and D. Zappalà, Phys. Rev. **D68** (2003) 065008.
9. P. Castorina, G. Riccobene and D. Zappalà, Phys. Rev. **D69** (2004) 105024.
10. P. Castorina, G. Riccobene and D. Zappalà, hep-th
11. J. Ambjorn, S. Catterall, Phys. Lett. **B549** (2002) 253; W. Bietenholz, F. Hofheinz, J. Nishimura, Nucl. Phys. Proc. Suppl. 119 (2003) 941; Fortsch. Phys. 51 (2003) 745; e-Print Archive: hep-th/0404020.
12. M. R. Douglas and N. A. Nekrasov, Rev. Mod. Phys. **73** (2001) 977; R.J. Szabo, Phys. Rep. **378** (2003) 207.

Perturbative and nonperturbative fields in SU(2) lattice gluodynamics

V.G. Bornyakov[*,†], M.I. Polikarpov[*], T. Suzuki[**] and S.N. Syritsyn[*]

[*]*ITEP, B. Cheremushkinskaya 25, Moscow, 117259, Russia*
[†]*Institute for High Energy Physics IHEP, 142284 Protvino, Russia*
[**]*Institute for Theoretical Physics, Kanazawa University, Kanazawa 920-1192, Japan*

Abstract. We discuss the separation of perturbative and nonperturbative fields in lattice SU(2) gauge theory fixing Maximal Abelian gauge and Landau U(1) gauge. We represent the link matrix as the product of diagonal monopole field matrix and the matrix which we call perturbative link field. It occurs that the perturbative link field correctly reproduces the Coulomb part of the nonabelian potential, and the monopole field reproduces the linear part of the confining potential. The monopole field is responsible for the nonperturbative contribution to plaquette action, proportional to the square of the lattice spacing as predicted by renormalon formalism.

Keywords: monopoles, confinement
PACS: 11.15.Ha,12.38.Gc

Are there Local Minima in the Magnetic Monopole Potential in Compact QED?

H. Bozkaya*, M. Faber*, P. Koppensteiner* and M. Pitschmann*

*Atominstitut der Österreichischen Universitäten, Technische Universität Wien, Wiedner Hauptstraße 8–10/142, A-1040 Wien, Austria

Abstract. In order to investigate the phase transition in compact QED (CQED) we examine the influence of the granularity of the lattice on the potential between monopoles in the strong coupling region of this theory. Using the flux definition of monopoles we introduce their centres of mass and are able to realize continuous shifts of the monopole positions. In this way we succeed in obtaining the monopole-antimonopole potential for continuous separations and find periodic deviations from the supposed $1/r$-behaviour. These deviations lead to local extrema and to metastabilities. We suppose that the latter may influence the order of the phase transition in compact QED.

Keywords: Lattice gauge theory; Quantum electrodynamics; Magnetic monopoles.
PACS: 11.15.Ha; 12.20.m; 14.80.Hv.

CP756, *Quark Confinement and the Hadron Spectrum VI*
edited by N. Brambilla, U. D'Alesio, A. Devoto, K. Maung, G.M. Prosperi and S. Serci
© 2005 American Institute of Physics 0-7354-0241-8/05/$22.50

π^0, $\eta \to \gamma\gamma$ at finite temperature and density

P. Costa*, M. C. Ruivo* and Yu. L. Kalinovsky[†]

*Centro de Física Teórica, Departamento de Física, Universidade, P3004-516 Coimbra, Portugal
[†]Université de Liège, Départment de Physique B5, Sart Tilman, B-4000, LIEGE 1, Belgium

Abstract. We investigate the anomalous decays π^0, $\eta \to \gamma\gamma$ in the framework of the three–flavor Nambu–Jona-Lasinio [NJL] model at finite temperature and density. The similarities and differences between both scenarios are discussed. It is found that the behavior of the relevant observables essentially reflects a manifestation of the partial restoration of chiral symmetry in non strange sector. In both cases the lifetimes of these mesons decrease significantly at the critical point, although this might not be sufficient to observe enhancement of these decays in heavy-ion collisions.

CP756, *Quark Confinement and the Hadron Spectrum VI*
edited by N. Brambilla, U. D'Alesio, A. Devoto, K. Maung, G.M. Prosperi and S. Serci
© 2005 American Institute of Physics 0-7354-0241-8/05/$22.50

The invariant measure for $SU(N)$

Michael Creutz

Physics Department, Brookhaven National Laboratory, Upton, NY 11973, USA

Abstract. I present a simple recursive relation between the invariant measure for integration over SU(N) and the product of the measure for a sphere S(2N-1) times the measure for SU(N-1). I also discuss the periodicity factor that appears in non-trivial mappings of S(2N-1) into SU(N).

Parametrizing an $SU(N)$ matrix in the form

$$U = \begin{pmatrix} 1 & \cdots \\ \vdots & g_{N-1} \end{pmatrix} g_s(\vec{z})$$

with $g_{N-1} \in SU(N-1)$ and $g_s \in SU(N)$ a standard form with given top row $\vec{z} \in S_{2N-1}$,

$$g_s = \begin{pmatrix} 1 & 0 & \cdots \\ 0 & 1 & \cdots \\ \vdots & \vdots & \ddots \\ & & & \prod_i p_i^* \end{pmatrix} \begin{pmatrix} c_1 & s_1 & 0 & \cdots \\ -s_1 & c_1 & 0 & \cdots \\ 0 & 0 & 1 & \cdots \\ \vdots & \vdots & \vdots & 1 \end{pmatrix} \begin{pmatrix} 1 & 0 & 0 & \cdots \\ 0 & c_2 & s_2 & \cdots \\ 0 & -s_2 & c_2 & \cdots \\ \vdots & \vdots & \vdots & 1 \end{pmatrix}$$
$$\cdots \begin{pmatrix} 1 & \vdots & \vdots & \vdots \\ \cdots & 1 & 0 & 0 \\ \cdots & 0 & c_{N-1} & s_{N-1} \\ \cdots & 0 & -s_{N-1} & c_{N-1} \end{pmatrix} \begin{pmatrix} p_1 & 0 & \cdots \\ 0 & p_2 & \cdots \\ \vdots & \vdots & \ddots \\ & & & p_N \end{pmatrix}$$

Then the invariant integration measure takes the form

$$dg_N = dS_{2N-1}\, dg_{N-1}$$

That is, the measure is uniform over an S_{2N-1} and $SU(N-1)$.

To construct $\Pi_{2N-1}(SU(N))$ via mapping an $S(2N-1)$ into the group non-trivially requires covering the above $S(2N-1)$, defined by $\vec{z}$, 4^{N-2} times due to singularities at the poles of the mapping when $s_i = 0$.

ACKNOWLEDGMENTS

This manuscript has been authored under contract number DE-AC02-98CH10886 with the U.S. Department of Energy. Accordingly, the U.S. Government retains a non-exclusive, royalty-free license to publish or reproduce the published form of this contribution, or allow others to do so, for U.S. Government purposes.

CP756, *Quark Confinement and the Hadron Spectrum VI*
edited by N. Brambilla, U. D'Alesio, A. Devoto, K. Maung, G.M. Prosperi and S. Serci
© 2005 American Institute of Physics 0-7354-0241-8/05/$22.50

SPINLESS SALPETER EQUATION
Some (Semi-) Analytical Approaches

Wolfgang Lucha* and Franz F. Schöberl†

*Institute for High Energy Physics, Austrian Academy of Sciences,
Nikolsdorfergasse 18, A-1050 Vienna, Austria
E-mail: wolfgang.lucha@oeaw.ac.at
†Institute for Theoretical Physics, University of Vienna,
Boltzmanngasse 5, A-1090 Vienna, Austria
E-mail: franz.schoeberl@univie.ac.at

The eigenvalue equation of a semirelativistic Hamiltonian composed of the relativistic kinetic term of spin-0 particles and static interactions is called spinless Salpeter equation. It is regarded as relativistic generalization of the nonrelativistic Schrödinger approach, or as approximation to the homogeneous Bethe–Salpeter equation in its instantaneous limit. The nonlocality inherent to this kind of operators makes hard to find analytical solutions. Nevertheless, rigorous analytical statements can be proved by sophisticated methods [1]:

- Combining minimum–maximum principle and suitable operator inequalities allows to derive "semianalytical" (or even analytical) upper bounds on all energy levels [2].
- Geometrical considerations summarized under the term "envelope technique" yield "semianalytical" expressions for both upper and lower bounds on eigenenergies [3]. For some particular interactions these bounds can be represented in analytical form.

Resulting eigenstates must be constructed numerically anyway, for instance, by scanning the Hilbert space variationally using standard Rayleigh–Ritz techniques or by integrating this equation of motion by conversion to some equivalent matrix eigenvalue problem [4]. The achieved accuracy of these approximate solutions can then be estimated by powerful criteria derived from generalized — in the present case relativistic — virial theorems [5].

REFERENCES

1. For some comprehensive review see, for instance, W. Lucha and F. F. Schöberl, arXiv:hep-ph/0408184.
2. W. Lucha and F. F. Schöberl, Phys. Rev. A **54** (1996) 3790; Int. J. Mod. Phys. A **14** (1999) 2309; Fizika B **8** (1999) 193; J. Math. Phys. **41** (2000) 1778.
3. R. L. Hall, W. Lucha, and F. F. Schöberl, J. Phys. A **34** (2001) 5059; J. Math. Phys. **42** (2001) 5228; Int. J. Mod. Phys. A **17** (2002) 1931; *ibid.* **18** (2003) 2657; J. Math. Phys. **43** (2002) 5913; *ibid.* **43** (2002) 1237; *ibid.* **44** (2003) 2724 (E); Phys. Lett. A **320** (2003) 127; J. Math. Phys. **45** (2004) 3086.
4. For all the details, consult, for example, W. Lucha, H. Rupprecht, and F. F. Schöberl, Phys. Rev. D **45** (1992) 1233; W. Lucha and F. F. Schöberl, Phys. Rev. D **50** (1994) 5443; Phys. Lett. B **387** (1996) 573; Phys. Rev. A **56** (1997) 139; Int. J. Mod. Phys. C **11** (2000) 485; as well as the references listed therein.
5. W. Lucha and F. F. Schöberl, Phys. Rev. Lett. **64** (1990) 2733; Mod. Phys. Lett. A **5** (1990) 2473; Phys. Rev. A **60** (1999) 5091; Int. J. Mod. Phys. A **15** (2000) 3221.

CP756, *Quark Confinement and the Hadron Spectrum VI*
edited by N. Brambilla, U. D'Alesio, A. Devoto, K. Maung, G.M. Prosperi and S. Serci
© 2005 American Institute of Physics 0-7354-0241-8/05/$22.50

PARTICIPANTS *(at 25 september, 2004)*

NAME	POST ADDRESS/EMAIL
Aguilar Arlene Cristina	c/Dr Molliner,50 — 46100 Valencia, SPAIN arlene.aguilar@uv.es
Alkofer Reinhard	Auf der Morgenstelle,14 - 72706 GERMANY reinhard.alkofer@uni-tuebingen.de
Andrianov Alexander	ul.Ulianovskaya -198504 Sankt-Petersburg, RUSSIA andrianov@bo.infn.it
Antinori Federico	CERN/CSI -Dept. Of Physics - 1211 Geneve,Switzerland federico.antinori@cern.ch
Awes Terry	Oak Ridge National Laboratory - 37831 Oak Ridge, TN, USA awes@bnl.gov
Baker Marshall	Dep. Of Physics Univ.of Washington — PO Box 351560 — Seattle, Washington - USA baker@phys.washington.edu
Bakulev Alexander	Joint institute for nuclear research, BLTPh, JINR - 141980 Dubna, RUSSIA bakulev@thsun1.jinr.ru
Bentz Wolfgang	Dept. Of Physics, school of science, Tokai University - 117 Kitakaname 259-1292 Hiratsuka-shi, JAPAN bentz@keyaki.cc.u-tokai.ac.jp
Beraudo Andrea	Univ.di Torino - Via Pietro Giuria 1 - 10100 Torino ITALIA beraudo@to.infn.it
Bodwin Geoffrey	Argonne National Laboratory - 9700 S.Cass Ave. - 60439 Argonne IL-USA gtb@hep.anl.gov
Boglione Mariaelena	Univ.di Torino, Via Pietro Giuria 1 - 10100 Torino ITALIA boglione@to.infn.it
Brambilla Nora	Univ.di Milano, Via Caloria 16 - 20133 MILANO ITALIA nora.brambilla@mi.infn.it
Casalbuoni Roberto	Sez. INFN e Dip. Di Fisica Univ. Di Firenze - via G. Sansone 1 50019 Sesto Fiorentino Firenze ITALIA casalbuoni@fi.infn.it

Castorina Paolo	Dip. Di Fisica di Catania - Via S. Sofia 64 - 95100 Catania ITALIA paolo.castorina@ct.infn.it
Chernodub Maxim	Institute of Theoretical and Experimental Physics - B.Cheremushkinskaja 25 117218 Moscow RUSSIA maxim.chernodub@itep.ru
Colangelo Gilberto	Theoretische Physik Universitaet Bern - Sidlerstr. 5 CH-3012 Bern SWISS gilberto@itp.unibe.ch
Corianò Claudio	Univ. Di Lecce - via Arnesano - 73100 Lecce ITALIA claudio.coriano@le.infn.it
Costa Pedro	Univ. De Coimbra Dep de Fisica da Univ de Coimbra - Rua Larga P-3004-516 Coimbra PORTUGAL pcosta@teor.fis.uc.pt
Creutz Michael	Brookhaven Lab - BNL 510 A - Upton,NY 11973 USA creutz@bnl.gov
Cucchieri Attilio	Institute de Fisica de Sao Carlos, Univ. De Sao Paulo - C.P.369 - 13560-970, Sao Carlos SP, BRAZIL attilio@if.sc.usp.br
D'Alesio Umberto	Dip. Di Fisica Univ. Di Cagliari Cittadella Univ. - S.P. Sestu km 0,700 09042 MONSERRATO,Cagliari ITALIA umberto.dalesio@ca.infn.it
De Fazio Fulvia	INFN Bari - via Orabona 4 - 70126 Bari ITALIA fulvia.defazio@ba.infn.it
Deldar Sedigheh (Mina)	Tehran Univ. - PO. Box 14395 - 547 Tehan IRAN sdeldar@khayam.ut.ac.ir
Di Giacomo Adriano	INFN - Dip.Fisica - Largo Pontecorvo 2 - 56127 Pisa ITALIA adriano.digiacomo@df.unipi.it
Droz-Vincent Philippe	LUTH Obserbatoire de Meudon - place Jules Janssen - 92195 Meudon FRANCE droz@mesiog.obspm.fr
Dzhunushaliev Vladimir	Mcr. Asanbai, d.25, kv.24, 720060, Bishkek, Kyrgyzstan dzhun@hotmail.kg
Eichten Estia	P.O.Box 500 60510 Battavia IL- USA eichten@fnal.gov

Engelfried Jurgen	Inst.de Fisica Univ. Autonoma de San Luis Potosi 78240 MEXICO jurgen@ifisica.uaslp.mx
Eugenio Paul	Florida State Univ. Dep.of physics 32306 Tallahassee Florida USA eugenio@fsu.edu
Ewerz Carlo	Univ. Di Milano - via Celoria 16 - 20133 Milano ITALIA carlo.ewerz@mi.infn.it
Faber Manfried	Vienna Univ. Of Techn.- Dept. of Physics – Vienna, AUSTRIA faber@kph.tuwien.ac.at
Faridi M. Ayub	CHEP Punjab Univ.,Lahore 54590 Lahore PAKISTAN ayubfaridi@yahoo.com
Feo Alessandra	INFN - Sez. di Parma Dipartimento di Fisica - Parco Area delle Scienze, 7/a 43100 PARMA feo@fis.unipr.it
Feuchter Claus	University of Tuebingen - Tiefenbachstrasse, 97 70329 STUTTGART - GERMANY Claus.feuchter@gmx.de
Fini Rosa Anna	INFN - Sez. di Bari - Via Orabona, 4 70126 BARI rosanna.fini@ba.infn.it
Fischer Christian	IPPP- Univ. Of Durham - South Road - DH1 3LE Durham (U.K.) christian.fischer@durham.ac.uk
Flierl Dominik	Univ. Of Frankfurt - August Euler Str., 6 60486 FRANKFURT (GERMANY) flierl@ikf.uni-frankfurt.de
Foka Panagiota	CERN/CSI Dept. Of Physics 1211 GENEVA - SWITZERLAND yiota.foka@cern.ch
Furui Sadataka	Teikyo Univ. 1-1 Toyosadotai 320-8551 UTSUNOMIYA (JAP) furui@umb.teikyo-u.ac.jp
Garcia Xavier	Univers. Of Barcelona Dept. ECM - Diagonal 647 08028 BARCELONA (SPAIN) xgarcia@ecm.ub.es
Gattullo Eulalia	Univ. Di Milano, Dipartimento di Fisica - via Celoria 16 20133 Milano ITALIA eulalia.gattullo@mi.infn.it
Gladilin Leonid	DESY - Zeus Exp. 22607 HAMBURG (GERMANY)

471

	gladilin@mail.desy.de
Goity Jose Luis	Hampton Univ. Jefferson Laboratori12000 - Jefferson Ave. 23606 NEWPORT (USA) goity@jlab.org
Green Anthony	Univ. Of Helsinky Institute of Physics - P.O. Box 64 00014 HELSINKY (FINLAND) anthony.green@helsinki.fi
Gromes Dieter	Inst.F.Theoretische Physik - Philosophenwek, 16 - 69120 HEIDELBERG (GERMANY) d.gromes@thphys.uni-heidelberg.de
Gutay Laszlo	Dept. Of Physics Purdue University 47907 WEST LAFAYETTE (USA) gutay@physics.purdue.edu
Horvath Ivan	Univ. Of Kentucky - 177, Chem/Phys Bldg. 40506 LEXINGTON (USA) horvath@pa.uky.edu
Huang Huan Zhong	UCLA Dept. Of Physics 90095-1547 LOS ANGELES (USA) huang@physics.ucla.edu
Kabana Sonia	University of Bern Cern 1211 GENEVA – SWITZERLAND sonia.kabana@cern.ch
Kalloniatis Alexander	CSSM - University of Adelaide 5005 ADELAIDE (AUSTRALIA) akalloni@physics.adelaide.edu.au
Karliner Marek	High Energy Physics, University of Cambridge, Cavendish Laboratory, Madingley Road Cambridge CB3 0HE, United Kingdom marek@proton.tau.ac.il
Kizilersu Ayse	CSSM University of Adelaide 5000 ADELAIDE (AUSTRALIA) akiziler@physics.adelaide.edu.au
Koch-Steinheimer Peter	Bitfabrik GmbH & Co.KG - Siedlerstrasse 84 63128 DIETZENBACH – GERMANY pks@bitfabrik.de
Konishi Kenichi	Univ. Di Pisa Dipart. Di Fisica - Via Buonarroti, 2 56127 PISA konishi@df.unipi.it
Kozlov Gennady	JINR – Juliot Curie st.6 141980 DUBNA MOSCOW - RUSSIA kozlov@jinr.ru

Krein Gastao	Instituto de Fisica Teorica - Rua Pamplona 145 01405-900 SAO PAOLO (BRAZIL) gkrein@ift.unesp.br
Langfeld Kurt	Institut fuer Theor.Physik, Univ.Tuebingen – Auf der Morgenstelle 14 72076 TUEBINGEN - GERMANY kurt.langfeld@uni-tuebingen.de
Leitch Michael	Los Alamos National Laboratory P-25, M.S. H846 87545 LOS ALAMOS (USA) leitch@lanl.gov
Loizides Constantinos	Univ. Of Frankfurt - August Euler Str., 6 60486 FRANKFURT (GERMANY) loizides@ikf.uni-frankfurt.de
Lucha Wolfgang	Inst. For High Energy Physics, Austrian Academy of Sciences – Nikolstorfergasse, 18 1050 VIENNA (AUSTRIA) wolfgang.lucha@oeaw.ac.at
Maas Axel	Inst. For Nuclear Physics, Darmstadt Univ. Of Techology - Schlosgardenstrasse, 9 64289 DARMSTADT (GERMANY) axel.maas@physik.tu-darmstadt.de
Marques Gonçalo	CFIF Instituto Superior Tecnico - Av. Rovisco Pais 01- 1049-001 LISBOA - PORTUGAL gmarques@cfif.ist.utl.pt
Martinelli Guido	Dipartimento di Fisica, Univ. La Sapienza – P.le A.Moro 2 – 00185 ROMA guido.martinelli@roma1.infn.it
Maung Khin	Department of Physics, Hampton University – 23668 HAMPTON, VIRGINIA, USA the5maungs@cox.net
Medinnis Michael	DESY - Notkestrasse, 85 22607 HAMBURG (GERMANY) michael.medinnis@desy.de
Megias Fernandez Eugenio	Departamento de Fisica Moderna, Universidad de Granada – Campus Fuentenueva s/n 18071 GRANADA - SPAIN emegias@ugr.es
Melikhov Dimitri	1) Institute fuer Teoretische Physik, Univ. Heidelberg – Philosophenweg 16 – 69120 HEIDELBERG GERMANY 2) Insitute of Nuclear Physics, Moscow State University – Leninskie Gory – 119992 – MOSCOW, RUSSIA melikhov@thphys.uni-heidelberg.de

Melis Stefano	Dip. Di Fisica Univ. Di Cagliari Cittadella Univ. S.P. Sestu km 0,700 - 09042 Monserrato, Cagliari ITALIA stefano.melis@ca.infn.it
Mendes Terza	IFSC-Univ. De Sao Paulo IFSC- USP - Caixa postal 369 13560-970 Sao Carlos SP BRAZIL mendes@if.sc.usp.br
Mevius Maaijke	Schermerhornstraat, 9/1 35115 UTRECHT (Netherlands) mevius@mail.desy.de
Muller-Preussker Michael	Dept. of Physics, Humboldt University - Newtonstr. 15 12489 BERLIN (GERMANY) mmp@physik.hu-berlin.de
Nakamura Atsushi	Hiroshima University RIISE - 1-7-1, Kagami-yama HIROSHIMA (JAPAN) nakamura@riise.hiroshima-u.ac.jp
Nakamura Yoshifumi	Kanazawa University - Kakuma - 920-1192 KANAZAWA (JAPAN) yoshi@hep.s.kanazawa-u.ac.jp
Napolitano Jim	Rensselaer Polytech Inst - 110 Eighth Street NEW YORK - USA napolj@rpi.edu
Nardulli Giuseppe	INFN - Sez. di Bari - Via Amendola, 173 70126 BARI giuseppe.nardulli@ba.infn.it
Nefediev Alexey	Institute of Theoretical and Experimental Physics, Theory Division – B.Cheremushkinskaya 25 – 117218 MOSCOW, RUSSIA nefediev@limon.itep.ru
Neubert Matthias	Cornell University - 318, Newman Laboratory 14850 N.Y. (USA) neubert@lepp.cornell.edu
Niemi Antti	Dept. Of Physics Uppsala University - P.O. Box 803 75108 UPPSALA (SWEDEN) niemi@teorfys.uu.se
Olejnik Stefan	Institute of Physics Slovak Academy of Sciences – Dubravska cesta 9 84511 BRATISLAVA, SLOVAKIA stefan.olejnik@savba.sk
Oliveira Orlando	Univ. Of Coimbra Dept. Of Physics 3004 - 516 COIMBRA (PORTUGAL)

	orlando@teor.fis.uc.pt
Pawlowski Jan Martin	ITP Universitat Tuebingen, Inst. Fur Theroetische Physik - Auf der Morgenstelle 14 72076 TUEBINGEN (GERMANY) j.pawlowski@thphys.uni-heidelberg.de
Peigne Stephane	CNRS - Chemin de Bellevue 74941 ANNECY-le-VIEUX, FRANCE peigne@lapp.in2p3.fr
Petrov Konstantin	Brookhaven Natl. Lab. - Bldg.510a, PO Box 5000 11973 UPTON – USA petrov@bnl.gov
Picariello Marco	Dipart. Di Fisica Univ. Di Milano - Via Celoria, 16 20133 MILANO marco.picariello@mi.infn.it
Piccinini Fulvio	INFN Sez. di Pavia - Via A.Bassi n.6 27100 PAVIA fulvio.piccinini@pv.infn.it
Polikarpov Mikhail	ITEP – B. Cheremushkinskaya 25 – 117259 MOSCOW, RUSSIA polikarp@itep.ru
Prosperi Giovanni	Univ. di Milano - Via Celoria 16 20133 MILANO giovanni.prosperi@mi.infn.it
Quandt Markus	Dept. of Physics University of Tuebingen - Auf der Morgenstelle, 14 72076 TUEBINGEN (GERMANY) quandt@tphys.physik.uni-tuebingen.de
Quercigh E.	NON COMUNICATO
Rafelski Johann	Dept. Of Physics Univ. Of Arizona - 6850 N Placita Sierra 85718 TUCSON - AZ (USA) johann.rafelski@cern.ch
Rak Jan	ISU 1005 – PHENIX 11973 UPTON - NY, (USA) janrak@bnl.gov
Reinhardt Hugo	Dept. Of Physics University of Tuebingen - Auf der Morgenstelle, 14 72076 TUEBINGEN (GERMANY) h.reinhardt@uni-tuebingen.de

Ribeiro Emilio	Dept. Of Physics Univ. Of Lisboa - Av. Rovisco Pais, 11049-001 LISBOA (PORTUGAL) emilioribeiro@netcabo.pt
Ruiz Arriola Enrique	Facultad de Ciencias Universidad de Granada - Av. Fuentenueva s/n 18071 GRANADA – SPAIN earriola@ugr.es
Rupp George	Centro de Fisica, Insituto Superior Tecnico – Av.Rovisco Pais 1 – 1049001 LISBOA george@ist.utl.pt
Salmela Antti	Dept. Of Physical Sciences Univ. Of Helsinki - P.O. Box 64 - 00014 HELSINKI (FINLAND) antti.salmela@helsinki.fi
Sandoval Andres	CERN/CSI Dept. Of Physics 1211 GENEVA - SWITZERLAND andres.sandoval@cern.ch
Sasai Yuji	Oshima National College of Maritime Technology - 1091-1 Komatsu, Oshima-cho, Oshima-gun 742-2193 YAMAGUCHI (JAPAN) sasai@oshima-k.ac.jp
Sazdjian Hagop	Institut de Physique Nucleare, Univ. Paris XI – 91406 ORSAY, FRANCE sazdjian@ipno.in2p3.fr
Scorzato Luigi	Dept. Of Physics Humboldt University - Newtonstr., 15 12489 BERLIN – GERMANY luigi.scorzato@physik.hu-berlin.de
Shnir Yakov	Institute of Physics Univ. Of Oldenburg – 26111 OLDENBURG, GERMANY shnir@marvin.physik.uni-oldenburg.de
Silva Paulo	Departamento de Fisica Universidade de Coimbra 3004-516 COIMBRA – PORTUGAL psilva@teor.fis.uc.pt
Simolo Claudia	V.le Papiniano 24 20123 MILANO claudiasimolo@hotmail.com
Simula Silvano	INFN - Sez. di Roma III - Via della Vasca Navale, 84 00146 ROMA simula@roma3.infn.it
Snellings Raimond	NIKHEF - Kruislaan, 409 1098SJ AMSTERDAM (The Netherlands) raimond.snellings@nikhef.nl

476

Sommer Rainer	DESY – Platanenellee 6 – 15738 ZEUTHEN, GERMANY rainer.sommer@desy.de
Soto Joan	Universitat de Barcelona - Diagonal 647 O8028 BARCELONA – SPAIN soto@ecm.ub.es
Sousa Cèlia	Departamento de Fisica Centro de Fisica Teorica 3004-516 COIMBRA – PORTUGAL celia@teor.fis.uc.pt
Stassart Pierre	University of Liège, Physique Theorique Batiment de Physique 4000 LIEGE - BELGIUM pierre.stassart@ulg.ac.be
Steinhauser Matthias	Univ. Hamburg - Luruper Chaussee 149 22761 HAMBURG – GERMANY matthias.steinhauser@desy.de
Sternbeck Andre	Humboldt University - Newtonstr. 15 12489 BERLIN – GERMANY andre.sternbeck@physic.hu-berlin.de
Stock Reinhard	August-Euler-Str. 6 60486 FRANKFURT – GERMANY stock@ikf.uni-frankfurt.de
Suganuma Hideo	Tokyo Institute of Technology 2-12-2 Ohokayama, Meguro 152-8551 TOKYO – JAPAN suganuma@th.phys.titech.ac.jp
Sugihara Taka	Brookhaven National Laboratory - RBRC, 510A, Brookhaven Ave 11973-5000 NEW YORK – USA sugihara@bnl.gov
Suzuki Tsuneo	Kanazawa University - Kaluma-machi 920-1192 KANAZAWA – JAPAN suzuki@hep.s.kanazawa-u.ac.jp
Szczepaniak Adam	Physics Department Indiana University 47405 BLOOMINGTON – USA aszczepa@indiana.edu
Toki Hiroshi	RCNP, Osaka University - Mihogaoka 10-1 567-5047 OSAKA – JAPAN toki@rcnp.osaka-u.ac.jp
Tornqvist Nils	Dept. Of Physical Sciences University of Helsinki - G. Hallstrom street 2 00014 HELSINKI – FINLAND

	nils.tornqvist@helsinki.fi
Turrisi Rosario	INFN Sez. di Padova - Via Marzolo, 8 35131 PADOVA rosario.turrisi@pd.infn.it
Vairo Antonio	Univ. Di Milano - Via Celoria, 16 20133 MILANO antonio.vairo@mi.infn.it
Vukotic Ilija	Hera-B Platanenalee 6 15738 ZEUTHEN – GERMANY vukotic@ifh.de
Watson Peter	Institute for Theoretical Physics Univ. Giessen - Heinrich-Buff-Ring 16 35392 GIESSEN – GERMANY peter.watson@theo.physik.uni-giessen.de
Wereszczynski Andrzej	Institute for Physics Jagiellonian Univ. - Reymont 4 - 30-059 KRAKOW – POLAND wereszcz@th.if.uj.edu.pl
Williams Anthony	CSSM - University of Adelaide 5005 ADELAIDE – AUSTRALIA awilliam@physics.adelaide.edu.au
Ynduráin Francisco Jose	Departamento de Fisica Teorica Universidad de Madrid – Canto Blanco – 28049 MADRID,Spain fjy@delta.ft.uam.es
Zakharov Valentin	MPI - Foehringer Ring 6 80805 MUENCHEN – GERMANY xxz@mppmu.mpg.de
Zheng Hanqing	Dept. Of Physics Peking University 100871 BEIJING – CHINA zhenghq@pku.edu.cn

479

481

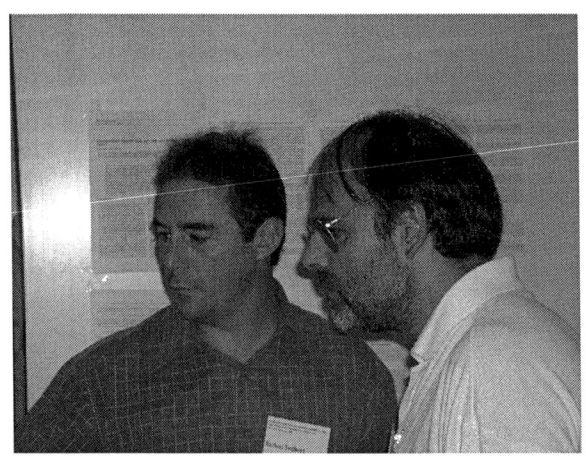

486